Basic Electrical Engineering

Third Edition

T.K. NAGSARKAR

Former Professor & Head
Department of Electrical Engineering
Punjab Engineering College
Chandigarh

M. S. SUKHIJA

Founder & Former Principal
Guru Nanak Dev Engineering College
Bidar, Karnataka

OXFORD

OXFORD
UNIVERSITY PRESS

Oxford University Press is a department of the University of Oxford.
It furthers the University's objective of excellence in research, scholarship,
and education by publishing worldwide. Oxford is a registered trade mark of
Oxford University Press in the UK and in certain other countries.

Published in India by
Oxford University Press
YMCA Library Building, 1 Jai Singh Road, New Delhi 110001, India

First Edition published in 2005
Third Edition published in 2017
Second Impression 2018

ISBN-13: 978-0-19-947936-8
ISBN-10: 0-19-947936-4

Typeset in Times New Roman
by Anvi Composers, New Delhi
Printed in India by Rakmo Press, New Delhi 110020

Cover image: NWM / Shutterstock

Dedicated to the readers – our inspiration

T.K. Nagsarkar

M.S. Sukhija

Preface to the Third Edition

The motivation to develop the *third edition* of our immensely popular textbook *Basic Electrical Engineering* comes from (i) an in-depth review by faculty from various institutions, undertaken by Oxford University Press (India), (ii) inputs from colleagues and students, and (iii) our own re-visits to stay abreast with the curricula of various universities/institutions.

New to the Third Edition

Based on the various inputs and in an effort to provide an all-in-one resource as a first course at the graduate level, the following additions are incorporated.

- More number of solved examples (Additional Examples) has been added to further strengthen the understanding and applications of various laws and principles.
- Short theoretical questions are included to help the readers to prepare for testing and evaluation.
- Chapter on Network Analysis and Network Theorems has been expanded by including (i) conversion of Thevenin and Norton equivalent networks, (ii) Compensation theorem, and (iii) Millman's theorem.
- New and special topics, such as Swinburne's test for testing dc machines, universal motors, tachometer generators, and synchros have been included.
- Electronic instruments have been added and Chapter 11 is renamed as Basic Analogue and Electronic Instruments.
- Solar power systems have been described to update the chapter on power systems.
- A new chapter on Illumination is added.

While incorporating the various inputs and evolving a textbook which has universal acceptance, the authors wish to state that the lucid style of writing and maintaining a smooth flow of the language has not been lost sight of.

Contents and Coverage

Chapter 1 - Introduction to Electrical Engineering - the fundamental laws of electrostatics, electromagnetism, and their applications in electrical engineering are explained. The principles and applications of various laws are further strengthened through solved examples.

Chapter 2 - Network Analysis and Network Theorems - after conceptualizing independent and dependent sources, the chapter defines and explains the application of various network theorems. Node voltage and mesh current methods for solving networks along with applications of super nodes and super meshes have also been explained.

Chapter 3 - Magnetic Circuits - beginning with the definition of the Biot-Savart law, the chapter explains the magnetic behaviour of materials, their properties and classification based on the dipole moment of electrons. It demonstrates the use of mesh analysis for solving a magnetic circuit. The dot convention to obtain the correct direction of the statically induced emf in a coupled circuit is also explained in detail.

Chapter 4 - Alternating Quantities - discusses the generation of ac quantities, their representation, manipulation, and application for analysing alternating networks.

Chapter 5 - Three-phase Systems - deals with the generation of three-phase voltages, current and power along with an analyses of three-phase circuits. Measurement of power and importance of power factor are also included.

Chapter 6 - Transformer Principles - proceeding with a description of the constructional features, the chapter details out the principle of operation and development of an equivalent circuit of a transformer. Application of oc and sc test results for computing efficiency, regulation, etc. of a transformer have also been demonstrated.

Chapter 7 - Synchronous Machines - describes the constructional features of a synchronous machine stator and rotor. The setting up of a synchronously rotating magnetic field has been conceptualized graphically and with the help of phasor diagrams. The concept of infinite bus and the advantages of operating generators in parallel have also been outlined.

Chapter 8 - Induction Motors - includes a detailed description of the construction and principle of operation of three-phase induction motors. Additional solved examples have been added to support the understanding of the operating principles.

Chapter 9 - Direct Current Machines - explains how voltage is induced and derives expressions for induced voltage and electromagnetic torque and discusses commutator action. It also elucidates armature reaction leading to cross and demagnetization mmf. Additionally, field applications of dc machines have also been included.

Chapter 10 - Single-phase Induction Motors and Special Machines - qualitatively discusses the working of fractional kilowatt motors such as single-phase induction motors, ac and dc servo motors, different types of stepper and hysteresis motors. The chapter also explains the working principle of universal motors, synchro systems, and tachometer generators.

Chapter 11 - Basic Analogue and Electronic Instruments - includes the principles of measurement of electrical quantities such as resistance, voltage, current, power and the working principles and calibration of measuring instruments.

Chapter 12 - Power Systems - describes the generation, transmission, and distribution systems and subsystems. Various types of domestic wiring, including staircase lighting and earthing systems are also included.

Chapter13 - Illumination - defines the different terminologies related to illumination such as luminous flux, candela, and luminous intensity and introduces the laws of illumination such as Proportionality law, Inverse Square law, and Lambert's Cosine law of incidence. The chapter also explains the application of the laws for computing luminosity through solved examples.

Acknowledgements

The authors would like to express their gratitude to the readers of their textbook, *Basic Electrical Engineering*, for being an uninterrupted source of inspiration. Also, no amount of appreciation would be enough for the editorial staff of OUP who always keep us on our toes by their feedback from the users and their constructive approach.

T.K. Nagsarkar

M.S. Sukhija

Preface to the First Edition

'Why another book on basic electrical engineering' was our initial thought when the idea of writing this book first came to mind. Basic electrical engineering is a core course offered to engineering students of all streams. It is extremely important to ensure that the fundamentals of the course are well understood by all engineering students since these have applications in all streams. In spite of a number of textbooks available in the subject, we felt that there was still a need for a book that would make the learning and understanding of the principles of electrical engineering an enjoyable experience.

The book is the outcome of our experience of over three decades of teaching both undergraduate and postgraduate courses. The initial draft of the chapters was written by one of us and then read thoroughly by the other. The content was also peer reviewed and suggestions of the reviewers incorporated. In this book, we have tried to ensure that students would easily grasp the basics of electrical engineering. We hope that students will discover that their learning and understanding of the subject progressively increases while using the book.

About the Book

The contents of this book have been designed, modelled, and written as per the AICTE's model curriculum and the syllabi of several universities. The book provides a comprehensive coverage of the different topics prescribed by various universities, thereby providing it with wide acceptability.

It is firmly believed that this book will help students to overcome their initial apprehensions and initiate a life-long affair with electrical engineering. Written in a simple, yet lucid style, the book presents a clear and concise exposition of the principles and applications of electrical engineering. Students will find the smooth flow of language an asset to quickly grasp the basic concepts and build a strong foundation in the subject.

Key Features

- Provides a chapter overview and recapitulation of important formulae in every chapter
- Includes a large number of illustrations to supplement the text
- Enhances the understanding of concepts with several worked examples
- Provides numerous chapter-end exercises with answers and multiple choice questions to stimulate student interest

Content and Coverage

The book introduces the fundamentals of electricity and electrical elements. It provides an exhaustive coverage of network theory and analysis, electromagnetic theory and energy conversion, alternating quantities, alternating and direct current machines, basic analog instruments, and power systems.

Finally, we hope that we have been able to make the subject of electrical engineering appealing not only for the students but also for the faculty. On the other hand, if you find that it falls short of expectations, please share your feedback with us to enable us to make improvements.

T.K. Nagsarkar

M.S. Sukhija

Contents

Symbols and Acronyms

ϕ	magnetic flux (Wb)		(T or Wb/m^2)
Φ	flux per pole (Wb)	B_m	maximum magnetic flux
ϕ_m	maximum magnetic flux		density (T or Wb/m^2)
γ	angle between two adjacent slots (degree electrical)	B_{ii}, B_{ij}	self and transfer suceptances (S)
		$[\boldsymbol{B}]$	susceptance matrix
ρ	specific resistance or resistivity (Ω m)	BCD	binary coded decimals
		BIS	Bureau of Indian Standards
u	velocity (m/sec)	C	capacitance (F)
μ	absolute permeability of the medium (H/m)	C_{eq}	equivalent capacitance (F)
μ_0	permeability of free space $= 4\pi \times 10^{-7}$ (Wb/ATm or H/m)	CT	current transformer
		CI	cast iron
μ_r	relative permeability of the medium	D	electric flux density (C/m^2)
σ	specific conductance or conductivity of a material (mho/m, or S/m)	dc/DC	direct current
		DOL	direct on-line starting
		DPIC	double-pole iron clad
α_t	temperature coefficient of resistance at temperature $t°C$	emf	electromotive force
		e	emf (V)
θ	angle (rad or deg electrical)	E	electric field intensity
ω	angular velocity $= 2\pi f$ (rad/sec)		(N/C or V/m)
η	efficiency	E	rms value of emf (V)
φ	phase or power factor angle	$\boldsymbol{E}$	phasor emf
β	short pitch angle	E_b	back emf (V)
τ	time constant of a circuit (sec); thickness of laminations (m)	E_c	rms value of emf induced in a coil (V)
		$\mathcal{E}$	electric field intensity
ε_0	permittivity of free space $= 8.85 \times 10^{-12}$ (F/m)		(N/C or V/m)
		ECC	earth continuity conductor
ε	permittivity of the medium (F/m)	EMEC	electromechanical energy conversion
ε_r	relative permittivity of the medium	$\mathcal{F}$	mmf (AT)
ψ	flux linkage ($N\phi$)	F	force of attraction or repulsion, between two charges (N)
$\parallel$	in parallel with		
a, A	cross-sectional area (m^2)	F	force (N)
a	number of parallel paths	F_b	backward revolving field
a	transformation ratio	F_f	forward revolving field
ac/AC	alternating current	f	frequency (cps or Hz)
B	magnetic flux density	f_r	frequency of rotor induced emf

G	conductance of a material (S)	MI	moving iron
$[G]$	conductance matrix	MNP	magnetic neutral plane
G_{ii}, G_{ij}	self and transfer conductances (S)	n	speed of revolution (rps); Steinmetz constant
g	conductance of a material (S)	N	speed of rotation (rpm); number of turns
GI	galvanized iron		
GNP	geometric neutral plane	N_S	synchronous speed (rpm)
H	magnetic field intensity or magnetizing force (AT/m)	NL	neutral link
		oc	open circuit
HV	high voltage	OCC	open-circuit characteristic
hp	horse power	P	number of poles; active power (W, kW, MW)
i	instantaneous value of current (A)	P_L	power delivered to load
I	rms value of current (A)	P_m	mechanical power output (W)
$\mathbf{I}$	current phasor	P_{ag}	power transferred from stator to rotor; air-gap power (W)
I_ϕ	rms value of magnetizing current (A)		
I_a	dc armature current (A); rms value of current in phase a (A); rms current in auxiliary winding	P_e	eddy current loss (W)
		P_h	hysteresis loss (W)
		p	instantaneous power (J/sec or W)
I_f	field current (A)		
I_m	rms current in main field winding (A)	$\mathcal{P}$	permeance which is reciprocal of reluctance $\mathcal{R}$ (Wb/AT)
I_{sc}	short-circuit current (A)		
I_{sh}	shunt field current (A)	pd	potential difference (V)
$\mathcal{J}$	current density (A/m^2)	pf	power factor of a circuit $= \cos \phi$
j	mathematical operator		
KCL	Kirchoff's current law	q	charge (C); number of phases
k_d	breadth or distribution factor	Q	reactive power or quadrature power (VAR, kVAR, MVAR)
k_p	pitch factor		
KVL	Kirchoff's voltage law	$\mathcal{R}$	reluctance (AT/Wb)
k_w	winding factor	R, r	resistance (Ω)
L	inductance (V sec/A or H)	r_a	armature resistance (Ω)
L_{eq}	equivalent inductance (H)	R_L	load resistance (Ω)
LV	low voltage	R_{TH}	Thevenin's equivalent resistance (Ω)
l	length (m)		
$\mathcal{M}$	magnetic potential drop (AT)	R_N	Norton's equivalent resistance (Ω)
mmf	magnetomotive force	rms	root mean square
M	mutual inductance (H)	S	complex power (VAR, kVAR, MVAR); total number of slots
m	number of slots/pole/phase		
max	maximum	s	slip in an induction motor
min	minimum	sc	short circuit
MCB	miniature circuit breaker	SCC	short-circuit characteristic
MDB	main distribution board		

SDB	sub-distribution board	V_p	phase voltage (V)
t	temperature (°C); unit of time (sec)	V_L	line voltage (V)
T	torque (N m); number of turns; time period (sec)	$\mathcal{V}$	volume (m^3)
		w	work done (N m);
T_c	controlling torque (N m)		Energy in joules (J)
T_d	deflecting torque (Nm)	X, x	reactance (Ω)
v	instantaneous voltage	X_L	inductive reactance (Ω)
$v_s(t)$	instantaneous time-varying voltage	X_C	capacitive reactance (Ω)
V_{oc}	open-circuit voltage	Y	complex admittance (mho)
V_s	dc voltage source	$[Y]$	admittance matrix
V_{TH}	Thevenin's voltage	Y-Δ	Star–delta
V_N	Norton's voltage	Z	complex impedance (Ω); number of conductors
V	rms voltage (V); dc voltage (V)		
		$[Z]$	impedance matrix
V	rms voltage phasor	Z_L	load impedance (Ω)
V_s	source voltage (V)		

INTRODUCTION TO ELECTRICAL ENGINEERING

Learning Objectives

This chapter will enable the reader to

- Understand the nature of structure of an atom and significance of free electrons
- Differentiate between conductors, semiconductors, and insulators based upon the energy levels of electrons
- Compute the resistance of a conductor from its physical dimensions at different temperatures
- Familiarize with electrostatic phenomena associated with electric charges and define electric field intensity, electric potential and potential difference, electric flux, and electric flux density
- Get familiar with basic electrical quantities: current, voltage, emf, and electric power
- Define Ohm's law for a resistor and compute the resistance of a conductor from its physical dimensions at different temperatures
- Compute the induced voltage due to varying current, power, and energy stored in an inductor
- Based upon an understanding of the charge storing nature of a capacitor, compute its capacitance, current, power, and energy stored
- Define Ampere's law and use it to estimate the force on a current carrying conductor when placed in a magnetic field, and use Fleming's left hand rule to determine direction of the force
- Use Faraday's laws of electromagnetic induction to compute the magnitude of dynamic or static induced voltage and apply Fleming's right hand rule or Lenz's law to determine the direction of the induced voltage
- Define Kirchhoff's voltage and current laws and apply these to compute currents and voltages in a circuit made up of resistors, inductors, and capacitors

---

1.1 ESSENCE OF ELECTRICITY

It is believed that electricity is present in nature. It is amazing how humankind has been able to put electricity to myriad uses for its own progress and comfort without having an exact knowledge of the nature of electricity. In fact, based on experimentation and observations, theories have been developed to explain the behaviour of electricity.

Electrical energy has been accepted as a form of energy that is most suited for transformation into other forms of energy, such as heat, light, mechanical energy, etc. Electricity can be converted into many different forms to bring about new and enabling technologies of high value. Conversion of electrical energy into pulses and electromagnetic waves has given rise to computers and communication systems. Its conversion into microwaves finds use in microwave ovens, industrial processes, and radars. Electricity in the arc form serves in arc furnaces and welding. Efficient lighting, lasers, visuals, sound, robots, medical tools are among many other examples of the use of electricity.

Electrical engineering deals with the generation, transmission, utilization, and control of electric energy. Electric energy is generated at electric power generating stations such as hydroelectric, thermal, and nuclear power stations. In a hydroelectric power station, the potential energy of the head of water stored in dams is converted into kinetic energy by regulating the flow of the stored water through turbines. This kinetic energy, in turn, gets transformed into electric energy by the process of electro-mechanical energy conversion. In a thermal station, the chemical energy of coal, oil, natural gas, and synthetic derivatives is converted by combustion into heat energy. Heat energy is also produced by nuclear fission of nuclear fuels in a nuclear reactor. It is then converted into mechanical energy, which in turn is transformed by electro-mechanical energy conversion to electric energy, through thermodynamic processes. Conversion of limitless energy from the sun into usable electric energy through photovoltaic energy conversion is achieved by using solar cells. Commercially, electricity is also being generated from renewable energy sources such as wind, biomass, and geothermal sources. Wind energy is converted into electrical form through a wind turbine coupled to an electrical generator. Geothermal power generation converts energy contained in hot rocks into electricity by using water to absorb heat from rocks and transport it to the earth's surface, where it is converted into electric energy through turbine generators. The majority of biomass electricity is generated using a steam cycle where biomass material is first converted into steam in a boiler; the resultant steam is then used to turn a turbine connected to a generator.

Electricity permits the source of generation to be remote from the point of application. Electric energy transmission systems are varied, such as power transmission systems and electronic communication systems. Electric energy for conversion into light energy, heat energy, and mechanical energy for use in industries, commercial establishments, and households would require bulk transmission of electric power from the source, which produces energy, to the load centre, where the electric energy is utilized. Electrical power transmission systems consist of chains of transmission towers on the earth's surface, from which the line conductors carrying current are suspended by porcelain insulators.

An electric system may be viewed as consisting of generating devices, transformers, and transmission systems which interconnect terminal equipment for converting electrical energy into light, heat, or mechanical energy and vice versa. All devices and equipment can be represented by idealized elements called circuit elements. These elements can be interconnected to form networks, which can be used for modelling and analysing the system behaviour. Conversely, networks may be designed to achieve the required performance from a system.

Electrical engineering is concerned with the study of all aspects of electric power, i.e., its generation, transmission, and utilization. Therefore, it is necessary to become familiar with the basic concepts and terms associated with electricity.

1.2 ATOMIC STRUCTURE AND ELECTRIC CHARGE

Atom is the smallest particle of an element. As per Bohr–Rutherford's planetary model of atom, the mass of an atom and all its positive charge is concentrated in a tiny nucleus, while negatively charged electrons revolve around the nucleus in elliptical orbits like planets around the sun (see Fig. 1.1). The nucleus contains protons and neutrons. A neutron carries no charge and its mass is 1.675×10^{-27} kg, while a proton carries a positive charge $+e$ and its mass is 1.672×10^{-27} kg. The electron carries a negative charge $-e = 1.602 \times 10^{-19}$ C and its mass is 9.109×10^{-31} kg. Thus an electron is lighter than a proton by a factor of about 1840. There are exactly as many protons in the nucleus of an atom as planetary electrons. Thus, the nucleus of an atom can be viewed as a core carrying a positive charge, and the negative charge of the encircling electrons is equal to the positive charge of protons.

An atom as a whole is electrically neutral. The orbits for the planetary electrons are called shells or energy levels. The electrons in successive shells named $K, L, M, N, O, P,$

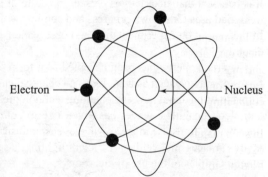

Fig. 1.1 Structure of an atom

and Q are at increasing distance outwards from the nucleus. Each shell has a maximum number of electrons for stability. For most elements, the maximum number of electrons in a filled inner shell equals $2n^2$, where n is the shell number in sequential order outward from the nucleus. Thus the maximum number of electrons in the first shell is 2, for the second shell is 8, for the third shell is 18, and so on. These values apply only to an inner shell that is filled with its maximum number of electrons. To illustrate this rule, a copper atom with 29 electrons is chosen. In this case, the number of electrons in the K, L, M, and N is 2, 8, 18, and 1, respectively.

1.3 CONDUCTORS, SEMICONDUCTORS, AND INSULATORS

As stated in the preceding section, electrons revolve in orbits around the nucleus. The electrons closer to the nucleus possess lower energies than those further from it, which is very much similar to a mass m possessing increasing potential energy as its distance above the earth's surface increases. Thus the position occupied by an electron in an orbit signifies a certain potential energy. Due to the opposite charge, there is a force of attraction between the electron and the nucleus. The closer an electron is to the nucleus, more strongly it is bound to the nucleus. Conversely, further away an electron is from the nucleus, lesser is the force of attraction between the electron and the nucleus. Since the bond between the outer electrons and the nucleus is weak, it is easy to detach such an electron from the nucleus.

When many atoms are brought close together, the electrons of an atom are subjected to electric forces of other atoms. This effect is more pronounced in the case of electrons in the outermost orbits. Due to these electric forces, the energy levels of all electrons are changed. Some electrons gain energy while others lose it. The outermost electrons suffer the greatest change in their energy levels. Thus the energy levels, which were sharply defined in an isolated atom, are now broadened into energy bands. Each band consists of a large number of closely packed energy levels. In general, two bands result, namely, the conduction band associated with the higher energy level and the valence band. A region called forbidden energy gap separates these two bands. Each material has its own band structure. Band structure differences may be used to explain the behaviour of conductors, semiconductors, and insulators.

In metals, atoms are tightly packed together such that the electrons in the outer orbits experience small, but significant, force of attraction from the neighbouring nuclei. The valence band and the conduction band are very close together or may even overlap. Consequently, by receiving a small amount of energy from external heat or electric sources the electrons readily ascend to higher levels in the conduction band and are available as electrons that can move freely within the metal. Such electrons are called free electrons and can be made to move in a particular direction by applying an external energy source. This movement of electrons is really one of negative electric charge and constitutes the flow of electric current. In metals the density of electrons in the conduction band is quite high. Such metals are categorized as conductors. In general metals are good conductors, with silver being the best and copper being the next best.

In semiconductors the valence and conduction bands are separated by a forbidden gap of sufficient width. At low temperatures, no electron possesses sufficient energy to occupy the conduction band and thus no movement of charge is possible. At room temperatures it is possible for some electrons to gain sufficient energy and make the transition to the conduction band. The density of electrons is not as high as in metals and thus cannot conduct electric current as readily as in conductors. Carbon, germanium, and silicon are semiconductors conducting less than the conductor but more than the insulators.

A material with atoms that are electrically stable, that is, with the outermost shell complete, is an insulator. In such materials the forbidden gap is very large, and as a result the energy required by the electron to cross over to the conduction band is impractically large. Insulators do not conduct electricity easily, but are able to hold or store electricity better than conductors. Insulating materials such as glass, rubber, plastic, paper, air, and mica are also called dielectric materials.

1.4 ELECTROSTATICS

Electrostatics is associated with materials in which electrical charge moves only slowly (insulating materials) and with electrically isolated conductors. Charges are static as insulation and isolation prevent easy migration of charge. Electrostatic phenomena arise from the forces that electric charges exert on each other. There are many examples as simple as the attraction of the plastic wrap to one's hand after it is removed from a package, to the operation of photocopiers.

Electrostatics involves the buildup of charge on the surface of objects due to contact with other surfaces. Although exchange of charge happens whenever any two surfaces contact and separate, the effects of charge exchange are usually noticed only when at least one of the surfaces has a high resistance to electrical flow. This is because the charges that transfer to or from the highly resistive surface are more or less trapped there for a long enough time for their effects to be observed. These charges then remain on the object until they either bleed off the ground or are quickly neutralized by a discharge.

The space surrounding a charged object is affected by the presence of the charge and an electric field is established in that space. A charged object creates an electric field—an alteration of the space or field in the region that surrounds it. Electric field is a vector quantity whose direction is defined as the direction in which a positive test charge would be pushed when placed in the field. Thus, the electric field direction about a positive source charge is always directed away from the positive source. And the electric field direction about a negative source charge is always directed toward the negative source.

1.4.1 Coulomb's Law

Coulomb's law states that the force of attraction or repulsion F, between two charges q_1 and q_2 coulombs, concentrated at two different points in a medium, is directly proportional to the product of their magnitudes and inversely proportional to the square of the distance r between them. Mathematically, it may be expressed as

$$F = \frac{1}{4\pi\varepsilon}\frac{q_1 q_2}{r^2} \quad \text{newton (or N)} \tag{1.1}$$

where ε is the absolute permittivity of the surrounding medium and is given by

$$\varepsilon = \varepsilon_0 \varepsilon_r \tag{1.2}$$

where ε_0 is the permittivity of free space and is equal to 8.84×10^{-12} F/m; ε_r is the relative permittivity of the medium.

If the charges are of like polarity, the force between them is repulsive, and if the charges are of opposite polarity, the force is attractive.

1.4.2 Electric Field Intensity

When a stationary electric charge is placed within an electrostatic field, it experiences a force of attraction or repulsion depending on the nature of the charge and its position in the field. The ratio of the force exerted on the charge to the magnitude of the charge is defined as the electric field intensity. Thus, if a charge of magnitude q coulomb, when placed within an electric field, experiences a force of F newton, then the electric field intensity E will be given by

$$E = \frac{F}{q} \quad \text{N/C or V/m} \tag{1.3}$$

The force F experienced by charge q_2 due to the presence of charge q_1 is given by Eq. (1.1). Hence, the field strength at the point where change q_2 is located will be [from Eq. (1.3)]

$$E = \frac{F}{q_2} \quad \text{N/C or V/m}$$

Substituting for F from Eq. (1.1) in the above equation yields

$$E = \frac{q_1 q_2}{4\pi\varepsilon r^2} \times \frac{1}{q_2} = \frac{q_1}{4\pi\varepsilon r^2} \quad \text{N/C or V/m} \tag{1.4}$$

Example 1.1 Find the force in free space between two like point charges of 0.1 C each and placed 1 m apart.

Solution Using Eq. (1.1), the force may be obtained as

$$F = \frac{0.1 \times 0.1}{4\pi \times 8.84 \times 10^{-12} \times 1} = 9 \times 10^7 \, \text{N}$$

It may be noted that the magnitude of the force is gigantic. This calculation shows that 0.1 C of electric charge is a very high value and is normally not encountered in engineering computations.

1.4.3 Electric Potential and Potential Difference

Moving a positive test charge against the direction of an electric field would require work by an external force. This work would in turn increase the potential energy of the charge. On the other hand, the movement of a positive test charge in the direction of an electric field would occur without the need for work by an external force. This motion would result in the loss of potential energy of the charge. Potential energy is the stored energy of position of a charge and it is related to the location of the charge within a field.

The above situation finds an analogy in mechanics where work has to be done against the gravitational force in raising a mass to some height above sea level. The greater the mass, the greater is the potential energy possessed by the mass.

While electric potential energy has a dependency upon the charge experiencing the electric field, electric potential is purely location dependent. It is the potential energy per charge.

The electric potential at any point within an electric field is defined as the amount of work done against the electric field (or the energy required) to bring a unit positive charge from infinity to that point, or alternatively, from a place of zero potential to the point. The unit of potential is volt, and 1 volt is equal to 1 joule/coulomb. An alternate name of this quantity, voltage, is named after the Italian physicist Alessandro Volta.

The potential difference between two points within an electric field is the work done by the field in shifting a unit positive charge from one point to the other. It is to be noted that positive charge always flows from higher potential point to lower potential point, whereas a negative charge flows from a lower potential point to higher potential point.

Both potential and potential difference are scalar quantities as these are position dependent in a field but are not dependent on the path by which the position is reached.

The total work per unit charge associated with the motion of charge between two points is called voltage. If v is the voltage in volts, w is the energy in joules, and q is the charge in coulombs, then

$$v = \frac{dw}{dq} \quad \text{J/C} \tag{1.5}$$

Example 1.2 Two charges $Q_1 = 2 \times 10^{-9}$ C and $Q_2 = 3 \times 10^{-9}$ C are spaced 6 m apart in air as shown in Fig. 1.2(a). Derive an expression for the net force on a unit positive charge Q at point A, located at x m from Q_1. If A and B are respectively located 1 m and 4 m away from the charge Q_1 as shown in Fig. 1.2(b), compute the voltage V_{AB} between the points A and B.

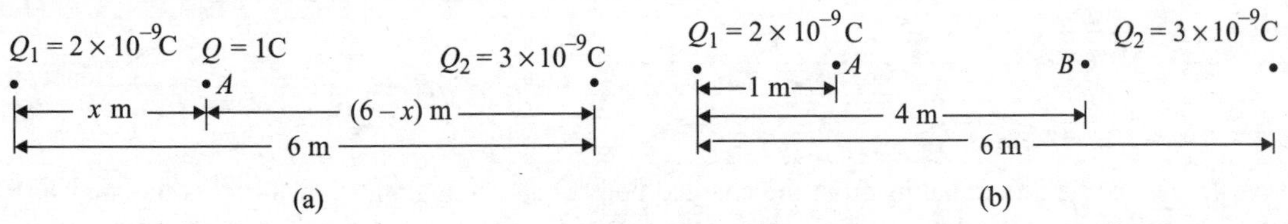

Fig. 1.2

Solution The force of repulsion between Q and Q_1 at point A, directed away from Q_1, is

$$\frac{1}{4\pi \times 8.84 \times 10^{-12}} \times \frac{2 \times 10^{-9}}{x^2} = \frac{18}{x^2} \text{ N}$$

Similarly, the force of repulsion between Q and Q_2 at point A, directed away from Q_2, is

$$\frac{1}{4\pi \times 8.84 \times 10^{-12}} \times \frac{3 \times 10^{-9}}{(6-x)^2} = \frac{27}{(6-x)^2} \text{ N}$$

The net force on Q, directed away from Q_2, is given by

$$F = \frac{27}{(6-x)^2} - \frac{18}{x^2} = 9 \left[\frac{3}{(6-x^2)} - \frac{2}{x^2} \right] \text{ N}$$

The work done in moving Q from point A to point B is given by

$$W_{BA} = \int_a^b F dx = 9 \int_1^4 \left[\frac{3}{(6-x)^2} - \frac{2}{x^2} \right] dx = 9 \left[\frac{3}{(6-x)} + \frac{2}{x} \right]_1^4 = -5.4 \text{ J}$$

Since voltage is defined as work done per unit charge, the voltage between points A and B is given by

$$V_{BA} = \frac{W_{BA}}{Q} = -5.4 \text{ J/C or V}$$

Then, $V_{AB} = - V_{BA} = 5.4$ V

1.4.4 Electric Flux

An electric field exists in space between a positively and a negatively charged body. The presence of an electric field is shown by certain imaginary lines through space. They are called *flux lines*. Conventionally, they radiate from a positive charge and converge on equal quantity of negative charge. The electric flux lines are not closed on themselves as a positive and negative charge cannot exist simultaneously. Electric flux lines of an isolated charged conductor are shown in Fig. 1.3.

Both electric charge q and flux ψ are measured in coulomb, and one coulomb of positive charge radiates one coulomb of flux.

1.4.5 Electric Flux Density

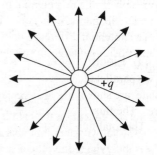

Fig. 1.3 Electric flux lines of an isolated charged conductor

Electric flux density D at any point in a medium is defined as the flux ψ (in coulomb) per unit area a (in m^2), at right angles to the direction of the flux. Thus,

$$D = \frac{\psi}{a} = \frac{q}{a} \text{ C/m}^2 \tag{1.6}$$

From Eq. (1.4), electric field intensity E at a distance r from the centre of a charged body of charge q is

$$E = \frac{q}{4\pi \varepsilon r^2}$$

or

$$\varepsilon E = \frac{q}{4\pi r^2} \tag{1.7}$$

Now, the electric flux radiating from the charged body is also q coulombs, and $4\pi r^2$ is the total surface area of the sphere, with the centre at the centre of the charged body and a radius of r. The electric flux density is given by

$$D = \frac{q}{4\pi r^2} \tag{1.8}$$

From Eqs. (1.7) and (1.8), we get the following relation:

$$D = \varepsilon E \tag{1.9}$$

1.4.6 Gauss' Law

Gauss' law states that the surface integral of the electric flux density over a closed surface enclosing a specific volume is equal to the algebraic sum of all the charges enclosed within the surface, i.e.,

$$\oint D \, ds = \Sigma q \tag{1.10}$$

1.4.7 Electric Field Due to a Long Straight Charged Conductor

A long conductor, having uniform charge q coulombs per metre, is shown in Fig. 1.4. The electric flux will be radial in all directions perpendicular to the conductor.

Let a point be chosen at a perpendicular distance r from the conductor. The total charge enclosed by an elementary cylindrical surface of length dl will be $q \times dl$, and the total flux ψ coming out of the cylindrical surface will be

$$\psi = q \times dl$$

and the flux density D on the cylindrical surface is

$$D = \frac{\psi}{2\pi r \, dl} = \frac{q \, dl}{2\pi r \, dl} = \frac{q}{2\pi r}$$

Then, the field intensity E is given by

$$E = \frac{D}{\varepsilon} = \frac{q}{2\pi \varepsilon r} \ \text{V/m} \tag{1.11}$$

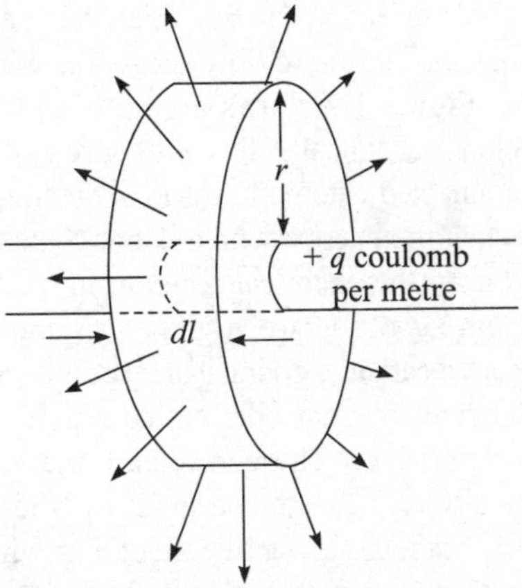

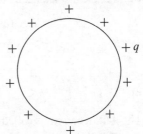

Fig. 1.4 Electric field around a long charged conductor

1.4.8 Electric Field Between Two Charged Parallel Plates

Two parallel plates, with charge $+q$ on one plate and charge $-q$ on the other plate, are shown in Fig. 1.5. The cross-sectional area of each plate is a metre^2. The flux lines in this case will be perpendicular to the charged plates.

The total flux $\psi = qC$ and the flux density inside the medium is

$$D = \frac{\psi}{a} = \frac{q}{a}$$

and the field intensity is

$$E = \frac{D}{\varepsilon} = \frac{q}{\varepsilon a} \ \text{V/m} \tag{1.12}$$

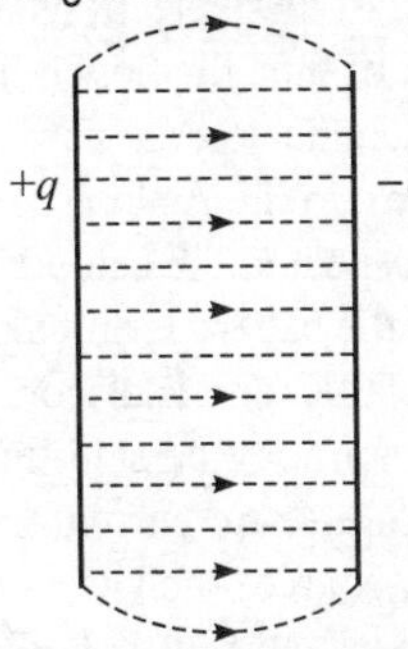

Fig. 1.5 Electric field of parallel charged conductors

1.4.9 Electric Field of a Uniformly Charged Sphere

A hollow metallic sphere with total charge q coulombs is shown in Fig. 1.6. Electric field intensity inside the hollow sphere is zero because of the fact that the electrical charge resides at the surface of the sphere only. Therefore, the electric field intensity outside the charged sphere is to be determined.

The total electric flux ψ going out of the charged sphere is $\psi = q$. The flux density D at a distance r from the centre of the sphere can be determined by considering a spherical shell of radius r with the same centre as the centre of the sphere. The surface area of this spherical shell is $4\pi r^2$ and the flux density will be

$$D = \frac{\psi}{4\pi r^2} = \frac{q}{4\pi r^2}$$

Fig. 1.6 A hollow metallic charged sphere

Also, the electric field intensity is

$$E = \frac{D}{\varepsilon} = \frac{q}{4\pi\varepsilon r^2} \text{ V/m} \tag{1.13}$$

1.5 ELECTRIC CURRENT

In an isolated metallic conductor, such as a length of copper wire, numerous free electrons exist in the conduction band and yet no current flows. Due to interactive forces between the free electrons themselves and with the positive ions the electrons are in motion which is essentially random in nature, and at any cross section of the copper wire the net movement of electrons is zero. In conductors, an orderly movement of electrons in a given direction can be achieved by applying an external energy source across the ends of the conductor. This makes current to flow across the wire/conductor.

Figure 1.7 shows an arrangement in which an electrochemical cell, commonly called a battery, is connected externally by a conducting wire. Initially, when the energy source is not connected externally, due to the chemical reaction in the battery a large number of electrons gather around one electrode, called cathode, giving it an excess of negative charge. The other electrode, called anode, has an excess of positively charged nuclei, thereby, charging it positively. The anode is at a higher potential than the cathode. When a conducting wire is connected externally to the battery terminals, electrons in the

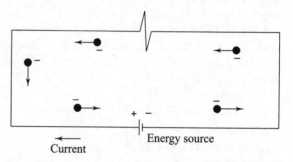

Fig. 1.7 Flow of electrons and current

conduction band are set in motion by the electric force due to accumulated charges at the battery terminals. The motion of these electrons is periodically interrupted by collisions with static atoms and ions. However, at any instant of time the flow of charge at the conductor cross section is constant. There is no accumulation of charge in the conductor; as many charges enter the cross section as leave it. The constant flow of charges constitutes electric current. As long as the chemical reactions in the battery maintain the anode terminal at a higher potential with respect to the cathode terminal, the flow of current continues. Further, the greater the potential difference across the battery terminals, the greater is the accumulated charge, the rate of flow of charge, and the current. If the metallic wire is disconnected from the battery terminals, its electrical neutrality is preserved.

Thus it may be said that the flow of electric current is associated with the movement of electric charge. Flow of electric current in a conductor is possible only when it is connected to the terminals of an electric energy source, such as a battery, and there exists a potential difference across its terminals.

Electric current is defined as the time rate change of charge passing through a cross-sectional area of a conductor. If Δq coulomb is the amount of charge flow in Δt seconds, then the average current i_{av} over a period of time, the instantaneous current i, and the charge q transferred from time t_0 to t_1 are given by

$$i_{av} = \frac{\Delta q}{\Delta t} \text{ C/sec (or amperes)} \tag{1.14}$$

$$i = \frac{dq}{dt} \text{ C/sec} \tag{1.15}$$

or $$q = \int_0^t i \, dt \text{ C} \tag{1.16}$$

The unit of current is called ampere, named after the French scientist Andre Marie Ampere. A current of 1 A means that the electric charge is flowing at the rate of 1 C/sec.

Currents have direction. In conductors the current consists of the movement of electrons. Conventionally, a positive current is taken to be a flow of positive charge in the direction of a reference arrow used to mark the direction of the current flow, as shown in Fig. 1.7. Thus, the positive direction of the flow of current is taken as opposite to that of the direction of the movement of electrons.

Example 1.3 In a metallic wire, 10^{19} electrons drift across a cross section per second. What is the average current flow in the wire?

Solution Charge on one electron $= 1.6 \times 10^{-19}$ C
From Eq. (1.14),

$$I_{av} = \text{Total charge movement per second} = 1.6 \times 10^{-19} \times 10^{19}$$
$$= 1.6 \text{ C/sec} = 1.6 \text{ A}$$

1.6 ELECTROMOTIVE FORCE

In an isolated metallic conductor, free electrons, which are loosely bonded with their nuclei, can be made to flow in a given direction by applying an electric pressure across the ends of the conductor. Such a pressure is provided by an external energy source, for example, a battery.

Due to the chemical reactions inside an electrochemical cell, commonly called a battery, separation of electric charges takes place. Negative charges accumulate at one terminal, the cathode, and positive charges accumulate at the other terminal, the anode. As the charges of unlike polarity attract each other, work has to be done by an external agency against these attractive forces to separate them. In the case of a battery, the work is done chemically. The greater the number of charges that are separated, the greater is the work that has to be done to achieve this separation and the greater is the potential energy of the separated charges.

The work done per unit charge is a measure of the amount of accumulated charge or a measure of the potential energy that has been established. The work done per unit charge in a battery is the potential difference (pd) between the terminals of the battery.

The pd between the battery terminals is known as the electromotive force, or emf. The emf represents the driving influence that causes a current to flow, and may be interpreted to represent the energy that is used during passing of a unit charge through the source. The term emf is always associated with energy conversion. The emf is usually represented by the symbol E and has the unit volt. When the battery is connected externally through a conductor to a load, energy transfer to the load commences through the conductor. The energy transfer due to the flow of unit charge between the two points in the circuit is termed as potential difference. When all the energy is transferred to the load unit, the pd across the load unit becomes equal to that of the battery emf.

In view of this discussion, it may be stated that both emf and pd are similar entities and have the same units. Thus emf is associated with energy while pd causes the passage of charge, or current. Both potential and potential difference are scalar quantities.

The emf and pd are represented in a diagram following certain conventions. Each is indicated by an arrow, as shown in Fig. 1.8. The arrowhead in each case points to a higher potential. It may be noted that the current leaves the source of emf at the positive terminal and therefore the direction of current flow is the same as that of the emf arrow. The current enters the load at the positive terminal, and thus the direction of current is opposite with respect to the pd arrow of the load.

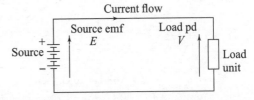

Fig. 1.8 Conventions of representing emf and pd

The unit of pd is volt and the symbol V is used to represent the pd. A volt is defined as the potential difference between two points of a conductor carrying a current of 1 A, when the power dissipated between the points is 1 W. As the pd is measured in volts, it is also termed as voltage drop.

1.7 ELECTRIC POWER

Power is defined as work done or energy per unit time. If force F newton acts for t seconds through a distance d metre along a straight line, then the work done is $F \times d$ joules. Then the power p either generated or dissipated by a circuit element can be represented by the following relationship:

$$p = \frac{F \times d}{t} = F \times u \tag{1.17}$$

where u is the velocity in m/sec.

In the case of rotating machines,

$$p = \frac{2\pi N_r T}{60} \tag{1.18}$$

where N_r is the speed of rotation of the machine in rpm (revolutions per minute) and T is the torque in N m.

$$\text{Power} = \frac{\text{work}}{\text{time}} = \frac{\text{work}}{\text{charge}} \times \frac{\text{charge}}{\text{time}} = \text{voltage} \times \text{current} \tag{1.19}$$

The unit of power is J/sec or watt (after the Scottish engineer, James Watt). The unit of energy is joule or watt-second. Commercially, the unit of energy is kilowatt-hour (kWh). It represents the work done at the rate of 1 kW for a period of 1 h. The electric supply authorities refer 1 kWh as one 'unit' for billing purposes.

Alternatively, if the current flowing between two points in a conductor is i and the voltage is v, then from the definitions of current and voltage given in Eqs (1.15) and (1.16), it is apparent that the product of current and voltage is power p dissipated between two points in the conductor carrying the current. Thus,

$$p = v \times i \tag{1.20}$$

$$= \frac{dw}{dq} \times \frac{dq}{dt}, \quad \frac{\text{joules}}{\text{coulomb}} \times \frac{\text{coulombs}}{\text{seconds}}$$

$$= \frac{dw}{dt}, \quad \frac{\text{joules}}{\text{second}} \quad \text{or} \quad \text{watts} \tag{1.20a}$$

Just like voltage, power is a signed quantity. Usually the electrical engineering community adopts the passive sign convention. As per this convention, if positive current flows into the positive terminal of an element, the power dissipated is positive, that is, the element absorbs power; while if the current leaves the positive terminal of an element, the power dissipated is negative, that is, the element delivers power.

Example 1.4 A circuit delivers energy at the rate of 30 W and the current is 10 A. Determine the energy of each coulomb of charge in the circuit.

Solution From Eq. (1.20)

$$v = \frac{p}{i} = \frac{30}{10} = 3 \text{ V}$$

Also, $$v = \frac{p}{i} = \frac{dw}{dt} \times \frac{dt}{dq} = \frac{dw}{dq}$$

$$\therefore \qquad dw = v \times dq$$

If $i = 10$ A, $dq = i \times dt = 10 \times 1 = 10$ C, then

$$dw = 3 \times 10 = 30 \text{ J}$$

Therefore, the energy of each coulomb of charge is 30/10 = 3 J.

Example 1.5 An electric motor is developing 15 kW at a speed of 1500 rpm. Calculate the torque available at the shaft.

Solution Substituting into Eq. (1.18), $15,000 \text{ W} = T \times 2\pi \times \dfrac{1500}{60}$

$\therefore \qquad T = 95.45 \text{ Nm}$

1.8 OHM'S LAW

This law is named after the German mathematician Georg Simon Ohm who first enunciated it in 1827. It states that *at constant temperature, potential difference V across the ends of a conductor is proportional to the current I flowing through the conductor*. Mathematically, Ohm's law can be stated as

$$V \propto I \quad \text{or} \quad V = R \times I$$

or $\qquad R = \dfrac{V}{I}$ (1.21)

In Eq. (1.21), R is the proportionality constant and is the resistance of the conductor. Its unit is ohm (Ω):

$\qquad 1 \, \Omega = 1 \text{ V/A}$ (1.22)

It may be noted that subsequently it was established that Ohm's law could not be applied to networks containing unilateral elements (such as diodes), or non-linear elements (such as thyrite, electric arc, etc.). A unilateral element is the one that does not exhibit the same *V-I* characteristic when the direction of the flow of current through it is reversed. Similarly, in non-linear elements the *V-I* characteristic is not linear.

Using a dc source, voltaic cell, Ohm achieved the experimental verification of his law. Later experiments with time-varying sources showed that this law is also valid when the potential difference applied across a linear resistance is time-varying. In this case, Eq. (1.21) is written as

$$v = R \times i$$ (1.23)

where v and i are instantaneous values of the potential difference and current, respectively.

1.9 BASIC CIRCUIT COMPONENTS

Resistor, inductor and capacitor are the three basic components of a network. A resistor is an element that dissipates energy as heat when current passes through it. An inductor stores energy by virtue of a current through it. A capacitor stores energy by virtue of a voltage existing across it. The behavior of an electrical device may be approximated to any desired degree of accuracy by a circuit formed by interconnection of these basic and idealized circuit elements.

1.9.1 Resistors

A resistor is a device that provides resistance in an electric circuit. As already stated in Section 1.5, ordinarily the free electrons in a conductor undergo random movement but the net movement of electrons is zero and hence this does not result in a net current flow. The free electrons in a conductor can be made to flow in a particular direction by applying an external voltage source. The application of the voltage source produces an electric field within the conductor, which produces a directed motion of free electrons. The motion of these free electrons is directed opposite to the electric field. During their motion these electrons collide with the fixed atoms in the lattice structure of the material of the conductor. Such collisions result in the production of irreversible heat loss. Thus resistance is the property of a circuit element which offers hindrance or opposition to the flow of current and in the process electric energy is converted into heat energy. Electric resistance is analogous to pipe friction in a hydraulic system and friction in a mechanical system. The resistance of a conductor opposes the current, pipe friction opposes the water flow through the pipe, and friction opposes the motion of a mechanical system, and the energy dissipated in overcoming this opposition appears as heat.

A physical device whose principal electrical characteristic is resistance is called *resistor*. A resistor is said to be linear if it satisfies Ohm's law, that is, the current through the resistor is proportional to the pd across it.

If the magnitude of resistance varies with the voltage or current, the resistor is said to be non-linear. Resistors made of semiconductor materials are non-linear resistors.

The resistance of a resistor depends on the material of which the conductor is made and the geometrical shape of the conductor. The resistance of a conductor is proportional to its length l and inversely proportional to its cross-sectional area a. Therefore, the resistance of a conductor can be written as

$$R \propto \frac{l}{a} \quad \text{or} \quad R = \frac{\rho \times l}{a} \qquad (1.24)$$

The proportionality constant ρ is called the specific resistance or resistivity of the conductor and its value depends on the material of which the conductor is made. Equation (1.24) is valid only if the current is uniformly distributed throughout the cross section of the conductor. In Eq. (1.24), if $l = 1$ m, $a = 1$ m^2, then $\rho = R$. Thus specific resistance is defined as the resistance of a conductor having a length of 1 m and a cross section of 1 m^2. The unit of resistivity can be obtained as under:

$$\rho = \frac{R \times a}{l}, \quad \frac{\text{ohm} \times \text{metre}^2}{\text{metre}} = \text{ohm-metre} \ (\Omega \, \text{m})$$

The inverse of resistance is called *conductance* and the inverse of resistivity is called *specific conductance* or *conductivity*. The symbol used to represent conductance is G and conductivity is σ. Thus, from Eq. (1.24), conductivity $\sigma = 1/\rho$, and its units are siemens per metre or mho.

$$G = \frac{1}{R} = \frac{a}{\rho l} = \frac{1}{\rho} \times \frac{a}{l} = \sigma \times \frac{a}{l} \ \text{mho} \qquad (1.25)$$

Example 1.6 Find the resistance of stranded annealed copper wire 200 m long and 25 mm^2 in cross section. Resistivity of copper is 1.72×10^{-8} Ω m.

Solution

$$R = \frac{\rho l}{a} = \frac{1.72 \times 10^{-8} \times 200}{25 \times 10^{-6}} = 0.1376 \, \Omega$$

Example 1.7 Find the resistance of the semicircular copper section, shown in Fig. 1.9, between the equipotential faces A and B. The inner radius is 6 cm, radial thickness 4 cm, and axial thickness 4 cm.

Solution The mean radius of the semicircular section is

$$6 + 2 = 8 \text{ cm} = 0.08 \text{ m}$$

Then, the mean length is $\quad l = \pi \times r = \pi \times 0.08$

Area of the cross section $\quad a = 0.04 \times 0.04 = 0.0016 \text{ m}^2$

Resistivity of copper $\quad \rho = 1.72 \times 10^{-8} \ \Omega \text{ m}$

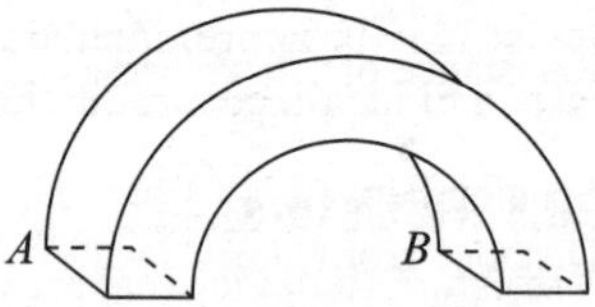

Fig. 1.9

Therefore, the resistance $\quad R = \dfrac{\rho \times l}{a} = \dfrac{1.72 \times 10^{-8} \times \pi \times 0.08}{0.0016} = 270.286 \times 10^{-8} \Omega = 2.703 \ \mu\Omega$

Example 1.8 A coil consists of 4000 turns of copper wire having a cross-sectional area of 0.8 mm^2. The mean length per turn is 80 cm. The resistivity of copper at normal working temperature is 0.02 m Wm. Calculate the resistance of the coil and the power dissipated when it is connected across a 230-V dc supply.

Solution

$$R = \frac{\rho \times l}{a} = \frac{0.02 \times 10^{-6} \times (4000 \times 80 \times 10^{-2})}{0.8 \times 10^{-6}} = 80 \, \Omega$$

Now, power dissipated $= V \times I = 230 \times 230/80 = 661.25$ W

Example 1.9   An aluminium wire 7.5 m long is connected in parallel with a copper wire 6 m long. When a current of 5 A is passed through the combination, it is found that the current in the aluminium wire is 3 A. The diameter of the aluminium wire is 1 mm. Determine the diameter of the copper wire. Resistivity of copper is 0.017 μΩ m and that of aluminium is 0.028 μΩ m.

Solution The resistance of aluminium wire is

$$R_{Al} = \frac{\rho_{Al} \times l_{Al}}{a_{Al}} = \frac{0.028 \times 10^{-6} \times 7.5}{\frac{\pi}{4}(10^{-3})^2} = 0.2675 \ \Omega$$

The potential drop across aluminium wire is $0.2675 \times 3 = 0.8025$ V. Then the potential drop across the copper wire is also 0.8025 V. Therefore,

$$\text{Resistance of copper wire} = \frac{0.8025}{2} = 0.40125 \ \Omega$$

The cross section of copper wire is

$$\frac{\rho_{Cu} \, l_{Cu}}{R_{Cu}} = \frac{0.017 \times 10^{-6} \times 6}{0.40125} = 0.2542 \times 10^{-6} \, m^2$$

Then $\frac{\pi}{4}(d_{Cu})^2 = 0.2542 \times 10^{-6}$

∴ $d_{Cu} = 0.569 \times 10^{-3} \, m = 0.569 \ mm$

Example 1.10   A porcelain cylinder 5 cm in diameter is wound with a bare high resistance wire having a resistance of 1 Ω m length and 1 mm^2 cross section. The distance between consecutive turns equals the diameter of the wire. If the external surface of the cylinder (excluding the ends) can dissipate 0.32 W/cm^2 at the permitted temperature rise, find the length of the cylinder and the diameter and length of wire for a loading of 100 W and a current of 1 A.

Solution The area required to dissipate 100 W = 100/0.32 = 312.5 cm^2

Let the length of the cylinder be L cm, length of the wire be l cm, and the diameter of the wire be d cm. Then

$$L = \frac{312.5}{\pi \times (\text{diameter of cylinder})} = \frac{312.5}{\pi \times 5} = 19.846 \ cm \approx 20 \ cm$$

$$\text{Resistance of the wire, } R = \frac{\text{Load, watts}}{(\text{current})^2} = \frac{100}{1^2} = 100 \ \Omega$$

Spacing between two consecutive turns = d cm
Distance along the axis of the cylinder between consecutive turns = $2d$ cm

$$\text{Therefore,} \quad \text{Number of turns} = \frac{L}{2d} = \frac{20}{2d} = \frac{10}{d}$$

Length of 1 turn of wire = $\pi \times 5$ cm

$$\text{Length of wire } l = \frac{10 \times 5 \times \pi}{d} = \frac{50\pi}{d} \ cm$$

Now, the resistance of wire of length 1 m and area of cross section 1 mm^2 is 1 Ω. Then,

$$\rho = \frac{1 \, \Omega \times 1 \, mm^2}{1 m} = \frac{1 \, \Omega \times 10^{-2} cm^2}{100 \ cm} = 10^{-4} \ \Omega cm$$

∴ $R = 100 = \dfrac{10^{-4} \times 50\pi \times 4}{d \times \pi d^2} = \dfrac{2 \times 10^{-2}}{d^3}$

Then, $d^3 = 2 \times 10^{-4}$ or $d = 0.058$ cm

and $l = \dfrac{50\pi}{d} = \dfrac{50\pi}{0.058} = 2700$ cm $= 27$ m

1.9.1.1 Temperature Coefficient of Resistance

All current-carrying conductors and resistors dissipate heat when carrying current. When V volts applied across a resistor of R ohm causes a current of I ampere to flow, the electrical energy absorbed by the resistor is at the rate of $V \times I$ or I^2R which is converted into heat, thereby causing a temperature rise in the resistor. When the resistor becomes warmer than its surrounding medium, it dissipates heat into the surrounding medium. Finally, when the release of heat energy is at the same rate as it receives electric energy, the temperature of the resistor no longer rises. All resistors have a power rating, which is the maximum power that can be dissipated without the temperature rise being damaging to the resistor. Thus a 4 W resistor of 100 Ω can pass a current of 20 mA, whereas a 1/4 W resistor of 100 Ω can allow only 50 mA. If the current level exceeds, the resistances are overheated and might burn.

The resistance of most conductors and all metals increases with increase in temperature. However, the resistance of carbon and insulating materials decreases with increase in temperature. Certain alloys such as constantan (60% copper and 40% nickel) and manganin (84% copper, 12% manganese, and 4% nickel) show no change in resistance for a considerable variation in temperature. This makes these alloys ideal for the construction of accurate resistances used in resistance boxes. Investigations reveal that a linear variation of resistance with temperature for copper prevails over a temperature range −50°C to 200°C. The change in resistance is usually proportional to the change in temperature. The temperature coefficient of resistance is the ratio of the change in resistance per degree change in temperature to the resistance at some definite (reference) temperature and is denoted by the Greek letter α.

Figure 1.10 shows the linear variation of the resistance of copper with the change in temperature. It may be seen from the graph that at −234.5°C its resistance becomes theoretically zero. If $R_0 = 1$ Ω is the resistance of copper at 0°C, then $R_{-234.5} = 0$ Ω at −234.5°C, and by definition the temperature coefficient of copper at 0°C, α_0 is given by

$$\alpha_0 = \dfrac{\dfrac{R_0 - R_{-234.5}}{0 - (-234.5)}}{R_0} \dfrac{\dfrac{\Omega}{\degree C}}{\Omega} = \dfrac{1}{234.5} = 0.004264/\degree C \tag{1.26}$$

In general, resistance R_2 at any temperature t_2 can be expressed in terms of resistance R_1 at temperature t_1 as

$$R_2 = R_1 [1 + \alpha_1 (t_2 - t_1) \tag{1.27}$$

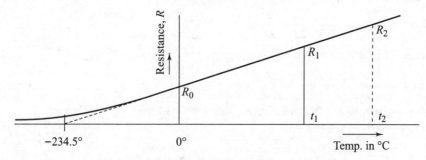

Fig. 1.10 Variation of resistance of copper with temperature

where α_1 is the temperature coefficient at temperature t_1. Suppose the reference temperature is taken as 0°C. Then

$$R_1 = R_0 (1 + \alpha_0 t_1)$$
$$R_2 = R_0 (1 + \alpha_0 t_2)$$

$\therefore \qquad R_2 = R_1 \dfrac{(1 + \alpha_0 t_2)}{(1 + \alpha_0 t_1)}$ \hfill (1.28)

Equating Eqs (1.27) and (1.28) and simplifying, the value of α_1 is given as

$$\alpha_1 = \frac{\alpha_0}{1 + \alpha_0 t_1} \qquad (1.29)$$

Similarly, the specific resistance ρ varies linearly with temperature. The expression for ρ_1, the resistivity at temperature t_1, in terms of ρ_0, the resistivity at 0°C, will be

$$\rho_1 = \rho_0 (1 + \alpha_0 t_1) \qquad (1.30)$$

Typical values of resistivity and temperature coefficients of resistances at 20°C are given in Table 1.1.

Table 1.1 Resistivity and temperature coefficient

Material	Resistivity at 20°C, Ω m	Temperature coefficient, α_{20}
Copper, annealed	1.69×10^{-8} to 1.74×10^{-8}	0.00393
Aluminium, hard drawn	2.80×10^{-8}	0.0039
Carbon	6500×10^{-8}	−0.000476
Tungsten	5.6×10^{-8}	0.0045
Manganin	48×10^{-8}	0
Constantan (Eureka)	48×10^{-8}	0

Example 1.11 A potential difference of 250 V is applied to a copper field coil at a temperature of 15°C and the current is 5 A. What will be the mean temperature of the coil when the current has fallen to 3.91 A, the applied voltage being the same as before? The temperature coefficient of copper at 0°C is 0.00426.

Solution At 15°C, $R_{15} = \dfrac{250}{5} = 50 \ \Omega$

At t°C, $R_t = \dfrac{250}{3.91} = 63.94 \ \Omega$

Then $\qquad \dfrac{R_t}{R_{15}} = \dfrac{1 + \alpha_0 \times t}{1 + \alpha_0 \times 15}$

or $\qquad \dfrac{63.94}{50} = \dfrac{1 + 0.00426 \times t}{1 + 0.00426 \times 15}$

Hence $t = 84.63$°C.

Example 1.12 If the resistance temperature coefficient of a conductor is α_1 at t_1°C, derive an expression for the temperature coefficient α_2 at t_2°C in terms of α_1 and the temperatures.

Solution From Eq. (1.29), it is seen that

$$\alpha_1 = \frac{\alpha_0}{1 + \alpha_0 t_1} \qquad \text{or} \qquad \alpha_0 = \frac{\alpha_1}{1 - \alpha_1 t_1}$$

Similarly, $\alpha_2 = \dfrac{\alpha_0}{1 + \alpha_0 t_2}$

Substitution of α_0 in the expression for α_2 results in

$$\alpha_2 = \frac{\left(\dfrac{\alpha_1}{1-\alpha_1 t_1}\right)}{1+\left(\dfrac{\alpha_1}{1-\alpha_1 t_1}\right)t_2} = \frac{\alpha_1}{1+\alpha_1(t_2-t_1)} = \frac{1}{\dfrac{1}{\alpha_1}+(t_2-t_1)}$$

1.9.2 Inductors

The electrical element that stores energy in association with a flow of current is called *inductor*. The idealized circuit model for the inductor is called an *inductance*. Practical inductors are made of many turns of thin wire wound on a magnetic core or an air core.

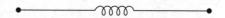

Fig. 1.11 Schematic representation of an inductor

A unique feature of the inductance is that its presence in a circuit is felt only when there is a changing current. Figure 1.11 shows a schematic representation of an inductor.

For the ideal circuit model of an inductor, the voltage across it is proportional to the rate of change of current in it. Thus if the rate of change of current is di/dt and v is the induced voltage, then

$$v \propto \frac{di}{dt}$$

or

$$v = L\frac{di}{dt} \text{ V} \tag{1.31}$$

In Eq. (1.31) the proportionality constant L is called inductance. The unit of inductance is henry, named after the American physicist Joseph Henry. Equation (1.31) may be rewritten as

$$L = \frac{v}{\dfrac{di}{dt}} = \frac{\text{volt·second}}{\text{ampere}} \quad \text{or} \quad \text{henrys (H)} \tag{1.32}$$

Equation (1.32) can be used to define inductance. If an inductor induces a voltage of 1 V when the current is uniformly varying at the rate of 1 A/sec, it is said to have an inductance of 1 H. Integrating Eq. (1.31) with respect to time t,

$$i = \frac{1}{L}\int_0^t v\,dt + i(0) \tag{1.33}$$

where $i(0)$ is the current at $t = 0$. From Eq. (1.33) it may be inferred that the current in an inductor cannot change suddenly in zero time.

Instantaneous power p entering the inductor at any instant is given by

$$p = vi = Li\frac{di}{dt} \tag{1.34}$$

When the current is constant, the derivative is zero and no additional energy is stored in the inductor. When the current increases, the derivative is positive and hence the power is positive; and, in turn, an additional energy is stored in the inductor. The energy stored in the inductor, W_L, is given by

$$W_L = \int_0^t vi\,dt = \int_0^t Li\frac{di}{dt}\times dt = L\int_0^t i\,di = \frac{1}{2}Li^2 \text{ joule} \tag{1.35}$$

Equation (1.35) assumes that the inductor has no previous history, that is, at $t = 0$, $i = 0$. The energy is stored in the inductor in a magnetic field. When the current increases, the stored energy in the magnetic field also increases. When the current reduces to zero, the energy stored in the inductor is returned to the source from which it receives the energy.

Example 1.13 A current having a variation shown in Fig. 1.12 is applied to a pure inductor having a value of 2 H. Calculate the voltage across the inductor at time $t = 1$ and $t = 3$ sec.

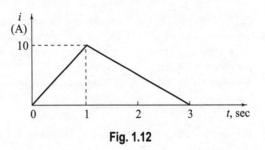

Fig. 1.12

Solution *For the period $0 \leq t \leq 1$ sec*

Current, $i = 10t$ A

Rate of change of current $\dfrac{di}{dt} = 10$ A/sec

Therefore, at $t = 1$ sec, voltage across the inductor is

$$L\frac{di}{dt} = 2 \times 10 = 20 \text{ V}$$

For the period $1 \leq t \leq 3$ sec

Rate of change of current $\dfrac{di}{dt} = -5$ A/sec

Therefore, at $t = 3$ sec, voltage across the inductor is

$$L\frac{di}{dt} = 2 \times -5 = -10 \text{ V}$$

Example 1.14 A voltage wave having the time variation shown in Fig. 1.13 is applied to a pure inductor having a value of 0.5 H. Calculate the current through the inductor at times $t = 1, 2, 3, 4, 5$ sec. Sketch the variation of current through the inductor over 5 sec.

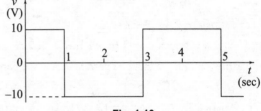

Fig. 1.13

Solution For the period $0 \leq t \leq 1$ sec, $v = 10$ V; $i(0) = 0$. The current i may be expressed using Eq. (1.33) as

$$i = \frac{1}{L}\int_0^t v \, dt + i\,(0) = \frac{1}{0.5}\int_0^t 10 \, dt = 20\int_0^t dt = 20\,t$$

Then at $t = 1$ sec, $i = 20 \times 1 = 20$ A

For the period $1 \leq t \leq 3$ sec, $v = -10$ V; $i(1) = 20$ A, then current

$$i = \frac{1}{L}\int_1^t v \, dt + i\,(1) = \frac{1}{0.5}\int_1^t -10 \, dt + 20 = -20\int_1^t dt + 20 = -20(t-1) + 20$$

Then at $t = 2$ sec, $\quad i = -20 \times (2-1) + 20 = -20 + 20 = 0$ A

And at $\quad t = 3$ sec, $\quad i = -20 \times (3-1) + 20 = -40 + 20 = -20$ A

For the period $3 \leq t \leq 5$ sec, $v = 10$ V; $i(3) = -20$ A,

$$i = \frac{1}{L}\int_3^t v \, dt + i(3) = \frac{1}{0.5}\int_3^t 10 \, dt - 20$$

$$= 20\int_3^t dt - 20 = 20\,(t-3) - 20$$

Then at $t = 4$ sec, $\quad i = 20 \times (4-3) - 20$

$$= 20 - 20 = 0 \text{ A}$$

And at $\quad t = 5$ sec, $\quad i = 20 \times (5-3) - 20$

$$= 40 - 20 = 20 \text{ A}$$

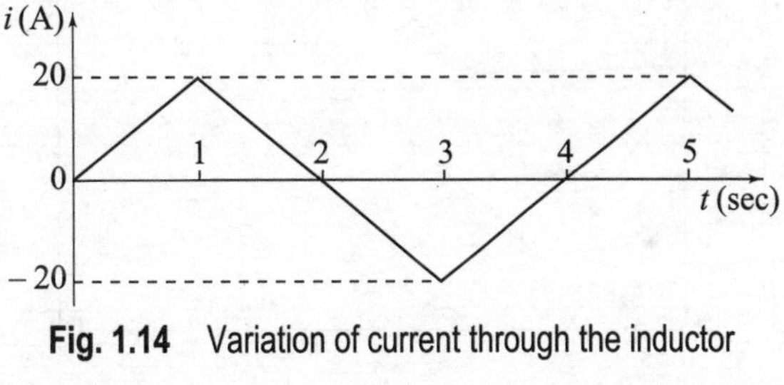

Fig. 1.14 Variation of current through the inductor

Example 1.15 The voltage waveform shown in Fig. 1.15 is applied across an inductor of 5 H. Derive an expression for current in the circuit and sketch the current and energy waveforms against time. Assume zero initial condition in the circuit.

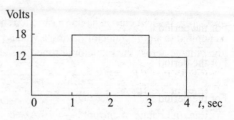

Fig. 1.15

Solution Using Eq. (1.31), the generalized relation for current through the inductor is written as $i = \dfrac{1}{5}\displaystyle\int_0^t v\,dt$

For the period $0 \le t \le 1$ sec, $v = 12$ V. Therefore, the current through the inductor is given by

$$i = \frac{1}{5}\int_0^t 12\,dt = 2.4t + i(0)$$

At $t = 0$, $i(0) = 0$.

Thus, $i = 2.4t$

For the period $1 \le t \le 3$ sec, $v = 18$ V. Therefore, the current through the inductor is given by

$$i = \frac{1}{5}\int_0^t 18\,dt = 3.6t + i(1.0)$$

At $t = 1.0$ sec, $i = 2.4$ A. Hence,

$$2.4 = 3.6 \times 1.0 + i(1.0) \text{ or } i(1.0) = -1.2 \text{ A}$$

The expression for the inductor current during the period $1 \le t \le 3$ sec is

$$i = 3.6t - 1.2$$

For the period $3 \le t \le 4$ sec, $v = 12$ V. Hence, the current through the inductor is expressed as

$$i = \frac{1}{5}\int_0^t 12\,dt = 2.4t + i(3.0)$$

At $t = 3.0$ sec, $i = 9.6$ A. Hence,

$$9.6 = 2.4 \times 3.0 + i(3.0) \text{ or } i(3.0) = 2.4 \text{ A}$$

$\therefore$ $i = 24t + 2.4$

Energy stored in the various periods is as follows.

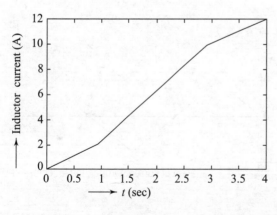

Fig. 1.16

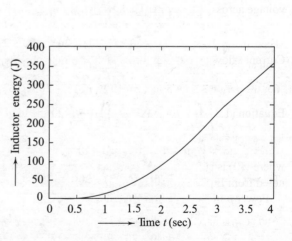

Fig. 1.17

For the period $0 \leq t \leq 1$ sec, $W_L = \dfrac{1}{2} \times 5 \times (2.4t)^2 = 14.4t^2$ J

For the period $1 \leq t \leq 3$ sec, $W_L = \dfrac{1}{2} \times 5 \times (3.6t - 1.2)^2 = 32.4t^2 - 21.6t + 3.6$ J

For the period $3 \leq t \leq 4$ sec, $W_L = \dfrac{1}{2} \times 5 \times (2.4t + 2.4)^2 = 14.4t^2 + 28.8t + 14.4$ J

The variation of inductor current and energy with time is sketched in Figs 1.16 and 1.17.

1.9.3 Capacitors

A *capacitor* is a device that can store energy in the form of a charge separation when it is suitably polarized by an electric field by applying a voltage across it. In the simplest form, a capacitor consists of two parallel conducting plates separated by air or any insulating material, such as mica. It has the characteristic of storing electric energy (charge), which can be fully retrieved, in an electric field. A significant feature of the capacitor is that its presence is felt in an electric circuit when a changing potential difference exists across the capacitor. The presence of an insulating material between the conducting plates does not allow the flow of dc current; thus a capacitor acts as an open circuit in the presence of dc current. Figure 1.18 shows the schematic representation of a capacitor.

The ability of the capacitor to store charge is measured in terms of capacitance C. *Capacitance* of a capacitor is defined as charge stored per volt applied and its unit is farad (F). However, for practical purposes the unit of farad is too large. Hence, microfarad (μF) is used to specify the capacitance of the components and circuits.

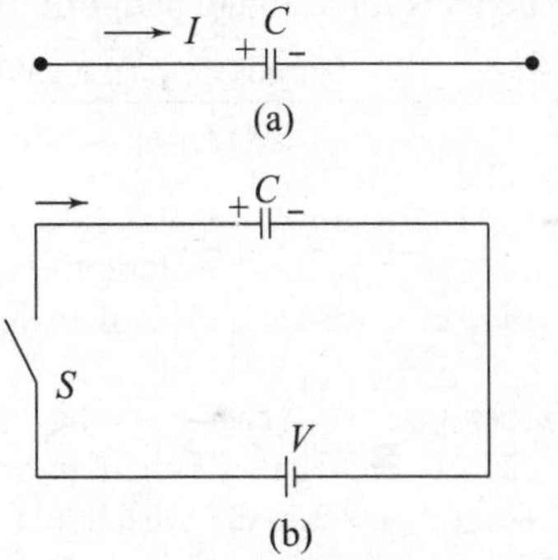

Fig. 1.18 (a) Schematic representation of a capacitor and (b) capacitor across a dc source

In Fig. 1.18(b), it is assumed that the charge on the capacitor at any time t after the switch S is closed is q coulombs and the voltage across it is v volts. Then by definition

$$C = \frac{q}{v} \text{ coulomb} \tag{1.36}$$

Current i flowing through the capacitor can be obtained as

$$i = \frac{dq}{dt} = C \frac{dv}{dt} \text{ ampere} \tag{1.37}$$

Equation (1.37) is integrated with respect to time to get the voltage across the capacitor as

$$v = \frac{1}{C} \int_0^t i \, dt + v \ (0) \tag{1.38}$$

where $v(0)$ is an integration constant which defines the initial voltage across the capacitor at $t = 0$. It may be noted from Eq. (1.38) that the voltage across a capacitor cannot change instantaneously, that is, in zero time.

Power p in the capacitor is given as

$$p = vi = Cv \frac{dv}{dt} \text{ watt} \tag{1.39}$$

Energy stored in the capacitor, W_C, is given by

$$W_C = \int p \, dt = C \int v \, dv = \frac{1}{2} Cv^2 \text{ joule} \tag{1.40}$$

From Eq. (1.40) it is evident that the energy stored in the capacitor is dependent on the instantaneous voltage and is returned to the network when the voltage is reduced to zero.

As stated earlier, a capacitor consists of two electrodes (plates) separated by an insulating material (dielectric). If the area of the plates is A m^2 and the distance between them is d m, it is observed that

$$C \propto A \quad \text{and} \quad C \propto \frac{1}{d}$$

$$\therefore \quad C = \frac{\varepsilon A}{d} \tag{1.41}$$

where ε is the absolute permittivity constant. The absolute permittivity constant depends on the type of dielectric employed in the capacitor. The ratio of the absolute permittivity constant of the dielectric ε to the permittivity constant of vacuum ε_0 is called relative permittivity ε_r, that is,

$$\varepsilon_r = \frac{\varepsilon}{\varepsilon_0}$$

Hence, $\varepsilon = \varepsilon_0 \varepsilon_r$.

The units for absolute permittivity ε can be established from Eq. (1.41) as under:

$$\varepsilon = \frac{C \text{ (farads)} \times d \text{ (metres)}}{A \text{ (metres)}^2} = \frac{C \times d}{A} \quad \text{farads/metre (F/m)}$$

Based on experimental results, the value of the permittivity constant of vacuum has been found to be equal to 8.84×10^{-12} F/m. Therefore, the value of ε_r for vacuum is 1.0 and for air is 1.0006. For practical purposes, the value of ε_r for air is also taken as 1.

Example 1.16 A voltage wave having a time variation of 20 V/sec is applied to a pure capacitor having a value of 25 μF. Find (a) the current during the period $0 \leq t \leq 1$ sec, (b) charge accumulated across the capacitor at $t = 1$ sec, (c) power in the capacitor at $t = 1$ sec, and (d) energy stored in the capacitor at $t = 1$ sec.

Solution (a) Current through the capacitor i may be obtained using Eq. (1.37) as

$$i = C\frac{dv}{dt} = 25 \times 10^{-6} \times 20 = 500 \text{ μA}$$

(b) At $t = 1$sec, $v = 20$V. Charge q at $t = 1$ sec may be obtained using Eq. (1.36) as

$$q = C v = 25 \times 10^{-6} \times 20 = 500 \text{ μC}$$

(c) At $t = 1$sec, power $p = v \times i = 20 \times 500 \times 10^6 = 1 \times 10^{-2}$ W

(d) At $t = 1$sec, energy stored in the capacitor, W_C, can be obtained using Eq. (1.40) as

$$W_C = \frac{1}{2}Cv^2 = \frac{1}{2} \times 25 \times 10^{-6} \times (20)^2 = 5 \times 10^{-3} \text{ J}$$

Example 1.17 A current having variation shown in Fig. 1.19 is applied to a pure capacitor having a value of 5 μF. Calculate the charge, voltage, power, and energy at time $t = 2$ sec.

Solution For the period $0 \leq t \leq 1$sec, $i = 100 \times 10^{-3} t = 0.1t$ A

At $t = 1$sec, $q = \int_0^t i\, dt = \int_0^t 0.1\, t\, dt = 0.1 \times \left[\frac{t^2}{2}\right]_{t=0}^{t=1} = 0.05[t^2]_{t=0}^{t=1} = 0.05[1-0] = 0.05$ C

$$v = \frac{q}{C} = \frac{1}{C}\int_0^t 0.1\, t\, dt = \frac{0.05\, t^2}{500 \times 10^{-6}}$$

$$= 100\, t^2 = 100 \text{ V}$$

where $t = 1$ sec,

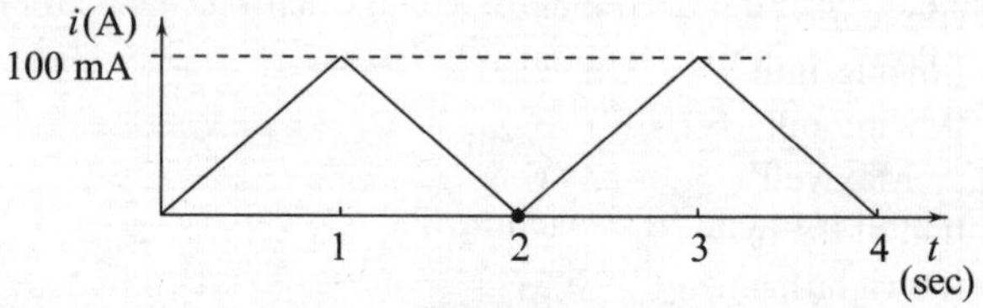

Fig. 1.19

$$p = v \times i = 100 \times 0.1 = 10 \text{ W}$$

$$W_C = \int_0^t vi\,dt = \int_0^1 100t^2 \times 0.01t \; dt = \int_0^1 t^3 dt = \left[\frac{t^4}{4}\right]_{t=0}^{t=1} = \frac{1}{4}[1-0] = 0.25 \text{ J}$$

For the period $0 \leq t \leq 1$ sec, $i = 0.2 - 0.1t$ A

At $t = 2$ sec, Charge $q = q_{t=1} + \int_1^t idt = 0.05 + \int_1^2 (0.2 - 0.1t)dt = 0.05 + \left[0.2t - 0.1 \times \frac{t^2}{2}\right]_{t=1}^{t=2}$

$$= 0.05 + [0.2(2-1) - 0.05(2^2 - 1^2)] = 0.05 + 0.05 = 0.1 \text{ C}$$

Voltage $\quad v = \dfrac{q}{C} = \dfrac{1}{C}\left[0.05 + \int_1^2 (0.2 - 0.1t)dt\right]$

$$= \frac{1}{500 \times 10^{-6}}[0.05 + 0.2t - 0.05t^2]_{t=1}^{t=2}$$

$$= \frac{10^6}{500}[0.05 + 0.2(2-1) - 0.05(2^2 - 1^2)] = \frac{10^6}{500} \times 0.1 = 200 \text{ V}$$

Power $\quad p = v \times i = 200 \times 0 = 0$ W

Energy $\quad W_C = W_{C.t=1} + \displaystyle\int_1^2 vi\,dt$

$$= 0.25 + \int_1^2 \left[\frac{1}{500 \times 10^{-6}}[0.05 + 0.2t - 0.05t^2] \times (0.2 - 0.1t)\right] dt$$

$$= 0.25 + \frac{10^6}{500}\int_1^2 [0.01 + 0.035t - 0.03t^2 + 0.005t^3]dt$$

Energy $\quad W_C = 0.25 + \dfrac{10^6}{500}\left[0.01t + 0.035 \times \dfrac{t^2}{2} - 0.03 \times \dfrac{t^3}{3} + 0.005 \times \dfrac{t^4}{4}\right]_{t=1}^{t=2}$

$$= 0.25 + \frac{10^6}{500} \times 0.01125 = 0.25 + 22.50 = 22.75 \text{ J}$$

1.10 ELECTROMAGNETISM RELATED LAWS

Some laws related to electromagnetism are discussed in this section.

1.10.1 Magnetic Field Due to Electric Current Flow

Hans Christian Oersted, a Danish physicist, in 1831 discovered that current flowing in a conductor generates a magnetic field all around it. He proved that the magnetic lines of force due to the current flow in a conductor were concentric circles closed on themselves as shown in Fig. 1.20. The direction of the magnetic lines of force depends upon the direction of current. The convention adopted to show the direction of the current flow is that current flowing into the plane of the paper is indicated by a cross sign and current flowing out of the plane of the paper is shown by a dot.

Maxwell's corkscrew rule is a convenient method of determining the direction of the magnetic field set up by a current-carrying conductor. It states that if a right-handed corkscrew is placed along the direction of the current flow,

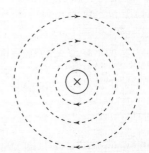

Fig. 1.20 Magnetic field due to current in a straight conductor

the direction of motion of the hand, which would advance the screw in the direction of the current, gives the direction of magnetic field as shown in Fig. 1.21(a).

Another method of determining the direction of the field is to employ the right hand rule. Imagine the current-carrying conductor to be held in the right hand with the thumb pointing in the direction of the current with the fingers wrapped around it. Then the direction of the fingers points in the direction of the magnetic field.

Magnetic field of a solenoid   When a current-carrying conductor is given the shape of circular coils of the conductor placed side by side and insulated from one another, it is called a *solenoid* [see Fig. 1.21(b)]. The magnetic field is represented by the dotted lines. If the fingers of the right hand are wrapped around the current-carrying conductor with the fingers pointing in the direction of the current, then the thumb outstretched parallel to the axis of the solenoid points in the direction of the magnetic field inside the solenoid.

If an iron rod is placed inside the solenoid coil, as shown in Fig. 1.21(b), and the coil is connected to a voltage source, the iron rod is magnetized and behaves like a magnet. The magnetic field becomes hundreds of times stronger.

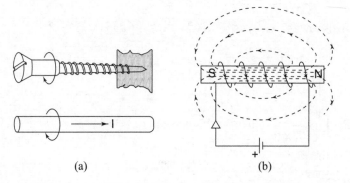

(a)   (b)

Fig. 1.21   (a) Maxwell's corkscrew rule and (b) solenoid with a magnetic core

1.10.2 Force on a Current-carrying Conductor Placed in a Magnetic Field

If a current-carrying conductor is placed at right angles to the lines of force of a magnetic field, a mechanical force will be exerted on the conductor. The magnitude of the mechanical force can be calculated by using *Ampere's law*.

1.10.2.1 Ampere's Law

When a straight elemental conductor of length l metres and carrying current I_2 amperes is placed in the same horizontal plane at a distance of r metres from a straight, long conductor carrying current I_1 amperes in the opposite direction to I_2, the small conductor experiences a force of repulsion, F, given by

$$F = \frac{\mu I_1}{2\pi r} I_2 l \;\; \text{newton} \tag{1.42}$$

$$= B I_2 l \;\; \text{newton} \tag{1.43}$$

where $B = \dfrac{\mu I_1}{2\pi r}$ tesla (T) $\tag{1.44}$

In Eq. (1.44) μ is a scalar constant of the medium (called permeability of the medium) and B is the magnetic flux density. From Eq. (1.43) it may be noted that the unit of flux density is taken as the density of the magnetic field such that a conductor carrying 1 A at right angles to that field experiences a force of 1 Nm. The unit is named tesla (T) after the famous electrical inventor Nikola Tesla. The units of B may also be found as follows:

$$B = \frac{newton}{ampere \times metre} = \frac{joule}{metre \times ampere \times metre}$$

$$= \frac{watt \times second}{ampere \times metre^2} = \frac{volt \times second}{metre^2} = \frac{weber}{metre^2} \quad or \quad Wb/m^2$$

From Faraday's law, it is shown later in Section 1.10.3 that weber = volt × second.

For a magnetic field having a cross-sectional area A m^2 and a uniform flux density B tesla, the total flux ϕ is given by

$$\phi = B \times A \tag{1.45}$$

and

$$B = \frac{\phi}{A} \ Wb/m^2 \tag{1.46}$$

Then, for $\phi = 1$ Wb,

$$A = 1 \ m^2, \ B = 1, \ T = 1 \ Wb/m^2$$

Figure 1.22(a) shows a part of a magnetic field with lines of force in the plane of the paper. Figure 1.22(b) shows a conductor arranged at right angles to the paper and carrying a current whose direction is inwards and produces a magnetic field around the conductor in the plane of the paper. Figure 1.22(c) shows the combined effect where the conductor is situated in the magnetic field of Fig. 1.22(a). On the right hand side of the conductor, the two fields, both due to the permanent magnet and the current-carrying conductor, are in the same direction, whereas on the left hand side they are in opposition. Thus the effect of the current is to transfer some of the lines of force from the left hand side to the right hand side of the conductor, resulting in bending of some of them, as shown. Since the flux lines behave like lastic cords and tend to return to the shortest path, the conductor experiences a mechanical force towards the left.

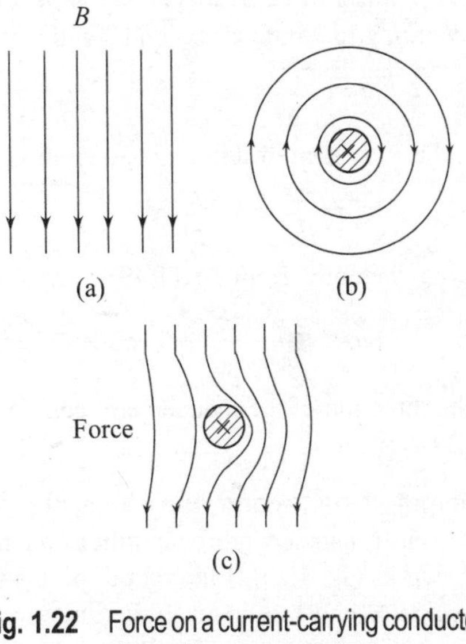

The magnitude of the force on the conductor, F, in the case of a conductor of length l metres arranged at right angles to the magnetic field B tesla or, Wb/m^2, and carrying current I is given by Eq. (1.43) as

$$F = BIl \ newton \tag{1.47}$$

Fig. 1.22 Force on a current-carrying conductor placed in a magnetic field

The direction of the force can be determined by Fleming's left hand rule, which is illustrated in Fig. 1.23. The rule states that if the thumb, forefinger, and the middle finger of the left hand are stretched and held at right angles to each other, with the middle finger pointing in the direction of current flow and the forefinger in the direction of the magnetic field, then the thumb will point in the direction of force on the conductor. As an aid to apply Fleming's left hand rule it may be remembered that the forefinger and field associate with each other, the middle finger has an **i** in middle, *i* also represents current, and finally thumb contains an **m** that is associated with motion.

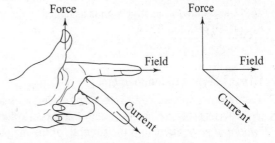

Fig. 1.23 Left-hand rule

1.10.3 Faraday's Laws of Electromagnetic Induction

Michael Faraday in 1831 experimentally demonstrated that a variable magnetic field could produce an electric field. He showed that when a conductor is moved in a stationary magnetic field, an emf is produced between the ends of the conductor; and if the ends of the conductor are joined by a wire, an induced current flows through the conductor. Alternatively, a stationary conductor when placed in a magnetic field that changes with time develops an emf across it. This phenomenon is called *electromagnetic induction*. Based on this, Faraday developed the following laws.

Faraday's first law   This law states that whenever the magnetic flux changes with respect to an electrical conductor or a coil, an emf is induced in the conductor.

Faraday's second law   This law states that the magnitude of the emf induced in the conductor or the coil by electromagnetic induction is directly proportional to the time rate of change of the flux linkages.

The term flux linkages merely means the product of flux in webers and the number of turns with which the flux is linked. Let the magnetic flux through a coil of N turns be increased by $\Delta\phi$ webers in Δt seconds, then according to Faraday's second law the magnitude of the induced emf, e, in the coil will be given by

$$e \propto \frac{N\Delta\phi}{\Delta t} \propto \frac{d\psi}{dt} \qquad (1.48)$$

In the SI system of units, e is given in volts and the constant of proportionality is unity. Hence,

$$e = \frac{d\psi}{dt} = N\frac{d\phi}{dt}\,\text{volt} \qquad (1.49)$$

It may be noted from Eq. (1.49)

$$\text{volt} = \frac{\text{weber}}{\text{second}} \quad \text{or} \quad \text{volt} \times \text{second} = \text{weber}$$

The direction of the induced emf can be determined in two ways, namely, (i) Fleming's right hand rule and (ii) Lenz's law.

Fleming's right hand rule   To apply the right hand rule, hold the thumb, forefinger, and second (or middle) finger at right angles to each other as shown in Fig. 1.24. If the thumb points in the direction of motion of the conductor and forefinger in the direction of the magnetic field, then the second finger gives the direction of the induced emf (voltage).

Lenz's law   German physicist Heinrich Lenz in 1834 enunciated a simple rule, presently known as Lenz's law. The law states that the direction of the induced emf is always such that it tends to establish a current which opposes the change of flux responsible for inducing that emf. In accordance with Lenz's law, a negative sign is assigned to the expression for emf and Eq. (1.49) get modified as

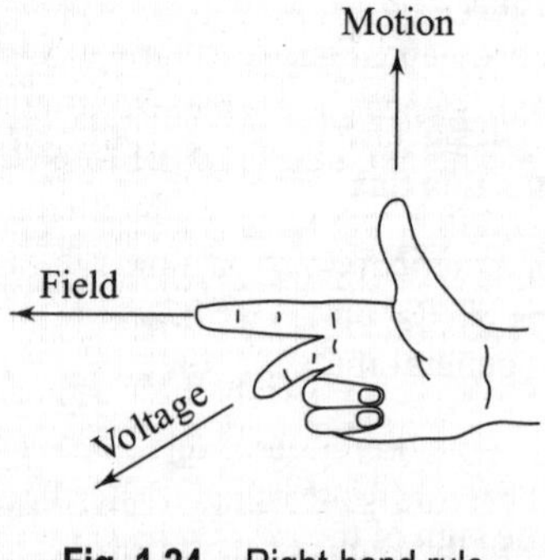

Fig. 1.24   Right-hand rule

$$e = -N\frac{d\phi}{dt} \qquad (1.50)$$

1.10.3.1 Types of Induced emf

There are two different ways of producing emf in accordance with Faraday's laws. First, emf may be produced by a relative movement of a conductor or a coil with respect to the magnetic field. Such an emf is called *dynamically induced emf (motional emf)*. Second, if the strength of the magnetic field is varied without changing its orientation and a conductor or a coil is placed in this field, an emf is induced in the conductor or the coil. This emf is called *statically induced emf (transformer emf)*.

Dynamically induced emf in a conductor   Figure 1.25 shows an isometric view of a pair of NS poles and the plan of a conductor AA placed in the air gap between the pair of poles. If the conductor length be l metres and current I amperes flows axially into the plane of the paper and B is the flux density in tesla, the conductor experiences a force BIl newtons tending to move the conductor to the left (Lenz's law). Thus a force of this magnitude is to be applied in the opposite direction to move the conductor from position A to position B through distance d metres. Work done, W, in moving the conductor is given by

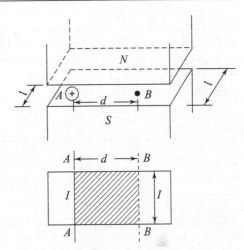

$$W = BIl \times d \quad \text{N m} \quad \text{or} \quad \text{J}$$

If the movement of conductor A from position A to B takes place at a uniform velocity in t seconds, so as to cut at right

Fig. 1.25   Current-carrying conductor moving across a magnetic field

angles the lines of force of the uniform magnetic field of the air gap, then a constant emf, say E volts, is induced in it. Electrical power generated due to the movement of the conductor is $E \times I$ watts and the corresponding energy produced is $E \times I \times t$ W/sec or joules. As the mechanical energy required for moving the conductor horizontally in the air gap is converted into electrical energy, the following equation results:

$$E \times I \times t = BIl \times d$$

or $\qquad E = Bl \times \dfrac{d}{t} = Blu$ $\qquad\qquad\qquad\qquad\qquad\qquad\qquad\qquad$ (1.51)

where u is the velocity in m/sec.

If the conductor velocity u makes an angle θ with the direction of the magnetic field then the emf induced is given by

$$E = Blu \sin \theta \qquad\qquad\qquad\qquad\qquad\qquad\qquad\qquad (1.52)$$

In Eq. (1.51), Bld is the total magnetic flux, ϕ webers, in the area shown shaded in Fig. 1.25. The conductor cuts this flux ϕ when it moves from AA to BB in t seconds. Thus

$$E \text{ volts} = \dfrac{\phi}{t} \quad \text{Wb/sec} \qquad\qquad\qquad\qquad\qquad\qquad (1.53)$$

In general if a conductor cuts a flux of $d\phi$ webers in dt seconds, then the generated emf e volts is given as

$$e = \dfrac{d\phi}{dt} \text{ V} \qquad\qquad\qquad\qquad\qquad\qquad\qquad\qquad (1.54)$$

The sign of the emf induced can be determined from the physical considerations. Therefore, the negative sign in Eq. (1.54) is left out.

Magnitude of induced emf in a coil   Assume a coil of N turns. The flux in it is increased by $d\phi$ webers in dt seconds by moving a permanent magnet towards the coil. Since there are N turns in the coil, all the turns link this flux. From Eq. (1.49) the emf developed, e, may be written

$$e = N\dfrac{d\phi}{dt} = \dfrac{d\psi}{dt} \qquad\qquad\qquad\qquad\qquad\qquad\qquad (1.55)$$

ψ is called the flux linkage and is equal to $N\phi$.

Example 1.18   A straight conductor 100 cm long and carrying a direct current of 50 A lies perpendicular to a uniform magnetic field of 1.5 Wb/m² (tesla). Find (a) the mechanical force on the conductor, (b) the mechanical power in watts to move the conductor against the force at a uniform speed of 5 m/sec, and (c) the electromotive force generated in the conductor.

Solution From Eq. (1.47), $\mathcal{F} = BIl$

(a) Force on the conductor is 1.5 [T] × 50 [A] × 1 [m] = 75 N

(b) The mechanical power to move the conductor against the force is $\mathcal{F} \times u$ = 75 [N] × 5 [m/sec] = 375 W

(c) From Eq. (1.51) $E = B \times l \times u$

The emf generated is 1.5 [T] × 1 [m] × 5 [m/sec] = 7.5 V

Example 1.19   A wire of length 50 cm moves in a direction at right angles to its length at 40 m/s in a uniform magnetic field of density 1.5 Wb/m². Calculate the electromotive force induced in the conductor when the direction of motion is (a) perpendicular to the field, (b) inclined at 30° to the direction of the field.

Solution From Eq. (1.52), $E = B \times l \times u \times \sin \theta$

When $\theta = 90°$, force on the conductor is 1.5 [T] × 0.5 [m] × 40 [m/sec] × sin 90° = 30 V

When $\theta = 30°$, force on the conductor is 1.5 [T] × 0.5 [m] × 40 [m/sec] × sin 30° = 15 V

1.11 KIRCHHOFF'S LAWS

Gustav Robert Kirchhoff (1824–1887), a German physicist, published the first systematic description of the laws of circuit analysis. These laws are known as *Kirchhoff's current law (KCL)* and *Kirchhoff's voltage law (KVL)*. His contribution forms the basis of all circuit analyses problems.

   Kirchhoff's current law states that the algebraic sum of the currents at a node (junction) in a network at any instant of time is zero. KCL may be expressed mathematically as

$$\sum_{j=1}^{n} i_j = 0 \tag{1.56}$$

where i_j represents current in the *j*th element and *n* is the number of elements connected to the node *k*. This means that the algebraic sum of the currents meeting at a junction is zero. If the currents entering the node are taken as positive, then the currents leaving the node are negative, or vice versa. The KCL may be thought of to be a consequence of the conservation of electric charge—charge cannot be created nor destroyed but must be conserved.

   As an example of KCL, consider the node *k* in Fig. 1.26 where currents i_1, i_2, i_3, i_4, and i_5 flowing in the five branches meet. For node *k*, KCL may written in the form

$$-i_1 + i_2 - i_3 + i_4 - i_5 = 0 \tag{1.57}$$

or $$i_2 + i_4 = i_1 + i_3 + i_5 \tag{1.58}$$

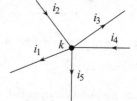

Fig. 1.26   Applications of KCL

   In Eq. (1.58) currents i_2 and i_4 are flowing towards node *k* and hence a positive sign is assigned to these currents while the currents i_1, i_3, and i_5, which leave node *k*, are negative.

Example 1.20   For the circuit shown in Fig. 1.27 determine the value of i_5 for the following values of voltages: $v_1 = 3 \sin t$, $v_2 = 10 \sin t$, $v_3 = 10 \cos t$, $i_4 = \cos t$.

Solution From Eq. (1.37)

$$i_1 = \frac{1}{3}\frac{d}{dt}(3 \sin t) = \cos t$$

Applying Ohm's law,

$$i_2 = \frac{10\sin t}{5} = 2\sin t$$

From Eq. (1.31)

$$10\cos t = 2\,\frac{di_3}{dt}$$

$$\therefore \qquad i_3 = 5\int \cos t\, dt = 5\sin t$$

Applying KCL to the node

$$-i_1 + i_2 - i_3 - i_4 + i_5 = 0$$

$$\therefore \qquad i_5 = i_1 - i_2 + i_3 + i_4$$

$$= \cos t - 2\sin t + 5\sin t + \cos t$$

$$= 3\sin t + 2\cos t$$

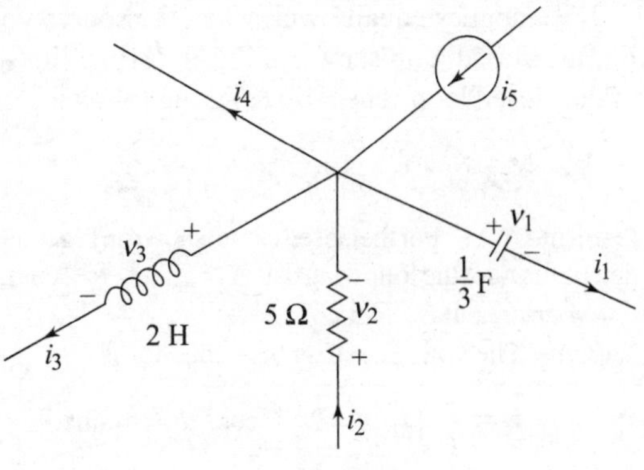

If $3 = K\cos\varphi$ and $2 = K\sin\varphi$, then $K = 5$ and $\varphi = \tan^{-1}(2/3)$.

Thus, $i_5 = 5\sin[t + \tan^{-1}(2/3)]$.

Fig. 1.27

Kirchhoff's voltage law states that at any instant of time the sum of voltages in a closed circuit is zero. KVL may be expressed mathematically as

$$\sum_{j=1}^{n} v_j = 0 \tag{1.59}$$

where v_j represents the individual voltage in the jth element around the closed circuit having n elements.

If the voltage drop from the positive polarity to the negative polarity is assigned a positive sign, then the voltage rise from the negative polarity to the positive polarity is assumed negative, or vice versa. KVL is a consequence of the fact that no energy is lost or created in an electric circuit. In other words, KVL states that in a closed loop, at any instant of time, the algebraic sum of the emfs acting around the loop is equal to the algebraic sum of the pds around the loop.

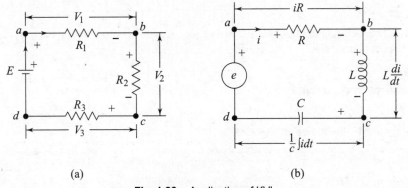

(a)   (b)

Fig. 1.28 Application of KVL

For the closed loop shown in Fig. 1.28(a) the dc source (battery) causes a constant current flow in the loop. Applying KVL, using the sign convention for the positive direction of voltage drop and emf, the following expression can be written:

$$E = V_1 + V_2 + V_3$$

or $\qquad V_1 + V_2 + V_3 - E = 0 \tag{1.60}$

KVL applies equally well when the source voltage is time-varying, causing a time-varying current flow. For the closed loop shown in Fig. 1.28(b) a time-varying voltage source e produces a time-varying current i in the closed loop. Then, by KVL the following expression can be written:

$$e = iR + L\,\frac{di}{dt} + \frac{1}{C}\int i\,dt \tag{1.61}$$

Example 1.21 For the closed circuit shown in Fig. 1.29 determine i_3 for the following data: $v_1 = 5\sin t$, $i_2 = 2\cos t$, $v_4 = 4\cos t$, and $i_5 = \sin t$.

Solution The voltage across the capacitor is

$$v_2 = \frac{1}{C}\int i_2\,dt = 2\int 2\,\cos t\,dt = 4\,\sin t,$$

and $v_5 = R \times i_5 = 4\sin t$

Applying KVL around the closed circuit shown in Fig. 1.29,

$$-v_1 + v_2 - v_3 - v_4 + v_5 = 0$$

or $v_3 = -v_1 + v_2 - v_4 + v_5 = -5\sin t + 4\sin t - 4\cos t + 4\sin t = 3\sin t - 4\cos t$

Now, $v_3 = L\dfrac{di_3}{dt} = \dfrac{1}{3}\dfrac{di_3}{dt}$

$$i_3 = 3\int (3\sin t - 4\cos t)\,dt = -9\cos t - 12\sin t = -(9\cos t + 12\sin t)$$

Suppose $12 = K\cos \varphi$ and $9 = K\sin \varphi$. Then $K = 15$ and $\varphi = \tan^{-1}(3/4)$. Therefore,

$$i_3 = -15(\sin \varphi \cos t + \cos \varphi \sin t) = -15\sin (t + \varphi)$$

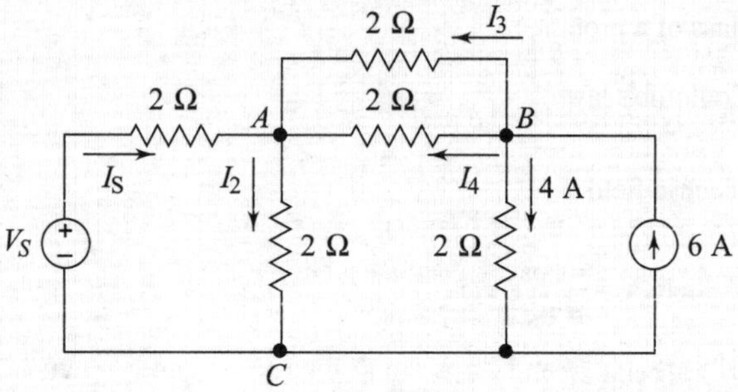

Example 1.22 For the circuit shown in Fig. 1.30 determine the values of I_2 and V_S.

Solution Let the node C be taken as the reference node in Fig. 1.30. Applying KCL to node B,

$$I_3 + I_4 = 6 - 4 = 2\,\text{A}$$

Now, $I_3 = I_4 = 1$ A, being current through two 2 Ω resistances in parallel across nodes A and B. Then the potential of node B with respect to node C is

$$V_B = 2 \times 4 = 8\,\text{V}$$

The voltage drop across nodes B and A,

$$V_{BA} = I_4 \times 2 = 1 \times 2 = 2\,\text{V}$$

Then the potential of node A with respect to node C is

$$V_A = V_B - V_{BA} = 8 - 2 = 6\,\text{V}$$

Therefore, $I_2 = \dfrac{V_A}{2} = \dfrac{6}{2} = 3\,\text{A}$

Applying KCL to node A,

$$I_S + I_3 + I_4 - I_2 = 0$$

$\therefore$ $I_S = 3 - 1 - 1 = 1\,\text{A}$

Then, $V_S = V_A + 2 \times I_S = 6 + 2 \times 1 = 8$ V.

Fig. 1.30

Recapitulation

Charge on an electron, $e = -1.602 \times 10^{-19}$ C

Mass of an electron $= 9.11 \times 10^{-31}$ kg

Charge on a proton $= 1.602 \times 10^{-19}$ C

Mass of a proton $= 1.673 \times 10^{-27}$ kg

Coulomb's law: $F = \dfrac{1}{4\pi\varepsilon} \dfrac{q_1 q_2}{r^2}$ newton (or N)

Electric field intensity, $E = \dfrac{F}{q}$ N/C or V/m

Voltage, $v = \dfrac{dw}{dq}$ J/C

Electric flux density, $D = \dfrac{q}{4\pi r^2} = \varepsilon E$

Gauss' law: $\oint D \, ds = \Sigma q$

Current, $I = \dfrac{Q}{t}$ C/sec or A

Quantity of electricity, $Q = I \times t$ C

Potential difference $V = \dfrac{P}{I} = \dfrac{W}{Q}$

Ohm's law: $I = \dfrac{V}{R}$

Resistance of a conductor, $R = \dfrac{\rho l}{a}$ Ω

Resistivity $\rho = \dfrac{R \times a}{l}$ Ω m

Resistance of a conductor at (t_1)°C, $R_1 = R_0 (1 + a_0 t_1)$

Temperature coefficient of resistance at t_1, $\alpha_1 = \dfrac{\alpha_0}{1 + \alpha_0 t_1}$

Temperature coefficient of resistivity, $\rho_1 = \rho_0 (1 + \alpha_0 t_1)$

Conductance of a conductor, $G = \dfrac{\sigma \times a}{l}$

Capacitance $C = \dfrac{q}{V}$ farad

Energy stored in a capacitor, $W_C = \dfrac{1}{2} CV^2$

Voltage induced in an inductor, $v = L\dfrac{di}{dt}$ V

Inductance of an inductor, $L = \dfrac{v}{di/dt}$ H

Energy stored in an inductor, $W_L = \dfrac{1}{2} Li^2$ J

Flux density, $B = \dfrac{\phi}{A}$ tesla or Wb/m^2

Instantaneous value of induced voltage,

$$e = -\dfrac{d(N\phi)}{dt} = -\dfrac{d\phi}{dt} = -N\dfrac{d\phi}{dt}$$ V

Assessment Questions

1. Write a short essay on the fundamental nature of electricity.
2. Describe the atomic structure and therefrom distinguish between (a) conductors, (b) semi-conductors, and (c) insulators.
3. Define Coulomb's law and electric field intensity.
4. Distinguish between (a) electric potential and potential difference and (b) electric flux and electric flux density.
5. Define Gauss' law and describe the electric field set up due to a long straight charged conductor.
6. Derive expressions for electric fields set up between (a) two charged parallel plates and (b) a uniformly charged sphere.
7. Describe the nature of current and express the co-relation between charge and current.
8. Explain (a) emf and (b) electric power. What is passive sign convention?
9. State and explain Ohm's law.
10. Enumerate the basic circuit elements and briefly describe their properties.
11. Specify the parameters that govern the resistance of a conductor. Distinguish between linear and non-linear resistors.
12. Explain what is meant by the temperature coefficient of resistance of a material.
13. State how the physical parameters of a capacitor are related to its capacitance and discuss the significance of the permittivity of the dielectric.
14. State and explain Ampere's law.
15. Define inductance and derive an expression for the energy stored in the inductor. State how the direction of induced emf is determined.

16. Define Kirchhoff's laws. What is the basis of these laws?
17. Apply Kirchhoff's laws and develop voltage and current equations for a hypothetical resistive circuit.

Problems

1.1 Plot the variation of charge when the electric current varies as shown in Fig. 1.31.

1.2 A pd of 1.5 V causes a current of 270 μA to flow in a conductor. Calculate the resistance of the conductor.
[5.56 kΩ]

1.3 What is the voltage across an electric heater of resistance 5 Ω through which passes a current of 46 A?
[230 V]

1.4 Calculate the current in a circuit due to a pd of 20 V applied to a 20 kΩ resistor. If the supply voltage is doubled while the circuit resistance is trebled, what is the new current in the circuit?
[1 mA, 0.67 mA]

1.5 A pd of 12 V is applied to a 4.7 kΩ resistor. Calculate the circuit current.
[2.55 mA]

1.6 A current in a circuit is due to a pd of 20 V applied to a resistor of resistance 200 Ω. What resistance would permit the same current to flow if the supply voltage were 200 V?
[2 kΩ]

1.7 A pd of 12 V is applied to a 7.5 Ω resistor for a period of 10 sec. Calculate the electric charge transferred during this time.
[16 C]

1.8 What is the charge transferred in a period of 8 sec by current flowing at the rate of 3.5 A?
[28 C]

1.9 For the assumed directions of current flows and voltages of the elements shown in Fig. 1.32. State whether the element is absorbing or dissipating power.

1.10 A dc motor connected to a 230 V supply, developing 20 kW at a speed of 1000 rpm, has an efficiency of 0.85. Calculate (a) the current and (b) cost of energy absorbed if the load is maintained constant for 12 h. Assume the cost of electrical energy to be Rs 2.50 per kWh. [(a) 102.3 A, (b) Rs 705.90]

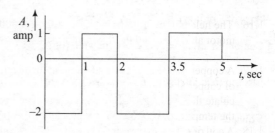

Fig. 1.31

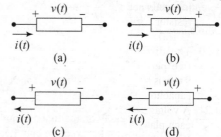

Fig. 1.32

1.11 An electric motor runs at 600 rpm when driving a load requiring a torque of 400 N m. If the motor input is 30 kW, calculate the efficiency of the motor and the heat lost per minute by the motor. Assume its temperature to remain constant. [83.8%, 292.8 kJ]

1.12 A conductor of length l and radius r has a resistance of $R\Omega$. If the volume of the conductor is V, show that

$$(i)\ r = \sqrt[4]{\frac{\rho V}{\pi^2 R}} \quad \text{and} \quad (ii)\ l = \sqrt{\frac{VR}{\rho}}$$

Assume that the resistivity of the conductor is ρ.

1.13 A voltage of 440 V is applied across a parallel plate liquid resistor. If the resistor absorbs 50 kW, calculate the distance between the plates. Assume a resistivity of 25 Ω-cm for the liquid and a current density of 0.30 A/cm². [58.67 cm]

1.14 A conductor has a resistance of R_1 ohms at t_1°C and is made of copper with a resistance-temperature coefficient α referred to 0°C. Find an expression for the resistance R_2 of the conductor at temperature t_2°C.

1.15 The resistance temperature coefficients of two conductors A and B, at a temperature of t °C are α_A and α_B respectively. The resistors are connected in series such that their resistances are in the ratio of $(R_A/R_B) = a$.

(a) Derive an expression for the resistance temperature coefficient α and the temperature t for the circuit. (b) If $a = 4$, $\alpha_A = 0.003/°C$, and $\alpha_B = 0.0003/°C$, determine the value of α and the temperature t. (c) What is the ratio of the resistors when $\alpha = 0.005/°C$, $\alpha_A = 0.0004/°C$, and $\alpha_B = 0.0025/°C$?

$$\left[\text{(a)} \; \frac{a\alpha_A - \alpha_B}{(a-1)}, \; \frac{1}{\alpha - (\alpha_A + \alpha_B)}, \text{(b)} \; 1666.7°C, \text{(c)} \; 1.84 \right]$$

1.16 The field coil of a motor has a resistance of 500 Ω at 15°C. By how much will the resistance increase if the motor attains an average temperature of 45°C when it is running? Take $\alpha = 0.00428$ per °C referred to 0°C.
[60.33 Ω]

1.17 A copper rod, 0.6 m long and 4 mm in diameter, has a resistance of 825 $\mu\Omega$ at 20°C. Calculate the resistivity of copper at that temperature. If the rod is drawn out into a wire having a uniform diameter of 0.8 mm, calculate the resistance of the wire when its temperature is 60°C. Assume the resistivity to be unchanged and the temperature coefficient of resistance of copper to be 0.00426 per °C.   [0.01727 $\mu\Omega$ m, 0.6035 Ω]

1.18 A coil of insulated copper wire has a resistance of 160 Ω at 20°C. When the coil is connected to a 240 V supply, the current after several hours is 1.35 A. Calculate the average temperature throughout the coil, assuming the temperature coefficient of resistance of copper at 20°C to be 0.0039 per °C.   [48.5°C]

1.19 The voltage waveform shown in Fig. 1.33 is applied across a parallel combination of a capacitor of 0.4 F and a resistor of 4 Ω. Plot the waveforms of the currents through the capacitor, resistor, and the total current and determine the (a) energy dissipated in the resistor, (b) maximum energy stored in the capacitor, (c) energy supplied by the source, (d) total charge flow through the resistor, and (e) average resistor voltage.
[(a) 66.67 J, (b) 80 J, (c) 66.67 J, (d) 5 C, (e) 13.33 V]

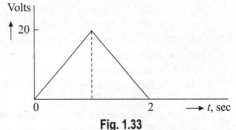

Fig. 1.33

1.20 The distance between the plates of a capacitor is 6 mm and its dielectric material has a relative permittivity of 3. Another sheet of dielectric material of relative permittivity ε_r and thickness 9 mm is inserted by moving the plates apart. If the capacitance of the composite capacitor is half of the original capacitor, determine the value of ε_r.   [4.5]

1.21 A voltage of 25 kV is applied to a parallel plate capacitor whose capacitance is 2.5×10^{-4} μF. If the area of each plate is 110 cm^2 and the plates are separated by a dielectric material of thickness 3 mm, calculate the (a) total charge in coulombs, (b) per sq m charge density, (c) relative permeability of the dielectric, and (d) potential gradient.   [(a) 6.25 μC, (b) 568.18 μC/m^2, (c) 7.7, (d) 83.33 kV/cm]

1.22 A parallel plate condenser has an area of A cm^2 and the distance between the plates is d mm. If the relative permittivity of the dielectric material is ε_r, determine how the energy stored in the capacitor will vary with each factor, when a voltage of V volts is applied across it.

$$\left[W_C \propto \frac{AV^2}{d} \right]$$

1.23 A conductor of l m is carrying a current of I A (whose direction is perpendicular to and coming out of the plane of the paper) and is placed at right angles to a magnetic field of magnitude B T. If the magnetic lines of force are in the plane of the paper and have a direction from top to bottom, determine the magnitude and direction of the force experienced by the conductor. How does the direction of the force change if (i) the direction of the conductor current is reversed, (ii) the direction of the magnetic field only is reversed, and (iii) the direction of the current flow and that of the magnetic field both are reversed? State the law used to determine the direction of the force.

1.24 A current-carrying conductor is situated at right angles to a uniform magnetic field having a density of 0.4 T. (a) Calculate the force (in N m length) on the conductor when the current is 100 A. (b) Calculate the current in the conductor when the force per metre length of the conductor is 25 N.   [(a) 40 Nm, (b) 62.5A]

1.25 The voltage across an inductor of 5 H varies as shown in Fig. 1.34. Determine the inductor current and energy at $t = 1, 2, 3,$ and 5 sec. Sketch the plot illustrating variation of current with time.

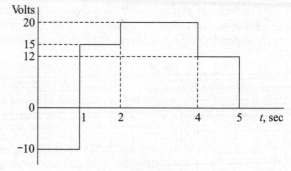

Fig. 1.34

1.26 The coil of a moving-coil loudspeaker has a mean diameter of 40 mm and is wound with 1000 turns. It is situated in a radial magnetic field of 0.4 T. Calculate the force on the coil, in newtons, when the current is 10 mA.   [0.5024N]

1.27 A square coil of side l cm and T number of turns revolves about its axis at right angles inside a magnetic field of density B Wb/m². If the speed of the coil is N rpm, derive an expression for the instantaneous value of the induced emf. If $l = 15$ cm, $B = 0.5$ Wb/m², and $N = 1200$ rpm, determine (a) maximum and (b) minimum values of the induced emf. (c) What are the respective angles made by the plane of the coil with the magnetic field? (d) Calculate the angle made by the plane of the coil with the magnetic field when the instantaneous value of the induced emf is 185 V.   [(a) 212.06 V, 90°, (b) 0 V, 0°, (c) 60.74]

1.28 A conductor, 750 mm long, is moved at a uniform speed at right angles to its length and in a uniform magnetic field having a density of 0.4 T. If the generated emf in the conductor is 3 V and the conductor forms part of a closed circuit having a resistance of 0.5 Ω, calculate: (a) the velocity of the conductor in m/sec, (b) the force acting on the conductor in newtons, (c) the work done in joules when the conductor has moved 500 mm.   [(a) 10 m/sec, (b) 1.8 N, (c) 0.9 J]

1.29 In a coil of 120 turns, the flux is varying with time as shown in Fig. 1.35. If $\phi_m = 0.025$ Wb and $T = 0.04$ sec, determine the value of the statically induced emf. Sketch to scale the waveform of the induced voltage.

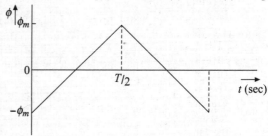

Fig. 1.35

1.30 The axle of a certain motorcar is 1.6 m long. Calculate the generated emf in the car when it is travelling at 120 km/h. Assume the vertical component of the earth's magnetic field to be 40 μT.   [2.13 mV]

1.31 A coil of 2500 turns gives rise to a magnetic flux of 5 mWb when carrying a certain current. If this current is reversed in 0.2 sec, what is the average value of the emf induced in coil?   [125 V]

1.32 A short coil of 500 turns surrounds the middle of a bar magnet. If the magnet sets up a flux of 60 μWb, calculate the average value of the emf induced in the coil when the latter is removed completely from the influence of the magnet in 0.04 sec.   [0.75 V]

1.33 For the circuit shown in Fig. 1.36 determine the value of the source current I_S for the following operating conditions:
(a) the source voltage $V_S = 12$ V, and $I_{AB} = 0$
(b) the source voltage $V_S = 15$ V, and $I_{AB} = 3$ A
   [(a) 4 A, (b) –2 A]

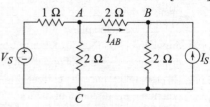

Fig. 1.36

1.34 In the circuit shown in Fig. 1.37, $v(t) = 3\,e^{-t}$. Use Kirchhoff's laws and the volt–ampere relations for te elements to determine the source current $i_S(t)$.   [$4.8e^{-t}$]

1.35 Repeat Problem 1.34 for $v(t) = 4 \sin t$.   [$3.2 \sin t + 9.6 \cos t$]

1.36 In the circuit shown in Fig. 1.37, $i(t) = -15e^{-2t}$. Use Kirchhoff's laws and the volt–ampere relations for the elements to determine the source voltage $v_S(t)$. $[22e^{-t}]$

1.37 Repeat Problem 1.36 for $i(t) = 20 \sin t$. $[40.67 \sin t + 43.33 \cos t]$

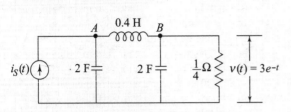

Fig. 1.37

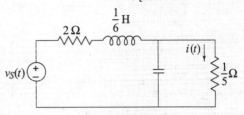

Fig. 1.38

Objective Type Questions

1. In Bohr's model of atomic structure, the nucleus consists of
   (i) electrons
   (ii) electrons and protons
   (iii) protons
   (iv) protons and neutrons

2. If 1 A current flows in a circuit, the number of electrons flowing in a circuit is
   (i) 0.625×10^{19}
   (ii) 1.6×10^{19}
   (iii) 1.6×10^{-19}
   (iv) 0.625×10^{-19}

3. The resistivity of a conductor depends on the
   (i) area of the conductor
   (ii) length of the conductor
   (iii) type of material
   (iv) none of these

4. Current flowing in a series circuit having four equal resistances is I amperes. What is the magnitude of the current if the four resistances are connected in parallel?
   (i) $0.25I$
   (ii) I
   (iii) $4I$
   (iv) $8I$

5. How many coulombs of charge flow through a circuit carrying a current of 10 A in 1 min?
   (i) 10
   (ii) 60
   (iii) 600
   (iv) 1200

6. Two parallel plates separated by a distance d are charged to V volts. The field intensity E is given by
   (i) $V \times d$
   (ii) V/d
   (iii) $V \times d^2$
   (iv) V^2/d

7. A capacitor carries a charge of 0.15 C at 10 V. Its capacitance is
   (i) 0.015 F
   (ii) 1.5 F
   (iii) 1.5 μF
   (iv) none of these

8. Four capacitors each of 20 μF are connected in parallel, the total capacitance is
   (i) 80 μF
   (ii) 5 μF
   (iii) 16 μF
   (iv) none of these

9. One farad is equal to
   (i) 1 Ω
   (ii) 1 V/C
   (iii) 1 C/V
   (iv) none of these

10. Point A has an absolute potential of 20 V and point B is at an absolute potential of − 5 V. V_{BA} has a value of
    (i) −25 V
    (ii) 15 V
    (iii) 25 V
    (iv) none of these

11. The unit of resistivity is
    (i) Ω
    (ii) Ω/m
    (iii) Ω/m^2
    (iv) $\Omega\,m$

12. Two resistors connected in parallel across a battery of 1 V draw a current of 1 A. When one of the resistors is disconnected, the current drawn is 0.2 A. The resistance of the disconnected resistor is
    (i) 1 Ω
    (ii) 1.25 Ω
    (iii) 5 Ω
    (iv) none of these

13. The effect of temperature on metals and insulating materials is that the
    (i) resistance of both increases
    (ii) resistance of both decreases
    (iii) resistance of metals decreases and that of insulating material increases
    (iv) resistance of metals increases and that of insulating materials decreases

14. Two resistors each of 100 Ω are rated at 100 W and 0.25 W. Which has a higher current rating?
    (i) 100 W
    (ii) 0.25 W
    (iii) both have the same rating
    (iv) none of these

15. The unit of inductance is henry. It is represented by
    (i) V/A
    (ii) V sec/A

(iii) V s   (iv) V/sec

16. Instantaneous power in an inductor is proportional to the
    (i) product of instantaneous current and rate of change of current
    (ii) square of instantaneous current
    (iii) the induced voltage
    (iv) none of these

17. The voltage induced in an inductor of L henry is represented by
    (i) Li (ii) $\dfrac{L}{i}$
    (iii) $L\dfrac{di}{dt}$ (iv) none of these

18. Absolute permittivity of a dielectric medium is represented by
    (i) $\dfrac{\varepsilon_0}{\varepsilon_r}$ (ii) $\dfrac{\varepsilon_r}{\varepsilon_0}$
    (iii) $\varepsilon_0\varepsilon_r$ (iv) none of these

19. A parallel plate capacitor has a capacitance of C farads. If one of the sides of the plates is doubled and the distance between them is halved, the capacitance of the capacitor is
    (i) $0.5C$ F   (ii) C F
    (iii) $2C$ F   (iv) $4C$ F

20. Which of these is not an expression for the energy stored in a capacitor?
    (i) $\dfrac{1}{2}CV^2$ (ii) $C\displaystyle\int vdv$
    (iii) $\displaystyle\int pdt$ (iv) QV^2

21. Magnetic flux has the unit of
    (i) newton   (ii) ampere turns
    (iii) coulomb   (iv) weber

22. A 1-m-long conductor carries a current of 50 A at right angles to a magnetic field of 100×10^{-3} T. The force on the conductor is
    (i) 5000 N   (ii) 500 N
    (iii) 50 N   (iv) 5 N

23. Whenever the magnetic flux changes with respect to an electrical conductor or a coil, an emf is induced in the conductor is Faraday's
    (i) first law   (ii) second law
    (iii) third law   (iv) none of these

24. A conductor of length 1 m is moving at right angles to a magnetic field of constant magnitude at a velocity of v m/sec. The magnitude of the induced emf is proportional to
    (i) $l \times v$ (ii) l/v
    (iii) v/l (iv) none of these

25. A coil wound around a magnetic ring is required to produce a flux of 800×10^{-6} Wb. What is the magnitude of the mmf required to set up the flux if the reluctance of the ring is 1.675×10^{-6} AT / Wb?
    (i) 13.4 AT   (ii) 134 AT
    (iii) 1340 AT   (iv) 1.34×10^6

Answers

1. (iv)	2. (i)	3. (iii)	4. (iii)	5. (iii)	6. (ii)	7. (i)
8. (i)	9. (iii)	10. (i)	11. (iv)	12. (ii)	13. (iv)	14. (i)
15. (ii)	16. (i)	17. (iv)	18. (iii)	19. (iv)	20. (iv)	21. (iv)
22. (iv)	23. (i)	24. (i)	25. (iii)			

NETWORK ANALYSIS AND NETWORK THEOREMS

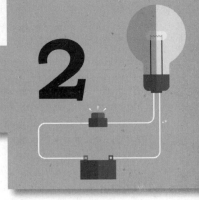

2

Learning Objectives

This chapter will enable the reader to

- Define and identify linear, non-linear, bilateral circuit elements, classify passive and active branches, and recognize nodes, loops, and meshes in networks
- Define independent, dependent, and practical voltage and current sources
- Learn to transform a voltage source into a current source and vice versa without modifying the response in the network
- Develop an ability to calculate equivalent resistance of series, parallel, and series-parallel connected resistances
- Learn to reduce series, parallel, and series-parallel combination of inductances as well as of capacitances
- Utilize star-delta or delta-star transformations for simplifying complex networks
- Employ mesh current and nodal voltage methods to circuits with independent and dependent sources for computing branch currents and node voltages
- Define and apply superposition theorem for analysing complex networks
- Define Thevenin and Norton theorems and be able to develop Thevenin and Norton equivalent circuits for analyses of intricate networks
- Understand and apply Compensation theorem to compute variations in voltages and currents in a circuit when resistance changes in one branch of a circuit.
- Recognize the use of Millman's theorem to analyse circuits where a single node is the focus of attention
- State maximum power transfer theorem and its utility in resistance matching
- Analyse transient or natural response of source-free *RL*, *RC*, and *RLC* circuits
- Analyse steady state or forced response of *RL*, *RC*, and *RLC* circuits when connected to dc source voltage

An electrical network comprises a number of basic elements such as resistors, inductors, capacitors, etc. interconnected in a desired manner and energized by one or more voltage/current sources. The elements of the network may be of different kinds, such as linear or non-linear, bilateral or unilateral, active or passive, lumped or distributed, time-dependent or time-invariant, and so on. This chapter deals with the analysis of linear, lumped parameter time-invariant circuits only.

Network analysis involves the calculation of voltages and currents in all the branches of a given network where the parameters of the elements constituting the network and the exciting sources are completely known. There are different approaches for analysis, but the solution is always unique.

2.1 BASIC DEFINITIONS OF SOME COMMONLY USED TERMS

Network elements The individual components such as a resistor, inductor, capacitor, diode, voltage source, current source, etc. that are used in a circuit are known as network elements.

Network A group of network elements interconnected in a manner so as to perform a desired function is known as a network (circuit).

Branch A branch is an element of a network having only two terminals.

Passive and active branches A branch that does not contain any source is known as a passive branch. When it contains one or more sources it is called an active branch.

Linear element An element is said to be linear when the current and voltage of the element are related through a linear equation, either of algebraic, differential, or integral type.

Bilateral and unilateral elements A bilateral element conducts equally well in either directions. Resistors and inductors are examples of bilateral elements. When the current–voltage relations are different for the two directions of current flow, the element is said to be unilateral. A diode is a unilateral element.

Node The junction point of two or more branches is known as a node.

Loop and mesh Any closed path formed by branches in a network is known as a loop. A mesh is a kind of loop which does not cross another mesh.

2.2 NETWORK SOURCES

Sources are circuit elements that supply energy. Sources are of two types, namely, voltage sources and current sources. These may further be classified as independent sources and dependent (controlled) sources. In this section representations of different types of sources are discussed.

2.2.1 Ideal Independent Voltage Sources

An ideal independent voltage source is an element capable of generating a specified voltage across its terminals irrespective of the magnitude and direction of current flowing through it. The circuit connected to it determines the amount of current supplied by the source. Such a source can deliver current and, consequently, energy continuously without any limit at constant voltage. The voltage of an ideal independent source may be constant (dc) or it may be a function of time. Figure 2.1 shows the symbolic representation and the *v-i* characteristic of an ideal independent

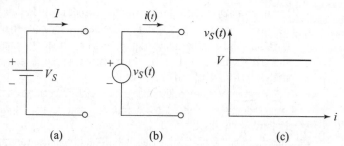

Fig. 2.1 Symbolic representation of an ideal independent voltage source: (a) dc voltage source, (b) time-dependent voltage source, and (c) *v-i* characteristics

voltage source. From the figure, it may be noted that the convention adopted is that the current flowing out of the positive terminal is positive. Further, the magnitude of the dc voltage source is represented by the upper case symbol V_S while the magnitude of the time-dependent voltage source is represented by the lower case symbol $v_S(t)$. From the *v-i* characteristic in Fig. 2.1(c), it is seen that the magnitude V of the voltage source is independent of the magnitude of the current supplied by it. Open-circuit and short-circuit conditions in a voltage source are identified respectively by $i_S = 0$ and $v_S = 0$.

2.2.2 Ideal Independent Current Source

An ideal independent current source is an element that can supply any amount of energy at a constant specified current to any circuit connected to it independent of the magnitude and direction of the voltage appearing across its terminals. Its terminal voltage is determined by the conditions of the network to which it is connected. Consequently no limit exists on the voltage and power delivered by a current source. The current source may be a dc source which delivers a constant dc current independent of time. Alternatively, it may be a time-dependent source. The output current waveform for such a source does not change with loading as the source itself adjusts its terminal voltage to keep the load current constant.

Figure 2.2 shows the symbolic representation and v-*i* characteristic of an independent current source. In this case also, the convention for the current flow is that it flows out of the positive terminal, lower case letter $i_S(t)$ is used to represent time-dependent current while upper case letter I_S is used for dc current. Correspondingly, open and short

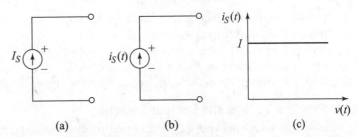

Fig. 2.2 Symbolic representation of an ideal independent current source: (a) dc current source, (b) time-dependent current source, and (c) v-*i* characteristics

circuit conditions in a current source are identified by $v_S = 0$ and $i_S = 0$, respectively.

It is a common practice to employ voltage sources in electric circuit analysis. Current sources find wider applications in electronic circuits. However, with the help of additional circuitry, it is feasible to transform a voltage source into a current source and vice versa. High-voltage dc transmission engineering is based on constant current flow through transmission lines.

Example 2.1 An ideal 10 V voltage source is connected to a 4 Ω resistor. (a) Find the current flowing in the circuit and the voltage drop across the resistor. (b) A 6 Ω resistor is connected in series with a 4 Ω resistor. Find the current flow and the voltage drop across each resistor. (c) If 10 V voltage source is replaced by 10 A current source repeat parts (a) and (b).

Solution (a) Current = $\dfrac{10\,\text{V}}{4\,\Omega} = 2.5\,\text{A}$

Voltage across the 4 Ω resistor = 10 V

(b) It may be noted that the voltage source makes its full voltage available to the circuit. Then current

$$I = \frac{10\,\text{V}}{(4+6)\,\Omega} = 1\,\text{A}$$

Voltage across the 4 Ω resistor, $V_4 = 1\,\text{A} \times 4\,\Omega = 4\,\text{V}$

Voltage across the 6 Ω resistor $V_6 = 1\,\text{A} \times 6\,\Omega = 6\,\text{V}$

(c) The current source supplies a constant current of 10 A in both the cases (a) and (b).

When only the 4 Ω resistor is connected in the circuit,

$$V_4 = 10\,\text{A} \times 4\,\Omega = 40\,\text{V}$$

When the 4 Ω resistor in series with the 6 Ω resistor is connected,

$$V_4 = 10\,\text{A} \times 4\,\Omega = 40\,\text{V} \quad \text{and} \quad V_6 = 10\,\text{A} \times 6\,\Omega = 60\,\text{V}$$

Therefore, the total voltage across the combined resistors is

$$V_4 + V_6 = 40 + 60 = 100\,\text{V}$$

2.2.3 Dependent Sources

The source whose output voltage or current is a function of the voltage or current in another part of a circuit is called a dependent source. Both controlled current and controlled voltage sources exist. A diamond shaped symbol is used to represent dependent sources as shown in Fig. 2.3. The four basic types of controlled sources with relationship between source voltage v_s or source current i_s and the voltage v_i or current i_i in any part of the circuit on which it depends are given below:

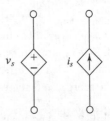

Fig. 2.3 Symbols for dependent sources

(a) Voltage-controlled voltage source (VCVS): $v_s = Av_i$
(b) Voltage-controlled current source (VCCS): $i_s = Av_i$
(c) Current-controlled current source (CCCS): $i_s = Ai_i$
(d) Current-controlled voltage source (CCVS): $v_s = Ai_i$

Dependent sources are very useful in modelling certain types of electronic circuits.

2.2.4 Practical Voltage and Current Sources

An ideal voltage source does not exist in practice. In all available practical voltage sources the terminal voltage is generally found to decrease linearly with the increase of current that flows through it. For example, the terminal voltage of an automobile battery falls with rise in load current. In case a practical voltage source is modelled as an ideal voltage source v_s as shown in Fig. 2.1 and a variable load resistance R_L is connected to it, then as R_L decreases the source current $i(t)$ increases and

$$i(t) = \frac{v_S}{R_L} \tag{2.1}$$

In the limit the load resistance approaches zero, the ideal voltage source is required to provide an infinite amount of current to the load. Naturally, this is impossible. This implies that there is a limit (a large one) to the amount of current a practical source can deliver to a load. Quite simply, exploiting the notion of the internal resistance of a source can approximate the limitations of practical sources. Figure 2.4 depicts a model for a practical voltage source, comprising an ideal voltage source v_S in series with a resistance r_S. The resisance r_S in effect poses a limit to the maximum current the voltage source can provide:

$$I_{S\max} = \frac{v_S}{r_S} \tag{2.2}$$

Typically r_S is small and its presence affects the voltage v_L across the load resistance. Now this voltage is no longer equal to the source voltage. Since the current provided by the source is

$$i_S = i_L = \frac{v_S}{r_S + R_L} \tag{2.3}$$

The load voltage can be determined to be

$$v_L = i_L R_L = \frac{v_S}{r_S + R_L} R_L \tag{2.4}$$

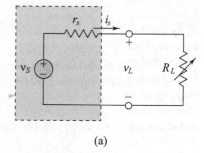

(a)

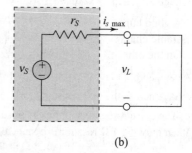

(b)

Fig. 2.4   Practical voltage source

Thus, in the limit as the source internal resistance r_S approaches zero, the load voltage v_L becomes exactly equal to the source voltage. Hence a desirable feature of an ideal voltage source is to have very small internal resistance. Often the effective internal resistance of a voltage source is quoted in the technical specifications for the source, so that the user can take this parameter into account.

Example 2.2   A practical 36 V independent source can supply a maximum current of 72 A. If this voltage source is connected across a variable load, (a) determine the internal resistance of the source and (b) write down the expression for the load voltage. (c) Plot the load characteristic when the load is varied from 0 Ω o 50 r_S Ω.

Solution Using Eq. (2.2), the internal resistance of the source is determined as follows:

$$r_S = \frac{36}{72} = 0.5\ \Omega$$

From Eq. (2.4), the load voltage is expressed as

$$v_L = \frac{36R_L}{0.5 + R_L}$$

By varying R_L from 0 Ω to 25 Ω, the load current is calculated from Eq. (2.3) and the load voltage is determined from the above equation. Figure 2.5 shows the plot of the load voltage against the load current.

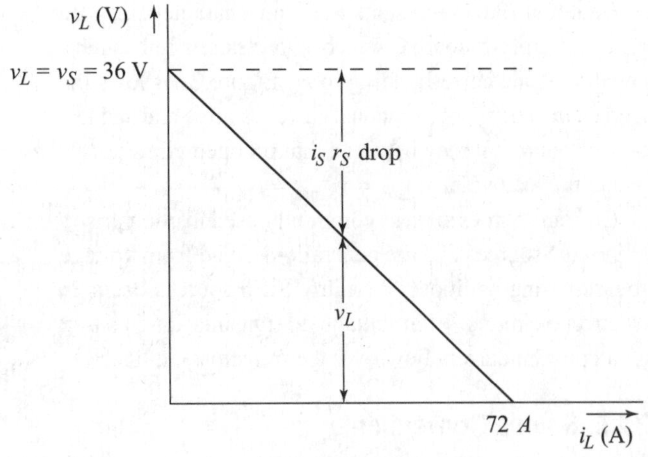

Fig. 2.5

It may be noted that when the load current is zero (open-circuit condition), the load voltage is equal to the source voltage, and for $R_L = 0$ (short-circuit condition), the maximum current $i_{S\,max}$ supplied by the voltage source depends on the value of r_S.

An ideal current source does not exist in practice. In all practical current sources the terminal supply current does change with change in load. A modification of the ideal current source model is useful to describe the behavior of a practical current source. The circuit illustrated in Fig. 2.6 depicts a simple representation of a practical current source consisting of an ideal source in parallel with a variable resistor load. It may be noted that as the load resistance approaches infinity (i.e., an open circuit), the output voltage of the current source approaches its limit,

$$v_{OC} = v_{S\,max} = i_S r_S \tag{2.5}$$

A good current source should be able to approximate the behavior of an ideal current source. Therefore, a desirable characteristic for the internal resistance of a current source is that it should be as large as possible.

On the other hand, when the load resistance becomes zero, the load current is the short-circuit current, i_{SC}, and $i_S = i_{SC}$. For any other value of the load resistance R_L, application of KCL at the positive terminal yields

$$i_S = \frac{v_L}{r_S} + i_L$$

or $\qquad v_L = i_S r_S - r_S i_L = v_{S\,max} - r_S i_L \tag{2.6}$

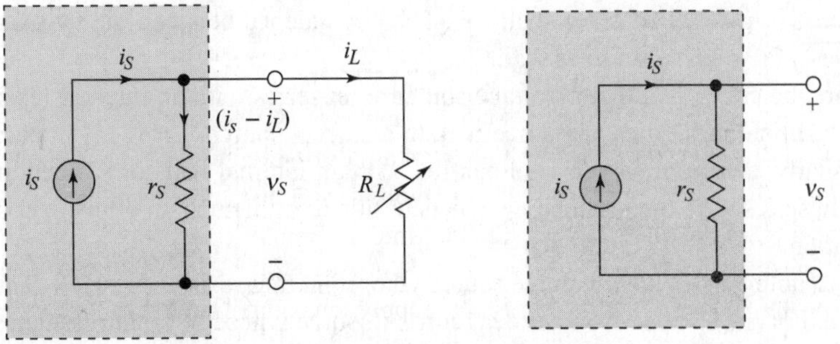

Fig. 2.6 Practical current source

Equation (2.6) represents the load characteristic of a practical current source, which is a straight line with a negative slope corresponding to r_S. Figure 2.7 shows the load characteristic of a practical current source and an ideal current source. It may be noted that on open circuit, $R_L = \infty$ and the load voltage $v_{OC} = v_{S\,max} = i_S r_S$.

Current sources are less commonly used in comparison to voltage sources and are generally derived from voltage sources using additional circuitry. Such sources occur in electronic circuits. High voltage dc transmission is based on a constant current flow over the transmission line.

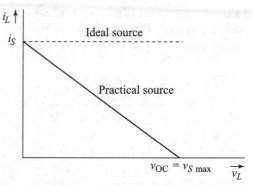

Fig. 2.7   Load characteristics of practical current source

2.2.5 Source Conversion

The classification of voltage and current sources is done depending upon their terminal characteristics. It is possible, for the sake of numerical computation, to replace a voltage source with an equivalent current source and vice versa without affecting the response in the rest of the network to which the source is connected. The equivalence between the two kinds of sources may be arrived at as explained next.

A practical voltage source, modelled as a combination of an ideal voltage source with voltage v_S in series with internal resistance r_S connected to a network N and delivering current i at a terminal voltage of v_L, as shown in Fig. 2.8(a), is considered. Due to a potential drop in the internal resistance, the voltage at the terminals of the source may be written as

$$v_L = v_S - i r_S \tag{2.7}$$

Dividing on both sides by r_S and assuming $r_S \neq 0$, the above equation may be written as

$$\frac{v_L}{r_S} = \frac{v_S}{r_S} - i = i_S - i \tag{2.8}$$

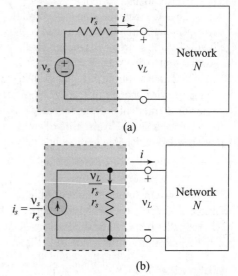

Fig. 2.8   Equivalence between voltage and current sources

Equation (2.8) may be regarded as the KCL equation for the network shown in Fig. 2.8(b), which consists of a current source i_S in parallel with resistance r_S. The equivalence between the two kinds of sources thus established is valid in the reverse sense as well.

It may, therefore, be concluded that a voltage source v_S in series with an internal resistance r_S is equivalent, as far as the terminal properties are concerned, to a current source $i_S = v_S/r_S$ in parallel with the same resistance r_S. Similarly, a current source i_S in parallel with an internal resistor r_S is equivalent to a voltage source $v_S = i_S \cdot r_S$ in series with a resistance r_S. The determination of such an equivalent source is known as source transformation.

It is important to note that an ideal voltage source whose internal resistance $r_S = 0$ cannot be transformed to a current source. For the same reason an ideal current source whose internal resistance is infinitely large cannot be converted into a voltage source.

Example 2.3 An independent voltage source has an open-circuit voltage of 24 V and an internal resistance of 0.75 Ω. Transform the voltage source into an equivalent current source and draw its equivalent circuit when it supplies a load of 12 Ω. Prove that the external behaviour of the two sources is the same.

Solution The magnitude of the equivalent source current is $i_S = \dfrac{24}{0.75} = 32$ A

Thus, the equivalent current source consists of a supply current of 32 A and a resistance of 0.75 Ω connected in parallel. The equivalent circuit of the transformation is shown in Fig. 2.9.

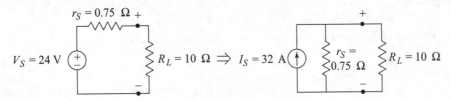

Fig. 2.9

In order to establish the external behaviour of the two types of sources, it is essential to calculate the power delivered to the load. When the source is represented as a voltage source, the current supplied to the load is

$$I_L = \left(\frac{24}{0.75 + 10} \right) = 2.23 \text{ A}$$

When the source is represented as a current, the load current is calculated as follows.
The voltage across the load terminals is $I_L R_L = I_L \times 10 = (32 - I_L) \times 0.75$ V. Thus,

$$10 I_L = (32 - I_L) \times 0.75$$

or $\qquad I_L = \dfrac{24}{10.75} = 2.23$ A

From the foregoing discussion, it may be noted that the current supplied at the terminals by the two sources and, therefore, the power supplied to the load is the same. Hence, the external behaviour of the two sources is the same.

Example 2.4 For the circuit shown in Fig. 2.10(a), $v_S = 12$ V, $r_S = 0.3$ Ω, and load current $I_L = 10$ A. Calculate the (a) total power supplied by the ideal source, (b) total power dissipated by the practical source, and (c) total power supplied by the practical source. Finally, convert the voltage source to a current source and draw the circuit with the current source.

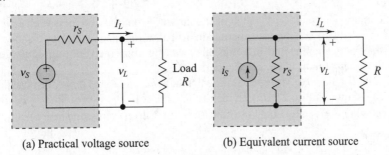

(a) Practical voltage source (b) Equivalent current source

Fig. 2.10

Solution (a) The total power supplied by the ideal source is

$$v_S \times I = 12 \text{ V} \times 10 \text{ A} = 120 \text{ W}$$

(b) The total power dissipated by the practical source is

$$I^2 \times r_S = 10^2 \times 0.3 = 30 \text{ W}$$

(c) The total power supplied by the practical source is

$$(v_S - I \times r_S) \times I = (12 - 10 \times 0.3) \times 10 = 90 \text{ W}$$

For the current source $i_S = \dfrac{v_S}{r_S} = \dfrac{12}{0.3} = 40 \text{ A}$

The circuit with an equivalent current source is shown in Fig. 2.10(b).

2.3 RESISTIVE NETWORKS

Practical electric circuits have complex performances. It is required to approximate them to simplify the analysis as combinations of circuit elements in series and parallel. Series and parallel circuits have a direct relationship with Kirchoff's laws. These circuits form the basis of all network analysis, and hence an overview of these circuits is taken up in the next section.

2.3.1 Series Resistors and the Voltage Divider Rule

Two or more resistances are said to be in series if the same current flows through all of them. Figure 2.11(a) shows resistances R_1, R_2, R_3, ..., R_n connected in series end to end across a dc voltage source of magnitude V volts.

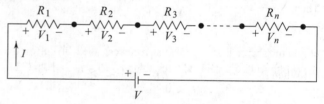

(a) Resistances in series

Applyng KVL around the closed loop network,

$$IR_1 + IR_2 + IR_3 + \cdots + IR_n - V = 0$$

or $\quad I(R_1 + R_2 + R_3 + \cdots + R_n) = V$

$\therefore \qquad I = \dfrac{V}{R_1 + R_2 + R_3 + \cdots + R_n}$ (2.9)

$\qquad\qquad = \dfrac{R_1}{R_{eq}}$ (2.9a)

where $\quad R_{eq} = R_1 + R_2 + R_3 + \cdots + R_n$ (2.10)

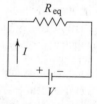

(b) Equivalent resistance

Fig. 2.11 Series connection

R_{eq} denotes the equivalent series resistance of the circuit. From the above, it is clear that the equivalent circuit shown in Fig. 2.11(b) can replace the original configuration of the circuit shown in Fig. 2.11(a). Thus, when resistances are connected in series they can be combined into an equivalent resistance by directly adding the values of individual resistances.

A concept very closely linked to series resistors is that of *voltage divider*. It may be observed that the source voltage in the circuit shown in Fig. 2.11(a) divides among the resistors R_1, R_2, R_3, ..., R_n in accordance with KVL. The series current I is given by Eq. (2.9)a, and the voltage drops across the resistances can be obtained as

$$V_1 = IR_1 = \frac{R_1}{R_{eq}} V, \quad V_2 = IR_2 = \frac{R_2}{R_{eq}} V,$$

$$V_3 = IR_3 = \frac{R_3}{R_{eq}} V, \quad \cdots, \quad V_n = \frac{R_n}{R_{eq}} V$$

(2.11)

That is, the voltage drop across each resistor in a series circuit is directly proportional to the ratio of its resistance to the total series resistance of the circuit. By using the voltage divider rule, the proportion in which the voltage drops are distributed around a circuit can be determined.

Example 2.5 For the voltage divider circuit shown in Fig. 2.12 it is desired to have $V_0 = 0.6\,V_S$. (a) Find R_2 when $R_1 = 10\,\text{k}\Omega$. (b) Find the percentage change in V_0 if a load resistance (i) $R_3 = 200\,\text{k}\Omega$, (ii) $R_3 = 20\,\text{k}\Omega$ is connected across R_2 in turn.

Solution

(a) $\dfrac{V_0}{V_S} = \dfrac{R_2}{R_1 + R_2} = 0.6 \quad \text{or} \quad 0.4R_2 = 0.6R_1$

$\therefore \qquad R_2 = \dfrac{0.6}{0.4}\,R_1 = \dfrac{6}{4}\times 10 = 15\,\text{k}\Omega$

(b) (i) When load $R_3 \parallel R_2$, the equivalent resistance

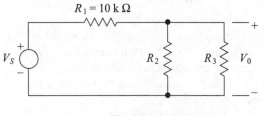

$$R_{eq} = \frac{200 \times 15}{200 + 15} = \frac{3000}{215} = 13.95\,\text{k}\Omega$$

Then, $\dfrac{V_0}{V_S} = \dfrac{R_{eq}}{R_1 + R_{eq}} = \dfrac{13.95}{10 + 13.95} = 0.5825$

$\therefore \qquad \%\ \text{change} = \dfrac{0.6 - 0.5825}{0.6}\times 100 = 2.91$

Fig. 2.12

(ii) When $R_3 = 20\,\text{k}\Omega$,

$$R_{eq} = \frac{20 \times 15}{20 + 15} = \frac{300}{35} = 8.57\,\text{k}\Omega$$

Then, $\dfrac{V_0}{V_S} = \dfrac{R_{eq}}{R_1 + R_{eq}} = \dfrac{8.57}{10 + 8.57} = 0.4615$

$\therefore \qquad \%\ \text{change} = \dfrac{0.6 - 0.4615}{0.6}\times 100 = 23.08$

It may be noted that as the load resistance decreases from 200 kΩ to 20 kΩ the per cent change in the output voltage rises from 2.91 to 23.08.

2.3.2 Parallel Resistors and the Current Divider Rule

Two or more resistances are said to be in parallel if an identical voltage appears across each element. Figure 2.13 shows parallel resistances $R_1, R_2, R_3, \ldots, R_n$ connected across a dc voltage source of magnitude V volts.

The currents $I_1, I_2, I_3, \ldots, I_n$ among the resistors $R_1, R_2, R_3, \ldots, R_n$, respectively, can be expressed as

$$I_1 = \frac{V}{R_1}, \quad I_2 = \frac{V}{R_2}, \quad I_3 = \frac{V}{R_3}, \quad \cdots, \quad I_n = \frac{V}{R_n} \tag{2.12}$$

Applying KCL to the junction at the top in Fig. 2.13(a)

$$I - I_1 - I_2 - I_3 - \cdots - I_n = 0$$

or $\qquad I = I_1 + I_2 + I_3 + \cdots + I_n \tag{2.13}$

$$= \frac{V}{R_1} + \frac{V}{R_2} + \frac{V}{R_3} + \cdots + \frac{V}{R_n}$$

$$= V \left[\frac{1}{R_1} + \frac{1}{R_2} + \frac{1}{R_3} + \cdots + \frac{1}{R_n} \right] \quad (2.14)$$

$$= V \frac{1}{R_{eq}} \quad (2.15)$$

where $\dfrac{1}{R_{eq}} = \dfrac{1}{R_1} + \dfrac{1}{R_2} + \dfrac{1}{R_3} + \cdots + \dfrac{1}{R_n}$ (2.16)

$$R_{eq} = \frac{1}{\dfrac{1}{R_1} + \dfrac{1}{R_2} + \dfrac{1}{R_3} + \cdots + \dfrac{1}{R_n}} \quad (2.17)$$

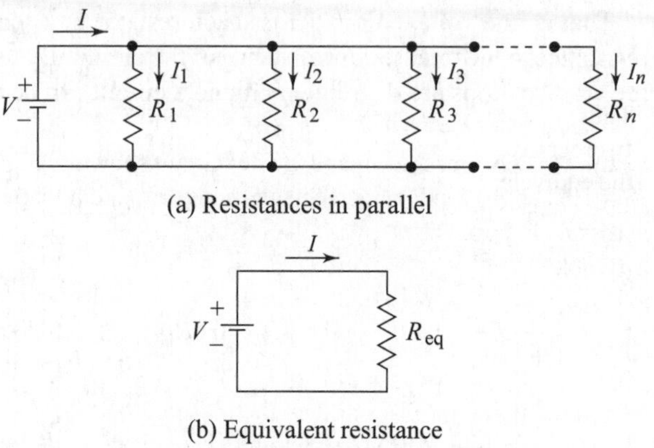

(a) Resistances in parallel

(b) Equivalent resistance

Fig. 2.13 Parallel connection

R_{eq} denotes equivalent resistance of the parallel combination of the resistances. From the above it is clear that the equivalent circuit shown in Fig 2.13(b) can replace the original configuration of the circuit shown in Fig. 2.13(a). Thus, the equivalent resistance of n parallel-connected resistances is the reciprocal of the sum of the reciprocals of individual resistances.

The parallel combination of two or more resistances is referred to, in the remainder of the book, with the following notation:

$$R_1 \parallel R_2 \parallel R_3 \parallel \cdots$$

where the symbol $\parallel$ means 'in parallel with'.

In terms of conductance, which is the reciprocal of resistance, Eq. (2.16) can be expressed as

$$G_{eq} = G_1 + G_2 + G_3 + \cdots + G_n = \sum_{k=1}^{n} G_k \quad \text{mho or siemens (S)}$$

where G_{eq} is the equivalent conductance and $G_1, G_2, G_3, \cdots, G_n$ are individual conductances.

A concept very closely linked to parallel resistors is that of *current divider*. From Eq. (2.15), the voltage across the parallel elements is obtained as $V = I R_{eq}$. Substituting this into Eq. (2.12), current in each branch may be expressed as

$$I_1 = \frac{R_{eq}}{R_1} I = \frac{1/R_1}{1/R_{eq}} I = \frac{G_1}{G_{eq}} I$$

$$I_2 = \frac{R_{eq}}{R_2} I = \frac{1/R_2}{1/R_{eq}} I = \frac{G_2}{G_{eq}} I$$

$$\vdots$$

$$I_n = \frac{R_{eq}}{R_n} I = \frac{1/R_n}{1/R_{eq}} I = \frac{G_n}{G_{eq}} I \quad (2.18)$$

Thus it can be seen that the current in a parallel circuit divides in inverse proportion to the resistances of the individual parallel resistances.

Example 2.6 For the circuit shown in Fig. 2.14, apply the current divider rule to compute the currents in the various resistors if the current flowing through the 2 Ω resistor is 2 A. Calculate the voltage V_S of the voltage source.
Solution The current I_3 by the application of the current divider rule yields

$$2 = \frac{4}{2+4} I_3 \quad \text{or} \quad I_3 = \frac{2 \times 6}{4} = 3 \text{ A}$$

Thus, $I_4 = I_3 - 2 = 1\text{A}$

In order to calculate I_2, Eq. (2.17) is employed to determine the equivalent parallel resistance, R_{eq}, as follows.

$$R_{eq} = \frac{1}{(1/2)+(1/4)+(1/6)} = 1.09 \, \Omega$$

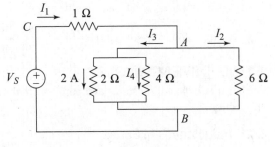

From Eq. (2.18), $I_4 = \frac{R_{eq}}{4} I_1$ and $I_2 = \frac{R_{eq}}{6} I_1$

Fig. 2.14

Dividing the two equations and substituting $I_4 = 1$ A, we get

$$I_2 = \frac{4}{6} \times 1 = 0.67 \text{ A}$$

Finally, $I_1 = \frac{6}{R_{eq}} I_2 = \frac{6}{1.09} \times 0.67 = 3.67 \text{ A}$

Tracing the circuit CABC, via the 6 Ω resistor, and application of Ohm's law yields

$$V_S = 3.67 \times 1 + 0.67 \times 6 = 7.67 \text{ V}$$

Example 2.7 For the resistive network shown in Fig. 2.15 find the equivalent resistance seen by the source.

Solution Combining series-parallel resistances,

$$R_{eq} = [\{\{(2+2+2) \parallel 6\} + 5 + 2\} \parallel 10] + 5$$

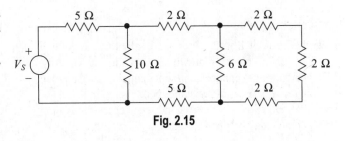

$$= \left[\left\{ \frac{6 \times 6}{6+6} + 7 \right\} \parallel 10 \right] + 5 = \left[\frac{10 \times 10}{10+10} \right] + 5$$

Fig. 2.15

$$= 5 + 5 = 10 \, \Omega$$

Example 2.8 In the resistive network shown in Fig. 2.16, find the (a) equivalent resistance between terminals a and b when (i) terminals c and d are open, (ii) terminals c and d are shorted together; (b) equivalent resistance between terminals c and d when (i) terminals a and b are open, (ii) if terminals a and b are shorted together.

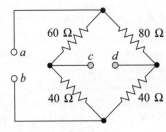

Solution (a) (i) when terminals c and d are open,

$$R_{ab} = (80+40) \parallel (60+40) = \frac{120 \times 100}{120+100} = 54.54 \, \Omega$$

Fig. 2.16

(ii) When terminals c and d are closed,

$$R_{ab} = (80 \parallel 60) + (40 \parallel 40) = \frac{80 \times 60}{80+60} + \frac{40 \times 40}{40+40} = 34.286 + 20 = 54.286 \, \Omega$$

(b) (i) When terminals a and b are open,

$$R_{cd} = (60+80) \parallel (40+40) = \frac{140 \times 80}{140+80} = 50.91 \, \Omega$$

(ii) When terminals a and b are closed,

$$R_{ab} = (60 \parallel 40) + (80 \parallel 40) = \frac{60 \times 40}{60 + 40} + \frac{80 \times 40}{80 + 40} = 24 + 26.67 = 50.67 \ \Omega$$

2.4 INDUCTIVE NETWORKS

The manner of handling series and parallel combinations of inductances in electric circuits is discussed in this section.

2.4.1 Inductances in Series

Figure 2.17(a) shows three inductances L_1, L_2, and L_3 connected in series to a time-varying voltage source $v_s(t)$ causing a changing current to flow through the inductances. Application of KVL to the closed circuit gives

$$v_s(t) = L_1 \frac{di}{dt} + L_2 \frac{di}{dt} + L_3 \frac{di}{dt}$$

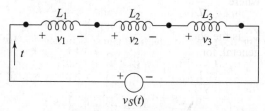

(a) Inductances in series

$$= (L_1 + L_2 + L_3) \frac{di}{dt} \qquad (2.19)$$

$$= L_{eq} \frac{di}{dt} \qquad (2.20)$$

where $\qquad L_{eq} = L_1 + L_2 + L_3 \qquad (2.21)$

On the basis of Eq. (2.20) the equivalent circuit shown in Fig. 2.17(a) can be redrawn as shown in Fig. 2.17(b). Thus inductances connected in series can be directly added in the same manner as resistances connected in series. In general, for n series-connected inductances the equivalent inductance may be obtained as

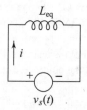

(b) Equivalent inductance

Fig. 2.17 Series connection

$$L_{eq} = L_1 + L_2 + L_3 + \cdots + L_n = \sum_{k=1}^{n} L_k \qquad (2.22)$$

2.4.2 Inductances in Parallel

Figure 2.18(a) shows three inductances L_1, L_2, and L_3 connected in parallel to the source $v_s(t)$. Applying KCL to the circuit yields

$$i = i_1 + i_2 + i_3 \qquad (2.23)$$

Assuming the initial current through each inductor to be zero, using Eq. (1.31), i_1, i_2, and i_3 may be expressed as

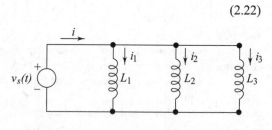

(a) Inductances in parallel

$$i_1 = \frac{1}{L_1} \int_0^t v_S(t) \ dt \qquad (2.24)$$

$$i_2 = \frac{1}{L_2} \int_0^t v_S(t) \ dt \qquad (2.25)$$

$$i_3 = \frac{1}{L_3} \int_0^t v_S(t) \ dt \qquad (2.26)$$

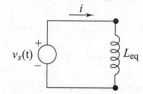

(b) Equivalent inductance

Fig. 2.18 Parallel connection

Substituting Eqs. (2.24)–(2.26) into Eq. (2.23),

$$i = \left(\frac{1}{L_1} + \frac{1}{L_2} + \frac{1}{L_3} \right) \int_0^t v_s(t) \, dt \qquad (2.27)$$

$$= \frac{1}{L_{eq}} \int_0^t v_s(t) \, dt \qquad (2.28)$$

where $\quad \dfrac{1}{L_{eq}} = \dfrac{1}{L_1} + \dfrac{1}{L_2} + \dfrac{1}{L_3} \qquad (2.29)$

L_{eq} is the equivalent inductance of the parallel combination of L_1, L_2, and L_3 and is shown in Fig. 2.18(b). In general, for n parallel-connected inductances the equivalent inductance may be obtained as

$$\frac{1}{L_{eq}} = \frac{1}{L_1} + \frac{1}{L_2} + \frac{1}{L_3} + \cdots + \frac{1}{L_n} = \sum_{k=1}^{n} \frac{1}{L_k} \qquad (2.30)$$

2.5 CAPACITIVE NETWORKS

Series and parallel combinations of capacitances occur quite often in electronic circuitry. Simplification of these circuits by equivalent circuits is discussed in this section.

2.5.1 Capacitances in Series

Three capacitors having capacitances C_1, C_2, and C_3 farad are connected in series as shown in Fig. 2.19(a) and the combination is connected to a time-varying voltage source $v_s(t)$ causing current i to flow through the capacitances. Applying KVL to the circuit,

$$v_s(t) = v_1 + v_2 + v_3 \qquad (2.31)$$

where v_1, v_2, and v_3 are voltages across C_1, C_2, and C_3, respectively.

Assuming initial capacitor voltages to be zero, using Eq. (1.3), v_1, v_2, v_3 may be expressed as

$$v_1 = \frac{1}{C_1} \int_0^t i \, dt \qquad (2.32)$$

$$v_2 = \frac{1}{C_2} \int_0^t i \, dt \qquad (2.33)$$

$$v_3 = \frac{1}{C_3} \int_0^t i \, dt \qquad (2.34)$$

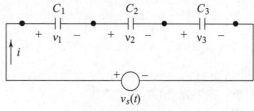

(a) Capacitances in series

Thus Eq. (2.31) becomes

$$v_s(t) = \frac{1}{C_1} \int_0^t i \, dt + \frac{1}{C_2} \int_0^t i \, dt + \frac{1}{C_3} \int_0^t i \, dt \qquad (2.35)$$

$$= \left(\frac{1}{C_1} + \frac{1}{C_2} + \frac{1}{C_3} \right) \int_0^t i \, dt \qquad (2.36)$$

$$= \frac{1}{C_{eq}} \int_0^t i \, dt \qquad (2.37)$$

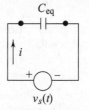

(b) Equivalent capacitance

Fig. 2.19 Combination of capacitances in series

where
$$\frac{1}{C_{eq}} = \frac{1}{C_1} + \frac{1}{C_2} + \frac{1}{C_3} \qquad (2.38)$$

C_{eq} is the equivalent capacitance of the series combination of C_1, C_2, and C_3 and is shown in Fig. 2.19(b). In general, for n series-connected capacitances, the equivalent capacitance may be obtained as

$$\frac{1}{C_{eq}} = \frac{1}{C_1} + \frac{1}{C_2} + \frac{1}{C_3} + \cdots + \frac{1}{C_n} = \sum_{k=1}^{n} \frac{1}{C_k} \qquad (2.39)$$

Thus, it can be seen that capacitances in series combine in the same manner as resistances or inductances connected in parallel.

2.5.2 Capacitances in Parallel

Figure 2.20(a) shows three capacitors of C_1, C_2, and C_3 farad connected in parallel across a voltage source $v_s(t)$. Applying KCL to the circuit gives

$$i = i_1 + i_2 + i_3 \qquad (2.40)$$

where i_1, i_2, i_3 are currents in C_1, C_2, C_3, respectively. Using Eq. (1.37), Eq. (2.40) may be written as

$$i = C_1 \frac{dv_s}{dt} + C_2 \frac{dv_s}{dt} + C_3 \frac{dv_s}{dt} \qquad (2.41)$$

$$= (C_1 + C_2 + C_3) \frac{dv_s}{dt} \qquad (2.42)$$

$$= C_{eq} \frac{dv_s}{dt} \qquad (2.43)$$

where $C_{eq} = C_1 + C_2 + C_3 \qquad (2.44)$

C_{eq} is the equivalent capacitance of the parallel combination of C_1, C_2, and $_3$ and is shown in Fig. 2.20(b).

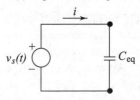

(a) Capacitances in parallel

(b) Equivalent capacitance

Fig. 2.20   Combination of capacitances in parallel

In general for n parallel-connected capacitances the equivalent capacitance may be obtained as

$$C_{eq} = C_1 + C_2 + C_3 + \cdots + C_n = \sum_{k=1}^{n} C_k \qquad (2.45)$$

Hence, it can be seen that capacitances connected in parallel can be directly added as in the case of resistances or inductances connected in series.

Example 2.9   For the circuit shown in Fig. 2.21, calculate the capacitance of the equivalent capacitor between terminals A and B.

Solution   Two capacitors, 1 μF each, in series between points D and E can be combined, using Eq. (2.39), to make one capacitor of 0.5 μF. Similarly, two sets of series capacitances of 2 μF each between points D and F and E and F combine to make 1 μF each. Figure 2.21 is redrawn replacing series-connected capacitors by their equivalents as shown in Fig. 2.22(a).

In Fig. 2.22(a), the two parallel capacitors of 0.5 μF and 0.5 μF between points D and E can be added [using Eq. (2.45)] to an equivalent capacitor of 1 μF, which in series with 1 μF capacitor between points E and F combine to make equivalent capacitor of 0.5 μF between points D and F. Again, two parallel capacitors of 1 μF and 0.5 μF between points D and F add to form equivalent capacitor of 1.5 μF. The new equivalent circuit is shown in Fig. 2.22(b). Equation (2.39) yields the equivalent capacitor of Fig. 2.22(b) as

$$C_{eq} = \frac{1}{(1/3) + (1/1.5) + (1/5)} = \frac{5}{6} = 0.83 \ \mu F$$

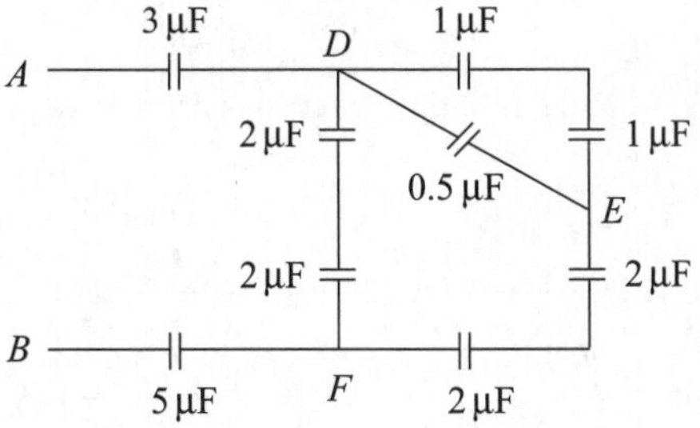

Fig. 2.21

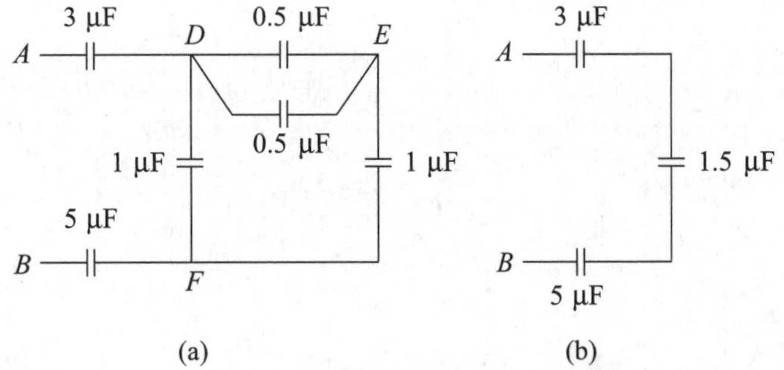

Fig. 2.22

(a)

(b)

2.6 SERIES-PARALLEL CIRCUITS

Practical electrical circuits are rarely composed of elements connected either in series or in parallel only. In many electrical circuits an element is in series with a parallel combination of other circuit elements. In the remainder of this chapter, only the resistance parameter is considered to make the analysis simple. The inductance and capacitance parameters involve current–voltage relationships that are dependant on time-varying voltages or currents applied to such elements as already seen in Sections 2.4 and 2.5. However, the methods of analysis and the network theorems discussed in the following sections for resistive circuits can be applied for circuits containing all the three circuit parameters, namely, R, L and C.

Figure 2.23(a) shows a series-parallel resistive circuit consisting of R_1 in series with a parallel combination of R_2 and R_3 connected to a dc source V_s. The circuit shown in Fig. 2.23(a) is simplified by combining the parallel resistances R_2 and R_3 into a single equivalent resistance, shown in Fig. 2.23(b), whose value may be obtained using Eq. (2.17) as

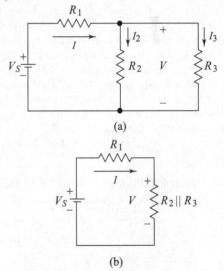

(a)

(b)

Fig. 2.23 Series-parallel circuit

$$R_2 \parallel R_3 = \frac{1}{1/R_2 + 1/R_3} = \frac{R_2 R_3}{R_2 + R_3} \qquad (2.46)$$

The equivalent circuit in Fig. 2.23 (b) can now represent the original circuit configuration. Then by KVL,

$$V_S = IR_1 + I \, (R_2 \parallel R_3) = IR_1 + I \, \frac{R_2 R_3}{R_2 + R_3}$$

$$= I \left[R_1 + \frac{R_2 R_3}{R_2 + R_3} \right] \qquad (2.47)$$

$$\therefore \qquad I = \frac{V_S}{R_1 + \dfrac{R_2 R_3}{R_2 + R_3}} \qquad (2.48)$$

Applying the current divider rule, the currents I_2 and I_3 through resistances R_2 and R_3, respectively, are computed as

$$I_2 = \frac{R_3}{R_2 + R_3} \, I, \quad I_3 = \frac{R_2}{R_2 + R_3} I \qquad (2.49)$$

The voltage V across the parallel resistances may be found by applying the voltage divider rule as

$$V = \frac{R_2 \parallel R_3}{R_1 + R_2 \parallel R_3} V_S \tag{2.50}$$

Example 2.10 In the circuit shown in Fig. 2.24 calculate (a) currents I, I_1, and I_2; (b) the power consumed by each resistor; (c) the voltage drop V_2 across the 2 Ω resistor.

Solution

(a) $I = \dfrac{12}{2 + (12 \parallel 24)} = \dfrac{12}{2 + \dfrac{12 \times 24}{12 + 24}} = \dfrac{12}{10} = 1.2\ \text{A}$

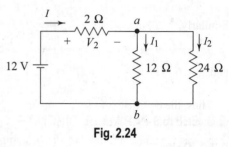

Fig. 2.24

$I_1 = I \times \dfrac{24}{24 + 12} = 1.2 \times \dfrac{2}{3} = 0.8\ \text{A}$

$I_2 = I \times \dfrac{12}{24 + 12} = 1.2 \times \dfrac{1}{3} = 0.4\ \text{A}$

(b) Power consumed in the 2 Ω resistor $= I^2 \times 2 = 1.2^2 \times 2 = 2.88\ \text{W}$

Power consumed in the 12 Ω resistor $= I_1^2 \times 12 = 0.8^2 \times 12 = 7.68\ \text{W}$

Power consumed in the 24 Ω resistor $= I_2^2 \times 24 = 0.4^2 \times 24 = 3.84\ \text{W}$

(c) Voltage drop $V_2 = I \times 2 = 1.2 \times 2 = 2.4\ \text{V}$

2.7 STAR–DELTA OR Y–Δ TRANSFORMATION

The methods of series, parallel, and series-parallel combination of elements do not always lead to simplification of networks. Such networks are handled by star–delta transformation. Figure 2.25(a) shows three resistances R_{ab}, R_{bc}, and R_{ca} connected in delta to three nodes A, B, and C. Figure 2.25(b) shows three resistances R_a, R_b, and R_c connected in star between the same three nodes A, B, and C and a common point n. If these two networks are to be equivalent, then the resistance between any pir of nodes of the delta-connected network of Fig. 2.25(a) must be the same as that between the same pair of nodes of the star-connected network of Fig. 2.25(b).

2.7.1 Star Resistances in Terms of Delta Resistances

Equating resistances between node pair AB,

$$R_a + R_b = R_b \parallel (R_{bc} + R_{ca}) = \frac{R_{ab}R_{bc} + R_{ab}R_{ca}}{R_{ab} + R_{bc} + R_{ca}} \tag{2.51}$$

Similarly, for the node pair BC,

$$R_b + R_c = R_{bc} \parallel (R_{ca} + R_{ab}) = \frac{R_{bc}R_{ca} + R_{bc}R_{ab}}{R_{ab} + R_{bc} + R_{ca}} \tag{2.52}$$

Similarly, for the node pair CA,

$$R + R_a = R_{ca} \parallel (R_{ab} + R_{bc}) = \frac{R_{ca}R_{ab} + R_{ca}R_{bc}}{R_{ab} + R_{bc} + R_{ca}} \tag{2.53}$$

Subtracting Eq. (2.52) from Eq. (2.53) gives

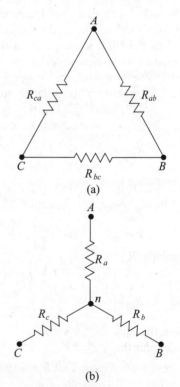

(a)

(b)

Fig. 2.25 (a) Delta-connected resistance, and (b) Star-connected resistances

$$R_a - R_b = \frac{R_{ca}R_{ab} - R_{bc}R_{ab}}{R_{ab} + R_{bc} + R_{ca}} \qquad (2.54)$$

Adding Eqs (2.51) and (2.54) and dividing by 2 results in

$$R_a = \frac{R_{ab}R_{ca}}{R_{ab} + R_{bc} + R_{ca}} \qquad (2.55)$$

Similarly, $\quad R_b = \dfrac{R_{bc}R_{ab}}{R_{ab} + R_{bc} + R_{ca}} \qquad (2.56)$

and $\quad R_c = \dfrac{R_{bc}R_{ab}}{R_{ab} + R_{bc} + R_{ca}} \qquad (2.57)$

Thus, the equivalent star resistance connected to a node is equal to the product of the two delta resistances connected to the same node divided by the sum of delta resistances.

2.7.2 Delta Resistances in Terms of Star Resistances

Dividing Eq. (2.55) by Eq. (2.56) gives

$$\frac{R_a}{R_b} = \frac{R_{ca}}{R_{bc}}$$

$\therefore \qquad R_{ca} = \dfrac{R_a R_{bc}}{R_b}$

Similarly, dividing Eq. (2.55) by Eq. (2.57) gives

$$\frac{R_a}{R_c} = \frac{R_{ab}}{R_{bc}}$$

$\therefore \qquad R_{ab} = \dfrac{R_a R_{bc}}{R_c}$

Substituting for R_{ab} and R_{ca} in Eq. (2.55) and simplifying gives

$$R_{bc} = R_b + R_c + \frac{R_c R_b}{R_a} \qquad (2.58)$$

Similarly, $R_{ca} = R_c + R_a + \dfrac{R_c R_a}{R_b} \qquad (2.59)$

and $\quad R_{ab} = R_a + R_b + \dfrac{R_a R_b}{R_c} \qquad (2.60)$

Thus, the equivalent delta resistance between two nodes is the sum of two star resistances connected to those nodes plus the product of the same two star resistances divided by the third star resistance.

Example 2.11 For the network shown in Fig. 2.26 calculate the equivalent resistance between (a) nodes A and B, and (b) nodes A and n.

Solution (a) For determining the equivalent resistance between nodes A and B the star-connected resistances at node n are converted into delta-connected resistances as shown in Fig. 2.26(b).

$$R_1 = 3 + 2 + \frac{3 \times 2}{6} = 6 \, \Omega$$

$$R_2 = 2 + 6 + \frac{2 \times 6}{3} = 12 \, \Omega$$

$$R_3 = 3 + 6 + \frac{3 \times 6}{2} = 18 \, \Omega$$

Then, $R_{ab} = [(R_1 \parallel 18) \parallel \{(R_2 \parallel 6) + (R_3 \parallel 12)\}]$

$$= [(6 \parallel 18) \parallel \{(12 \parallel 6) + (18 \parallel 12)\}]$$

$$= 4.5 \parallel (4 + 7.2) = \frac{4.5 \times 11.2}{4.5 + 11.2} = 3.21 \, \Omega$$

(b) For determining the equivalent resistance between nodes Δ and n, the star-connected resistances at node C are converted into delta-connected resistances as shown in Fig. 2.26 (c).

$$R_4 = 12 + 6 + \frac{12 \times 6}{6} = 30 \, \Omega$$

$$R_5 = 6 + 6 + \frac{6 \times 6}{12} = 15 \, \Omega$$

$$R_6 = 6 + 12 + \frac{6 \times 12}{6} = 30 \, \Omega$$

$$R_{an} = [(R_6 \parallel 3) \parallel \{(R_5 \parallel 2) + (R_4 \parallel 18)\}]$$

$$= [(30 \parallel 18) \parallel \{(15 \parallel 6) + (30 \parallel 12)\}]$$

$$= (11.25) \parallel (4.29 + 8.57) = (11.25) \parallel (12.86) = 6 \, \Omega$$

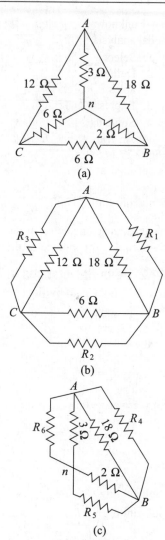

Fig. 2.26

2.8 NODE VOLTAGE ANALYSIS METHOD

Node voltage analysis is one of the general methods for analysis of electrical networks. The node voltage method is based on defining the voltage at each node as an independent variable. One of the nodes is selected as a reference node (usually, but not necessarily, ground), and each of the other node voltages is referenced to this node. Once node voltages are defined, the current flowing in each branch between two adjacent nodes may be determined by applying Ohm's law. The branch currents are expressed in terms of one or more node voltages; thus, currents do not explicitly enter into the equations. Once each branch current is defined in terms of the node voltage, Kirchhoff's current law is applied at each of the nodes except the reference node.

The systematic application of this method to a circuit with n nodes would lead to writing $n-1$ linear equations. However, one of the node voltages is the reference voltage and is therefore already known, since it is usually assumed to be zero. Thus, $n-1$ independent linear equations may be written in terms of the $n-1$ independent node voltage variables. The solution of $n-1$ independent linear equations provides the values of

$n - 1$ unknown node voltages. The branch voltages or currents may be determined from nodal voltages. The nodal analysis method, defined as a sequence of steps, is outlined below:

1. Select a reference node (usually ground).
2. Define the remaining $n - 1$ nodal voltages, with respect to this reference node, as the independent variables.
3. Express each current in terms of the adjacent node voltages. Apply KCL at each of the $n - 1$ nodes.
4. Solve the linear system of $n - 1$ equations in $n - 1$ unknown node voltages.

As an illustration of the method, the circuit shown in Fig. 2.27 is considered. The bottom node c of Fig. 2.27(a) is selected as the reference node and is assumed to be at zero potential, that is, $V_c = 0$. This node is designated as 0 node in Fig. 2.27(b). The direction of the current flow is selected arbitrarily (assuming that I_S is a positive current). The application of KCL at node a yields

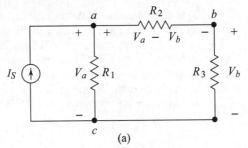

$$I_S - I_1 - I_2 = 0 \qquad (2.61)$$

and, at node b

$$I_2 - I_3 = 0 \qquad (2.62)$$

Now, in applying the node voltage method, the currents I_1, I_2, and I_3 are expressed as functions of V_a, V_b, and V_c, the independent variables. The application of Ohm's law gives

$$I_1 = \frac{V_a - V_c}{R_1} = \frac{V_a}{R_1} \qquad (2.63)$$

$$I_2 = \frac{V_a - V_b}{R_2} \qquad (2.64)$$

$$I_3 = \frac{V_b - V_c}{R_3} = \frac{V_b}{R_1} \qquad (2.65)$$

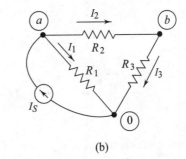

Fig. 2.27   Illustration of node voltage analysis

Substituting the expressions for the three currents in the nodal equation, i.e., Eq. (2.61), the following relationship is obtained:

$$I_S - \frac{V_a}{R_1} - \frac{V_a - V_b}{R_2} = 0$$

or

$$\left(\frac{1}{R_1} + \frac{1}{R_2} \right) V_a + \left(-\frac{1}{R_2} \right) V_b = I_s \qquad (2.66)$$

or

$$(G_1 + G_2)V_a + (-G_1)V_b = I_s \qquad (2.67)$$

where the terms G_1 and G_2 are conductances of resistances R_1 and R_2, respectively. Similarly substituting Eqs (2.63)–(2.65) into Eq. (2.62) and simplifying gives

$$\left(-\frac{1}{R_2} \right) V_a + \left(\frac{1}{R_2} + \frac{1}{R_3} \right) V_b = 0 \qquad (2.68)$$

or

$$(-G_1)V_a + (G_1 + G_3)V_b = 0 \qquad (2.69)$$

When the network parameters and the source current are known, Eqs (2.67) and (2.69) can be solved as simultaneous equations to calculate the unknown node voltages.

Equations (2.67) and (2.69) may be represented in matrix form as

$$\begin{bmatrix} G_1 + G_2 & -G_1 \\ -G_1 & G_2 + G_3 \end{bmatrix} \begin{bmatrix} V_a \\ V_b \end{bmatrix} = \begin{bmatrix} I_s \\ 0 \end{bmatrix} \tag{2.70}$$

or

$$\begin{bmatrix} G_{11} & G_{12} \\ G_{21} & G_{22} \end{bmatrix} \begin{bmatrix} V_a \\ V_b \end{bmatrix} = \begin{bmatrix} I_s \\ 0 \end{bmatrix} \tag{2.71}$$

or $\quad [G][V] = [I]$ (2.72)

where $[G]$ is the conductance matrix of the two-node network, $[V]$ is the column vector of the node voltages, and $[I]$ is the column vector of the algebraic sum of currents of all sources entering the node.

The matrix formulation of the nodal equation [Eq. (2.72)] for an *n*-node system can be made easily by the inspection of the network. The diagonal element G_{ii} is equal to the sum of the conductances connected to the node, the off-diagonal element G_{ij} is negative of the sum of all the conductances connected between the *i*th and *j*th nodes. $[G]$ matrix is symmetric.

The solution for node voltages in Eq. (2.71) by Cramer's rule gives the following expressions:

$$V_a = \frac{\begin{vmatrix} I_S & G_{12} \\ 0 & G_{22} \end{vmatrix}}{\begin{vmatrix} G_{11} & G_{12} \\ G_{21} & G_{22} \end{vmatrix}} = \frac{G_{22}}{D} I_S \tag{2.73}$$

$$V_b = \frac{\begin{vmatrix} G_{11} & I_S \\ G_{21} & 0 \end{vmatrix}}{\begin{vmatrix} G_{11} & G_{12} \\ G_{21} & G_{22} \end{vmatrix}} = -\frac{G_{21}}{D} I_S \tag{2.74}$$

where D is the determinant of the matrix $[G]$ of the two-node network.

Example 2.12 Using the node voltage analysis for the circuit shown in Fig. 2.28(a), find all the node voltages and currents in 1/2 Ω, 1/4 Ω, and 1/3 Ω resistances.

Solution The circuit shown in Fig. 2.28(a) is redrawn in Fig. 2.28(b) with two nodes 1, 2, while the third node is 0 node, the reference node, whose voltage is assumed to be zero. The resistances (in ohm) are converted into conductances (in siemen) as indicated in Fig. 2.28(b).

Applying KCL to node 1 gives

$$2 - 2V_1 - 4(V_1 - V_2) = 0$$

or $\qquad 6V_1 - 4V_2 = 2$ (I)

Applying KCL to node 2 gives

$$-3 - 3V_2 - 4(V_2 - V_1) = 0$$

or $\qquad -4V_1 + 7V_2 = -3$ (II)

Equations (I) and (II) may be written in matrix form as

$$\begin{bmatrix} 6 & -4 \\ -4 & 7 \end{bmatrix} \begin{bmatrix} V_1 \\ V_2 \end{bmatrix} = \begin{bmatrix} 2 \\ -3 \end{bmatrix}$$

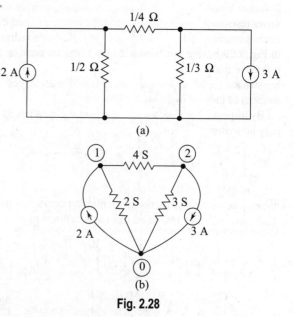

(a)

(b)

Fig. 2.28

Using Cramer's rule $V_1 = \dfrac{\begin{vmatrix} 2 & -4 \\ -3 & 7 \end{vmatrix}}{\begin{vmatrix} 6 & -4 \\ -4 & 7 \end{vmatrix}} = \dfrac{14-12}{42-16} = \dfrac{2}{26} = 0.0769$ V

Similarly, $V_2 = \dfrac{\begin{vmatrix} 6 & 2 \\ -4 & -3 \end{vmatrix}}{26} = \dfrac{-18+8}{26} = \dfrac{-10}{26} = -0.3846$ V

and the current in $1/2\ \Omega$ resistance is $2 \times 0.0769 = 0.154$ A, in $1/4\ \Omega$ resistance is $4 \times (0.0769 + 0.3846) = 1.846$ A, and in $1/3\ \Omega$ resistance is $3 \times (-0.3846) = -1.154$ A.

The negative current in $1/3\ \Omega$ resistance indicates that the current flows from node 0 to node 2 of magnitude 1.154 A.

2.8.1 Nodal Analysis with Voltage Sources

It may appear that the node voltage method is easily applicable when current sources are present, since current sources can be directly accounted for while applying KCL at a node. In case an ideal voltage source is present in a circuit and is connected between the reference and non-reference nodes, the voltage of the latter node is held constant at the voltage of the ideal source. To that extent, the nodal analysis is simplified since the number of unknown node voltage equations required to be solved is one less than the number of equations when a circuit contains current sources only. In case of a practical voltage source, it can be converted into a current source as discussed in Section 2.2.5. A numerical example to illustrate such a case follows.

Example 2.13 Using the node voltage analysis for the circuit shown in Fig. 2.29(a), find all the node voltages and currents in $1/3\ \Omega$, $1/5\ \Omega$, and $1/6\ \Omega$ resistances.

Solution Since two of the sources are voltage sources with series resistances, these two sources are first converted into current sources. The circuit shown in Fig. 2.29(a) is redrawn in Fig. 2.29(b) with two nodes 1 and 2 while 0 node is the reference node whose voltage is assumed to be zero. The resistances (in ohm) are converted into conductances (in siemen) as indicated in Fig. 2.29(b).

By inspection, the node voltage equations in matrix form may be written as follows:

$$\begin{bmatrix} 2+3+5 & -5 \\ -5 & 5+6+4 \end{bmatrix} \begin{bmatrix} V_1 \\ V_2 \end{bmatrix} = \begin{bmatrix} 20+5 \\ 20-5 \end{bmatrix}$$

or $\begin{bmatrix} 10 & -5 \\ -5 & 15 \end{bmatrix} \begin{bmatrix} V_1 \\ V_2 \end{bmatrix} = \begin{bmatrix} 25 \\ 15 \end{bmatrix}$

Using Cramer's rule the node voltages are obtained as

$$V_1 = \dfrac{\begin{vmatrix} 25 & -5 \\ 15 & 15 \end{vmatrix}}{\begin{vmatrix} 10 & -5 \\ -5 & 15 \end{vmatrix}} = \dfrac{375+75}{150-25} = \dfrac{450}{125} = 3.6\ V$$

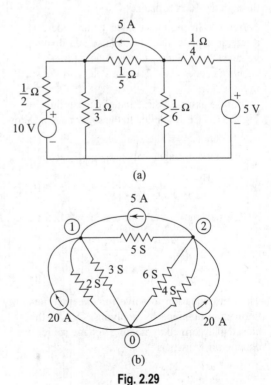

(a)

(b)

Fig. 2.29

Similarly, $V_2 = \dfrac{\begin{vmatrix} 10 & 25 \\ -5 & 15 \end{vmatrix}}{125} = \dfrac{150+125}{125} = \dfrac{275}{125} = 2.2$ V

and the current in $1/3$ Ω resistance is $3 \times 3.6 = 10.8$ A, in $1/5$ Ω resistance is $5 \times (3.6 - 2.2) = 7$ A, and in $1/6$ Ω resistance is $6 \times 2.2 = 13.2$ A.

2.8.2 Circuit Analysis with Super Nodes

When neither terminal of a voltage source is connected to the reference node, it is called a floating voltage source. Such types of circuits require special techniques for undertaking nodal analysis. Since the voltage in a voltage source is defined as being independent of the current and the current in the voltage source branch is also unknown, the direct application of the KCL is not possible. Such a difficult situation is avoided by creating a *super node*, which is formed by enclosing the terminal nodes of the voltage source within a curve and treating it as a single node, that is, applying KCL in the same manner as it is applied to a regular node. Thus, the application of KCL to a super node requires that the sum of the currents entering/leaving a closed curve is zero, which is to be expected since the law of conservation of charge for a super node must also hold true. The technique for applying the nodal voltage analysis to a circuit with a super node is demonstrated through an example given below.

Example 2.14 A 24 V voltage source is connected between nodes 1 and 2 as shown in the circuit shown in Fig. 2.30. Determine (a) the voltages of nodes 1 and 3 and (b) the current through the 16 Ω resistor.

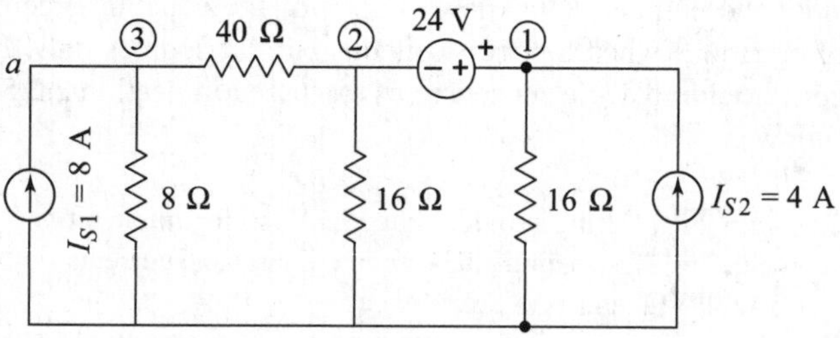

Solution Since neither node 1 nor node 2 is a reference node, the 24 V source is a floating voltage source. Hence, a super node is formed by enclosing nodes 1 and 2 within a curve as shown in Fig. 2.31.

For the assumed direction of current flow in Fig. 2.31, KCL is applied to the super node a follows:

$$I_1 + I_{S2} = I_3 + I_4 \qquad \text{(I)}$$

In terms of the node voltages, Eq. (I) becomes

or $\quad \dfrac{V_3 - V_2}{40} + 4 = \dfrac{V_2}{16} + \dfrac{V_1}{16} \qquad$ (II)

KCL is next applied to node 3 as shown below:

$$I_1 + I_2 = I_{S1} \quad \text{or} \quad \dfrac{V_3 - V_2}{40} + \dfrac{V_3}{8} = 8$$

or $\quad -V_2 + 6V_3 = 320 \qquad$ (III)

The third equation to solve for the three unknowns is obtained by applying KVL to the closed path formed by floating voltage source and the reference node. Thus, starting with the reference node, the equation is written as

$$-V_1 + 24 + V_2 = 0 \quad \text{or} \quad V_1 - V_2 = 24 \qquad \text{(IV)}$$

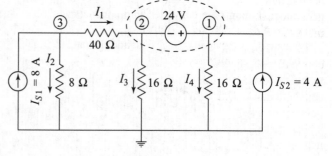

Fig. 2.30

Fig. 2.31

Writing Equations (II), (III), and (IV) in matrix form leads to

$$\begin{bmatrix} 5 & 7 & -2 \\ 0 & -1 & 6 \\ 1 & -1 & 0 \end{bmatrix} \begin{bmatrix} V_1 \\ V_2 \\ V_3 \end{bmatrix} = \begin{bmatrix} 320 \\ 320 \\ 24 \end{bmatrix}$$

or

$$\begin{bmatrix} V_1 \\ V_2 \\ V_3 \end{bmatrix} = \begin{bmatrix} 5 & 7 & -2 \\ 0 & -1 & 6 \\ 1 & -1 & 0 \end{bmatrix}^{-1} \begin{bmatrix} 320 \\ 320 \\ 24 \end{bmatrix} = \begin{bmatrix} 50.29 \\ 26.29 \\ 57.71 \end{bmatrix} \text{ V}$$

Thus, the voltages of nodes 1 and 3 are, respectively, 50.29 and 57.71 V and the current through the 16 Ω resistor is $I_3 = 26.29/16 = 1.64$ A.

The computations are easily verified when it is seen that $50.29 - 26.29 = 24$ V and the current at node 3 is $(57.71 - 26.29)/40 + (57.71/8) = 8$ A

2.9 MESH CURRENT ANALYSIS METHOD

Analysis by mesh currents consists of defining the currents around the individual meshes as independent variables. The direction of the current flow around a mesh may be assigned either a clockwise or a counterclockwise sense. To avoid confusion, mesh currents will be defined exclusively clockwise when using this method. A simple application of KVL around each mesh provides the desired equations. The number of equations is equal to the number of meshes in the circuit. All branch currents and voltages may subsequently be obtained from mesh currents. The procedure used in applying the mesh current method to a linear circuit may be outlined as follows:

1. Define each mesh current in a clockwise direction.
2. Apply KVL around each mesh expressing voltage across each element in terms of one or more mesh currents flowing through it.
3. Solve the resulting linear equations with mesh currents as independent variables.

The number of independent mesh equations, m, is related to branches b and nodes n in an electrical network by the equation

$$m = b - n + 1 \tag{2.75}$$

To illustrate the mesh current method, a simple two-mesh circuit shown in Fig. 2.32 is considered. This circuit will be used to generate two equations in the two unknown mesh currents I_1 and I_2.

In applying KVL, the proper sign convention explained in Section 1.11 is to be followed. According to the sign convention the voltage drop from a positive polarity to a negative polarity is assigned a positive sign and the voltage rise from a

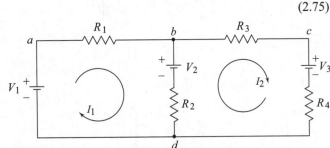

Fig. 2.32 Illustration of mesh current analysis method

negative polarity to a positive polarity is assigned a negative sign. KVL is a consequence of the fact that no energy is lost or created in an electric circuit.

By applying KVL to mesh 1 along with the sign convention of taking voltage drops as positive and voltage rise as negative, the following expression may be obtained:

$$R_1 I_1 + R_2 (I_1 - I_2) - V_1 + V_2 = 0 \tag{2.76}$$

or

$$(R_1 + R_2) I_1 - R_2 I_2 = V_1 - V_2 \tag{2.77}$$

Similarly, applying KVL to mesh 2 the following expression results:

$$R_2(I_2 - I_1) + R_3I_2 + R_4I_2 - V_2 + V_3 = 0 \tag{2.78}$$

or $\quad -R_2I_1 + (R_2 + R_3 + R_4)I_2 = V_2 - V_3 \tag{2.79}$

Equations (2.77) and (2.79) may be solved simultaneously to obtain the desired solution, namely, the mesh currents I_1 and I_2. Knowledge of the mesh currents permits the determination of all the other voltages and currents in the circuit.

Equations (2.77) and (2.79) may be written in matrix form as

$$\begin{bmatrix} R_1 + R_2 & -R_2 \\ -R_2 & R_1 + R_2 + R_3 \end{bmatrix} \begin{bmatrix} I_1 \\ I_2 \end{bmatrix} = \begin{bmatrix} V_1 - V_2 \\ V_2 - V_3 \end{bmatrix} \tag{2.80}$$

or $\quad \begin{bmatrix} R_{11} & R_{12} \\ R_{21} & R_{22} \end{bmatrix} \begin{bmatrix} I_1 \\ I_2 \end{bmatrix} = \begin{bmatrix} V_1 - V_2 \\ V_2 - V_3 \end{bmatrix} \tag{2.81}$

or $\quad [R][I] = [V] \tag{2.82}$

where $[R]$ is the resistance matrix of the two-node network, $[V]$ is the column vector of the algebraic sum of all the source voltages around the mesh, $[I]$ is the column vector of mesh currents.

The matrix formulation of the nodal equation [Eq. (2.82)] for an n-mesh system can be made easily by inspection of the network. The diagonal element R_{ii} is equal to the sum of all the resistances connected in mesh 1, the off-diagonal element R_{ij} is negative of the sum of all resistances common between the ith and jth meshes. $[R]$ matrix is symmetric. Equation (2.81) may be solved by Cramer's rule to obtain the values of mesh currents I_1 and I_2.

Example 2.15 Using the mesh current analysis, find the voltage v across the 3 Ω resistor.

Solution Thee independent mesh currents I_1, I_2, and I_3 are first identified. Then, for the three mshes, KVL is applied. By inspection the mesh equation in matrix form may be written as

$$\begin{bmatrix} 1+2+3 & -2 & -3 \\ -2 & 2+1+2 & -1 \\ -3 & -1 & 1+1+3 \end{bmatrix} \begin{bmatrix} I_1 \\ I_2 \\ I_3 \end{bmatrix}$$

$$= \begin{bmatrix} 20 \\ -10 \\ 0 \end{bmatrix}$$

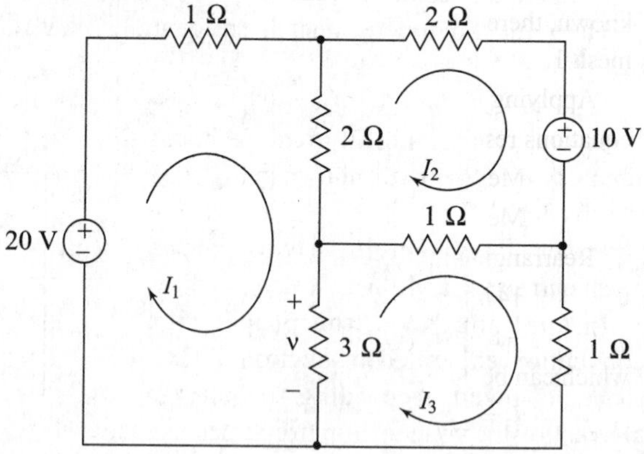

or $\quad \begin{bmatrix} 6 & -2 & -3 \\ -2 & 5 & -1 \\ -3 & -1 & 5 \end{bmatrix} \begin{bmatrix} I_1 \\ I_2 \\ I_3 \end{bmatrix} = \begin{bmatrix} 20 \\ -10 \\ 0 \end{bmatrix}$ (A)

Fig. 2.33

Solving equation (A) by Cramer's rule,

$$I_1 = \frac{\begin{vmatrix} 20 & -2 & -3 \\ -10 & 5 & -1 \\ 0 & -1 & 5 \end{vmatrix}}{\begin{vmatrix} 6 & -2 & -3 \\ -2 & 5 & -1 \\ -3 & -1 & 5 \end{vmatrix}} = \frac{20\begin{vmatrix} 5 & -1 \\ -1 & 5 \end{vmatrix} + 10\begin{vmatrix} -2 & -3 \\ -1 & 5 \end{vmatrix}}{6\begin{vmatrix} 5 & -1 \\ -1 & 5 \end{vmatrix} + 2\begin{vmatrix} -2 & -3 \\ -1 & 5 \end{vmatrix} - 3\begin{vmatrix} -2 & -3 \\ 5 & -1 \end{vmatrix}}$$

$$= \frac{20(25-1)+10(-10-3)}{6(25-1)+2(-10-3)-3(2+15)} = \frac{350}{67} = 5.224 \text{ A}$$

Similarly,

$$I_2 = \frac{\begin{vmatrix} 6 & 20 & -3 \\ -2 & -10 & -1 \\ -3 & 0 & 5 \end{vmatrix}}{67} = \frac{250}{67} = 0.7463 \text{ A}$$

and $$I_3 = \frac{\begin{vmatrix} 6 & -2 & 20 \\ -2 & 5 & -10 \\ -3 & -1 & 0 \end{vmatrix}}{67} = \frac{220}{67} = 3.28$$

The voltage across the 3 Ω resistor, $v = (I_1 - I_3) \times 3 = (5.224 - 3.28) \times 3 = 5.832$ V.

2.9.1 Mesh Analysis with Current Sources

Mesh analysis is particularly simple when applied to circuits having voltage sources only. However, when a current source is present in the circuit, the current source must be converted into an equivalent voltage source as discussed in Section 2.2.5. A numerical example to illustrate such a case follows.

Example 2.16   Using mesh analysis find the mesh currents in Fig. 2.34.

Solution   For mesh 1 it can be inferred that the current source forces the mesh current $I_1 = 1$ A. As I_1 is already known, there is no need to write the KVL equation for mesh 1.

Applying KVL to mesh 2 and mesh 3 the following equations result:

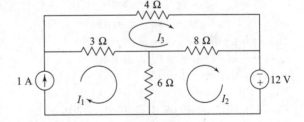

Fig. 2.34

Mesh 2 : $6(I_2 - I_1) + 8(I_2 - I_3) - 12 = 0$
Mesh 3 : $3(I_3 - I_1) + 4I_3 + 8(I_3 - I_2) = 0$

Rearranging these equations and substituting $I_1 = 1$ A, two equations with two unknowns are obtained as

$$14I_2 - 8I_3 = 18$$
$$-8I_2 + 15I_3 = 3$$

which can be solved to obtain

$$I_3 = 2.013 \text{ A}, \quad I_2 = 1.273 \text{ A}$$

2.9.2 Circuit Analysis with Super Meshes

Difficulties arise in writing mesh equations when an independent current source or a number of sources are common to two or more meshes. In such a case, it is difficult to assign a voltage to the independent current source(s) since a current source will supply rated current at all voltages. In order to get around such a tricky situation, a *super mesh* is created by including the meshes on the two sides of the current source and any resistance connected in series with it.

The following example shows the formation of a super mesh and the development of mesh current equations along with an auxiliary equation, which correlates the current source output with the adjoining mesh currents.

Example 2.17 For the circuit shown in Fig. 2.35, use mesh current analysis technique to compute (a) the voltage at node D, (b) current in the 4 Ω resistor, and (c) the power supplied by the 18 V source. All data is shown in the circuit diagram.

Solution The 12 A independent current source is first transformed in to an equivalent voltage source of $12 \times (3/4)$ = 9 V and a $(3/4)$ Ω resistor connected in series. The transformation, along with the assumed mesh currents, is shown in Fig. 2.36.

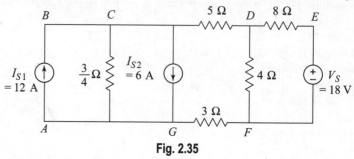

Fig. 2.35

Next, since the 6 A current source is common to meshes 1 and 2, a super mesh is formulated by including the two meshes and the current source as shown in Fig. 2.36. To analyse the circuit commence by writing the mesh current equation for the mesh ABCDFGA as follows:

$-9 + I_1 \times (3/4) + I_2 \times 5 +$
$$(I_2 - I_3) \times 4 + I_2 \times 3 = 0$$
or $\quad 3I_1 + 48I_2 - 16I_3 = 36 \quad$ (I)

Similarly, the mesh current for FDEF is given by

$$(I_3 - I_2) \times 4 + I_3 \times 8 + 18 = 0$$
or $\quad -4I_2 + 12I_3 = -18 \quad$ (II)

The auxiliary mesh equation is obtained from the mesh currents I_1, I_2 and the output of the current source is

$\quad I_1 - I_2 = 6 \quad$ (III)

Substitution of Eq. (III) in (I) gives $51I_2 - 16I_3 = 18 \quad$ (IV)

Multiplying Eq. (II) by 4 and Eq. (IV) by 3 and adding the two resultant equations, we get

$\quad 137\,I_2 = -18$
or $\quad I_2 = -18/137 = -0.13$ A

From Eq. (III), $I_1 = 6 - 0.13 = 5.87$ A

and Eq. (II) yields $I_3 = (-18 - 4 \times 13)/12 = -1.54$ A

Voltage at node D = $18 - 1.54 \times 8 = 5.68$ V

Current in the 4 Ω resistor = $5.86/4 = 1.42$ A

Power supplied by the 18 V source = $18 \times 1.54 = 27.72$ W

2.10 NODAL AND MESH ANALYSIS WITH DEPENDENT SOURCES

Solution methods that allow for the presence of dependent sources will be particularly useful in the study of circuits with active elements. When a dependent source is present in a circuit to be analysed by node or mesh analysis, one can initially treat it as an ideal source and write the node or mesh equations accordingly. In addition to this, there will also be an equation relating the dependent source to one of the circuit voltages or currents. This constraint equation can then be substituted in the set of equations obtained by the techniques of nodal and mesh analyses, and the equations can subsequently be solved for unknowns. It is important to note that once the constraint equation has been substituted in the initial system of equations, the number of unknowns remains unchanged.

Figure 2.37(a) shows a circuit with a dependent source. The dependent source voltage $V_S = kV_2$, where V_2 is the voltage across R_3 and k is a constant whose value is known. It is desired to find node voltages.

Since one of the sources is a voltage source V_S with a series resistance R_1, it is first converted into a current source as I_1

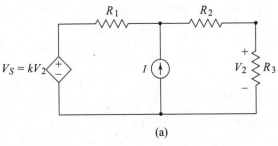

where

$$I_1 = \frac{V_S}{R_1} = \frac{kV_2}{R_1}$$

The circuit is redrawn in Fig. 2.37(b), where three nodes are identified and labeled as 1, 2, and ground node 0. Applying KCL to node 1, the following equations result:

$$I + \frac{kV_2}{R_1} - \frac{V_1}{R_1} - \frac{V_1 - V_2}{R_2} = 0 \qquad (2.83)$$

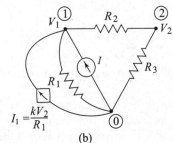

$$\text{or} \quad \left(\frac{1}{R_1} + \frac{1}{R_2}\right) V_1 + \left(-\frac{k}{R_1} - \frac{1}{R_2}\right) V_2 = I \qquad (2.84)$$

Fig. 2.37 Circuit with dependent source

Applying KCL to node 2, the following expression is obtained:

$$\frac{V_1 - V_2}{R_2} - \frac{V_2}{R_3} = 0 \qquad (2.85)$$

$$\text{or} \quad \left(-\frac{1}{R_2}\right) V_1 + \left(\frac{1}{R_2} + \frac{1}{R_3}\right) V_2 = 0 \qquad (2.86)$$

The node voltages V_1 and V_2 may be obtained by solving Eqs (2.84) and (2.86) as simultaneous equations.

Example 2.18   The dependent current source I_d is related to the voltage V_{ab} in Fig. 2.38 through the relation $I_d = 0.4V_{ab}$. Find the current through the 8 Ω resistor by nodal analysis.

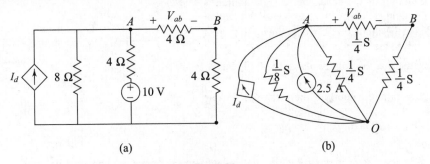

Fig. 2.38

Solution   First of all, a 10 V source in series with 4 Ω resistance is converted into a current source $I_S = 2.5$ A in parallel with conductance of 1/4 S. The circuit shown in Fig 2.3(a) is redrawn showing two nodes A ad B and the reference node 0 and the resistance of each branch replaced by its conductance as shown in Fig. 2.38(b).

Nodal equation for node A:

$$0.4\left(V_A - V_B\right) + 2.5 - \left(\frac{1}{8} + \frac{1}{4}\right)V_A - \frac{1}{4}\left(V_A - V_B\right) = 0$$

or $\dfrac{9}{40}V_A - \dfrac{3}{20}V_B = 2.5$ (I)

Nodal equation for node B:

$$-\frac{1}{4}V_A + \frac{1}{2}V_B = 0$$ (II)

Solving Eqs (I) and (II),

$V_A = 8.33$ V, $V_B = 4.17$ V

Therefore, current through the 8 Ω resistor is 1.04 A.

Example 2.19   In the circuit shown in Fig. 2.39 the voltage magnitude of the dependent source $V_d = 2I_8$, where I_8 is the current flowing through the 8 Ω resistance. Find the current through the 4 Ω resistance by using the mesh current method.

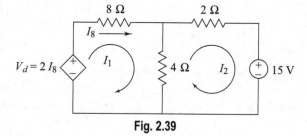

Fig. 2.39

Solution   From Fig. 2.39 it may be noted that $I_8 = I_1$, hence $V_d = 2I_1$. The mesh equations are:

For mesh 1, $12I_1 - 4I_2 - V_d = 0$

or $10I_1 - 4I_2 = 0$ (I)

For mesh 2, $-4I_1 + 6I_2 = -15$ (II)

Solving Eqs (I) and (II),

$I_1 = -1.363$ A and $I_2 = -3.40$ A

Therefore, current through the 4 Ω resistance is $I_1 - I_2 = 2.037$ A.

Example 2.20   Calculate (a) the current supplied by the dependent current source and (b) the power supplied by the independent voltage source in the circuit shown in Fig. 2.40. The output of the dependent current source is reliant upon the voltage across the 2 Ω resistor. All other data is provided in the circuit diagram. Use the mesh current technique to analyse the circuit.

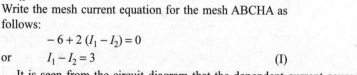

Fig. 2.40

Solution   Assume that the voltage across the 2 Ω resistor is V_X volts.

Write the mesh current equation for the mesh ABCHA as follows:

$$-6 + 2\left(I_1 - I_2\right) = 0$$

or $I_1 - I_2 = 3$ (I)

It is seen from the circuit diagram that the dependent current source is connected between meshes 2 and 3. Therefore, to formulate the mesh current equation super mesh GFEDCHG is employed as follows:

$$4 \times I_3 + 3 \times I_2 + 4 \times I_2 + V_X + 6 \times I_3 = 0$$

or $\qquad 7I_2 + 10I_3 + V_X = 0 \qquad\qquad\qquad\qquad\qquad\qquad$ (II)

It may be noted that $V_X = 2(I_2 - I_1)$. Substitution for V_X in Eq. (II) results in

$$-2I_1 + 9I_2 + 10I_3 = 0 \qquad\qquad\qquad\qquad\qquad\qquad\text{(III)}$$

In order to obtain the third auxiliary equation, the co-relation between the output of the depnent current source and mesh currents I_2 and I_3 is used. Thus,

$$I_3 - I_2 = 2V_X = 2(I_2 - I_1)$$

or $\qquad 2I_1 - 3I_2 + I_3 = 0 \qquad\qquad\qquad\qquad\qquad\qquad$ (IV)

Use of Cramer's rule leads to the mesh currents as follows:

$$I_1 = \frac{\begin{vmatrix} 3 & -1 & 0 \\ 0 & 9 & 10 \\ 0 & -3 & 1 \end{vmatrix}}{\begin{vmatrix} 1 & -1 & 0 \\ -2 & 9 & 10 \\ 2 & -3 & 1 \end{vmatrix}} = \frac{117}{17} = 6.89 \text{ A}$$

$$I_2 = \frac{\begin{vmatrix} 1 & 3 & 0 \\ -2 & 0 & 10 \\ 2 & 0 & 1 \end{vmatrix}}{\begin{vmatrix} 1 & -1 & 0 \\ -2 & 9 & 10 \\ 2 & -3 & 1 \end{vmatrix}} = \frac{66}{17} = 3.89 \text{ A}$$

$$I_3 = \frac{\begin{vmatrix} 1 & -1 & 3 \\ -2 & 9 & 0 \\ 2 & -3 & 0 \end{vmatrix}}{\begin{vmatrix} 1 & -1 & 0 \\ -2 & 9 & 10 \\ 2 & -3 & 1 \end{vmatrix}} = \frac{-36}{17} = -2.12 \text{ A}$$

Current supplied by the dependent source $= 2V_X = 2 \times 2(I_2 - I_1) = -12 \text{ A}$

Power supplied by the voltage source $= 6 \times I_1 = 41.29 \text{ W}$

2.10.1 Node-Voltage versus Mesh-Current Techniques

Nodal analysis is used when voltage sources are present and mesh current technique is used when current sources are present. Generally, however, the choice of the technique depends on the number of unknowns and hence the number of simultaneous equations required to be solved.

In the case of nodal analysis, the number of simultaneous equations required to be solved is equal to the number of nodes minus one. Further, every voltage source connected to the reference decreases the number of simultaneous equations by unity.

Mathematically, the number of simultaneous equations, N_{eq}, may be expressed as

$$N_{eq} = N_n - N_V - 1 \qquad\qquad\qquad\qquad\qquad\qquad\text{(2.87a)}$$

where $\quad N_n$ = number of nodes in the circuit

$\qquad\quad N_V$ = number of voltage sources connected to the reference node

In the case of mesh current analysis, the number of unknown current variables and, therefore, the number of simultaneous equations required is equal to the number of meshes reduced by the number of current sources. Thus,

$$N_{eq} = N_m - N_C \qquad (2.87b)$$

where N_m = number of meshes in the circuit

N_C = number of current sources

It may also be noted that when a circuit is made up of crossing elements, that is, the circuit is non-planar, it is not feasible to employ mesh current technique. Hence, nodal voltage analysis is required to be employed.

Another important point to be noted is that the presence of every super node/super mesh increases the number of equations by one. Finally, the technique that leads to a lesser number of simultaneous equations decides the method of circuit analysis.

2.11 NETWORK THEOREMS

A systematic procedure to write a set of equations that describe a given network completely has been studied so far. Branch currents and voltages can be determined from the solution of these equations. Many network problems involve only a restricted analysis, namely, the determination of current and voltage of one branch only, or computing voltage and current supplied to a load, or finding maximum power that can be transferred from a network to a load, etc. Such network analysis problems can be expeditiously solved by the application of certain network theorems. Some network theorems will be discussed in the following sections.

2.11.1 Superposition Theorem

The superposition theorem states that in a linear bilateral network containing N sources, each branch current (branch voltage) is the algebraic sum of N currents (branch voltages), each of which is determined by considering one source at a time and removing all other sources. In removing the sources, voltage sources are short-circuited (or replaced by resistances equal to their internal resistances for non-ideal sources), while the ideal current sources are open-circuited.

Te application of the superposition theorem is illustrated with the following example.

Example 2.21   Determine the current flowing in the 8 Ω resistor in the circuit shown in Fig. 2.41.

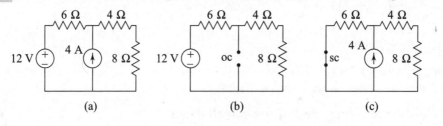

(a)   (b)   (c)

Fig. 2.41

Solution   Assuming the bottom node to be the reference node of the circuit, first the current source is open-circuited (oc) as shown in Fig. 2.41(b). The circuit becomes a simple series circuit consisting of a 12 V source in series with three resistances in series. Then the current through the 8 Ω resistor is $\dfrac{12}{6+4+8} = 0.667$ A

Next the voltage source is short-circuited (sc) as shown in Fig. 2.41(c). The resultant circuit has two parallel branches—the branch at the centre is a 4 A current source, with a 6 Ω resistance on the left, and on the right a total resistance of $(4 + 8) = 12\ \Omega$. The current of 4 A is distributed in the parallel branches as per the current divider rule.

Then the current through the 8 Ω resistor is $4 \times \dfrac{6}{6+12} = 1.333$ A

Thus the total current through the 8 Ω resistor is $0.667 + 1.333 = 2.0$ A.

2.11.2 Thevenin's Theorem

Thevenin's theorem states that any pair of terminals AB of a linear active network may be replaced by an equivalent voltage source V_{TH}, in series with an equivalent resistance R_{TH}. The value of V_{TH} (called the Thevenin voltage) is equal to pd between the terminals AB when they are open-circuited, and R_{TH} is the equivalent resistance looking into the network at AB with the independent active sources set to zero, i.e., with all the independent voltage sources short-circuited and all the independent current sources open-circuited; of course, leaving behind their internal resistances.

The computation of V_{TH} and R_{TH} is now illustrated by considering the terminal pair AB of the circuit shown in Fig. 2.42(a). To compute V_{TH} the resistance R_L is disconnected from AB and the circuit configuration with R_L removed is as shown in Fig. 2.42(b). As no current flows through R_3, the open-circuit voltage V_{oc} must be equal to the voltage cross R_2. Since in the circuit shown in Fig. 2.42(b) R_1 and R_2 connected across V_S form a closed mesh, the voltage across R_2 may be obtained by applying the voltage divider rule as

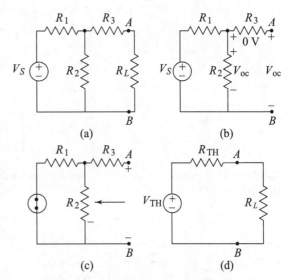

Fig. 2.42   Illustration of Thevenin's theorem

$$V_{oc} = V_{R_2} = \frac{R_2}{R_1 + R_2} V_S \tag{2.88}$$

and $\qquad V_{TH} = V_{oc} \tag{2.89}$

In order to compute R_{TH}, resistance R_L is removed from the circuit and the independent voltage source is replaced by a short circuit as shown in Fig. 2.42(c). Then R_{TH}, the total resistance between the terminals A and B, can be determined as

$$R_{TH} = R_3 + R_1 \parallel R_2 = R_3 + \frac{R_1 R_2}{R_1 + R_2} \tag{2.90}$$

Figure 2.42(d) shows the Thevenin equivalent circuit of the network shown in Fig. 2.42(a).

It should be noted that the dependent current sources are left unaltered and they are not set equal to zero while finding out R_{TH}. In case dependent sources are present, then R_{TH} is determined using the following relationship:

$$R_{TH} = \frac{V_{TH}}{I_{sc}} \tag{2.91}$$

where I_{sc} is the current through the short circuit between the terminals A and B.

Thevenin's theorem is sometimes referred to as Helmholtz's theorem. Thevenin's theorem is a handy tool to simplify computations in networks with varying loads. If R_L connected to the terminals A and B in

Fig. 2.42 indicates varying load, then the response can be easily determined reducing the network to its Thevenin equivalent. However, the information with regard to currents and voltages within the network is lost once Thevenin's equivalent is formed.

Example 2.22   Use Thevenin's theorem to find current through the 6 Ω resistor in the circuit shown in Fig. 2.43(a).

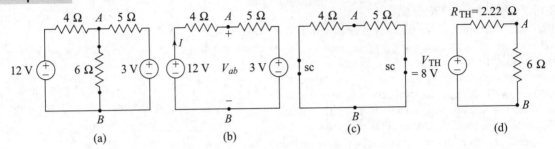

Fig. 2.43

Solution   First the 6 Ω resistor across *AB* is disconnected, as shown in Fig. 2.43(b). By applying KVL to the circuit shown in Fig. 2.43(b),

$$I(4+5) + 3 - 12 = 0$$

∴ $I = 1$ A

Then, $V_{TH} = V_{AB} = 12 - I \times 4 = 12 - 1 \times 4 = 8$ V

To determine R_{TH}, the voltage sources are short-circuited as shown in Fig. 2.43 (c). The resistance seen at the terminals *AB* is equal to the parallel combination of 4 Ω and 5 Ω resistances. Then

$$R_{TH} = 4 \parallel 5 = \frac{4 \times 5}{4 + 5} = 2.22 \ \Omega$$

and the current through the 6 Ω resistor is $\dfrac{V_{TH}}{R_{TH} + 6} = \dfrac{8}{2.22 + 6} = 0.973$ A

Example 2.23   Use (a) Thevenin's theorem and (b) the principle of superposition to find the current when a 2 Ω resistor is connected between terminals *X* and *Y* in the circuit shown in Fig. 2.44(a).

Solution   (a) First step is to determine the open-circuit voltage V_{TH} across *XY*. Applying the esh analysis to the circuit shown in Fig. 2.44(a), I_1 and I_2 are the two mesh currents. Applying KVL to the two meshes, the following equations can be written in matrix form by inspection:

$$\begin{bmatrix} 14 & -12 \\ -12 & 16 \end{bmatrix} \begin{bmatrix} I_1 \\ I_2 \end{bmatrix} = \begin{bmatrix} 6 \\ -12 \end{bmatrix}$$

Solving this equation for I_1 and I_2 using Cramer's rule,

$$I_1 = \frac{\begin{vmatrix} 6 & -12 \\ -12 & 16 \end{vmatrix}}{\begin{vmatrix} 14 & -12 \\ -12 & 16 \end{vmatrix}} = \frac{98 - 144}{80} = \frac{-48}{80} = -0.6 \text{ A}$$

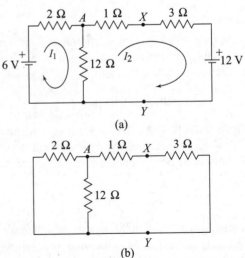

(a)

(b)

Fig. 2.44

and $I_2 = \dfrac{\begin{vmatrix} 14 & 6 \\ -12 & -12 \end{vmatrix}}{80} = \dfrac{-168 + 72}{80} = \dfrac{-96}{80} = -1.2 \text{ A}$

The negative sign indicates that the actual flow of currents is opposite to the assumed direction of flow. Thus,

$$V_{TH} = 12 - 1.2 \times 3 = 8.4 \text{ V}$$

To determine R_{TH} the voltage sources are short-circuited and the circuit redrawn as shown in Fig. 2.44(b). Then

$$R_{TH} = (1 + 2 \parallel 12) \parallel 3 = \left(1 + \frac{2 \times 12}{2 + 12}\right) \parallel 3$$

$$= \frac{19}{7} \parallel 3 = \frac{\frac{19}{7} \times 3}{\frac{19}{7} + 3} = \frac{57}{40} = 1.425 \ \Omega$$

Then the current through the 2 Ω resistor connected across XY is

$$I_{XY} = \frac{V_{TH}}{R_{TH} + 2} = \frac{8.4}{1.425 + 2} = 2.45 \text{ A}$$

(b) Before applying the superposition theorem, the voltage sources of the circuit shown in Fig. 2.44(a) are first converted to current sources. The 6 V source in series with the 2 Ω resistor is converted to a 3 A current source and the 12 V source is converted to a current source of 4 A in parallel with the 3 Ω resistor as shown in Fig. 2.45(a). In Fig. 2.45(a) the resistance values are replaced by their conductance values in siemens (S). The circuit has three nodes: A, X, and Y. The node Y is taken as the reference node, and V_A and V_X are the node voltages of nodes A and X, respectively.

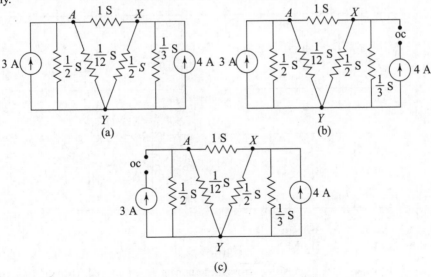

Fig. 2.45

For using the superposition theorem first the 4 A current source is open-circuited, as shown in Fig. 2.45(b), and the node voltage of node X is determined by the nodal analysis. By inspection of Fig. 2.45(b), the nodal equations in matrix form can be written as

$$\begin{bmatrix} \dfrac{1}{2}+\dfrac{1}{12}+1 & -1 \\ -1 & \dfrac{1}{3}+\dfrac{1}{2}+1 \end{bmatrix} \begin{bmatrix} V_A \\ V_X \end{bmatrix} = \begin{bmatrix} 3 \\ 0 \end{bmatrix} \quad \text{or} \quad \begin{bmatrix} \dfrac{19}{12} & -1 \\ -1 & \dfrac{11}{6} \end{bmatrix} \begin{bmatrix} V_A \\ V_X \end{bmatrix} = \begin{bmatrix} 3 \\ 0 \end{bmatrix}$$

Solving this for V_X using Cramer's rule,

$$V_X = \frac{\begin{vmatrix} 19/12 & 3 \\ -1 & 0 \end{vmatrix}}{\begin{vmatrix} 19/12 & -1 \\ -1 & 11/6 \end{vmatrix}} = \frac{3}{(19/12) \times (11/6) - 1} = \frac{3}{137/72} = 1.57 \text{ V}$$

Next, the 3 A current source is open-circuited, as shown in Fig. 2.45(c), and the node voltage of node X is determined by the nodal analysis. By inspection of Fig. 2.45(c), the nodal equations in matrix form can be written as

$$\begin{bmatrix} \dfrac{1}{2}+\dfrac{1}{12}+1 & -1 \\ -1 & \dfrac{1}{3}+\dfrac{1}{2}+1 \end{bmatrix} \begin{bmatrix} V_A \\ V_X \end{bmatrix} = \begin{bmatrix} 0 \\ 4 \end{bmatrix}$$

Solving the above for V_X using Cramer's rule, $V_X = 3.33$

Using the superposition theorem, voltage $V_X = 1.57 + 3.33 = 4.90$, and the current flowing through the 2 Ω resistor connected between XY, $I_X = 4.9/2 = 2.45$ A.

Example 2.24   For the circuit shown in Fig. 2.46(a) determine the current through the 5 Ω resistor connected across the terminals A and B by using Thevenin's theorem.

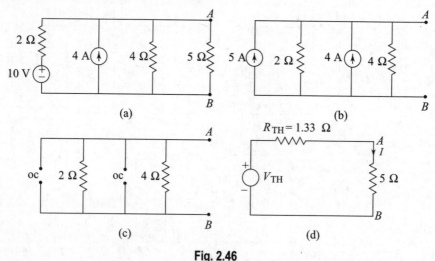

Fig. 2.46

Solution   To determine V_{TH} first the 5 Ω resistor across the terminals AB is removed. Then the 10 V voltage source in series with the 2 Ω resistor is converted to a (10/2 = 5 A) current source in parallel with the 2 Ω resistor and the circuit shown in Fig. 2.46(a) is redrawn as shown in Fig. 2.46(b). The node voltage at B is chosen as the reference.

Applying KCL at node A, $5 - \dfrac{V_A}{2} + 4 - \dfrac{V_A}{4} = 0$

$\therefore \qquad V_A = 12$ V

To find R_{TH}, the two current sources in Fig. 2.46(b) are open-circuited. The circuit with current sources removed is shown in Fig. 2.46(c). It may be seen that $R_{TH} = 2\ \Omega \parallel 4\ \Omega = 1.33\ \Omega$. Finally, the Thevenin's equivalent circuit is drawn as shown in Fig. 2.46(d). From the equivalent circuit, the current I through the 5 Ω resistor can be computed as

$$I = \frac{V_{TH}}{R_{TH} + 5} = \frac{12}{1.33 + 5} = 1.896\ A$$

Example 2.25   Find Thevenin's equivalent for the circuit shown in Fig. 2.47(a).

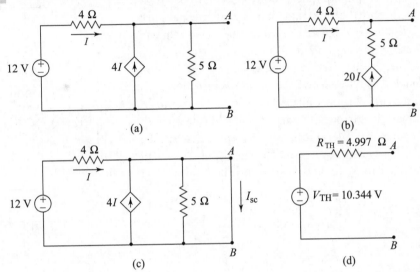

Fig. 2.47

Solution   First the current source, $4I$, in parallel with the 5 Ω resistor is converted into a voltage source, $4I \times 5 = 20I$, in series with the 5 Ω resistor as shown in Fig. 2.47(b). For the circuit in Fig. 2.47(b), applying KVL,

$$-12 + 4I + 5I + 20I = 0$$

∴ $I = 0.414\ A$

$V_{oc} = V_{TH} = 12 - 4 \times I = 12 - 4 \times 0.414 = 10.344\ V$

As a dependent current source is present in the circuit in Fig. 2.47(a), R_{TH} cannot be found by short-circuiting the voltage source. For this, terminals A and B are short-circuited as shown in Fig. 2.47(c). Then applying KCL to node A gives

$$I_{sc} = I + 4I = 5I = 2.07\ A$$

Thevenin's equivalent circuit is shown in Fig. 2.47(d). Using Eq. (2.91),

$$R_{TH} = \frac{V_{oc}}{I_{sc}} = \frac{10.344}{2.07} = 4.997\ A$$

2.11.3 Norton's Theorem

Norton's theorem states that any pair of terminals AB of a network with linear passive and active elements may be replaced by an equivalent current source I_N, in parallel with an equivalent resistance, R_N. The value of I_N (called the Norton current) is equal to the current that would flow from A to B when the terminals A and B are short-circuited, and R_N is the equivalent resistance looking into the network at AB with the independent active sources set to zero, i.e., with all the voltage sources shorted and all the current sources open-circuited leaving behind their internal resistances.

From the statement of Norton's theorem it may be noted that $R_N = R_{TH}$. The computation of I_N will now be illustrated by considering the terminal pair AB of the network shown in Fig. 2.48(a), which is the same network as that in Fig. 2.42(a) taken for the computation of V_{TH} and R_{TH}. To compute I_N the resistance R_L is replaced by a short-circuit as shown in Fig. 2.48(b). When the terminals A and B are shorted, the short-circuit current I_{sc} is equal to the Norton current I_N since all the source current in the circuit in Fig. 2.48(b) must flow through the short circuit. To determine I_{sc} the network shown in Fig. 2.48(a) seen from the terminals AB to the left is replaced by Thevenin's equivalent voltage V_{TH} in series with Thevenin's equivalent resistance R_{TH} as shown in Fig. 2.48(c). Then

$$I_{sc} = I_N = \frac{V_{TH}}{R_{TH}} \qquad (2.92)$$

where $\quad V_{TH} = \dfrac{R_2}{R_1 + R_2}\ V, \quad R_{TH} = R_3 + \dfrac{R_1 R_2}{R_1 + R_2} = R_N \quad (2.93)$

Thus the Norton's equivalent circuit can be drawn as shown in Fig. 2.48(d)

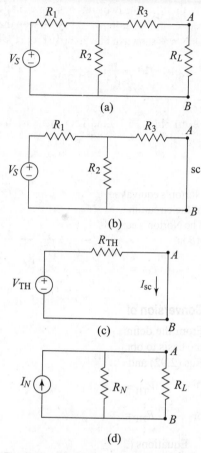

(a)

(b)

(c)

(d)

Fig. 2.48 Illustration of Norton's theorem

Example 2.26 Determine the current in the 4 Ω resistor of the circuit shown in Fig. 2.49(a) using Norton's theorem

Solution The 5 A current source with the 2 Ω resistance in parallel in the circuit in Fig. 2.49(a) is converted to a voltage source with voltage $= 5 \times 2 = 10$ V in series with the 2 Ω resistance as shown in Fig. 2.49(b): the 4 Ω resistor across AB is removed and the terminals A and B are short-circuited. Then applying KVL to the closed circuit in Fig. 2.49(b) $I_{sc} = I_N = \dfrac{12}{2+2} = 3$ A

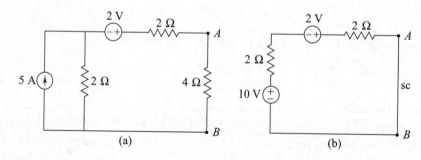

(a)

(b)

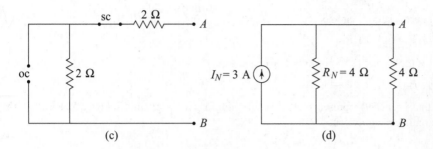

Fig. 2.49

Norton's equivalent resistance may be obtained by short-circuiting the voltage source and opening the current source of the circuit in Fig. 2.49(a). The corresponding circuit is shown in Fig. 2.49(c), from which $R_N = 4\ \Omega$

The Norton's equivalent circuit is shown in Fig. 2.49(d). Then the current through the 4 Ω resistor connected across AB is

$$I_{AB} = \frac{4}{4+4} \times 3\ \text{A} = 1.5\ \text{A}$$

Conversion of Thevenin Equivalent into Norton Equivalent and Vice Versa

From the definition of Norton's theorem, it may be stated that $R_N = R_{TH}$. Equations (2.92) and (2.93) form the basis to obtain a Norton equivalent circuit from a Thevenin equivalent network and vice versa. Rewriting Eqs (2.92) and (2.93) yields

$$V_{TH} = R_{TH}I_N \tag{2.94}$$

or $\qquad R_{TH} = \dfrac{V_{TH}}{I_N} \quad$ for $\quad I_N \neq 0 \tag{2.95}$

Equations (2.94) and (2.95) form the basis of source conversion.

Example 2.27 Using the principle of superposition prove Norton theorem.

Solution Figure 2.50 shows a linear circuit excited by a voltage source of V volts.

Assume that the circuit in Fig. 2.51 is the Norton equivalent of the circuit in Fig. 2.50. In order to establish the equivalence it is necessary to prove that the conditions at the terminals A-B in the two circuits are identical.

Application of the superposition principle to the current flowing into the circuit in Fig. 2.50 is given by

$$I = \underbrace{c_0 V}_{\substack{\text{contribution to } I \text{ due} \\ \text{to external source } v}} + \underbrace{d_0}_{\substack{\text{contribution to } I \text{ due} \\ \text{to internal source}}} \tag{2.96}$$

When terminals A-B are shorted, $V = 0$, from Eq. (2.96) $I = d_0 = -I_{SC}$ where I_{SC} is the short circuit current out of terminal A which in turn is the Norton current I_N. Hence,

$$d_0 = -I_N \tag{2.97}$$

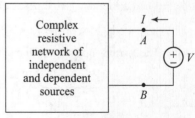

Fig. 2.50

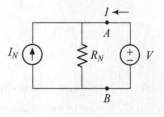

Fig. 2.51

When all the internal sources are de-activated $d_0 = 0$, thus from Eq. (2.96)

$$c_0 = I/V = G_{EQ} = 1/R_{TH} \tag{2.98}$$

Substituting Eqs (2.97) and (2.98) in Eq. (2.96) results in

$$I = V/R_{TH} - I_N \tag{2.99}$$

It can be seen that Eq. (2.99) represents the terminal conditions in the circuits in Figs 2.50 and 2.51, thereby proving that the two circuits are equivalent.

Example 2.28   Determine the Norton equivalent for the circuit shown in Fig. 2.52.

Solution

Step 1: Determine I_N by short circuiting terminal $A - B$ as shown in Fig. 2.53

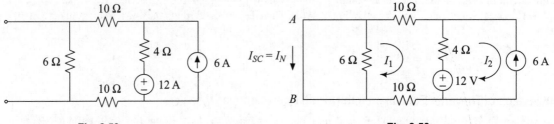

Fig. 2.52   **Fig. 2.53**

For the assumed directions of mesh currents, the mesh equations are written as

$$24I_1 - 4I_2 = -12 \tag{2.100}$$

$$I_2 = -6 \tag{2.101}$$

Substituting Eq. (2.101) in Eq. (2.100) gives $I_1 = -36/24 = -1.5 = -i_{SC} = -I_N$ A

Step 2: To compute R_N, the voltage source is shorted and the current source is open circuited as shown in Fig. 2.54. Looking into the terminals A-B, R_N is given by

$$R_N = 6 \parallel (10+4+10) = (6 \times 24)/(6+24) = 4.8\,\Omega$$

Step 3: The Norton equivalent is drawn in Fig. 2.55.

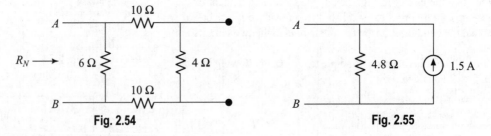

Fig. 2.54   **Fig. 2.55**

2.11.4 Maximum Power Transfer Theorem

The maximum power transfer theorem states that the maximum power transferred to a load resistance R_L, connected across a voltage source V_S and internal resistance r_S, is maximum when the value of the load resistance is equal to the internal resistance of the voltage source.

Figure 2.56 shows a practical voltage source V_S in series with r_s connected to a variable load R_L. Then the power P_L delivered by the source to the load is given by

$$P_L = I_L^2 R_L \qquad (2.102)$$

where the load current I_L is given by

$$I = \frac{V_S}{r_s + R_L} \qquad (2.103)$$

Combining Eqs (2.102) and (2.103),

$$P = \frac{V_S^2}{(r_s + R_L)^2} R_L \qquad (2.104)$$

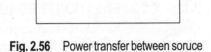

Fig. 2.56 Power transfer between soruce and load

Assuming V_S and r_s to be constant, power P_L is to be maximized by varying R_L.

The condition for maximization is

$$\frac{dP_L}{dR_L} = 0 \qquad (2.105)$$

or

$$\frac{dP_L}{dR_L} = \frac{V_S^2 (R_L + r_s)^2 - 2V_S^2 R_L (r_s + R_L)}{(r_s + R_L)^2} = 0 \qquad (2.106)$$

This leads to the following expression:

$$(R_L + r_s)^2 - 2R_L (r_s + R_L) = 0 \qquad (2.107)$$

Solving Eq. (2.107) gives the condition

$$R_L = r_s \qquad (2.108)$$

Thus in order to transfer maximum power to a load, the load resistance must be equal to the source resistance. This condition is referred to as resistance matching. Under this condition the maximum power transfer, $P_{L\,max}$, to the load is given by substituting $R_L = r_s$ into Eq. (2.104). Then

$$P_{L\,max} = \frac{V_S^2}{(r_s + R_L)^2} R_L = \frac{V_S^2}{4R_L} \qquad (2.109)$$

Now, the power lost in the internal resistance is the same as $P_{L\,max}$ and the total input power P_0 may be obtained as

$$P_0 = V_S I_L = \frac{V_S^2}{2R_L} \qquad (2.110)$$

Thus, under the condition of maximum power transfer the efficiency η is $\quad \dfrac{P_{L\,max}}{P_0} = 50\%$.

When the source is a current source with the internal resistance parallel to it, then also the condition of maximum power transfer remains the same as the current source can always be transformed into a voltage source in series with the same value of internal resistance.

The maximum power transfer theorem can be applied to any network connected to a load because the network may be replaced by the Thevenin equivalent voltage V_{TH} in series with Thevenin's equivalent resistance R_{TH}. The maximum power transfer to load R_L takes place when $R_{TH} = R_L$.

Resistance matching is quite important in the operation of electronics and communication circuits. The power involved is sufficiently small in such circuits compared to that in power systems. In communication and electronic circuits the source resistance is relatively high and it is desirable to transfer the largest possible power from the source to the load. In such cases a low value of the maximum efficiency of 50% is tolerable. However, this 50% loss of power in electronic and communication circuits must be suitably dissipated to limit the temperature rise. In power systems, the sources such as generators and batteries have a low internal source resistance and the load resistance is many times greater than the source resistance. Further, in power systems efficiency is of prime importance and the power loss and voltage drops are kept as minimum as possible. Hence the power circuits are never operated under the maximum power transfer condition.

2.11.5 COMPENSATION THEOREM

Compensation theorem is a convenient means to determine changes in voltages and currents in a circuit when resistance is changed in one of the branches of a circuit. Such a situation may arise, when a resistor, whose exact value is not known due to component tolerance, is to be employed in a circuit. The Compensation theorem is derived from the principle of superposition and is stated as follows.

In a linear network, the change in voltage and current in a circuit, when resistance in a branch varies by ΔR, is equal to that produced by an opposing voltage source of magnitude $V_C = I\Delta R$ where I is the current flowing through the branch prior to the variation in R.

Consider the Thevenin equivalent circuit shown in Fig. 2.57(a) which is connected across a load R_L.

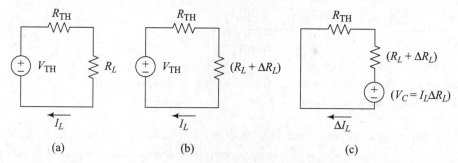

(a)   (b)   (c)

Fig. 2.57   Illustration of the Compensation theorem (a) Thevenin equivalent circuit supplying a load R_L; (b) Thevenin equivalent supplying a load $(R_L + \Delta R_L)$; (c) Compensation theorem circuit.

From Fig. 2.57 (a), load current is given by

$$I_L = \frac{V_{TH}}{(R_{TH} + R_L)} \tag{2.111a}$$

Load current from Fig. 2.57(b) due to a change in load resistance R_L

$$I_L' = \frac{V_{TH}}{(R_{TH} + R_L + \Delta R_L)} \tag{2.111b}$$

Subtracting Eq. (2.111b) from Eq. (2.111a) leads to

$$\Delta I_L = I_L' - I_L = \frac{V_{TH}}{(R_{TH} + R_L + \Delta R_L)} - \frac{V_{TH}}{(R_{TH} + R_L)}$$

$$= -\frac{V_{TH}}{(R_{TH} + R_L)}\left[\frac{\Delta R_L}{(R_{TH} + R_L + \Delta R_L)}\right] \tag{2.111c}$$

Substitution of Eq. (2.111a) in Eq. (2.111c) leads to

$$\Delta I_L = -\frac{V_C}{\left(R_{\mathrm{TH}} + R_L + \Delta R_L\right)} \qquad (2.112)$$

From the circuit in Fig. 2.57(c) it can be seen that

$$\Delta I_L = -\frac{V_C}{\left(R_{\mathrm{TH}} + R_L + \Delta R_L\right)} \qquad (2.113)$$

Comparing Eqs (2.112) and (2.113) it is seen that

$$V_C = I\Delta R_L \qquad (2.114)$$

Example 2.29   A 230 V, 200 W resistor has a tolerance of 10%. Apply the Compensation theorem to calculate the change in load current when it is connected across a Thevenin equivalent circuit of $V_{\mathrm{TH}} = 300$ V, $R_{\mathrm{TH}} = 100\ \Omega$. Comment on the result.

Solution   The resistance and rated current of the resistor are computed from its heat rating and the voltage in the following way:

$$R = \frac{V^2}{W} = \frac{(230)^2}{200} = 264.5\ \Omega \ \text{ and } \ I = \frac{V}{R} = \frac{230}{264.5} = 0.8696\ \text{A}$$

Figure 2.58 shows the resistor connected across the Thevenin equivalent circuit.
From the Thevenin equivalent circuit load current is calculated as given below.

$$I_L = \frac{v_{\mathrm{TH}}}{\left(R_{\mathrm{TH}} + R\right)} = \frac{300}{(100 + 264.5)} = 0.8230\ \text{A}$$

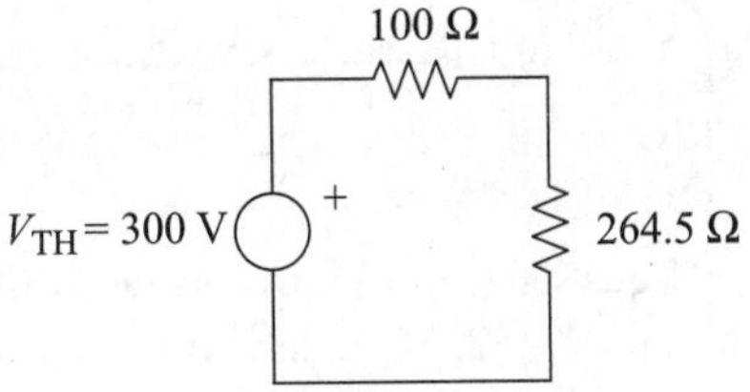

Fig. 2.58

Tolerance in a resistor implies that its resistance value would lie within the specified tolerance. If the upper tolerance is assumed then, $\Delta R = 0.1 \times 264.5 = 26.45\ \Omega$. Using Eqs (2.113) and (2.114) in turn provides

$$V_C = I_L\Delta R = 0.8230 \times 26.45\ \text{V}$$

and $\qquad \Delta I_L = -\dfrac{V_C}{\left(R_{\mathrm{TH}} + R + \Delta R\right)} = -\dfrac{0.8230 \times 26.45}{(100 + 264.5 + 26.45)} = -0.0557\ \text{A}$

The load current, if the resistance value is at the upper tolerance level is given by

$$I_L' = I_L + \Delta I_L = 0.8230 - 0.0557 = 0.7674\ \text{A} \qquad (2.115)$$

On the other hand, for the lower tolerance, $\Delta R = -0.1 \times 264.5 = -26.45\ \Omega$, and the use of Eqs (2.113) and (2.114) will lead to

$$V_C = -0.8230 \times 26.45\ \text{V}$$

and $\qquad \Delta I_L = -\dfrac{(-0.8230 \times 26.45)}{(100 + 264.5 - 26.45)} = 0.0644\ \text{A}$

The load current, if the resistance value is at the lower tolerance level, is given by

$$I_L' = I_L + \Delta I_L = 0.8230 + 0.0644 = 0.8874\ \text{A}$$

From Eq. (2.115) it is seen that if the resistance value is accepted at the upper tolerance level, the load current is below the rated current of 0.8696 A. On the other hand, at the lower level of tolerance, the current exceeds the rated current.

2.11.6 Millman's Theorem

Millman's theorem states that when a number of linear resistors meet at a common node point O', and the potentials from a point of reference (say, the ground) to the free ends of the resistors are known, then the potential at the node point O' is the ratio of the sum of the products of potentials and conductances, and the sum of conductances.

Millman's theorem is explained by considering Fig. 2.59 in which linear conductances $G_1, G_2, \cdots, G_n$, meet at a common node point O' and if the voltages at free points $1, 2, 3, \cdots, n$ with respect to the ground are $V_1, V_2, \cdots, V_n$, then by Millman's theorem, the voltage $V_{O'}$ at node O', with respect to the ground, can be expressed as

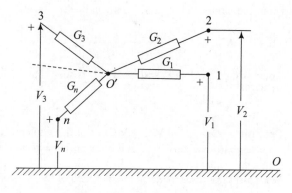

Fig. 2.59 Millman's theorem

$$V_{O'} = \frac{V_1 G_1 + V_2 G_2 + \cdots + V_n G_n}{G_1 + G_2 + \cdots + G_n} = \frac{\sum_{k-1}^{n} V_k G_k}{\sum_{k-1}^{n} G_k} \tag{2.116}$$

Equation (2.116) provides the basis for transforming a circuit containing several voltage sources of magnitudes $V_1, V2, \cdots, Vn$ having respective internal conductances $G_1, G_2, \cdots, G_n$, into an equivalent generator of magnitude V_{Eq} volts and equivalent internal resistor R_{EQ} ohms as follows

$$V_{\text{Eq}} = \frac{\sum_{k-1}^{n} V_k G_k}{\sum_{k-1}^{n} G_k} = \frac{\sum_{k-1}^{n} V_k G_k}{G_{\text{eq}}} \text{ V} \tag{2.117a}$$

and $\quad R_{\text{eq}} = \dfrac{1}{G_{\text{eq}}} = \dfrac{1}{\sum_{k-1}^{n} G_k} \ \Omega \tag{2.117b}$

Example 2.30 For the circuit shown in Fig. 2.60 use Millman's theorem to determine $V_{0'-0}$, and there from draw the circuit for the equivalent generator.

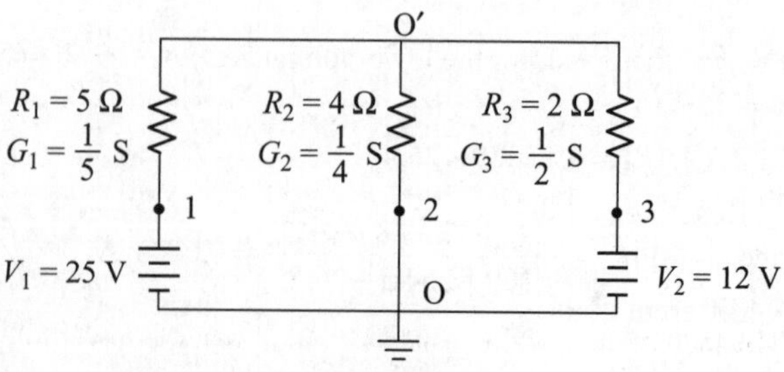

Solution The voltages at points 1, 2, and 3 are respectively $V_1 = 25$ V, $V_2 = 0$ V, and $V_3 = 12$ V. The equivalent conductances are shown in the circuit diagram. Using Eq. (2.116) gives

$R_1 = 5\,\Omega$
$G_1 = \dfrac{1}{5}$ S

$R_2 = 4\,\Omega$
$G_2 = \dfrac{1}{4}$ S

$R_3 = 2\,\Omega$
$G_3 = \dfrac{1}{2}$ S

$V_1 = 25$ V

$V_2 = 12$ V

Fig. 2.60

$$V_{0'-0} = \frac{25 \times \dfrac{1}{5} + 0 \times \dfrac{1}{4} + 12 \times \dfrac{1}{2}}{\left(\dfrac{1}{5} + \dfrac{1}{4} + \dfrac{1}{2}\right)} = 11.5789 \text{ V}$$

$$G = \left(\frac{1}{5} + \frac{1}{4} + \frac{1}{2}\right) = 0.95 \text{ S}$$

From Eqs (2.117), the Millman equivalent generator voltage magnitude and internal resistance is obtained as under

$$V_m = V_{0'-0} = 11.5789 \text{ V and } R_m = \frac{1}{G} = \frac{1}{0.95} = 1.0526 \,\Omega$$

The Millman equivalent generator circuit is drawn in Fig. 2.61.

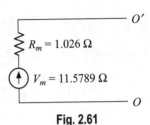

$R_m = 1.026 \,\Omega$

$V_m = 11.5789 \,\Omega$

Fig. 2.61

2.12 TRANSIENT ANALYSIS

The response of a circuit that contains resistances, inductances, capacitances, voltage and current sources and switches to the sudden application of a voltage or current is called transient response. The most common instance of a transient response in a circuit occurs when a switch is turned on or off. Transient response is a natural result of the presence of inductance and capacitance, or in general the energy-storage elements, in circuits. Equations for the circuits containing these elements contain derivatives and integrals of currents/voltages, the solutions of which are functions of time and not constant as in the case of purely resistive circuits. Thus the equations that result from applying Kirchhoff's laws are differential equations and not algebraic equations as obtained for resistive circuits. The solution of the differential equation gives the response of the circuit.

A circuit may be source free, or may be driven by an independent source. For circuits that are source free, the response depends upon the type of elements and their interconnections, and it is called a natural response or natural solution. However, practical energy storage elements cannot store energy indefinitely, as the stored energy is dissipated as heat in the resistances associated with inductors and capacitors. The response thus dies out and hence is termed as the transient response.

When an independent source is present in a circuit, a part of the response depends upon the nature of the source (forcing function), and this part of the response is called the steady state solution, or the forced response. The complete response of the circuit under such conditions is given by the sum of the transient response (natural response) and the steady state response (forced response).

The mathematical solution of the linear differential equation will have two parts: the complementary function (or the homogeneous solution) and the particular integral. For the circuit, the transient response corresponds to complimentary function and the steady state response refers to the particular integral.

In this section, circuits constituted of energy-storing elements are studied for two types of responses, namely, (i) source free or natural response, and (ii) forced response due to an independent dc voltage source.

2.12.1 Natural or Transient Response of Source-free Circuits

Reactive circuits of various configurations such as RL, RC, and RLC in series/parallel can be analysed for their natural response by the application of Kirchhoff's laws and specifying the initial conditions. The response of different combinations for the given initial conditions will be taken up in subsequent sections.

2.12.1.1 Series *RL* Circuit

Consider a series *RL* circuit shown in Fig. 2.62. Let $i(t)$ represent the time varying current flowing in the circuit, and at $t = 0$, the initial current $i(0) = I_0$ is assumed to be known.

Application of KVL around the closed circuit of Fig. 2.62 gives

$$v_R(t) = v_L(t) \tag{2.118}$$

But $\quad v_R(t) = -R\,i(t) \quad \text{and} \quad v_L = L\dfrac{di}{dt}$

Substitution of $v_R(t)$ and $v_L(t)$ in Eq. (2.118) yields

$$R\,i(t) + L\frac{di(t)}{dt} = 0$$

or $\quad \dfrac{di(t)}{dt} + \dfrac{R}{L}i(t) = 0 \tag{2.119}$

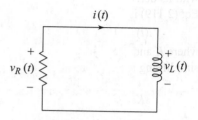

Fig. 2.62　RL series circuit

It may be recalled that Eq. (2.119), in mathematics, is categorized as a homogeneous linear differential equation with constant coefficients. The solution may be obtained by different ways. Since RHS of Eq. (2.119) is zero (source-free or zero-forcing function), mathematically, complimentary function is its solution. In Eq. (2.119), the variables are separable, and the solution is obtained by separating the variables and integrating. Thus,

$$\int \frac{di}{i(t)} = -\int \frac{R}{L}dt + K$$

or $\quad \ln i(t) = -\dfrac{R}{L}t + K \tag{2.120}$

where K is the integration constant. The value of K can be determined from the initial conditions. At $t = 0$, $i = I_0$, and Eq. (2.120) becomes

$$\ln I_0 = K$$

Substitution of the value of K in Eq. (2.120) yields

$$\ln i(t) = \frac{R}{L}t + \ln I_0$$

or $\quad i(t) = I_0 e^{-\left(\frac{R}{L}\right)t} = I_0 e^{-\frac{t}{\tau}} \quad \text{for } t > 0 \tag{2.121}$

where, $\tau = L/R =$ the time constant of the *RL* circuit in seconds $\tag{2.122}$

On the basis of Eq. (2.122), the variation in current against time is shown in Fig. 2.63. Substituting $t = \tau$ in Eq. (2.122),

$$i = I_0 e^{-1} = 0.368 I_0$$

$$= 36.8\% \text{ of the initial value}$$

Similarly, if $t = 5\tau$, then

$i = I_0 e^{-5} = 0.0067 I_0$. In other words, the time taken for the current to reduce practically to zero is five times the time constant.

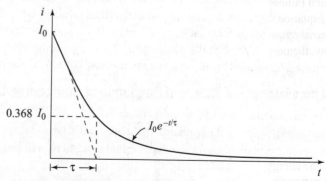

Fig. 2.63　Decay of current in RL circuit

Generalized method of solution   The above method of solution is used when the variables are separable. In case the variables are not separable, the generalized method of solution is used, wherein a solution is assumed based on experience, and then it is tested by substitution in the differential equation. The only function whose linear combination with its derivative can be zero is an exponential function of the form Ke^{st}. Thus, assume that the solution of Eq. (2.119) is represented by

$$i(t) = Ke^{st} \tag{2.123}$$

where K and s are constants, which are required to be determined. Substituting $i(t) = Ke^{st}$ in Eq. (2.119) and simplifying it, we get

$$sKe^{st} + \frac{R}{L}Ke^{st} = 0$$

or

$$\left(s + \frac{R}{L}\right)Ke^{st} = 0 \tag{2.124}$$

In order that Eq. (2.123) is a solution of Eq. (2.119), either

$$Ke^{st} = 0 \tag{2.125a}$$

or

$$\left(s + \frac{R}{L}\right) = 0 \tag{2.125b}$$

For Eq. (2.125) to be equal to zero, either $K = 0$ or $s = -\infty$. In both cases, it leads to an inconsequential solution, that is, the response of the circuit is zero. Hence, a realistic solution is obtained by putting Eq. (2.125b) to be equal to zero, which gives

$$s = -\frac{R}{L} \tag{2.125c}$$

Substitution of the value of s from Eq. (2.125c) into Eq. (2.123) yields

$$i(t) = Ke^{-\left(\frac{R}{L}\right)t}, \quad \text{for } t > 0 \tag{2.126}$$

The constant K is determined by substituting the initial condition, $t = 0$, $i = I_0$, in Eq. (2.126), which gives

$$K = I_0$$

Thus, the natural response of the above circuit is represented by the equation

$$i(t) = I_0 e^{-\left(\frac{R}{L}\right)t} = I_0 e^{-\frac{t}{\tau}}, \quad \text{for } t > 0 \tag{2.127}$$

which indicates the response to be a damped exponential.

Equation (2.125b), that is, $s + (R/L) = 0$, is called the *characteristic equation* and $s = -R/L$ is called the *natural frequency* of the series RL circuit.

Now, the power dissipated in the resistor, p_R, can be written as

$$p_R = [i(t)^2]\,R = I_0^2\,Re^{-(2R/L)t} \tag{2.128a}$$

and the total energy W_R that is turned into heat is found by integrating p_R from $t = 0$ to $t = \infty$ as

$$W_R = \int_0^\infty p_R dt = I_0^2 R \int_0^\infty e^{-(2R/L)\,t} dt = I_0^2 R \times \frac{-L}{2R} \times e^{-(2R/L)t}\Big|_0^\infty = \frac{1}{2}LI_0^2 \tag{2.128b}$$

The energy stored in the inductor at $t = 0$ is $W_L = (1/2)LI_0^2$ J [see Eq. (1.35)]. This energy is dissipated in the resistance at infinite time as at $t = \infty$, the current $I(\infty)$ has decreased to zero and the energy stored in the inductor also reduces to zero.

The time constant of series RL circuit $\tau = (L/R)$ has the unit of seconds, where L is in henries and R is in ohms. It can be shown that henries/ohm is seconds. This can be deduced as follows:

Voltage V_L developed across L is given by

$$V_L = L\frac{di}{dt}$$

So, $L = \dfrac{V_L}{di/dt} = \dfrac{\text{volts} \times \text{seconds}}{\text{amperes}}$

and $\tau = \dfrac{L}{R} = \dfrac{\text{volts} \times \text{seconds}}{\text{amperes} \times \text{ohms}} = \text{seconds}$

2.12.1.2 Series *RC* Circuit

Consider a series RC circuit shown in Fig. 2.64. It is assumed that at $t = 0$, the initial voltage across capacitor plates $v_C(0) = V_0$.

Application of KVL around the closed circuit of Fig. 2.64 gives

$$v_R(t) = v_C(t) \qquad (2.129)$$

However, $v_R(t) = -R\, i_C(t) = -RC\dfrac{dv_C(t)}{dt}$

Substitution of $v_R(t)$ in Eq. (2.129) and dividing by RC yields

$$\frac{dv_C(t)}{dt} + \frac{v_C(t)}{RC} = 0 \qquad (2.130)$$

Following the same procedure as adopted for solving Eq. (2.119), the solution of Eq. (2.130) can be obtained as

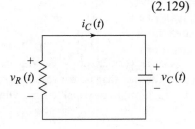

Fig. 2.64   *RC* Series circuit

$$v_C(t) = V_0\, e^{-\frac{t}{RC}} = V_0 e^{-\frac{t}{\tau}} \quad \text{for } t > 0 \qquad (2.131)$$

and

$$i_C(t) = C\frac{dv_C}{dt} = -\frac{V_0}{R} e^{-\frac{t}{RC}} = -\frac{V_0}{R} e^{-t/\tau} \quad \text{for } t > 0$$

$$(2.132)$$

where $\tau = RC =$ the time constant of RC circuit in seconds (2.133)

The voltage and current of the capacitor are exponentially decaying with respect to time with initial values of V_0 and V_0/R respectively. In this case, the voltage and current of the capacitor become zero at $t = \infty$, and the total energy stored in the capacitor, $\mathcal{W}_C = (1/2)(CV^2)$ J [see Eq. (1.40)] is dissipated through the resistor R. Figure 2.65 shows the plot of decay of voltage and current in the charged capacitor with respect to time. From Fig. 2.65, it can be seen that the time constant of

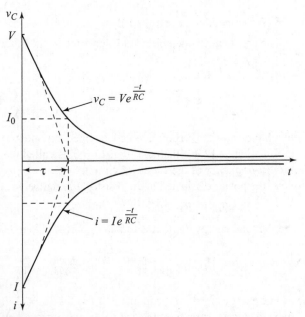

Fig. 2.65   Decay of voltage and current in a charged capacitor

the network τ in seconds is the time taken for the voltage v_C or the current i to decrease to zero value if the rates of change of these quantities are maintained at their initial values at time $t = 0$. Further, in Eqs (2.131) and (2.132), substituting $t = RC$, values of v_C and i become

$$v_C = Ve^{-1} = 0.368V \quad \text{and} \quad i = Ie^{-1} = 0.368\,I$$

Alternately, the time constant τ in seconds can be defined as the time taken for the voltage across or the current through the capacitor to decrease to 36.8% of its initial values.

It is left to the reader, as a tutorial exercise, to verify that the time constant of series RC circuit has the unit of seconds.

2.12.1.3 Significance of Time Constant (τ)

As seen from Eqs (2.127) and (2.131), the natural (or transient) responses of both series RL and series RC source-free circuits are similar. In the former, it is an exponentially decaying current in the inductor and in the latter, it is an exponentially decaying voltage of the capacitor. The rate of decay in both the cases is dependent on the time constants τ of the respective circuits. Figure 2.66 shows the variation of $i(t)/I_0$ in RL circuits and the variation of $v_C(t)/V_0$ in RC circuits as the variable with respect to time, for different values of time constant τ. It may be noted from Fig. 2.66 that the transient response of a series RL or RC circuit is wholly governed by the time constant τ—the smaller the time constant faster is the response (the decay of the variable) and vice versa. For $\tau = 1$ s, the response at $t = 1, 2, 3, 4, 5$ s is consecutively 36.79%, 13.53%, 4.98%, 1.83%, and 0.67%.

At any time t, the rate of decay of current $i(t)$ in a RL circuit is obtained from Eq. (2.127) and the rate of decay of voltage $v_C(t)$ is given by Eq. (2.131) as follows:

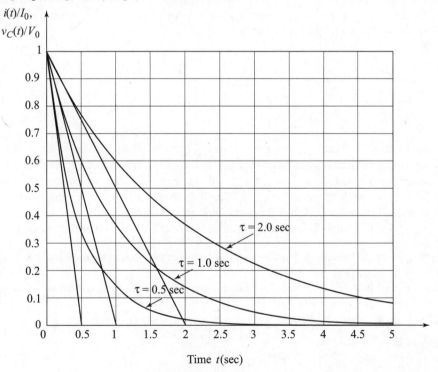

Fig. 2.66 Natural response of series *RL/RC* circuit

$$\frac{d(i)}{dt} = -\frac{I}{\tau} e^{-\frac{t}{\tau}}$$

(2.134a)

and $\quad \dfrac{d(v)}{dt} = -\dfrac{V}{\tau} e^{-\frac{t}{\tau}}$

(2.134b)

The initial rate of decay of current in a *RL* circuit is obtained at $t = 0$ as follows:

$$\frac{di(0)}{dt} = -\frac{I_0}{\tau}$$

(2.135)

and the initial rate of decay of current in a *RC* circuit is obtained a $t = 0$ as

$$\frac{dv_C(0)}{dt} = -\frac{V_0}{\tau}$$

(2.136)

It may be noted from Fig. 2.66 that if the initial rate of decay is continued, the straight line meets the *x*-axis at $t = \tau$, that is, the response decays to zero. Further, it is also seen that the current for *RL* series circuit (or voltage for *RC* series circuit) reduces to 36.88% of the initial value in a time equal to the time constant of the circuit. Hence, the time constant of a circuit is also defined as the time taken for the current (or voltage) to reduce to 36.88% of its initial value.

2.12.1.4 Natural Frequency

Natural frequency of a circuit is its natural mode of oscillations. For a first-order circuit, for example an *RL* or *RC* series circuit having a response proportional to e^{-st}, the coefficient *s* in the exponent is the natural frequency. The unit of natural frequency is second^{-1}.

Thus, from Eqs (2.122) and (2.133),

$$s = -\frac{1}{\tau} = -\frac{R}{L} \qquad \text{for } RL \text{ circuit}$$

(2.137a)

$$s = -\frac{1}{\tau} = -\frac{1}{RC} \qquad \text{for } RC \text{ circuit}$$

(2.137b)

Computation of natural frequency   For a source-free *RL* or *RC* circuit, all resistors and inductors (or capacitors) are combined separately to reduce the circuit to the simple form of one equivalent resistor and one equivalent inductor (or capacitor) as shown in Fig. 2.62 [or Fig. 2.64]. Natural frequency is then obtained from Eq. (2.137a) for a series *RL* circuit and from Eq. (2.137b) for a series *RC* circuit. However, if energy sources are present, all voltage sources are short-circuited and all current sources are open-circuited and then the circuit is simplified to a single equivalent resistor and one equivalent inductor (or capacitor).

Example 2.31   Determine the natural frequency of the series parallel *RL* circuit shown in Fig. 2.67. The circuit is energized by a voltage and a current source.

Solution   Since natural frequency is required to be calculated, the voltage source is short-circuited and the current source is open-circuited as shown in Fig. 2.68

Equivalent resistance, $R_{eq} = [(6 + 6) \parallel 4] + [1 \parallel 2] = 3 + 0.6667 = 3.6667\ \Omega$

Natural frequency from Eq. (2.137a) is

$$s = -\frac{3.6667}{4} = -0.91667 \text{ sec}^{-1}$$

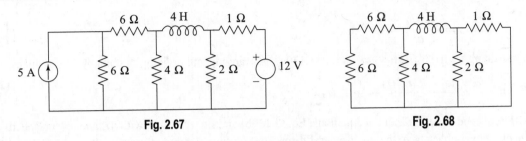

Fig. 2.67   **Fig. 2.68**

Example 2.32   For the circuit shown in Fig. 2.69, determine (a) the natural frequency and (b) the time constant.

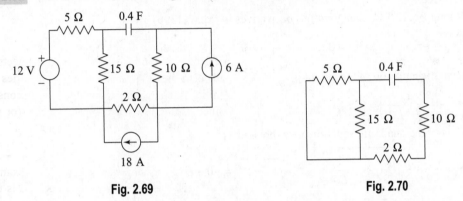

Fig. 2.69   **Fig. 2.70**

Solution   The simplified circuit with voltage sources short circuited and the current source open circuited is shown in Fig. 2.70.

Equivalent resistance of the circuit, $R_{eq} = [10 + 2 + (5 \parallel 15)] = 15.75 \, \Omega$

(i) Natural frequency, $s = -\dfrac{1}{(15.75 \times 0.4)} = -0.15873 \text{ second}^{-1}$.

(ii) Time constant, $\tau = 15.75 \times 0.4 = 6.3$ sec.

2.12.1.5 Series *RLC* Circuit

The series *RLC* circuit shown in Fig. 2.71 contains no energy source. Applying KVL to the closed circuit shown in Fig. 2.71,

$$v_R(t) + v_L(t) + v_C(t) = 0 \tag{2.138}$$

Choosing current $i(t)$ as the circuit variable and expressing $v_R(t)$, $v_L(t)$, and $v_C(t)$ in terms of $i(t)$ yields

$$R\,i(t) + L\frac{di(t)}{dt} + \frac{1}{C}\int i(t)dt = 0 \tag{2.139a}$$

Differentiating Eq. (2.139a), rearranging and dividing through by L, we get

$$\frac{d^2 i(t)}{dt^2} + \frac{R}{L}\frac{di(t)}{dt} + \frac{1}{LC}i(t) = 0 \tag{2.139b}$$

If $v_C(t)$ is chosen as the circuit variable and expressing $v_R(t)$, $v_L(t)$, and $v_C(t)$ in terms o $v_C(t)$ gives

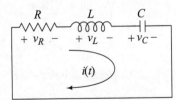

Fig. 2.71   *RLC* series circuit

$$R\,i(t) + L\frac{di(t)}{dt} + v_C = 0 \tag{2.139c}$$

Substituting for $i(t) = C[dv_C(t)/dt]$ in Eq. (1.139c), rearranging and dividing through by LC produces

$$\frac{d^2v_C(t)}{dt^2} + \frac{R}{L}\frac{dv_C(t)}{dt} + \frac{1}{LC}v_C(t) = 0 \tag{2.139d}$$

The second-order differential equations, Eq. (2.139b) and Eq. (2.139d), with $i(t)$ and $v_C(t)$ respectively as the circuit variables, have the same general form. The solution in terms of variable $i(t)$ will be taken up now. Assuming the solution of Eq. (2.139b) to be

$$i(t) = Ke^{st} \tag{2.140}$$

and substituting Eq. (2.134) along with its derivatives in Eq. (2.139b) gives

$$\left(s^2 + \frac{R}{L}s + \frac{1}{LC}\right)K\,e^{st} = 0 \tag{2.141}$$

The characteristic equation of the circuit is

$$\left(s^2 + \frac{R}{L}s + \frac{1}{LC}\right) = 0 \tag{2.142}$$

The two roots of the quadratic equation are expressed as

$$s_1 = -\frac{R}{2L} + \sqrt{\left(\frac{R}{2L}\right)^2 - \frac{1}{LC}} = -\sigma + \sqrt{\sigma^2 - \omega_0^2} = -\sigma + \beta \tag{2.143a}$$

$$s_2 = -\frac{R}{2L} - \sqrt{\left(\frac{R}{2L}\right)^2 - \frac{1}{LC}} = -\sigma - \sqrt{\sigma^2 - \omega_0^2} = -\sigma - \beta \tag{2.143b}$$

where $\sigma = R(2L)$, $\beta = \sqrt{\sigma^2 - \omega_0^2}$, and $\omega_0 = 1/\sqrt{LC}$

In Eq. (2.143), σ is called the damping constant and ω_0 is the resonant or natural frequency in radians/second. The natural response of the *RLC* series circuit may now be written as

$$i(t) = K_1 e^{s_1 t} + K_2 e^{s_2 t} \tag{2.144}$$

where K_1 and K_2 are constants that are determined from the initial conditions of the circuit. The nature of response depends upon the relative magnitudes of σ and ω_0. The radical that appear in the expressions of s_1 and s_2 will be real when $\sigma > \omega_0$, zero when $\sigma = \omega_0$, and imaginary when $\sigma < \omega_0$.

Case 1: Overdamped case

When $\left(\dfrac{R}{2L}\right)^2 > \dfrac{1}{LC}$, that is $\sigma > \omega_0$, both σ and β are real positive numbers. Then

$$i(t) = K_1 e^{(-\sigma+\beta)t} + K_2 e^{(-\sigma-\beta)t} = e^{-\sigma t}(K_1 e^{\beta t} + K_2 e^{-\beta t}) \tag{2.145}$$

Case 2: Critically damped case

When $\left(\dfrac{R}{2L}\right)^2 = \dfrac{1}{LC}$, that is, $\sigma = \omega_0$, the solution takes the form

$$i(t) = e^{-\sigma t}(K_1 + K_2 t) \tag{2.146}$$

Case 3: Underdamped or oscillatory case

When $\left(\dfrac{R}{2L}\right)^2 < \dfrac{1}{LC}$, that is, $\sigma < \omega_0$, then the roots are

$$s_1 = -\sigma + j\sqrt{\omega_0^2 - \sigma^2} = -\sigma + j\omega_d$$

(2.147)

$$s_2 = -\sigma - j\sqrt{\omega_0^2 - \sigma^2} = -\sigma - j\omega_d$$

where ω_d is called the damped resonant frequency and is given by

$$\omega_d = \sqrt{\omega_0^2 - \sigma^2} = \sqrt{\dfrac{1}{LC} - \left(\dfrac{R}{2L}\right)^2} \text{ rad/s}$$

Substitution of Eq. (2.147) in Eq. (2.140) gives

$$i(t) = e^{-\sigma t}\left(K_1 e^{j\omega_d t} + K_2 e^{-j\omega_d t}\right)$$

or $\quad i(t) = e^{-\sigma t}\left[K_1\left(\cos\omega_d t + j\sin\omega_d t\right) + K_2\left(\cos\omega_d t - j\sin\omega_d t\right)\right]$

or $\quad i(t) = e^{-\sigma t}\left[\left(K_1 + K_2\right)\cos\ \omega_d t + j\left(K_1 - K_2\right)\sin\omega_d t\right]$

or $\quad i(t) = e^{-\sigma t}\left[A\cos\omega_d t + B\sin\omega_d t\right]$

(2.148)

where $A = (K_1 + K_2)$ and $B = j(K_1 - K_2)$

2.12.2 Forced Response of Reactive Circuits

A forcing function in a reactive circuit produces in addition to the transient or natural response a response that is dependent on the forcing function called the forced response. This section describes the response of reactive circuits to dc sources.

2.12.2.1 Inductor

It was shown in Section 1.9.2 [see Eq. (1.33)] that current in an inductor cannot change suddenly as such a change would require infinite voltage. Therefore, an ideal inductor (zero resistance) at $t = 0^+$ behaves like a perfect current source of capacity $i_L(0^+) = i_L(0^-)$. If $i_L(0^-) = 0$, the inductor behaves like an open circuit. Symbolic representation of the conditions is shown in Fig. 2.72.

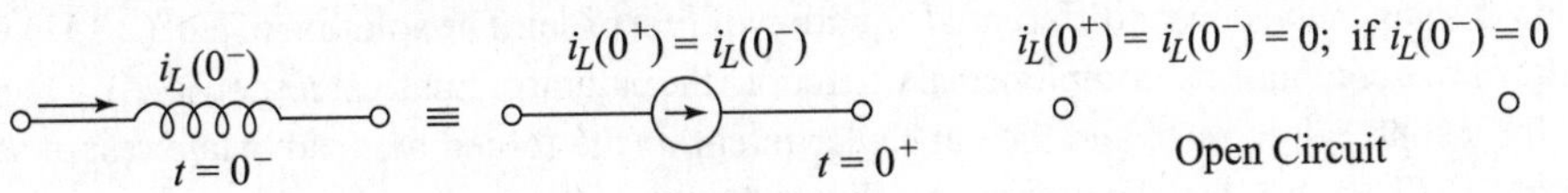

Fig. 2.72   Behaviour of an inductor at $t = 0^+$

If an inductor is connected to a dc excitation then at $t = \infty$, the current I_L flowing through the inductor will acquire a constant value and the voltage across the inductor is given by

$$v_L = \dfrac{dI_L}{dt} = 0$$

(2.149)

Equation (2.149) indicates that in the steady state condition, an ideal inductor behaves like a short circuit. However, an inductor is built from conductor wire and possesses some resistance. Therefore, at $t = \infty$, the steady state current in an inductor is limited by its resistance.

2.12.2.2 Capacitor

It was shown in Section 1.9.3 that the voltage across a capacitor [see Eq. (1.38)] cannot change suddenly. Hence, a capacitor at $t = 0^+$ sec behaves like an ideal voltage source whose magnitude $v_C(0^+) = v_C(0^-)$. At $t = 0^-$, if $v_C(0^-) = 0$, the capacitor would act as a short-circuited voltage source. Representation of the capacitor under the above condition is symbolically shown in Fig. 2.73.

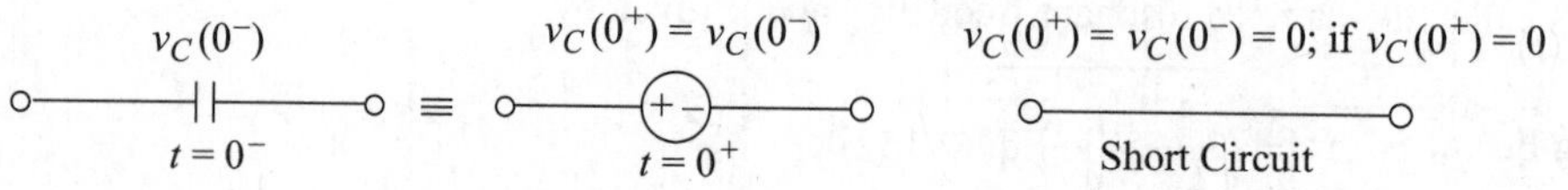

$$v_C(0^-) \qquad v_C(0^+) = v_C(0^-) \qquad v_C(0^+) = v_C(0^-) = 0; \text{ if } v_C(0^+) = 0$$

$$t = 0^- \qquad \equiv \qquad t = 0^+ \qquad \text{Short Circuit}$$

Fig. 2.73 Behaviour of a capacitor at $t = 0^+$

A capacitor also attains a steady state voltage, V_C, which is constant at $t = \infty$. Thus, the current through the capacitor at $t = \infty$ is given by

$$i_C = C \frac{dV_C}{dt} = 0 \tag{2.150}$$

From Eq. (2.150) it is seen that in the steady state current flowing through the capacitor is zero, and the capacitor behaves like an open circuit. However, the voltage across a capacitor is determined by its conductance, which is always present in a practical capacitor.

2.12.2.3 Series *RL* Circuit

Figure 2.74 shows a series *RL* circuit connected across a dc source of voltage V_S through a switch S. As the switch S is in the open position at the instant $t = 0^-$, $i(0^-) = 0$. The dc voltage of V_S volts is applied to the circuit by closing the switch S at $t = 0$.

For $t > 0$, application of KVL to the circuit of Fig. 2.74 leads to the following equation:

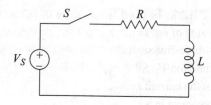

Fig. 2.74 Series *RL* circuit

$$Ri(t) + L \frac{di(t)}{dt} = V_S \tag{2.151}$$

which is a non-homogeneous linear differential equation of first order. The solution of Eq. (2.151) will give $i(t)$, which consists of two components: complimentary function (the natural or transient response, i_n), which will satisfy Eq. (2.119) for a source-free case, and the particular integral (the forced or steady state response, i_f), which should satisfy Eq. (2.151). Thus, the complete solution may be written as

$$i(t) = i_n(t) + i_f(t) \tag{2.152}$$

Equation (2.121) provides the natural response and is reproduced below:

$$i_n(t) = K e^{-\left(\frac{R}{L}\right)t} = K e^{-t/\tau} \tag{2.153}$$

Next step is to determine the particular integral of Eq. (2.151), which gives the forced response. The forcing function is constant voltage, V_S, and it is intuitively expected that the steady state response $i_f(t) = I = a$ constant. Equation (2.151) can now be written as

$$RI + L\frac{dI}{dt} = V_S \tag{2.154}$$

since I = constant, $L\dfrac{dI}{dt} = 0$

$$i_f(t) = I = \frac{V_S}{R} \tag{2.155}$$

Substituting Eqs (2.153) and (2.155) in Eq. (2.152) yields the solution of Eq. (2.151) as

$$i(t) = Ke^{-t/\tau} + \frac{V_S}{R} = Ke^{-t/\tau} + I \tag{2.156}$$

K is determined from the initial condition. Substitution of the initial $i(0^-) = i(0^+) = 0$ in Eq. (2.156) gives

$$K = -\frac{V_S}{R} = -I \tag{2.157}$$

Hence, the complete solution of Eq. (2.151) is given by

$$i(t) = \frac{V_S}{R}\left(1 - e^{-\frac{R}{L}t}\right) = I\left(1 - e^{-t/\tau}\right) \quad \text{for } t > 0 \tag{2.158}$$

It may be noted from Eq. (2.158) that the current in the circuit increases exponentially and reaches a final value of I at $t = \infty$. The exponentially decaying component of the response is similar to the one obtained for a source-free series RL circuit except that it has an amplitude that is dependent on the magnitude of the forcing function V_S. Since the exponential component of current decays to zero as $t \rightarrow \infty$, the steady state component of the current represented by $I = V_S/R$ is due to the forcing function. Thus, it may be concluded that the forced response in a circuit has the same form as the forcing function.

The rate of increase of current in the circuit is obtained from Eq. (2.158) as follows:

$$\frac{di(t)}{dt} = \frac{I}{\tau}e^{-t/\tau} \tag{2.159}$$

The initial rate of rise at $t(0^+) = 0$ is equal to I/τ A/sec.

From Fig. 2.75, it is seen that the rate of increase of current in a series RL circuit, due to a step voltage forcing function, increases exponentially and reaches a steady state value $I = V_S/R$ at $t = \infty$. It may also be noted that in time $t = \tau$ sec, the current in the circuit attains 0.632 times the final steady state value. Therefore, the time constant (τ) may also be defined as the time in which the current in a series RL circuit attains 63.2% of the final steady state value.

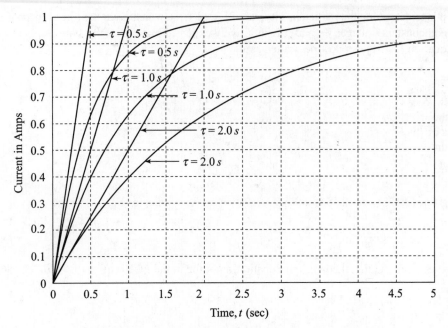

Fig. 2.75   Response of series *RL* circuit to dc source voltage

Example 2.33   A coil of resistance 30 Ω and inductance 0.6 H is switched on to a 240 V supply. (a) Calculate the rate of change of current (i) at the instant of closing of the switch when $t = 0$, (ii) at time $t = 2$ (L/R); (b) the magnitude of the final steady state current.

Solution   (a) From Eq. (2.158),

$$i = \frac{V}{R}(1 - e^{-(R/L)t}) = \frac{240}{30}(1 - e^{-(30/0.6)t}) = 8\,(1 - e^{-50t})$$

Then,

$$\frac{di}{dt} = 400\,e^{-50t}$$

(i)   At $t = 0$, $\dfrac{di}{dt} = 400$ A/s

(ii)   At $t = 2$ (L/R) $= 2 \times 0.6/30 = 0.04$ sec

$$\frac{di}{dt} = 400\,e^{-50 \times 0.04} = 400e^{-2} = 400 \times 0.1353 = 54.13 \text{ A/sec}$$

(b) The final steady state current $= \dfrac{V}{R} = \dfrac{240}{30} = 8$ A

Example 2.34   In the circuit shown in Fig. 2.65, the switch is initially in position 1. (a) For $R = 500$ Ω, find the voltage across the field coil at the instant at which the switch is changed to position 2. (b) Calculate the value of R for the voltage across the coil to be 120 V at the instant of switching. (c) With R of the value found in (b), find the time taken to dissipate 95% of the stored energy.

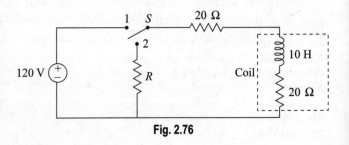

Fig. 2.76

Solution (a) From Eq. (2.158)

$$i = I(1 - e^{-t/\tau})$$

Since the switch has been in position 1 for a long time, then

$$i = I_0 = \frac{V}{R} = \frac{120}{40} = 3 \text{ A}$$

When the switch is moved to position 2, at $t = 0$, $i(0) = I_0 = 3$ A, since the circuit has reached a steady state condition and the energy stored in the inductor $= (1/2) LI_0^2$, and this energy will be dissipated in the circuit resistances by driving a current in the closed circuit. This current will cause a voltage drop in the circuit. Applying KVL to the closed circuit containing the coil, the 20 Ω resistor, and $R = 500$ Ω, the coil voltage is

$$I_0 (500 + 20) = 3 \times 520 = 1560 \text{ V}$$

(b) In order to reduce the voltage across the coil to 120 V at the time of switching, R can be found as follows:

$$I_0 (R + 20) = 120 \text{ V}$$

where $I_0 = 3$ A. Therefore,

$$R = \frac{120 - 60}{3} = 20 \text{ }\Omega$$

(c) When $R = 20$ Ω, the total resistance in the circuit is equal to $20 + 20 + 20 = 60$ Ω.

The time constant of the circuit,

$$\tau = \frac{L}{R} = \frac{10}{60} = 0.1667 \text{ sec}$$

Then the current from Eq. (2.126),

$$i = I_0 e^{-t/\tau} = 3e^{-t/0.1667} = 3e^{-6t} \tag{I}$$

The total energy stored in the coil at the instant of switching is

$$\frac{1}{2} \times 10 \times 3^2 = 45 \text{ J}$$

Balance energy $= 0.05 \times 45 = 2.25$ J

Since the stored energy in a coil at a particular instant is proportional to the square of the current flow at that instant, when the stored energy is 2.25 J the current i in the coil is given by

$$\frac{1}{2} \times L \times i^2 = 2.25 \text{ J}$$

$$\therefore \qquad i = \sqrt{0.45} = 0.6708 \text{ A}$$

The time taken by current i to decay from 3 A to 0.6708 A is obtained from Eq. (I) as

$$3e^{-6t} = 0.6708 \quad \text{or} \quad e^{-6t} = 0.2236 \quad \text{or} \quad -6t = \ln(0.2236) = -1.5$$

$$\therefore \qquad t = 0.25 \text{ sec}$$

Example 2.35 In the *RL* circuit shown in Fig. 2.77, the switch S has been in position 'a' for a long time. At $t = 0$ sec, S is quickly moved to position 'b'. For the values of the resistors and the inductor shown in the figure, calculate the voltages across (i) the nodes 1 and 2 and (ii) the current through the 4 Ω resistor for $t \geq 0$. Employ the mesh current technique for determining the voltages.

Solution Assume the independent mesh currents to be i_1 and i_2 and clockwise as son in Fig. 2.67(a). The mesh current equations are formulated by applying KVL to the two meshes as follows:

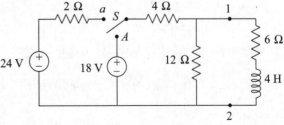

Fig. 2.77

$$4i_1 + 12(i_1 - i_2) = 18 \quad \text{or} \quad 8i_1 - 6i_2 = 9 \tag{I}$$

and $$12(i_2 - i_1) + 6i_2 + 4\frac{di_2}{dt} = 0 \quad \text{or} \quad -6i_1 + 9i_2 + 2\frac{di_2}{dt} = 0 \tag{II}$$

Substitution of $i_1 = \dfrac{(9 + 6i_2)}{8}$ from Eq. (I) in Eq. (II) produces

$$\frac{-6 \times (9 + 6i_2)}{8} + 9i_2 + 2\frac{di_2}{dt} = 0 \quad \text{or} \quad i_2 + \frac{4}{9} \times \frac{di_2}{dt} = 1.5$$

or $$\frac{di_2}{1.5 - i_2} = \frac{9dt}{4} \tag{III}$$

Integration of Eq. (III) leads to

$$\frac{9t}{4} = -\ln(1.5 - i_2) + \ln(K)$$

or $$\frac{1.5 - i_2}{K} = e^{-\frac{9}{4}t}$$

or $$i_2 = 1.5 - Ke^{-\frac{9}{4}t} \tag{IV}$$

The integration constant K is computed from the initial condition in the circuit shown in Fig. 2.78(b). Noting that the circuit is in the steady-state condition, the inductor behaves like a short circuit and the current flowing through it at $t = 0$ sec is given by

$$i_2 = \left[\frac{24}{2 + 4 + (12 \times 6)/18} \right] \times \frac{12}{18} = 1.6 \text{ A}$$

Thus, substitution of K in Eq. (IV) leads to

$$K = 1.5 - 1.6 = -0.1$$

Therefore, the current through the inductor for $t \geq 0$ sec is

$$i_2 = 1.5 + 0.1e^{-2.25t} \text{ A}$$

(i) Voltage between nodes $1 - 2$ is

$$V_{1-2} = 6\left(1.5 + 0.1e^{-2.25t}\right) - 4\left(2.25 \times 0.1e^{-2.25t}\right) \text{ V}$$

$$= 9.0 - 0.3e^{-2.25t} \text{ V}$$

(ii) The current through the 4 Ω resistor is i_1 and is obtained by substituting for i_2 in Eq. (II) as under

$$i_1 = \frac{9}{8} + \frac{6}{8}\left(1.5 + 0.1e^{-2.25t}\right) = 2.25 + 0.075e^{-2.25t} \text{ A}$$

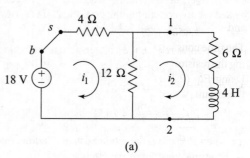

(a)

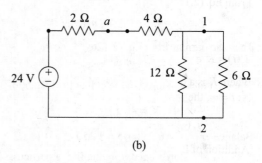

(b)

Fig. 2.78

Example 2.36 A dc source of voltage V_S is applied to the circuit shown in Fig. 2.79 by closing the switch S at $t = 0$. Determine the response of the circuit.

Solution The equivalent circuit obtained by short circuiting the voltage source is shown in Fig. 2.80(a). For the circuit

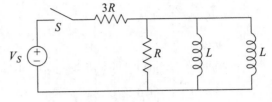

Fig. 2.79

$$R_{eq} = 3R \parallel R = \frac{3R \times R}{3R + R} = \frac{3R}{4}$$

and $\quad L_{eq} = L \parallel L = L/2$

The time constant of the circuit is obtained by combining the resistors and inductors as shown in Fig. 2.80(b). Using Eq. (2.151) to write the response equation of the circuit, we get

$$\frac{d\,i(t)}{dt} + \left(\frac{R_{eq}}{L_{eq}}\right) i(t) = \frac{V_S}{L_{eq}} \quad \text{for } t > 0$$

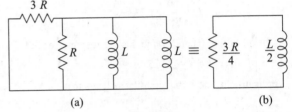

(a)　　　　　　　　　(b)

Fig. 2.80

From Eq. (2.153), the natural response for the circuit is

$$i_n(t) = K\,e^{-\left(\frac{R_{eq}}{L_{eq}}\right)t} = K\,e^{-\left(\frac{3R}{2L}\right)t} = K\,e^{-t\big/\left(\frac{2L}{3R}\right)} = K e^{-\frac{t}{\tau}} \qquad \text{(I)}$$

where K is a constant to be determined from the initial condition.

At $t = \infty$, the inductor behaves like a short circuit. Hence,

$$i_f(\infty) = \frac{V_S}{3R} \qquad \text{(II)}$$

Addition of Equations (I) and (II) yields the response of the circuit as follows:

$$i(t) = i_n(t) + i_f(\infty) = K e^{-\frac{t}{\tau}} + \frac{V_S}{3R} \qquad \text{(III)}$$

At $t = 0$, $i(0^-) = i(0^+) = 0$. Substitution of the initial condition in Eq. (III) gives

$$K = -\frac{V_S}{3R}$$

Thus, the response of the circuit is given by

$$i(t) = \frac{V_S}{3R}\left[1 - e^{-\frac{t}{2L/(3R)}}\right]$$

Example 2.37 Switch S in the circuit shown in Fig. 2.81 is moved to position 2 at $t = 0$ sec, after it has been in position 1 for a long time. (a) Derive an expression for the current $i(t)$ for $t > 0$. (b) Determine the initial rate of current at $t = (0^+)$ and the voltage across the inductor at $t = (0^-)$ and $t = (0^+)$ sec. All data is shown in the figure.

Solution At the instant the switch is moved to position 2, the circuit is in the steady state. At $t = 0$, the initial values of the circuit current and voltage across the inductor are

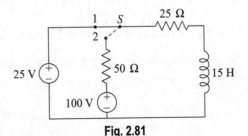

Fig. 2.81

$$i(0^-) = \frac{25}{25} = 1 \text{ A}$$

$$v_L(0^-) = 0$$

Time constant of the circuit with switch at position 2 is

$$\tau = \frac{15}{(50+25)} = 0.2 \text{ sec}$$

Using Eqs (2.148) and (2.150), we get

$$i_n(t) = K\, e^{-t/0.2} = K\, e^{-5t} \tag{I}$$

$$i_f(t) = I = \frac{100}{75} = 1.333\,\text{A} \tag{II}$$

Addition of Eqs (I) and (II) yields the response of the circuit as under

$$i(t) = Ke^{-5t} + 1.333$$

Application of the initial conditions leads to

$$K + 1.333 = 1.0$$

or $K = -0.333$

(a) Current through the circuit is

$$i(t) = 1.333 - (0.333)e^{-5t}, \qquad \text{for } t > 0 \tag{III}$$

(b) At $t = 0^+$ sec, from Eq. (III),

$$i(0^+) = 1.333 - 0.333 = 1.0\,\text{A}$$

Application of KVL to the circuit at $t = 0^+$ sec gives

$$1.0 \times (50 + 25) + v_L(0^+) = 100$$

So, $v_L(0^+) = 25$ V

Initial rate of rise of current is

$$\frac{di(0^+)}{dt} = 5 \times (0.333) = 1.665 \text{ A/sec}$$

2.12.2.4 Series *RC* Circuit

Figure 2.82 shows a series *RC* circuit connected across a dc voltage source of V_S through a switch S. Let it be assumed that at the instant $t = 0^-$, just before the switch S is closed, the capacitor is charged to a voltage V_0. After the switch S is closed, application of KVL to the series *RC* circuit yields the following equation:

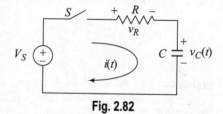

Fig. 2.82

$$V_S - R\, i\,(t) - v_C(t) = 0 \tag{2.160}$$

For the analysis of the circuit of Fig. 2.82, the capacitor voltage $v_C(t)$ is chosen as the variable. Substituting for $i(t) = C\dfrac{dv_C(t)}{dt}$ in Eq. (2.160) and rearranging gives

$$RC\frac{dv_C(t)}{dt} + v_C(t) = V_S, \qquad \text{for } t > 0 \tag{2.161}$$

Equation (2.161), like Eq. (2.151), is a first-order, non-homogeneous linear differential equation and, therefore, its solution, similar to Eq. (2.151), can be written as

$$v_C(t) = Ke^{-t/t} + V_S \tag{2.162}$$

In Eq. (2.162), the time constant $\tau = RC$ and the constant K is determined from the history of the circuit. At $t = 0^+$, $v_C(0^-) = v_C(0^+) = V_0$. Substitution of the initial condition in Eq. (2.162) leads to

$$K = V_0 - V_S$$

Substitution of the value of K in Eq. (2.162) and its simplification gives

$$v_C(t) = V_0 e^{-\frac{t}{\tau}} + V_S\left(1 - e^{-\frac{t}{\tau}}\right) \text{V}, \quad \text{for } t > 0 \tag{2.163}$$

The expression for the current in the circuit is given by

$$i(t) = C\frac{dv_C(t)}{dt} = C\left[\left(-\frac{1}{\tau}V_0 e^{-\frac{t}{\tau}}\right) + \left(\frac{1}{\tau}V_S e^{-\frac{t}{\tau}}\right)\right] = \frac{C}{RC}(V_S - V_0)e^{-\frac{t}{\tau}}$$

or

$$i(t) = \left(\frac{V_S - V_0}{R}\right)e^{-\frac{t}{\tau}} \tag{2.164}$$

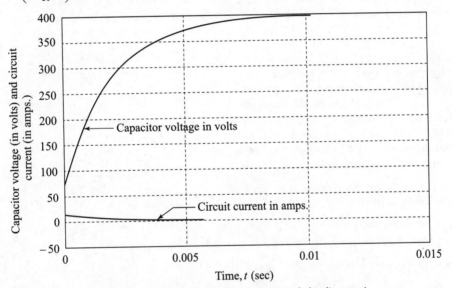

Fig. 2.83 Variation of capacitor voltage and circuit current

For fictitious values of $\tau = 0.002$ sec, $V_S = 400$ V, $V_0 = 75$ V, and $R = 25$ Ω, plots of capacitor voltage $v_C(t)$ and circuit current $i(t)$ are shown against time in Fig. 2.83.

Example 2.38 A resistance R and a 2 μF condenser are connected in series across a 200 V dc supply as shown in Fig. 2.84. Across the condenser is a neon lamp that strikes at 120 V. Calculate R to make the lamp strike 5 sec. after the switch has been closed. If $R = 5$ MΩ, how long will it take the lamp to strike.

Solution The time constant $\tau = RC = 2R$ seconds as R is in megaohms and C is in microfarads. In order that the lamp strikes at $t = 5$ sec, the condenser must acquire a voltage of 120 V. From Eq. (2.163) the voltage at time t is given by

$$v = V(1 - e^{-t/\tau})$$

Substituting $v = 120$, $V = 200$, $t = 5$ sec, and $\tau = 2R$ in the above equation,
$$120 = 200\,(1 - e^{-5/2R})$$

or $e^{-5/2R} = 1 - 0.6 = 0.4$

or $e^{2.5/R} = \dfrac{1}{0.4} = 2.5$

or $\dfrac{2.5}{R} = \ln 2.5 = 0.916$

$\therefore$ $R = 2.72$ MΩ

When $R = 5$ MΩ and $\tau = 2 \times 5 = 10$ sec,
$$120 = 200\,(1 - e^{-t/10})$$

or $1 - e^{-0.1t} = 0.6$

or $e^{2.5/R} = 0.4$

or $0.1\,t = \ln 2.5 = 0.916$

$\therefore$ $t = 9.16$ sec

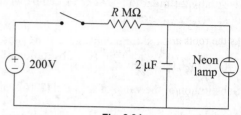

Fig. 2.84

2.12.2.5 Series *RLC* Circuit

The mathematical simulation of the response of an *RLC* series circuit to a dc voltage produces a second-order differential equation with constant coefficients. Hence, the analysis of such a circuit can best be explained by numerical example.

Example 2.39   Figure 2.85 shows a series *RLC* circuit, in which the switch S has been in position 1 for time long enough for the circuit to acquire steady state conditions. At $t = 0$, S is moved to position 2. Determine and plot (a) the current response for $t > 0$ and (b) the voltage across the capacitor for $t > 0$. Compute (c) the rate of voltage increase across the inductor at $t = 0^+$ and (d) the initial rate of energy consumption in the resistor.

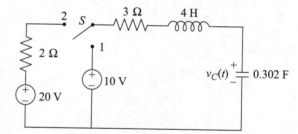

Fig. 2.85

Solution   In the steady state condition with switch in position 1, the voltage across the capacitor is 10 V. Hence, the initial conditions are
$$v_C(0^+) = 10 \text{ V} \quad \text{and} \quad i(0^+) = 0$$
The circuit response equation at any time t after S is moved to position 2 is given by
$$(3 + 2) \times i(t) + 4\frac{di(t)}{dt} + v_C(t) = 20 \tag{I}$$

Also $i(t) = C\dfrac{dv_C(t)}{dt} = 0.302 \times \dfrac{dv_C(t)}{dt}$

or $\dfrac{di(t)}{dt} = 0.302 \times \dfrac{d^2 v_C(t)}{dt^2} \tag{II}$

Substituting Eq. (II) in Eq. (I) and rearranging leads to
$$\frac{d^2 v_C(t)}{dt} + \frac{2.5}{2}\frac{dv_C(t)}{dt} + 0.828 v_C = 16.55 \tag{III}$$

and the characteristic equation is
$$(s^2 + 1.25s + 0.828) = 0 \tag{IV}$$

The roots of Eq. (IV) are

$$s_1 = -0.6250 + j0.6614 = -\alpha + j\omega_d$$
$$s_2 = -0.6250 - j0.6614 = -\alpha - j\omega_d$$

As the roots are complex conjugate, the response is under-damped and is given by

$$v_{C\,n}(t) = K_1 e^{(-0.6250 + j0.6614)\,t} + K_2 e^{(-0.6250 - j0.6614)\,t}$$

or $\qquad v_{Cn}(t) = e^{-0.6250t}\left[A_1 \cos(0.6614t) + A_2 \sin(0.6614t)\right]$ $\qquad$ (V)

In order to obtain the forced response, the particular integral (*PI*) is obtained by recalling that the response has the same form as the excitation and that at $t = \infty$, the inductor behaves like a short circuit and the capacitor is an open circuit. Hence,

$$v_{Cf}(t) = 20 \qquad\qquad\qquad (VI)$$

Hence, the solution of Eq. (II) is given by

$$v_C(t) = v_{Cn}(t) + v_{Cf}(t)$$
$$= e^{-0.6250t}[A_1 \cos(0.6614t) + A_2 \sin(0.6614t)] + 20 \qquad (VII)$$

The constants in Eq. (VII) are determined from the initial conditions, that is,

$$10 = A_1 + 20 \quad\text{or}\quad A_1 = -10$$

Differentiating of Eq. (VII) gives

$$\frac{dv_C(t)}{dt} = e^{-0.6250t} \times 0.6614 \times \left[-A_1 \sin(0.6614t) + A_2 \cos(0.6614t)\right]$$

$$+ (-0.6250)e^{-0.6250t}\left[A_1 \cos(0.6614t) + A_2 \sin(0.6614t)\right] \qquad (VIII)$$

Application of the second initial condition yields

$$i(0^+) = 0 = 0.302 \times \frac{dv_C(0^+)}{dt} = (0.302) \times \left[A_2 \times 0.6614 - 0.6250 \times A_1\right]$$

or $\qquad A_2 = \dfrac{0.625 \times -10}{0.6614} = -9.4496$

(a) The complete response of the capacitor is obtained as follows:

$$v_C(t) = 20 - e^{-0.6250\,t}\left[10\cos(0.6614t) + 9.4496\sin(0.6614t)\right]\text{V}$$

(b) The circuit current is obtainted as follows:

$$i(t) = 0.302 \times \frac{dv_C(t)}{dt} = 0.302 \times (0.6250)e^{-0.6250\,t}\left[10\cos(0.6614t) + 9.4496\sin(0.6614t)\right]$$

$$- (0.302) \times (0.6614)e^{-0.6250\,t}\left[-10\sin(0.6614t) + 9.4496\cos(0.6614t)\right]$$

$$= e^{-0.6250\,t}\left[-2 \times 10^{-5} \times \cos(0.6614t) + 6.26 \times \sin(0.6614t)\right]\text{A} \qquad (IX)$$

The coefficient of the cosine term in Eq. (IX) is very small. Therefore, the cosine term can be neglected without affecting the accuracy of the result. Thus,

$$i(t) = e^{-0.6250\,t}\left[6.26 \times \sin(0.6614t)\right]\text{ A} \qquad\qquad (X)$$

The time response of the circuit current and voltage is shown in Fig. 2.86.

$$\frac{di(t)}{dt} = (-0.625)e^{-0.625t}\left[6.26\sin(0.6614t)\right] + 6.26e^{-0.625t}\left[\cos(0.6614t) \times 0.6614\right]$$

$$= 0.626e^{-0.635t}\left[0.6614 \cos(0.6614t) - 0.625 \sin(0.6614t)\right]$$

$$\frac{di(0^+)}{dt} = \left[(-0.625) \times (-2.0 \times 10^{-5}) + (0.6614) \times 6.26\right] = 4.1404 \text{ A/sec}$$

(c) Rate of initial voltage increase across the inductor is

$$v_L(0^+) = 4 \times \frac{di(0^+)}{dt} = 4 \times 4.1404 = 16.5616 \text{ V/sec}$$

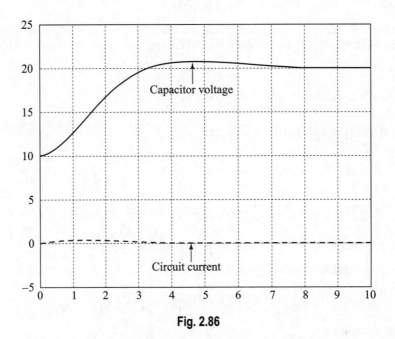

Fig. 2.86

(d) The current in the circuit at $t = 1.5$ sec is obtained by substituting $t = 1.5$ in Eq. (IX) and it works out to be 2.0523 A.

Energy consumed in the resistors $= (2.0523)^2 \times 5 = 21.06$ W.

Example 2.40   The switch S in the *RLC* circuit in Fig. 2.87 has been in position 1 for an infinitely long time. At $t = 0$ sec, S is moved to position 2. Derive expressions (a) for the discharge of capacitor voltage v_C and (b) the capacitor current for $t > 0$ when $R_1 = $ (i) 6 Ω, (ii) 4 Ω, and (iii) 0.5 Ω. Plot graphs to show variation of v_C and i_C with time.

Solution   The initial conditions in the circuit are determined by ascertaining the capacitor voltage and circuit current just before the position of S is changed to 2. Since the circuit is in the steady state, prior to the change in position of S, the capacitor is fully charged to a voltage of 18 V and the current through the circuit is zero.

Assume that the voltage across the capacitor at any time $t > 0$ sec is v_C volt and the circuit current is i ampere. Application of KVL around the closed circuit is given by

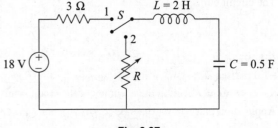

Fig. 2.87

$$Ri + L\frac{di}{dt} + \frac{1}{C}\int i\,dt = 0$$

Substitution of $i = C\,(dv_C/dt)$ in the above equation results in

$$LC\frac{d^2v_C}{dt^2} + R_1 C\frac{dv_C}{dt} + v_C = 0$$

or $\qquad \dfrac{d^2v_C}{dt^2} + \dfrac{R_1}{L}\dfrac{dv_C}{dt} + \dfrac{1}{LC}v_C = 0 \qquad\qquad\qquad\qquad\qquad$ (I)

It may be noted that Eq. (I) is a second-order differential equation with constant coefficients. Thus, its solution may be assumed to be

$$v_C = Ke^{st} \qquad\qquad\qquad\qquad\qquad \text{(II)}$$

Substitution of Eq. (II) along with its derivatives in Eq. (I) leads to

$$\left(s^2 + \frac{R_1}{L}s + \frac{1}{LC}\right)Ke^{st} = 0$$

The characteristic equation, which provides a meaningful solution, is written as

$$\left(s^2 + \frac{R_1}{L}s + \frac{1}{LC}\right) = 0 \qquad\qquad\qquad\qquad \text{(III)}$$

Clearly, the nature of the roots of Eq. (III) determines the response of the circuit. Substitution of the given inductor and capacitor values produces

$$(s^2 + 0.5\,R_1 s + 1) = 0$$

(i) When $R_1 = 6\ \Omega$, the roots of the characteristic equation are given by $s_1 = -0.38$ rad/sec and $s_2 = -2.62$ rad/sec. The variation of the capacitor voltage is given by

$$v_C = K_1 e^{-0.38t} + K_2 e^{-2.62t}$$

The values of the constants K_1 and K_2 are determined from the initial conditions as follows.
When $t = 0$, $v_C = 18$. Thus,

$$K_1 + K_2 = 18 \qquad\qquad\qquad\qquad\qquad \text{(IV)}$$

The circuit current is given by

$$i = C\frac{dv_C}{dt} = C\left[-0.38K_1 e^{-0.38t} - 2.62K_2 e^{-2.62t}\right]$$

Substitution of $i = 0$ at $t = 0$ produces

$$0.38\,K_1 + 2.62\,K_2 = 0 \qquad\qquad\qquad\qquad \text{(V)}$$

Simultaneous solution of Eqs (IV) and (V) yields $K_1 = 21.05$ and $K_2 = -3.05$. The voltage across the capacitor is

$$v_C = 21.05e^{-0.38t} - 3.05e^{-2.62t}$$

The circuit current is expressed as

$$i = 0.5\frac{dv_C}{dt} = 0.5\left[-0.38\times 21.05e^{-0.38t} + 2.62\times 3.05e^{-2.62t}\right] = 4\left[e^{-2.62t} - e^{-0.38t}\right]$$

(ii) When $R_1 = 4\ \Omega$, the roots of the characteristic equation are identical, real and are given by $s_1 = s_2 = -1.0$ rad/sec. The variation of the capacitor voltage is given by

$$v_C = [K_1 + K_2 t]e^{-t}$$

Application of the initial voltage condition gives $K_1 = 18$

and $\qquad i_C = 0.5\dfrac{dv_C}{dt} = 0.5\left[K_2(1-t) - K_1\right]e^{-t}$

When the initial current condition is applied to the above equation, it gives $K_2 = 18$. The transient response of the circuit is then expressed as

$$v_C = 18e^{-t}[1 + t]$$

and $i_C = -9te^{-t}$

(iii) When $R_1 = 2\ \Omega$, the roots of the characteristic equation are complex with negative real parts and are given by $s_1 = -0.5 + j\,0.866$ and $s_2 = -0.5 - j\,0.866$ rad/sec. The variation of the capacitor voltage is written as

$$v_C = \left[K_1 e^{j0.866t} + K_2 e^{-j0.866t}\right]e^{-0.5t} = \left[A\cos(0.866t) + B\sin(0.866t)\right]e^{-0.5t}$$

where $A = K_1 + K_2$ and $B = j(K_1 - K_2)$.

Again constants A and B are computed from the initial conditions as shown below.

Application of the voltage condition gives $A = 18$

The circuit current is expressed as

$$i_C = 0.5\frac{dv_C}{dt} = 0.5 \times 0.866\left[-A\sin(0.866t) + B\cos(0.866t)\right]e^{-0.5t}$$
$$- 0.5 \times 0.5\left[A\cos(0.866t) + B\sin(0.866t)\right]e^{-0.5t}$$

Substitution of the initial condition for circuit current results in

$$0.433B - 0.25A = 0$$

or $B = 0.25 \times 18/0.433 = 10.39$

Therefore, the capacitor voltage is given by

$$v_C = e^{-0.5t}[18\cos(0.866t) + 10.39\sin(0.866t)]$$

and the circuit current is

$$i_C = 0.433\left[-18\sin(0.866t) + 10.39\cos(0.866t)\right]e^{-0.5t} - 0.25\left[18\cos(0.866t) + 10.39\sin(0.866t)\right]e^{-0.5t}$$
$$= -e^{-0.5t}\left[0.0011\cos(0.866t) + 10.39\sin(0.866t)\right]$$

Figures 2.88 and 2.89 respectively show the plot of capacitor voltage and circuit current against time.

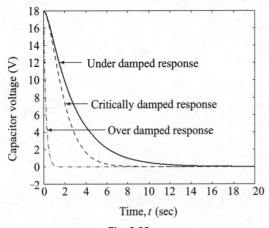

Fig. 2.88

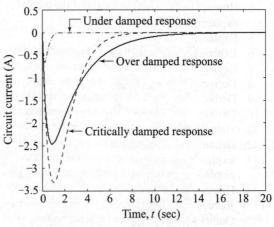

Fig. 2.89

Recapitulation

In a series circuit, $V = V_1 + V_2 + V_3 + \cdots + V_n$ volts

For resistances, $R_{eq} = R_1 + R_2 + R_3 + \cdots + R_n$ ohms

For inductances, $L_{eq} = L_1 + L_2 + L_3 + \cdots + L_n$ ohms

For capacitances, $\dfrac{1}{C_{eq}} = \dfrac{1}{C_1} + \dfrac{1}{C_2} + \dfrac{1}{C_3} + \cdots + \dfrac{1}{C_n}$

Voltage divider rule for n resistances, $V_n = \dfrac{R_n}{R_{eq}} V$ volts

In a parallel circuit, $I = I_1 + I_2 + I_3 + \cdots + I_n$ amperes

For resistances, $\dfrac{1}{R_{eq}} = \dfrac{1}{R_1} + \dfrac{1}{R_2} + \dfrac{1}{R_3} + \cdots + \dfrac{1}{R_n}$ ohms

For inductances, $\dfrac{1}{L_{eq}} = \dfrac{1}{L_1} + \dfrac{1}{L_2} + \dfrac{1}{L_3} + \cdots + \dfrac{1}{L_n}$

For capacitances, $C = C_1 + C_2 + C_3 + \cdots + C_n$, farads

Current division rule for n resistances, $I_n = \dfrac{R_{eq}}{R_n} I$ amperes

For delta–star transformation,

$$R_a = \dfrac{R_{ab} R_{ac}}{R_{ab} + R_{bc} + R_{ca}} \text{ ohms}$$

For star–delta transformaion, $R_{ab} = R_a + R_b + \dfrac{R_a R_b}{R_c}$

For maximum power transfer, $R_S = R_L$

The time constant of an RL circuit, $\tau = L/R$ seconds

Current rise in an R and L series circuit with dc supply,
$$i = I_0 (1 - e^{-t/\tau}) \text{ amperes}$$

Current decay in an RL circuit, $i = I_0 e^{-t/\tau}$ amperes

Voltage rise in R and C series circuit with dc supply,
$$v_C = V(1 - e^{-t/\tau})$$

Charging current, $i = I_0 e^{-t/\tau}$

Time constant of RC circuit, $\tau = RC$ seconds

Capacitor voltage when C discharges through R,
$$v_C = V e^{-t/RC}$$

Discharging current, $i = C \dfrac{dv_C}{dt} = -\dfrac{V}{R} e^{-t/RC} = -I_0 e^{-t/RC}$

Assessment Questions

1. Differentiate between (i) network, (ii) network elements, (iii) passive and active branch, (iv) bilateral and unilateral elements, and (vi) linear element.
2. What are ideal independent voltage and current sources? Sketch the symbolic representations of each and draw their *v-i* characteristics.
3. What are dependent sources? Enumerate the different types of dependent sources and sketch their symbolic diagrams.
4. How are practical (a) voltage and (b) current sources represented? Draw the load characteristic of each and explain their behaviour.
5. Explain source conversion.
6. Explain series connection of resistors and the voltage divider rule.
7. Explain parallel connection of resistors and the current divider rule.
8. Derive expressions for (a) inductors and (b) capacitors connected in series.
9. Derive expressions for (a) inductors and (b) capacitors connected in parallel.
10. Derive expressions for Y- Δ / Δ – Y transformations.
11. State the rule for Y- Δ / Δ – Y transformations by inspection of such a circuit.
12. Define a node. Enumerate the various steps to write the node-voltage equations for analysing a n-node circuit.
13. Explain the circumstances when it becomes necessary to create a super node in a circuit. Assume a fictitious circuit and show how a super node is created and treated.
14. Discuss the modifications in the node voltage method when (a) an independent and (b) a dependent voltage source are present.
15. Outline the steps required to analyse a circuit by applying the mesh current method.
16. Describe the modifications in the mesh current method when (a) an independent and (b) a dependent current source are present.
17. Explain, with the help of a circuit, how a super mesh is created and the method used to analyse such a circuit.
18. Describe matrix formulation by inspection, for (a) node voltage and (b) mesh current analysis of circuits.
19. Compare the node voltage and mesh current methods of circuit analysis.
20. State the basis for selecting between node voltage and mesh current techniques for circuit analysis.

21. Define the superposition theorem and list its limitations.
22. State the superposition theorem and explain its application by assuming an arbitrary circuit consisting of independent voltage and current sources. Can the superposition theorem be applied to power computations? Justify your answer.
23. Define (a) Thevenin and (b) Norton theorems. Explain how Norton equivalent is obtained from Thevinin equivalent.
24. State and explain maximum power transfer theorem. What is the efficiency under maximum power transfer condition? State in which types of applications such a level of efficiency is acceptable and why.
25. Define compensation theorem and state its limitations.
26. State Millman's theorem and develop expressions for an equivalent Millman generator.
27. Write an essay on the significance of the time constant of a circuit.
28. What is natural frequency of a circuit and how is it determined.
29. Discuss the behaviour of an ideal (a) inductor and (b) capacitor when connected across a dc excitation. Show the symbolic representation of their behaviour.

Problems

2.1 A 6 V battery having an internal resistance of 0.25 Ω is connected to a load resistance $R_L = 2\,\Omega$. Determine the: (a) total power supplied to the load, (b) power dissipated and lost within the voltage source, (c) load voltage, (d) terminal voltage and power supplied to the circuit when the voltage source is replaced by an ideal source with the same magnitude of source voltage. [(a) 14.22 W, (b) 1.78 W, (c) 5.33 V, (d) 6 V, 18 W]

2.2 The voltage at the terminals of a particular power supply on no load is 12.5 V. When a 5 W load is attached, the voltage drops to 12 V. (a) Determine V_S and R_S for this practical source. (b) What would be the terminal voltage of the source when a 10 Ω load resistor is connected? (c) How much current could be drawn from this power supply under short-circuit conditions? [(a) 12.5 V, 1.2 Ω, (b) 11.16 V, (c) 7.833 A]

2.3 An incandescent light bulb rated at 100 W dissipates 100 W as heat and light when connected across a 230 V ideal voltage source. If four such bulbs are connected in series across the same source, determine the power each bulb will dissipate. [6.34 W]

2.4 Determine the equivalent resistance seen by the source of the circuit shown in Fig. 2.90 by combining resistors in series and in parallel. Find V_1, I_1, I_2.
[$R_{eq} = 6.05\,\Omega$, $V_1 = 5.2$ V, $I_1 = 0.72$ A, $I_2 = 0.29$ A]

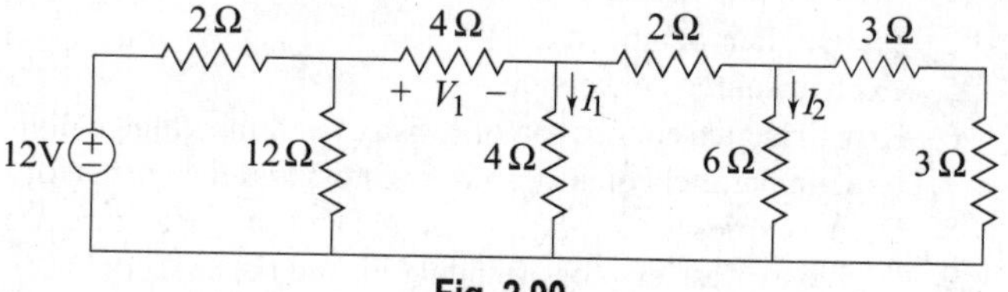

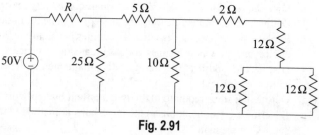

Fig. 2.90

2.5 For the circuit shown in Fig. 2.91 the power consumed by the 25 Ω resistor is 25 W. Find R. [7.96 W]

2.6 Find the currents I_1 and I_2, the power delivered by the 5 A current source and by the 20 V voltage source, and the total power dissipated by the circuit shown in Fig. 2.92. [$I_1 = 4.5$ A, $I_2 = 0.5$ A, 812.5 W, 44 W, 856.5 W]

Fig. 2.91

2.7 In the bridge circuit in Fig. 2.93, the terminals P and Q are shorted and $R_1 = 1.1$ kΩ, $R_2 = 20$ kΩ, $R_3 = 4.7$ kΩ, and $R_4 = 6.6$ kΩ. Determine the (a) equivalent resistance of the bridge as seen at the battery terminals and (b) voltage between the nodes P and Q when the short circuit is removed. [(a) 3.78 kΩ, (b) 6.75 V]

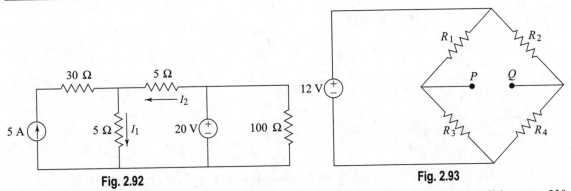

Fig. 2.92   **Fig. 2.93**

2.8 The capacitance values of three capacitors are 20, 40, and 60 µF. If these are placed in parallel across a 230 V source, find (a) the equivalent capacitance, (b) total charge residing on the capacitors, and (c) charge on each capacitor.   [(a) 120 µF, (b) 27.5 mC, (c) 4.6 mC, 9.2 mC, 13.8 mC]

2.9 The three capacitors of Problem 2.8 are placed in series across a 440 V source. (a) Compute the equivalent capacitance. (b) Find the charge on each capacitor. (c) Determine the voltage drop across each capacitor. (d) Compute the total stored energy.   [(a) (120/11) µF; (b) 4.8×10^{-3} C; (c) 240 V, 120 V, 80 V; (d) 0.132 J]

2.10 Three series-connected inductor coils have voltages of 20, 30, and 50 V appearing across their terminals when the circuit current is changing at a rate of 100 A/sec. Determine the equivalent series inductance.   [1.0 H]

2.11 An inductor $L_1 = 0.02$ H is connected in series with a parallel combination of two inductors $L_2 = 0.01$ H and $L_3 = 0.05$ H. (a) Find the equivalent inductance L_{eq} of the combination. (b) Determine the value of the emf of self-inductance in coil L_2 when the current in coil L_1 is changing at a rate of 1500 A/sec.
[(a) 0.02833 H, (b) 12.5 V]

2.12 Determine the resistance between the points A and B of the network shown in Fig. 2.94.   [2 Ω]

2.13 Determine the resistance between the points A and B of the network shown in Fig. 2.95.   [2.88 Ω]

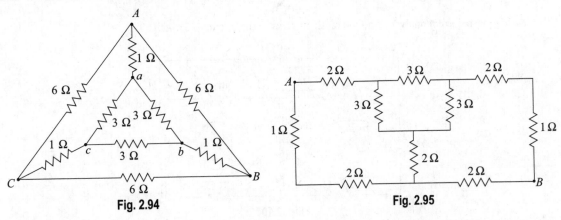

Fig. 2.94   **Fig. 2.95**

2.14 Using node voltage analysis in the circuit shown in Fig. 2.96, find the voltage V across the 8 S conductance.   [0.34 V]

2.15 Using node voltage analysis in the circuit shown in Fig. 2.97, find current I through the voltage source.   [29.5 A]

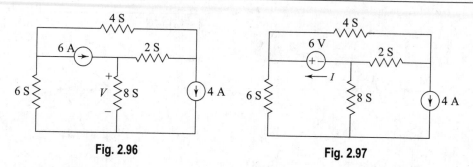

Fig. 2.96   **Fig. 2.97**

2.16 Using node voltage analysis in the circuit shown in Fig. 2.98, find the three indicated node voltages.
$$[V_1 = 157.5 \text{ V}, V_2 = 120.51 \text{ V}, V_3 = -437.3 \text{ V}]$$

2.17 Calculate the node voltages and the power consumed in the 4 Ω resistor in the circuit which contains an independent and a dependent voltage source as shown in Fig. 2.99. All data is provided in the figure.
$$[V_A = 5.20 \text{ V}, V_B = 10.74 \text{ V}, V_C = 4.41 \text{ V}, V_D = 0.20 \text{ V}, 7.69 \text{ W}]$$

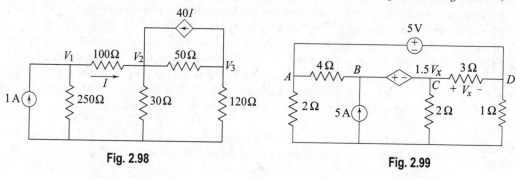

Fig. 2.98   **Fig. 2.99**

2.18 Using mesh current analysis, find the mesh currents in the circuit of Fig. 2.100.

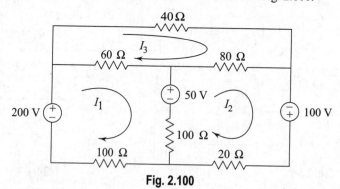

Fig. 2.100

$$[I_1 = 1.864 \text{ A}, I_2 = 2.35 \text{ A}, I_3 = 0.384 \text{ A}]$$

2.19 Using mesh current analysis, find the current I through the voltage source in the circuit shown in Fig. 2.97.
[29.51 A]

2.20 Using mesh current analysis, find the current I through the 150 W resistor in the circuit shown in Fig. 2.98.
[0.37 A]

2.21 Compute (i) the current in the 5 Ω resistor and (ii) the power supplied by the 12 V source in the circuit shown in Fig. 2.101.
[(i) 1.49 A, (ii) 35.31 W]

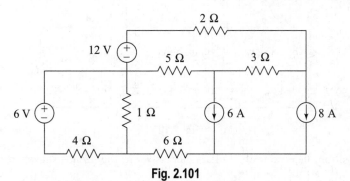

Fig. 2.101

2.22 Find the Thevenin equivalent circuit as seen by the 3 Ω resistor for the circuit shown in Fig. 2.102.
[4.8 V, 4.4 Ω]

2.23 Compute the voltage *V* across the 4 Ω resistor in the circuit shown in Fig. 2.103 using the Thevenin equivalent circuit.
[– 4.33 V]

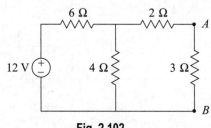

Fig. 2.102

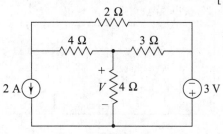

Fig. 2.103

2.24 Find the Thevenin equivalent to the left of *AB* for the circuit shown in Fig. 2.104.   [13.333 V, 0.667 Ω]

2.25 Find the Thevenin equivalent to the left of *AB* for the circuit shown in Fig. 2.105.   [– 4.8 V, 0 Ω]

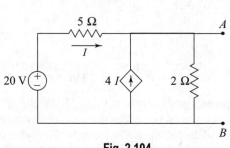

Fig. 2.104

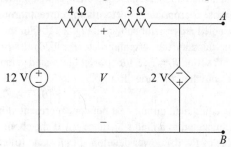

Fig. 2.105

2.26 Find the Norton equivalent to the left of *AB* of the circuit shown in Fig. 2.102.   [1.09 A, 4.4 Ω]

2.27 Find the Norton equivalent to the left of *AB* of the circuit shown in Fig. 2.104.   [20 A, 0.6675 Ω]

2.28 The 5 Ω resistor in the circuit shown in Fig. 2.106 increases by 5%. Verify the compensation theorem.

[*Hint*: Apply a test voltage of 1.0 V at terminals *AB*].

2.29 (a) Using Millman's theorem, develop the equivalent generator at terminals *AB*, for the circuit given in Fig. 2.107; (b) calculate the current through the load resistance and power dissipated. Verify the computations by any suitable method.

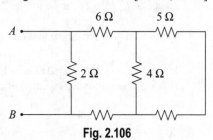

Fig. 2.106

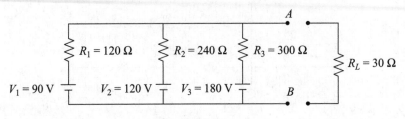

Fig. 2.107

2.30 Using superposition, determine the voltage across R_2 in the circuit shown in Fig. 2.108 in which $V_{S1} = V_{S2} = 120$ V. [40 V]

2.31 With reference to Fig. 2.109, using superposition, determine the component of the current through the 8 Ω resistor that is due to V_{S2}. The battery voltage $V_{S1} = V_{S2} = 24$ V. [4.118 A]

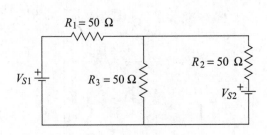

Fig. 2.108

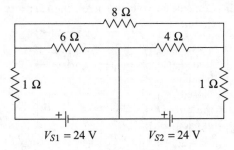

Fig. 2.109

2.32 Between (i) node voltage, (ii) mesh current, (iii) superposition, and (iv) Thevenin equivalent, which method will be most appropriate for determining current through the variable load resistor shown in Fig. 2.110. For the given data, calculate the current through the load resistor when $R_L = 1$ Ω and $R_L = 3$ Ω. Plot power dissipated versus load current when R_L varies from 0.5 Ω to 5 Ω.
[0.77 A, 0.46 A]

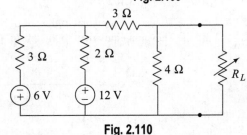

Fig. 2.110

2.33 The equivalent circuit of a network represented by Thevenin's equivalent with $V_{TH} = 12$ V and $R_{TH} = 8$ Ω is connected to a load resistance R_L. If the conditions for maximum power transfer exist, determine the (a) value of R_L, (b) power developed in R_L, and (c) efficiency of the circuit, that is, the ratio of power absorbed by the load to power supplied by the source. [(a) 8 Ω, (b) 4.5 W, (c) 0.5]

2.34 An inductor of inductance L H and resistance of r Ω is connected in series with a resistor of R Ω across a voltage source of V_s volts as shown in Fig. 2.111. The switch S, which has been in the open position for a long time, is closed at $t = 0$ sec. Show that for $t > 0$, the current through the circuit is given by $i = \dfrac{V_s}{r}\left(1 - \dfrac{R}{R+r}e^{-\frac{r}{L}t}\right)$ A.

If $L = 5$ H, $r = 5$ Ω, $R = 10$ Ω, and $V_s = 72$ V, determine the circuit current and energy at (a) $t = 0.25$ sec and (b) $t = 5.0$ sec. Plot the variation of circuit current against time and comment on the result.
[(a) 119.84 J (b) 513.75 J]

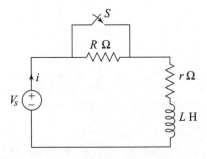

Fig. 2.111

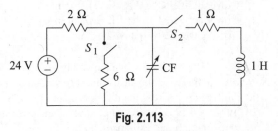

Fig. 2.112

2.35 At $t = 0$ sec, the switch S in the circuit shown in Fig. 2.112 is closed. For the data shown in the figure, determine the time required at which the battery is supplying a current of 20 mA. After the circuit has attained steady state, the switch S is again opened. Compute the capacitor voltage and charge at t equal to τ.

[0.086 sec, 4.41 V, 529.20 μC]

2.36 The switch S_1 in the circuit in Fig. 2.113 is in closed position and switch S_2 is in open position. After the circuit has reached steady state condition, at $t = 0$ sec, S_1 is opened while S_2 is closed at the same time. For the data shown in the circuit diagram, derive expressions for the capacitor voltage and current for $t > 0$ when (i) $C = 2$ F, (ii) $C = 4$ F, and (iii) $C = 8$ F. Plot the capacitor voltage and current curves against time.

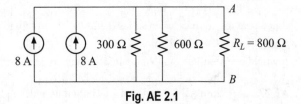

Fig. 2.113

$$[\text{(i)} \ v_C = 18e^{-0.5t}[\cos(0.5t) + \sin(0.5t)], \ i = -36e^{-0.5t}\sin(0.5t)$$
$$\text{(ii)} \ v_C = 9e^{-0.5t}(2 + t), \ i = -9te^{-0.5t}$$
$$\text{(iii)} \ v_C = 25.71e^{-0.15t} - 7.71e^{-0.35t}, \ i = 5.40e^{-0.35t} - 7.71e^{-0.15t}]$$

Additional Examples

AE 2.1 A voltage source when tested produced (a) 120 V and 10 A, and (b) 100 V and 20 A at its terminals. Calculate source (i) voltage and (ii) internal resistance.

Solution Assume the source voltage and internal resistance to be V_S volt and r_S ohm respectively. Application of Ohm's law leads to the following simultaneous equations.

$$V_S - 10 \times r_S = 120 \quad V_S - 20 \times r_S = 100$$

Solution of the above equations yields (i) $V_S = 140$ V and (ii) $r_S = 2 \ \Omega$

AE 2.2 Two independent current sources of 4 A and 8 A and internal resistances of 300 Ω and 600 Ω each are connected in parallel as shown in Fig. AE 2.1. Compute (a) open circuit voltage, (b) load current if $R_L = 800 \ \Omega$. What is the power supplied to the load?

Solution Combining the two internal resistances gives

$$R_S = \frac{300 \times 600}{900} = 200 \ \Omega$$

Fig. AE 2.1

(a) Open circuit voltage $V_{A-B} = (4 + 8) \times 200 = 2400$ V

(b) Load current $I_L = 2400/800 = 3$ A

Power supplied to the load $= 3^2 \times 800 = 7200$ W

AE 2.3   An 8 Ω resistance is connected across a 24 V ideal voltage source. Determine (a) circuit current and (b) voltage drop across the resistance. If a 4 Ω resistance is connected in series with the 8 Ω resistance, calculate (c) circuit current, (d) voltage drop across each resistor, and (e) power delivered by the source in each case.
Solution
   (a) Circuit current $= 24/8 = 3$ A
   (b) Since the full voltage of the voltage source is applied across the 8 Ω resistance, voltage across it is 24 V.
   (c) In this case, full voltage of the source is applied across the $(8 + 4)$ Ω resistances in series. Thus, current $I = 24/(8 + 4) = 2$ A
   (d) Voltage across 8 Ω resistance $2 \times 8 = 16$ V
        Voltage across 4 Ω resistance $2 \times 4 = 8$ V
   (e) Power delivered to the 8 Ω resistor $= V \times I = 24 \times 3 = 72$ W
        Power delivered to the 8 Ω resistor in series combination $= V \times I = 16 \times 2 = 32$ W
        Power delivered to the 4 Ω resistor in series combination $= V \times I = 8 \times 2 = 16$ W
        Total power delivered by the source $= 32 + 16 = 48$ W

AE 2.4   Repeat AE 2.3 when the voltage source is replaced by an ideal current source of 24 A.
Solution   In this case, the ideal current source supplies a constant current of 24 A under both types of connections.
   (a) Voltage drop across the 8 Ω resistance $= 24 \times 8 = 192$ V
   (b) When a 4 Ω resistance is connected in series, the constant current of 24 A from the current source flows through both the resistances.
   (c) Voltage drop across 8 Ω resistance $= 24 \times 8 = 192$ V
        Voltage drop across 4 Ω resistance $= 24 \times 4 = 96$ V
   (d) Voltage drop across the series combination $= 24 \times (4 + 8) = 288$ V
   (e) Power delivered to the 8 Ω resistor $= V \times I = 192 \times 24 = 4608$ W
        Power delivered to the 8 Ω resistor in series combination $= V \times I = 192 \times 24 = 4608$ W
        Power delivered to the 4 Ω resistor in series combination $= V \times I = 96 \times 24 = 2304$ W
        Total power delivered by the source $= V \times I = 288 \times 24 = 6912$ W
        From the foregoing it can be concluded that there is no limit on the voltage and power delivered by a current source.

AE 2.5   The branch shown in Fig. AE 2.2 below, has an independent voltage source of 12 V connected in series with a dependent CCVS of $v_s = 10i_1$. Determine the voltage v when $i_1 = 1.5$ A.
Solution   For $i_1 = 1.5$ A, the voltage output from the dependent CCVS
$$v_s = 10 \times 1.5 = 15 \text{ V}$$
Since both the sources are in series (additive) the voltage
$$v = 10 + 12 = 27 \text{ V}$$

Fig. AE 2.2

It would be useful to note that the multiplying constant of 10 of the CCVS has the unit of ohm.

AE 2.6   Figure AE 2.3 shows a CCCS and an independent current source connected to a node. Determine the magnitude and unit of the multiplying constant A for the KCL to be applicable. All data is provided in the figure. What happens when the output current of the independent current source is changed to 18 A?

Solution Application of KCL at the node point yields

$$12 + A \times 15 = 15 \quad \text{or} \quad A = (15 - 12)/15 = 0.2$$

The multiplying constant is dimensionless.

When the output current of the independent current source is changed to 18 A, the expression for KCL becomes

$$12 + A \times 15 = 15 \quad \text{or} \quad A = (15 - 18)/15 = -0.2$$

Negative multiplying constant means that the direction of the CCCS should be reversed.

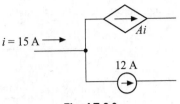

Fig. AE 2.3

AE 2.7   In the circuit shown in Fig. AE 2.4, the practical current source has a capacity of 10 A and an internal resistance of 100 Ω. Plot the v–i characteristics of the current source when load resistance is equal to (a) 10 Ω, (b) 50 Ω, and (c) 100 Ω.

Solution   From Fig. AE 2.4, the following data is available: $i_S(t) = 10$ A and $r_S = 100$ Ω.

On OC $(R_L = \infty)$ $i_L(t) = 0$, and $v_L(t) = 10 \times 100 = 1000$ V

(a) For $R_L = 10$ Ω, $i_L(t) = 10 \times \dfrac{100}{(100+10)} = 0.091$ A, and $v_L(t) = 9.091 \times 10 = 90.91$ V

(b) For $R_L = 50$ Ω, $I_L(t) = 10 \times \dfrac{100}{(100+50)} = 6.667$ A, and $v_L(t) = 6.667 \times 50 = 333.333$ V

(c) For $R_L = 100$ Ω, $i_L(t) = 10 \times \dfrac{100}{(100+100)} = 5.0$ A, and $v_L(t) = 5.0 \times 100 = 500$ V

The v–i characteristics are plotted in Fig. AE 2.5.

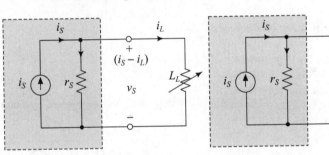

Fig. AE 2.4

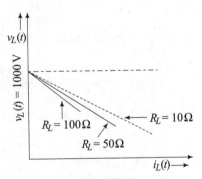

Fig. AE 2.5

AE 2.8   Use source transformation technique to compute the voltage across the 2 Ω load resistor in the circuit in Fig. AE 2.5. All data is shown in the circuit in the figure.

Solution   As a first step the voltage source is transformed into a current source and the circuit is redrawn as shown in Fig. AE 2.7 below.

The current sources are combined and the circuit is modified as shown in Fig. AE 2.8 below.

Assume the current through the load to be I_L. The voltage across points *A-B* is given by

$$(4 + 2 + 3) \times I_L = (12 - I_L) \times 6$$

or $I_L = 72/15 = 4.8$ A

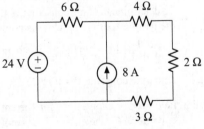

Fig. AE 2.6

Voltage across the 2 Ω load resistor = 4.8 × 2 = 9.6 V

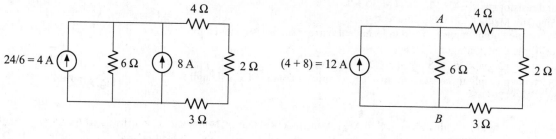

Fig. AE 2.7   **Fig. AE 2.8**

AE 2.9   Figure AE 2.9 shows a *RL* circuit supplied by a dc voltage source of 6 V. At *t* = 0, due to a power surge the source voltage jumps to 12 V. Calculate the current through the 40 Ω resistor at (a) *t* = 0⁻ and (b) *t* = 0⁺. What is the time constant of the circuit.

Solution   Figure. AE 2.10 shows the circuit at *t* = 0⁻, that is, just before the power surge.

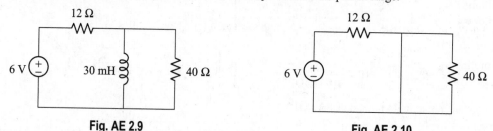

Fig. AE 2.9   **Fig. AE 2.10**

(a) Since the inductor behaves like a short circuit, current through it = 6/12 = 0.5 A and current through the 40 Ω is 0 A.

Thus, at *t* = 0⁺, 0.5 A current flows through the inductor and the equivalent circuit takes the form as shown in Fig. AE 2.11.

If the voltage at node point A is *V* volts and the current through the 40 Ω resistor is *I* A, application of KCL yields

$$\frac{12-V}{12} = 0.5 + \frac{V}{40}$$

or *V* = 4.615 V

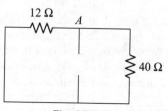

Fig. AE 2.11

(b) Thus, at *t* = 0⁺, current through the 40 Ω resistor *i* = 4.617/40 = 0.115 A
To determine the time constant τ of the circuit, the Thevenin resistance at the terminals of the inductor is computed from Fig. AE 2.12 as follows

$$R_{TH} = (40 \times 12)/(40 + 12) = 9.231 \, \Omega$$

And time constant τ = 30/9.231 = 3.25 ms

AE 2.10   Derive expressions for current through the inductor and the 40 Ω resistor for *t* ≥ 0 for the circuit in Example AE 2.9.

Fig. AE 2.12

Solution Mesh current technique is employed to determine inductor and resistor currents as follows.

Step 1: Mesh equation for resistor current i_R for $t \geq 0$ is given by

$$12(i_R + i_L) + 40i_R = 12$$

or, $i_R = 0.231(1 - i_L)$ (AE.2.10.1)

Mesh equation for inductor current i_L for $t \geq 0$ is written as

$$12(i_R + i_L) + 0.03\frac{di_L}{dt} = 12$$

(AE.2.10.2)

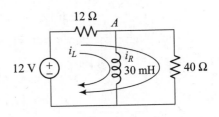

Fig. AE 2.13

Substituting for i_R from Eq. (AE.2.10.1) in Eq. (AE.2.10.2) and simplifying leads to

$$\frac{di_L}{dt} + 307.6i_L = 307.6$$

and its solution gives $i_L = \left(1 + Ke^{-307.6t}\right)$ A   (AE.2.10.3)

Substituting the initial condition, that is, at $t = 0^+$, $i_L = 0.5$ A in Eq. (AE.2.10.3) results in the expression for the inductor current as under

$$i_L = \left(1 - 0.5e^{-307.6t}\right) \text{A}$$

(AE.2.10.4)

Substituting for i_L in Eq. (AE.2.10.1) gives

$$i_R = 0.231\left(1 - \left[1 - 0.5e^{-307.6t}\right]\right) = 0.115e^{-307.6t} \text{ A}$$

(AE.2.10.5)

AE 2.11 Calculate the current in the 40 Ω resistor (a) immediately after the surge has passed and (b) at $t = \infty$.

Solution In the present case when the surge has passed at $t = 5$ ms, the source voltage immediately drops to 6 V, however, the circuit is still in transient condition. In order to compute the resistor current at $t = 5$ ms, it is necessary to calculate the current through the inductor at $t = 5^+$ ms. Substituting $t = 5$ ms in Eq. (AE.2.10.4) yields $i_L = 0.893$ A. The equivalent circuit at $t = 5^+$ ms is shown in Fig. AE 2.14.

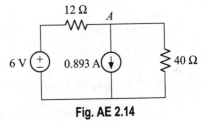

Fig. AE 2.14

If the voltage at node point A is V volts and the current through the 40 Ω resistor is I A, application of KCL yields

$$\frac{6 - V}{12} = 0.893 + \frac{V}{40} \quad \text{or} \quad V = -3.628 \text{ V}$$

Current through the resistor at $t = 5$ ms, $i_R = -3.628/40 = -0.91$A

As earlier, expression for transient current through the resistor for $t \geq 5$ ms, is expressed as under

$$i_R = -0.91e^{-307.6(t - 0.005)} \text{ A}$$

AE 2.12 Use Superposition theorem to compute output voltage v_0 for the circuit shown in Fig. AE 2.15,

Solution
Step 1: Deactivate the independent current source. Fig. AE 2.16 shows the re-drawn circuit.
Step 2: The voltage v_0' due to the 12 V source is calculated by applying voltage divider rule as under

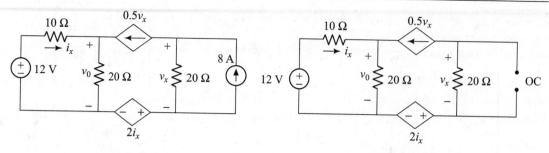

Fig. AE 2.15

Fig. AE 2.16

$$v_0' = \frac{20}{30} \times 12 = 8 \text{ V} \qquad (AE\ 2.12.1)$$

Step 3: With the independent voltage source short-circuited the circuit is re-drawn in Fig. AE 2.17.
For easy understanding the nodes have been numbered and a reference node added.

Step 4: Application of KCL at nodes 1 and 2 leads to

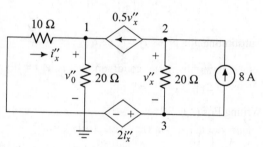

Fig. AE 2.17

Node 1: $\dfrac{v_0''}{10} + \dfrac{v_0''}{20} = 0.5 v_x''$, or, $0.15 v_0'' - 0.5 v_x'' = 0$ (AE 2.12.2)

Node 2: $0.5\ v_x'' + \dfrac{v_2'' - 2\ i_x''}{20} = 8$, or,

$$0.5\ v_x'' + 0.05 v_2'' - 0.1\ i_x'' = 8 \qquad (AE\ 2.12.3)$$

From Fig. AE 2.17, it is seen that $v_2'' = v_x'' + 2\ i_x''$. Substituting for v_2'' in Eq. (2.12.3) yields

$$0.5 v_x'' + 0.05(v_0'' + 2\ i_x'') - 0.1\ i_x'' = 8, \text{ or, } v_x'' = 8/0.55 = 14.545 \text{ V}$$

Substituting for v_x'' in Eq. (AE. 2.12.2) gives

$$v_0'' = (0.5 \times 14.545)/0.15 = 48.46 \text{ V} \qquad (AE\ 2.12.4)$$

Step 5: Adding Eqs (AE 2.12.1) and (2.12.4) results in

$$v_0 = v_0' + v_0'' = 8 + 27.778 = 35.778 \text{ V}$$

Again it may be noted that while analysing circuits using superposition theorem the dependent sources are *never* deactivated.

AE 2.13 Use superposition principle to calculate i_x in the circuit shown in Fig. AE 2.18.
Solution
Step 1: Fig. AE 2.19 shows the given circuit with the voltage source deactivated and the dependent source retained.
Step 2: For the assumed directions, the mesh current equations are written as follows

For mesh 1: $i_1 = -5$ (AE 2.13.1)

For mesh 2: $-2i_1 + 6i_2 - i_3 - 6i_x' = 0$ (AE 2.13.2)

For mesh 3: $-5i_1 - i_2 + 10i_3 + 6i_x' = 0$ (AE 2.13.3)

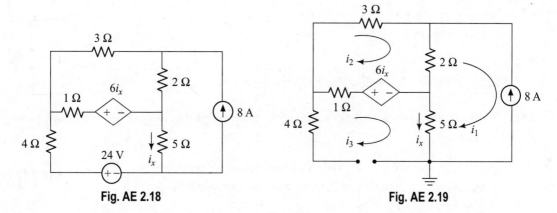

| **Fig. AE 2.18** | **Fig. AE 2.19** |

Substituting $i'_x = i_3 - i_1$ in Eqs (AE 2.13.2) and (AE 2.13.3) and simplifying leads to

$$4i_1 + 6i_2 - 7i_3 = 0 \qquad \text{(AE 2.13.4)}$$

$$-11i_1 - i_2 + 16i_3 = 0 \qquad \text{(AE 2.13.5)}$$

Writing Eqs (AE 2.13.1), (AE 2.13.4), and (AE 2.13.5) in matrix form gives

$$\begin{bmatrix} 1 & 0 & 0 \\ 4 & 6 & -7 \\ -11 & -1 & 16 \end{bmatrix} \begin{bmatrix} i_1 \\ i_2 \\ i_3 \end{bmatrix} = \begin{bmatrix} -5 \\ 0 \\ 0 \end{bmatrix}, \quad \text{then,} \quad \begin{bmatrix} i_1 \\ i_2 \\ i_3 \end{bmatrix} = \begin{bmatrix} 1 & 0 & 0 \\ 4 & 6 & -7 \\ -11 & -1 & 16 \end{bmatrix}^{-1} \begin{bmatrix} -5 \\ 0 \\ 0 \end{bmatrix} \text{A}$$

or

$$\begin{bmatrix} i_1 \\ i_2 \\ i_3 \end{bmatrix} = \begin{bmatrix} 1 & 0 & 0 \\ 0.1461 & 0.1798 & 0.0787 \\ 0.6966 & 0.0112 & 0.0674 \end{bmatrix} \begin{bmatrix} -5 \\ 0 \\ 0 \end{bmatrix} \text{A}$$

Thus, $\quad i'_x = -3.483 + 5.0 = 1.517$ A $\qquad\qquad$ (AE 2.13.6)

Step 3: Figure AE 2.20 shows the given circuit with the current source de-activated.

Step 4: For the assumed directions, the mesh current equations are written as under

$$6\,i_1 - i_2 - 6\,i''_x = 0 \qquad \text{(AE 2.12.7)}$$

$$-i_1 + 10\,i_2 + 6\,i''_x = 24 \qquad \text{(AE 2.12.8)}$$

Substituting $i''_x = i_2$ in Eqs (AE 2.12.7) and (AE 2.12.8) and combining the like terms results in

$$6\,i_1 - 7\,i_2 = 0 \qquad \text{(AE 2.12.9)}$$

$$-i_1 + 16\,i_2 = 24 \qquad \text{(AE 2.12.10)}$$

Writing Eqs (AE 2.12.9) and (AE 2.12.10) in matrix form gives

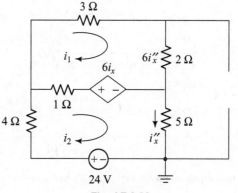

Fig. AE 2.20

$$\begin{bmatrix} 6 & -7 \\ -1 & 16 \end{bmatrix} \begin{bmatrix} i_1 \\ i_2 \end{bmatrix} = \begin{bmatrix} 0 \\ 24 \end{bmatrix}, \text{ or,} \quad \begin{bmatrix} i_1 \\ i_2 \end{bmatrix} = \begin{bmatrix} 6 & -7 \\ -1 & 16 \end{bmatrix}^{-1} \begin{bmatrix} 0 \\ 24 \end{bmatrix} \text{A}$$

or $\begin{bmatrix} i_1 \\ i_2 \end{bmatrix} = \begin{bmatrix} 0.18 & 0.0112 \\ 0.0786 & 0.0674 \end{bmatrix} \begin{bmatrix} 0 \\ 24 \end{bmatrix} = \begin{bmatrix} 0.2688 \\ 1.6176 \end{bmatrix}$ A

Thus, $i_x'' = i_2 = 1.6176$ A $\hspace{5cm}$ (AE 2.12.11)

Step 5: Addition of Eqs (AE 2.12.6) and (AE 2.12.11) gives $i_x = 1.517 + 1.6876 = 3.1346$ A

AE 2.14 Use the superposition principle to determine the power supplied by each source in the circuit shown in Fig. AE 2.21.
Solution

Step 1: Figure AE 2.21 shows the circuit with 18 V and 5 A sources deactivated.

Step 2: From Fig. AE 2.21 it is seen that the three resistors are in series. Thus,

$i_1' = -i_2' = i_3' = 24/18 = 1.333$ A $\hspace{4cm}$ (AE 2.14.1)

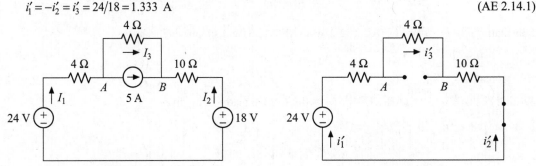

Fig. AE 2.21 $\hspace{5cm}$ **Fig. AE 2.21**

Step 3: Figure AE 2.22 shows the circuit with 24 V and 5 A sources deactivated.

Step 4: From Fig. AE 2.22 it is seen that the three resistors are in series. Thus,

$-i_1'' = i_2'' = -i_3'' = 18/18 = 1.0$ A $\hspace{4cm}$ (AE 2.14.2)

Step 5: Figure AE 2.23 shows the circuit with 24 V and 18V sources deactivated.

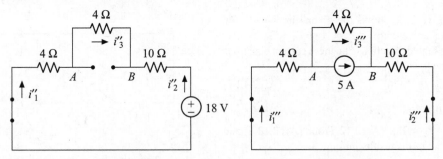

Fig. AE 2.22 $\hspace{5cm}$ **Fig. AE 2.23**

Step 6: Observe from Fig. AE 2.23 that resistors $(4 + 10)\Omega \parallel 4\Omega$ resistor. Thus,

$i_1''' = -i_2''' = \dfrac{4}{(4+10+4)} \times 5 = 1.111$ A and $i_3''' = -(5-1.111) = -3.889$ A $\hspace{2cm}$ (AE 2.14.3)

Step 7: Various components from Eqs (AE 2.14.1) to (AE 2.14.3) to get the circuit currents as under

$I_1 = i_1' + i_1'' + i_1''' = 1.333 - 1.0 + 1.111 = 1.444$ A

$$I_2 = i_2' + i_2'' + i_2''' = -1.333 + 1.0 - 1.111 = -1.444 \text{ A}$$

$$I_3 = i_3' + i_3'' + i_3''' = 1.333 - 1.0 - 3.889 = -3.556 \text{ A}$$

Step 8:   Power supplied by the sources are determined as follows

Power by 24 V source $= 24 \times 1.4444 = 34.67 \text{ W}$

Power by 18 V source $= 18 \times (-1.4444) = -26.00 \text{ W}$

Power by 5 A source $= I_3 \times V_{BA} = 5 \times (4 \times 3.556) = 71.12 \text{ W}$

Verification of the above computations may be done as given below:

Total power supplied by two voltage sources and current source $= 34.67 - 26.00 + 71.12 = 79.79$

Total power consumed by the resistances of the circuit $= (1.444)^2 \times 4 + (-1.444)^2 \times 10 + (-3.556)^2 \times 4 = 79.79 \text{ W}$

Thus the total power supplied by two voltage sources and current source equals the power consumed by the resistances of the circuit.

Objective Type Questions

1. A branch in a network is said to be active when it contains a
    (i) resistor   (ii) inductor
    (iii) capacitor   (iv) source

2. Any closed path formed by the branches in a network is called a
    (i) loop   (ii) mesh
    (iii) node   (iv) none of these

3. Dependent voltage and current sources are
    (i) unidirectional
    (ii) output is dependent on input
    (iii) independent of any other network variable
    (iv) all of these

4. As the current supplied by a practical voltage source increases, its voltage
    (i) remains constant
    (ii) increases linearly
    (iii) decreases linearly
    (iv) decreases exponentially

5. When resistances are connected in parallel, the voltage across the resistors is
    (i) equal
    (ii) unequal
    (iii) proportional to the square of the current
    (iv) none of these

6. The total equivalent resistance of a network when a number of resistances are connected in series is obtained by
    (i) directly adding the conductance of all the resistances
    (ii) multiplying the magnitudes of the resistances

    (iii) directly adding resistances
    (iv) the ratio of the sum of the conductances of all resistances to the sum of all resistances

7. When a number of resistances are connected in parallel, the current divides
    (i) in inverse proportion to the resistances of the individual parallel resistances
    (ii) in direct proportion to the resistances of the individual parallel resistances
    (iii) in inverse proportion to the conductance of the individual parallel conductance
    (iv) none of these

8. Three capacitors, each of C microfarad are first connected in series and then in parallel. The equivalent capacitance
    (i) is greater in the series combination
    (ii) is greater in the parallel combination
    (iii) is the same in the two combinations
    (iv) none of these

9. Two resistances of 5 Ω and 20 Ω are connected in parallel. The parallel combination is connected in series with a 1 Ω resistance and this series parallel combination is connected across a dc source of 100 V. The current supplied by the source is
    (i) 25 A   (ii) 20 A
    (iii) 5 A   (iv) 4 A

10. In Question 9, the current through the 5 Ω resistance is
    (i) 16 A   (ii) 4 A
    (iii) 20 A   (iv) none of these

11. Three resistances of 3 Ω each are connected in delta. The value of the resistances in the equivalent star is
    (i) 27
    (ii) 9 Ω
    (iii) 1.5 Ω
    (iv) 1 Ω

12. In a network having N nodes, the number of independent equations required to solve the network, with ground as the reference node, is
    (i) $2N$
    (ii) $2N - 1$
    (iii) N
    (iv) $N - 1$

13. In the mesh current method of network analysis, the independent variable is
    (i) node current   (ii) mesh current
    (iii) node voltage   (iv) none of these

14. Nodal and mesh methods of network analyses can be applied to
    (i) independent current sources
    (ii) independent voltage sources
    (iii) dependent current and voltage sources
    (iv) all of these

15. Superposition theorem can be employed to
    (i) voltages only   (ii) currents only
    (iii) power only   (iv) all of these

16. Thevenin's theorem can be applied to networks containing
    (i) passive elements only
    (ii) active elements only
    (iii) linear elements only
    (iv) all of these

17. Which of the following theorems helps in simplifying computations when the load across a circuit is varying?
    (i) superposition
    (ii) Norton's
    (iii) Thevenin's
    (iv) maximum power transfer

18. Which of the following forms the basis of Compensation theorem?
    (i) Millman's theorem
    (ii) Superposition theorem
    (iii) Thevenin's theorem
    (iv) Norton's theorem

19. In a linear circuit, the voltage at a single node is the point of interest. Which of the following theorems will be suitable?
    (i) Compensation theorem
    (ii) Superposition theorem
    (iii) Millman's theorem
    (iv) Maximum power transfer

20. When maximum power transfer takes place, the efficiency of power transfer of the circuit is
    (i) 100%
    (ii) 75%
    (iii) 50%
    (iv) 25%

21. Equations obtained by applying Kirchoff's laws to a circuit to study its transient behaviour are
    (i) linear algebraic (ii) non-linear algebraic
    (ii) differential   (iv) none of these

22. Time constant of a circuit containing a resistance of R Ω in series with an inductance of L H is equal to
    (i) $R \times L$
    (ii) $\dfrac{R}{L}$
    (iii) $\dfrac{L}{R}$
    (iv) none of these

23. The natural frequency in a source-free reactive circuit is repreented by
    (i) $-\dfrac{1}{\tau}$
    (ii) $\dfrac{1}{\tau}$
    (iii) τ
    (iv) none of these

24. An RLC series circuit is excited by a voltage source of V volt. Which of the following denotes the steady state current in the circuit?
    (i) $\dfrac{V}{R}$
    (ii) $\dfrac{V}{L}$
    (iii) $\dfrac{V}{C}$
    (iv) $\dfrac{V}{R+L+C}$

25. Which of the following correctly simulates the response of a reactive circuit?
    (i) Homogenous   (ii) Integro-differential
    (iii) Linear   (iv) All of these

Answers

1. (iv)	2. (i)	3. (iv)	4. (iii)	5. (i)	6. (iii)	7. (i)
8. (ii)	9. (ii)	10. (i)	11. (iv)	12. (iii)	13. (ii)	14. (iv)
15. (iv)	16. (iv)	17. (iii)	18. (ii)	19. (iii)	20. (iii)	21. (iii)
22. (iii)	23. (i)	24. (i)	25. (iv)			

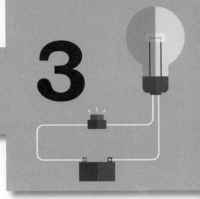

MAGNETIC CIRCUITS

3

Learning Objectives

This chapter will enable the reader to
- Understand and differentiate between linear and non-linear magnetic circuits
- Define the laws of electromagnetism and apply them for analysing magnetic circuits
- Recognize the analogy between electric and magnetic circuits and apply the laws of the former to solve the latter types of circuits
- Classify magnetic materials based on their properties
- Be able to identify and understand the significance of the various parts of the magnetization characteristic of ferromagnetic materials
- Understand the cause of hysteresis and eddy current losses in magnetic cores
- Develop expressions for self-inductance and mutual inductance based on Faraday's laws of electromagnetic induction
- Apply dot convention for developing equations for coupled circuits
- Derive expressions for energy stored in linear magnetic circuits and utilize the stored magnetic energy to do mechanical work

3.1 INTRODUCTION

Electromagnetic field is the coupling medium for most of the electric apparatuses; for example, in generators and motors the interchange of energy from the electrical system to the mechanical system and vice versa may be explained using the concept of electromagnetic field. For transformers magnetic field is the coupling medium for the interchange of energy from one electrical system to two or more electrical systems. Practically all electric machinery and transformers use magnetic materials for shaping and directing the magnetic field. Thus the analysis of magnetic field quantities is needed for understanding the operation of these devices. In this chapter some of the basic terms associated with the analysis of magnetic field systems, such as magnetic flux, magnetic flux density, magnetic field intensity, permeability and reluctance, are defined. A brief introduction to the properties of practical magnetic materials is given. The concepts of a magnetic circuit that simplify the calculations involved in the analysis of magnetic systems are presented.

3.2 MAGNETIC CIRCUITS

In Section 1.10 the fundamental laws regarding electromagnetism have been discussed. It has been seen that the natural phenomenon of a current flowing through a conductor sets up a magnetic field around the conductor and when a current-carrying conductor is placed in another magnetic field, a force is exerted on the conductor (motor action). Alternatively, when a conductor is moved inside a uniform magnetic field, an electromotive force is induced in the moving conductor (generator action). It has also been observed that the lines of magnetic flux emanate in a non-magnetic medium from the north pole to the south pole and the magnetic flux forms closed loops.

A closed restricted path in which magnetic flux is established is known as a *magnetic circuit*. The magnetic circuit consists of a structure composed, for the most part, of magnetic material of high permeability. The high-permeability material causes the magnetic flux to be confined to the paths defined by the magnetic structure. This is analogous to the electric circuit where current is confined to the conducting path of the circuit.

The simplest form of a magnetic circuit is a ring made of a ferromagnetic material such as iron, around which is wound a current-carrying conductor, as shown in Fig. 3.1. The iron ring is called the core. The current flowing through the coil sets up magnetic flux within the ring. The flux lines are shown by dotted lines in Fig. 3.1. The direction of the magnetic flux is given by the right hand rule. The magnetic flux so set up follows, by and large, a circular path along the ferromagnetic core and is confined almost entirely to the core. The path of the field lines along the core constitutes the magnetic circuit. A small amount of flux may take a path through the air. This flux is called *leakage flux*. For convenience in simplifying the analysis, leakage flux is neglected.

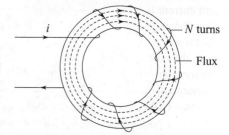

Fig. 3.1   A magnetic circuit

3.3 BIOT–SAVART LAW

Biot–Savart law is the basic law providing a relation between cause and effect in electromagnetism. In electrostatics, Coulomb's law defines the relation between a point charge (cause) and the electric field (effect) that the charge produces in its surrounding. Similarly, Biot–Savart's law describes the relation between a current element or moving charge (cause) and the magnetic field (effect) that the current element or moving charge produces in its surrounding.

Biot–Savart law states that if I ampere is the current flowing through an elemental length of wire dl metre, the magnetic field produced by the elemental current Idl at a perpendicular distance of r metre from it is given by

$$dB = \frac{\mu}{4\pi} \times \frac{I\,\vec{dl} \times \vec{r}}{r^3}$$

or
$$B = \frac{\mu_0 \mu_r}{4\pi} \int_l \frac{I\,\vec{dl} \times \vec{r}}{r^3} \tag{3.1}$$

In Eq. (3.1), μ is the absolute permeability constant of the medium (discussed in detail in a later section in the chapter, vector $\vec{r}$ points from the short elemental current towards the point at which the magnetic field is to be determined and the direction of dl is in the direction of the current as shown in Fig. 3.2

The total magnetic field is obtained by integration of Eq. (3.1) over the length of the conductor. Evaluation of the line integral over any random length of a conductor, however, is not easy to perform. Fortunately, magnetic field integrations for conductors in practical use, i.e., infinitely long wire and circular loop, can be conveniently performed.

In the case of an infinitely long current-carrying conductor, the field at a distance d metre is expressed as

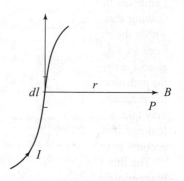

Fig. 3.2   Biot–Savart law

$$B = \frac{\mu_0 I}{2\pi d} \text{ Wb/m}^2$$

Similarly, the field at the centre of a circular loop of a wire of diameter d metre is given by $B = \frac{\mu_0 I}{d}$ Wb/m^2.

3.4 MAGNETIC FIELD STRENGTH

From Section 1.10.2, flux density B may be defined as the force experienced by an elementary conductor per unit current length when placed in a magnetic field produced by a straight, long conductor carrying current. Now the magnetic field may embrace different media with varying permeabilities. The same elementary conductor when placed in different media experiences different forces even though placed at the same distance from the current-carrying conductor. This is because the flux density is a function of μ, the permeability of the medium, as seen from Eq. (1.44). Thus, to calculate the force, irrespective of the medium, a new term *magnetic field intensity H* is introduced. The magnetic field intensity or the magnetizing force H is defined as the ratio of the magnetic flux density B and the absolute permeability of the medium, μ. Thus,

$$H = \frac{B}{\mu} \tag{3.2}$$

From Eq. (1.34) $B = \frac{\mu I}{2\pi r}$ \qquad (3.3)

Then \quad $H = \frac{I}{2\pi r}$ A/m \qquad (3.4)

where I is the current in the straight, long conductor and r is the distance of the point at which the magnetic field intensity is measured from the axis of the conductor, as shown in Fig. 3.3. The magnetizing force is tangential all along the flux lines. The magnetizing force H has the same circular locus as the flux density B. The direction of the flux is given by the right-hand rule.

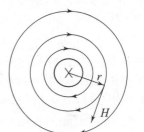

Fig. 3.3 Flux and magnetic field intensity

3.5 MAGNETOMOTIVE FORCE

A steel ring of mean radius r metres and a circular cross section of diameter d metres, with a coil of N turns wound around it through which a current of I amperes flows, is shown in Fig. 3.4. The current flowing through the coil sets up a magnetic flux within the ring. The flux lines are shown by dotted lines in Fig. 3.4. The magnetic flux is confined almost entirely to the core and the field lines follow the path along the ferromagnetic core and constitute the magnetic circuit. A very small amount of flux may take a path through the air. This flux is called leakage flux. The leakage flux is not taken into account in this chapter.

The flux set up in the ferromagnetic core of the magnetic circuit shown in Fig. 3.4 is due to the current of I amperes flowing through N turns of the

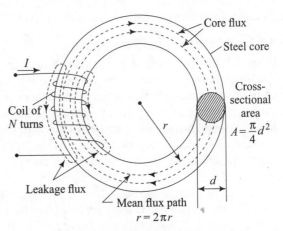

Fig. 3.4 Ring-shaped magnetic circuit

coil. All the flux lines enclose the N turns of the coil. The product NI is defined as magnetomotive force. It may be said that magnetic flux is produced in the coil due to the presence of a magnetomotive force (mmf). This is analogous to the current flow in a closed electric circuit which is caused due to the presence of one or more electromotive forces (emf). The unit of mmf in engineering practice is ampere-turns (AT). The mmf is like magnetic potential difference and is represented by $\mathcal{F}$. Hence, $\mathcal{F} = NI$ AT $\hspace{2cm}$ (3.5)

The magnetic field intensity or magnetizing force H is defined as the magnetomotive force per metre length of the magnetic circuit. The magnetic circuit shown in Fig. 3.4 is homogeneous and has a uniform cross section. Thus, by definition

$$H = \frac{NI}{l} = \frac{\mathcal{F}}{l} \text{ AT/m} \hspace{2cm} (3.6)$$

where l is the length of the mean flux path and is equal to $2\pi r$. Substituting $2\pi r$ for l in Eq. (3.6), the magnetizing force becomes

$$H = \frac{NI}{2\pi r} \text{ AT/m} \hspace{2cm} (3.7)$$

From the above, it can be seen that the unit of magnetic intensity or magnetizing force H is ampere-turns per metre (AT/m).

3.5.1 Ampere's Circuital Law

Electric current flowing through a number of turns of a conductor produces a magnetic field in its vicinity. The collective effect of all the turns will equal the line integral of the magnetic field intensity around a close magnetic path.

Thus, $\hspace{1cm} NI = \mathcal{F} = \oint H \, dl = Hl \hspace{2cm} (3.8)$

Equation (3.8) gives the relationship between H and $\mathcal{F}$, which is same as that obtained in Eq. (3.6). The number of turns is a dimensionless quantity and the unit of $\mathcal{F}$ is ampere. However, it is customary to write the unit of $\mathcal{F}$ as ampere-turns (AT) in order to have a better perception.

3.6 PERMEABILITY

Permeability is a measure of the receptiveness of a material to having magnetic flux set up in it. In this section, permeability of free space and relative permeability of magnetic materials are discussed.

3.6.1 Permeability of Free Space

Figure 3.5 shows the cross section of a single turn of a conductor placed in vacuum. At C a current of 1 A flows into the plane of the paper and the return path of the current is located at a considerable distance away from C such that the magnetic flux set up by the return current does not affect the magnetic flux set up by the current flowing into the plane of the paper. The current flowing into the plane of the paper at C sets up magnetic flux lines as concentric circles and the direction of the flux is clockwise as per the corkscrew rule. C_1 represents the path of the flux line having a radius of 1 m.

Since it is a single-turn coil, $N = 1$, $I = 1$ A, and $R = 1$ m, using Eq. (3.7), the magnetic field intensity H is given by

$$H = \frac{1 \times 1}{2\pi \times 1} = \frac{1}{2\pi} \text{ AT/m} \hspace{2cm} (3.9)$$

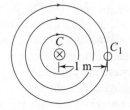

Fig. 3.5 Magnetic field due to current in a single-turn long straight conductor

Assuming that the flux density in the region of the flux line C_1 is B tesla, the force per metre on a conductor placed at C_1 may be obtained from Eq. (1.43), taking $I = 1$ A and $l = 1$ m, as

$$\text{Force per metre} = B \times I \times l = B\,[\text{tesla}] \times 1\,[\text{A}] \times 1\,[\text{m}] = B \text{ newton} \qquad (3.10)$$

Using the definition of ampere as that current which if maintained in two straight parallel conductors of infinite length, of negligible circular cross section, and placed 1 m apart in vacuum, would produce between these conductors a force of 2×10^{-7} N/m. Thus,

$$\text{Force per metre} = 2 \times 10^{-7} \text{ N} \qquad (3.11)$$

Comparing Eqs (3.10) and (3.11), flux density may be obtained as

$$B = 2 \times 10^{-7} \text{ T or Wb/m}^2 \qquad (3.12)$$

Dividing Eq. (3.12) by Eq. (3.9),

$$\frac{B}{H} = \frac{2 \times 10^{-7}}{1/2\pi} = 4\pi \times 10^{-7} \quad \frac{\text{Wb}}{\text{ATm}} \qquad (3.13)$$

From Eq. (3.13) it can be seen that the flux density B set up by the magnetic field intensity H is dependent on the medium. For free space or a non-magnetic medium, the ratio B/H is called *permeability of free space* and is represented by μ_0. Thus,

$$\mu_0 = \frac{B}{H} = 4\pi \times 10^{-7} \quad \text{Wb/ATm} \qquad (3.14)$$

The permeability of free space is a universal constant. For all practical purposes the absolute permeability of air and other non-magnetic materials is taken to be the same as μ_0. The unit of permeability in SI units (Standard International system of units) is henry/metre (H/m). This unit of permeability follows from Eq. (1.42), which is written below taking the medium a free space:

$$\mathcal{F} = \frac{\mu_0 I_1}{2\pi r} I_2 l \text{ newton}$$

Thus, $\quad \mu_0 \propto \dfrac{\text{newtons}}{\text{ampere}^2}$

But, newton-metre = joule = volt × ampere × second. Therefore,

$$\mu_0 \propto \frac{\text{volt} \times \text{ampere} \times \text{second}}{\text{ampere}^2 \times \text{metre}} = \frac{\text{volt} \times \text{second}}{\text{ampere} \times \text{metre}}$$

Now (volt × second)/ampere is the unit of inductance expressed in henry. Thus permeability is expressed in henry/metre or H/m.

3.6.2 Relative Permeability

The magnetic flux inside a coil is increased manifold when a core of ferromagnetic material such as iron core is inserted. The ratio of the flux density produced in a material to the flux density produced in vacuum by the same magnetic field strength is termed as *relative permeability* and is denoted by μ_r. For air, $\mu_r = 1$; for some form of nickel–iron alloy the value of μ_r can be 10^5.

From Eq. (3.2), it can be seen that

$$B = \mu H \qquad (3.15)$$

where μ is the absolute permeability of the medium. Magnetic materials such as iron, cast steel, etc. have a higher permeability and for a given H would set up a much higher flux density. The absolute permeability of the medium is constituted of two parts, that is,

$$\mu = \mu_0 \mu_r H \qquad\qquad (3.16)$$

and $\quad B = \mu H = \mu_0 \mu_r H \text{ weber/m}^2 \qquad\qquad (3.17)$

For magnetic materials μ_r has a range that varies between 600 and 10,000 and $\mu_r = 1$ for non-magnetic materials.

3.6.3   Classification of Materials Based on Magnetic Behaviour

Dipole moment of electrons in materials is largely due to (i) the orbital movement of the electrons around their nucleus and (ii) due to the spin around their own axis. The orbital movement and spin give rise to circulating currents and, therefore, set up an mmf. This mmf, however, is neutralized by an opposite one in most of the molecules. But in magnetic materials such as iron and steel there are a number of unneutralized orbits such that a resultant axis of mmf exists which produces a magnetic dipole. In un-magnetized specimens, because of mutual attraction and repulsion among dipoles, the molecular mmf axes lie along continuous closed paths and no external magnetic effect can be detected. In magnetized specimens, the dipoles line up parallel with the exciting mmf and when the exciting mmf is removed, a number of dipoles may remain aligned in the direction of the external field and thus exhibit permanent magnetism. The readiness of a material to accept magnetism is expressed by its permeability. For all materials, except a few magnetic materials, permeability is constant and is equal to the permeability of free space, μ_0. For magnetic materials, permeability μ equals $\mu_0 \mu_r$.

Magnetic properties of materials are characterized by their relative permeabilities. In accordance with the value of relative permeability, materials may be divided into three broad classes.

Ferromagnetic materials   The presence of magnetic dipole moment, due to both the orbital movement and spin of the electrons, in this category of materials ensures a relative permeability μ_r whose values typically vary from a few hundred to tens of thousands. As such, ferromagnetic materials show a great affinity to magnetic fields and exhibit residual magnetic properties even when the external field is removed. Due to the residual magnetic property, ferromagnetic materials are employed to fabricate permanent magnets, in addition to their applications as electromagnets in electrical machines. Iron, steel, nickel, cobalt, etc. are commonly known ferromagnetic materials.

Paramagnetic materials   Such types of materials exhibit weak magnetic properties due to a resultant spin magnetic moment produced by the existence of a few un-paired electrons, which align themselves in the direction of the applied external magnetic field. When the applied external field is removed, these materials do not exhibit any magnetic properties. In this category, the relative permeability of materials is slightly greater than unity and they are considered to be non-magnetic in comparison with ferromagnetic materials. Due to sensitivity of the paramagnetic materials to temperature, such materials become more magnetic at lower temperatures. Examples of materials that are classified as paramagnetic are aluminium, platinum, uranium, magnesium, and titanium.

Diamagnetic materials   These materials set up small opposing persistent currents, when subjected to a strong external field, by rearranging the orbital motion of their electrons. Therefore, materials in this category have a relative permeability, which is less than unity ($\mu_r < 1$), and do not show any magnetic properties when the external field is removed. Diamagnetic materials find applications in the process of magnetic levitation, i.e., they are employed to float an object in a strong magnetic field. Bismuth, copper, silver, gold, diamond, silicon, etc. are some examples of diamagnetic materials.

Table 3.1 presents a summary of the properties of different categories of magnetic materials.

Ferromagnetic materials are of great importance in electrical engineering. Ferromagnetic materials are of two types: (a) Materials that are easily magnetized are called *soft ferromagnetic* materials. Examples are iron and its various alloys with nickel, cobalt, aluminium, and tungsten. (b) Materials that retain their magnetism over a long period of time after the removal of magnetizing force are called *hard ferromagnetic* materials.

Carbon steel, cobalt steel, alnico are some examples of such materials. Silicon steel and cast steel are the most widely used ferromagnetic materials in transformers and electrical machines.

Table 3.1 Properties of magnetic materials

Property	Diamagnetic	Paramagnetic	Ferromagnetic
Main magnetism due to	Orbital movement	Spin	Orbital and spin
Relative permeability	$\mu_r < 1$	$\mu_r > 1$	$\mu_r \gg 1$
Relative direction of B and H	In opposition	Supportive	Hysteresis loop
Residual magnetism	Nil	Weak	Very strong
Examples	Copper, silver, gold, silicon, lead, mercury, etc.	Aluminium, calcium, platinum, tungsten, etc.	Iron, steel, nickel, cobalt, their alloys, etc.

3.7 RELUCTANCE

The ferromagnetic ring wound with N turns of a conductor carrying a current of I ampere, shown in Fig. 3.4, is considered. Assuming that the mean circumference of the ring is l metre and cross-sectional area is A m^2, the flux ϕ in the steel core is

$$\phi = B \times A \quad \text{weber} \tag{3.18}$$

Substituting $B = \mu\mu_0 H$ as per Eq. (3.17), flux ϕ may be obtained as

$$\phi = \mu_0 \mu_r H \times A \quad \text{weber}$$

Substituting for H from Eq (3.6), the flux becomes

$$\phi = \mu_0 \mu_r \times \frac{\mathcal{F}}{l} \times A \quad \text{weber}$$

$$\therefore \quad \phi = \frac{\mathcal{F}}{\dfrac{l}{\mu_0 \mu_r A}} = \frac{\mathcal{F}}{\mathcal{R}} \quad \text{weber} \tag{3.19}$$

$\mathcal{R}$ is called the reluctance of the magnetic circuit and is given by

$$\mathcal{R} = \frac{l}{\mu_0 \mu_r A} \tag{3.20}$$

The units of $\mathcal{R}$ can be obtained from Eq. (3.19). Thus,

$$\mathcal{R} = \frac{\mathcal{F}}{\phi} \quad \text{AT/Wb} \tag{3.21}$$

The reciprocal of reluctance $\mathcal{R}$ is called *permeance* of the magnetic circuit and is denoted by $\mathcal{P}$.

For a given magnetic circuit the reluctance $\mathcal{R}$ is a constant and its value depends on the geometrical dimensions and permeability of the magnetic circuit. From Eq. (3.19) it may be seen that the flux produced in a magnetic circuit is proportional to the magnetomotive force producing it. Thus,

Magnetic fux $\propto$ mmf = constant $\times$ mmf

$$= \frac{1}{\text{reluctance}} \times \text{mmf}$$

This expression is similar to Ohm's law for electric circuits.

Equation (3.20) is similar to Eq. (1.24) for an electric circuit. Equation (1.24) expresses

$$R = \frac{V}{I} = \frac{\rho l}{A}$$

where R is the resistance of the conductor; V is the applied voltage; I is the current through the conductor; l, A, and ρ are length, cross-sectional area, and the specific resistivity of the conductor, respectively. The expression for $\mathcal{R}$ in Eq. (3.20) is similar to that of R given above except that the absolute permeability $\mu_0 \mu_r$ corresponds to the reciprocal of the specific resistivity ρ.

3.8 ANALOGY BETWEEN ELECTRIC AND MAGNETIC CIRCUITS

In all electrical machines and transformers the magnetic field problems are concerned with the calculations of quantities such as flux and flux density in space. In electrical machines the flux passes through both stationary and rotating masses of ferromagnetic materials as well as very small air gaps in between the two masses because of high permeability of the ferromagnetic materials. The space occupied by the magnetic field and the space occupied by the ferromagnetic materials are almost the same. The flux is confined to the ferromagnetic materials in the same way as a current flow is restricted in the conductor. This gives rise to the concept of *magnetic circuit*. Thus the magnetic circuit consists of mainly iron paths with possibly air gaps in between, of given geometry, to keep the flux confined to it.

The concept of magnetic circuit simplifies the analysis of magnetic structures. The simplification is based on the following assumptions.

1. All the magnetic flux is linked by all the turns of the coil.
2. Flux is confined only within the magnetic core.
3. Flux density is uniform across the cross section of the core.

The first of these assumptions might not hold true near the ends of the coil, but that it might be more reasonable if the coil is tightly wound. The second assumption is equivalent to stating that the relative permeability of the core is infinitely higher than that of the air surrounding the core. Some of the flux in a structure would thus not be confined within the core but takes a path through the air (this is usually referred to as leakage flux). Finally, the assumption that the flux is uniform across the core cannot hold for a finite-permeability medium, but it is very helpful in giving an approximate mean behaviour of the magnetic circuit. The magnetic circuit analogy is therefore far from being exact. However, it is the most convenient tool at the engineer's disposal for the analysis of magnetic structures.

A typical magnetic circuit consisting mainly of an iron core with a very small air gap is shown in Fig. 3.6(a). The mmf produced by the current I flowing through N turns of the coil produces flux ϕ. Usually, the flux produced due to the mmf is not entirely restricted to the core, a small part of the flux lines complete their paths through the air. Since the length l_a of the air gap is small compared to the dimensions of the core faces, most of the flux lines take the shortest path to cross from one core face to the other. However, some flux lines tend to bulge out, as shown in Fig. 3.6(a). The bulging of the lines in Fig. 3.6(a) is called *fringing* and it has a tendency to increase the air gap area A_a. When fringing is to be considered, its effect is accounted for by increasing the effective

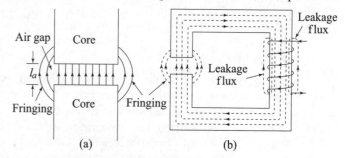

Fig. 3.6 Fringing and leakage flux

area by 10 to 12%. The magnitude of the leakage flux shown in Fig. 3.6(b) is small since the reluctance of the core is much less than the reluctance of the air. Therefore, the leakage flux can be neglected and it can be assumed that the entire flux is confined to the core and to that part of the air gap which has effectively the same cross-sectional area as the iron core. This is similar to a dc emf which when applied to an electric circuit causes a current to flow in the circuit with the current being confined to the conductor constituting the circuit. This similarity between a magnetic circuit and an electric circuit reveals that a single-line equivalent circuit, as depicted in Fig. 3.7(b), may replace the magnetic circuit shown in Fig. 3.7(a). Analysing Eqs (3.19) and (3.21)

suggests that the equivalent circuit consists of the magnetomotive force F driving the flux ϕ through two series-connected reluctances R_e, the reluctance of the iron core, and R_a, the reluctance of the air gap. This analogy of the magnetic circuit with the electric circuit carries through in many other respects. Current in an electric circuit and flux in a magnetic circuit are analogous since the former is set up by a potential difference and the latter is set up by a magnetic potential difference. The resistance of a conductor depends on the geometrical parameters of the electric circuit and is equal to $(\rho l)/A$; the reluctance of a magnetic circuit also depends on the geometrical configuration of the magnetic circuit and is equal to $l/(\mu A)$.

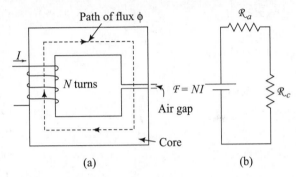

Path of flux ϕ

N turns
$F = NI$
Air gap
Core

(a)
(b)

Fig. 3.7 (a) Magnetic system and (b) equivalent dc circuit analog

The representation of a distributed magnetic circuit by a lumped reluctance is similar to the representation of a dc electric circuit by a lumped resistance. Above all, both Kirchoff's current law (KCL) and Kirchoff's voltage law (KVL) are equally applicable to magnetic circuits. Thus the algebraic sum of fluxes at a junction is zero (KCL) The analogous Kirchhoff's current law (KCL) for magnetic circuits can be mathematically stated as

$$\sum_{k=1}^{k=n} \phi_k = 0 \qquad (3.22)$$

where k represents the number of magnetic core branches.
Consider the magnetic circuit shown in Fig. 3.8. Use of Eq. (3.22) at the T junction in Fig. 3.8 leads to the following expression:

$$\phi_1 - \phi_2 - \phi_3 = 0 \quad \text{or} \quad \phi_1 = \phi_2 + \phi_3$$

Similarly, the analogous Kirchhoff's voltage law (KVL) is mathematically written as

$$\sum F_k = \sum \phi_k R_k \qquad (3.23)$$

Table 3.2 compares the properties of electric and magnetic circuits.

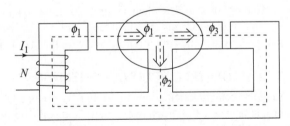

Fig. 3.8 Application of KCL to the T-joint in the magnetic circuit

It would be useful to note the dissimilarities between electric and magnetic circuits. Following are the distinctions between the two:

(i) The flow of flux ϕ is perpendicular to the current flow.
(ii) At a given temperature, electric resistance is constant.
(iii) Magnetic reluctance is variable since the magnitude of permeability depends on the magnetic field B and flux intensity H.

(iv) Flow of current in an electric circuit leads to dissipation of heat energy. For flux to be generated in a magnetic circuit, electric energy is required to be input.

(v) In a magnetic circuit, energy is stored in the magnetic field, while in an electric circuit, energy is released in the form of heat.

Table 3.2 Comparison of electric and magnetic circuits

Electric circuits				Magnetic circuits			
Quantity	Symbol	Formula	Units	Quantity	Symbol	Formula	Units
emf	E or V		V	mmf	$\mathcal{F}$		AT
Electric field strength	$\mathcal{E}$		V/m	Magnetic field strength	H	B/μ	AT/m
Current	I or i	V/R	A	Magnetic flux	ϕ	$B \times A$	Wb
Current density	J	I/A	A/m^2	Magnetic flux density	B	ϕ/A	Wb/m^2
Resistance	R	$(\rho l)/A$	Ω	Reluctance	$\mathcal{R}$	$l/(\mu A)$	AT/Wb
Conductivity	σ		$1/\Omega$ m	Permeability	μ		Wb/Am
Voltage drop		$R \times I$	V	mmf drop	$\phi \times \mathcal{R}$		AT

3.9 MAGNETIC POTENTIAL DROP

When a magnetic flux ϕ is established in a magnetic circuit of reluctance $\mathcal{R}$, there is always a magnetic potential drop $\mathcal{M}$ along the direction of the flux. The magnetic potential drop is the product of the flux and the reluctance. Thus,

$$\mathcal{M} = \phi \times \mathcal{R} \tag{3.24}$$

Equation (3.24) is similar to the potential drop (voltage drop) in an electric circuit expressed as $V = I \times R$, where V is the electric potential drop, R is the resistance, and I is the current. Also, from Eqs (3.8), (3.21), and (3.24) the following relationship results:

$$\mathcal{F} = \phi \times \mathcal{R} = H \times l = \mathcal{M} \tag{3.25}$$

So it can be said that when an mmf is applied to a magnetic circuit it creates a flux in the magnetic circuit, which in turn causes a magnetic potential drop along the path of the flux. The magnetic potential drop along the path of the magnetic circuit is the product of the magnetic field intensity and the length of the magnetic circuit.

3.10 MAGNETIC CIRCUIT COMPUTATIONS

Magnetic circuit calculations involving ferromagnetic materials are of two types. The first type deals with problems where the structure geometry and coil parameters, such as the number of turns and current flowing through it, are known. It is required to calculate the magnetic flux in the structure. This involves computation of the mmf, determination of the length and cross section of the magnetic path for each leg of the path, calculation of the equivalent reluctance of the leg, and drawing of the equivalent magnetic circuit. Next, the calculation of the total equivalent reluctance and computation of the flux, flux density, and magnetic field intensity is done as needed. In the second type the value of the flux and the structure geometry are known. The number of turns of the coil and the current flowing through it are to be determined. This requires the calculation of the total equivalent reluctance of the structure from the desired flux, generation of the equivalent magnetic circuit diagram, determination of the mmf to establish the required flux, and choosing the coil current and the number of turns required to establish the desired mmf.

3.10.1 Series Magnetic Circuit

A closed magnetic path consisting of different sections of varying permeabilities and physical dimensions but having the same magnetic flux is called a *series magnetic circuit*. It may contain one or more coils at different locations which impart the mmfs required for creating the flux.

Figure 3.9(a) shows a closed magnetic circuit consisting of three magnetic cores of lengths l_1, l_2, l_3; cross-sectional areas A_1, A_2, A_3, and absolute permeabilities μ_1, μ_2, μ_3, respectively. There are also two coils with the number of turns N_1 and N_2 carrying current I_1 and I_2, respectively, placed at two different positions around the core.

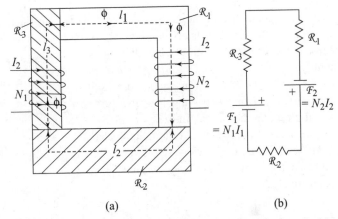

(a) (b)

Fig. 3.9 (a) Series magnetic circuit and (b) Electric circuit equivalent

The total magnetomotive force applied in the closed magnetic path is $F = F_1 + F_2 = N_1 I_1 \pm N_2 I_2$. The terms will be additive when the fluxes produced due the two mmfs are in the same direction, and subtractive when the two mmfs produce fluxes in the opposite direction. In other words, this depends on the direction of winding (clockwise or counterclockwise) of the two coils and the direction of currents in the coils. The two mmfs will produce a resultant flux ϕ in the core, which will cause a magnetic potential drop in three different portions of the core. Thus,

$$N_1 I \pm N_2 I_2 = \mathcal{M}_1 + \mathcal{M}_2 + \mathcal{M}_3$$

$$= H_1 l_1 + H_2 l_2 + H_3 l_3$$

$$= \frac{B_1}{\mu_1} l_1 + \frac{B_2}{\mu_2} l_2 + \frac{B_3}{\mu_3} l_3$$

$$= (B_1 A_1)\left(\frac{l_1}{A_1 \mu_1}\right) + (B_2 A_2)\left(\frac{l_2}{A_2 \mu_2}\right) + (B_3 A_3)\left(\frac{l_3}{A_3 \mu_3}\right)$$

But $\qquad \phi = B_1 A_1 = B_2 A_2 = B_3 A_3$

Then, $\quad N_1 I_1 \pm N_2 I_2 = \phi \mathcal{R}_1 + \phi \mathcal{R}_2 + \phi \mathcal{R}_3$

$$= \phi(\mathcal{R}_1 + \mathcal{R}_2 + \mathcal{R}_3) \qquad (3.26)$$

$$= \phi \times \mathcal{R} \qquad (3.27)$$

$$\therefore \qquad \mathcal{R} = \mathcal{R}_1 + \mathcal{R}_2 + \mathcal{R}_3 \qquad (3.28)$$

where H_1, H_2, H_3 are magnetic field strengths; B_1, B_2, B_3 are the flux densities; and $\mathcal{R}_1$, $\mathcal{R}_2$, $\mathcal{R}_3$ are the reluctances, respectively, of the three different magnetic cores constituting the closed magnetic circuit. Equation (3.28) shows that for a series magnetic circuit the total reluctance of the magnetic circuit is the sum of the individual reluctances of the different portions of the circuit. The magnetic circuit analogous to the series electrical circuit is shown in Fig. 3.9(b), where the mmfs, reluctances, and fluxes are analogous to emfs, resistances, and currents, respectively, for a series electric circuit.

In a magnetic series circuit, sometimes a portion contains a non-magnetic substance or air. In either case the absolute permeability becomes equal to the permeability of free space, μ_0. This is true, of course, if we neglect leakage or fringing of flux.

3.10.2 Parallel Magnetic Circuit

For certain geometrical configurations of a magnetic core, the flux created by the magnetomotive force acting in the magnetic circuit gets divided into two or more branches/portions of the magnetic core. The fluxes in these branches may be different, but the magnetic potential drops across the branches remain the same. Such a magnetic circuit is called *parallel magnetic circuit*.

Figure 3.10(a) shows a magnetic core excited by a current of I ampere passing through N turns of a coil placed in the central limb. The mmf $\mathcal{F} = NI$ creates a flux of ϕ weber in the central limb of length l_c metre. The flux gets divided into two fluxes ϕ_1 and ϕ_2 in the outer limbs of lengths l_1 and l_2 metre, respectively. Thus,

$$\phi = \phi_1 + \phi_2 \tag{3.29}$$

The equivalent circuit of the magnetic circuit shown in Fig. 3.10(a) is given in Fig 3.10(b), which can be reduced to the circuit shown in Fig. 3.10(c). From the equivalent circuit, the total reluctance of the magnetic pah $\mathcal{R}_{total}$ may be computed as

$$\mathcal{R}_{total} = \mathcal{R}_c + \mathcal{R}_1 \parallel \mathcal{R}_2 = \mathcal{R}_c + \frac{\mathcal{R}_1\mathcal{R}_2}{\mathcal{R}_1 + \mathcal{R}_2} \tag{3.30}$$

and the flux ϕ may be computed as

$$\phi = \frac{mmf}{reluctance} = \frac{NI}{\mathcal{R}_{total}} = \frac{NI}{\mathcal{R}_c + \dfrac{\mathcal{R}_1\mathcal{R}_2}{\mathcal{R}_1 + \mathcal{R}_2}} \tag{3.31}$$

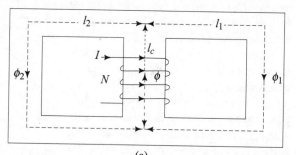

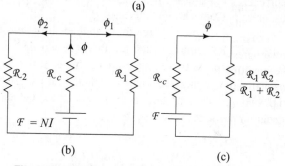

Fig. 3.10 Parallel magnetic circuit and its equivalent

where $\mathcal{R}_1$ and $\mathcal{R}_2$ are the reluctances of the two outer limbs and $\mathcal{R}_c$ is the reluctance of the central limb. From Fig. 3.10(b), equating the magnetic potential drops across the parallel branches,

$$NI - H_c l_c = H_1 l_1 = H_2 l_2 \tag{3.32}$$

where H_1 and H_2 are the magnetic field strengths of the outer limbs and H_c is the magnetic field strength of the central limb. From Eq. (3.25) the magnetic potential drop in a branch is equal to the product of the flux and reluctance. Hence, rearranging Eq. (3.32), the mmf is given by

$$\mathcal{F} = NI = H_c l_c + H_1 l_1 = H_c l_c + H_2 l_2 \tag{3.33}$$

The mmf $\mathcal{F}$ acting in the central limb will cause magnetic potential drops $H_c l_c + H_1 l_1$ and $H_c l_c + H_2 l_2$ in the two closed-loop magnetic circuits, each loop consisting of a central limb and one outer limb. The fluxes in the outer limbs are given by

$$\phi_1 = \frac{H_1 l_1}{\mathcal{R}_1} = \frac{NI - H_c l_c}{\mathcal{R}_1} \tag{3.34}$$

$$\phi_2 = \frac{H_2 l_2}{\mathcal{R}_2} = \frac{NI - H_c l_c}{\mathcal{R}_2} \tag{3.35}$$

$$\phi = \phi_1 + \phi_2 = \frac{H_1 l_1}{\mathcal{R}_1} + \frac{H_2 l_2}{\mathcal{R}_2} = \frac{NI - H_c l_c}{\mathcal{R}_1} + \frac{NI - H_c l_c}{\mathcal{R}_2}$$

$$= (NI - H_c l_c)\left(\frac{1}{\mathcal{R}_1} + \frac{1}{\mathcal{R}_2}\right) = \frac{NI - H_c l_c}{\mathcal{R}_{eq}}$$

where $\quad \dfrac{1}{\mathcal{R}_{eq}} = \dfrac{1}{\mathcal{R}_1} + \dfrac{1}{\mathcal{R}_2}$ (3.36)

From Eq. (3.36) it is evident that the formula for determining the equivalent reluctance $\mathcal{R}_{eq}$ of a number of reluctances in parallel is analogous to that of resistances connected in parallel.

3.11 MAGNETIZATION CHARACTERISTICS OF FERROMAGNETIC MATERIALS

Ferromagnetic materials are characterized by high values of relative permeability. Iron, nickel, cobalt, and gadolinium are ferromagnetic elements. Ferromagnetic materials composed of iron and alloys of iron with cobalt, tungsten, nickel, aluminium, and other metals are most common magnetic materials.

If an iron specimen, e.g., a closed ring, is subjected to a progressively increasing magnetizing force H and the resulting flux density B is plotted against H, a magnetization curve will be obtained as shown in Fig. 3.11.

Since $\quad B = \mu_0 \mu_r H$

then $\quad \mu_r = \dfrac{B}{\mu_0 H}$

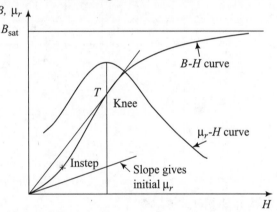

Fig. 3.11 Characteristics of ferromagnetic materials

Using this relationship the curve of relative permeability μ_r against H may be obtained from the plotted magnetization characteristics. The value of the relative permeability, μ_r, varies considerably with the magnetic field intensity H that establishes the operating flux density. The process of magnetization up to a high value of B can be conveniently divided into three stages at two points, which are called *instep* and *knee* of the curve.

Initial stage Near the origin, the graph is a straight line through the origin, and the slope gives the initial permeability. The value of H at which the graph begins to bend over varies over very wide limits from one material to another.

Middle portion, between the instep and the knee The slope is the greatest and consequently the point of maximum permeability is within this region. If curves for decreasing, as well as increasing, values of H are obtained, it is found that the values of B differ markedly, and it is within this region that these irreversible changes are the greatest. These changes and the associated losses in energy are discussed in Section 3.11.1.

Upper portion beyond the knee At point T, at which the tangent to the magnetization curve passes through the origin, μ_r has its maximum value. For increases in H beyond this particular value, there is a progressive but non-linear decrease in μ_r.

With the present state of the development of the theory of magnetism, it is not possible to describe analytically the characteristics of a magnetic material even though the exact composition of the material is known. It is customary to get the relevant information about magnetic characteristics of a sample from the manufacturer's bulletin where such information, based on measured magnetic quantities of the test samples of materials using prescribed standard testing methods, is given. Normally, these magnetic characteristics of various iron and steel samples are presented in a graphical form showing the variation of flux density B as a function of the magnetic field intensity H, as shown in Fig. 3.11. The characteristic of B versus H is a universal plot for a given material and it can be used for any geometrical shape, cross-sectional area, and length.

Typical magnetization characteristics for a variety of commonly used magnetic materials are shown in Fig. 3.12.

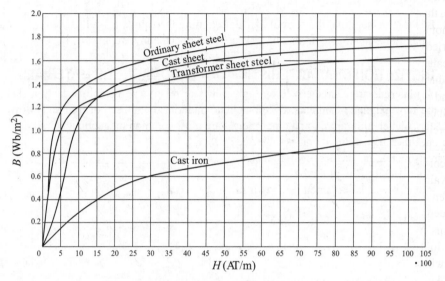

Fig. 3.12 Magnetization curves

3.11.1 Hysteresis Loop

The most common curve used to describe a magnetic material is the *B-H* curve or hysteresis loop. If the magnetizing force applied to a specimen of iron is increased from zero to some maximum value H_{max} and the magnetizing force is gradually reduced again to zero, it is found that the material opposes demagnetization and accordingly the new *B-H* curve for decreasing values of H lies above the original curve *OA*. Furthermore, it is seen that when H is returned to zero value again, B is no longer zero but has a finite value. Since the values of B apparently lag behind those of H, this effect is called *hysteresis* (the word 'hysteresis' is derived from the Greek word 'hysterein' meaning

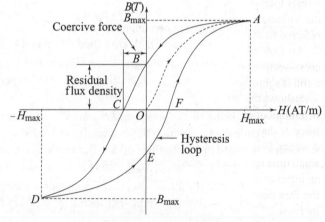

Fig. 3.13 Hysteresis loop

to lag or to be behind). The finite value of B, OB in Fig. 3.13, when H is zero, is a measure of the residual magnetism or retentivity. It is necessary to apply a negative magnetizing force represented by OC to fully demagnetize the iron sample. This negative force is called *coercive force*. Now, if H is increased in the negative direction to $-H_{max}$, the curve will reach a point D, at which the flux density is equal to the previous maximum value, B_{max}, but the flux will be in the opposite direction. If, finally, H is gradually reduced to zero, reversed, and increased to its maximum value H_{max} in the original direction, the curve $DEFA$ will be traced out. The complete curve forming a closed loop $ABCDEFA$ is called a *hysteresis loop*. The loop is always traversed in the direction indicated by arrows. If these alternations are maintained, the value of B_{max} being the same during each cycle, the iron sample will continue to go through the same series of changes. In general, the sample of iron does not follow a closed hysteresis cycle if the cycle is only carried out once. The iron sample must be brought to what is called the cyclic state, by subjecting it to a series of cycles, during which the maximum values of H, both positive and negative, are kept constant.

The unit of retentivity has the dimensions Wb/m^2 and that of coercive force has the dimensions AT/m.

From the hysteresis loops shown in Fig. 3.13 it is evident that corresponding to a fixed value of magnetizing force the value of flux density is not single-valued but lies between certain limits. For magnetic calculations, a single-valued curve called *normal magnetization curve* is used. This curve is obtained by joining the tips of the hysteresis loops generated in the cyclic magnetization condition with different maximum limits of H, as shown in Fig. 3.14.

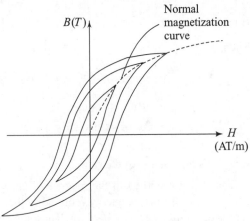

Fig. 3.14 Normal magnetization curve from hysteresis loops

3.11.2 Hysteresis Losses

In each cycle of the hysteresis loop, a certain amount of energy is expended to reverse the magnetic fields, which is represented by the area enclosed by the hysteresis loop. This energy appears in the form of heat in the volume of the magntic material. These losses are referred to as *hysteresis losses*.

An iron ring of mean circumference l metre and a cross-sectional area A wound uniformly with N turns of a coil is subjected to cyclic flux density variations. Figure 3.15 shows the hysteresis loop obtained for the iron ring. Instantaneous current i in the coil causes a magnetizing force H that produces a flux density B Wb/m^2 and flux ϕ weber. For an increase of current from i to $(i + di)$ in a small time interval dt, the magnetizing force H undergoes an infinitesimal change by PQ as shown in Fig. 3.15 and the flux density increases by dB with a corresponding increase of flux by $d\phi = AdB$. As per Lenz's law, the change in flux sets up an instantaneous emf e induced in the coil, which is given by Eq. (1.50). Thus,

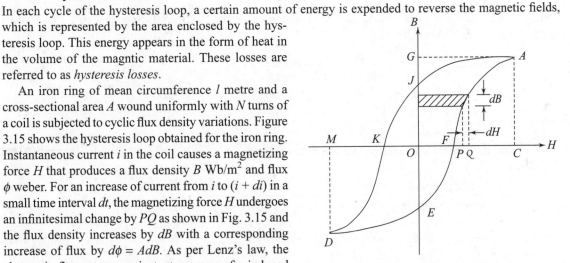

Fig. 3.15 Hysteresis loss

$$e = -N \frac{d\phi}{dt} = -NA \frac{dB}{dt} \qquad (3.37)$$

and the component of applied voltage to neutralize the emf equals e. Therefore, the instantaneous power supplied to the magnetic field is

$$p = ei = NiA \frac{dB}{dt} \text{ watt} \qquad (3.38)$$

The energy dw supplied to the magnetic field during the time interval dt is given by

$$dw = NiAdB \text{ joule} \qquad (3.39)$$

But $\qquad OP = H = \dfrac{Ni}{l} \quad \text{or} \quad i = \dfrac{OP \times l}{N} \qquad (3.40)$

Then,

$$dw = N \frac{OP \times l}{N} A \times dB = OP \times lA \times dB \text{ joule}$$

$$= v \times OP \times dB \quad \text{joule}$$

$$= v \times \text{area of shaded strip joule} \qquad (3.41)$$

$$= \text{area of shaded strip J/m}^3 \qquad (3.42)$$

where v is the volume of the ring in m^3.

The energy supplied to the magnetic field when H is increased from zero to OC is equal to the area $EFAGE$ J/m^3. Similarly, energy returned to the source from the magnetic field as H is reduced from OC to zero is equal to the area $AGJA$ J/m^3. Thus, the net energy absorbed by the magnetic field is equal to the area $EFAJE$ J/m^3, and the hysteresis loss for one complete cycle is equal to the area of loop $DEFAJKD$ J/m^3. Thus the hysteresis loss w_h is the energy loss per cycle and can be expressed as

$$w_h = [\text{area of hysteresis loop}] \frac{J}{m^3 \times \text{cycle}} \qquad (3.43)$$

It may be noted that the unit associated with the product of B and H is J/m^3. In Section 1.10.2.1 it is shown that the unit of B is N/Am. Then

$$\text{units of } HB = \frac{\text{ampere}}{\text{metre}} \times \frac{\text{newton}}{\text{ampere} \times \text{metre}} = \frac{\text{newton}}{\text{metre}^2}$$

But $\qquad$ newton-metre = joule

Hence, $\quad$ units of $HB = \dfrac{\text{joule}}{\text{metre}^3} \quad \text{or} \quad \text{J/m}^3$

Normally, it is desirable to express hysteresis loss in watts. This can be obtained directly from Eq. (3.43) as follows:

$$w_h = \frac{\text{energy}}{\text{volume} \times \text{cycles}} = \frac{\text{power} \times \text{time}}{\text{volume} \times \text{cycles}} = \frac{\text{power}}{\text{volume} \times \text{cycles/second}}$$

$$= \frac{P_h}{vf} \qquad (3.44)$$

and $\qquad P_h = w_h vf \qquad (3.45)$

where P_h is hysteresis power loss in watts and f is the frequency of variations of H in hertz or cycles/second.

Based on experimental work, Steinmetz developed an empirical relation to determine the power loss due to hysteresis. This relation is given by

$$P_h = K_h v f B_{max}^n \quad \text{watt} \tag{3.46}$$

where K_h is a constant whose value depends on the material of the core and range of flux density, f is the frequency of alternating magnetization in hertz, v is the volume of the magnetic core in m^3, B_{max} is the maximum value of flux density in Wb/m^2, and n is the Steinmetz constant. The value of n varies between 1.5 and 2.0 depending on the quality of steel and the range of operating flux density. The typical value of n is usually taken as 1.6.

3.11.3 Eddy Current Losses

Ferromagnetic material of the core subjected to time-varying flux induces emf in the core, which in turn produces circulating currents in the core. The currents so generated are called eddy currents. The effect of eddy currents is to dissipate energy in the form of heat. If the magnetic circuit is composed of solid iron, the power loss in the core would be considerable owing to the very low resistance of the core. If the core is made of laminations insulated from one another, the resistance to the path of eddy current is increased due to decrease in the cross-sectional area, and the flow of currents being confined to the respective lamination reduces eddy current loss.

As induced emf and currents are proportional to the frequency and flux density, it is reasonable to expect the power loss due to eddy currents to be proportional to the square of the frequency and the square of the flux density. An empirical formula for eddy current loss is

$$P_e = k_e f^2 B_{max}^2 \tau^2 v \quad \text{watt} \tag{3.47}$$

where K_e is characteristic constant of the core material, f is the frequency of variation of flux, B_{max} is the maximum flux density (Wb/m^2), τ is the thickness of lamination (m), and v is the volume of material (m^3). The sum total of hyteresis and eddy current losses constitutes core losses.

Example 3.1 A mild steel ring has a mean diameter of 20 cm and cross-sectional area of 400 mm^2. Calculate the (a) reluctance of the ring and (b) mmf to produce a flux of 500 μWb. Assume the relative permeability of mild steel to be 1000.

Solution (a) Mean length of the ring, $l = \pi \times$ mean diameter of the ring $= \pi \times 0.2$ m

Cross-sectional area of the ring, $A = 400$ mm$^2 = 400 \times 10^{-6}$ m

Relative permeability of steel = 1000

Therefore, reluctance of the steel ring

$$\mathcal{R} = \frac{l}{A \mu_0 \mu_r} = \frac{0.2\pi}{400 \times 10^{-6} \times 4\pi \times 10^{-7} \times 1000}$$

$$= 1.25 \times 10^6 \text{ AT/Wb}$$

(b) Flux $\phi = 500$ μWb $= 500 \times 10^{-6}$ Wb

The magnetomotive force, $\mathcal{F} = \phi \times \mathcal{R}$

$$= 500 \times 10^{-6} \times 1.25 \times 10^6 = 625 \text{ AT}$$

Example 3.2 A steel magnetic circuit has a uniform cross-sectional area of 4 cm^2 and a length of 50 cm. A coil of 250 turns is wound uniformly over the magnetic circuit. When the current in the coil is 1.5 A, the total flux is 0.25 mWb; when the current is 5 A, the total flux is 0.6 mWb. For each value of current, calculate the (a) magnetic field strength and (b) relative permeability of the steel.

Solution Mean length of the magnetic circuit, $l = 50$ cm $= 0.50$ m

Cross-sectional area of the magnetic circuit, $A = 4$ cm$^2 = 4 \times 10^{-4}$ m^2

Number of turns, $N = 250$

(a) When current I is 1.5 A, the flux is

$$\phi = 0.25 \text{ mWb} = 0.25 \times 10^{-3} \text{ Wb}$$

Flux density, $B = \dfrac{\phi}{A} = \dfrac{0.25 \times 10^{-3}}{4 \times 10^{-4}} = 0.625 \text{ Wb/m}^2$

Magnetomotive force, $\mathcal{F} = N \times I = 250 \times 1.5 = 375 \text{ AT}$

Magnetic field strength,

$$H = \frac{NI}{l} = \frac{375}{0.5} = 750 \text{ AT/m}$$

Now, $B = \mu H = \mu_0 \mu_r H$

$\therefore \qquad \mu_r = \dfrac{B}{\mu_0 H} = \dfrac{0.625}{4\pi \times 10^{-7} \times 750} = 663$

(b) When current I is 5 A, the flux is

$$\phi = 0.6 \text{ mWb} = 0.6 \times 10^{-3} \text{ Wb}$$

Flux density, $\quad B = \dfrac{\phi}{A} = \dfrac{0.6 \times 10^{-3}}{4 \times 10^{-4}} = 1.5 \text{ Wb/m}^2$

Magnetomotive force, $\mathcal{F} = N \times I = 250 \times 5 = 1250 \text{ AT}$

Magnetic field strength,

$$H = \frac{NI}{l} = \frac{1250}{0.5} = 2500 \text{ AT/m}$$

Now, $B = \mu H = \mu_0 \mu_r H$

$\therefore \qquad \mu_r = \dfrac{B}{\mu_0 H} = \dfrac{1.5}{4\pi \times 10^{-7} \times 2500} = 477.5$

Example 3.3 A steel ring, 20 cm mean diameter and circular cross section of diameter 2.5 cm, has an air gap of 1 mm. The ring is uniformly wound with 500 turns of copper wire carrying a current of 3 A. Calculate the (a) magnetomotive force, (b) magnetic flux, (c) flux density and (d) reluctance. Neglect magnetic leakage and fringing. Assume that the steel ring takes 30% of the total magnetomotive force.

Solution Mean length of the steel ring, $l_s = \pi \times 0.2 - 1 \times 10^{-3} = 0.627 \text{ m}$

Length of the air gap, $l_a = 0.001 \text{ m}$

Cross-sectional area of the magnetic circuit,

$$A = \pi \times (1.25 \times 10^{-2})^2 = 4.91 \times 10^{-4} \text{ m}^2$$

(a) Magnetomotive force, $\mathcal{F} = N \times I = 500 \times 3 = 1500 \text{ AT}$

Mmf of the steel ring, $\mathcal{F}_s = 0.3 \times 1500 = 450$

Mmf of the air gap, $\mathcal{F}_a = 0.7 \times 1500 = 1050 \text{ AT}$

(b) Reluctance of air gap,

$$\mathcal{R}_a = \frac{l_a}{\mu_0 A} = \frac{0.001}{4\pi \times 10^{-7} \times 4.91 \times 10^{-4}} = 1.6207 \times 10^6 \text{ AT/Wb}$$

Thus, $\quad \phi = \dfrac{\mathcal{F}_a}{\mathcal{R}_a} = \dfrac{1050}{1.6207 \times 10^6} = 6.47868 \times 10^{-4} \text{Wb}$

(c) Flux density,

$$B = \frac{\phi}{A} = \frac{6.47868 \times 10^{-4}}{4.91 \times 10^{-4}} = 1.3195 \text{ Wb/m}^2$$

(d) Reluctance of steel ring,

$$\mathcal{R}_s = \frac{\mathcal{F}_s}{\phi} = \frac{450}{6.47868 \times 10^{-4}} = 0.6956 \times 10^6 \text{ AT/Wb}$$

Example 3.4 A cast steel magnet has an air gap of length 2 mm and an iron path of 30 cm. Find the number of ampere turns necessary to produce a flux density of 1.2 Wb/m². The relative permeability of cast steel is 900. Neglect leakage and fringing.

Solution Length of the air gap, $l_a = 2 \times 10^{-3}$ m
Length of the cast steel core, $l_s = 0.3$ m
Required flux density, $B = 1.2$ Wb/m²
  Now for air gap, $B = \mu_0 H_a$

$$\therefore \qquad H_a = \frac{B}{\mu_0} = \frac{1.2}{4\pi \times 10^{-7}} = 0.955 \times 10^6 \text{ AT/m}$$

Then, magnetomotive force for the air gap,

$$\mathcal{F}_a = H \times l = 0.955 \times 10^6 \times 2 \times 10^{-3} = 1910 \text{ AT}$$

For cast steel ring, $B = \mu_0 \mu_r H_s$

$$\therefore \qquad H_s = \frac{B}{\mu_0 \mu_r} = \frac{1.2}{4\pi \times 10^{-7} \times 900} = 1061.03 \approx 1061 \text{ AT/m}$$

Then, magnetomotive force for the air gap,

$$\mathcal{F}_s = H_s \times l_s = 1061 \times 0.3 = 318.3 \text{ AT}$$

Then, total mmf is $\mathcal{F}_a + \mathcal{F}_s = 1910 + 318.3 = 2228.3$ AT

Example 3.5 An iron ring of mean length 30 cm has an air gap of 2 mm and a winding of 200 turns. If the permeability of the iron core is 300 when a current of 1 A flows through the coil, find the flux density.
Solution The magnetic circuit with its analogue is shown in Fig. 3.16.
Assume the area of the cross section of the ring and the air gap to be A m².

Total mmf, $\mathcal{F} = 200 \times 1 = 200$ AT

Reluctance of air gap $\mathcal{R}_a = \dfrac{2 \times 10^{-3}}{\mu_0 \times A}$ AT/Wb

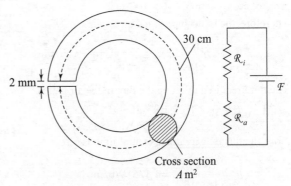

Fig. 3.16 Magnetic circuit with its analogue

Reluctance of iron core, $\mathcal{R}_i = \dfrac{30 \times 10^{-2}}{300 \times \mu_0 \times A} = \dfrac{10^{-3}}{\mu_0 \times A}$ AT/Wb

Reluctance of the magnetic circuit, $\mathcal{R} = \mathcal{R}_a + \mathcal{R}_i = \dfrac{10^{-3}}{\mu_0 \times A}(2+1) = \dfrac{3 \times 10^{-3}}{\mu_0 \times A}$ AT/Wb

Total flux, $\phi = \dfrac{\mathcal{F}}{\mathcal{R}} = \dfrac{200 \times \mu_0 \times A}{3 \times 10^{-3}}$

Therefore, flux density, $B = \dfrac{\phi}{A} = \dfrac{200 \times 4\pi \times 10^{-7}}{3 \times 10^{-3}} = 83.77$ mWb/m^2

Example 3.6 A 600-turn coil is wound on the central limb of the cast steel frame shown in Fig. 3.17(a). A total flux of 1.8 mWb is required in the gap. Find what current is required. Assume that the gap density is uniform and that all flux lines pass straight across the gap. Dimensions given are in centimetres.

Solution The electrical equivalent of the magnetic circuit is shown in Fig. 3.17(b). The method of solution can be easily ascertained from the equivalent circuit. From the knowledge of the air gap, flux ϕ_a, which is the flux in the centre limb as well, the mmf required to set up the flux through the air gap hg and the central limb gf can be determined. The flux ϕ_a gets divided equally into the two paths *fabh* and *fedh* as these have the same reluctance. Thus the flux in each of the outer limbs is $(\phi_a)/2$. The mmf needed to maintain $(\phi_a)/2$ in the outer limb can be found by multiplying ϕ_a by $\mathcal{R}$. Thus, from Fig. 3.17(b), for the closed loop *hgfabh*,

$$\mathcal{F} = \mathcal{F}_a + \mathcal{F}_c + \mathcal{F}_{fabh} = \phi_a \times (\mathcal{R}_a + \mathcal{R}_c) + (\phi_a/2) \times \mathcal{R}$$

Flux in the air gap $\phi_a = 1.8 \times 10^{-3}$ Wb

Length of the air gap $= 0.1 \times 10^{-2}$ m

Area of cross-section $= 16 \times 10^{-4}$ m^2

Therefore, the reluctance of the air gap is

$$\mathcal{R}_a = \frac{0.1 \times 10^{-2}}{4\pi \times 10^{-7} \times 16 \times 10^{-4}} = 4972 \times 10^2 \text{ AT/Wb}$$

The mmf required to set up the flux in the air gap is

$$\mathcal{F}_a = \phi_a \times \mathcal{R}_a = 1.8 \times 10^{-3} \times 4972 \times 10^2$$
$$= 894.96 \text{ AT} \approx 895 \text{ AT}$$

Now, the flux density in the central limb is

$$B = \frac{1.8 \times 10^{-3}}{16 \times 10^{-4}} = 1.125 \text{ Wb/m}^2$$

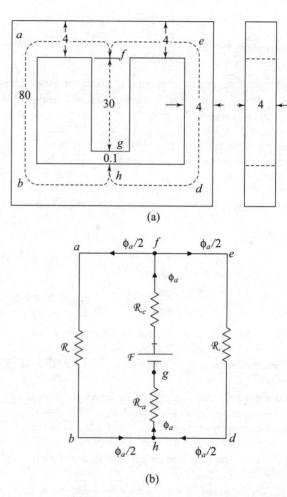

(a)

(b)

Fig. 3.17 (a) Magnetic circuit and (b) analogue of magnetic circuit

From the magnetization curve (*B-H* curve) of cast steel (shown in Fig. 3.12), to set up a flux density $B = 1.125$ Wb/m^2, the field intensity required is $H_c = 1000$ AT/m.
The length of the central limb is 30×10^{-2} m.
The mmf drop in the central limb is

$$\mathcal{F}_c = H_c \times l_c = 1000 \times 0.3 = 300 \text{ AT}$$

The flux density in the parallel path *fabh* is $(\phi_a)/2 = 0.5625$ Wb/m^2 and the field intensity H is 625 AT/m.
The length of the parallel path is 80×10^{-2} m.
Hence mmf for the parallel path, $\mathcal{F}_{fabh} = 625 \times 80 \times 10^{-2} = 500$ AT

Therefore, the total mmf required is

$$\mathcal{F} = \mathcal{F}_a + \mathcal{F}_c + \mathcal{F}_{fabh} = 895 + 300 + 500 = 1695 \text{ AT}$$

$$\text{Required current} = \frac{\text{mmf}}{\text{number of turns}} = \frac{1695}{600} = 2.825 \text{ A}$$

Example 3.7 A magnetic core made from transformer sheet steel and wound with 500 turns of wire is shown in Fig. 3.18. Calculate the current required to set up a flux of 1.4 mWb in the air gap. Use the *B-H* curve given in Fig. 3.12 and neglect fringing. All dimensions, in millimetre, are shown in the figure.

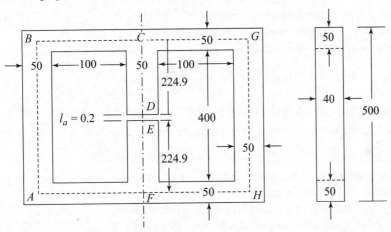

Fig. 3.18

Solution The equivalent electrical analogue (magnetic circuit) is shown in Fig. 3.19.

Flux in the air gap, $\phi_g = 1.4 \times 10^{-3}$ Wb

Area of the air gap $= 50 \times 10^{-3} \times 40 \times 10^{-3} = 0.002$ m^2

Flux density in the air gap $B_a = \dfrac{1.4 \times 10^{-3}}{0.0020} = 0.7$ Wb/m^2

Magnetic field in the air gap, $H_a = \dfrac{0.7}{4\pi \times 10^{-7}} = 55.70$ AT/m

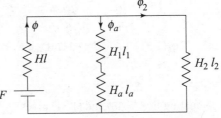

Fig. 3.19

Magnetomotive force (mmf) required for the air gap

$$= H_a l_a = 55.7 \times 10^4 \times 0.2 \times 10^{-3} = 111.41 \text{ AT} \qquad \text{(I)}$$

Since the flux density in the central limb is also 0.7 Wb/m^2, the corresponding field strength is obtained from Fig. 3.12 as $H_1 = 250$ AT/m. Thus, the mmf for the magnetic central limb is given by

$$\text{mmf} = H_1 l_1 = 250 \times (450 - 0.2) \times 10^{-3} = 112.45 \text{ AT} \qquad \text{(II)}$$

Total mmf required for the central limb is obtained by adding (I) and (II) as

$$\text{mmf}_{CDEF} = 111.41 + 112.45 = 223.86 \text{ AT}$$

Mean length of the magnetic core *CGHF*, $l_2 = (2 \times 150 + 450) \times 10^{-3} = 0.75$ m

Since the central limb and the magnetic core *CGHF* are in parallel, the two paths have the same mmf, that is, 223.86 AT. Hence, the magnetic intensity in *CGHF* is

$$H_2 = \frac{223.86}{0.75} = 298.48 \text{ AT}$$

Using Fig. 3.12, the flux density is obtained as $B_2 = 0.7$ Wb/m^2. Therefore,

$$\phi_2 = 0.7 \times 50 \times 40 \times 10^{-6} = 0.0014 \text{ Wb}$$

Total flux in the magnetic core *FABC*, $\phi = \phi_a + \phi_2 = 0.0014 + 0.0014 = 0.0028$ Wb

Flux density in the magnetic core *FABC*, $B = \dfrac{0.0028}{50 \times 40 \times 10^{-6}} = 1.4$ Wb/m^2

From Fig. 3.12, the corresponding magnetic intensity, $H = 3000$ AT/m

Total mmf required for the magnetic core *FABC* $= Hl = 3000 \times 0.75 = 2250$ AT

Total mmf required for the magnetic core $= 223.86 + 2250 = 2473.86$ AT

The required current to set up the flux in the air gap $= \dfrac{2473.86}{500} = 4.95$ A

3.12   SELF INDUCTANCE AND MUTUAL INDUCTANCE

In this section, self inductance and mutual inductance are discussed.

3.12.1   Self Inductance

In Section 1.9.2 an inductor, which induced a voltage of 1 V with the current varying at the rate of 1 A/sec, is said to have an inductance of 1 H. Figure 3.20 shows a coil of *N* turns, carrying a time-varying current of *i* ampere. The current sets up a flux ϕ, which links with all the *N* turns of the coil. Due to the flux in the coil, voltage *e* will be induced, which as per Lenz's law has a direction opposite to the flow of current that sets up the flux. Thus,

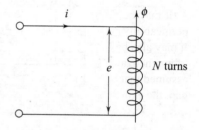

Fig. 3.20   Self-inductance of a coil

$$e = -N\frac{d\phi}{dt} = -N\frac{d\phi}{di}\frac{di}{dt} \tag{3.48}$$

Using Eq. (1.31), the induced emf *e* may be written as

$$e = L\frac{di}{dt}$$

Comparing Eqs (3.48) and (1.31) gives

$$L = N\frac{d\phi}{di} = N\frac{d\phi}{di} = \frac{d\psi}{di} \tag{3.49}$$

where *L* is called the self inductance of the coil and $\psi = N\phi$ denotes the total flux linkages. The sign of the induced voltage can be determined from the physical considerations; therefore the negative sign in Eq. (3.48) is left out. When the reluctance of the magnetic circuit of the coil is not constant, the magnetization characteristic (*B-H* curve) is non-linear. Under these conditions, *L* is not constant but changes with the value of magnetizing force *H*. The value of *H* may be obtained as a value corresponding to the incremental changes of *B* and *H* around an operating point on the curve. However, when the reluctance of the magnetic circuit of the coil is constant, such as an air-cored coil, the flux is proportional to the current. Hence the magnetization characteristic (*B-H* curve) is linear and *L* is a constant and may be expressed as

$$L = \frac{N\phi}{i} = \frac{\psi}{i} \text{ Henry} \tag{3.50}$$

Therefore, self inductance can be defined as (self) flux linkage per ampere. Neglecting the leakage flux, Eq. (3.50) can be used to compute the self inductance of the coil shown in Fig. 3.1.

$$L = \frac{N\phi}{i} = \frac{N^2\phi}{Ni} \text{ Henry} \tag{3.51}$$

Substituting $\phi = B \times A$ into Eq. (3.51) and using Eq. (3.6), self inductance may be expressed as

$$L = \frac{N^2 \times B \times A}{H \times l} \text{ Henry} \tag{3.52}$$

Substituting $B = \mu_0\mu_r H$ from Eq. (3.17) into Eq. (3.52), L becomes

$$L = \frac{N^2 \times \mu_0\mu_r \times A}{l} = \frac{N^2}{\mathcal{R}} = \mathcal{P} \times N^2 \text{ H} \tag{3.53}$$

where $\quad \mathcal{R} = \dfrac{1}{\mathcal{P}} = \dfrac{l}{\mu_0\mu_r A}$

It may be noted from Eq. (3.53) that the self-inductance is a function of the number of turns in the coil, permeability of the magnetic material, and geometry of the core or the reluctance of the core since $\mathcal{R} = l(\mu_0\mu_r A)$. It may also be observed that due to the non-linearity of the permeability μ_r, which is dependent on the extent of magnetization of the core, it is difficult to employ Eq. (3.53) to numerical computations. However, if it is assumed that there exists a linear relationship between flux ϕ and the varying current i or a pre-dominant air gap, the expression for the self-inductance takes the form as given below:

$$L = \frac{\mu_0 N^2 A}{l} \text{ H} \tag{3.54}$$

3.12.2 Mutual Inductance

When two coils are placed in close proximity, a change in current in the first coil produces a change in magnetic flux, which links not only the coil itself but also the second coil as well. According to Faraday's laws, the change in flux induces an emf in the second coil. This emf is called mutually induced emf and the two coils are said to have mutual inductance.

Figure 3.21(a) shows coil 1 having N_1 turns placed on a common magnetic core near coil 2 with N_2 turns. The voltage induced in coil 2 is proportional to the rate of change of current i_1 in coil 1. If di_1 is the increase of current in coil 1 in dt seconds, then the emf e_2 induced in coil 2 may be expressed as

$$e_2 \propto \frac{di_1}{dt}$$

or $\quad e_2 = -M_{12}\dfrac{di_1}{dt} \tag{3.55}$

where M_{12} is called mutual inductance between coil 1 and coil 2. The unit of mutual inductance is the same as for self inductance, namely, henry. Two coils have mutual inductance of 1 H if an emf of 1 V is induced in one coil when the current in the other coil varies uniformly at the rate of 1 A/sec.

If $d\phi_{12}$ weber is the increase in the mutual flux in coil 2 due to the increase of di_1 ampere in the coil 1, then the emf induced in coil 2 is

$$e_2 = -N_2 \frac{d\phi_{12}}{dt} \tag{3.56}$$

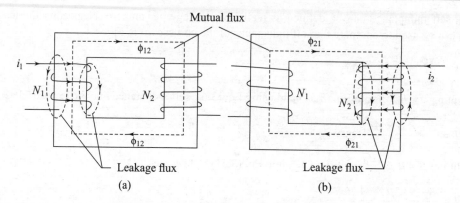

Fig. 3.21 Two coils on a common magnetic core

From expressions (3.55) and (3.56),

$$M_{12}\frac{di_1}{dt} = N_2\frac{d\phi_{12}}{dt} = N_2\frac{d\phi_{12}}{di_1}\frac{di_1}{dt}$$

$$\therefore \qquad M_{12} = N_2\frac{d\phi_{12}}{di_1} \qquad\qquad\qquad (3.57)$$

Thus mutual inductance M_{12} is equal to the ratio of the change of flux linkages of coil 2 to the change of current in coil 1.

If the relative permeability of the magnetic core remains constant, the ratio $d\phi_{12}/di_1$ is constant and is equal to flux per ampere. Then

$$M_{12} = \frac{N_2\phi_{12}}{i_1} = \frac{\psi_{12}}{i_1} \qquad\qquad\qquad (3.58)$$

where ϕ_{12} and ψ_{12} are, respectively, the flux and flux linkages of coil 2 due to current i_1 in coil 1.

Similarly, if ϕ_{21} and ψ_{21} are, respectively, the flux and flux linkages of coil 1 due to current i_2 in coil 2, as shown in Fig. 3.21(b), the mutual inductance M_{21} between coil 2 and coil 1 can be expressed as

$$M_{21} = \frac{N_1\phi_{21}}{i_2} = \frac{\psi_{21}}{i_2} \qquad\qquad\qquad (3.59)$$

Dot Convention In Fig. 3.22(a), two windings on a common magnetic core are shown. When a varying current i_1 flows through coil 1, a flux ϕ_1 is set up, which induces a varying emf in coil 2. Now if the switch is closed, a current i_2 begins to circulate in coil 2. The direction of the current i_2, as per Lenz's law, should be such that the flux ϕ_2 set up by it should be in the direction opposite to ϕ_1. After the switch is closed, application of the right hand rule with the thumb pointing in the direction of ϕ_2 leads to the fingers pointing to the direction of current i_2 as indicated in Fig. 3.22(a). Since the direction of i_2 has been established the polarity of the induced emf in coil 2 is easily identified and shown in Fig.3.22(a).

From the previous discussion, it is clear that the direction of the induced emf of a mutual inductance depends upon the direction of the windings of the coupled coils. However, it is difficult to determine the directions of the coupled coils since the equipment is sealed by the manufacturer. Therefore, a dot convention is adopted to indicate symbolically a pictorial representation of the instantaneous positive polarity of the terminals. Figure

3.22(b) may be taken to represent the equivalent circuit of Fig. 3.22(a).

Dot convention is employed in the following manner to determine the sign of the mutual inductance.

(i) When the assumed currents in a mutually coupled pair of coils enter or leave the dotted terminals, the M terms will have the same sign as the L terms.

(ii) When one current enters a dotted terminal and the other current leaves the other dotted terminal, the M terms will have the opposite signs as the L terms.

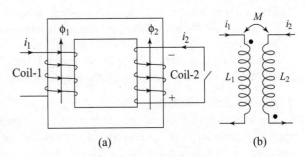

(a)   (b)

Fig. 3.22   Dot convention

3.12.2.1 Coupling Coefficient

Figure 3.21 shows two coils placed on a common magnetic core. A part of the total flux (called leakage flux) produced by one coil links only the coil producing the flux, while the balance flux (called mutual flux) links the other coil. If a fraction k of total flux ϕ_1 produced by current i_1 in coil 1 links coil 2, the mutual flux is $\phi_{12} = k\phi_1$. Similarly, $\phi_{21} = k\phi_2$, where ϕ_2 is the total flux produced by current i_2 in coil 2, linking coil 1. Then

$$\phi_1 = \frac{N_1 i_1}{\mathcal{R}} \quad \text{and} \quad \phi_2 = \frac{N_2 i_2}{\mathcal{R}} \tag{3.60}$$

where $\mathcal{R}$ is the reluctance of the magnetic circuit.

Substituting the values of ϕ_{12} and ϕ_{21} in Eqs (3.58) and (3.59) gives

$$M_{12} = k\,\frac{N_1 N_2}{\mathcal{R}} \text{ henry} \quad \text{and} \quad M_{21} = k\,\frac{N_1 N_2}{\mathcal{R}} \text{ henry} \tag{3.61}$$

Hence, for a bilateral circuit, $M_{12} = M_{21} = M$ \hfill (3.62)

Thus, $M^2 = M_{12} \times M_{21} = k^2 \left(\dfrac{N_1 N_2}{\mathcal{R}}\right)^2$ \hfill (3.63)

Using Eq. (3.53), the self inductances of coil 1 and coil 2, respectively, are expressed as

$$L_1 = \frac{N_1^2}{\mathcal{R}} \text{ Henry} \tag{3.64}$$

$$L_2 = \frac{N_2^2}{\mathcal{R}} \text{ Henry} \tag{3.65}$$

Multiplying Eqs (3.64) and 3.65)

$$L_1 L_2 = \frac{N_1^2 N_2^2}{\mathcal{R}^2} \tag{3.66}$$

or $\sqrt{L_1 L_2} = \dfrac{N_1 N_2}{\mathcal{R}}$ \hfill (3.67)

Substituting Eq. (3.67) into Eq. (3.63) gives

$$M = k\sqrt{L_1 L_2} \tag{3.68}$$

The term k is called the coefficient of coupling. When 100% of the flux lines link each coil, then $k = 1$. The term coupling coefficient is widely used in radio networks to denote the degree of coupling between two coils. If the two coils are close together, most of the flux produced by current in one coil passes through the other coil, and the coils are said to be tightly coupled. If the coils are well apart, a small part of the flux produced by current in one coil passes through the other coil and the coils are said to be loosely coupled.

3.12.2.2 Coupled Circuits

Figure 3.23 shows a bilateral circuit ($M_{12} = M_{21} = M$) with coils 1 and 2 with self inductances L_1 and L_2 and carrying currents i_1 and i_2, respectively. As per the dot convention, since both currents i_1 and i_2 are entering the dots, both the self-inductances L_1 and L_2 and the mutual inductance M have the same sign. The total flux linkages with coils 1 and 2 are given as

$$\psi_1 = L_1 i_1 + M i_2 \tag{3.69}$$

$$\psi_2 = M i_1 + L_2 i_2 \tag{3.70}$$

The induced voltage in each coil can be written as

$$e_1 = \frac{d\psi_1}{dt} = L_1 \frac{di_1}{dt} + M \frac{di_2}{dt} \tag{3.71}$$

$$e_2 = \frac{d\psi_2}{dt} = M \frac{di_1}{dt} + L_2 \frac{di_2}{dt} \tag{3.72}$$

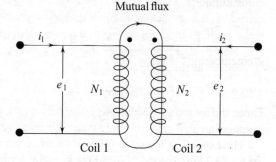

Fig. 3.23   Coupled circuits

The circuit used to represent the magnetic coupling between two coils with current in both the coils is shown in Fig. 3.24. The equivalent network representation of Eqs (3.71) and (3.72) is shown in Fig. 3.25. The veracity of the representation can be verified by developing mesh equations for the network.

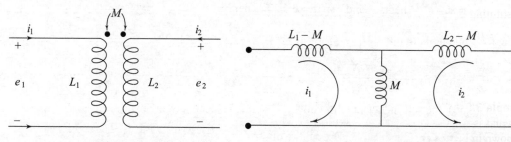

Fig. 3.24   Circuit representation of mutually coupled coils

Fig. 3.25   The equivalent network representation of mutually coupled coils

3.13   ENERGY IN LINEAR MAGNETIC SYSTEMS

In Section 1.9.2, it is shown that energy W [see Eq. (1.35)] stored in a magnetic field is given by

$$W = \frac{1}{2} L i^2 \text{ joule} \tag{3.73}$$

Substituting the value of L from Eq. (3.50), the expression of W becomes

$$W = \frac{1}{2} \psi i = \frac{1}{2} N\phi i \text{ joule} \tag{3.74}$$

From Eq. (3.8), $Ni = H \times l$ and from Eq. (3.18), $\phi = B \times A$. Substituting these in Eq. (3.74) gives

$$W = \frac{1}{2}(H \times l)(B \times A) = \frac{1}{2}BHV \text{ joule} \tag{3.75}$$

where V is the volume of the magnetic material/core.

The energy density of the magnetic field can be expressed per unit volume as

$$\frac{W}{V} = \frac{1}{2}BH \text{ J/m}^3 \tag{3.76}$$

Substituting for L from Eq. (3.53) in Eq. (3.73),

$$W = \frac{1}{2}\frac{N^2 i^2}{\mathcal{R}} \text{ joule} \tag{3.77}$$

Remembering that mmf $\mathcal{F} = \phi \times \mathcal{R}$, then

$$W = \frac{1}{2}\phi^2 \times \mathcal{R} \text{ joule} \tag{3.78}$$

Since the air gap has a much higher reluctance, most of the energy would be stored in the air gap if the core contains an air gap.

The expression for energy stored in two excited and mutually coupled coils can be derived with the help of Eq. (3.74) and takes the following general form:

$$W = \int_0^{\psi_1} i_1 d\psi_1 + \int_0^{\psi_2} i_2 d\psi_2 \tag{3.79}$$

Differentiating Eqs (3.69) and (3.70) with respect to current gives

$$d\psi_1 = L_1 di_1 + M di_2$$

$$d\psi_1 = M di_1 + L_2 di_2$$

Substituting for $d\psi_1$ and $d\psi_2$ in Eq. (3.79), the energy stored

$$W = L_1 \int_0^{i_1} i_1 di_1 + L_2 \int_0^{i_2} i_2 di_2 + M \int_0^{i_2}(i_1 di_1 + i_2 di_2)$$

$$= \frac{1}{2}L_1 i_1^2 + \frac{1}{2}L_2 i_2^2 + M i_1 i_2 \text{ joule} \tag{3.80}$$

Example 3.8 A magnetic circuit contains three air gaps and is wound with 150 and 200 turns of wire as shown in Fig. 3.26. If the relative permeability of the material approaches infinity, calculate (a) the self-inductance and mutual inductance, (b) the mutual inductance between the coils when the air gap l_3 is closed, and (c) the energy stored in the magnetic system when each coil is carrying a current of 1 A. The lengths of the air gaps are shown in the figure and the area of each gap is 200 mm². Neglect fringing and assume $\mu_0 = 4\pi \times 10^{-7}$ Wb/ATm.

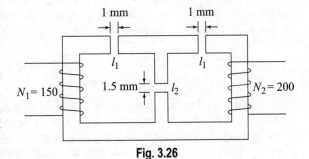

Fig. 3.26

Solution Reluctance of air gaps 1 and 2, $\mathcal{R}_1 = \mathcal{R}_2 = \dfrac{1 \times 10^{-3}}{\mu_0 \times 200 \times 10^{-6}} = 3.98 \times 10^6$ AT/Wb

Reluctance of air gap 3, $\mathcal{R}_3 = \dfrac{1.5 \times 10^{-3}}{\mu_0 \times 200 \times 10^{-6}} = 5.97 \times 10^6$ AT/Wb

The equivalent electrical analogue of the magnetic circuit is shown in Fig. 3.27.

(a) Assume that a current of 1 A flows through the 150-turns coil. For the assumed directions of fluxes, application of the mesh current method leads to the following matrix equations:

$$\begin{bmatrix} (\mathcal{R}_1 + \mathcal{R}_3) & -\mathcal{R}_3 \\ -\mathcal{R}_3 & (\mathcal{R}_2 + \mathcal{R}_3) \end{bmatrix} \begin{bmatrix} \phi_1 \\ \phi_2 \end{bmatrix}$$

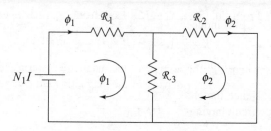

Fig. 3.27

$$= \begin{bmatrix} 9.95 & -5.97 \\ -5.97 & 9.95 \end{bmatrix} \begin{bmatrix} \phi_1 \\ \phi_2 \end{bmatrix} \times 10^6 = \begin{bmatrix} 150 \\ 0 \end{bmatrix}$$

or $$\begin{bmatrix} \phi_1 \\ \phi_2 \end{bmatrix} = \left\{ \begin{bmatrix} 9.95 & -5.97 \\ -5.97 & 9.95 \end{bmatrix} \times 10^6 \right\}^{-1} \begin{bmatrix} 150 \\ 0 \end{bmatrix} = \begin{bmatrix} 2.36 \\ 1.41 \end{bmatrix} \times 10^{-5} \text{ Wb}$$

The self-inductance and mutual inductance are obtained as follows:

$$L_1 = \frac{N_1 \phi_1}{i_1} = \frac{150 \times 2.36 \times 10^{-5}}{1} = 3.5 \text{ mH} \quad \text{and} \quad M_{12} = \frac{N_2 \phi_2}{i_1} = \frac{200 \times 1.41 \times 10^{-5}}{1} = 2.8 \text{ mH}$$

Next assume a current of 1 A flowing through the 200-turns coil. The self-inductance of the coil is determined in the same manner. The steps are shown below.

$$\begin{bmatrix} 9.95 & -5.97 \\ -5.97 & 9.95 \end{bmatrix} \begin{bmatrix} \phi_1 \\ \phi_2 \end{bmatrix} \times 10^6 = \begin{bmatrix} 0 \\ 200 \end{bmatrix} \quad \text{or} \quad \begin{bmatrix} \phi_1 \\ \phi_2 \end{bmatrix} = \begin{bmatrix} 1.89 \\ 3.14 \end{bmatrix} \times 10^{-5} \text{ Wb}$$

Hence, the self-inductance and mutual inductance are compute as follows:

$$L_2 = \frac{N_2 \phi_2}{i_2} = \frac{200 \times 3.14 \times 10^{-5}}{1} = 6.3 \text{ mH} \quad \text{and} \quad M_{21} = \frac{N_1 \phi_1}{i_2} = \frac{150 \times 1.89 \times 10^{-5}}{1} = 2.8 \text{ mH}$$

It may be noted that the computations are verified since $M_{12} = M_{21}$.

(b) When the air gap l_3 is closed, the reluctance of the limb is zero since the permeability of the magnetic material is infinity. Thus, the limb behaves like a short circuit and the entire flux ϕ_1 passes through it. Thus, the flux linking the 200-turns coil is zero and the mutual inductance, therefore, is also zero.

(c) Use of Eq. (3.80) for determining the energy in the magnetic system leads to

$$W = (3.5/2) + (6.3/2) + 2.8 = 7.7 \times 10^{-3} \text{ J}$$

3.14   COILS CONNECTED IN SERIES

Coils 1 and 2 having self inductances L_1 and L_2 henry, respectively, and mutual inductance M henry are wound coaxially on an insulating cylinder, with terminals 1′ and 2 joined together as shown in Fig. 3.28(a). When current i flows through the two coils, as shown in Fig. 3.28(a), as per the dot convention, the fluxes produced in the coils are in the same direction, and the coils are said to be cumulatively coupled.

If the current increases by di amperes in dt seconds, then

Emf induced in coil 1 due to its self inductance $= -L_1 \dfrac{di}{dt}$

Emf induced in coil 2 due to its self inductance $= -L_2 \dfrac{di}{dt}$

Emf induced in coil 1 due to increase of current in coil 2 $= -M\dfrac{di}{dt}$

Emf induced in coil 2 due to increase of current in coil 1 $= -M\dfrac{di}{dt}$

Hence the emf induced in the whole circuit regarded as a single circuit having an equivalent self inductace of L_A henry is

$$-(L_1 + L_2 + 2M)\frac{di}{dt} = -L_A\frac{di}{dt}$$

and $L_A = L_1 + L_2 + 2M$ (3.81)

If the direction of current in coil 2 is reversed relative to that in coil 1 by joining together terminals 1′ and 2′, as in Fig. 3.28(b), as per the dot convention, the fluxes produced in the coils are in opposition, and the coils are said to be differentially coupled. With this coupling, the emf, $M(di/dt)$, induced in coil 1 due to an increase di in current in time dt in coil 2 is in the same direction as the current and is therefore in opposition to the emf induced in coil 1 due to its self inductance. Similarly, the emf induced in coil 2 by mutual inductance is in opposition to that induced by the self inductance of coil 2. Hence, the total emf induced in coils 1 and 2 when differentially coupled is

$$= -(L_1 + L_2 - 2M)\frac{di}{dt} = -L_B\frac{di}{dt}$$

and $L_B = L_1 + L_2 - 2M$ (3.82)

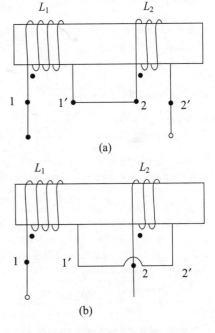

Fig. 3.28   Coils connected in series

Therefore, the total inductance of inductively coupled circuits is

$$L_1 + L_2 \pm 2M$$

The positive sign is used if the coils are cumulatively coupled and the mutual inductance is then regarded as positive. But, the negative sign is used if the coils are differentially coupled.

From expressions (3.81) and (3.82), the value of M may be obtained as

$$M = \frac{L_A - L_B}{4} \qquad (3.83)$$

Example 3.9   The mean diameter of a steel ring is 40 cm and a flux density of 0.8 Wb/m² is produced by 50 AT/cm. If the cross section of the ring is 25 cm² and the number of turns is 800, find: (a) the inductance in henry; (b) the exciting current and inductance in henry when a gap of 1 cm is cut in the ring. The flux density is 1 Wb/m². Ignore leakage and fringing.

Solution   (a) Flux density, $B = 0.8$ Wb/m²

Cross-sectional area of the ring, $A = 25 \times 10^{-4}$ m²

Flux, $\phi = B \times A = 0.8 \times 25 \times 10^{-4} = 20 \times 10^{-4}$ Wb

Field intensity, $H = 50$ AT/cm $= 50 \times 100$ AT/m $= 5000$ AT/m

Mean length of the steel ring, $l = \pi \times 40 \times 10^{-2} = 0.4\pi$ m

Magnetomotive force, $\mathcal{F} = H \times l = 5000 \times 0.4\pi = 2000\pi$ AT

Number of turns $N = 800$

Exciting current is $i = \dfrac{\mathcal{F}}{N} = \dfrac{2000\,\pi}{800} = 7.854$ A

Using Eq. (3.50), self inductance is $L = \dfrac{N\phi}{i} = \dfrac{800 \times 20 \times 10^{-4}}{7.854} = 0.2037$ H

(b) Flux in the air gap: $\phi_{air} = 20 \times 10^{-4}$ Wb

Reluctance of the air gap is $\mathcal{R}_{air} = \dfrac{1 \times 10^{-2}}{4\pi \times 10^{-7} \times 25 \times 10^{-4}} = 3.1831 \times 10^{6}$ AT/Wb

The mmf required for the air gap, $\mathcal{F}_{air} = \phi_{air} \times \mathcal{R}_{air} = 20 \times 10^{-4} \times 0.3183 \times 10^{8} = 6366$ AT

The mmf for the core: $\mathcal{F}_{core} = 5000 \times (0.4\pi - 0.01) = 6233.18$ AT

Total mmf, $\mathcal{F} = \mathcal{F}_{air} + \mathcal{F}_{core} = 6366 + 6233.18 = 12599.18 \approx 12600$ AT

Exciting current is $i = \dfrac{\mathcal{F}}{N} = \dfrac{12600}{800} = 15.75$ A

Self inductance $L = \dfrac{N\phi}{i} = \dfrac{800 \times 20 \times 10^{-4}}{15.75} = 0.10158$ H

(*Note*:   It may be seen that if a gap of 1 cm is cut in the steel ring the self inductance of the coil is reduced to almost half the value of inductance without the air gap.)

Example 3.10   Two identical 2000-turn coils X and Y are in parallel planes such that 75% of the magnetic flux produced by one coil links the other. A current of 10 A in X produces a flux of 0.05 mWb in it. If the current in X changes from + 10 A to −10 A in 0.01 sec, what will be the magnitude of the electromotive force induced in Y? Calculate the self inductance of each coil and the mutual inductance.

Solution   Self inductance of coil X

$$L_X = \frac{N\phi}{i} = \frac{2000 \times 0.05 \times 10^{-3}}{10} = 10 \ mH$$

Since the coils are identical, the self inductance of coil Y is $L_Y = 10$ mH.

Flux linking coil Y due to current in coil $X = 0.75 \times 0.05 \times 10^{-3}$ Wb

Flux linkages in coil Y, $\psi_{YX} = 2000 \times 0.75 \times 0.05 \times 10^{-3}$ Wb

Then, the mutual inductance

$$M = \frac{\psi_{YX}}{i_X} = \frac{2000 \times 0.75 \times 0.05 \times 10^{-3}}{5} = 15 \ mH$$

The rate of change of current:

$$\frac{di}{dt} = \frac{+10 - (-10)}{0.01} = 2000 \ A/sec$$

The induced emf in the coil Y,

$$e_Y = M \frac{di}{dt} = 15 \times 10^{-3} \times 2000 = 30 \ V$$

Example 3.11   The combined inductance of two coils connected in series is 1.2 H or 0.2 H, depending on the relative directions of the currents in the coils. If one of the coils, when isolated, has a self inductance of 0.4 H, calculate (a) the mutual inductance and (b) the coupling coefficient.

Solution   When the currents in the coils are in the same direction, the flux produced is additive. On the other hand, when the currents in the coils are in opposite directions, the flux produced is subtractive. Therefore, when the currents are in the same direction, the total inductance is

$$L_T = L_1 + L_2 + 2M$$

and when the currents are in the opposite directions,

$$L_T = L_1 + L_2 - 2M$$

For the data given in this problem,

$$L_1 + L_2 + 2M = 1.2 \qquad\qquad\qquad \text{(I)}$$

and $\qquad L_1 + L_2 - 2M = 0.2 \qquad\qquad\qquad \text{(II)}$

Subtracting (II) from (I),

$$4M = 1.0 \quad \text{or} \quad M = 0.25 \text{ H}$$

Substituting the values of $L_1 = 0.4$ H and $M = 0.25$ H in (I),

$$0.4 + L_2 + 2 \times 0.25 = 1.2$$

$\therefore \qquad L_2 = 0.3$ H

Using the general form of Eq. (3.68), the coefficient of coupling

$$k = \frac{0.25}{\sqrt{0.4 \times 0.3}} = 0.72168$$

Example 3.12 An iron ring 25 cm in diameter and 10 cm^2 cross section is wound with 250 turns of wire. For a flux density of 1 Wb/m^2 and a permeability of 800, find (a) the exciting current, the inductance and the stored energy; (b) the corresponding quantities when there is a 1 mm air gap.

Solution (a) The flux $\phi = B \times A = 1 \times 10 \times 10^{-4} = 10^{-3}$ Wb

The reluctance of the core without the air gap is

$$\mathcal{R} = \frac{\pi \times 25 \times 10^{-2}}{800 \times 4\pi \times 10^{-7} \times 10 \times 10^{-4}} = \frac{25 \times 10^6}{32} = 0.78125 \times 10^6 \text{ AT/Wb}$$

The magnetomotive force is

$$\mathcal{F} = \phi \times \mathcal{R} = 781.25 \text{ AT}$$

The exciting current required is

$$i = \frac{\mathcal{F}}{N} = \frac{781.25}{250} = 3.125 \text{ A}$$

The self inductance of the coil is

$$L = \frac{N\phi}{i} = \frac{250 \times 10^{-3}}{3.125} = 0.08 \text{ H}$$

The energy stored is given by

$$w = \frac{1}{2} Li^2 = \frac{1}{2} \times 0.08 \times (3.125)^2 = 0.391 \text{ J}$$

(b) With an air gap of 1 mm in the iron ring:

The reluctance of the air gap is

$$\mathcal{R}_{air} = \frac{1 \times 10^{-3}}{4\pi \times 10^{-7} \times 10^{-3}} = \frac{1}{4\pi \times 10^{-7}} \text{ AT/Wb}$$

The magnetomotive force for the air gap:

$$\mathcal{F}_{air} = \phi \times \mathcal{R}_{air} = 10^{-3} \times \frac{1}{4\pi \times 10^{-7}} = 795.7 \text{ AT}$$

The reluctance of the core is

$$\mathscr{R}_{core} = \frac{25\pi \times 10^{-2} - 1 \times 10^{-3}}{800 \times 4\pi \times 10^{-7} \times 10^{-3}} = \frac{25\pi - 0.1}{32\pi \times 10^{-6}}$$

The magnetomotive force for the core is given as

$$\mathscr{F}_{core} = \phi \times \mathscr{R}_{core} = 10^{-3} \times \frac{25\pi - 0.1}{32\pi \times 10^{-6}} = \frac{78.44}{32\pi \times 10^{-3}}$$

$$= 780.2 \approx 780 \ \text{AT}$$

The magnetomotive force for the magnetic circuit is given as

$$\mathscr{F} = \mathscr{F}_{air} + \mathscr{F}_{core} = 795.7 + 780 = 1575.7$$

The exciting current is

$$i = \frac{\mathscr{F}}{N} = \frac{1575.7}{250} = 6.30 \ \text{A}$$

The self inductance of the coil with the air gap is

$$L = \frac{N \times \phi}{i} = \frac{250 \times 10^{-3}}{6.3} = 0.0397$$

The energy stored in the coil is given as

$$w = \frac{1}{2}L \times i^2 = \frac{1}{2} \times 0.0397 \times (6.3)^2 = 0.7878 \ \text{J}$$

3.15 ATTRACTING FORCE OF ELECTROMAGNETS

The stored energy in the magnetic field is capable of doing mechanical work. When a magnetic substance is held
near an electromagnet and the coil of the electromagnet
is excited by a current, it exerts a force of attraction on the
substance. This force of attraction pulls the substance
and draws it nearer to the pole face of the electromagnet.
When this force causes a displacement of the substance,
a mechanical work is done at the expense of the stored
energy of the field.

An electromagnet with a pole face of A m^2 and a
magnetic substance placed at a distance l metre from the
pole face is shown in Fig. 3.29. The magnetic circuit of
the electromagnet is completed through the ferromagnetic medium of the electromagnet, its iron frame, and
the magnetic material placed under the electromagnet.
The only non-magnetic medium is the air gap of l metre.

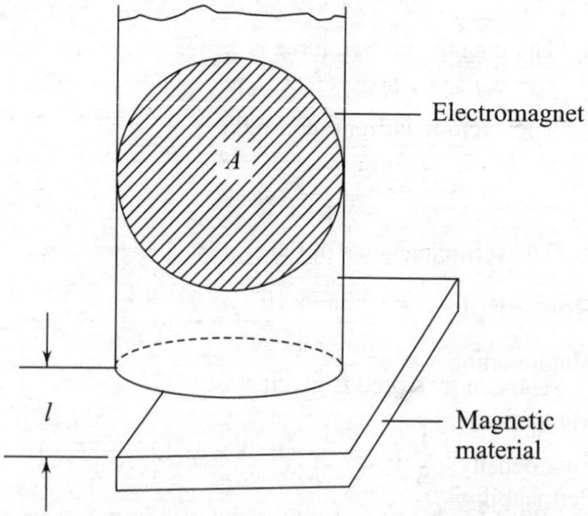

Now, as the field intensity in a ferromagnetic medium

Fig. 3.29 Attractive force of an electromagnet

is many times less than the field intensity in an air gap, it may be assumed that the whole stored energy of the
magnetic field is concentrated in the air gap only. So the work done in attracting the magnetic substance is equal
to the stored energy of the air gap volume. From Eq. (3.75), the stored energy is given by

$$W = \frac{1}{2}BHV = \frac{1}{2}BH \ (A \times l) \ \text{joule} \tag{3.84}$$

The work done in attracting the magnetic material is equal to the energy stored in the magnetic field. Then
the force of attraction F is obtained as

$$F = \frac{\text{work done}}{l} = \frac{BHAl}{2l} = \frac{1}{2} BHA \text{ newton} \tag{3.85}$$

But $H = B/\mu_0$. Therefore,

$$F = \frac{B^2 A}{2\mu_0} \text{ newton} \tag{3.86}$$

and Eq. (3.84) becomes

$$W = \frac{B^2 (A \times l)}{2\mu_0} \text{ joule} \tag{3.87}$$

Example 3.13 A smooth core armature working in a two-pole field magnet has an air gap (from iron to iron) of 1 cm. The area of the surface of each pole is 1000 cm². The flux from each pole is 0.1 Wb. Find (a) the mechanical force exerted by each pole on the armature; (b) the energy (in joules) stored in the two air gaps.

Solution   (a) Flux per pole, $\phi = 0.1 = 10^{-1}$ Wb
Cross-sectional area of the pole, $A = 1000 \times 10^{-4} = 10^{-1}$ m²
Flux density,

$$B = \frac{\phi}{A} = \frac{10^{-1}}{10^{-1}} = 1 \text{ Wb/m}^2$$

Then force is given by

$$F = \frac{B^2 A}{2\mu_0} = \frac{1^2 \times 10^{-1}}{2 \times 4\pi \times 10^{-7}} = 39,788.7 \text{ N}$$

(b) The length of the air gap is $l = 10^{-2}$ m. Using Eq. (3.87) the stored energy in the two air gaps can be obtained as

$$W = \frac{B^2 (A \times l)}{2\mu_0} \times (\text{number of gaps})$$

$$= \frac{1^2 \times 10^{-1} \times 10^{-2}}{2 \times 4\pi \times 10^{-7}} \times 2 = 795.77 \text{ J}$$

Recapitulation

Magnetomotive force, $F = NI$ AT

Magnetic field strength, $H = \dfrac{F}{l} = \dfrac{NI}{l}$ AT/m

Flux density, $B = \mu H$ Wb/m

Permeability, $\mu = \mu_0 \mu_r$

Permeability of free space, $\mu_0 = 4\pi \times 10^7$ Wb/ATm

Flux, $\phi = \dfrac{F}{R}$ weber

Reluctance, $R = \dfrac{l}{\mu \times A}$ AT/Wb

Self inductance of a coil,

$$L = \frac{N\phi}{I} = \frac{N^2 \mu\, l}{A} = \frac{N^2}{R} = P \times N^2 \text{ henry}$$

Mutual inductance for a bilateral circuit,

$$M_{12} = \frac{\phi_{12} N_2}{i_2} = \frac{N_1 N_2}{R} \text{ henry}$$

Coupling coefficient of a mutual inductor, $k = \dfrac{M}{\sqrt{L_1 L_2}}$

Effective inductance of a mutual inductor,

$$L = L_1 + L_2 \pm 2M \text{ henry}$$

Energy stored in magnetic field, $W = \dfrac{1}{2}\phi^2 R$ joule

Energy stored in excited and mutually coupled coils,

$$W = \frac{1}{2} L_1 i_1^2 + \frac{1}{2} L_2 i_2^2 + M i_1 i_2 \text{ joule}$$

Attracting force of electromagnets, $F = \dfrac{B^2 A}{2\mu_0}$ Newton

Assessment Questions

1. (a) What is a magnetic circuit? Explain the role of magnetic materials in a magnetic circuit.
   (b) Explain the significance of permeability and distinguish between (i) permeability of free space and (ii) relative permeability.
2. (a) Define and explain (i) magnetic field strength and (ii) magnetomotive force.
   (b) State and explain Ampere's circuital law.
3. (a) Explain how magnetism occurs in materials.
   (b) Classify magnetic materials based on their relative permeability. Give at least two examples of materials in each category and state the order of magnitudes of relative permeability in each case.
4. Explain the term reluctance and identify the factors on which its magnitude depends.
5. (a) Develop the analogy between a magnetic circuit and an electric circuit.
   (b) State and explain the equivalent KCL and KVL for magnetic circuits.
6. (a) Discuss the various stages of a *B-H* curve of a magnetic material when it is subjected to a continuous increasing magnetizing force *H*.
   (b) Explain how a hysteresis loop is obtained. Identify and explain the significance of the critical points on the curve.
7. (a) Prove that the area within the loop of a *B-H* curve represents the hysteresis loss.
   (b) Derive the relationship for determining the hysteresis loss of a magnetic material. Write down Steinmetz formula for calculating the hysteresis loss and give the ranges of the empirical constants.
   (c) Write down the empirical formula for calculating the eddy current loss.
8. (a) Define self-inductance and mutual inductance.
   (b) Derive the formula in terms of self-inductance and mutual inductance for determining the coefficient of coupling.
   (c) Discuss the significance of coefficient of coupling.
9. (a) State the dot convention and explain why it is necessary.
   (b) Apply the dot convention in writing down the induced voltage equations for the circuit shown in Fig. 3.30.

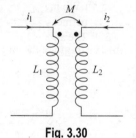

Fig. 3.30

Problems

Note: Data of *B-H*, when not given in the problem, should be taken from Fig. 3.12.

3.1 For the magnetic circuit shown in Fig. 3.31 draw the equivalent electrical analogue and write the equations to obtain the flux flows in the air gaps.

3.2 A cast steel ring has a mean diameter of 200 mm and a cross-sectional area of 250 mm². Calculate: (a) the mmf to produce a flux of 300 μWb and (b) the corresponding values of the reluctance of the ring and of the relative permeability.

$$[(a)\ 785\ \text{AT/m; (b)}\ 2.617 \times 10^6\ \text{AT/Wb, 764}]$$

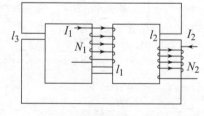

Fig. 3.31

3.3 (a) A circular wooden ring of cross section 600 sq mm and mean circumference 500 mm is wound with a coil of 250 turns. If the coil carries a current of 4 A, compute (i) magnetic field intensity, (ii) flux density, and (iii) flux in the ring. (b) If the wooden ring is replaced by an iron ring of relative permeability 600, calculate the new values of magnetic field intensity, flux density, and flux in the ring.

$$[(a)\ 2000\ \text{A/m, 0.0025 T, 1.51 } \mu\text{Wb, (b)}\ 2000\ \text{A/m, 1.51 T, 904.78 } \mu\text{Wb}]$$

3.4 A steel, magnetic circuit has a uniform cross-sectional area of 4 cm^2 and a length of 30 cm. A coil of 150 turns is wound uniformly over the magnetic circuit. When the current in the coil is 1.5 A, the total flux is 0.35 mWb and when the current is 5 A, the total flux is 0.65 mWb. For each value of current, calculate: (a) the magnetic field strength and (b) the relative permeability of the steel.

[(a) 750 AT/m, 2500 AT/m; (b) 928, 538]

3.5 A mild steel ring has a mean circumference of 600 mm and a uniform cross-sectional area of 250 mm^2. Calculate the mmf required to produce a flux of 450 μWb. An air gap, 1.0 mm in length, is now cut in the ring. Determine the flux produced if the mmf remains constant. Assume the relative permeability of the mild steel to remain constant at 220. [3906 AT/m, 329 μWb]

3.6 Schematic representation of a synchronous machine is shown in Fig. 3.32. The length of the air gap between the rotor and the stator is 0.75 cm each and the rotor pole face area is 2500 cm^2. The winding on the rotor has 1500 turns and carries a current of 12 A. If both the rotor and the stator are assumed to have negligible reluctance and there is no fringing, (a) draw the equivalent magnetic circuit and calculate (b) mmf, (c) reluctance of each air gap, (d) magnetic flux, and (e) magnetic flux density in each air gap.

[(b) 18,000 AT, (c) 2.39 × 10^4 AT/Wb, (d) 0.38 Wb, (e) 1.51 T]

3.7 A steel ring has a mean diameter of 20 cm, a cross section of 25 cm^2, and a radial air gap of 0.5 mm cut in it. The ring is uniformly wound with 2000 turns of insulated wire and a magnetizing current of 1 A produces a flux of 1.25 mWb in the air gap. Neglecting the effect of magnetic leakage and fringing, calculate: (a) the reluctance of the magnetic circuit and (b) the relative permeability of the steel.

[(a) 1.6 × 10^6 AT/Wb, (b) 138]

3.8 A magnetic core made of cast steel is shown in Fig. 3.33. The core is symmetrical about the Y-axis and its central limb carries a coil of 600 turns. Compute the magnitude of the exciting current to produce a flux of 25 μWb in either of the side limbs of the core. Assume a square cross section of 1.0 cm × 1.0 cm and a relative permeability of 200 for cast steel. [7.0 A]

3.9 The air gap in a magnetic circuit is 1.5 mm long and 2500 mm^2 in cross section. Calculate: (a) the reluctance of the air gap and (b) the mmf to send a flux of 800 μWb across the air gap.

[(a) 0.477 × 10^6 AT/Wb, (b) 382 AT]

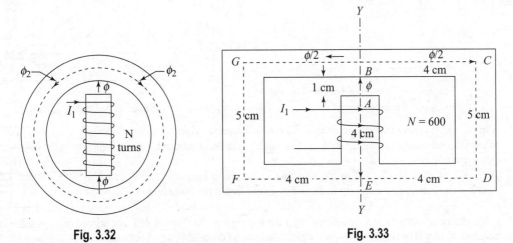

Fig. 3.32   **Fig. 3.33**

3.10 A magnetic circuit made from ferromagnetic material has a core length of l_c metre and an air gap of length l_a metre. Derive an expression for the ratio of the core to air gap reluctance. Therefrom, show that in the presence of an air gap, the magnetic core can be assumed to be behaving like a short circuit even though $l_c \gg l_a$.

3.11 A magnetic circuit consists of a cast steel yoke which has a cross-sectional area of 300 mm^2 and a mean length of 160 mm. There are two air gaps, each 0.2 mm long. Calculate: (a) the mmf required to produce a flux of

0.06 mWb in the air gaps and (b) the value of the relative permeability of cast steel at this flux density. The magnetization curve for cast steel is given by the following

B (T)	0.1	0.2	0.3	0.4	
H (A/m)	170	300	380	460	[(a) 112 AT, (b) 530]

3.12 An electromagnet has a magnetic circuit comprising three parts in series: part A of length 100 mm and cross-sectional area 80 mm^2; part B of length 80 mm and cross-sectional area 70 mm^2; part C, an air gap of length 1 mm and cross-sectional area 80 mm^2. Parts A and B are made of a material having magnetic characteristics given by the following table:

H (A/m)	100	210	340	400	500	560	800	1280	1500	1800
B (T)	0.2	0.4	0.6	0.7	0.8	0.85	1.0	1.15	1.2	1.25

Determine the current necessary in a coil of 5000 turns wound on part B to produce a flux density of 0.8T in the air gap. Magnetic leakage may be neglected. [145.4 mA]

3.13 A magnetic circuit made of transformer sheet steel is arranged as shown in Fig. 3.34. The centre limb has a cross-sectional area of 1000 mm^2 and each of the side limbs has a cross-sectional area of 600 mm^2. Calculate the mmf required to produce a flux of 1.2 mWb in the centre limb. Assume the magnetic leakage to be negligible. [1841 AT]

3.14 A magnetic core made of cast steel has the dimensions shown in Fig. 3.35. There is an air gap 1 mm long in one side limb and a coil of 400 turns is wound on the centre limb. The cross-sectional area of the centre limb is 1200 mm^2 and that of each side limb is 800 mm^2. Calculate the exciting current required to produce a flux of 600 μWb in the air gap. Neglect any magnetic leakage and fringing. [1.34 A]

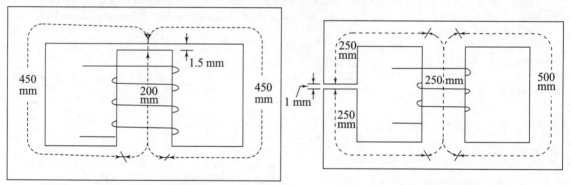

Fig. 3.34 Magnetic circuit   **Fig. 3.35** Magnetic circuit

3.15 A toroid of 300 mm mean diameter and 650 mm^2 circular cross section is wound uniformly with 1200 turns of wire. Calculate: (a) the inductance of the toroid; (b) the emf induced when the current in the coil increases at the rate of 250 A/sec. [(a) 1.248 mH, (b) 0.312 V]

3.16 Two coils of 1500 and 200 turns are wound on a common magnetic circuit of reluctance 250,000 AT/Wb. Find mutual inductance when (a) leakage flux is neglected (b) leakage flux is 10% of total flux. [(a) 1.2 H, (b) 1.08 H]

3.17 Calculate the inductance of a ring-shaped coil having a mean diameter of 400 mm wound on a wooden core of diameter 40 mm. The winding is evenly wound and has 1000 turns. If the wooden core is replaced by a ferromagnetic core that has a relative permeability of 600 when the current is 5 A, calculate the new value of inductance. [1.257 mH, 0.754 H]

3.18 A large electromagnet is wound with 2000 turns. A current of 1 A in this winding produces a flux of 0.01 Wb through the coil. Calculate the inductance of the electromagnet. If the current in the coil is reduced from 1 A to zero in 0.01 sec, what average emf will be induced in the coil? Assume that there is no residual flux. [20 H, 2000 V]

3.19 If two coils have a mutual inductance of 300 µH, calculate the emf induced in one coil when the current in the other coil varies at a rate of 4000 A/sec. [1.2 V]

3.20 Two similar coils have a coupling coefficient of 0.2. When they are connected in series cumulatively, the total inductance is 120 mH. Calculate: (a) the self inductance of each coil, (b) the total inductance when the coils are connected in series differentially, and (c) the total magnetic energy due to a current of 3 A when the coils are connected in series (i) cumulatively and (ii) differentially.   [(a) 50 mH; (b) 80 mH; (c) 540 mJ, 360 mJ]

3.21 Determine the force (in kilogram) necessary to separate two surfaces of contact area of 200 cm^2 when the flux density normal to the surface is 1.2 Wb/m^2. [1.146×10^4 N]

Additional Examples

AE 3.1   A 1.2 metre long air cored solenoid has a diameter of 1.2 cm. If the inductance of the solenoid is 0.2 mH, calculate the number of turns in the coil.

Solution   Cross sectional area $A = \pi \times \left(\dfrac{1.2}{2 \times 100} \right)^2 = 1.131 \times 10^{-4}$ m^2

Number of turns is calculated by using Eq. (3.54) as follows

$$N = \sqrt{\frac{0.2 \times 10^{-3} \times 1.2}{4\pi \times 10^{-7} \times 1.131 \times 10^{-4}}} = 1300 \text{ turns}$$

AE 3.2   Two coils having 750 and 1200 turns respectively are wound on a common non-magnetic core. The leakage flux and mutual flux, due to a current of 7.5 A in coil 1, is 0.25 mWb and 0.75 mWb respectively. Calculate (a) self inductance of the coils, (b) mutual inductance and (c) coefficient of coupling.

Solution   From the data it is seen that $\Phi_{11} = 0.25$ mWb and $\Phi_{12} = 0.75$ mWb. Therefore, total flux linking coil 1 is $\Phi_1 = \Phi_{11} + \Phi_{12} = 1.0$ mWb.

(a) From Eq. (3.50), the self inductance of the coil is computed as follows

$$L_1 = \frac{750 \times 1.0 \times 10^{-3}}{7.5} = 100 \text{ mH}$$

To compute the mutual inductance, Eq. (3.58) is employed in the following manner,

$$M = M_{12} = \frac{1200 \times 0.75 \times 10^{-3}}{7.5} = 120 \text{ mH}$$

Coefficient of coupling is calculated as

$$k = \frac{\Phi_{12}}{\Phi_1} = \frac{0.75}{1.0} = 0.75$$

Using Eq. (3.68) leads to the self inductance L_2. Thus,

$$L_2 = \frac{M^2}{k^2 L_1} = \frac{\left(120 \times 10^{-3} \right)^2}{(0.75)^2 \times \left(100 \times 10^{-3} \right)} = 256 \text{ mH}$$

AE 3.3   Two inductors of L_1 and L_2 H and mutual inductance of M H are connected in series as shown in Fig. AE 3.1. Apply dot convention and derive expressions for the equivalent inductance.

Solution   Assume that the equivalent inductance of the series combination is L_{eq} and a voltage v is applied across the terminals A-B. As per the dot convention M is aiding in Fig. AE 3.1(a) and it is opposing in Fig. AE 3.1(b).

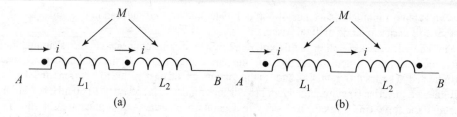

(a)   (b)

Fig. AE 3.1

When M is positive, $L_{eq} \dfrac{di}{dt} = L_1 \dfrac{di}{dt} + M \dfrac{di}{dt} + L_2 \dfrac{di}{dt} + M \dfrac{di}{dt}$ or $L_{eq} = (L_1 + L_2 + 2M)$ H

When M is negative, $L_{eq} \dfrac{di}{dt} = L_1 \dfrac{di}{dt} - M \dfrac{di}{dt} + L_2 \dfrac{di}{dt} - M \dfrac{di}{dt}$ or $L_{eq} = (L_1 + L_2 - 2M)$ H

AE 3.4 Determine the equivalent inductance of the three coupled inductors shown in Fig. AE 3.2. Assume the following mutual inductances between the coils: $M_{12} = 1$ H, $M_{23} = 2$ H and $M_{31} = 3$ H.

Solution As per the dot convention, $M_{12} = 1$ H, $M_{23} = -2$ H, and $M_{31} = -3$ H. Making use of the expressions in example AE 3.3, the equivalent inductance is given by

$$L_{eq} = (4 + 5 + 6 + 2 \times 1 - 2 \times 2 - 2 \times 3) = 7 \ \text{H}$$

Fig. AE 3.2

AE 3.5 Two mutually coupled inductors are connected in parallel across a voltage v. If the mutual inductance between the coils is M H, derive an expression for the equivalent inductance of the parallel combination. Assume that mutual flux is aiding the inductors.

Solution The connection diagram is shown in Fig. AE 3.3.

Assume the equivalent inductance to be L_{eq} H. Using Eq. (1.31) gives

$$v = L_{eq} \dfrac{di}{dt} \tag{AE 3.5.1}$$

Also it is seen that $di/dt = di_1/dt + di_2/dt$ (AE 3.5.2)

The voltage across each of the inductors is written as

$$v = L_1 \dfrac{di_1}{dt} + M \dfrac{di_2}{dt} = M \dfrac{di_1}{dt} + L_2 \dfrac{di_2}{dt} \tag{AE 3.5.3}$$

or $(L_1 - M) \dfrac{di_1}{dt} = (L_2 - M) \dfrac{di_2}{dt}$ or $\dfrac{di_1}{dt} = \dfrac{(L_2 - M)}{(L_1 - M)} \dfrac{di_2}{dt}$ (AE 3.5.4)

Substitution of Eq. (AE 3.5.4) in Eq. (AE 3.5.2) eliminates di_1/dt as follows

$$\dfrac{di}{dt} = \left[\dfrac{(L_2 - M)}{(L_1 - M)} + 1 \right] \dfrac{di_2}{dt} \tag{AE 3.5.5}$$

Equating Eqs (AE 3.5.1) and (AE 3.5.2) it is seen that

Fig. AE 3.3

$$L_{eq}\frac{di}{dt} = L_1\frac{di_1}{dt} + M\frac{di_2}{dt}$$

Substituting from Eq. (AE 3.5.3) in the above expression results in

$$\frac{di}{dt} = \frac{1}{L_{eq}}\left[L_1\frac{(L_2 - M)}{(L_1 - M)} + M\right]\frac{di_2}{dt}$$ (AE 3.5.6)

Equating Eqs (AE 3.5.5) and (AE 3.5.6) and simplifying gives

$$L_{eq} = \frac{L_1 L_2 - M^2}{L_1 + L_2 - 2M} \text{ H}$$ (AE 3.5.7)

Substituting $M = -M$ in Eq. (AE 3.5.7) leads to the equivalent inductance when the mutual inductance is opposing the inductors, that is,

$$L_{eq} = \frac{L_1 L_2 - M^2}{L_1 + L_2 + 2M} \text{ H}$$

AE 3.6   Two coils have inductances of $L_1 = 40$ mH, and $L_2 = 65$ mH. If the coefficient of coupling $k = 0.45$, compute the maximum total inductance when the coils are connected in (a) series and (b) parallel.

Solution   From Eq. (3.68) the mutual inductance $M = 0.45\sqrt{40 \times 65} = 22.946$ mH

(a) In series connection, maximum inductance occurs when M is positive. Thus,

$$L_T = 40 + 65 + 2 \times 22.946 = 150.891 \text{ mH}$$

(b) In parallel connection, maximum inductance occurs when M is negative. Thus,

$$L_T = \frac{40 \times 65 - (22.946)^2}{40 + 65 - 2 \times 22.946} = 35.079 \text{ mH}$$

Objective Type Questions

1. Electromagnetic field is employed as a medium for the operation of
   - (i) motors
   - (ii) generators
   - (iii) transformers
   - (iv) all of these

2. When a conductor is moved inside a uniform magnetic field,
   - (i) it experiences a force
   - (ii) an emf is induced
   - (iii) a current flows through it
   - (iv) a flux is produced

3. The unit of magnetic field intensity is
   - (i) ampere turns
   - (ii) weber
   - (iii) ampere per metre
   - (iv) weber/metre^2

4. In a magnetic circuit
   - (i) flux is analogous to current
   - (ii) reluctance is analogous to resistance
   - (iii) magnetomotive force is analogous to potential difference
   - (iv) all of these

5. Which of the following are not representative of the magnetic field intensity?
   - (i) H
   - (ii) E
   - (iii) magnetomotive force per metre
   - (iv) ampere turns per metre

6. The circuital law which states that an electric current flowing through a number of turns of a conductor produces a magnetic field in its vicinity was enunciated by
   - (i) Faraday
   - (ii) Lenz
   - (iii) Ohm
   - (iv) none of these

7. When a conductor of length l m is moved perpendicular to a magnetic field of intensity B tesla, at a velocity of v m/sec, the magnitude of the induced voltage is given by

(i) Blv (ii) BIl

(iii) $Blv \sin \theta$ (iv) none of these

8. The permeability of a non-magnetic medium is
   (i) 6×10^{-12} (ii) $4\pi \times 10^{-7}$
   (iii) 8.85×10^{-12} (iv) 1.0

9. The relative permeability for air is
   (i) more than 10,000
   (ii) between 10,000 and 4000
   (iii) equal to unity
   (iv) none of these

10. When compared with an electric circuit which of the following parameters in a magnetic circuit are analogous?
    (i) current and flux
    (ii) electric field strength and magnetic field strength
    (iii) voltage drop and mmf drop
    (iv) all of these

11. Magnetic circuits obey
    (i) Kirchoff's laws
    (ii) Thevenin's theorem
    (iii) Norton's theorem
    (iv) none of these

12. Saturation of a magnetic circuit implies that with an increase in field intensity the flux density
    (i) decreases
    (ii) increases proportionally
    (iii) increases marginally
    (iv) none of these

13. 'Retentivity' and 'coercive force' are used in connection with
    (i) hysteresis loop   (ii) loop current
    (iii) current cycle   (iv) none of these

14. A magnetic circuit consists of an iron core and an air gap. Most of the energy is stored in the air gap because the air gap has
    (i) a much higher reluctance
    (ii) a much lower reluctance
    (iii) zero reluctance
    (iv) none of these

15. When two coils are placed close together and there is no flux leakage, the coupling coefficient k is
    (i) zero   (ii) unity
    (iii) infinity   (iv) none of these

16. The self inductance of a coil is dependent on
    (i) the number of turns in the coil only
    (ii) the geometry of the core only
    (iii) the number of turns in the coil and geometry of the core
    (iv) none of these

17. The dot convention is used to define the sign of
    (i) self inductance of the coils
    (ii) mutual inductance of the coils
    (iii) direction of current flow in the coils
    (iv) none of these

18. The energy stored in a linear magnetic circuit is given by
    (i) $\frac{1}{2} Li^2$ (ii) $\frac{1}{2} BHV$
    (iii) $\frac{1}{2} \phi^2 \times \mathcal{R}$ (iv) all of these

19. Two coils having self inductances of L_1 and L_2 henry are differentially coupled. If the mutual inductance between the coils is M, the total inductance of the coils is
    (i) $L_1 + L_2 + 2M$ (ii) $L_1 + L_2 - 2M$
    (iii) $L_1 - L_2 + 2M$ (iv) $L_1 - L_2 - 2M$

20. The force of attraction in an electromagnet is due largely to
    (i) energy stored in the air gap between the magnet and the non-magnetic material
    (ii) energy stored in the ferromagnetic material
    (iii) energy stored in the non-magnetic material
    (iv) none of these

21. What is the effect of fringing on the area of an air gap?
    (i) Increases it   (ii) Decreases it
    (iii) Keeps it same   (iv) None of these

22. The mutual inductance between two closely coupled coils is 0.5 H. The coils are rewound to reduce the number of turns in one coil 1/3 and to increase in the other by three times. Which of the following represents the mutual inductance of the coils?
    (i) 0.17 H   (ii) 0.5 H
    (iii) 0.75 H   (iv) 1.5 H

23. How should the current in a magnetic circuit be changed to produce the same flux, if the reluctance is halved?
    (i) Doubled   (ii) No change
    (iii) Halved   (iv) One-fourth
24. The magnitude of the coefficient of coupling is dependent on
    (i) L_1
    (ii) L_2

(iii) Proximity of the coils
(iv) All of these
25. Which of the following represents the energy stored in two mutually coupled coils?
    (i) $(L_1 i_1^2 + L_2 i_2^2 + M i_1 i_2)/2$
    (ii) $(L_1 i_1^2 - L_2 i_2^2 + M i_1 i_2)/2$
    (iii) $(L_1 i_1^2 + L_2 i_2^2 - M i_1 i_2)/2$
    (iv) None of these

Answers

1. (iv)	2. (ii)	3. (iii)	4. (iv)	5. (ii)	6. (iv)	7. (i)
8. (ii)	9. (iii)	10. (iv)	11. (i)	12. (iii)	13. (i)	14. (i)
15. (ii)	16. (iii)	17. (ii)	18. (iv)	19. (ii)	20. (i)	21. (i)
22. (ii)	23. (iii)	24. (iv)	25. (i)			

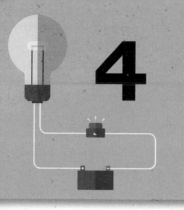

4 ALTERNATING QUANTITIES

Learning Objectives

This chapter will enable the reader to

- Grasp the importance and origin of sinusoidal voltages and currents
- Understand the process of generation of ac voltages
- Define a periodic function, waveform, cycle, time period, and frequency
- Correlate speed with frequency
- Compute average and rms values of different types of waveforms
- Represent sinusoidal currents and voltages as phasor quantities
- Perform addition/subtraction, multiplication/division of phasor quantities by representing them in rectangular, trigo-nometric, and polar forms
- Analyse series, parallel, and series-parallel RL, RC, and RLC circuits when subjected to alternating voltage and current sources
- Draw phasor diagrams of currents and voltages of all types of circuits
- Grasp the concept of power and its exchange in reactive circuits and perform power calculations for different types of circuits
- Understand series and parallel resonance and compute Q-factor and bandwidth
- Proficiently apply Y-Δ and Δ-Y transformations and network theorems

4.1 INTRODUCTION

In the preceding chapter on network analysis only dc sources have been considered. Another type of sources, alternating (ac) sources, in which voltage or current varies in a repetitive manner as a function of time is considered in this chapter. Though the term 'alternating quantity' has a much broader meaning, it is generally used in a restricted sense to mean a sinusoidal quantity. Usually, alternating current (referred to as ac current) or alternating voltage (referred to as ac voltage) means a sinusoidally varying current or voltage. Almost all electrical power supply systems involve sinusoidal ac current, which is derived from sinusoidal ac voltages.

The electric power industry began by the end of the nineteenth century. Two competitive systems emerged in generation and transmission, namely, dc and ac systems. The dc system was strongly pursued by Thomas A. Edison (1876) as it was simple to visualize, and only real quantities such as voltage, current, and resistance were involved. The ac system was transformed into a practical scheme with Sir Nikola Tesla's invention of transformer (1884). George Westinghouse implemented a practical ac system in 1886. The ac system was widely accepted due to the technical feasibility of stepping up/down the ac voltage by transformers, which solved the problem of long-distance transmission of power.

Today the ac supply system is universally used. In order to understand ac systems, it is necessary to be familiar with the generation of sinusoidally varying ac voltage, which is discussed in this chapter. For the analysis of ac networks, new terms related to alternating waveforms are used. Some of these terms are defined in this chapter. The concepts of average and root mean square values are introduced. The representation of a sinusoidal quantity by a phasor, which simplifies the analysis of ac circuits, is discussed. The analysis of series and parallel ac circuits containing resistances, inductances, and capacitances and the analytical techniques that greatly simplify the solution of the circuit with sinusoidal excitation are presented in detail. The phenomenon of resonance in series as well as parallel ac circuits is also discussed in this chapter. Star–delta transformation and network theorems applied to ac circuits are given.

4.2 GENERATION OF AC VOLTAGE

The generation of ac voltage may be explained by considering a rectangular coil AA' whose conductor length is l metre and width is b metre. The coil is mounted on a spindle DD' as shown in Fig. 4.1. The coil is rotated in the anticlockwise direction with angular velocity ω rad/sec in a uniform magnetic field having a flux density of B Wb/m^2. The ends of the coil are connected to two slip rings S and S' which are rigidly fixed to but insulated from the shaft DD' which is used to rotate the coil. The slip rings maintain a sliding contact with carbon brushes C and C', which are connected to a resistor R.

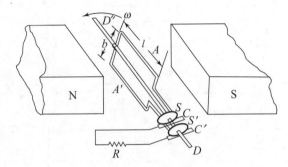

Fig. 4.1 Coil rotating in a uniform permanent magnetic field

The principle of the generation of ac voltage in coil AA' can be explained with the help of Fig. 4.2, which shows different positions of this coil in one complete rotation in the anticlockwise direction in a magnetic field. In position I, shown in Fig. 4.2(a), the plane of coil AA' is vertical, the sides A and A' of the coil are moving parallel to the magnetic flux and thus do not cut any flux; therefore, the voltage generated in the coil is zero. As the coil rotates to position II, shown in Fig. 4.2(b), the conductor starts cutting the flux and a voltage is generated in the two sides of the coil. Both sides of the coil, A and A', move in the counter-clockwise direction. Applying Fleming's right hand rule to the coil sides A and A', it is found that the direction of emf generated in side A is opposite to that generated in side A', and is shown as dot and cross notation. However, around the coil loop the two emfs aid each other in the loop and the total emf that appears at the brushes is equal to twice the magnitude of the emf generated by each coil side. As the coil rotates to position III, shown in Fig. 4.2(c), the maximum flux is being cut and the emf induced in the coil is maximum. As the coil further rotates from position III to position IV, shown in Fig. 4.2(d), the voltage generated reduces from maximum to zero since the flux being cut reduces from maximum to zero. It may be noted that in Fig. 4.2(d) the sides of the coil have undergone a rotation of 180° or half a cycle. As the coil takes position V, shown in Fig. 4.2(e), the voltage generated is equal in magnitude to that of the voltage generated in position II [Fig. 4.2(b)], but has the opposite polarity since sides A and A' of the coil have now interchanged positions. The same observation holds good when the coil takes position VI, shown in Fig. 4.2(f). Thus, it is seen that as the coil rotates from 0° to 180°, the voltage generated increases from zero to maximum to zero. When the coil rotates through 180° to 360°, the voltage generated again increases from zero to maximum to zero, but is of opposite polarity to that generated when the coil was moving through 0° to 180°. The brush in touch with slip ring S is supposed to be positive with respect to S' to start with and S' is chosen as the reference.

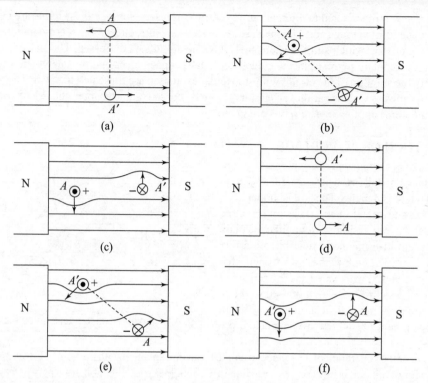

Fig. 4.2 Generation of emf in a rotating coil (a) Position I: Zero voltage, (b) Position II, (c) Position III: Peak positive voltage, (d) Position IV: Zero voltage, (e) Position V, (f) Position VI: Peak negative voltage

Let it be assumed that the coil has been rotating with an angular velocity ω rad/sec. Then in t seconds the coil rotates through angle $\omega t = \theta$ rad from the vertical position, the position of zero emf shown in Fig. 4.3(a), and the coil sides A and A' have a tangential velocity of u m/sec. The horizontal component of the velocity tangential, H, can be found out from Fig. 4.3(b) as

$$H = u \sin\theta = u \sin\omega t$$

The voltage induced in each of the conductor sides A and A' is due to the component of the velocity perpendicular to the magnetic field. From Eq. (1.52) the emf induced in one side of the coil is

$$Blu \sin\theta \text{ volt}$$

and the total emf induced in the coil is

$$e = 2Blu \sin\theta \text{ volt} \tag{4.1}$$

From Eq. (4.1) it may be stated that the generated emf is proportional to $\sin\theta$. It follows that when $\theta = 90°$ the generated emf is at its maximum value $E_m = 2Blu$ vot.

Thus, Eq. (4.1) may be written as

$$e = E_m \sin\theta = E_m \sin\omega t \tag{4.2}$$

where $E_m = 2Blu$.

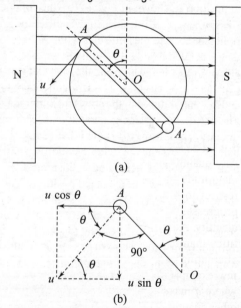

Fig. 4.3 Computation of the magnitude of the generated voltage

If n is the speed of rotation of the coil in revolutions per second, then

$$u = \pi bn \text{ m/sec}$$

and

$$E_m = 2Bl \times \pi bn \text{ volt}$$

$$= 2\pi BAn \text{ volt} \qquad (4.3)$$

where $A = l \times b$ is the area of the coil.

For a coil having T number of turns in series, each turn having an area of A m^2, the maximum value of the generated emf is given by

$$E_m = 2\pi BAnT \text{ volt} \qquad (4.4)$$

and the instantaneous value of the generated voltage is given by

$$e(t) = 2\pi BAnT \sin \theta \text{ volt}$$

$$= E_m \sin \theta = E_m \sin \omega t \qquad (4.5)$$

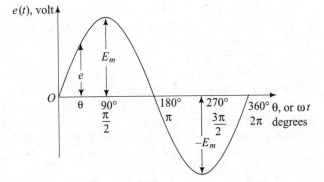

Fig. 4.4 Waveform of generated voltage

The instantaneous value of the generated voltage can be represented as a function of θ as shown in Fig. 4.4, where $e(t)$ represents the value of voltage when the coil has rotated through an angle θ from the position of zero emf and E_m is the maximum value of induced voltage.

Example 4.1 A square coil of 20 cm side and with 500 turns is rotated at a uniform speed of 1000 rev/min. about an axis at right angles to a uniform magnetic field having a flux density of 0.2 Wb/m^2. Calculate the instantaneous value of the induced electromotive force when the plane of the coil is (a) at right angles to the field, (b) at 60° to the field, and (c) in the plane of the field.

Solution The value of the instantaneous value of electromotive force induced can be calculated using Eq. (4.5). Data provided are $B = 0.2$ Wb/m^2, $A = 4 \times 10^{-2}$ m^2, $n = 1000/60$ rev/sec, and $T = 500$.

(a) When the coil is at right angles to the field, that is, $\theta = 0°$,

$$e(t) = 0 \text{ V}$$

(b) When the coil is at 30° to the field, that is, $\theta = 60°$,

$$e(t) = 2\pi \times 0.2 \times 4 \times 10^{-2} \times \frac{1000}{60} \times 500 \times \sin 60° = 362.76 \text{ V}$$

(c) When the coil is in the plane of the field, that is, $\theta = 90°$,

$$e(t) = 2\pi \times 0.2 \times 4 \times 10^{-2} \times \frac{1000}{60} \times 500 \times \sin 90° = 418.87 \text{ V}$$

4.3 WAVEFORMS AND BASIC DEFINITIONS

Any quantity, voltage or current, whose magnitude has alternating values with respect to time repetitively as the time progresses is said to be an alternating quantity. A system in which such quantities manifest themselves is called an alternating system. Figure 4.5 shows a few examples of alternating waveforms, where q may be voltage, current, or any other quantity.

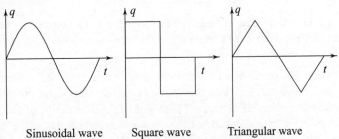

Sinusoidal wave   Square wave   Triangular wave

Fig. 4.5 Examples of alternating waveforms

The figure shows that in all the waveforms the quantities are varying in time and the cycle of variation repeats. Thus, a periodic quantity that has identical positive and negative half-cycles is known as an alternating quantity.

Though the term 'alternating quantity' has a much broader meaning, it is generally used in a restricted sense to mean a sinusoidal quantity. Usually alternating current (ac current) or alternating voltage (ac voltage) means a sinusoidally varying current or voltage. For such a function the sum or the difference, the derivative or the integral, are all sinusoidal functions of the same frequency. Thus, if a sinusoidal excitation is applied to a linear network, it is found that the response under steady state would be again a sinusoid of the same frequency at all the points in the network. This is observed only for sinusoidal excitation and is not found with any other form of excitation. Almost all electrical power supply systems involve sinusoidal ac currents, which is derived from sinusoidal ac voltages.

In order to be able to analyse the effect of alternating voltages and currents in electrical networks, it is necessary to introduce and define the various terms commonly used to study their behaviour. Some of the terms are introduced in this section.

Periodic function A function $f(t)$ is said to be a periodic function in time t with a period of T seconds if the same pattern of values of the function repeats after every T seconds. Mathematically, for the function which satisfies the condition

$$f(t) = f(t + T) \qquad (4.6)$$

for all values of t is said to be periodic with a period of T seconds.

Waveform The shape of the curve obtained when the instantaneous values of a periodic variable are plotted along the ordinate with the time as the abscissa is known as the waveform of the variable.

Cycle Each repetition of the set of values of a periodic function in equal intervals is termed a cycle.

Frequency (f) The number of cycles described by a periodic function per second is termed as frequency. It is measured in cycles per second or hertz (Hz). In India, the standard frequency of voltages and currents is 50 Hz, while in USA it is 60 Hz.

Period (T) The time duration in seconds of one cycle is termed as its period. The period T in seconds is related to frequency f by the relation

$$T = \frac{1}{f} \qquad (4.7)$$

Angular velocity (ω) Since a sinusoidal quantity recurs at an interval of 2π radians, the duration T for one complete cycle of a periodic variable is assumed to correspond to a time angle of 2π electrical radians. The angular velocity of tracing the time angle is therefore

$$\omega = \frac{\text{variation in angle in one cycle}}{\text{time taken to complete a cycle}}$$

$$= \frac{2\pi}{T} = 2\pi f \text{ electrical radians/sec} \qquad (4.8)$$

Instantaneous value The instantaneous value of an alternating quantity is its magnitude at any instant of time. Lowercase letters v and i are used in this book to denote instantaneous values of the ac voltage and ac current, respectively.

Phase The phase of an alternating variable at any instant is the time interval that has elapsed since the instantaneous value of the variable last passed through zero from negative to positive direction. This interval is measured either (i) in terms of a fraction of the time period of the waveform as $\varphi = t_1/T$, where φ is the phase and t is the time elapsed; or (ii) more commonly in terms of the time angle $f = 2\pi\varphi = 2\pi(t_1/T) = \omega t_1$, where φ is the phase and ω is the angular frequency of the supply.

Phase difference The phase difference between two periodic variables of the same frequency is defined as the difference in phase between the two variables measured at any instant of time. If the two waveforms pass through their maximum values at the same instant, the phase difference between them is zero. When the peak values of two periodic variables do not occur simultaneously, the variable that attains its peak earlier in time is said to *lead* the other, or conversely the second variable is said to *lag* behind the former by the angle which is termed as the phase difference between the two variables. Figure 4.6 shows two waveforms a and b of the same frequency and having phase angles φ_a and φ_b, at time t_0 corresponding to angle ωt_0. The phase difference between a and b is φ with the waveform a leading b or b lagging behind a.

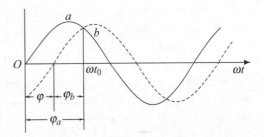

Fig. 4.6 Periodic waveforms with a phase difference

Peak value The maximum magnitude of the instantaneous value is defined as its peak value.

Peak-to-peak value The maximum variation between the maximum positive instantaneous value and the maximum negative instantaneous value of an alternating quantity is the peak-to-peak value. The peak-to-peak ac voltages and ac currents are denoted as V_{pp} and I_{pp}. For pure ac, the peak-to-peak value is equal to two times the peak value.

Peak amplitude The maximum instantaneous value measured from the mean value of a waveform is the peak amplitude.

Figure 4.7 graphically shows the relationship between various terms.

Average value The average value of an alternating quantity is defined as the time average of its instantaneous values over one cycle.

Root mean square (rms) or effective value The root mean square value of an alternating quantity is defined as the square root of the average of the squares of its instantaneous values over one cycle.

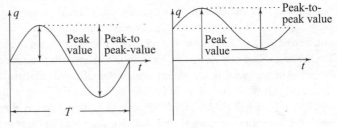

Fig. 4.7 Peak values

4.4 RELATIONSHIP BETWEEN FREQUENCY, SPEED, AND NUMBER OF POLES

In Section 4.2 it is seen that the waveform of the emf developed in an ac generator goes through one complete cycle of variation when the conductors move past one north pole and one south pole. The shape of the waveform over the negative half-cycle is exactly similar to that over the positive half-cycle. In a two-pole machine, as the ac generator rotates through one revolution the voltage generated also completes one cycle of variation. Thus, for P poles, when the ac generator is driven at a speed N rpm, $P/2$ cycles of variation of emf are produced in one revolution of the generator. The frequency f of the induced emf is given by

f = number of cycles per second

= number of cycles per revolution × number of revolution per second

$$= \frac{P}{2} \times \frac{N}{60}$$

$\therefore \qquad f = \dfrac{PN}{120}$ cycles/sec or Hz $\qquad\qquad\qquad (4.9)$

Hence, for a two-pole machine, the speed at which the machine must be driven to generate emf at 50 Hz frequency is given by Eq. (4.9) as

$$N = \frac{120 \times f}{P} = \frac{120 \times 50}{2} = 3000 \text{ rpm}$$

As the minimum number of poles that a machine can have is 2, the highest speed at which 50 Hz ac generator can be operated is 3000 rpm. In practice, this is the operating speed of steam turbine driven generating stations. For hydroelectric generators speeds are lower and, correspondingly, the number of poles may be 4, 8, 10, 12, or 16.

4.5 ROOT MEAN SQUARE AND AVERAGE VALUES OF ALTERNATING CURRENT AND VOLTAGE

An alternating quantity, voltage or current, varies from instant to instant and it is necessary to quantify the strength of such a time-varying quantity. Most commonly, ac voltage and ac current are measured by root mean square or effective values and average values. These terms are discussed in the following sections.

4.5.1 Root Mean Square or Effective Values

As seen in the preceding section, an alternating quantity such as current or voltage is not steady, but varies from instant to instant. The effectiveness of such currents and voltages in performing power transfer cannot be readily assessed. In addition, the average value of an alternating quantity over a cycle is zero and the maximum amplitudes do not reflect power accurately. Hence an alternative specification of expressing the strength of alternating current or voltage is necessary. The most convenient basis for this purpose is the comparison of the heating effect in a resistance by both direct and alternating currents. The derivation is as follows.

Let it be assumed that a steady-state alternating current flows through a certain resistance for a certain time. A steady-state current is that whose maximum value/amplitude and waveform remain constant. A certain amount of heat will be produced. Now suppose that a steady direct current is passed through the same resistance for the same duration of time and that the value of this current is such that exactly the same amount of heat is produced for the given time. Then this particular value of the steady direct current is taken as a measure of the alternating current. This value is called the virtual or effective value.

Figure 4.8(a) shows a waveform of an alternating current, which cannot be represented by a simple mathematical expression. It may be noted that the heating effect, being proportional to the square of the current, is not directional. Since the negative half-wave is a repetition of the positive half-wave, therefore, as far as numerical values are concerned, one half-wave only needs to be considered. In the positive half-cycle of the waveform in Fig. 4.8(a), $i_1, i_2, i_3, \ldots, i_n$ are taken as n equidistant mid-ordinates. If the current, represented by the waveform of Fig. 4.8(a), is passed through a resistance of R ohm, the heating effect of i_1 is $i_1^2 R$, i_2 is $i_2^2 R$, and so on, as shown in Fig. 4.8(b). The variation of the heating effect in the negative half-cycle is the same as during the positive half-cycle.

$$\text{Average heat produced} = \frac{i_1^2 R + i_2^2 R + i_3^2 R + \cdots + i_n^2 R}{n}$$

If I is the value of the direct current through the same resistance R to produce a heating effect equal to the average heating effect of the alternating current, then

$$I^2 R = \frac{i_1^2 R + i_2^2 R + i_3^2 R + \cdots + i_n^2 R}{n}$$

$$\therefore \quad I = \sqrt{\frac{i_1^2 + i_2^2 + i_3^2 + \cdots + i_n^2}{n}} \tag{4.10}$$

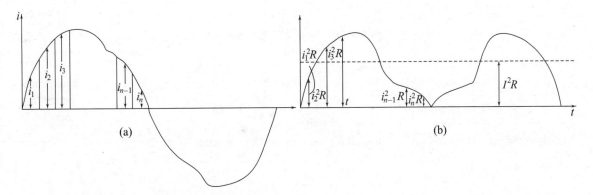

Fig. 4.8 (a) Non-sinusoidal alternating current wave (b) Square of alternating current wave

The quantity I is termed as the effective value of the current. From the form of the expression, this current I is also termed as the root mean square (rms) value of the current.

It can be seen that the effective value or the rms value of an alternating current may be defined as that value of current which when flowing through a resistance produces the same amount of heat as a direct current flowing through that resistance. Thus, the rms or effective value of an alternating quantity can be defined as the square root of the average of the square of the instantaneous values of the alternating quantity. Normally I and V designate rms values of the alternating current and voltage, respectively.

Alternatively, the average heating effect can be expressed as follows:

$$\begin{Bmatrix} \text{Average heating effect} \\ \text{over half-cycle} \end{Bmatrix} = \frac{\begin{Bmatrix} \text{area enclosed by } i^2R \text{ curve} \\ \text{over half-cycle} \end{Bmatrix}}{\{\text{length of the base}\}} \tag{4.11}$$

This method of determining the rms value is more convenient when dealing with sinusoidal waves.

4.5.1.1 Root Mean Square Values of Sinusoidal Current and Voltage

The waveform of a sinusoidally varying current passing through a resistance R is shown in Fig. 4.9. The instantaneous value of the current can be expressed as

$$i = I_m \sin \theta = I_m \sin \omega t$$

where I_m is the maximum value of the current, θ is the angle in radians from the instant of zero current, and ω is the angular velocity in rad/sec. Consider an elementary strip of magnitude i ampere and width $d\theta$ or $d(\omega t)$ rad. The variation of i^2R during one complete cycle is also shown in Fig. 4.9. During interval $d\theta$ rad the heat produced is $i^2R \times d\theta$ watt-radian. The area of an elementary strip of the square of the current of

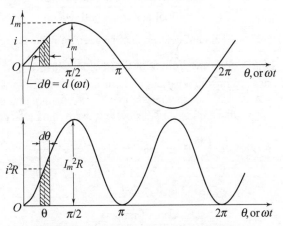

Fig. 4.9 Full-cycle sinusoidal current wave

magnitude i ampere and width $d\theta$ rad is $i^2R\, d\theta = I_m^2R \sin^2 \omega t\, d(\omega t)$ and is represented by the area of the shaded strip. Thus the heat generated during the first half-cycel is the area enclosed by the i^2R curve and is given by

$$\int_0^\pi i^2R\, d(\omega t) = I_m^2R \int_0^\pi \sin^2 \omega t\, d(\omega t) = \frac{I_m^2R}{2} \int_0^\pi (1 - \cos 2\omega t)\, d(\omega t)$$

$$= \frac{I_m^2 R}{2} \left[\omega t - \frac{1}{2}\sin 2\omega t \right]_0^\pi = \frac{\pi}{2} I_m^2 R \quad \text{watt-radians}$$

By definition, from expression (4.11),

$$\text{Average heating effect} = \frac{\dfrac{\pi}{2} I_m^2 R}{\pi} \frac{\text{watt-radians}}{\text{radians}} = \frac{I_m^2 R}{2} \quad \text{watts} \tag{4.12}$$

If the value of the direct current I through R produces the same heating effect, then

$$I^2 R = \frac{1}{2} I_m^2 R$$

$$\therefore \qquad I = \frac{I_m}{\sqrt{2}} = 0.707 I_m \tag{4.13}$$

Since the voltage across the resistance is proportional to the current, it follows that the rms value of the voltage is $V = V_m / \sqrt{2} = 0.707 V_m$, where V_m is the peak value of the sinusoidal voltage wave.

4.5.2 Average Value

The average value of an alternating quantity is defined as the average over the positive half-cycle of the waveform.

For the non-sinusoidal alternating waveform shown in Fig. 4.8(a) the average value of the current I_{av} over half a cycle is

$$I_{av} = \frac{i_1 + i_2 + i_3 + \cdots + i_n}{n} \tag{4.14}$$

Or, alternatively, the average current I_{av} is given by

$$I_{av} = \frac{\text{area enclosed under half-cycle}}{\text{length of base over half-cycle}} \tag{4.15}$$

This method of determining the average value is more convenient when dealing with sinusoidal waves.

4.5.2.1 Average Values of Sinusoidal Current and Voltage

For the sinusoidal variation of current shown in Fig. 4.9,

Area of the elementary strip = $i \times d(\omega t)$ ampere radian

The area enclosed by the current wave in a half-cycle is

$$\int_0^\pi i \, d(\omega t) = \int_0^\pi I_m \sin \omega t \, d(\omega t) = -I_m [\cos \omega t]_0^\pi$$

$$= -I_m[-1-1] = 2I_m \quad \text{ampere-radians}$$

The average value over half a cycle is

$$I_{av} = \frac{2I_m}{\pi} \frac{\text{ampere-radians}}{\text{radians}}$$

$$= 0.637 I_m \quad \text{A} \tag{4.16}$$

Similarly, the average value of the voltage is $V_{av} = 0.637 V_m$.

4.6 FORM FACTOR AND PEAK FACTOR

The form factor of a wave is defined as the ratio of its rms value to its average value. The peak or crest factor of a wave is the ratio of its peak value to its rms value. From Eqs (4.13) and (4.16),

$$\text{Form factor} = \frac{0.707 I_m}{0.637 I_m} = 1.11 \qquad (4.17)$$

$$\text{Peak factor} = \frac{I_m}{0.707 I_m} = 1.414 \qquad (4.18)$$

If a wave is not sinusoidal, the form factor will depend on the shape. Thus, for a rectangular wave, it is unity; while for a triangular wave, it is 1.15. In general, it can be said that the more pointed or peaking the wave shape, the greater will be the form factor.

Sinusoidal Voltage and Current Sources   The generated voltage being sinusoidal, its magnitude not only varies from instant to instant with respect to time but also repeats after an interval of 2π radians or after every time period of T sec. Mathematically, the instantaneous value v of a sinusoidal voltage source is represented as

$$v = V_m \sin(\omega t + \varphi)\text{V} \qquad (4.19)$$

It may be noted that at a given frequency f Hz, Eq. (4.19) is described by two factors, namely, maximum amplitude V_m and phase angle φ, which represents the phase of the sinusoid with respect to the reference voltage v. The time period of the voltage source is specified by T and is equal to $2\pi/\omega = 1/f$ sec and the angular speed in rad/sec is denoed by ω. Figure 4.10 shows the plot of Eq. (4.19) with respect to ωt.

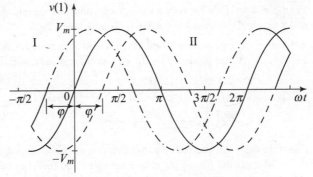

It may be observed that at $t = 0$ sec, the sinusoidal I has a positive value and the sinusoidal II has a negative value with respect to the reference, which has a zero value. Thus, sinusoidal I

Fig. 4.10   Variation of sinusoidal voltage with time

is said to lead the reference sinusoidal by an angle $+\varphi$ while sinusoidal II lags the reference sinusoidal by an angle $-\varphi$ since it has a negative value.

Example 4.2   Calculate (i) time period T, (ii) frequency f, and (iii) rms value of the sinusoidal voltage represented by $v = 155 \sin(377t + \varphi)$V. Determine the phase angle φ if the instantaneous value of the voltage at $t = 0$ sec is -77.5 V. Also compute the time when the instantaneous value of the voltage is zero when t is getting close to zero from the negative time side.

Solution   Data: $V_m = 155$ V and $\omega = 377$ rad/sec

  (i) Time period, $T = 2\pi/\omega = 2\pi/377 = 0.0167$ sec
 (ii) Frequency, $f = 1/T = 60$ Hz
(iii) The rms value, $V = 155/\sqrt{2} = 109.60$ V

At $t = 0, -77.5 = 155 \sin \varphi$

or, $\varphi = \sin^{-1}(-77.5/155) = -0.5236 \text{ rad} = -30.0°$

The time, just before $t = 0$, when the instantaneous voltage is zero,

$$\varphi/\omega = -0.5236/377 = -0.0014 \text{ sec}$$

Example 4.3 An alternating current is expressed as $i = 14.14 \sin 314t$. Determine: (a) rms current, (b) frequency, (c) instantaneous current when $t = 0.02$ ms.

Solution (a) The current is normally expressed as $i = I_m \sin \omega t$. Thus by comparing

$$I_m = 14.14 \text{ A} = \sqrt{2} I$$

Therefore, $\quad I = \dfrac{14.14}{\sqrt{2}} = 10 \text{ A}$

(b) Also by comparison,

$$\omega = 314 \text{ rad/sec} = 2\pi f$$

$\therefore \qquad f = \dfrac{314}{2\pi} = 50 \text{ Hz}$

(c) When $t = 2$ ms $= 2 \times 10^{-3}$ sec,

$$i = 14.4 \sin (314 \times 2 \times 10^{-3}) = 14.14 \sin(0.628)$$

$$= 14.14 \times 0.5875 = 8.30 \text{ A}$$

Example 4.4 Find the root mean square value of the resultant current in a wire that carries a dc current of 20 A and a sinusoidal alternating current with peak value 20 A.

Solution Let I be the rms value of the resultant current and R be the resistance of the wire. The heat produced by I is the sum of heat produced by the dc current and heat produced by the ac current,

$$I^2 R = 20^2 R + (20/\sqrt{2})^2 R$$

or $\qquad I^2 = 400 + 200$

$\therefore \qquad I = \sqrt{600} = 24.495 \text{ A}$

Example 4.5 An alternating current varying sinusoidally with a frequency of 50 Hz has a rms value of 10 A. Write down the equation for the instantaneous value and find this value for (a) 0.0025 sec, (b) 0.0125 sec after passing through a positive maximum value. At what time measured from a positive maximum value will the instantaneous current be 7.07 A?

Solution The instantaneous value of current, i, in terms of the maximum value of current, I_m, can be written as

$$i = I_m \sin \theta = I_m \sin \omega t$$

The rms value of current, $I = 0.707 \, I_m$,

$\therefore \qquad I_m = \dfrac{I}{0.707}$

and $\qquad \theta = \omega t = 2\pi f \times t = 2 \times 3.14 \times 50 \times t = 314t$

$$i = \dfrac{I}{0.707} \sin 314t$$

For $t = 0.0025$ sec,

$$i = \dfrac{10}{0.707} \sin (314 \times 0.0025) = 10 \text{ A}$$

The maximum positive value of current occurs when $\sin (314t) = 1$. Then,

$$314t = \sin^{-1}(1) = \dfrac{\pi}{2}$$

$\therefore \qquad t = \dfrac{\pi}{2 \times 314} = 0.005 \text{ sec}$

Hence at $t = 0.005 + 0.0125 = 0.0175$ sec,

$$i = \frac{10}{0.707} \sin (314 \times 0.0175) = -10 \text{ A}$$

No when $i = 7.07$ A the tie at which it occurs can be found from

$$7.07 = \frac{10}{0.707} \sin 314t$$

or $\quad \sin 314t = \dfrac{7.07 \times 0.707}{10} = 0.4998$

$\therefore \quad t = \dfrac{\sin^{-1} (0.4998)}{314} = 0.001667 \text{ or } 0.00833 \text{ sec}$

So in the positive half of the cycle, the instantaneous value of 7.07 A occurs at 0.00167 and 0.00833 sec. Hence the time at which the instantaneous value is 7.07 A after the positive maximum value is at time
$$t = 0.00833 - 0.005 = 0.0033 \text{ sec}$$

Example 4.6 Calculate the root mean square value, the form and peak factors of a periodic voltage having the following values for equal time intervals, changing suddenly from one value to the next: 0, 5, 10, 20, 40, 50, 40, 20, 10, 5, 0, −5, −10 V, etc. What would be the root mean square value of a sine wave having the same peak value?

Solution Figure 4.11 shows the periodic variation of the voltage graphically.

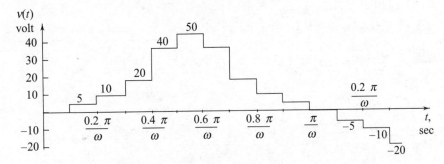

Fig. 4.11 Perodicvariation of voltage with time

$$V = \sqrt{\frac{0.1\pi}{\omega} \left[\frac{0^2 + 5^2 + 10^2 + 20^2 + 40^2 + 50^2 + 40^2 + 20^2 + 10^2 + 5^2}{\pi/\omega} \right]}$$

$$= \sqrt{675} = 25.98 \text{ V}$$

$$V_{av} = \frac{0.1\pi}{\omega} \left[\frac{0 + 5 + 10 + 20 + 40 + 50 + 40 + 20 + 10 + 5}{\pi/\omega} \right] = 20 \text{ V}$$

$$\text{Form factor} = \frac{\text{rms value}}{\text{average value}} = \frac{25.98}{20} = 1.299 \approx 1.3$$

$$\text{Peak factor} = \frac{\text{maximum value}}{\text{rms value}} = \frac{50}{25.98} = 1.9245$$

The rms value for a sine wave with the same peak value is $0.707 \times 50 = 35.35$ V.

Example 4.7 An alternating voltage wave increases uniformly from zero at 0° to 100 V at α, remains constant at 100 V from α to $(\pi - \alpha)$ and decreases uniformly from 100 V at $(\pi - \alpha)$ to zero at π. Calculate the rms value and average value of this wave for one half-cycle. Evaluate rms and average values of voltage for $\alpha = \pi/2$.

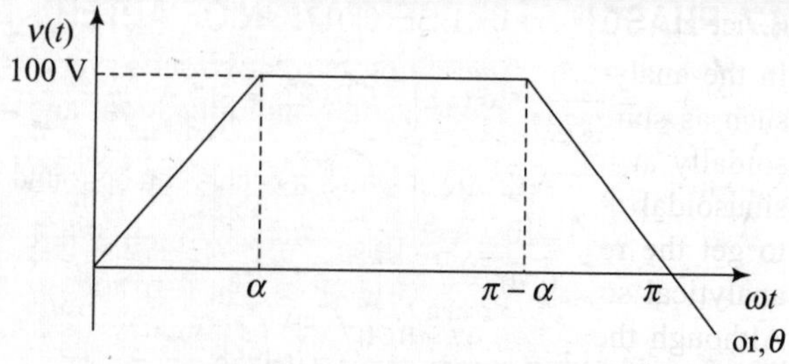

Solution Figure 4.12 shows the plot of the alternating current wave.

Fig. 4.12 Plot of the half-cycle of the alternating current wave

The half-cycle of the alternating current wave can be written mathematically as

$$\frac{100}{\alpha}\theta \qquad \text{for } 0° \le \theta \le \alpha$$

$$v = 100 \qquad \text{for } \alpha \le \theta \le \pi - \alpha$$

$$v = -\frac{100}{\alpha}\theta + \frac{100}{\alpha}\pi \qquad \text{for } \pi - \alpha \le \theta \le \pi$$

Then, $I_{av} = \frac{1}{\pi}\left[\int_0^\alpha \frac{100}{\alpha}\theta\, d\theta + \int_\alpha^{\pi-\alpha} 100\, d\theta + \int_{\pi-\alpha}^\pi \frac{100}{\alpha}(\pi - \theta)\, d\theta\right]$

$= \frac{100}{\pi}\left[\frac{1}{\alpha}\left(\frac{\theta^2}{2}\right)_0^\alpha + \theta\Big|_\alpha^{\pi-\alpha} + \frac{1}{\alpha}\left(\pi\theta - \frac{\theta^2}{2}\right)_{\pi-\alpha}^\pi\right]$

$= \frac{100}{\pi}\left[\frac{1}{\alpha}\left(\frac{\alpha^2}{2}\right) + (\pi - 2\alpha) + \frac{1}{\alpha}\left(\pi^2 - \frac{\pi^2}{2} - \pi(\pi - \alpha) + \frac{(\pi - \alpha)^2}{2}\right)\right]$

$= 100\left[\frac{1}{2\pi}\alpha + 1 - \left(\frac{2\alpha}{\pi}\right) + \frac{1}{2\pi}\alpha\right]$

$= 100\left(1 - \frac{\alpha}{\pi}\right)$

$I = \sqrt{\frac{1}{\pi}\left[\int_0^\alpha \frac{100^2}{\alpha^2}\theta^2\, d\theta + \int_\alpha^{\pi-\alpha} 100^2\, d\theta + \int_{\pi-\alpha}^\pi \frac{100^2}{\alpha^2}(\pi - \theta)^2 d\theta\right]}$

$= 100\sqrt{\frac{1}{\pi}\left[\frac{1}{\alpha^2}\left(\frac{\theta^3}{3}\right)_0^\alpha + \theta\Big|_\alpha^{\pi-\alpha} + \frac{1}{\alpha^2}\left[-\frac{(\pi - \theta)^3}{3}\right]_{\pi-\alpha}^\pi\right]}$

$= 100\sqrt{\frac{1}{\pi}\left(\pi - \frac{4\alpha}{3}\right)} = 100\sqrt{\left(1 - \frac{4\alpha}{3\pi}\right)}$

On substituting various values of $\alpha = \pi/2$ in I_{av} and I,

$$I_{av} = 100\left(1 - \frac{\pi/2}{\pi}\right) = 50 \text{ V}$$

$$I = 100\sqrt{\left(1 - \frac{4 \times \pi/2}{3\pi}\right)} = 100\frac{1}{\sqrt{3}} = 57.735 \text{ V}$$

4.7 PHASOR REPRESENTATION OF ALTERNATING QUANTITIES

In the analysis of alternating current circuits, very often one has to perform basic arithmetic operations such as summation, subtraction, multiplication, and division on two or more quantities that vary sinusoidally at the same frequency. One way of carrying out the mathematical operations is to plot each sinusoidal quantity and then make point-by-point summation, subtraction, etc. of the sinusoidal waves to get the result graphically. However, such a method is cumbersome and very time consuming. The analytical solution using trigonometric identities is an alternative to the graphical method of solution. Although the analytical procedure requires less time compared to the graphical method, it is still cumbersome and laborious particularly when more than two such quantities are involved. A simple and more direct method of dealing with sinusoidal quantities is to use the phasor representation. Charles P. Steinmetz introduced, in 1893, the concept of using a constant-amplitude line rotating at a frequency ω to represent a sinusoid.

A phasor is a complex number that contains the amplitude and phase angle information of a sinusoidal function. The concept of phasors can be developed using Euler's identity which relates the exponential function to the trigonometric functions as

$$e^{\pm j\omega t} = \cos\omega t \pm j\sin\omega t \qquad (4.20)$$

From Eq. (4.20) the sine and cosine functions can be represented as

$$\sin\omega t = \text{Im}\{e^{j\omega t}\} \qquad (4.21)$$

$$\cos\omega t = \text{Re}\{e^{j\omega t}\} \qquad (4.22)$$

where Im stands for 'imaginary part of' and Re stands for 'real part of'. The exponential function $e^{j\omega t}$ may be treated as a unit rotational operator. Its amplitude is always unity and the sine and the cosine terms vary as time progresses. Thus, a sinusoidal voltage function $V_m \sin\omega t$ and $V_m \cos\omega t$ can be written in the form suggested by Eq. (4.22) as

$$v = V_m \sin\omega t = \text{Im}\{V_m e^{j\omega t}\} \qquad (4.23)$$

$$v = V_m \cos\omega t = \text{Re}\{V_m e^{j\omega t}\} \qquad (4.24)$$

where v is the instantaneous value and V_m is the maximum value of the sinusoidal voltage.

Equations (4.23) and (4.24) indicate the representation of the sine and cosine voltage functions in terms of the scaled unit rotational operator $e^{j\omega t}$. This is depicted graphically in Fig. 4.13, where the line length OP is the voltage phasor and is drawn to scale to represent the maximum value V_m. The line OP is rotated in the anti-clockwise direction about O at a uniform angular velocity ω rad/sec. This is purely a conventional direction, which has been adopted universally. An arrowhead is drawn at the outer end of the phasor to indicate the exact length of the phasor. Figure 4.13(a) shows that the phasor OP is moved from the horizontal position when the voltage was passing through its zero value to an angle θ in time t seconds, then $\theta = \omega t$. If PA and PB are drawn perpendicular from P to the horizontal and vertical axes, the vertical component OB of the phasor OP is

$$OB = AP = OP \sin\theta = OP \sin\omega t$$

$$= V_m \sin\omega t$$

$= v$, the instantaneous value of voltage at time t seconds.

Hence the projection of V_m along the vertical axis represents to scale the instantaneous value of the voltage. When $\omega t = 0°$, the position of OP is horizontal directed towards the right. The vertical component of OP is zero at this instant, and hence the instantaneous value of the voltage $v = 0$. When $\omega t = 90°$, $v = V_m$; when $\omega t = 180°$, $v = 0$; when $\omega t = 210°$, $v = -1/2\,V_m$; when $\omega t = 270°$, $v = -V_m$; and when $\omega t = 360°$, $v = 0$. Thus OP rotates through one revolution or 2π rad in one cycle of sinusoidal variation of the voltage wave.

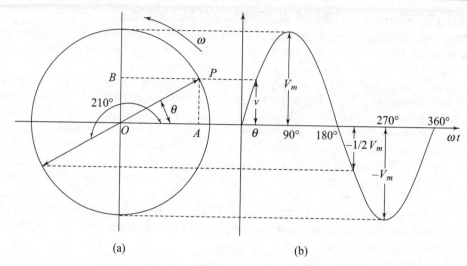

(a)　　　　　　　　　　　(b)

Fig. 4.13 Phasor representation of an alternating quantity

If f is the frequency in cycles per second, the phasor revolves through $2\pi f$ rad in 1 sec. Then the angular displacement in t second is $\theta = 2\pi ft$ rad. Then the instantaneous value of voltage is given by

$$v = V_m \sin\theta = V_m \sin\omega t = V_m \sin 2\pi ft \text{ volt}$$ (4.25)

Thus, a general sinusoidal function $V_m \sin(\omega t + \varphi)$ may be represented as

$$V_m \sin(\omega t + \varphi) = \text{Im}\{V_m e^{j(\omega t + \varphi)}\} = \text{Im}\{V_m e^{j\varphi} e^{j\omega t}\}$$ (4.26)

In Eq. (4.26) $V_m e^{j\varphi}$ is a complex number that carries the amplitude and phase angle of the given sinusoidal function. This complex number, by definition, is the phasor representation of the given sinusoidal function.

Comparing the representation of $V_m \sin\omega t$ and $V_m \sin(\omega t + \varphi)$ as per Eqs (4.23) and (4.26), respectively, it may be noticed that the factor $e^{j\omega t}$ is common to both and it contains the information about the angular frequency. Once the sinusoids of the same frequency are involved, this information is redundant. Then the representations of $V_m \sin\omega t$ and $V_m \sin(\omega t + \varphi)$ are simplified, by dropping the term $e^{j\omega t}$, to the following form:

$$V_m \sin \omega t = V_m e^{j0} \text{ or simply } V_m \angle 0°$$ (4.27)

$$V_m \sin(\omega t + \varphi) = V_m e^{j\varphi} \text{ or simply } V_m \angle \varphi°$$ (4.28)

It has been shown in Section 4.5.1.1 that the rms value of a sinusoidal waveform is 0.707 times its maximum value. In the steady-state analysis of ac circuits, the rms values are generally used instead of peak values. Also most measuring instruments such as voltmeters, ammeters, etc. are calibrated to measure rms values. Thus the length of the phasor may be made equal to the rms value instead of the maximum values. The rms value of voltage $V = V_m/\sqrt{2}$, then the sinusoidal voltage of Eq. (4.28) may be written as

$$V_m \sin(\omega t + \varphi) = \sqrt{2} \ V e^{j\varphi} = \sqrt{2} V$$ (4.29)

where $V = V e^{j\varphi} = V \angle \varphi°$ is the phasor. In the text that follows hereon, the boldfaced capital letter will represent the phasor so as to distinguish them from the complex numbers. The rms values of voltage and current would be represented by capital letters.

4.7.1 Phasor Representation of Quantities with a Phase Difference

Figure 4.14(b) shows the representation of sinusoidal voltage and current waves with the voltage leading the current by angle φ. In Fig 4.14(a) OA represents the maximum value of voltage V_m and OB the maximum value of current I_m. The angle between OA and OB must be the same angle φ. When OA is along the horizontal axis, the voltage at that instant is zero, while the current has a negative value represented by the projection of OB on the vertical axis. This corresponds to instant O in Fig. 4.14(b).

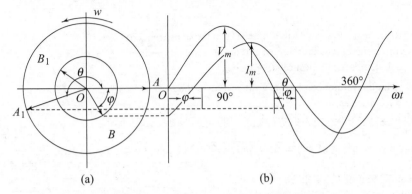

Fig. 4.14   Representation of phasors having a phase difference

After the phasors have rotated through an angle θ, they occupy the positions OA_1 and OB_1, respectively, and the instantaneous voltage and current are given by the projections of OA_1 and OB_1 on the vertical axis as shown by the horizontal dotted lines.

If the instantaneous value of voltage is represented by

$$v = V_m \sin\omega t$$

then the instantaneous value of current is represented by

$$i = I_m\sin(\omega t - \varphi)$$

The current is said to lag the voltage by an angle φ. The phase difference φ remains constant irrespective of the position.

4.7.2 Addition and Subtraction of Phasor Quantities

Figure 4.15 shows two phasors $OA\angle\theta_1$ and $OB\angle\theta_2$ representing to scale the maximum values of, say, two alternating voltages of the same frequency and differing in angle by φ. The two phasors are to be added and subtracted. The addition and subtraction of the two phasors are shown in Figs 4.15(a) and 4.15(b), respectively. First the parallelogram $OACB$ and its diagonal OC are drawn as shown in Fig. 4.15(a). Then the projections of OA, OB, and OC on the vertical axis give their instantaneous values. Thus,

Instantaneous value of $OA = OA_1$

Instantaneous value of $OB = OB_1$

Instantaneous value of $OC = OC_1$

Now, $BC \parallel OA$ and $BC = OA$. Then the projections on the vertical axis $OA_1 = B_1C_1$,

$$OC_1 = OB_1 + B_1C_1 = OB_1 + OA_1$$

Thus the instantaneous value of OC equals the sum of the instantaneous values of OA and OB, and the phasor OC represents to scale the maximum value of the resultant voltage. Figure 4.15(b) shows the subtraction of the phasor quantities

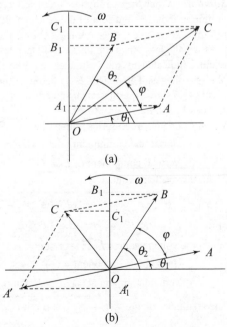

Fig. 4.15   (a) Addition of phasor quantities and (b) subtraction of phasor quantities

OA and *OB*. *OA'* is drawn in the direction opposite to *OA* and equal in magnitude to *OA*. The parallelogram *OA'CB* is drawn. It can then be proved that the phasor *OC* is equal to the phasor difference of the phasors *OA* and *OB*.

Example 4.8 Two voltages $V_{AB} = 60$ V and $V_{BC} = 90$ V at the same frequency and with a phase difference between them of 60° act in series in a circuit. Find voltage V_{AC} and its phase position relative to V_{BC}.

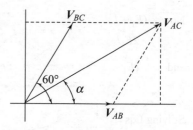

Fig. 4.16

Solution Let V_{AB} be the reference phasor. Voltage V_{AC} is the sum of two voltages V_{AB} and V_{BC} in series.

The phasor diagram is drawn as shown in Fig. 4.16. From the phasor diagram

$$V_{AC} = V_{AB} + V_{BC}$$

Now, the horizontal component of $V_{AB} = 60$ V and the vertical component of $V_{AB} = 0$ V.

The horizontal component of $V_{BC} = 90 \cos 60° = 45$ V

The vertical component of $V_{BC} = 90 \sin 60° = 77.94$ V

$$\text{Resultant } V_{AC} = \sqrt{(60+45)^2 + (77.94)^2} = \sqrt{17,099.64} = 130.76 \text{ V}$$

The phase angle α of the phasor is given by

$$\alpha = \tan^{-1} \frac{77.94}{105} = \tan^{-1} 0.7423 = 36.59°$$

Therefore, the phase position of V_{AC} with respect to $V_{BC} = 60° - 36.59° = 23.41°$.

Example 4.9 Two sinusoidal sources of emf have rms values E_1 and E_2. When connected in series, with displacement α, the resultant voltage read on an electrodynamic voltmeter is 76.4 V, and with one source reversed is 26.7 V. When the phase displacement is made zero, a reading of 80 V is obtained. Calculate E_1, E_2, and α.

Solution When the two sources are in series, the electrodynamic voltmeter indicates the sum of the phasor voltages and when one resource is reversed, the voltmeter shows the difference of the phasor voltages. When the phase displacement is made zero, the reading indicated by the voltmeter is the arithmetic sum of the phasor voltages. The phasor E_1 is taken as the reference phasor. Figure 4.17 shows the phasor sum and difference of the two voltage sources E_1 and E_2.

When E_1 and E_2 are in series with a phase difference α,

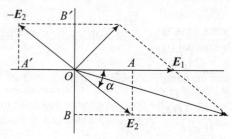

Fig. 4.17

Horizontal component of $E_2 = OA = E_2 \cos \alpha$

Vertical component of $E_2 = OB = - E_2 \sin \alpha$

Then, the resultant voltage

$$\sqrt{(E_1 + E_2 \cos\alpha)^2 + (-E_2 \sin\alpha)^2} = 76.4$$

or $\quad E_1^2 + 2E_1 E_2 \cos \alpha + E_2^2 = 5836.96 \quad\quad\quad\quad\quad\quad\quad$ (I)

When the voltage source E_2 is reversed

Horizontal component $= OA' = -E_2 \cos \alpha$

Vertical component $= OB' = E_2 \sin \alpha$

The resultant voltage $\sqrt{(E_1 - E_2 \cos \alpha)^2 + (E_2 \sin \alpha)^2} = 26.7$

or $\quad E_1^2 - 2E_1 E_2 \cos \alpha + E_2^2 = 712.89 \quad\quad\quad\quad\quad\quad\quad$ (II)

When the phase displacement is made zero,

$$E_1 + E_2 = 80 \tag{III}$$

Adding Eqs (I) and (II), $E_1^2 + E_2^2 = \dfrac{1}{2} \times 6549.85 = 3274.92$ (IV)

Now, $(E_1 + E_2)^2 - (E_1^2 + E_2^2) = 2\,E_1 E_2$

$\therefore \qquad 2E_1 E_2 = (6400 - 3274.92) = 3125.08$ (V)

and $\qquad (E_1 - E_2) = \sqrt{(E_1 + E_2)^2 - 4E_1 E_2} = \sqrt{80^2 - 2 \times 3125.08}$

$$= \sqrt{6400 - 6250.16} = \sqrt{149.84} = 12.24 \tag{VI}$$

Solving Eqs (III) and (VI), $E_1 = 46.12$ V, $E_2 = 33.88$ V

From Eq. (I), (IV), and (V) $\cos\alpha = \dfrac{5836.96 - 3274.92}{3125.08} = 0.8198$

$\therefore \qquad \alpha = 34.93°$

4.8 THE j OPERATOR AND PHASOR ALGEBRA

The use of a phasor diagram for solving alternating current problems is a very convenient tool. However, the method works satisfactorily for simple problems. For more involved problems the calculations become simpler when complex algebra is used, which enables the alternating sinusoidal voltages, currents, and their phase angles to be represented by a simple algebraic form. Here $j = \sqrt{-1}$. To represent an imaginary number, mathematicians use i while electrical engineers use j.

4.8.1 Resolution of Phasors

A complex phasor can be represented by a complex number, which is a result of locating a line in a plane. As discussed in Section 4.7, a phasor $\boldsymbol{OA}_1 = OA_1 \angle \theta_1$ can be located by the magnitude OA_1 and its displacement from the horizontal axis θ_1, as shown in Fig. 4.18. However, $\boldsymbol{OA}_1$ can also be located in terms of a horizontal (real axis or x-axis) component $a_1 = OA_1 \cos\theta_1$ and a vertical (imaginary or j-axis) component $b_1 = OA_1 \sin\theta_1$. An operator symbol j is introduced to denote that the component b_1 is at right angle to the component a_1 of the phasr $\boldsymbol{OA}_1$ and the phasor can be represented as

$$\boldsymbol{OA}_1 = a_1 + jb_1 = OA_1(\cos\theta_1 + j\sin\theta_1) \tag{4.30}$$

Hence, the various phasors shown in Fig. 4.18 can be written in their resolved form as shown below.

$$\boldsymbol{OA}_2 = -a_2 + jb_2 = OA\angle\theta_2$$

where $a_2 = OA_2 \cos\theta_2$ and $b_2 = OA_2 \sin\theta_2$.

$$\boldsymbol{OA}_3 = -a_3 - jb_3 = OA\angle-\theta_3$$

where $a_3 = OA_3 \cos(-\theta_3)$ and $b_3 = OA_3 \sin(-\theta_3)$

$$\boldsymbol{OA}_4 = a_4 - jb_4 = OA\angle-\theta_4$$

where $a_4 = OA_4 \cos(-\theta_4)$ and $b_4 = OA_4 \sin(-\theta_4)$

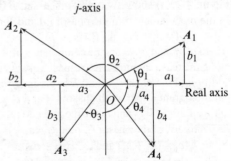

Fig. 4.18   Resolution of phasors

4.8.2 The j Operator

The multiplication of any phasor by the operator j causes the phasor to rotate through 90° in the anticlockwise direction without affecting its magnitude. Thus if a phasor A is multiplied by the operator j, then jA represents a phasor which is at 90° to A. When phasor jA is again multiplied by the operator j, it causes a displacement of 180° from the original position of the phasor A. The result can be written as

$$j(jA) = j^2A = -A$$

Hence, $j^2 = -1$ or $j = \sqrt{-1}$

Similarly, $j(j^2A) = j^3A = -jA$

$$j(j^3A) = j^4A = A$$

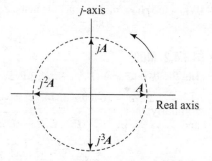

Fig. 4.19 Application of the j operator

Figure 4.19 shows the effect of the operator j when applied consecutively to phasor A.

Thus, alternating voltage and current phasors can be resolved along the real axis and j-axis in the complex plane with the help of the j operator.

4.8.3 Representation of Phasors in the Complex Plane

Phasors in the complex form can be represented in the following forms.
  (i) *Rectangular or Cartesian form* If a and b are the horizontal and vertical components along the real axis and j-axis, respectively, of a phasor A with a phase displacement of θ with respect to the real axis, then

$$A = a + jb, \quad \theta = \tan^{-1}(b/a) \tag{4.31}$$

 (ii) *Trigonometric form*

$$A = Ae^{j\theta} = A(\cos\theta + j\sin\theta) \tag{4.32}$$

(iii) *Polar form*

$$A = A\angle\theta \tag{4.33}$$

4.8.4 Phasor Algebra

In order to manipulate phasor quantities, the four basic arithmetic operations, namely, addition, subtraction, multiplication, and division of phasors, are described in the following sections.

4.8.4.1 Addition and Subtraction of Phasors

Figure 4.20(a) shows two phasors A and B that are to be added. In the rectangular form the phasors A and B can be written as

$$A = a_1 + ja_2 \text{ and } B = b_1 + jb_2$$

where a_1, a_2 and b_1, b_2 are the components of A and B of the phasors along the real axis and j-axis, respectively. As per Section 4.7.2 the addition of the phasors A and B is given by the diagonal C of the parallelogram drawn with phasors A and B as its sides, as shown in Fig. 4.20(a). If c_1 and c_2 are the real and imaginary components, respectively, of the phasor C, then $C = c_1 + jc_2$. From Fig. 4.20(a)

$$c_1 = (a_1 + b_1) \text{ and } c_2 = (a_2 + b_2)$$

$$\begin{aligned} C &= (a_1 + b_1) + j(a_2 + b_2) \\ &= (a_1 + ja_2) + (b_1 + jb_2) \\ &= A + B \end{aligned}$$

The phasor difference of the phasors A and B is given by the resultant phasor D, as shown in Fig. 4.20(b). The subtraction of the phasors A and B is given by the diagonal D of the parallelogram drawn with the phasors $-A$ and B as its sides, as shown in Fig. 4.20(b). If d_1 and d_2 are the real and imaginary components, respectively, of the phasor D, then $D = d_1 + j\,d_2$. Then, from Fig. 4.20(b), if $b_1 < a_1$ then d_1 is negative, and

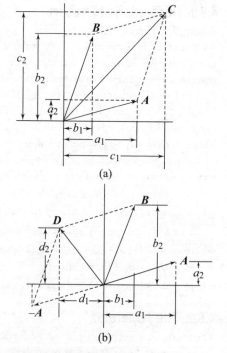

Fig. 4.20 (a) Addition of phasors
(b) Subtraction of phasors

$$D = -d_1 + j\, d_2 = -a_1 + b_1 + j\,(b_2 - a_2)$$
$$= (b_1 + j\, b_2) - (a_1 + j\, a_2) = B - A.$$

4.8.4.2 Multiplication and Division of Phasors

Let the two phasors be $A = Ae^{j\theta}$ and $B = Be^{j\varphi}$. Then the product of the two phasors can be writen as

$$A \cdot B = Ae^{j\theta} \times Be^{j\varphi} = AB\, e^{j(\theta + \varphi)}$$

Similarly, the division of the two phasors can be written as

$$\frac{A}{B} = \frac{Ae^{j\theta}}{Be^{j\varphi}} = \frac{A}{B} e^{j(\theta - \varphi)}$$

From the above it is clear that for adding and subtracting phasors it is expeditious to represent complex quantities by the rectangular form, while for multiplication and division of phasors it is more convenient to represent complex quantities by the trigonometric or polar form.

Example 4.10 Given phasor $A = 6.34 + j\, 13.59$ and phasor $B = 20 \angle -35°$. Write down the following in (i) rectangular form and (ii) polar form

(a) $A + B$
(b) A/B
(c) $(A + B)/(A - B)$
(d) $(A + B) \times B/(A - B) \times A$

Solution In polar form, $A = \sqrt{(6.34)^2 + (13.59)^2} \angle \tan^{-1}(13.59/6.34) = 15\angle 65°$

In rectangular form, $B = 20\left[\cos\left(-35°\right) + j\sin\left(-35°\right)\right] = 16.38 - j11.47$

Having expressed phasors A and B in both forms, the desired operations are performed by employing the appropriate form of representation.

(a) $A + B = 6.34 + j\, 13.59 + 16.38 - j\, 11.47 = 22.72 + j\, 2.12$

$$= \sqrt{(22.72)^2 + (2.12)^2} \angle \tan^{-1}(2.12/22.72) = 22.82\angle 5.33°$$

(b) $A/B = (15\angle 65°/20\angle -35°) = 0.75\angle 100° = 0.75\,[\cos(100°) + j\sin(100°)] = -0.13 + j0.74$

(c) Expressing $(A - B)$ in polar form gives

$$(A - B) = (-10.04 + j\, 25.06) = 27\angle 111.84°$$

Thus, $(A + B)/(A - B) = (22.8\angle 5.33°/27.0\angle 111.84°) = 0.85\angle -106.51° = -0.24 - j0.81$

(d) $(A + B) \times B/(A - B) \times A$

$$0.85\angle -106.51°/0.75\angle 100° = 1.13\angle 153.50° = -1.01 + j0.50$$

Example 4.11 Find the sum of the currents $i_1 = 20\sin(314t + \pi/3)$, $i_2 = -10\sin 314t$, and $i_3 = 15\sin(314t - \pi/4)$.

Solution The three currents are at the same frequency and may be represented as three phasors as

$$I_1 = \frac{20}{\sqrt{2}} e^{j\frac{\pi}{3}} = 14.144\left(\cos\frac{\pi}{3} + j\sin\frac{\pi}{3}\right) = (7.072 + j12.25)\,\text{A}$$

$$I_2 = -\frac{10}{\sqrt{2}} e^{j0} = -7.072(\cos 0 + j\sin 0) = (-7.072 + j)\,\text{A}$$

$$I_3 = \frac{15}{\sqrt{2}} e^{j\frac{\pi}{4}} = 10.607\left[\cos\left(-\frac{\pi}{4}\right) + j\sin\left(-\frac{\pi}{4}\right)\right] = (7.5 - j7.5)\,\text{A}$$

Then, the sum of the currents

$$I = I_1 + I_2 + I_3 = (7.5 + j4.75) = \sqrt{7.5^2 + 4.75^2} \ \angle\tan^{-1}\frac{4.75}{7.5}$$

$$= 8.8776 \ \angle 0.633 \text{ rad} = 8.8776 \ \angle 32.34°$$

The instantaneous value of the resultant current i is then given by the expression

$$i = \sqrt{2} \times 8.8776 \sin(314t + 0.633) = 12.5548 \sin(314t + 32.34°)$$

4.9 ANALYSIS OF AC CIRCUITS WITH SINGLE BASIC NETWORK ELEMENT

The steady-state response of the basic network elements such as resistance, inductance, and capacitance when excited with a sinusoidal source of voltage is presented in this section.

4.9.1 Resistive Circuit

Figure 4.21 shows a pure resistance of R ohm connected across a sinusoidal ac voltage source. The instantaneous value v of the source voltage at the instant of time t second is given by

$$v = V_m \sin \omega t = \text{Im}\{V_m e^{j\omega t}\} \qquad (4.34)$$

where V_m = the maximum value of voltage in volts

$\omega = 2\pi f$ rad/sec

f = frequency in cycles per seconds = 50 Hz

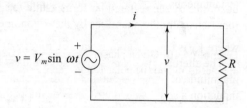

Fig. 4.21 Resistive circuit with a sinusoidal voltage source

As the resistance is a linear circuit element the steady-state response of the circuit must also be a sinusoid of the same frequency as the source voltage. At any instant of time t second, if the voltage is v volt, then by applying KVL to the circuit of Fig. 4.21 the current i at that instant can be determined from

$$v = iR$$

and
$$i = \frac{v}{R} = \frac{V_m \sin \omega t}{R} = \text{Im}\left\{\frac{V_m}{R}e^{j\omega t}\right\} \qquad (4.35)$$

$$= I_m \sin \omega t = \text{Im}\{I_m e^{j\omega t}\} \qquad (4.36)$$

where $I_m = \dfrac{V_m}{R}$ $\qquad (4.37)$

From Eq. (4.36), it can be seen that the current in the circuit has the same waveform an angular speed ω as the voltage waveform. Figure 4.22(a) shows the voltage and current waveforms.

As already discussed with reference to the phasor representation of the sinusoids, it is desirable to deal with rms or effective values of voltage and current rather than the maximum values. Thus Eq. (4.37) may be expressed as

$$I_m = \sqrt{2}I = \frac{\sqrt{2}V}{R} \qquad (4.38)$$

where V and I are the rms values of the sinusoidal voltage and the current, respectively. Thus

$$I = \frac{V}{R} \text{ ampere} \qquad (4.39)$$

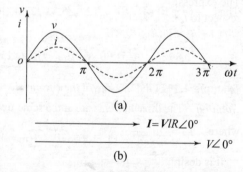

Fig. 4.22 (a) Response of a resistive circuit to ac voltage input (b) Phasor diagram

From Eqs (4.34) and (4.36) it can be seen that both the voltage and current phasors have similar waveforms and angular speeds. The two phasors are actually coincident or in phase. Thus, if voltage is taken as the reference phasor, then the current phasor also has the same position, as shown in Fig. 4.22(b). Usually the voltage and the current phasors of a resistive circuit are drawn slightly apart so that each may be clearly recognized.

4.9.2 Purely Inductive Circuits

A purely inductive circuit is that which possesses inductance only and has no resistance or capacitance. The nearest to such a circuit is obtained by winding a coil of heavy-section copper wire on a laminated iron core. Such a coil is called a choke coil, inductor, or reactor.

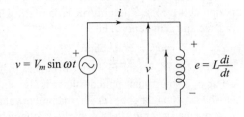

Figure 4.23 shows a pure inductor of L henry with negligible resistance connected across a sinusoidal ac voltage source. The instantaneous value v of the source voltage is given by

Fig. 4.23 Sinusoidal voltage applied to pure inductance

$$v = V_m \sin \omega t = \text{Im} \{V_m e^{j\omega t}\} \qquad (4.40)$$

where V_m is the maximum value of voltage in volt, $\omega = 2\pi f$ rad/sec, and $f = 50$ Hz.

The alternating current flowing through the inductance coil will set up alternating magnetic field; hence, the magnitude of the magnetic field will be changing at every instant. According to the laws of electromagnetic induction a self-induced emf, e, is developed across the inductance coil. Since there is no resistance, the applied voltage has to oppose the self-induced emf only. Hence, the applied voltage is equal and opposite to the self-induced emf at every instant. But, the circuit will take a fraction of the time to settle the transients which occur when the circuit is first switched on. The following relations have been obtained on the assumption that the transients have died down.

$$v = -e = L \frac{di}{dt} \qquad (4.41)$$

Substituting Eq. (4.40) into Eq. (4.41)

$$V_m \sin \omega t = L \frac{di}{dt}$$

or,

$$di = \int \frac{V_m}{L} \sin \omega t \, dt$$

The expression for the circuit current in the steady state is obtained by integrating the above equation with respect to t, as an indefinite integral. Thus the expression for current becomes

$$i = -\frac{V_m}{\omega L} \cos \omega t = \frac{V_m}{\omega L} \sin \left(\omega t - \frac{\pi}{2} \right) = \text{Im} \left\{ \frac{V_m}{\omega L} e^{j[\omega t - (\pi/2)]} \right\} \qquad (4.42)$$

$$= I_m \sin \left(\omega t - \frac{\pi}{2} \right) = \text{Im} \left\{ I_m e^{j[\omega t - (\pi/2)]} \right\} \qquad (4.43)$$

where

$$I_m = \frac{V_m}{\omega L} \qquad (4.44)$$

It is desirable to deal with rms or effective values of voltage and current rather than the maximum values. Thus Eq. (4.44) may be expressed as

$$I_m = \sqrt{2} \, I = \frac{\sqrt{2} \, V}{\omega L} \qquad (4.45)$$

where V and I are the rms values of the sinusoidal voltage and current, respectively.

Thus $\quad I = \dfrac{V}{\omega L} \quad$ ampere $\hspace{2cm}$ (4.46)

The term ωL is called the inductive reactance and is denoted by X_L. The unit of X_L is ohm, when ω is in rad/sec and L in henry. Thus

$$X_L = \omega L = 2\pi f L \quad \text{ohm} \hspace{2cm} (4.47)$$

From Eqs (4.40) and (4.43) it can be seen that the current set up by a sinusoidal applied voltage is also sinusoidal and has the same angular speed ω. The maximum value of current I_m is reached $\pi/2$ rad after the applied voltage has attained maximum value of V_m. Thus the current phasor lags the applied voltage phasor by $\pi/2$ rad. In other words, the current in an inductance lags the voltage by $\pi/2$ rad. Figure 4.24(a) shows voltage and current waveforms for a purely inductive circuit.

If the rms value of the source voltage is taken to be the reference phasor, that is, $V = V\angle 0°$, then the current phasor I in the pure inductive circuit may be represented as

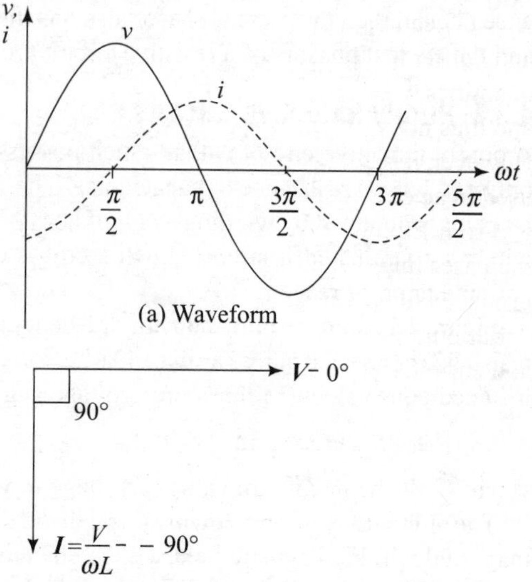

$$\boldsymbol{I} = \dfrac{V}{\omega L} \angle{-90°} = \dfrac{V}{\omega L} e^{-j\frac{\pi}{2}} = \dfrac{V}{\omega L\, e^{j\frac{\pi}{2}}}$$

$$= \dfrac{V}{\omega L\left[\cos\left(\dfrac{\pi}{2}\right) + j\sin\left(\dfrac{\pi}{2}\right)\right]} = \dfrac{V}{\omega L(0+j)} = \dfrac{V}{j\omega L} = \dfrac{V}{jX_L}$$

$$\hspace{12cm} (4.48)$$

(a) Waveform

(b) Phasor diagram

Fig. 4.24 Response of a purely inductive circuit to ac voltage input

Figure 4.24(b) shows the phasor diagram in which the rms values of voltage and current have been used.

Alternatively, employing the j operator to represent the phasor voltage and current in the complex plane, the voltage phasor V and the current phasor I may be written as

$$V = V + j0 \quad \text{and} \quad \boldsymbol{I} = 0 - jI$$

Hence $\quad \dfrac{V}{I} = \dfrac{V + j0}{0 - jI} = \dfrac{V}{-jI} = \dfrac{V}{-jI} \times \dfrac{j}{j} = j\dfrac{V}{I} = j\omega L = jX_L = \boldsymbol{X}_L \hspace{1.5cm} (4.49)$

since from Eq. (4.46) the ratio $V/I = \omega L$.

Alternatively, employing the j operator to represent the phasor voltage and current in the complex plane, the voltage phasor V and the current phasor I may be written as

$$V = V + j0 \quad \text{and} \quad \boldsymbol{I} = 0 - jI$$

Hence

$$\dfrac{V}{I} = \dfrac{V + j0}{0 - jI} = \dfrac{V}{-jI} = \dfrac{V}{-jI} \times \dfrac{j}{j} = j\dfrac{V}{I} = j\omega L = jX_L = \boldsymbol{X}_L \hspace{1.5cm} (4.49)$$

since from Eq. (4.46) the ratio $V/I = \omega L$.

It may be noted from Eq. (4.49) that the ratio of the phasor voltage V across an inductance and the phasor current I flowing through it gives the complex inductive reactance X_L. Thus, the inductive reactance

$X_L = jX_L$ can be represented in the complex plane. The phasors **V** and **I** differ from X_L, in that they are associated with time-varying quantities, whereas X_L is a complex number independent of time and thus not a phasor in the same sense as **V** and **I** are phasors. However, it is a common practice to refer to all complex numbers used in ac circuit computations as phasors.

From Eq. (4.49) it can be seen that the inductive reactance increases linearly with the increase in frequency and the current produced for a given voltage varies inversely with the frequency variation. Figure 4.25 shows the variation of inductive reactance and current with frequency.

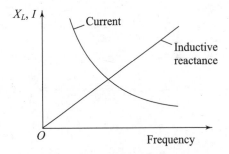

Fig. 4.25 Variation of inductive reactance and current with frequency

4.9.3 Purely Capacitive Circuit

A purely capacitive circuit is that which possesses capacitance only and no resistance. This can be realized when no heat is produced while current flows through the capacitor. However, in commercial capacitors a very small amount of loss takes place in the dielectric in addition to a very minute i^2R loss due to the flow of current in plates of the capacitor having a definite ohmic resistance.

A capacitor of C farad is connected across an ac voltage source, as shown in Fig. 4.26. The instantaneous value v of the source voltage is given by

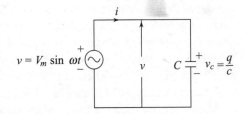

Fig. 4.26 Capacitor connected across ac voltage source

$$v = V_m \sin \omega t = \text{Im}\{V_m e^{j\omega t}\} \tag{4.50}$$

where V_m is the maximum value of voltage in volt, $\omega = 2\pi f$ rad/sec and $f = 50$ Hz.

Assume that the charge on the capacitor at any instant of time t is q coulomb. The potential difference, say v_c, across C due to the charge can be written as

$$v_c = \frac{q}{C} \text{ volt}$$

Applying KVL around the closed circuit after all transients have died down,

$$v - v_c = 0 \quad \text{or} \quad v_c = v$$

Then, $\quad \dfrac{q}{C} = V_m \sin \omega t \tag{4.51}$

Differentiating Eq. (4.51) gives the current flowing in the circuit as

$$i = \frac{dq}{dt} = C \frac{d}{dt}(V_m \sin \omega t) = \omega C V_m \cos \omega t$$

$$= \frac{V_m}{1/\omega C} \cos \omega t = \frac{V_m}{X_C} \sin\left(\omega t + \frac{\pi}{2}\right) \tag{4.52}$$

$$= I_m \sin\left(\omega t + \frac{\pi}{2}\right) = I_m e^{j(\omega t + \pi/2)} \tag{4.53}$$

The term $1/(\omega C)$ is called the capacitive reactance and is denoted by X_C. The unit of X_C is ohm, when ω is in rad/sec and C in farad. Thus

$$X_C = \frac{1}{\omega C} = \frac{1}{2\pi fC} \text{ ohm} \qquad (4.54)$$

The maximum current I_m is given by

$$I_m = \sqrt{2}I = \frac{V_m}{1/\omega C} = \frac{V_m}{X_C} = \frac{\sqrt{2}V}{X_C} \qquad (4.55)$$

and $$I = \frac{V}{X_C} \qquad (4.56)$$

where V and I are the rms values of voltage and current, respectively.

From Eqs (4.50) and (4.52) it can be seen that the current set up by a sinusoidal applied voltage is also sinusoidal and has the same angular speed ω. The maximum value of current I_m is reached $\pi/2$ rad before the applied voltage has attained the maximum value of V_m. Thus the current phasor leads the applied voltage phasor by $\pi/2$ rad. In other words, the current flowing in the capacitance leads the voltage applied to the capacitance by $\pi/2$ rad. Figure 4.27(a) shows voltage and current waveforms for a purely capacitive circuit.

If the rms value of the source voltage is taken to be the reference phasor, that is, $V = V\angle 0°$, then the current phasor I in a purely capacitive circuit may be represented as

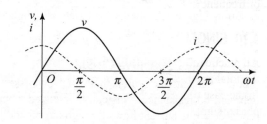

(a) Waveforms

$I = \dfrac{V}{1/\omega c}\angle 90°$

(b) Phasor diagram

Fig. 4.27   Response of a capacitive circuit to ac voltage input

$$I = \frac{V}{(1/\omega C)}\angle 90° = \frac{V}{(1/\omega C)}e^{j\frac{\pi}{2}} = \frac{V}{(1/\omega C)e^{-j\frac{\pi}{2}}}$$

$$= \frac{V}{\dfrac{1}{\omega C}\left[\cos\left(-\dfrac{\pi}{2}\right)+j\sin\left(-\dfrac{\pi}{2}\right)\right]} = \frac{V}{\dfrac{1}{\omega C}(0-j)} \qquad (4.57)$$

$$= \frac{V}{-\dfrac{j}{\omega C}} = \frac{V}{-jX_C} = \frac{V}{-jX_C}\times\frac{j}{j} = j\omega CV \qquad (4.58)$$

The phasor diagram showing the relationship between the rms values of the applied voltage and current phasors is given in Fig. 4.27(b). The current phasor will lead the applied voltage by $\pi/2$. Alternatively, employing the j operator to represent the phasor voltage and current in the complex plane, the voltage phasor V and the current phasor I may be written as

$$V = V + j0 \quad \text{and} \quad I = 0 + jI$$

Hence $\dfrac{V}{I} = \dfrac{V+j0}{0+jI} = \dfrac{V}{jI} = \dfrac{V}{jI}\times\dfrac{j}{j} = -\dfrac{j}{\omega C} = -\dfrac{j}{2\pi fC} = -jX_C = X_C \qquad (4.59)$

Equation (4.59) provides the representation of capacitive reactance $X_C = -jX_C$ in the complex plane. From Eq. (4.54) it is evident that the value of the capacitive reactance X_C is inversely proportional to frequency f. When $f = 0$, $X_C = \infty$ and when $f = \infty$, $X_C = 0$. From Eq. (4.58) it can be seen that the current produced in a capacitance when an ac voltage is applied to it is proportional to frequency. Figure 4.28 shows the variation of the capacitive reactance X_C and current I for different values of frequency.

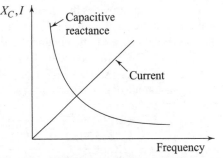

Fig. 4.28 Variation of capacitive reactance and current with frequency

4.10 SINGLE-PHASE SERIES CIRCUITS

All practical circuits consist of two or more of the elements R, L, and C. In the case of dc circuits, as per KVL, the total applied pd is equal to the algebraic sum of the voltage drops in the individual parts of the circuit. In the case of ac circuits under steady state, KVL is applicable with the exception that the sum must be the phasor sum; thus the applied voltage is equal to the sum of the phasors representing the voltage drops in the individual elements constituting the circuit. In a series circuit current is common to all elements of the circuit. Hence it is convenient to use the current phasor as the reference phasor.

4.10.1 Resistance and Inductance in Series

A series combination of a resistance of R ohm and an inductance of L henry is connected across an ac voltage source as shown in Fig. 4.29. The resistance R is assumed to include the resistance of the inductance coil.

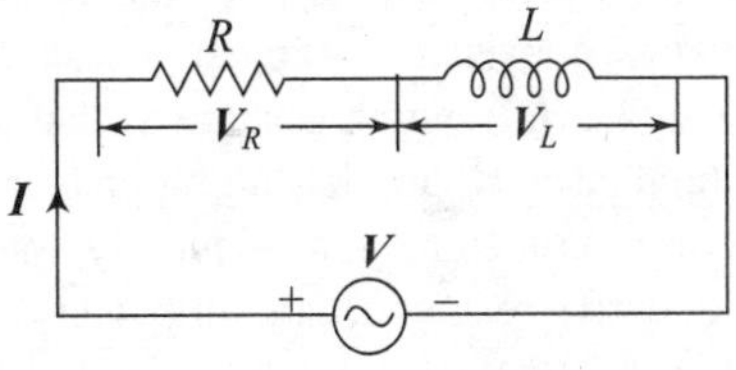

As per the KVL, for ac networks the phasor sum of voltages in a closed network is zero. Thus,

$$V - V_R - V_L = 0 \qquad (4.60)$$

Fig. 4.29 *RL series network*

where V is the source voltage phasor while V_R and V_L are the phasors representing the voltage drops across R and ωL, respectively. Suppose that the current flowing through the circuit is I ampere. In a series circuit the current is chosen as the reference phasor, thus $I = Ie^{j0} = I\angle 0°$. The voltage drops may be found as

$$V_R = IR$$

where V_R is in phase with I and is equal to $V_R \angle 0°$.

$$V_L = jIX_L$$

where V_L leads I by 90° and is equal to $V_L \angle 90°$.

The presence of the rotational operator j in the term jIX_L indicates that the quantity IX_L is to be rotated in the counterclockwise (or positive) direction through 90°. Then Eq. (4.60) may be written as

$$V = V_R + V_L = IR + jIX_L = I(R + jX_L) = IZ_L \qquad (4.61)$$

Therefore,

$$Z_L = V/I \qquad (4.62)$$

where $Z_L = (R + jX_L)$ $\qquad (4.63)$

The term Z_L is known as the complex inductive impedance of the RL circuit.

From Eq. (4.62) it may be observed that the complex impedance Z_L is the ratio between the voltage and current of an ac circuit. This is similar to Ohm's law in a dc circuit which gives the relation $V = IR$. For a given frequency, the impedance Z_L is constant and $V = IZ_L$. Hence Ohm's law applies to ac circuits as well.

Figure 4.30(a) provides the phasor representation of the voltages and the current phasor. From the phasor diagram the source voltage V can be obtained as the complex sum of V_R and V_L. The arithmetic sum of V_R and V_L is incorrect and gives too large a value.

The magnitude of the source voltage V can be found out from the phasor diagram as

$$V = \sqrt{V_R^2 + V_L^2} = \sqrt{I^2 R^2 + I^2 X_L^2}$$

$$= I\sqrt{R^2 + X_L^2}$$

Thus $\quad V = IZ_L$ \hfill (4.64)

where Z_L is the magnitude of the complex impedance Z_L and is given by

$$Z_L = \sqrt{R^2 + X_L^2} = \sqrt{R^2 + \omega^2 L^2} \qquad (4.65)$$

Figure 4.30(b) shows the waveforms of v and i for an RL circuit excited by an ac source. The current lags the voltage by φ. The value of angle φ depends on the ratio of resistance

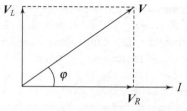

(a) Phasor diagram

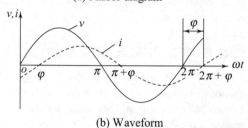

(b) Waveform

Fig. 4.30 Phasor diagram and waveforms of v and i of an RL series network

and inductive reactance of the circuit. Greater is the value of this ratio, lesser is the value of φ. For a finite value of resistance, φ must be less than 90°. This can be shown with the help of the phasor diagram given in Fig. 4.31(a), which is same as that given in Fig. 4.30(a). In Fig. 4.31(a) the phasor V_L is drawn at the tip of the phasor V_R and is leading phasor I by 90° to form the summation triangle. Each side of the triangle has a common factor I. Consequently, the same triangle can represent to some other scale the value of resistance, inductive reactance, and impedance. Such a triangle is named impedance triangle and is shown separately in Fig. 4.31(b).

From Fig. 4.31 angle φ can be determined using trigonometrical relationships, as the triangles are right-angled. Thus

$$\varphi = \tan^{-1} \frac{V_L}{V_R} = \tan^{-1} \frac{IX_L}{IR} = \tan^{-1} \frac{X_L}{R} = \tan^{-1} \frac{\omega L}{R} \qquad (4.66)$$

To show that the current lags the voltage, normally a negative sign is assigned to the angle φ, or the word lag is attached to it.

Alternatively, the phase angle φ may be determined as follows:

$$\varphi = \cos^{-1} \frac{V_R}{V} = \cos^{-1} \frac{R}{Z_L} = \cos^{-1} \frac{R}{\sqrt{R^2 + \omega^2 L^2}} \qquad (4.67)$$

Both the magnitude ratio and the phase shift between the voltage and current may be expressed in terms of a complex impedance Z_L of the circuit. Complex impedance Z_L may be expressed in the polar form as $Z_L = Z_L \angle \varphi$. The phase angle φ represents the angle by which the voltage leads the current and is known as the impedance angle.

A complex impedance $Z_L = Z_L \angle \varphi$ can be converted to the Cartesian form as follows:

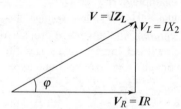

(a) Phasor voltage triangle

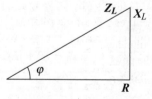

(b) Impedance triangle

Fig. 4.31 Voltage and impedance triangle

$$Z_L = Z_L \angle \varphi = Z_L e^{j\varphi} = Z_L(\cos \varphi + j \sin \varphi) \qquad (4.68)$$

Then $\quad R = Z_L \cos \varphi \quad$ and $\quad X_L = Z_L \sin \varphi \qquad (4.68a)$

The reciprocal of impedance is admittance and is denoted by Y. The unit of Y is mho and is written as

$$V_L = \frac{1}{Z_L} = \frac{1}{R + jX_L} = \frac{1}{R + jX_L} \times \frac{R - jX_L}{R - jX_L}$$

$$= \frac{R}{R^2 + X_L^2} - j\frac{X_L}{R^2 + X_L^2}$$

$$= G_L - jB_L \qquad (4.69)$$

where $\quad G_L = R/(R^2 + X_L^2) \quad$ and $\quad B_L = X_L/(R^2 + X_L^2) \qquad (4.70)$

The term G_L is called the conductance and the term B_L is the susceptance of the admittance.

Example 4.12 A resistance of 10 Ω is connected in series with a 50 mH inductance across a 230-V, 50-Hz supply. Calculate: (a) the current flowing in the circuit, (b) the phase angle of current.

Solution (a) Let the supply voltage V be assumed as the reference phasor. Then $V = 230\angle 0°$.

$$X_L = 2\pi fL = 2\pi \times 50 \times 50 \times 10^{-3} = 15.71 \ \Omega$$
$$\mathbf{Z} = R + jX_L = 10 + j15.71$$

$\therefore \qquad Z = \sqrt{10^2 + 15.71^2} = \sqrt{346.8} = 18.62 \ \Omega$

$$I = \frac{V}{Z} = \frac{230}{18.62} = 12.35 \ \text{A}$$

(b) $\varphi = -\tan^{-1} \dfrac{X_L}{R} = -\tan^{-1} \dfrac{15.71}{10} = -57.52° \ \text{lag or} \ \angle -57.52°$

Example 4.13 A resistance of 50 Ω is connected in series with a pure inductor of 250 mH. The circuit is connected to a 50 Hz sinusoidal supply and the voltage across the resistance is 150 V. Calculate the supply voltage.

Solution The current in the series circuit

$$I = \frac{V_R}{R} = \frac{150}{50} = 3 \ \text{A}$$

Now, $\quad X_L = 2\pi fL = 2\pi \times 50 \times 250 \times 10^{-3} = 78.5 \ \Omega$

$\qquad V_L = I \times X_L = 3 \times 78.5 = 235.5 \ \text{V}$

Then $\quad V = \sqrt{V_R^2 + V_L^2} = \sqrt{150^2 + 235.5^2} = \sqrt{77,960.25} = 279.21 \ \text{V}$

Example 4.14 A supply of 200 V, 50 Hz is connected to a 20 Ω resistor in series with a choke coil. The reading of the voltmeter across the resistor is 100 V and across the coil is 144 V. Calculate: (a) the power factor of the circuit, (b) the power consumed in the resistance, (c) the power consumed in the coil.

Solution (a) The network and phasor diagrams for this example are shown in Fig. 4.32.

The phasor diagram has been drawn with the current phasor chosen as the reference phasor. The voltage V_R across the resistor is in phase with the current phasor, while the voltage across the coil leads the current by the impedance angle θ of the coil.

$$\text{Current } I = \frac{V_R}{R} = \frac{100}{20} = 5 \ \text{A}$$

In the phasor diagram shown in Fig. 4.32, $V = 200$ V, $V_L = 150$ V, and $V_R = 100$ V. Then,

$$(V_R^2 + V_L \cos \theta)^2 + (V_L \sin \theta)^2 = V^2$$

or $\quad(100 + 150 \cos \theta)^2 + (150 \sin \theta)^2 = 200^2$

$$\cos \theta = \frac{200^2 - 100^2 - 150^2}{2 \times 100 \times 150} = \frac{7500}{30,000} = 0.25$$

and $\quad \theta = 75.52°$

The power factor

$$\cos \varphi = \frac{V_R + V_L \cos \theta}{V} = \frac{100 + 150 \times 0.25}{200} = 0.6875$$

(b) Power consumed in the resistance $= 5^2 \times 20 = 500$ W

(c) Power consumed in the choke coil $= V_L I \cos \theta = 150 \times 5 \times 0.25$
$$= 187.5 \text{ W}$$

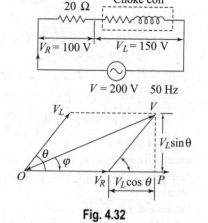

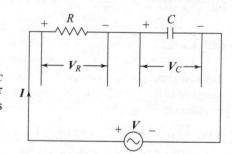

Fig. 4.32

4.10.2 Resistance and Capacitance in Series

The treatment of an *RC* series network connected across an ac voltage source is exactly similar to the *RL* series network as discussed in Section 4.10.1. Figure 4.33 shows a resistance of *R* ohm in series with a capacitor of *C* farad connected across a voltage source of *V* volt.

As per the KVL, for ac networks the phasor sum of voltages in a closed network is zero. Thus,

$$V - V_R - V_C = 0 \tag{4.71}$$

where *V* is the source voltage phasor while V_R and V_C are the phasors representing the voltage drops across *R* and $1/\omega C$, respectively. Suppose that the current flowing through the circuit is *I* ampere. In a series circuit, the current is chosen as the reference phasor, thus $I = Ie^{j0} = I\angle 0°$. The voltage drops may be found as

$$V_R = IR$$

where V_R is in phase with *I* and is equal to $V_R \angle 0°$

$$V_C = -jIX_C$$

where V_C lags *I* by 90° and is equal to $V_C \angle -90°$.

The presence of the rotational operator $-j$ in the term $-jIX_C$ indicates that the quantity IX_C is to be rotated in the clockwise (or negative) direction through 90°. Then Eq. (4.71) may be written as

$$V = V_R + V_C = IR - jIX_C = I(R - jX_C) = IZ_C \tag{4.72}$$

Therefore,

$$Z_C = V/I \tag{4.73}$$

where $\quad Z_C = R - jX_C \tag{4.74}$

The term Z_C is known as the complex capacitive impedance of the *RC* circuit.

Fig. 4.33 *RC* series network

Figure 4.34(a) provides the phasor representation of the voltages and the current phasor. From the phasor diagram, the source voltage *V* can be obtained as the complexor sum of V_R and V_C.

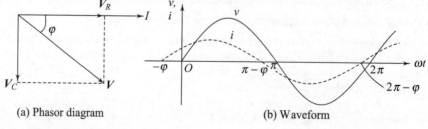

(a) Phasor diagram

(b) Waveform

Fig. 4.34 Phasor diagram and waveforms of *v* and *i* of an *RC* series network

The magnitude of the source voltage V can be found out from the phasor diagram as

$$V = \sqrt{V_R^2 + V_C^2} = \sqrt{I^2 R^2 + I^2 X_C^2}$$
$$= I \sqrt{R^2 + X_C^2}$$

Thus $\quad V = I Z_C$ $\qquad\qquad$ (4.75)

where Z_C is the magnitude of the complex capacitive impedance $\mathbf{Z}_C$ and is given by

$$Z_C = \sqrt{R^2 + X_C^2} = \sqrt{R^2 + \frac{1}{\omega^2 C^2}} \qquad\qquad (4.76)$$

Figure 4.34(b) shows the waveforms of v and i for the RC circuit excited by an ac source. The current leads the voltage by an angle φ. The value of angle φ depends on the ratio of resistance and capacitive reactance of the circuit. Greater is the value of the ratio, lesser is the value of angle φ. For a finite value of resistance, φ must be less than 90°. This can be shown with the help of the phasor diagram given in Fig. 4.35(a), which is same as that given in Fig. 4.34(a). In Fig. 4.35(a) the phasor $\mathbf{V}_C$ is drawn at the tip of the phasor $\mathbf{V}_R$ and is lagging phasor $\mathbf{I}$ by 90° to form the summation triangle. Each side of the triangle has a common factor $\mathbf{I}$. Consequently the same triangle can represent to some other scale the value of resistance, capacitive reactance, and impedance. Such a triangle is named capacitive impedance triangle and is shown separately in Fig. 4.35(b).

From Fig. 4.35 the angle φ can be determined using trigonometrical relationships because the triangles are right-angled. Thus

$$\varphi = \tan^{-1} \frac{V_C}{V_R} = \tan^{-1} \frac{IX_C}{IR} = \tan^{-1} \frac{X_C}{R} = \tan^{-1} \frac{1}{\omega CR}$$

$\qquad\qquad$ (4.77)

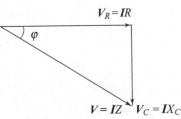

(a) Phasor voltage triangle

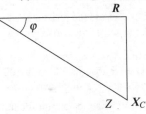

(b) Impedance triangle

Fig. 4.35 Voltage and impedance triangle

To show that the current leads the voltage, normally a positive sign is assigned to the angle φ, or the word lead is attached to it.

Alternatively, the phase angle φ may be determined as follows:

$$\varphi = \cos^{-1} \frac{V_R}{V} = \cos^{-1} \frac{R}{Z_C} = \cos^{-1} \frac{R}{\sqrt{R^2 + \frac{1}{\omega^2 C^2}}} \qquad\qquad (4.78)$$

Both the magnitude ratio and the phase shift between the voltage and current may be expressed in terms of a complex capacitive impedance Z_C of the circuit. Complex impedance Z_C may be expressed in the polar form as $\mathbf{Z}_C = Z_C \angle -\varphi$. The phase angle φ represents the angle by which the voltage lags the current and is known as the capacitive impedance angle.

A complex impedance $\mathbf{Z}_C = Z_C \angle -\varphi$ can be converted into the Cartesian form as follows:

$$\mathbf{Z}_C = Z_C \angle -\varphi = Z_C e^{-j\varphi} = Z_C (\cos \varphi - j\sin\varphi) \qquad\qquad (4.79)$$

The reciprocal of impedance Z_C is admittance and is denoted by Y_C. The unit of Y_C is mho and is written as

$$Y_C = \frac{1}{Z_C} = \frac{1}{R - jX_C} = \frac{1}{R - jX_C} \times \frac{R + jX_C}{R + jX_C}$$

$$= \frac{R}{R^2 + X_C^2} + j \frac{X_C}{R^2 + X_C^2} = G_C + jB_C \qquad\qquad (4.80)$$

where

$$G_C = R/(R^2 + X_C^2) \quad \text{and} \quad B_C = X_C/(R^2 + X_C^2) \tag{4.81}$$

The term G_C is called conductance and the term B_C is the susceptance of the admittance.

Example 4.15 A 10 Ω resistor is connected in series with a 100 μF capacitor to a 230-V, 50-Hz supply. Find: (a) the impedance, (b) current, (c) power factor, (d) phase angle, (e) voltage across the resistor and the capacitor.

Solution   (a) Capacitive reactance

$$X_C = \frac{1}{\omega C} = \frac{1}{314 \times 100 \times 10^{-6}} = 31.85 \ \Omega$$

Impedance $Z_C = 10 - j31.85 = 33.38 \ \angle{-72.6°}$

(b) Current $I = \dfrac{V}{Z_C} = \dfrac{230}{33.38} = 6.89 \ A$

(c) Power factor $\cos\varphi = \dfrac{R}{Z} = \dfrac{10}{33.38} = 0.2995 \approx 0.3$

(d) Phase angle $\varphi = \cos^{-1} 0.3 = 72.6°$

(e) Voltage across the resistor $R \times I = 10 \times 6.89 = 68.9 \ V$

Voltage across the capacitor $= X_C \times I = 31.85 \times 6.89 = 219.45 \ V$

Example 4.16 A resistor R in series with a capacitor C is connected to a 230-V, 50-Hz supply. Find (a) the value of C so that R absorbs 500 W at 100 V, (b) the maximum voltage across the capacitor, (c) the phase angle between the current and the supply voltage.

Solution   The series circuit diagram for the example is given in Fig. 4.36. The voltage V_R across R is in phase with the current I while the voltage V_C across C lags I by 90°. The supply voltage is the phasor sum of V_R and V_C.

(a) From the phasor diagram, $V^2 = V_R^2 + V_C^2$.

$$(230)^2 = (100)^2 + V_C^2$$

$$V_C = 207.12 \ V$$

The current through R is $\dfrac{500 \ W}{100 \ V} = 5 \ A$

Now, $I = \omega C V_C = 2\pi f C V_C$

or $5 = 2 \times \pi \times 50 \times C \times 207.12$

∴ $C = 76.88 \ \mu F$

(b) Maximum voltage across C is

$$\sqrt{2} \times V_C = \sqrt{2} \times 207.12 = 292.91 \ V$$

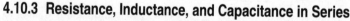

Fig. 4.36   (a) Circuit diagram for Example 4.13 (b) Phasor diagram

(c) Phase angle

$$\varphi = \cos^{-1} \frac{V_R}{V} = \cos^{-1}\left(\frac{100}{230}\right) = \cos^{-1}(0.4348) = 64.2°$$

4.10.3 Resistance, Inductance, and Capacitance in Series

Figure 4.37 shows the series combination of a pure resistance of R ohm, an inductance of L henry, and a capacitance of C farad connected to a sinusoidal ac voltage source of V volt (rms) and frequency f hertz. Let the rms current in the series circuit be I ampere.

Applying KVL around the network yields

$$V_R + V_L + V_C = V \tag{4.82}$$

In the series circuit the current is chosen as the reference phasor, thus $\boldsymbol{I} = Ie^{j0} = I\angle 0°$. The voltage drops may be found as

$$V_R = \boldsymbol{I}R$$

where V_R is in phase with $\boldsymbol{I}$ and is equal to $V_R \angle 0°$

$$V_L = j\boldsymbol{I}X_L$$

where V_L leads $\boldsymbol{I}$ by 90° and is equal to $V_L \angle 90°$.

$$V_C = -j\boldsymbol{I}X_C$$

where V_C lags $\boldsymbol{I}$ by 90° and is equal to $V_C \angle -90°$

Then Eq. (4.82) may be written as

$$V = V_R + V_R + V_C = \boldsymbol{I}R + j\boldsymbol{I}X_L - j\boldsymbol{I}X_C$$

$$= \boldsymbol{I}\,[R + j\,(X_L - X_C)] = \boldsymbol{I}Z_C \tag{4.83}$$

Therefore, $Z = \dfrac{V}{\boldsymbol{I}}$ \hfill (4.84)

where $Z = R + j(X_L - X_C)$ \hfill (4.85)

The term Z is known as the complex impedance of the *RLC* circuit.

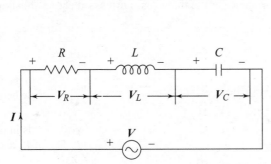

Fig. 4.37   *RLC* series network

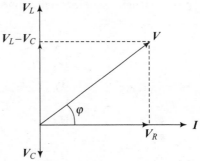

Fig. 4.38   Phasor diagram for *RLC* series network

Figure 4.38 provides the phasor representation of the voltages and the current phasor. The phasors V_L and V_C are opposite to each other and the phasor diagram is for the case when $V_L > V_C$. The source voltage V can be obtained from the phasor diagram as the complex sum of V_R, V_L, and V_C.

Then $V = \sqrt{V_R^2 + (V_L^2 - V_C^2)} = \sqrt{I^2 R^2 + I^2(X_L^2 - X_C^2)}$

$$= I\sqrt{R^2 + (X_L^2 - X_C^2)}$$

Thus $V = IZ$ \hfill (4.86)

where Z is the magnitude of the total complex impedance Z of the circuit and is given by

$$Z = \sqrt{R^2 + (X_L^2 - X_C^2)} = \sqrt{R^2 + \left(\omega^2 L^2 - \frac{1}{\omega^2 C^2}\right)} \tag{4.87}$$

From Fig. 4.38 the angle φ can be determined as

$$\varphi = \tan^{-1} \frac{V_L - V_C}{V_R} = \tan^{-1}\frac{I(X_L - X_C)}{IR} = \tan^{-1}\frac{X_L - X_C}{R}$$

$$= \tan^{-1} \frac{\omega L - \dfrac{1}{\omega C}}{R} \tag{4.88}$$

Alternatively, the phase angle φ may be determined as follows:

$$\varphi = \cos^{-1} \frac{V_R}{V} = \cos^{-1} \frac{R}{Z} = \cos^{-1} \frac{R}{\sqrt{R^2 + \left(\omega^2 L^2 - \dfrac{1}{\omega^2 C^2}\right)}} \tag{4.89}$$

If the inductive reactance is greater than the capacitive reactance, then φ is positive and the current lags the voltage by φ.

From Eq. (4.88) it may be observed that the phase angle φ is dependent on the relative values of X_L and X_C of the series circuit. There are three cases to the general *RLC* series circuit as follows:

(i) $X_L > X_C$, then the total reactance of the circuit is inductive, tan φ is positive, and the current lags the supply voltage by an angle φ.

(ii) $X_L < X_C$, then the total reactance of the circuit is capacitive, tan φ is negative, and the current leads the supply voltage by an angle φ.

(iii) $X_L = X_C$, then the total reactance being zero, the circuit is resistive, tan φ is zero, and the angle $\varphi = 0$. The current and the supply voltage are in phase. This is the condition of resonance in a series circuit and is discussed separately later in this chapter.

The phasor voltage triangle and the impedance triangle for the three cases discussed above are shown in Fig. 4.39.

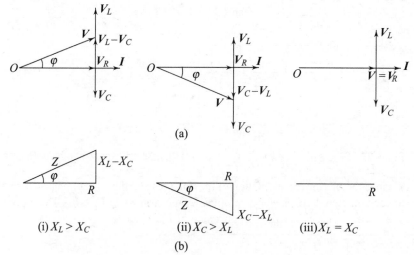

Fig. 4.39   (a) Phasor voltage triangles, (b) impedance triangles

Similarly, the admittance Y of the *RLC* series circuit can be written as

$$Y = \frac{1}{Z} = \frac{1}{R + j(X_L - X_C)}$$

$$= \frac{R}{R^2 + (X_L - X_C)^2} - j\frac{X_L - X_C}{R^2 + (X_L - X_C)^2} \tag{4.90}$$

$$= G + jB$$

where $$G = \frac{R}{R^2 + (X_L - X_C)^2} \tag{4.91}$$

$$B = -\frac{(X_L - X_C)}{R^2 + (X_L - X_C)^2} \tag{4.92}$$

The admittance Y, which is complex, has the unit of mho ($\mho$) or siemen (S) while G is called conductance and B is designated as susceptance.

Example 4.17 A choke coil of resistance 8 Ω and inductance 0.15 H is connected in series with a capacitor of capacitance 125 μF across a 230-V, 50-Hz supply. Calculate: (a) the inductive reactance, (b) capacitive reactance, (c) impedance, (d) current, (e) voltage across the coil and the capacitor, respectively, (f) phase difference between the current and the supply voltage.

Solution (a) Inductive reactance

$$X_L = 2\pi f L = 2 \times \pi \times 50 \times 0.15 = 314 \times 0.15 = 47.1 \ \Omega$$

(b) Capacitive reactance

$$X_C = \frac{1}{2\pi fC} = \frac{1}{314 \times 125 \times 10^{-6}} = \frac{10^6}{314 \times 125} = 25.48 \ \Omega$$

(c) Complex impedance

$$Z = R + j(X_L - X_C) = 8 + j(47.1 - 25.48) = 8 + j21.62$$

Then $Z = \sqrt{8^2 + 21.62^2} = \sqrt{64 + 467.42} = \sqrt{531.42} = 23.05$

Impedance angle

$$\varphi = \tan^{-1}\frac{(X_L - X_C)}{R} = \tan^{-1}\frac{21.62}{8} = \tan^{-1} 2.7025 = 69.7°$$

Thus $Z = 23.05 \angle 69.7°$

(d) Let the voltage phasor be chosen as the reference. Then, $V = 230\angle 0°$.

Current $I = \dfrac{V}{Z} = \dfrac{230\angle 0°}{23.05\angle 69.7°} = 9.98\angle -69.7°$

(e) Voltage across the coil

$$(R + jX_L) \times I = (8 + j47.1) \times 9.98\angle -69.7°$$
$$= 44.77\angle 80.36° \times 9.98\angle -69.7° = 446.8\angle 10.66° \ V$$

Voltage across the capacitor

$$-jX_C \times I = -25.48\angle -90° \times 9.98\angle -69.7° = -254.29 \angle -159.7 \ V$$

(f) Phase difference between the supply voltage V and the current I is 69.7° lag.

Example 4.18 In the arrangement shown in Fig. 4.40, the condenser C has a capacitance of 50 μF and the current flowing through the circuit is 2.355 A. If the voltages are as indicated, find the applied voltage, frequency, and the loss in the iron-cored choking coil L.

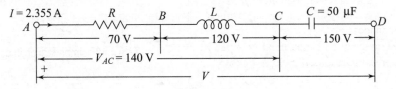

Fig. 4.40

Solution The phasor diagram for the circuit shown in Fig. 4.40, choosing the current phasor I as the reference, is shown in Fig. 4.41.

The phasor sum of V_R and V_L is the phasor OC; the applied voltage V that is the phasor sum of V_C and OC is represented by the phasor OV. Let θ to be the impedance angle of the RL combination. From the right-angled triangle OCD, the angle θ can be determined as follows:

$$(V_R + V_L \cos\theta)^2 + (V_L \sin\theta)^2 = OC^2 = (V_{AC})^2$$

$$(70 + 120 \cos \theta)^2 + (120 \sin \theta)^2 = 150^2$$
$$16{,}800 \cos \theta = 22{,}500 - 14{,}400 - 4900$$

$$\cos\theta = \frac{3200}{16{,}800} = 0.1905$$

$\therefore \qquad \theta = 79°$

Impendance of the coil $Z_L = \dfrac{V_L}{I} = \dfrac{120}{2.355} = 50.95 \ \Omega$

Inductance of the coil $X_L = Z_L \sin \theta = 50.95 \times \sin 79° = 50.01 \ \Omega$

Resistance of the coil $r = Z_L \cos\theta = 50.95 \times \cos 79° = 9.71 \ \Omega$

Voltage drop in X_L $IX_L = 2.355 \times 50.01 = 117.77 \ \text{V} = DC$

Voltage drop in r $\quad Ir = 2.335 \times 9.71 = 22.87 \ \text{V} = BD$

From the right-angled triangle *ODV*,

$$V^2 = (70 + 22.87)^2 + (150 - 117.77)^2 = 9663.61$$
$\therefore \qquad V = 98.3 \ \text{V}$

Capacitive reactance $X_C = \dfrac{150}{2.355} = 63.7 \ \Omega$

Then $\qquad \dfrac{1}{2\pi f \times 50 \times 10^{-6}} = 63.7$

$\qquad\qquad f = 49.995 \approx 50 \ \text{cycles/sec}$

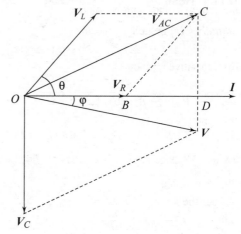

Fig. 4.41

The power loss in the iron-cored choke is $(2.355)^2 \times 9.71 = 53.85 \ \text{W}$.

4.10.4 Impedances in Series

The series circuits considered till now consisted of simple combinations of basic network elements. Quite often several impedances may be connected in series in ac circuits. The computations of currents in such cases become simpler by replacing the impedances connected in series by a single equivalent impedance. The law of combining the impedances connected in series is derived as follows.

Figure 4.42 shows complex impedances Z_1, Z_2, and Z_3 connected in series across an ac voltage source V, and a current I passes through the impedances. Let it be assumed that V_1, V_2, and V_3 are the voltage drop phasors across impedances Z_1, Z_2, and Z_3, respectively, caused by the current I flowing through the impedances.

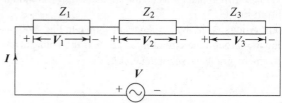

Fig. 4.42 Impedances in series

Applying KVL to the circuit shown in Fig. 4.42, the following expression results: $V = V_1 + V_2 + V_3$

Now, $\quad V_1 = Z_1 \times I, \ \ V_2 = Z_2 \times I, \ \ V_3 = Z_3 \times I$

Then $\quad V = (Z_1 + Z_2 + Z_3) \times I = Z_{eq} \times I$

where $\quad Z_{eq} = Z_1 + Z_2 + Z_3$ \hfill (4.93)

Thus the equivalent complex impedance for a series circuit may be obtained by the addition of all the complex impedances connected in series.

4.11 SINGLE-PHASE PARALLEL CIRCUITS

All parallel circuits consist of two or more series circuits connected to a common voltage source. Thus each branch can be analysed separately as a series circuit. The combined effect of each series circuit can be found out by applying KCL to the common node of the parallel branches. In applying KCL, the phasor addition of all the branch currents is found out to determine the total current supplied by the source. As all individual branches are connected to the common voltage source, it is convenient to use the supply voltage phasor as the reference phasor.

4.11.1 Resistance and Inductance in Parallel

Figure 4.43 shows a resistance of R ohm and a pure inductance of L henry connected in parallel to an ac voltage source of V volt and frequency f hertz. The inductive reactance $X_L = \omega L = 2\pi f L$ ohm.

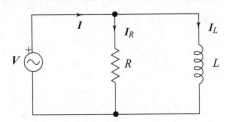

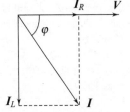

Fig. 4.43 *RL parallel network*   **Fig. 4.44** Phasor diagram

As per the KCL, for ac networks the phasor sum of currents at a node is zero.

Thus, $I = I_R + I_L$ (4.94)

where I_R and I_L are the currents (in ampere) through the resistance and inductance, respectively, and I is the source current (in ampere).

Let the voltage phasor V be the reference phasor since it is common to both the resistance and the inductance. Hence, $V = V\angle 0°$. Then

$$I_R = \frac{V\angle 0°}{R} = \frac{V}{R} \tag{4.95}$$

$$I_L = \frac{V\angle 0°}{j\omega L} = -j\frac{V}{\omega L} = \frac{V}{\omega L}\angle -90° \tag{4.96}$$

From Eqs (4.95) and (4.96) it is clear that the current I_R is in phase with the applied voltage V while the current I_L lags the supply voltage V by $\pi/2$ rad or 90°. The phasor diagram for a parallel RL circuit is shown in Fig. 4.44. Substituting Eqs (4.95) and (4.96) into Eq (4.94),

$$I = \frac{V}{R} - j\frac{V}{\omega L}$$

$$= V\left(\frac{1}{R} - j\frac{1}{\omega L}\right) = \frac{V}{Z} = VY \tag{4.97}$$

where $Y = \frac{1}{Z} = \left(\frac{1}{R} - j\frac{1}{\omega L}\right) = G - jB$ (4.98)

and $G = \frac{1}{R}$ and $B = \frac{1}{\omega L}$ (4.99)

From the phasor diagram it is observed that the source current I lags the applied voltage V by an angle φ. The angle φ is given by

$$\tan\varphi = \frac{I_L}{I_R} = \frac{V/\omega L}{V/R} = \frac{R}{\omega L} = \frac{R}{X_L} = \frac{B}{G} \tag{4.100}$$

Equation (4.97) expresses source current I in the complex form. The source current in the polar form is given as $I = Ie^{-j\varphi}$. The magnitude of current is given by

$$I = \sqrt{\left(\frac{V}{R}\right)^2 + \left(\frac{V}{\omega L}\right)^2} = V\sqrt{\frac{1}{R^2} + \frac{1}{\omega^2 L^2}} = V\sqrt{\frac{1}{R^2} + \frac{1}{X_L^2}} \qquad (4.101)$$

Example 4.19 A resistance of 20 Ω in parallel with a pure inductance of 200 mH is connected to a 230-V, 50-Hz supply. Calculate the total current drawn from the supply and the phase angle between the voltage and the total current.

Solution   The parallel *RL* circuit is shown in Fig. 4.45(a).
Let the supply voltage be chosen as the reference phasor. Then
$V = V\angle 0° = 230\angle 0°$.

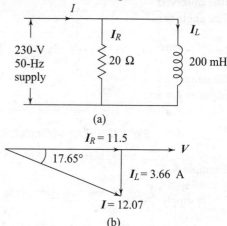

$$R = 20\ \Omega$$
$$X_L = j\omega L = j314 \times 200 \times 10^{-3} = j62.8\ \Omega = 62.8\angle 90°\ \Omega$$
$$I_R = \frac{230\angle 0°}{20} = 11.5\angle 0° = 11.5 + j0\ \text{A}$$
$$I_L = \frac{230\angle 0°}{62.8\angle 90°} = 3.66\angle -90° = 0 - j3.66\ \text{A}$$

Then $I = I_R + I_L = 11.5 - j3.66 = \sqrt{11.5^2 + 3.66^2} \angle \tan^{-1}\dfrac{-3.66}{11.5}$

$\qquad = 12.07\angle -17.65°\ \text{A} = 12.07\ \text{A},\ 17.65°$ lagging
The sketch of the phasor diagram is shown in Fig. 4.45(b).

Fig. 4.45

4.11.2 Resistance and Capacitance in Parallel

The treatment of a resistance and capacitance in parallel across an ac voltage source is exactly similar to the resistance and inductance in parallel as discussed in Section 4.11.1. Figure 4.46 shows the *RC* network in parallel across an ac voltage source of *V* volt. The current flow in the network is indicated in the figure.

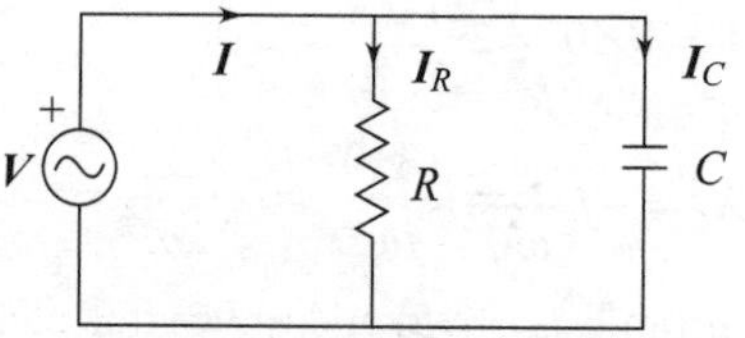

Applying the KCL for the circuit shown in Fig. 4.46, the following equation may be written:

$$I = I_R + I_C \qquad (4.102)$$

Fig. 4.46   *RC parallel network*

where I_R and I_C are the phasor currents (in ampere) through the resistance and capacitance, respectively, and I is the source current (in ampere). Let the voltage phasor V be the reference phasor. Then, $V = V\angle 0°$ and

$$I_R = \frac{V\angle 0°}{R} = \frac{V}{R} \qquad (4.103)$$

$$I_C = \frac{V\angle 0°}{-jX_C} = j\frac{V}{X_C} = j\frac{V}{1/(\omega C)} = jV\omega C = V\omega C\angle 90° \qquad (4.104)$$

It is clear from Eq. (4.103) that the current I_R is in phase with the applied voltage V while from Eq. (4.104) it may be seen that the current I_C leads supply voltage V by $\pi/2$ rad or 90°. The phasor diagram for the parallel *RC* circuit is shown in Fig. 4.47. Substituting Eqs (4.103) and (4.104) into Eq. (4.102),

$$I = \frac{V}{R} + jV\omega C$$

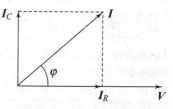

Fig. 4.47   Phasor diagram

$$= V\left(\frac{1}{R} + j\omega C\right) = \frac{V}{Z} = VY \tag{4.105}$$

where $\quad Y = \dfrac{1}{Z} = \left(\dfrac{1}{R} + j\omega C\right) = G + jB \tag{4.106}$

and $\quad G = \dfrac{1}{R}\quad$ and $\quad B = \omega C = \dfrac{1}{X_C} \tag{4.107}$

It is observed from the phasor diagram that the source current I leads the applied voltage V by an angle φ. The angle φ is given by

$$\tan\varphi = \frac{I_C}{I_R} = \frac{V\omega C}{V/R} = R\omega C = \frac{R}{X_C} = \frac{B}{G} \tag{4.108}$$

Equation (4.105) expresses the source current I in the complex form. The source current in the polar form is given as $I = Ie^{-j\varphi}$. The magnitude of current is given by

$$I = \sqrt{\left(\frac{V}{R}\right)^2 + (V\omega C)^2} = V\sqrt{\frac{1}{R^2} + \omega^2 C^2} = V\sqrt{\frac{1}{R^2} + \frac{1}{X_C^2}} \tag{4.109}$$

Example 4.20 A circuit consisting of resistance of 100 Ω in parallel with a pure capacitance of 50 µF is connected to a 230-V, 50-Hz supply. Calculate: (a) the branch currents and the supply current, (b) the circuit phase angle, and (c) the circuit impedance.

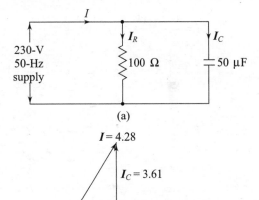

(a)

Solution The parallel *RC* circuit is shown in Fig. 4.48(a). Let the supply voltage be chosen as the reference phasor. Then $V = V\angle 0° = 230\angle 0°$.

$$R = 100\ \Omega$$

$$X_C = -j\frac{1}{\omega C} = \frac{-j}{314 \times 50 \times 10^{-6}} = -j63.7\ \Omega = 63.7\angle -90°.$$

$$I_R = \frac{230\angle 0°}{100} = 2.3\angle 0° = 2.3 + j0\ \text{A}$$

$$I_C = \frac{230\angle 0°}{63.7\angle -90°} = 3.61\angle 90° = j3.61\ \text{A}$$

Then $\quad I = I_R + I_C = 2.3 + j3.61$

$$= \sqrt{2.3^2 + 3.61^2}\ \angle \tan^{-1}\frac{3.61}{2.3}$$

$$= 4.28\ \angle\ 57.5°\ \text{A} = 4.28\ \text{A},\ 57.5°\ \text{leading}$$

The sketch of the phasor diagram is shown in Fig. 4.48(b).

(b)

Fig. 4.48

4.11.3 Resistance, Inductance, and Capacitance in Parallel

A circuit with three elements—resistance of R ohm, inductance of L henry, and capacitance of C farad—connected in parallel across an ac voltage source of V volt is shown in Fig. 4.49, with the current distribution indicated therein.

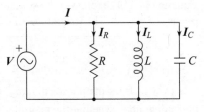

Fig. 4.49 Parallel *RLC* circuit

Applying the KCL for the circuit shown Fig. 4.49, the following equation may be written:

$$I = I_R + I_L + I_C \tag{4.110}$$

where I_R, I_L, and I_C are the phasor currents (in ampere) through the resistance, inductance, and capacitance, respectively, and I is the source current (in ampere). Let the voltage phasor V be the reference phasor. Then, $V = V\angle 0°$ and

$$I_R = \frac{V\angle 0°}{R} = \frac{V}{R} \tag{4.111}$$

$$I_L = \frac{V\angle 0°}{j\omega L} = -j\frac{V}{\omega L} = \frac{V}{\omega L}\angle -90° \tag{4.112}$$

$$I_C = \frac{V\angle 0°}{-jX_C} = j\frac{V}{X_C} = j\frac{V}{1/\omega C} = jV\omega C = V\omega C\angle 90° \tag{4.113}$$

It is evident that I_R is in phase with the voltage V, I_L lags the voltage V by 90°, while I_C leads the voltage V by 90°. The phasor diagram for the parallel *RLC* circuit when $I_L > I_C$, that is, $X_C > X_L$, is shown in Fig. 4.50. Substituting Eqs (4.111)–(4.113) into Eq. (4.110),

$$I = \frac{V}{R} - j\frac{V}{\omega L} + jV\omega C$$

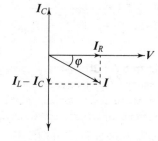

$$= V\left[\frac{1}{R} + j\left(\omega C - \frac{1}{\omega L}\right)\right] = \frac{V}{Z} = VY \tag{4.114}$$

Fig. 4.50 Phasor diagram

where

$$Y = \frac{1}{Z} = \left[\frac{1}{R} + j\left(\omega C - \frac{1}{\omega L}\right)\right] = G + j(B_C - B_L) \tag{4.115}$$

and

$$G = \frac{1}{R}, \quad B_L = \frac{1}{\omega L} = \frac{1}{X_L}, \quad B_C = \omega C = \frac{1}{X_C} \tag{4.116}$$

It is observed from the phasor diagram that the source current I lags the applied voltage V by an angle φ. The angle φ is given by

$$\tan\varphi = \frac{I_C - I_L}{I_R} = \frac{V[\omega C - (1/\omega L)]}{V/R} = R\left(\omega C - \frac{1}{\omega L}\right)$$

$$= R\left(\frac{1}{X_C} - \frac{1}{X_L}\right) = \frac{B_C - B_L}{G} \tag{4.117}$$

Equation (4.114) expresses the source current I in the complex form. The source current in the polar form is given as $I = Ie^{-j\varphi}$. The magnitude of current is given by

$$I = \sqrt{\left(\frac{V}{R}\right)^2 + \left(V\omega C - \frac{V}{\omega L}\right)^2} = V\sqrt{\frac{1}{R^2} + \left(\omega C - \frac{1}{\omega L}\right)^2} \tag{4.118}$$

It is evident from the phasor diagram that the supply current phasor I will lag or lead the voltage phasor V depending upon whether the magnitude of current phasors $I_L > I_C$ or $I_C > I_L$, respectively. In case $I_L = I_C$, the current phasor I will be in phase with the supply voltage phasor V. This is the condition of parallel resonance, which is discussed later in this chapter.

Example 4.21 Three elements, a resistance of 100 Ω, an inductance of 0.1 H, and a capacitance of 150 μF, are connected in parallel to a 230-V, 50-Hz supply. Calculate: (a) the current in each element, (b) the supply current, (c) the phase angle between the supply voltage and the supply current.

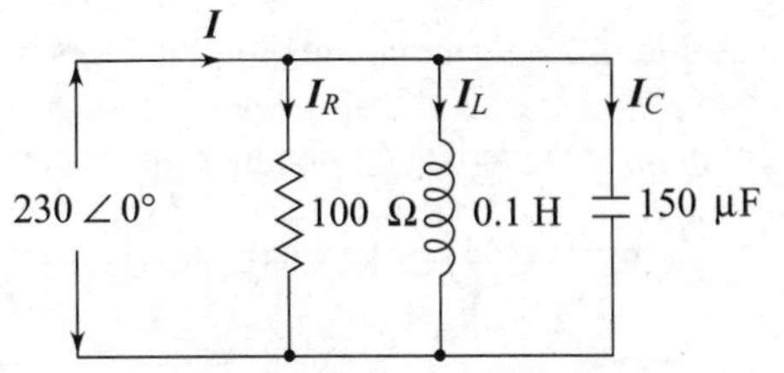

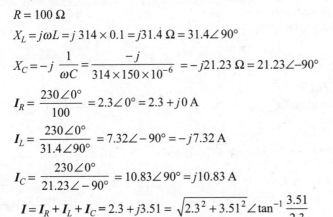

Fig. 4.51   Circuit diagram

Solution   The circuit diagram for the parallel branches is shown in Fig. 4.51. Let the supply voltage be chosen as the reference phasor. Then, $V = 230\angle 0°$.

$$R = 100\ \Omega$$

$$X_L = j\omega L = j\,314 \times 0.1 = j31.4\ \Omega = 31.4\angle 90°$$

$$X_C = -j\,\frac{1}{\omega C} = \frac{-j}{314 \times 150 \times 10^{-6}} = -j21.23\ \Omega = 21.23\angle -90°$$

$$I_R = \frac{230\angle 0°}{100} = 2.3\angle 0° = 2.3 + j0\ \text{A}$$

$$I_L = \frac{230\angle 0°}{31.4\angle 90°} = 7.32\angle -90° = -j7.32\ \text{A}$$

$$I_C = \frac{230\angle 0°}{21.23\angle -90°} = 10.83\angle 90° = j10.83\ \text{A}$$

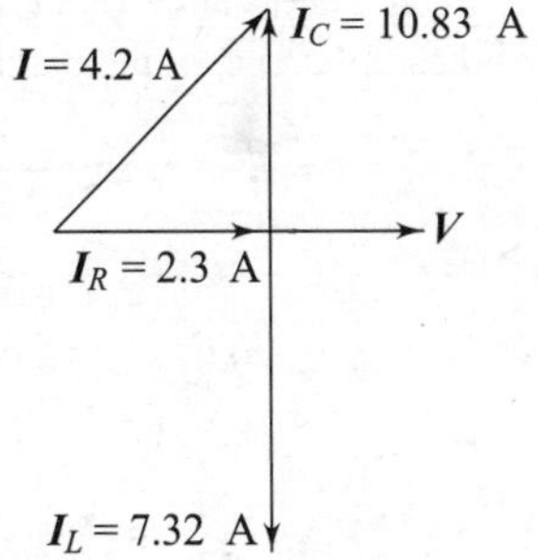

Fig. 4.52   Phasor diagram

Then $$I = I_R + I_L + I_C = 2.3 + j3.51 = \sqrt{2.3^2 + 3.51^2}\,\angle\tan^{-1}\frac{3.51}{2.3}$$

$$= 4.196\angle 56.76°\ \text{A} = 4.196\ \text{A},\ 56.76°\ \text{leading}$$

The sketch of the phasor diagram is shown in Fig. 4.52.

4.11.4 Impedances in Parallel

Figure 4.53 shows complex impedances Z_1, Z_2, and Z_3 connected in parallel across an ac voltage source V. These deliver a total current I to the impedances. Let it be assumed that I_1, I_2, and I_3 are representing the phasors of currents passing through the impedances Z_1, Z_2, and Z_3, respectively, caused by the voltage V connected across the impedances.

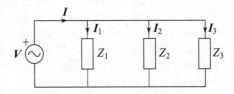

Fig. 4.53   Impedances in parallel

Applying KCL to the circuit shown in Fig. 4.53, the following expression results:

$$I = I_1 + I_2 + I_3$$

Now, $$I_1 = Y_1 \times V,\quad I_2 = Y_2 \times V,\quad I_3 = Y_3 \times V$$

where $Y_1 = 1/Z_1$, $Y_2 = 1/Z_2$, and $Y_3 = 1/Z_3$ are the complex admittances of the three impedances, respectively. Then

$$I = (Y_1 + Y_2 + Y_3) \times V$$

or $$I = Y_{eq} \times V$$

where $$Y_{eq} = Y_1 + Y_2 + Y_3 \tag{4.119}$$

and $$Z_{eq} = 1/Y_{eq} \tag{4.120}$$

Thus the equivalent complex admittance for a parallel circuit may be obtained by addition of all the complex admittances connected in parallel, and the reciprocal of the equivalent complex admittance is the equivalent complex impedance.

Example 4.22 In the circuit shown in Fig. 4.54 determine the equivalent impedance that appears across the terminals *AC*.

Solution Converting the complex impedances into the polar form

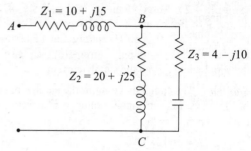

$$Z_1 = 10 + j15 = 18.03 \angle 56.31°$$
$$Z_2 = 20 + j25 = 32.02 \angle 51.34°$$
$$Z_3 = 4 - j10 = 10.77 \angle 56.31°$$

To combine the impedances Z_2 and Z_3 connected in parallel across the terminals *B* and *C*, it is easier to work first in terms of the admittance of the parallel combination. Thus,

Fig. 4.54

$$Y_{BC} = \frac{1}{Z_{BC}} = \frac{1}{Z_2} + \frac{1}{Z_3} = \frac{1}{32.02 \angle 56.31°} + \frac{1}{10.77 \angle -68.2°}$$

$$= 0.031 \angle -56.31° + 0.093 \angle 68.2°$$
$$= 0.0172 - j0.0258 + 0.0345 + j0.0863$$
$$= 0.0517 + j0.0605 = 0.0796 \angle 49.5°$$

$$\therefore \quad Z_{BC} = \frac{1}{Y_{BC}} = \frac{1}{0.0796 \angle 49.5°} = 12.563 \angle -49.5° = 8.159 - j9.553$$

Thus the total impedance between the terminals *A* and *C* is given by

$$Z_{AC} = Z_1 + Z_{BC} = 10 + j15 + 8.159 - j9.553$$
$$= 18.159 + j5.447 = 18.96 \angle 16.7°$$

4.12 SERIES PARALLEL COMBINATION OF IMPEDANCES

Quite often in ac electrical circuits several impedances are connected in series parallel combinations and computations of voltages and currents become complicated. The computations can be simplified by determining the equivalent impedance of the series parallel combination of impedances. The complex impedances are expressed in both rectangular and polar complex forms. The rectangular form is used for the addition/subtraction of impedances, while the polar form is conveniently used for the multiplication/division of impedances. Since the impedances and phasor voltages and currents are

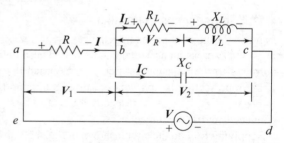

Fig. 4.55 Series parallel impedance circuit

expressed as complex numbers, the algebraic manipulations of complex numbers are extensively undertaken.

The computations of voltages and currents in series parallel combinations of impedances connected to the ac source may be accomplished by using the phasor diagram.

Figure 4.55 shows a resistance of R_1 ohm connected in series with an inductance coil having complex impedance $Z_L = Z_L \angle \theta = (R_L + jX_L)$ ohm and a capacitive reactance of X_C ohm connected in parallel. The series parallel combination is connected to an ac supply. The assumed current distributions in the circuit and the voltage drops across the series impedance and the parallel combination are indicated in Fig. 4.55.

By applying KVL around the network *abcdea* and KCL at the junction *b*,

$$V = V_1 + V_2 \tag{4.121}$$

$$I = I_L + I_C \tag{4.122}$$

In order to determine the supply current I and branch currents I_L and I_C and the voltage drops V_1 and V_2 from the phasor diagram, it is important to designate a reference phasor. In the present case V_2 is chosen as the reference phasor. Thus, $V_2 = V_2 + j0$. Then the current I_C will lead V_2 by 90°, that is, $I_C = jI_C$. The current I_L will lag V_2 by the impedance angle θ, where $\theta = \tan^{-1}(X_L/R)$. The current I_L may be resolved along the reference phasor V_2 and along a direction perpendicular to it. Thus

$$I_L = I_L (\cos\theta - j\sin\theta) \tag{4.123}$$

The phasor sum of I_L and I_C gives the supply current I. Then

$$V_1 = IR \tag{4.124}$$

$$V_2 = I_C (jX_C) = I_L Z_L \tag{4.125}$$

Equations (4.121)–(4.125) can be employed to obtain the various currents and voltage drops.

Figure 4.56 shows the phasor diagram for the circuit shown in Fig. 4.55. The voltage drop across the parallel branch V_2 is the reference phasor and is represented by OV_2. The current through the capacitance I_C leads V_2 by 90° and is OI_C in the phasor diagram. The current through the inductance I_L that lags V_2 by angle θ is shown as OI_L. The phasor sum of I_L and I_C is I and is marked as OI. The voltage drop V_R across R is in phase with the current I and is shown as OV_1. The phasor sum of V_1 and V_2 is the applied voltage and is drawn as OE.

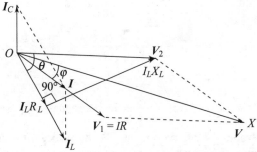

Fig. 4.56 Phasor diagram for series parallel circuit of Fig. 4.55

Example 4.23 A coil of resistance 25 Ω and inductance 0.159 H is in parallel with a circuit having 60 Ω resistor and 125 μF capacitor. This parallel circuit is connected to a 230-V, 50-Hz supply. Calculate: (a) the supply current, (b) the equivalent circuit impedance, resistance, and reactance.

Solution Figure 4.57 shows the circuit and its phasor diagram. The supply voltage is chosen as the reference phasor.

(a) $X_L = 2\pi f L = 2\pi \times 50 \times 0.159 = 50\ \Omega$

$$Z_1 = \sqrt{R_1^2 + X_L^2} = \sqrt{25^2 + 50^2} = 55.9\ \Omega$$

$$I_1 = \frac{V}{Z_1} = \frac{230}{55.9} = 4.11\ \text{A}$$

$$\varphi_1 = \cos^{-1}\frac{R_1}{Z_1} = \cos^{-1}\frac{25}{55.9} = \cos^{-1} 0.4472 = 63.43°\ \text{lag}$$

$$I_1 = 4.11\ \angle - 63.43°\ \text{A}$$

$$X_C = \frac{1}{2\pi f C} = \frac{1}{2\times\pi\times 50\times 125\times 10^{-6}} = 25.46\ \Omega$$

$$Z_2 = \sqrt{R_2^2 + X_C^2} = \sqrt{60^2 + 25.46^2} = 65.18\ \Omega$$

$$I_2 = \frac{V}{Z_2} = \frac{230}{65.18} = 3.53\ \text{A}$$

$$\varphi_2 = \cos^{-1}\frac{R_2}{Z_2} = \cos^{-1}\frac{60}{65.18} = \cos^{-1} 0.4472 = 23°\ \text{lead}$$

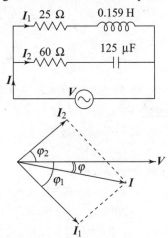

Fig. 4.57 The circuit and the phasor diagram

$I_2 = 3.53 \angle 23°$ A

$I = I_1 + I_2$

$I \cos \varphi = I_1 \cos \varphi_1 + I_2 \cos \varphi_2$

$\qquad = 4.11 \cos(-63.43°) + 3.53 \cos 23° = 5.088$A

$I \sin \varphi = I_1 \sin \varphi_1 + I_2 \sin \varphi_2$

$\qquad = 4.11 \sin(-63.43°) + 3.53 \sin 23° = -2.30$ A

$I = \sqrt{(I \cos \varphi)^2 + (I \sin \varphi)^2} = \sqrt{5.935^2 + 1.477^2} = 5.58$ A

(b) $\qquad Z = \dfrac{V}{I} = \dfrac{230}{5.58} = 41.20 \ \Omega$

$\qquad R = Z \cos \varphi = Z \times \dfrac{I \sin \varphi}{I} = 41.20 \times \dfrac{5.088}{5.58} = 37.57 \ \Omega$

$\qquad X = Z \sin \varphi = Z \times \dfrac{I \sin \varphi}{I} = 41.20 \times \dfrac{-2.30}{5.58} = -16.98 \ \Omega$

$I \sin \varphi$ is negative, hence the reactance must be inductive.

Example 4.24 Two circuits, the impedances of which are given by $Z_1 = 15 + j20 \ \Omega$ and $Z_2 = 8 - j10 \ \Omega$ are connected in parallel. If the total current supplied is 20 A, calculate (a) the currents in each circuit, (b) the power taken by each branch.

Solution Figure 4.58 shows the circuit and its phasor diagram. The supply voltage is chosen as the reference phasor.

$Z_1 = Z_1 \angle \theta_1 = 15 + j20 = 25 \angle 53.13°$

$Z_2 = Z_2 \angle \theta_2 = 8 - j10 = 12.81 \angle -51.34°$

$V = V \angle 0°$, reference phasor

$I_1 = \dfrac{V \angle 0°}{Z_1 \angle \theta_1} = \dfrac{V}{Z_1} \angle -\theta_1 = I_1 \angle -\theta_1,$

$I_2 = \dfrac{V \angle 0°}{Z_2 \angle \theta_1} = \dfrac{V}{Z_2} \angle -\theta_2 = I_2 \angle -\theta_2$

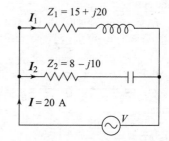

From the phasor diagram, $I = I_1 + I_2 = I \angle \varphi = 20 \angle \varphi$ A

Supply voltage, $V = I_1 Z_1 = I_2 Z_2$.

Then $\qquad 25 I_1 = 12.81 I_2$

$\therefore \qquad I_2 = 1.95 I_1$

From the phasor diagram,

$I^2 = (I_1 \cos\theta_1 + I_2 \cos\theta_2)^2 + (I_2 \sin\theta_2 - I_1 \sin\theta_1)^2$

$20^2 = I_1^2 + I_2^2 + 2I_1 I_2 (\cos\theta_1 \cos \theta_2 - \sin\theta_1 \sin\theta_2)$

$400 = I_1^2 + I_2^2 + 2I_1 I_2 \cos(\theta_1 + \theta_2)$

$\qquad = I_1^2 [1 + 1.95^2 + 2 \times 1.95 \times \cos(53.13° - 51.34°)]$

$\qquad = 8.7 \times I_1^2$

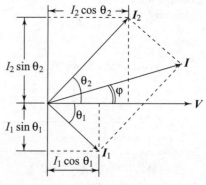

Fig. 4.58 Circuit and the phasor diagram

$\therefore \qquad I_1 = \sqrt{\dfrac{400}{8.7}} = 6.78 \text{ A}$

$\qquad I_2 = 1.95 I_1 = 1.95 \times 6.78 = 13.22 \text{ A}$

The power loss in $Z_1 = I_1^2 \times 15 = 6.78^2 \times 15 = 689.5 \text{ W}$

The power loss in $Z_2 = I_1^2 \times 8 = 13.22^2 \times 8 = 1398 \text{ W}$

Example 4.25 A circuit has an inductive reactance of 25 Ω at 50 Hz, its resistance being 20 Ω. For an applied voltage of 230 V at 50 Hz, calculate (a) the value of the current, (b) the angle of phase difference between current and voltage, (c) the value of the shunting capacitance to bring the resultant current into phase with the applied voltage, and (d) the resultant current in case (c).

Solution Figure 4.59 shows the circuit and its phasor diagram. The supply voltage is chosen as the reference phasor.

(a) Let the supply voltage be the reference phasor. Then, $V = V \angle 0°$.

$$Z_1 = Z_1 \angle \varphi = R_1 + jX_L = 25 + j20 = \sqrt{25^2 + 20^2} \angle \tan^{-1}\dfrac{20}{25}$$

$$= 32 \angle 38.66°$$

$$I_1 = \dfrac{V\angle 0°}{Z_1 \angle \varphi} = \dfrac{230 \angle 0°}{32 \angle 38.66} = 7.19 \angle -38.66°$$

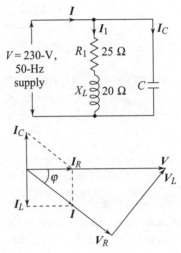

(b) The current lags the voltage by an angle $\varphi = \angle -38.66°$

(c) $\qquad I_R = I_1 \cos\varphi = 7.19 \times \cos(-38.66°) = 5.614 \text{ A}$

$\qquad I_L = I_1 \sin\varphi = 7.19 \times \sin(-38.66°) = -4.491 \text{ A}$

It can be observed from the phasor diagram that the current I_C drawn by C equals the current component I_L. Thus

$$I_C = I_L = 4.491 \text{ A}$$

$$I_C = \dfrac{V}{X_C} = V \times 2\pi \times 50 \times C$$

Thus $\qquad 4.491 = 230 \times 100\pi \times C$

$\therefore \qquad C = 62.15 \text{ μF}$

Fig. 4.59 Circuit and phasor diagram

(d) Resultant current is $I_R = 5.614 \text{ A}$.

Example 4.26 For the circuit shown in Fig. 4.60 calculate the impedance between the terminals A and B and the phase angle between applied voltage and circuit current.

$$Z_2 = 10 + j15 \ \Omega$$

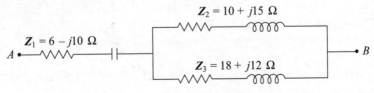

$$Z_1 = 6 - j10 \ \Omega$$

$$Z_3 = 18 + j12 \ \Omega$$

Fig. 4.60 Series parallel circuit

Solution $Z_1 = 6 - j10 = 11.66 \angle -59.04°$

$\qquad Z_2 = 10 + j15 = 18.03 \angle 56.31°$

$\qquad Z_3 = 18 + j12 = 21.63 \angle 33.69°$

$$\boldsymbol{Z}_{AB} = Z_1 + Z_2 \| Z_3 = Z_1 + \frac{Z_2 Z_3}{Z_2 + Z_3}$$

$$= 6 - j10 + \frac{18.03\angle56.31° \times 21.63\angle33.69°}{10 + j15 + 18 + j12}$$

$$= 6 - j10 + \frac{18.03 \times 21.63\angle(56.31 + 33.69)}{28 + j27}$$

$$= 6 - j10 + \frac{18.03 \times 21.63\angle56.31 + 33.69}{38.9\angle43.96}$$

$$= 6 - j10 + 10.025\angle(90 - 43.96)°$$

$$= 6 - j10 + 10.025 \angle46.04° = 6 - j10 + 6.96 + j7.22$$

$$= 12.96 - j2.78 = 13.25\angle-12.11°$$

Phase angle between applied voltage and circuit current is − 12.11°.

4.13 POWER IN AC CIRCUITS

Electric energy is delivered from an ac supply to resistive and reactive loads. For resistive loads the energy is dissipated in the same way as a direct current dissipates energy in a resistor. The power that gives rise to energy dissipation in a resistor is called active power. The power associated with energy transfer from the electrical system to another system, such as heat, light, or mechanical drives, is termed active power. The average power given by I^2R watt is the active power of the arrangement, where I ampere is the rms value of current that flows through a resistor of R ohm.

For pure inductive/capacitive reactance loads, energy is delivered to the load and then returned to the source. The power describing the rate of energy moving in and out of a reactance is called reactive power. In circuits containing both a resistance and a reactance, there is a mixture of active and reactive powers. The concept of the power factor is also introduced in this section.

4.13.1 Power in Resistive Circuits

Figure 4.61 shows the waveforms of current i, voltage v, and the product vi for a resistive circuit connected to an ac source. Since the instantaneous values of vi represent the instantaneous power p, it follows that the waveform of vi represents the power wave. As power is continually fluctuating, the power in an ac circuit is taken to be the average value of the vi wave.

The power in the resistance at any instant can be written as

$$p = vi \text{ watt}$$

$$= V_m \sin \omega t \, I_m \sin \omega t = V_m I_m \sin^2 \omega t$$

$$= V_m I_m \left(\frac{1 - \cos 2\omega t}{2} \right) \qquad (4.126)$$

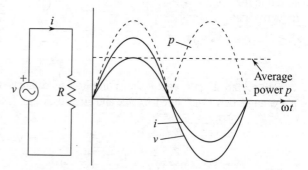

Fig. 4.61   Waveforms of i, v, and p for a resistive circuit

It is clear from Eq. (4.126) that the power wave has a frequency which is twice that of the voltage and current wave. However, the waveform of p is sinusoidal although it is displaced from the horizontal axis as shown in Fig. 4.61.

The average power can be derived from the power waveform as follows:

$$P = \frac{\omega}{2\pi} \int_0^{2\pi/\omega} V_m I_m \left(\frac{1 - \cos 2\omega t}{2} \right) dt$$

$$= V_m I_m \frac{\omega}{2\pi} \left[\frac{t}{2} - \frac{\sin 2\omega t}{4\omega} \right]_0^{2\pi/\omega}$$

$$= V_m I_m \frac{\omega}{2\pi} \frac{2\pi}{2\omega} = \frac{V_m I_m}{2}$$

$$= VI \text{ watt} \tag{4.127}$$

4.13.2 Power in a Purely Inductive Circuit

Figure 4.62 shows the waveforms of current i, voltage v, and the product vi for an inductance coil, having negligible resistance, connected to an ac source. Since the resistance is very small, the heating effect and, therefore, the active power are also very small, even though the voltage and the current may be large. Since the instantaneous values of vi represent the instantaneous power p, it follows that the waveform of vi represents the power wave. It may be observed that during the interval 0 to $\pi/2$ the applied voltage is positive, but the current is negative, so that the power is negative. It is also observed that during the interval $\pi/2$ to π, both the current and the voltage are positive, so that the power is positive. Thus the power changes sign after every quarter of a cycle. The power curve is found to be symmetrical about the horizontal axis. Consequently, the mean value of the power over the complete cycle is zero.

The physical meaning of this is made clear by the graph of power shown in Fig. 4.62. It may be observed that the power is negative during the period 0 to $\pi/2$, the first quarter-cycle of the applied voltage, the reason being that the current is decreasing and the flux in the coil is therefore collapsing. The energy stored in the magnetic field of the inductance coil is thus being returned to the source. Thus the flow of power is from the inductance to the source, and consequently must be reckoned as negative. The same applies in the third quarter-cycle, the period π to $3\pi/2$, the current decreases and the flux therefore collapses. The fact that the current decreases in the first and third quarter-cycles makes no difference, the negative sign of the power flow being decided by the fact that the stored energy from the magnetic field is being returned to the circuit in both cases. In the second and fourth quarter-cycles the current is increasing, the flux is increasing, the energy stored in the magnetic field is increasing. Thus energy is now being transferred from the supply circuit to the inductive coil and the power flow is thus positive.

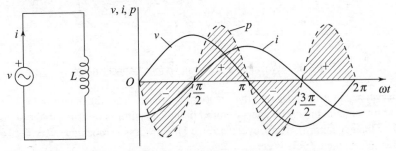

Fig. 4.62 Waveforms of i, v, and p for an inductive circuit

It can be proved analytically that inductive circuits do not dissipate power. The power in an inductance at any instant can be written as

$$p = vi \text{ watt}$$

$$= V_m \sin \omega t \, I_m \sin (\omega t - \pi/2) = -V_m I_m \sin \omega t \cos \omega t$$

$$= V_m I_m \left(\frac{-\sin 2\omega t}{2} \right) \tag{4.128}$$

The average power over one cycle is given by

$$P = \frac{\omega}{2\pi} \int_0^{2\pi/\omega} V_m I_m \left(\frac{-\sin 2\omega t}{2} \right) dt$$

$$= V_m I_m \frac{\omega}{2\pi} \left[\frac{\cos 2\omega t}{4\omega} \right]_0^{2\pi/\omega} = 0$$

From the power waveform, it can be seen that the average value of power in a pure inductive circuit is zero. Positive power implies that the coil is drawing power from the source to set up the magnetic field. The maximum energy stored in the magnetic field is $(1/2)(L I_m^2)$. Negative power means that the inductance is acting as a generator and returning the energy stored in the magnetic field to the voltage source. Thus, it can be seen that during each half-cycle of the voltage or current waveform there is a continuous exchange of energy between the inductance and the voltage source.

4.13.3 Power in Purely Capacitive Circuits

In purely capacitive circuits, the current leads the applied voltage by a quarter of a cycle, as shown in Fig. 4.63, and by multiplying v and i, the instantaneous values of the voltage and current, respectively, the variation of the power curve can be derived. During the interval 0 to $\pi/2$, the voltage and current are both positive, so that the power is positive, that is, the power is being supplied from the source to the capacitor. The shaded area enclosed by the power curve during the interval 0 to $\pi/2$ represents the value of the electrostatic energy stored in the capacitor at the instant $\omega t = \pi/2$.

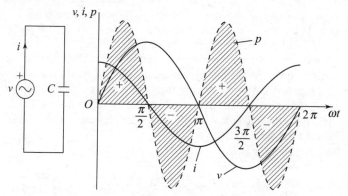

Fig. 4.63 Waveforms of i, v, and p for a capacitive circuit

During the interval $\pi/2$ to π, the pd across the capacitor decreases from the maximum value V_m at the instant $\omega t = \pi/2$ to zero at $\omega t = \pi$, and the entire energy stored in the capacitor is returned to the source. Thus the net energy absorbed by the capacitor in a half-cycle is zero. This process is repeated every half-cycle. Hence the average power over a complete cycle is zero.

It can be proved analytically that capacitive circuits do not dissipate power. The power in a capacitance at any instant can be written as

$$p = vi \text{ watt}$$

$$= V_m \sin\omega t \, I_m \sin(\omega t + \pi/2) = V_m I_m \sin \omega t \cos \omega t$$

$$= V_m I_m \left(\frac{\sin 2\omega t}{2} \right)$$

$$\tag{4.129}$$

The average power over one cycle is given by

$$P = \frac{\omega}{2\pi} \int_0^{2\pi/\omega} V_m I_m \left(\frac{\sin 2\omega t}{2} \right) dt \text{ watt}$$

$$= V_m I_m \frac{\omega}{2\pi} \left[-\frac{\cos 2\omega t}{4\omega} \right]_0^{2\pi/\omega} = 0$$

As in the case of a purely inductive network, the frequency of the power waveform is twice that of either the applied voltage or the current waveform. In addition, the average power of the waveform is zero, which implies that during a quarter of a cycle of the voltage wave, the capacitor draws power (positive power) from the voltage source to build the charge, and during the next quarter of the cycle, it returns an equal amount of power (negative power) to the source, since it is discharging. Thus, there is a continuous exchange of power between the source and the capacitor. The maximum energy stored in a capacitor is $(1/2)(CV_m^2)$.

4.13.4 Power in a Circuit with Resistance and Reactance

It is intended to develop a general expression for power in an ac circuit. Let it be assumed that the current lags the applied voltage by an angle φ. The instantaneous values of voltage and current may be expressed as

$$v = V_m \sin \omega t \qquad (4.130a)$$
$$i = I_m \sin(\omega t - \varphi) \qquad (4.130b)$$

The waveforms of v, i, and power $p = v \times i$ are shown in Fig. 4.64.

It may be observed from Fig. 4.64 that p is negative for the periods 0 to φ and π to $(\pi + \varphi)$, as shown by the shaded negative area, indicating that the circuit is returning power to the source during these intervals. The shaded positive areas during the periods φ to π and $(\pi + \varphi)$ to 2π indicate that the energy is supplied to the circuit from the source of supply. The difference between the positive and negative areas in a cycle represents the net energy absorbed by the circuit. As the angle φ is made smaller, the negative areas become smaller and the average power increases. As φ is made zero, that is, v and i are brought in phase, there are no negative areas and the circuit absorbs all the power.

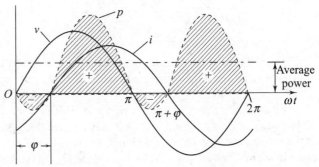

Fig. 4.64 Waveforms of i, v, and p for a resistive–reactive circuit

4.13.4.1 Active and Reactive Powers

The instantaneous value of power p is given by

$$p = V_m I_m \sin \omega t \sin(\omega t - \varphi) \qquad (4.131)$$

$$= \frac{1}{2} V_m I_m [\cos \varphi - \cos(2\omega t - \varphi)]$$

$$= \frac{V_m}{\sqrt{2}} \frac{I_m}{\sqrt{2}} [\cos \varphi - \cos(2\omega t - \varphi)]$$

$$= VI \cos \varphi - VI(\cos 2\omega t \cos \varphi + \sin 2\omega t \sin \varphi)$$

$$= VI \cos \varphi (1 - \cos 2\omega t) - VI \sin 2\omega t \sin \varphi \text{ watts} \qquad (4.132)$$

From Eq. (4.132), it may be noted that for a fixed value of φ the instantaneous power consists of two terms. The first term

$$p_1 = VI \cos \varphi(1 - \cos 2\omega t) \qquad (4.133)$$

has two components—a constant part and a time-varying part. The time-varying part has a frequency that is twice that of the voltage and the current, and this component does not contribute towards the average value of power taken from the source, as the average value of a cosine curve over one complete cycle is zero. Therefore, the first term p_1 has an average value $p_{av} = VI \cos\varphi$, with a minimum value of zero and a maximum value of $2VI \cos\varphi$, and varies about the average value sinusoidally at twice the supply frequency. This power is termed the instantaneous real power, the active power, or simply the power.

The second term of Eq. (4.132),

$$p_2 = -VI \sin 2\omega t \sin \varphi \qquad (4.134)$$

has a peak value of $q = VI \sin\varphi$ and varies about its mean value of zero sinusoidally at twice the supply frequency. This power is referred to as the instantaneous reactive power. This is a component of power that oscillates between the source and the energy-storing elements of the circuit, that is, the reactive elements of the load.

Hence, the average real power P is defined as the average value of the instantaneous power p over one cycle and is obtained as

$$P = \frac{\omega}{2\pi} \int_0^{2\pi/\omega} p \, dt = \frac{\omega}{2\pi} \int_0^{2\pi/\omega} VI \cos \varphi(1 - \cos 2\omega t) \, dt$$

$$= VI \cos\varphi \, \frac{\omega}{2\pi} \left(t - \frac{\sin 2\omega t}{2\omega} \right)_0^{2\pi/\omega}$$

$$= VI \cos\varphi \quad \text{watt} \qquad (4.135)$$

The unit of measurement of active power is the watt. The reactive power, also known as reactive volt-ampere or quadrature power, is defined as the average value of the instantaneous reactive power p_2 and may be written as

$$Q = VI \sin\varphi \qquad (4.136)$$

The unit of measurement of reactive power is VAR (volt-ampere reactive).

From the expressions for P and Q, it may be seen that for a resistive circuit with $\varphi = 0°$, for a given value of the voltage and current, P is maximum and equal to VI, while Q is zero. Similarly, for purely inductive and capacitive circuits with $\varphi = \pm 90°$, $P = 0$, while Q is maximum and equal to VI.

The reactive power in ac systems is important due to the basic fact that it accounts for the energy storage and the release of energy from the magnetic field produced in the inductive elements and the electric field produced in the capacitive elements that may be present in the electric circuits. Conventionally, reactive power is considered to be positive in inductive circuits and negative in capacitive circuits.

4.13.4.2 Volt-Ampere and Complex Power

The product of the voltage and current in an ac circuit is termed the apparent power. It is usually denoted by the symbol S. Thus,

$$S = VI \text{ volt-amperes (VA)} \qquad (4.137)$$

Equations (4.135) and (4.136) may now be written as

$$P = VI \cos\varphi = S \cos\varphi \quad \text{watt} \qquad (4.138)$$

$$Q = VI \sin\varphi = S \sin\varphi \quad \text{VAR} \qquad (4.139)$$

The $$S = \sqrt{P^2 + Q^2} \quad \text{VA} \qquad (4.140)$$

This suggests that power can be interpreted by means of the phasor diagram shown in Fig. 4.65(a). The current component $I \cos\varphi$ in phase with the voltage V supplies active power, whereas the current component

$I \sin\varphi$ in quadrature with the voltage V supplies reactive power. The component $I \cos\varphi$ is termed the active power component of the current, while the component $I \sin\varphi$ is termed the reactive power component of the current. The complex power S is defined as a complex number whose real part represents active power and imaginary part represents reactive power.

The complex power phasor diagram is drawn in Fig. 4.65(b). From the phasor diagram of Fig. 4.65(b), the complex power S can be written in the form

$$S = P + jQ \qquad (4.141)$$

where Q is positive for inductive load and negative for capacitive load. Complex power in terms of V and I can be expressed as

$$S = VI^* \qquad (4.142)$$

where I^* is the complex conjugate of I. Let $V = Ve^{j0}$, $I = Ie^{j(0-j)}$, and I lag V by j.

Then, $S = VI^* = Ve^{j0}Ie^{-j(0-\varphi)}$

$\qquad = VIe^{j\varphi} = VI (\cos\varphi + j \sin\varphi)$

$\qquad = P + jQ = S\angle\varphi = S(\cos\varphi + j \sin\varphi) \qquad (4.143)$

While dealing with large powers, the unit of active power is the kilowatt or megawatt, that of reactive power is the kVAR or MVAR, and that of complex power is kVA or MVA.

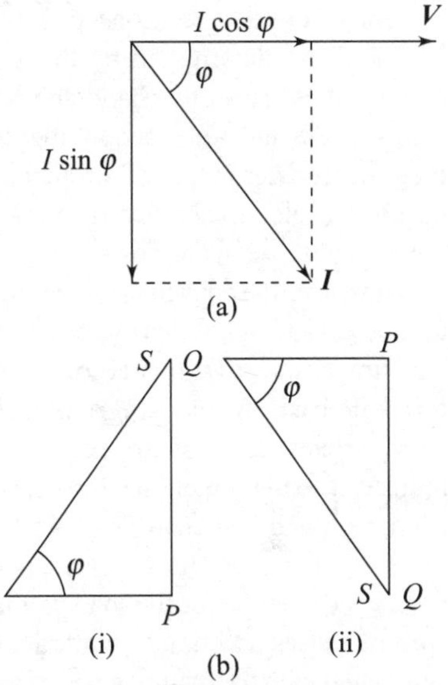

Fig. 4.65 (a) In-phase and quadrature components of current. (b) Complex power triangle: (i) inductive load, (ii) capacitive load

4.13.4.3 Power Factor and Reactive Factor

It is seen that the active power $P = VI \cos\varphi$, where V and I are rms values of the voltage and current, respectively, and φ is the phase angle difference between the voltage and the current. The value of $\cos\varphi$ must lie between 0 and 1. Thus, the value of power P can at the most be equal to VI volt-ampere. The term $\cos\varphi$ is called the power factor. Thus,

$$\text{Power factor} = \frac{\text{active power } P \ (\text{in watts})}{\text{apparent power } S \ (\text{in volt-amperes})} \qquad (4.144)$$

The power factor (pf) is considered to be lagging or leading depending, respectively, on whether the current lags or leads the voltage.

Similarly, the reactive power $Q = VI \sin\varphi$, and the term $\sin\varphi$ is called the reactive factor. The value of $\sin\varphi$ lies between -1 and 1. Thus, the reactive factor varies from $+1$, for a purely inductive circuit, to -1, for a purely capacitive circuit. The power factor is introduced in ac circuits to specify the amount of real power when the apparent power is VI.

4.13.5 Need for Power Factor Improvement

In Section 4.13.4.2, it is shown that a current I, lagging a voltage V by an angle φ, can be resolved into two components—the in-phase component $I \cos\varphi$, in phase with V, and the quadrature component $I \sin\varphi$, lagging V by 90°. The quadrature component can contribute nothing to the active power, as all the active power is conveyed by the in-phase component. When $\cos\varphi = 1$, for given values of V and I, the active power P is maximum and numerically equal to volt-ampere VI. The value of active power decreases with decreasing power factor.

Normally, ac apparatus, such as generators, transformers, transmission lines, and cables, is generally rated in kVA or MVA rather than the active power output in kilowatts or megawatts. The allowable output of the device is limited by heating caused by the losses in the device; these losses are in turn determined by the operating voltage and current of the device. Consequently, the capacity of electrical equipment installed to supply a given load is essentially determined by the kVA or MVA of that load rather than by the load power in kilowatts or megawatts. On the other hand, the boiler and turbine sizes and fuel requirements in a thermal generating station (turbine size and water requirements in a hydroelectric station) are determined essentially by the kilowatt or megawatt power output of the generator and not by its kVA or MVA rating. Thus, it is quite logical that electric utilities fix electrical rate structures depending mainly on active power and on either volt-ampere or reactive power. All three quantities are of economic importance.

Thus, from the point of view of a supply company, a low power factor is a very serious matter, and since the value of the power factor is decided by the nature of the load, that is, by the consumer devices, such as motors, etc., the company encourages the consumer to make his power factor as high as possible. The usual method is to frame the tariff on a kVA basis by imposing a high charge for the kVA and a low charge for the kilowatts of energy consumed.

Most electric loads are reactive in nature and have lagging power factors less than unity. Particularly, industrial loads which are mostly induction motor drives have low power factors. The power factor may even be much less than 0.8 if induction motors are not fully loaded and in the presence of arc furnaces. Transmission lines, transformers, and generators of the electric power utility have to carry the lagging reactive power requirement of the load so that their full active power capability is not exploited. In addition, reactive current causes additional ohmic losses and large voltage drops. These factors create operational difficulties associated with the maintenance of the required voltage at the alternator terminals or at the end of a transmission line and cause financial loss to the utility. The utility, therefore, induces its industrial consumers to improve the power factor of the load by installing static shunt capacitors which draw leading current to neutralize the lagging current drawn by the low power factor of the load. If the industries continue to draw lagging power factor, then a penalty is imposed through enhanced tariff based on the reactive component of the consumer's load. Industrial consumers, thus, find it economical to improve the power factor of their individual motors and/or the total installation by installing static shunt capacitors. The limit to which the power factor must be improved is dictated by the balance of the yearly tariff saving against the yearly interest and depreciation cost of installing the capacitors.

Summarizing, it is seen that a low power factor necessitates a high capital cost for alternators, switchgear, transformers, and cables. The disadvantages of a low power factor do not end here, however, since there are operational difficulties such as low voltage at the consumer end.

Example 4.27 A coil having a resistance of 8 Ω and an inductance of 0.05 H is connected across a 230-V, 50-Hz supply. Calculate (a) the current, (b) the phase angle between the current and supply voltage, (c) the apparent power, (d) the active power, and (e) the reactive power.

Solution   (a) The inductive reactance

$$X_L = 2\pi f L = 2 \times 3.14 \times 50 \times 0.05 = 15.7 \ \Omega$$

The impedance of the coil

$$Z_L = 8 + j15.7 = \sqrt{8^2 + 15.7^2} \angle \tan^{-1}\frac{15.7}{8}$$

$$= 17.62 \ \angle 63° \ \Omega$$

Let the supply voltage be the reference phase, i.e., $V = 230\angle 0°$. The current

$$I = \frac{V}{Z_L} = \frac{230\angle 0°}{17.62\angle 63°} = 13.05 \ \angle -63° \ \text{A}$$

(b) The phase angle between the supply voltage and current, $\varphi = -63°$, that is, I lags V by 63°.

(c) The apparent power = $VI = 230 \times 13.05 = 3001.5$ VA, or 3.0015 kVA.

(d) The active power = $VI \cos \varphi = 3.0015 \cos 63° = 1.363$ kW.
(e) The reactive power = $VI \sin \varphi = 3.0015 \sin 63° = 2.674$ kVAR.

Example 4.28 A single-phase motor operating at 230 V, 50 Hz is developing 7.5 kW at a power factor of 0.7 lagging with an efficiency of 80%. Calculate (a) the input apparent power, (b) the active and reactive components of current, (c) the reactive power, and (d) the capacitance required in parallel with the motor to improve the power factor to 0.95 lagging.

Solution   (a) Let I_m be the motor current drawn from the supply.

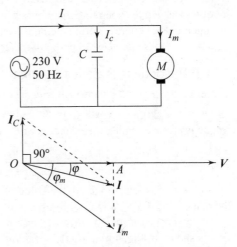

$$\text{Efficiency} = \frac{\text{output in watts}}{\text{input in watts}} = \frac{7500}{VI_m \cos\varphi} = \frac{7500}{VI_m \times 0.7}$$

or $0.8 = \dfrac{7500}{VI_m \times 0.7}$

$VI_m = 13{,}392.85$ VA ≈ 13.393 kVA

Thus,   Input = 13.393 kVA

(b) Current $I_m = \dfrac{VI_m}{V} = \dfrac{13{,}393}{230} = 58.23$ A

The power factor $\cos \varphi_m = 0.7$

The reactive factor $\sin \varphi_m = \sqrt{1 - \cos^2 \varphi_m} = \sqrt{1 - 0.7^2} = \sqrt{0.51}$
   $= 0.7141$

The active component of the motor current

   $I_m \cos \varphi_m = 58.23 \times 0.7 = 40.76$ A

The reactive component of the motor current

   $I_m \sin \varphi_m = 58.23 \times 0.7141 = 41.58$ A

(c) The reactive power = $VI_m \sin\varphi = 13.393 \times 0.7141 = 9.56$ kVAR.

(d) The circuit and phasor diagram of the motor M taking a current $I_m = 58.23$ A from the supply is shown in Fig. 4.66. The capacitor C connected across the motor takes a current I_C that leads the voltage by 90°. The combined current $I = I_m + I_C$ lags the voltage by an angle φ, so that $\cos \varphi = 0.95$.

Then, $I \cos \varphi = I \times 0.95$

Now, $I \cos \varphi = I_m \cos \varphi_m = OA$ of the phasor diagram

$$I = \frac{I_m \cos \varphi_m}{0.95} = \frac{40.76}{0.95} = 42.905 \text{ A}$$

The reactive component of the total current

   $I \sin\varphi = 42.905 \times \sqrt{1 - \cos^2 \varphi} = 42.905 \times \sqrt{1 - 0.95^2} = 13.40$

Then, $I_C = I_m \sin\varphi_m - I \sin\varphi = 41.58 - 13.40 = 28.18$ A
   $= 2\pi f C V$

or $28.18 = 2\pi \times 50 \times C \times 230$
   $C = 390.2$ μF

4.14 RESONANCE IN AC CIRCUITS

Resonance is a phenomenon found in all engineering systems, such as electrical, mechanical, hydraulic, and pneumatic, involving two independent energy-storing elements. The term resonance refers to a condition existing in any physical system when a fixed amplitude sinusoidal-forcing function produces a response of maximum amplitude and the resultant response is in time phase with the source function. For resonance, two types of independent energy-storing elements capable of interchanging energy between one another must be present; for example, inductance and capacitance for electrical systems or mass and spring for mechanical systems.

In ac circuits, the supply to a consumer is at constant voltage and frequency. However, most communication systems involve circuits in which the supply voltage operates with a varying frequency, or a number of signals operate together, each with its own frequency. In power application, the speed control of motors also involves varying the operating frequency. Thus, the effects of varying the frequency in series as well as parallel arrangements of energy-storing elements L and C are important from the point of view of understanding the operation of these devices. In particular, the condition known as resonance must be investigated.

4.14.1 Resonance in Series Circuits

The circuit of Fig. 4.67 shows a series *RLC* circuit connected to a voltage source that operates with constant voltage magnitude V and varying frequency. The complex impedance Z of the circuit is

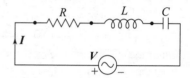

$$Z = \frac{V}{I} = R + j\left(2\pi fL - \frac{1}{2\pi fC}\right) = R + j(X_L - X_C)$$

Fig. 4.67 Series resonant *RLC* circuit

As the frequency f is varied, the value of Z changes. Let us consider the behaviour of R, L, and C with respect to a change in f.

The resistance R may be regarded as constant and, consequently, the graph of R against f is a horizontal line as shown in Fig. 4.68. Strictly speaking, the resistance of a conductor to ac is higher than that to dc; the difference increases as the frequency increases. This phenomenon is not associated in principle with the resonance phenomenon under discussion. So it is justified to consider R to remain constant.

In Section 4.9.2, it was shown that the inductive reactance X_L is directly proportional to f and increases linearly with frequency. The graph of X_L against f is a straight line through the origin, as shown in Fig. 4.68. It was also shown that the change in capacitive reactance is hyperbolic. As the capacitive reactance is regarded as negative, the hyperbola is drawn below the f-axis in Fig. 4.68.

The total reactance of the circuit is numerically equal to $X = X_L - X_C$, and therefore the variation of $X_L - X_C$ with f is hyperbolic as shown in the figure. The graph of $X_L - X_C$ crosses the f-axis at f_0. At the frequency f_0, $X_L = X_C$.

The magnitude of the impedance Z of the circuit is given by

$$Z = \sqrt{R^2 + \left(X_L - X_C\right)^2}$$

The magnitude of impedance depends upon the difference between the inductive and the capacitive reactance. Z is always positive and its graph against f is shown in Fig. 4.68. At f_0, the magnitude of Z is R.

It can be seen from Fig. 4.68 that when the frequency is zero, then $X_L = 2\pi fL = 0$ and $X_C = 1/2\pi fC = \infty$. As f is increased from zero, the value of X_L increases and that of X_C decreases. The net reactance $X = X_L - X_C$ is capacitive and the current I leads the applied voltage V, indicating the power factor to be leading. This continues

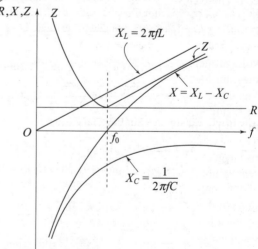

Fig. 4.68 Variation of R, X_L, X_C, $(X_L - X_C)$, and Z with respect to f for an *RLC* series circuit

as long as $f < f_0$. When $f > f_0$, then $X_L > X_C$, the net reactance is inductive, and the current I lags the applied voltage V, indicating the power factor to be lagging. At the frequency f_0, $X_L = X_C$; thus the net reactance of the circuit, $X = 0$; the impedance of the circuit Z exhibits a minimum value equal to R. In this condition, the current I attains its maximum possible value, equal to V/R, and is in phase with the voltage V; that is, the power factor is unity. In such a situation, the series electrical circuit is said to be in series resonance. Thus, for resonance,

$$X_L = X_C \tag{4.145}$$

From the graph of reactance against frequency of Fig. 4.68, the condition $X_L = X_C$ occurs at a frequency f_0. Equation (4.145) may be written in terms of f_0 as

$$2\pi f_0 L = \frac{1}{2\pi f_0 C}$$

$$f_0 = \frac{1}{2\pi\sqrt{LC}} \tag{4.146}$$

The frequency f_0 is termed the resonant frequency. Its value is specified entirely in terms of the parameters of the energy-storing elements. The current in the circuit in the resonant condition is limited only by the resistance R of the circuit, as the impedance of the circuit $Z = R$, and is given by

$$I_0 = \frac{V}{R} \tag{4.147}$$

During resonance, the current is the highest. The power dissipation $I_0^2 R$ is also the highest, while the peak rates of energy storage in the reactances X_L and X_C are maximum and equal. The resistance is usually very small in a resonant circuit and the current has a large value. This large current causes potential drops across the inductive reactance, $V_L = 2\pi f L I_0$, and the capacitive reactance, $V_C = I_0/2\pi f C$. Depending upon the values of L and C, the potential drops across the inductor and capacitor may attain very high values, which may damage the insulation of the circuit. These drops, however, have equal magnitudes but opposite phases, so that the net voltage across the inductor and capacitor is zero. At resonance, the voltage across the resistor is always equal to and in phase with the supply voltage. The entire supply voltage therefore appears across the resistance R. The phasor diagram for the voltages and currents of the circuit is shown in Fig. 4.69.

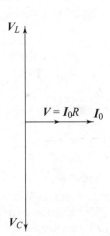

Fig. 4.69 Phasor diagram for an RLC series circuit in resonance frequency

Figure 4.70 shows the plot of the current against frequency for two different values of circuit resistance R_1 and R_2, where $R_1 > R_2$. The curve for R_2 is sharper for the simple reason that the current at resonance is larger in this case. For a very low value of R, the current exhibits a very pronounced peak at resonance. Since $X_L = X_C$ and the current is in phase with the voltage at resonance, it may be thought that there is no reactive power. Actually, the energy stored by each of the reactances is constant and equal to $I^2 X$. The energy oscillates between electromagnetic energy stored in the inductance and electrostatic energy stored in the capacitance. The energy remains with the reactances only, and the oscillations of the energy make the circuit appear resistive. When the peak rate of energy storage in each of the reactances is at least 10 times the resistive power $I^2 R$ of the

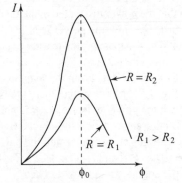

Fig. 4.70 Variation of current with RLC frequency

network, the network is said to be in resonance; otherwise the power factor of the circuit is adjusted to unity.

4.14.1.1 *Q*-Factor

It may be observed from the plot of frequency against current that when R is small the current falls off sharply on either side of the resonant frequency. This property of the circuit is of particular significance in electronics and communications engineering, since it indicates the capability of the circuit to accept current and power at resonant frequency. Such a circuit is called an acceptor (receiver) circuit, as it allows greater current flow when the signal frequency is close to f_0, but restricts signals having frequencies different from f_0 by offering high impedance.

Higher the reactive power, $I^2 X_L$ or $I^2 X_C$, of oscillation, better is the network's ability to receive current or power signals. The ratio of the reactive power to the resistive power is used to determine the quality of the network and is denoted by the Q-factor. Therefore, by definition,

$$Q\text{-factor} = \frac{\text{reactive power}}{\text{resistive power}} = \frac{I^2 X_L}{I^2 R} \quad \text{or} \quad \frac{I^2 X_C}{I^2 R}$$

$$= \frac{\omega_0 L}{R} \tag{4.148}$$

where $\omega_0 = 2\pi f_0$. The Q-factor may also be defined as the voltage magnification at resonance. Thus,

$$Q\text{-factor} = \frac{\text{voltage across } L \text{ or } C}{\text{supply voltage}}$$

$$= \frac{I X_L}{IR} \quad \text{or} \quad \frac{I X_C}{IR} = \frac{2\pi f_0 L}{R} \tag{4.149}$$

Substituting Eq. (4.146) in Eq. (4.149),

$$Q\text{-factor} = \frac{1}{R}\sqrt{\frac{L}{C}} \tag{4.150}$$

4.14.1.2 Bandwidth

The term bandwidth (BW) is used to describe the width of the resonance curve. It is defined as the range of frequencies for which the power delivered to the resistor R is equal to or greater than half the power delivered at resonance, and which are easily passed by the circuit. It is thus considered to be the range of frequencies over which the current is equal to or greater than $I_0/\sqrt{2}$ or $0.707 I_0$, so that the power dissipated in the resistance is equal to or greater than $V^2/2R$; i.e., the power is at least half of the maximum that would be dissipated in the resistance at resonance. The frequencies at which the power is just half of that at resonance are termed half-power points. From the shape of the resonance curve shown in Fig. 4.71, it is clear that there are two frequencies f_1 and f_2 for which the power delivered to R is half the power at resonance.

Let ω_0 and ω denote the angular frequencies and I_0 and I denote the currents at resonance and the half-power point, respectively. Then, $I_0 = V/R$ and $I = V/Z$, where Z is the series circuit impedance at the angular frequency ω. Then,

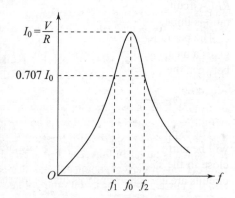

Fig. 4.71 Resonance curve of current and half-power points

$$\frac{V}{\sqrt{R^2 + \left(\omega L - \dfrac{1}{\omega C}\right)^2}} = \frac{V}{\sqrt{2}R}$$

or

$$\left(\omega L - \frac{1}{\omega C}\right) = \pm R \tag{4.151}$$

This is a quadratic equation in ω and may be solved to obtain two positive values of ω. The values are

$$\omega_1 = \sqrt{\frac{R^2}{4L^2} + \frac{1}{LC}} + \frac{R}{2L} \tag{4.152a}$$

$$\omega_2 = \sqrt{\frac{R^2}{4L^2} + \frac{1}{LC}} - \frac{R}{2L} \tag{4.152b}$$

Normally, in a tuned circuit, $R^2/4L^2 \ll 1/LC$. Therefore, neglecting the smaller terms, Eqs (4.152a) and (4.152b) become

$$\omega_1 = \omega_0 + \frac{R}{2L} \tag{4.153a}$$

$$\omega_2 = \omega_0 - \frac{R}{2L} \tag{4.153b}$$

Then, $\quad \omega_{BW} = \omega_2 - \omega_1 = \dfrac{R}{L}$ $\tag{4.154}$

Therefore, the bandwidth is given by

$$f_{BW} = f_2 - f_1 = \frac{\omega_2 - \omega_1}{2\pi} = \frac{R}{2\pi L} \tag{4.155}$$

where f_1 and f_2 are the lower and upper half-frequency points. It may be noted that the bandwidth of a series *RLC* circuit depends solely upon the R/L ratio. The individual values of R and L are not important but their ratio is important, while the value of C does not affect the bandwidth.

The per unit bandwidth is defined as the bandwidth expressed as a fraction of the resonance frequency. It gives a measure of the selectivity of the circuit. The quality factor Q of an *RLC* circuit is defined as the ratio between the resonance frequency and the bandwidth. Then,

$$Q = \frac{f_0}{f_2 - f_1} = \frac{f_0}{R/2\pi L} = \frac{2\pi f_0 L}{R} = \frac{\omega_0 L}{R} \tag{4.156}$$

The expression of Q in Eq. (4.156) is same as that given in Eq. (4.148). If the Q factor is high, the pass-band will be narrow, which means that the circuit will be highly selective and will allow only signal frequencies close to the resonance frequency to pass through. Frequencies away from resonance will be cut off. Such circuits are also referred to as tuned circuits.

Example 4.29 A series circuit has a resistance of 2 Ω, an inductance of 0.25 H, a variable capacitance, connected across a 230-V, 50-Hz supply. Calculate (a) the value of the capacitance for resonance, (b) the voltage across the inductance and the capacitance at resonance, and (c) the Q factor for the circuit.

Solution (a) For resonance, $X_L = X_C$ or $2\pi f L = 1/2\pi f C$. Then,

$$C = \frac{1}{(2 \times 3.14 \times 50)^2 \times 0.25} = 40.57 \text{ μF}$$

(b) The current at resonance, $I = \frac{V}{R} = \frac{230}{2} = 115 \text{ A}$

The voltage across the inductance $V_L = 2\pi \times 50 \times 0.25 \times 115 = 9027.5 \text{ V}$

The voltage across the capacitance $V_C = \frac{115 \times 10^6}{2 \times 3.14 \times 50 \times 40.57} = 9027.5 \text{ V} = V_L$

(c) Q factor $\frac{\omega L}{R} = \frac{2 \times 3.14 \times 50 \times 0.25}{2} = 39.25$

Example 4.30 A coil of inductance 10 Ω and resistance 100 Ω in series with a condenser is supplied a constant voltage from a variable frequency source. If the maximum current is 1 A at 100 Hz, find the frequencies when the current is 0.5 A.

Solution Since the current in the series *RLC* circuit is maximum at 100 Hz, the circuit is in resonance. The resonant frequency

$$f_0 = \frac{1}{2\pi\sqrt{LC}}$$

Then, $C = \dfrac{1}{4 \times 3.14^2 \times 100^2 \times 10} = 0.25 \text{ μF}$

At resonance, the voltage $V = IR = 1 \times 100 = 100 \text{ V}$. When the current is 0.5 A,
impedance $\quad Z = V/I = 100/0.5 = 200 \text{ Ω}$.
Now, $\quad\quad\quad Z = 100 + jX$
where $\quad\quad\quad X = X_L - X_C$

Therefore, $X = \sqrt{200^2 - 100^2} = 173.2 \text{ Ω}$

Let the angular frequency be ω when the current in the circuit is 0.5 A.

$$173.2 = \omega \times 10 - \frac{10^6}{\omega \times 0.25}$$

or $\quad 2.5\omega^2 - 173.2 \times 0.25\omega - 10^6 = 0$

or $\quad 2.5\omega^2 - 43.3\omega - 10^6 = 0$

Therefore, $\omega = 641.1 \text{ rad/sec}, -623.85 \text{ rad/sec}$

However, in engineering, negative angular speed is not possible and is hence ignored.

Then, $\quad f_0 = \dfrac{641.1}{2\pi} = 102.0342 \text{ Hz}$

$$f_1 = f_0 - \frac{R}{4\pi L} = 102.0342 - \frac{100}{2\pi \times 10} = 101.2385 \text{ Hz}$$

$$f_2 = f_0 + \frac{R}{4\pi L} = 102.0342 + \frac{100}{2\pi \times 10} = 102.8300 \text{ Hz}$$

4.14.2 Resonance in Parallel Circuits

The circuit of Fig. 4.72 shows a pure inductance of L henry, a pure capacitance of C farads, and a resistance of R ohms connected in parallel across an ac voltage source of V volt. The total current I drawn from the source is a sum of the currents I_R, I_L, and I_C drawn by the individual branches containing R, L, and C, respectively. Taking the supply voltage as the reference phasor, i.e., $V = V\angle 0° = V + j0$, the expression for the total current is

$$I = VY = V(G + jB) \tag{4.157}$$

and

$$Y = Y_R + Y_L + Y_C$$

$$= \frac{1}{R} + \frac{1}{j\omega L} - \frac{1}{j/(\omega C)} = \frac{1}{R} + j\left(\omega C - \frac{1}{\omega L}\right) \tag{4.158}$$

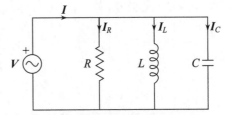

Fig. 4.72 Circuit for parallel resonance

Then, $\quad Y = G + j(B_L + B)$ \hfill (4.159)

where $\quad G = \dfrac{1}{R}, \; B_L = -\dfrac{1}{\omega L} = -\dfrac{1}{2\pi L}, \; B_C = \omega C$ \hfill (4.160)

Then, $\quad I = V\left[\dfrac{1}{R} + j\left(\omega C - \dfrac{1}{\omega L}\right)\right]$ \hfill (4.161)

The phase angle of the total current with respect to the applied voltage

$$\varphi = \tan^{-1} \frac{\omega C - 1/\omega L}{1/R} = \tan^{-1} \frac{B_L + B_C}{G} \tag{4.162}$$

The graphs of G, B, and Y are plotted against frequency f in Fig. 4.73. From this figure, it may be seen that B_L varies inversely with frequency, and its graph is a rectangular hyperbola, while B_C varies linearly with f, and its graph is a straight line passing through the origin. The value of conductance being constant, the graph of G is a horizontal straight line. The variation of the magnitude of the admittance $Y = (G^2 + B^2)^{1/2}$ is also shown in the figure. When the supply voltage is constant, the magnitude of the total current $I = VY$; thus the current I is proportional to Y, and as the frequency is changed, the variation of I is similar to that of Y. Thus, the graph showing the variation of Y will represent I as well but will have a different scale. If $G = 0$, then the variation of Y with f is as shown by the dotted line in Fig. 4.73.

For fixed supply voltage, the current can be determined from Eq. (4.161). As the current $I = VY$, with V having a constant value, the variation of current is the same as the variation of Y. It may be noted that for small values of ω, the susceptance B is large, so the current I is large; as ω increases, B decreases, I decreases; when B equals zero, V has its minimum value. At this point, the current I has its minimum value and is in time phase with the voltage, and the circuit is said to be in parallel resonance. The frequency at which resonance occurs is found from

$$\frac{1}{\omega_0 L} = \omega_0 C \tag{4.163}$$

or

$$\omega_0 = \frac{1}{\sqrt{LC}} \tag{4.164}$$

and the corresponding frequency f_0 is given as

$$f_0 = \frac{1}{2\pi\sqrt{LC}} \tag{4.165}$$

At the frequency f_0, the admittance Y has the minimum possible value of $G\,(= 1/R)$. The phasor diagram for the resonant condition is shown in Fig. 4.74.

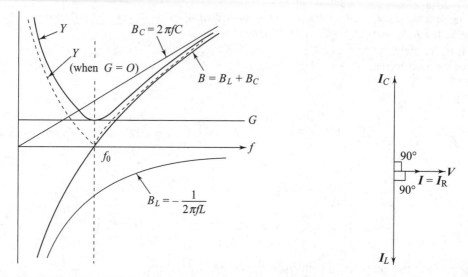

Fig. 4.73   Variation of circuit parameters with frequency   **Fig. 4.74**   Phasor diagram for parallel resonance

Just as there occurs a rise in voltage associated with L and C during the condition of series resonance, so also in the case of parallel resonance, there occurs a rise in current in the elements L and C during resonant conditions. The value of the phasor voltage V across the parallel branch is IR. Since the same voltage appears across L and C, the currents I_{L0} and I_{C0} through L and C, respectively, may be obtained as $I_{L0} = V/(j\omega_0 L)$ $= IR/(j\omega_0 L)$ and $I_{C0} = V/(1/j\omega_0 L) = jIR\omega_0 C$.

If the circuit of Fig. 4.72 contains L and C only but no resistance, then at resonant frequency f_0 the current I will be zero. The branch currents flowing in L and C are exactly equal in magnitude but in phase opposition; their phasor sum, which gives the total current I, therefore becomes zero. This phenomenon is called parallel resonance. Hence, under the condition of resonance, the circuit behaves like an open circuit network, that is, $I =$ 0. It may be noted that the active power is zero. However, the reactive power in L and C manifests itself in the form of electric and magnetic energy being continually exchanged between the capacitance and the inductance.

4.14.2.1 Practical *LC* Parallel Circuit

In actual practice, an inductance coil always has some resistance. If the inherent resistance of the inductance coil is r in the LC parallel circuit as shown in Fig. 4.75, taking the voltage phasor as reference, the current phasor I_L through the inductance coil is given by

$$I_L = \frac{V}{r + j\omega L} = \frac{V \angle 0°}{\sqrt{r^2 + \omega^2 L^2} \angle \tan^{-1}(\omega L / r)}$$

$$= \frac{V}{\sqrt{r^2 + \omega^2 L^2}} \angle - \tan^{-1}(\omega L / r) \qquad (4.166)$$

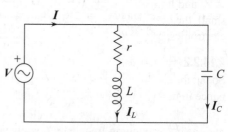

Fig. 4.75   Coil and capacitor in parallel

As $r \ll \omega L$, $\varphi \approx 90°$. Similarly, the current through the capacitance $I_C = \omega CV$, leading the applied voltage phasor V by 90°.

In the resonance condition, the total current $I = I_L + I_C$ will be in phase with the applied voltage V as shown in the phasor diagram in Fig. 4.76. Then

$$I_C = I_L \sin \varphi \qquad (4.167)$$

Now at resonant frequency f_0, $\omega = \omega_0 = 2\pi f_0$,

$$\sin\varphi = \frac{2\pi f_0 L}{\sqrt{r^2 + (2\pi f_0 L)^2}}$$

Then, Eq. (4.167) becomes

$$2\pi f_0 CV = \frac{V}{\sqrt{r^2 + \omega_0^2 L^2}} \frac{2\pi f_0 L}{\sqrt{r^2 + \omega_0^2 L^2}} = \frac{2\pi f_0 LV}{r^2 + 4\pi^2 f_0^2 L^2}$$

Therefore,

$$f_0 = \frac{1}{2\pi}\sqrt{\frac{1}{LC} - \frac{r^2}{L^2}} \qquad (4.168)$$

In communication circuits, $r \ll \omega L$; thus, $r^2/(\omega L)^2 \approx 0$. Then,

$$f_0 = \frac{1}{2\pi\sqrt{LC}} \qquad (4.169)$$

This is the same as that obtained for the series resonance condition. From the phasor diagram of Fig. 4.76, it can be seen that the capacitive current is several times the total circuit current.

The Q factor in the case of parallel resonance is defined as the ratio of the capacitive current to the network current. Hence,

$$Q \text{ factor} = \frac{I_C}{I} = \frac{I_L \sin\varphi}{I_L \cos\varphi} = \frac{\sin\varphi}{\cos\varphi} = \tan\phi = \frac{2\pi f_0 L}{r} \qquad (4.170)$$

The total current during parallel resonance is in phase with the applied voltage. The impedance Z_0 of the network in resonance can be obtained from the phasor diagram of Fig. 4.76 as

$$Z_0 = \frac{V}{I} = \frac{V}{I_C \cot\varphi} = \frac{V}{I_C}\tan\varphi = \frac{1}{2\pi f_0 C}\frac{2\pi f_0 L}{r} = \frac{L}{Cr} \qquad (4.171)$$

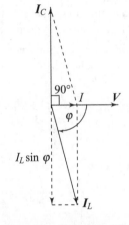

Fig. 4.76 Phasor diagram at resonance

The impedance Z_0 is called the dynamic impedance of the network and is a non-reactive resistor. It may be noted that smaller the value of r, higher the value of Z_0 and lower the resultant network current. When the resistance is negligibly small, the current I is zero. Such a network is referred to as a rejector network at that particular frequency in electronics and communications engineering.

4.14.2.2 Resonance by Varying *L* and *C*

So far, the effect of varying f on both series and parallel circuits has been considered. The conditions for resonance in both cases, assuming negligible resistance, is that

$$\omega L = \frac{1}{\omega C} \text{ (numerically)}$$

Clearly, this condition can be achieved by keeping the frequency constant and varying either L or C. The dependence of resonance on frequency is not usually of importance in circuits or apparatus operating at power frequency, unless there is such a distortion of the voltage wave from the sinusoidal form that there exists a higher order harmonic that gives rise to the resonant conditions. This sometimes happens with insulated cables. The creation of resonance by adjustment of L or C is, of course, the normal adjustment in transistor receiver sets, in which the detector circuit is 'tuned' (adjusted) to the frequency of the station it is desired to receive.

It may be useful to compare the properties of networks in series and parallel resonance. Table 4.1 compares the responses of series and parallel networks in resonance.

Table 4.1 Comparison of series and parallel networks in resonance

S.No	Parameter	Series resonance	Parallel resonance
1.	Current magnitude	Maximum	Minimum
2.	Power factor	Unity	Unity
3.	Power	Low	High
4.	Impedance	Low	High
5.	Q factor	Depends on voltage amplification	Depends on current amplification
6.	Type	Acceptor	Rejecter

Example 4.31 A broadcasting station transmits on a wavelength of 3000 m. A capacitor of 0.0005 μF is connected in parallel to a variable inductor to form a closed oscillating circuit. To what value must the inductor be adjusted in order to tune in to the particular station?

Solution   Electromagnetic waves are propagated with a velocity of almost exactly 3×10^8 m/sec. The relationship between the velocity u, wavelength λ, and frequency f is

$$u = \lambda f$$

$$f = \frac{u}{\lambda} = \frac{3 \times 10^8}{3000} = 10^5 \text{ Hz}$$

The condition for resonance is $2\pi f_0 L = \dfrac{1}{2\pi f_0 C}$

Therefore,

$$L = \frac{1}{4\pi^2 f_0^2 C} = \frac{1}{4\pi^2 \times 10^{10} \times 0.0005 \times 10^{-6}} = 5.07 \text{ mH}$$

Example 4.32 A coil of 1.5 kΩ resistance and 0.2 H inductance is connected in parallel with a variable capacitor across a 1.5-V, 15-kHz ac supply. Calculate (a) the value of the capacitor when the supply current is minimum, (b) the effective impedance of the network, and (c) the current drawn from the supply.

Solution   (a) When the circuit is in resonance, the supply current is minimum.

$$f_0 = \frac{1}{2\pi}\sqrt{\frac{1}{LC} - \frac{r^2}{L^2}} = \frac{1}{2\pi}\sqrt{\frac{1}{0.2 \times C} - \frac{(1.5 \times 10^3)^2}{(0.2)^2}}$$

Then, $\dfrac{1}{0.2C} = 4 \times \pi^2 \times (15 \times 10^3)^2 + \dfrac{1.5^2 \times 10^6}{0.2^2} = 8838.4 \times 10^6$

$$C = 55.935 \times 10^{-9} \text{ F}$$

(b) The effective impedance of the network

$$Z = \frac{L}{CR} = \frac{0.2}{55.935 \times 10^{-9} \times 1.5 \times 10^3} = 2.384 \times 10^3 = 2.384 \text{ k}\Omega$$

(c) The current drawn from the supply

$$I = \frac{V}{Z} = \frac{1.5}{2.384 \times 10^3} = 0.63 \text{ mA}$$

4.15 STAR–DELTA OR Y-Δ TRANSFORMATION

The methods of series, parallel, and series–parallel combination of elements do not always lead to the simplification of networks. Such networks are handled by star–delta transformations, as already discussed in Section 2.7. Figure 4.77(a) shows three impedances Z_{ab}, Z_{bc}, and Z_{ca} connected in delta to three nodes A, B, and C. Figure 4.77(b) shows three impedances Z_a, Z_b, and Z_c connected in star between the same three nodes A, B, and C and a common point n. If these two networks are to be equivalent, then the impedances between any two nodes of the delta-connected network of Fig. 4.77(a) must be the same as that between the same pair of nodes of the star-conneced network of Fig. 4.77(b).

Star impedances in terms of delta impedances

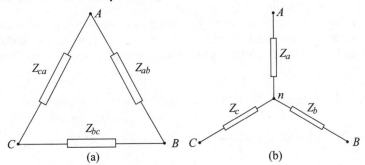

Fig. 4.77 (a) Delta network and (b) star network

$$Z_a = \frac{Z_{ab}Z_{ac}}{Z_{ab} + Z_{bc} + Z_{ca}} \tag{4.172}$$

$$Z_b = \frac{Z_{bc}Z_{ba}}{Z_{ab} + Z_{bc} + Z_{ca}} \tag{4.173}$$

and $$Z_c = \frac{Z_{ca}Z_{cb}}{Z_{ab} + Z_{bc} + Z_{ca}} \tag{4.174}$$

Thus, the equivalent star impedance connected to a node is equal to the product of the two delta impedances connected to the same node divided by the sum of the delta impedances.

Delta impedances in terms of star impedances

$$Z_{bc} = Z_b + Z_c + \frac{Z_c Z_b}{Z_a} \tag{4.175}$$

$$Z_{ca} = Z_c + Z_a + \frac{Z_c Z_a}{Z_b} \tag{4.176}$$

and $$Z_{ab} = Z_a + Z_b + \frac{Z_a Z_b}{Z_c} \tag{4.177}$$

Thus, the equivalent delta impedance between two nodes is the sum of the two star impedances connected to those nodes plus the product of the same two star impedances divided by the third star impedance.

4.16 NODAL VOLTAGE AND MESH CURRENT ANALYSIS OF AC NETWORKS

AC electrical networks are solved by reducing them to series, parallel, or series–parallel combinations of elements. Often, such networks cannot be reduced to a simple form, as discussed earlier in Sections 4.10–

4.12. In such cases, the mesh and nodal methods introduced for dc networks in Sections 2.8 and 2.9 may be employed with the following modifications:

  (i)   Resistance and conductance in dc circuits are replaced by complex impedance and admittance in ac circuits.
 (ii)   Current and voltage phasors are used in place of scalar values of dc current or voltage.
(iii)   KCL and KVL equations in phasor form are used in ac networks.
 (iv)   The mesh currents or nodal voltage phasors are treated as independent variables and are chosen exactly in the same manner as in dc circuits.
  (v)   In the case of ac networks, equilibrium equations are a set of linear simultaneous complex equations instead of real algebraic equations for dc networks.

  Mesh or nodal equations are written, as before, by inspecting the network only when mutual inductors are absent. When mutual inductors are present, the network equilibrium equations should be obtained from first principles by applying KCL or KVL. The mesh and nodal methods of solution are illustrated below using simple examples.

Example 4.33 For the circuit shown in Fig. 4.78, determine the values of the none phasor voltages V_1 and V_2.

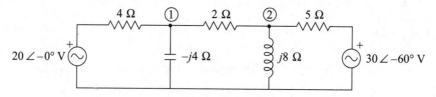

Fig. 4.78

Solution   The voltage sources of Fig. 4.78 are converted to the current sources and the resistances and reactances are converted to conductances and susceptances, respectively. With these changes incorporated, the network changes to that shown in Fig. 4.79. Node 0 is taken as the reference node.

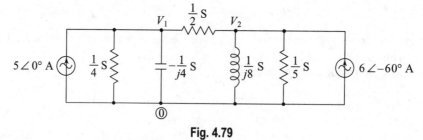

Fig. 4.79

  By inspection, the nodal equations in matrix form may be written as

$$\begin{bmatrix} \dfrac{1}{4}+\dfrac{1}{2}+j\dfrac{1}{4} & -\dfrac{1}{2} \\[2mm] -\dfrac{1}{2} & \dfrac{1}{5}+\dfrac{1}{2}-j\dfrac{1}{8} \end{bmatrix}\begin{bmatrix} V_1 \\[2mm] V_2 \end{bmatrix} = \begin{bmatrix} 5\angle 0° \\[2mm] 6\angle -60° \end{bmatrix}$$

Then, $V_1 = \dfrac{\Delta_1}{\Delta}$ and $V_2 = \dfrac{\Delta_2}{\Delta}$

where $\Delta = \begin{vmatrix} 0.75 + j0.25 & -0.5 \\ -0.5 & 0.7 - j0.125 \end{vmatrix} = 0.306 + j0.081 = 0.3165 \angle 14.83°$

$\Delta_1 = \begin{vmatrix} 5 & -0.5 \\ 3.0 - j5.196 & 0.7 - j0.125 \end{vmatrix} = 5.0 - j3.223 = 5.95 \angle -32.8°$

$\Delta_2 = \begin{vmatrix} 0.75 + j0.25 & 5 \\ -0.5 & 3.0 - j5.196 \end{vmatrix} = 6.049 - j3.147 = 6.8 - \angle -27.48°$

$V_1 = \dfrac{\Delta_1}{\Delta} = \dfrac{5.95 \angle -32.8°}{0.3165 \angle 14.83°} = 18.79 \angle -47.63° \text{ V}$

$V_2 = \dfrac{\Delta_2}{\Delta} = \dfrac{6.82 \angle -27.48°}{0.3165 \angle 14.83°} = 21.55 \angle -42.30° \text{ V}$

Example 4.34 For the circuit shown in Fig. 4.80, determine the current I through the impedance Z_3 using mesh analysis.

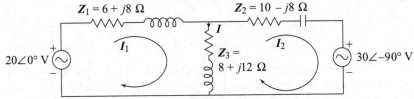

Fig. 4.80

Solution

$Z_1 = 6 + j8 = 10 \angle 53.1°$
$Z_2 = 10 - j8 = 12.806 \angle -38.66°$
$Z_3 = 6 + j12 = 13.416 \angle 63.43°$

The independent mesh currents are I_1 and I_2. By inspection, the following mesh equations in matrix form may be written

$$\begin{bmatrix} 14 + j20 & -8 - j12 \\ -8 - j12 & 18 + j6 \end{bmatrix} \begin{bmatrix} I_1 \\ I_2 \end{bmatrix} = \begin{bmatrix} 20 \angle 0° \\ -30 \angle \square 90° \end{bmatrix} = \begin{bmatrix} 20 \\ j30 \end{bmatrix}$$

Then, $I_1 = \dfrac{\Delta_1}{\Delta}$ and $I_2 = \dfrac{\Delta_2}{\Delta}$

where $\Delta = \begin{vmatrix} 14 + j20 & -8 - j12 \\ -8 - j12 & 18 + j6 \end{vmatrix} = 212 + j252 = 329.31 \angle 49.93°$

$\Delta_1 = \begin{vmatrix} 20 & -8 - j12 \\ j30 & 18 + j6 \end{vmatrix} = 0 + j360 = 360 \angle 90°$

$\Delta_2 = \begin{vmatrix} 14 + j20 & 20 \\ -8 - j12 & j30 \end{vmatrix} = -440 + j660 = 793.22 \angle 123.69°$

$I = \dfrac{\Delta_1}{\Delta} = \dfrac{360 \angle 90°}{329.31 \angle 49.93°} = 1.093 \angle 40.07° = 0.836 + j0.703 \text{ A}$

$I_2 = \dfrac{\Delta_2}{\Delta} = \dfrac{793.22 \angle 123.69°}{329.31 \angle 49.93°} = 2.409 \angle 73.76° = 0.674 + j2.313 \text{ A}$

$I = I_1 - I_2 = 0.163 - j1.61 = 1.62 \angle -84.21° \text{ A}$

4.17 NETWORK THEOREMS FOR AC NETWORKS

All the theorems discussed in Section 2.11 are applicable to ac networks as well with phasor voltages/currents and complex impedances/admittances. The application of the superposition, Thevenin, Norton, and maximum power transfer theorems is illustrated here through examples.

4.17.1 Superposition Theorem

Example 4.35 Determine the currents in the network shown in Fig. 4.81(a).

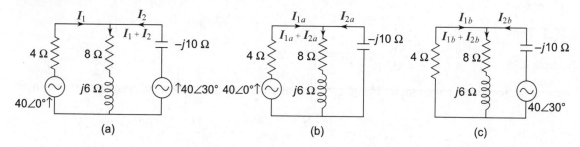

Fig. 4.81

Solution To apply the superposition theorem, two separate networks are drawn; each network has only one source. Figure 4.81(b) shows the network with only the source $40\angle0°$ V in the circuit, while the other source $40\angle30°$ V is replaced by a short circuit. Figure 4.81(c) shows the network with only the source $40\angle30°$ V in the circuit, while the other source $40\angle0°$ V is replaced by a short circuit. From Fig. 4.81(b), the total impedance of the circuit

$$Z = 4 + \frac{(8+j6)(0-j10)}{8+j6+0-j10} = 4 + \frac{60-j80}{8-j4}$$

$$= 14 - j5 = 14.87\angle-19.65°$$

Then, $I_{1a} = \dfrac{40\angle0°}{14.87\angle-19.65°} = 2.69\angle19.65° = 2.533 + j\,0.904$ A

$$I_{2a} = -2.69\angle19.65° \times \frac{8+j6}{8-j4} = -2.69\angle19.65° \times \frac{10\angle36.87°}{8.94\angle-26.56°}$$

$$= -2.69\angle(19.65+36.87+26.56)° = -2.69\angle83.08°$$

$$= -0.324 - j2.67 \text{ A}$$

Therefore, $I_{1a} + I_{2a} = (2.533 + j0.904) + (-0.324 - j2.67) = 2.209 - j1.766$

$$= 2.828\angle-38.64° \text{ A}$$

From Fig. 4.81(c), the total impedance of the circuit

$$Z = 0 - j10 + \frac{(8+j6)(4+j0)}{8+j6+4+j0} = -j10 + \frac{32+j24}{12+j6}$$

$$= 2.93 - j9.47 = 9.91\angle-72.81°$$

Then, $I_{1b} = \dfrac{40\angle30°}{9.91\angle-72.81°} = 4.036\angle102.81° = -0.895 + j3.935$ A

$$I_{2b} = -4.036\angle102.81° \times \frac{8+j6}{12+j6} = -4.036\angle102.81° \times \frac{10\angle36.87°}{13.416\angle26.56°}$$

$$= -3.008\angle(102.81 + 36.87 - 26.56)° = -2.69\angle113.12°$$

$$= 1.056 - j2.474 \text{ A}$$

Therefore, $I_{1b} + I_{2b} = (-0.895 + j3.935) + (1.056 - j2.474) = 0.161 + j1.461$

$$= 1.47\angle83.71° \text{ A}$$

The total currents are

$$I_1 = I_{1a} + I_{1b} = (2.533 + j0.904) + (-0.895 + j3.935) = 1.638 + j4.839 = 5.11\angle71.3° \text{ A}$$
$$I_2 = I_{2a} + I_{2b} = (-0.324 - j2.67) + (1.056 - j2.474) = 0.732 - j5.144 = 5.196\angle-81.9° \text{ A}$$
$$I_1 + I_2 = (1.638 + j4.839) + (0.732 - j5.144) = 2.37 - j0.305 = 2.39\angle-7.33° \text{ A}$$

4.17.2 Thevenin's Theorem

Example 4.36 Determine the current through the load impedance for the network shown in Fig. 4.82(a) using Thevenin's theorem.

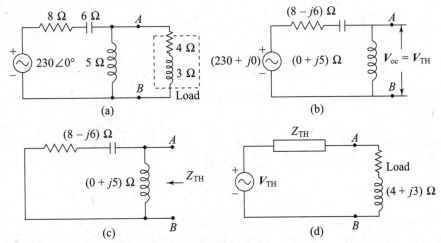

Fig. 4.82 (a) Network of Examples 4.36, (b) determination of VTH, (c) determination of ZTH, (d) Thevenin's equivalent circuit with load

Solution With the terminals *AB* open, the network s shown in Fig. 4.82(b). Then,

$$\frac{V_{oc}}{V} = \frac{0 + j5}{(8 - j6) + (0 + j5)} = \frac{j5}{8 - j} = \frac{j5 \times (8 + j)}{65} = -0.077 + j0.6154$$

Therefore,

$$V_{oc} = V_{TH} = (-17.71 + j141.54) = 142.64\angle97.13° \text{ V}$$

For the determination of Z_{TH}, the voltage source is short-circuited as shown in Fig. 4.82(c).

$$Z_{TH} = \frac{(8 - j6)(0 + j5)}{(8 - j6) + (0 + j5)} = \frac{30 + j40}{8 - j} = \frac{(30 + j40)(8 + j)}{8^2 + 1^2} = \frac{200 + j350}{65}$$

$$= 3.077 + j5.384 = 6.20\angle60.25°$$

Thevenin's equivalent circuit along with the load is shown in Fig. 4.82(d). The current through the load is given by

$$I = \frac{V_{TH}}{Z_{TH} + Z_{load}} = \frac{-17.71 + j141.54}{(3.077 + j5.384) + (4 + j3)} = \frac{142.64\angle97.13°}{7.077 + j8.384}$$

$$= \frac{142.64\angle97.13°}{10.97\angle49.83°} = 13\angle47.3° \text{ A} = 8.817 + j9.55 \text{ A}$$

4.17.3 Norton's Theorem

Example 4.37 Determine the current through the load impedance for the network shown in Fig 4.83(a) using Norton's theorem.

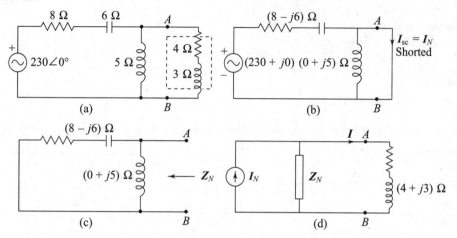

Fig. 4.83   (a) Network of Example 4.37, (b) determination of I_N, (c) determination of Z_N, (d) Norton's equivalent circuit with load

Solution   With the terminals *AB* shorted, the network is shown in Fig. 4.83(b). Then,

$$I_{sc} = \frac{230 + j0}{8 - j6} = \frac{230 \times (8 + j6)}{(8 - j6) \times (8 + j6)} = \frac{1840 + j1380}{100}$$

Therefore, $I_{sc} = I_N = 18.4 + j13.8 = 23\angle 36.87°$ A

For the determination of Z_N, the voltage source is short-circuited as shown in Fig. 4.83(c).

$$Z_N = \frac{(8 - j6)(0 + j5)}{(8 - j6) + (0 + j5)} = \frac{30 + j40}{8 - j} = \frac{(30 + j40)(8 + j)}{8^2 + 1^2} = \frac{200 + j350}{65}$$

$$= 3.077 + j5.384 = 6.20\angle 60.25°$$

Norton's equivalent circuit along with the load is shown in Fig. 4.83(d). The current through the load is given by

$$I = I_N \times \frac{Z_N}{Z_N + Z_{load}} = 23\angle 36.87° \frac{6.2\angle 60.25°}{(3.077 + j5.384) + (4 + j3)}$$

$$= \frac{23 \times 6.2\angle (60.25 + 36.87)°}{7.077 + j8.384} = \frac{142.6\angle 97.12°}{10.97\angle 49.83°} = 13\angle 47.29°\ \text{A}$$

4.17.4 Maximum Power Transfer Theorem

The statement of this theorem is slightly different than in the case of a dc network. The maximum power transfer theorem for the sinusoidal steady-state condition states that the power transferred to an impedance load will be maximum when the load impedance is equal to the complex conjugate of the internal impedance of the source. The theorem can be proved as follows.

Figure 4.84 shows a circuit consisting of a sinusoidal voltage source *V* in series with an internal impedance $Z_S = R_S + jX_S$, connected to a load impedance $Z_L = R_L + jX_L$. If the source voltage is taken as the reference phasor, the current through the load is given by

$$I = \frac{V}{Z_S + Z_L} = \frac{V + j0}{(R_S + R_L) + j(X_S + X_L)}$$

and the rms current

$$I = \frac{V}{\sqrt{(R_S + R_L)^2 + (X_S + X_L)^2}} \qquad (4.178)$$

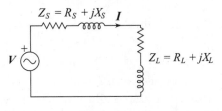

Fig. 4.84 Sinusoidal voltage source connected to a load

The power dissipated P in the load resistance is given by

$$P = |I|^2 R_L = \frac{|V|^2 R_L}{(R_S + R_L)^2 + (X_S + X_L)^2} \qquad (4.179)$$

From Eq. (4.179), it may be observed that for fixed values of source voltage magnitude and source impedance, the power P is a function of the load parameters R_L and X_L. So, for maximum power transfer, the following conditions must be satisfied:

$$\frac{\partial P}{\partial R_L} = 0 = \frac{|V|^2 [R_S + R_L]^2 + (X_S + X_L)^2 - 2R_L(R_S + R_L)}{[(R_S + R_L)^2 + (X_S + X_L)^2]^2} \qquad (4.180)$$

$$\frac{\partial P}{\partial X_L} = 0 = \frac{|V|^2 [-2R_L(X_S + X_L)]}{[(R_S + R_L)^2 + (X_S + X_L)^2]^2} \qquad (4.181)$$

Equations (4.180) and (4.181) upon simplification give, respectively, the following conditions:

$$R_L = R_S \qquad (4.182)$$
$$X_L = -X_S \qquad (4.183)$$

Thus, for maximum power transfer,

$$Z_L = Z_S^* \qquad (4.184)$$

where Z_S^* is the complex conjugate of the source impedance Z_S.

When a network is represented by its Thevenin equivalent loaded at the terminals by a load impedance, the maximum power transfer theorem may be applied. In such a case, Thevenin's equivalent impedance Z_{TH} replaces the source impedance Z_S. The maximum power is transferred to the load when

$$Z_L = Z_{TH}^* \qquad (4.185)$$

This method of transferring maximum power to the load is known as impedance matching and is widely used in electronic circuits. However, the maximum efficiency of power transfer is only 50%. In power systems, the power transfer to the load has to be done with minimum losses, hence the condition of operation is far from the condition of impedance matching.

Example 4.38 For the circuit shown in Fig. 4.85(a), determine the impedance of the load that will dissipate maximum power, and also the maximum power.

Solution The equivalent Thevenin impedance is given by

$$Z_{TH} = \frac{(3 + j4)(0 - j2)}{(3 + j4) + (0 - j2)} = \frac{8 - j6}{3 + j2} = \frac{(8 - j6)(3 - j2)}{3^2 + 2^2}$$

$$= \frac{12 - j34}{13} = 0.923 - j2.615 \ \Omega$$

The open circuit or Thevenin voltage across the load terminals is given by

$$V_{TH} = (15 + j0) \times \frac{0 - j2}{(3 + j4) + (0 - j2)} = \frac{0 - j30}{3 + j2} = \frac{-j30(3 - j2)}{13}$$

$$= -4.615 - j6.923 \text{ V}$$

For maximum power, $Z_L = Z_{TH}^*$

Thus, $Z_L = 0.923 + j2.615$

The total impedance of the circuit

$$Z_L + Z_{TH} = (0.923 + j2.615) + (0.923 - j2.615) = 1.846 \ \Omega$$

Then the current in load is given by

$$I = \frac{V_{TH}}{Z_{TH} + Z_L} = \frac{-4.615 - j6.923}{1.846} = -2.50 - j3.75$$

$$I = 4.51 \text{ A}$$

The power consumed by the load is given by

$$P_L = I^2 R_L = 4.51^2 \times 0.923 = 18.77 \text{ W}$$

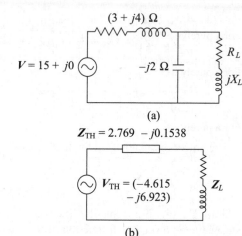

$$Z_{TH} = 2.769 - j0.1538$$

Fig. 4.85   (a) Circuit diagram for Example 4.38
(b) Thevenin's equivalent circuit for Example 4.38

Recapitulation

Any quantity (voltage or current) whose direction changes and whose magnitude varies with time, repetitively, is said to be an alternating quantity.

All electrical machines (except direct current machines) are designed to generate sinusoidal voltages and currents. Even in electrical communication systems, sinusoidal voltages and currents are of importance.

Any quantity having a magnitude and a direction is said to be a phasor quantity. Thus, alternating quantities are phasor quantities.

The j operator when applied to a phasor can be interpreted to rotate it through 90° in the anticlockwise direction without altering its magnitude.

For adding and subtracting phasors, it is expeditious to use the rectangular form for representing complex quantities. For multiplication and division of phasors, it is more convenient to employ the trigonometric or polar form of representation.

The network theorems that were used in dc networks are applicable for ac networks as well through the use of complex notation.

Time period $T = 1/f$ sec

Angular velocity $\omega = 2\pi f$ rad/sec

$$f = (NP/120) \text{ Hz or cycles/sec}$$

Average value of a rectified sine wave, $I_{av} = 0.637 I_m$ A

Root mean square value of a sine wave, $I = 0.707 I_m$ A

Form factor $= (0.707 I_m)/(0.637 I_m) = 1.11$

Peak factor $= I_m/(0.707 I_m) = 1.414$

The instantaneous values of voltage and current are given by

$$v = V_m \sin \theta = V_m \sin 2\pi f t \text{ V}$$
$$i = I_m \sin \theta = I_m \sin 2\pi f t \text{ A}$$

Properties of the j operator:

$$j^2 = -1 \text{ or } j = \sqrt{-1}$$
$$j^3 = -j$$
$$j^4 = 1$$

Representation of phasors in Cartesian or rectangular form: $A = a + jb$ and $\tan \theta = b/a$

Representation of phasors in trigonometric form:

$$A = Ae^{j}\theta = A(\cos \theta + j \sin \theta)$$

Representation of phasors in polar form:

$A = A\angle\theta$

For a purely resistive circuit,

$$V = IR$$

For a purely inductive circuit,

$$V = IX_L; \; X_L = 2\pi f L = \omega L$$
$$v = 2\pi f L I_m \sin(2\pi f t - \pi/2)$$

For a purely capacitive circuit,

$$V = IX_C; \; X_C = \frac{1}{2\pi f C} = \frac{1}{\omega C}$$

$$v = \frac{1}{2\pi f C} \, I_m \sin(2\pi f t + \pi/2)$$

For an RL series circuit,

$$V = IZ$$

$$Z = \sqrt{R^2 + \omega^2 L^2} = \sqrt{R^2 + X_L^{\,2}}$$

For an RC series circuit,

$$Z = \sqrt{R^2 + \frac{1}{\omega^2 C^2}} = \sqrt{R^2 + X_C^{\,2}}$$

For an RLC series circuit,

$$Z = \sqrt{R^2 + \left(\omega^2 L^2 - \frac{1}{\omega^2 C^2}\right)} = \sqrt{R^2 + (X_L - X_C)^2}$$

For R and L in parallel, $I = I_R + I_L$

For R and C in parallel, $I = I_R + I_C$

Power in ac circuit:

Active power $P = VI\cos\varphi$ watt

Reactive power $Q = VI\sin\varphi$ VAR

Apparent power $S = VI$ volt-amperes

Power factor $\cos\varphi = P/S$

Reactive factor $\sin\varphi = Q/S$

$$S^2 = P^2 + Q^2$$

Series resonance in an RLC series circuit: $X_L = X_C$

Resonant frequency $f_0 = \dfrac{1}{2\pi\sqrt{LC}}$

Q factor $= \omega_0 L / R$

Resonance in a practical LC parallel circuit:

Resonant frequency,

$$f_0 = \frac{1}{2\pi}\sqrt{\frac{1}{LC} - \frac{r^2}{L^2}}$$

$$Z_0 = \frac{L}{Cr}$$

For maximum power transfer, $Z_L = Z_{TH}^*$

Assessment Questions

1. (a) Show that when a coil of length l metre and width b metre is rotated at a speed of n revolutions per second inside a magnetic field of intensity B Wb/m^2, the generated voltage is alternating and its variation is sinusoidal. What is the magnitude of the maximum voltage?
   (b) Derive an expression for the frequency of the ac generated voltage.
2. (a) Define (i) periodic function, (ii) cycle, (iii) time period, (iv) frequency, and (v) angular velocity.
   (b) What is angular velocity? How is angular velocity related to time period and frequency?
3. (a) Use an arbitrary periodic waveform to show (i) instantaneous value, (ii) phase, and (iii) phase difference.
   (b) Graphically show the relationship between (i) peak value and (ii) peak-to-peak value.
4. (a) Define and derive mathematical expressions for (i) average and (ii) effective values of a periodic waveform.
   (b) Define (i) form factor and (ii) peak factor. What is the significance of the form factor?
5. (a) Explain the various forms of phasor representation.
   (b) Discuss the various forms of phasor representation that can be conveniently employed for (i) addition/subtraction and (ii) multiplication/division of phasor quantities.
6. (a) Derive an expression for the average power consumed in a resistive circuit.
   (b) A purely reactive component does not consume power. Justify the statement with the help of appropriate wave forms and explain how the exchange of power takes place between the source and the reactive element.
7. Distinguish between (i) instantaneous power, (ii) real power, (iii) reactive power, and (iv) apparent power.
8. (a) Explain the power triangle and there from define (i) power factor, and (ii) reactive factor.
   (b) Discuss the importance of power factor and the need for its improvement.

9. (a) Explain the phenomenon of resonance in an electric circuit.
    (b) Compare the characteristics of an acceptor circuit with a rejecter circuit.
10. (a) Explain resonance in a realistic parallel *LC* circuit. Draw the phasor diagram when the circuit is in resonance and show that the resonance frequency is equal to the series resonance frequency.
    (b) Explain dynamic impedance and outline its characteristics.
11. (a) Derive the condition for transfer of maximum power when an alternating sinusoidal voltage source, having a source impedance of ZS, is supplying a load ZL.
    (b) What is the modification in the condition for maximum power transfer when a network is modelled by its Thevenin equivalent?

Problems

4.1 A coil wound with 500 turns on a square former having sides 10 cm in length is rotated at 1000 rpm in a uniform magnetic field of density 0.75 T. Calculate (a) the maximum value of the emf generated in the coil, (b) the frequency of this emf, (c) the instantaneous value of the induced emf when the coil is at 60° after passing its maximum induced voltage, (d) the root mean square value of the induced emf, and (e) the emf at an instant 0.002 sec after the plane of the coil has been perpendicular to the field.
$$[(a)\ 392.5\ V,\ (b)\ 16.67\ Hz,\ (c)\ 196.25\ V,\ (d)\ 277.5\ V,\ (e)\ 81.6\ V]$$

4.2 Find the average and rms values of the wave shape shown in Figs 4.86(a) and 4.86(b).
$$[(a)\ 0,\ I_m;\ (b)\ 0.5I_m,\ 0.707I_m]$$

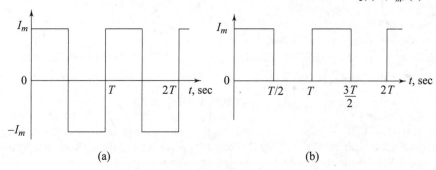

(a)   (b)

Fig. 4.86

4.3 Compute the average and effective values for the waveshape. shown in Figs 4.87(a) and 4.87(b).
$$[(a)\ 0.5I_m,\ 0.577I_m;\ (b)\ 0,\ 0.577I_m]$$

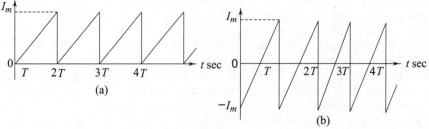

(a)   (b)

Fig. 4.87

4.4 A triangular voltage wave has a periodic time of 0.05 sec. For the first 0.03 sec of each cycle, it increases uniformly at the rate of 2000 V/sec, while for the last 0.02 sec, it falls away uniformly to zero. Find its (a) average value, (b) rms value, and (c) form factor.
$$[(a)\ 30\ V,\ (b)\ 34.64\ V,\ (c)\ 1.1547]$$

4.5 Figures 4.88(a) and 4.88(b) show the sketches of sine waves or parts thereof for a full-wave and a half-wave rectifier, respectively. Find the average and effective voltages for each waveform.

$$[(a)\ 0.637V_m,\ 0.707V_m;\ (b)\ 0.3185V_m,\ 0.5V_m]$$

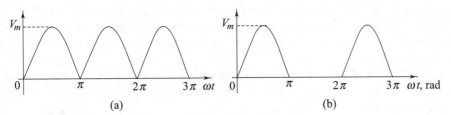

Fig. 4.88

4.6 A voltage of 200 sin(314t) V is maintained across a circuit consisting of a half-wave rectifier in series with a 25 Ω resistor. The resistance of the rectifier may be assumed to be negligible in the forward direction and infinity in the reverse direction. Calculate the average and the rms value of the current. [2.548 A, 4.0 A]

4.7 A moving-coil ammeter used for measuring direct current and a moving-iron ammeter used for measuring the rms value of alternating current are connected in series with a rectifier across a 230-V (rms) ac supply. The total resistance of the circuit in the conducting direction is 50 Ω and that in the reverse direction may be taken to be infinity. Assuming the waveform of the supply voltage to be sinusoidal, calculate from first principles the reading on each ammeter. [2.07 A, 3.25 A]

4.8 An ac voltage v is represented as $v = 100$ sin(1256t) A. Determine (a) the frequency, (b) the period, (c) the time taken from $t = 0$ for the current to reach a value of 60 V for a first and second time, and (d) the energy dissipated when the current flows through a 25 W resistor for 30 min.

$$[(a)\ 200\ Hz;\ (b)\ 0.005\ sec;\ (c)\ 0.000512\ sec,\ 0.005512\ sec;\ (d)\ 36\ kW]$$

4.9 Two sinusoidal emfs of peak values 150 V and 120 V, respectively, but differing in phase by 50° are induced in a circuit. Draw the phasor diagram, and find the peak and rms values of the resultant emf and the phase angle of the resultant with respect to the two emfs.

$$[245.03\ V,\ 173.24\ V,\ \angle 22.03°\ \text{with peak emf of 150 V}]$$

4.10 Two circuits connected in parallel take alternating currents which can be expressed trigonometrically as $i_1 = 6$ sin(314t) A and $i_2 = 8$ sin(314$t + \pi/4$) A. Determine the resultant of these currents using a phasor diagram and express it in trigonometric form. Give the rms value and the frequency of the resultant current.

$$[12.98\angle 25.89°,\ 12.96\ \sin(314t + 25.89°),\ 9.164\ A,\ 50\ Hz]$$

4.11 Four emfs, $e_1 = 100$ sinωt, $e_2 = 80$ sin($\omega t - \pi/3$), $e_3 = 90$ sin($\omega t + \pi/4$), and $e_4 = 120$ cos ωt, are induced in four coils connected in series. Find the resultant emf and its phase difference with (a) e_1 and (b) e_2.

$$[e = 233.54\ \sin(\omega t + 29.32°),\ (a)\ 29.32°,\ (b)\ 89.32°]$$

4.12 A voltage $v = 330$ sin(314$t + 60°$) V is maintained across a coil having a resistance of 30 Ω and an inductance of 0.2 H. Determine the rms values of the voltage and current phasors in (a) rectangular notation and (b) polar notation. Draw the phasor diagram.

$$[(a)\ 116.67 + j202.08\ V,\ 3.343 - j0.261\ A;\ (b)\ 233.34\angle 60°\ V,\ 3.353\angle -4.46°\ A]$$

4.13 A voltage $v = 330$ sin(314$t + 60°$) V is maintained across a circuit consisting of a 30 Ω nonreactive resistor in series with a loss-free 150 μF capacitor. Determine the rms value of the current phasor in (a) rectangular notation and (b) polar notation. Draw the phasor diagram. [(a) $-0.584 + j6.323$ A, (b) $6.35\angle 95.28°$ A]

4.14   A sinusoidal source $v(t) = 330 \sin(314t)$ is applied to an RL circuit. It is found that the circuit absorbs 800 W when an effective current of 10 A flows. (a) Find the power factor of the circuit, (b) compute the value of the impedance, and (c) calculate the inductance of the circuit in henry. [(a) 0.3428, (b) 23.334 W, (c) 0.0698 H]

4.15   A coil of inductance 0.15 H and negligible resistance is connected in series with a 30 Ω resistor across a 230-V, 50-Hz supply. Calculate (a) the current in the circuit, (b) the pd across the coil, (c) the pd across the resistor, and (d) the phase angle of the circuit. Draw to scale a phasor diagram representing the current and the component voltages.   [(a) $4.12\angle{-57.5°}$ A, (b) $194.05\angle{-32.5°}$ V, (c) $123.6\angle{-57.5°}$ V, (d) $\angle57.5°$]

4.16   When a sinusoidal voltage of 230 V rms is applied to a series RL circuit, it is found that there occurs a power dissipation of 2300 W and a current flow given by $i(t) = 28.28 \sin(314t - \theta)$. Find (a) the circuit resistance in ohms and (b) the circuit inductance in henry.   [(a) 5.75 Ω, (b) 0.0317 H]

4.17   The potential difference measured across a coil is 30 V when a direct current of 5 A is passed through it. With an alternating current of 5 A at 40 Hz, the pd across the coil is 140 V. If the coil is connected to a 230-V, 50-Hz supply, calculate (a) the current, (b) the active power, and (c) the power factor.
[(a) 6.63 A, (b) 263.74 W, (c) 0.164 lagging]

4.18   In an RC series circuit supplied by a voltage of $283 \sin \omega t$, it is found that a steady-state current flows leading the voltage by 60°. Find the effective voltage drops across the resistive and reactive elements.
[100 V, 173.2 V]

4.19   A circuit is composed of a resistance of 12 Ω and a series capacitive reactance of 16 Ω. A voltage $v(t) = 282.4 \sin(314t)$ is applied to the circuit. Determine (a) the complex impedance, (b) the effective and instantaneous values of the current, (c) the power delivered to the circuit, and (d) the value of the capacitance in farads.
[(a) $12 - j16$ Ω; (b) $10\angle53.3°$ A, $14.14 \sin(\omega t + 53.3°)$; (c) 1.2 kW; (d) 199 μF]

4.20   A 100 Ω resistor and a 50 μF capacitor are connected in parallel across a 230-V, 50-Hz supply. Calculate (a) the current in each branch, (b) the resultant current, (c) the phase difference between the resultant current and the applied voltage, (d) the active power, and (e) the power factor. Sketch the phasor diagram.
[(a) 2.3 A, 3.61 A; (b) 4.28 A; (c) $\angle57.5°$; (d) 528.92 W; (e) 0.5373 leading]

4.21   In order to use three 110-V, 25-W lamps on a 230-V, 50-Hz supply, they are connected in parallel, and a capacitor is connected in series with the group. Find (a) the capacitance required to give the correct voltage across the lamps and (b) the power factor of the network. (c) If one of the lamps is removed, to what value will the voltage across the remaining two rise, assuming that their resistances remain unchanged?
[(a) 0.537 μF, (b) 0.6911 leading, (c) 188.67 V]

4.22   A series circuit consists of a non-inductive resistor of 10 Ω, an inductor having a reactance of 50 Ω, and a capacitor having a reactance of 30 Ω. It is connected to a 230-V ac supply. Calculate (a) the current and (b) the voltage across each component. Draw to scale a phasor diagram showing the supply voltage and current and the voltage across each component.   [(a) $10.286\angle{-63.43°}$ A; (b) $102.86\angle{-63.43°}$ V,
$514.3\angle26.57°$ V, $308.58\angle{-153.43°}$ V]

4.23   Three impedances Z_1, Z_2, and Z_3 are connected across an alternating voltage source of V volt. Derive expressions for the voltages across the impedance and therefrom show that the series circuit can be employed as a voltage divider.

4.24   A voltage $v(t) = 300 \sin(100t)$ is applied across a series RLC circuit, where $R = 40 \Omega$, $L = 0.25$ H, and $C = 200 \mu$F. Determine (a) the rms value of the steady-state current, (b) the expression for the instantaneous voltage appearing across the capacitor terminals, (c) the expression for the instantaneous voltage appearing across the inductor terminals, (d) the rms value of the voltages appearing across L and C, (e) the power supplied by the source, (f) the reactive power supplied by the source, (g) the reactive power of the capacitor, (h) the reactive power of the inductor, and (i) the power factor of the circuit. Draw the complete phasor diagram showing all voltage components and the current.   [(a) $4.5\angle32°$ A, (b) $\sqrt{2} \times 225 \sin(100t - 58°)$,(c) $\sqrt{2} \times 112.5 \sin(100t + 122°)$, (d) 225 V,
(e) 112.5 V, (f) 505.85 W, (g) 1012 VAR, (h) 506.25 VAR]

4.25 A coil of resistance 25 Ω and inductance 0.05 H is connected in parallel with a branch made up of a 50 μF capacitor in series with a 40 Ω resistor; this arrangement is connected to a 230-V, 50-Hz supply. Calculate (a) the current drawn by each branch, (b) the total current taken from the supply, and (c) the phase angle of the total current. Draw the complete phasor diagram.
$$[(a)\ 7.79\angle-32.13°\ A,\ 3.058\angle57.87°;\ (b)\ 8.368\angle-10.7°;\ (c)\ \angle-10.7°]$$

4.26 The current in a circuit is $6 + j14$ A when the applied voltage is $120 + j180$ V. Determine (a) the complex expression for the impedance, stating whether it is inductive or capacitive, (b) the active power, and (c) the phase angle between the voltage and the current. $\quad$ [(a) $14.008 - j2.342$, capacitive; (b) 3.25 kW; (c) 9.49°]

4.27 Impedance $Z_1, Z_2, ... Z_n$ are connected across an alternating current source of I ampere. Prove that the current flowing through each impedance is directly proportional to its admittance.

4.28 A sinusoidal voltage is applied to three parallel branches, yielding branch currents as follows:
$$i_1 = 14.14 \sin(\omega t - 45°),\ i_2 = 28.3 \cos(\omega t - 60°),\ i_3 = 7.07 \sin(\omega t + 60°)$$
(a) Find the complete time expression for the source current. (b) Draw the phasor diagram in terms of the effective values. Use the voltage as reference. $\qquad$ [$39.39 \sin(\omega t + 15.11°)$]

4.29 A sinusoidal voltage $V_m \sin \omega t$ is applied to three parallel branches. Two of the branch currents are given by $i_1 = 14.14 \sin(\omega t - 37°)$ and $i_2 = 28.28 \sin(\omega t - 143°)$. The source current is found to be $63.8 \sin(\omega t + 12.8°)$. Determine (a) the effective value of the current in the third branch and (b) the complete time expression for the instantaneous value of the current in part (a). Draw the phasor diagram showing the source current and the three branch currents. Use the voltage as the reference phasor.
$$[(a)\ 58.496\angle27.31°\ A;\ (b)\ 82.726 \sin(\omega t + 27.31°)\ A]$$

4.30 A coil having a resistance of 10 Ω and an inductance of 0.2 H is connected in series with a 100 μF capacitor across a 230-V, 50-Hz supply. Calculate (a) the active and reactive components of the current and (b) the voltage across the coil. Sketch the phasor diagram. $\qquad$ [(a) 2.173 A, 6.728 A; (b) $448.19\angle8.82°$ V]

4.31 A network consists of three branches in parallel. Branch A is a resistance of 20 Ω, branch B is a series circuit consisting of a non-inductive resistor of 10 Ω and an inductor having a reactance of 15 Ω, and branch C is a series circuit consisting of a non-inductive resistor of 5 Ω and a capacitor having a reactance of 30 Ω. The network is connected to a 230-V ac supply. Calculate (a) the current in each branch, (b) the voltage across each component, and (c) the total current drawn from the supply. Sketch the phasor diagram showing the supply voltage, currents, and the voltage across each component.
$$[(a)\ 11.5\ A,\ 12.758\angle-56.31°\ A,\ 7.562\angle80.54°\ A;$$
$$(b)\ 127.58\ V,\ 191.37\ V,\ 37.81\ V,\ 226.86\ V;\ (c)\ 20.07\angle9.05°\ A]$$

4.32 A coil having a resistance of 20 Ω and an inductance of 0.05 H is connected in parallel with a circuit consisting of a 150 μF capacitor in series with a 10 Ω resistor. The arrangement is connected to a 230-V, 50-Hz supply. Determine (a) the current in each branch, (b) the total supply current, and (c) the total active and reactive power. Sketch the phasor diagram. $\qquad$ [(a) $9.046\angle-38.13°$ A, $9.8\angle64.78°$ A; (b) $11.52\angle11.42°$ A;
$$(c)\ 2.597\ kW,\ 524.6\ VAR]$$

4.33 A circuit consists of a 20 Ω non-reactive resistor in series with a coil having an inductance of 0.15 H and a resistance of 10 Ω. A loss-free capacitor of 80 μF is connected in parallel with the coil. The network is connected across a 230-V, 50-Hz supply. Calculate the value of the current in each branch and its phase relative to the supply voltage.
$$[(a)\ 4.922\angle-91.3°\ A,\ 5.953\angle76.7°\ A;\ (b)\ \angle-91.3°,\ \angle76.7°]$$

4.34 An impedance of $6 + j8$ Ω is connected in series with a parallel combination of the impedances of $10 + j6$ Ω and $8 - j10$ Ω, and is connected to a 230-V, 50-Hz supply. Calculate the magnitude and power factor of the main current. $\qquad$ [14.725 A, 0.897 lagging]

4.35 The pd across and the current in a circuit are $(220 + j50)$ V and $(10 - j4)$ A, respectively. (a) Calculate the active power and the reactive power. Is the reactive power leading or lagging? (b) If the current is $(10 + j4)$ A, calculate the active power and the reactive power. State whether the reactive power is leading or lagging.
$$[(a)\ 2\ kW,\ 1.38\ kVAR\ lagging,\ (b)\ 2.4\ kW,\ 0.38\ kVAR\ leading]$$

4.36 A single-phase motor takes 10 A at a power factor of 0.866 lagging when connected to a 230-V, 50-Hz supply. A capacitance bank is now connected in parallel with the motor to raise the power factor to unity. What is the value of the capacitance?   [69.23 μF]

4.37 A 10-kVA, single-phase motor has a power factor of 0.8 lagging. A capacitor of 4 kVA rating is connected for power-factor correction. Calculate the input apparent power in kVA taken from the mains and its power factor when the motor is on (a) half load and (b) full load. Sketch a phasor diagram for each case.
[(a) 4.123 kVA, 0.97 lagging; (b) 8.246 kVA, 0.97 lagging]

4.38 The load taken from an ac supply consists of (a) a heating load of 25 kVa, (b) a motor load of 50 kVA at 0.6 power factor lagging, and (c) a load of 30 kW at 0.8 power factor lagging. Calculate the total load from the supply (in kW) and its power factor. What should be the kVAR rating of a capacitor being used to bring the power factor to unity and how should the capacitor be connected?   [79 kW, 0.806, 58 KVAR]

4.39 A series *RLC* circuit has the following parameter values: $R = 10 \, \Omega$, $L = 0.1$ H, $C = 5 \, \mu$F. Compute (a) the resonant frequency in rad/sec, (b) the quality factor of the circuit, (c) the bandwidth, and (d) the lower and upper half-frequency points of the bandwidth. (e) If a signal $e(t) = 10 \sin(1415t)$ is applied to this series *RLC* circuit, calculate the maximum value of the voltage appearing across the capacitor terminals.
[(a) 225.19 Hz; (b) 14.15; (c) 15.91 Hz;(d) 1365 rad/sec, 1465 rad/sec; (e) 141.34 V]

4.40 A voltage of $e(t) = 10 \sin \omega t$ is applied to a series *RLC* circuit. At the resonant frequency of the circuit, the maximum voltage across the capacitor is found to be 400 V. Moreover, the bandwidth is known to be 500 rad/sec and the impedance at resonance is 50 Ω. Find (a) the resonant frequency, (b) the upper and lower limits of the bandwidth, and (c) the value of L and C for this circuit.
[(a) 3184.7 Hz; (b) 3105.08 Hz, 3264.32 Hz; (c) 0.1 H, 0.025 μF]

4.41 A series circuit comprises an inductor, of resistance 10 Ω and inductance 100 μH, and a variable capacitor connected to a 10 mV sinusoidal supply of frequency 1.2 MHz. What value of capacitance will result in resonant conditions and what will be the current then? For what values of capacitance will the current at this frequency be reduced to 10% of its value at resonance?   [175.9 pF, 1 mA, 155.5 pF]

4.42 A circuit consists of a 5 Ω resistor, a 20 mH inductor, and a 10 μF capacitor, and is supplied by a 15-V, variable-frequency source. Find the frequency for which the voltage developed across the capacitor is a maximum and calculate the magnitude of this voltage.   [356 Hz, 134.19 V]

4.43 A coil, of resistance R and inductance L, is connected in series with a capacitor C across a variable-frequency source. The voltage is maintained constant at 500 mV and the frequency is varied until a maximum current of 5 mA flows through the circuit at 10 kHz. If, under these conditions, the Q factor of the circuit is 105, calculate (a) the voltage across the capacitor and (b) the values of R, L, and C.
[(a) 52.53 V; (b) 100 Ω, 0.1672 H, 1515 pF]

4.44 A current source is applied to a parallel arrangement of R, L, and C, where $R = 10 \, \Omega$, $L = 0.1$ H, and $C = 10$ μF. Compute (a) the resonant frequency, (b) the quality factor, (c) the value of the bandwidth, and (d) the lower and upper half-frequency points of the bandwidth.
[(a) 159.23 Hz; (b) 10; (c) 15.92 Hz; (d) 950 rad/sec, 732 rad/sec]

4.45 A coil of resistance 15 Ω and inductance 0.15 H is connected in parallel with a 80 μF capacitor to a 50-V, variable-frequency supply. Calculate the frequency at which the circuit will behave as a non-reactive resistor, and also the value of the dynamic impedance. Draw for this condition the complete phasor diagram.
[43.12 Hz, 125 Ω]

4.46 A coil has a resistance of 200 Ω and an inductance of 152 μH. Find the capacitance of a capacitor which when connected in parallel with the coil will produce resonance with a supply frequency of 1.5 MHz. If a second capacitor of capacitance 50 pF is connected in parallel with the first capacitor, find the frequency at which resonance will occur.   [72.65 pF, 1.147 MHz]

4.47 Determine the star equivalent network for the circuit shown in Fig. 4.89.
[$\mathbf{Z}_a = 6.488 + j9.888 \, \Omega$, $\mathbf{Z}_b = 5.472 - j8.56 \, \Omega$, $\mathbf{Z}_c = 9.04 + j0.9533 \, \Omega$]

4.48 Determine the delta equivalent network for the circuit shown in Fig. 4.90.
[$\mathbf{Z}_{bc} = 30.615 - j12.923 \, \Omega$, $\mathbf{Z}_{ca} = 1.71 + j18.634 \, \Omega$, $\mathbf{Z}_{ab} = 15.29 - j10.295 \, \Omega$]

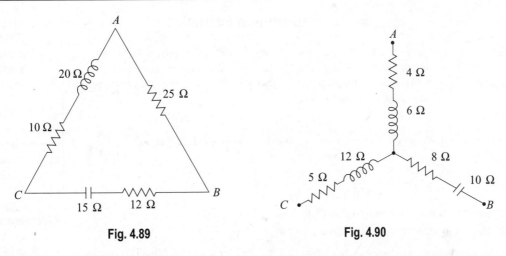

Fig. 4.89

Fig. 4.90

4.49 For the circuit shown in Fig. 4.91, find the current that flows through the impedance Z_3 using the nodal method of analysis. [$1.81\angle-24.1°$ A]

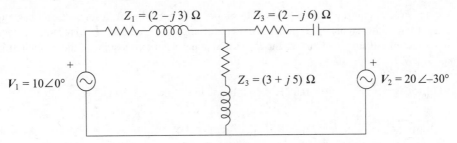

Fig. 4.91

4.50 For the circuit shown in Fig. 4.91, find the current that flows through the impedance Z_3 using the mesh method of analysis. [$1.8093\angle-24.2°$ A]

4.51 For the circuit shown in Fig. 4.91, find the current that flows through the impedance Z_3 using the superposition theorem. [$1.8096\angle-24.15°$ A]

4.52 For the circuit shown in Fig. 4.92, determine the current flowing in the capacitor connected across AB using Thevenin's theorem. [$0.902\angle88.01°$ A]

4.53 For the circuit shown in Fig. 4.92, determine the current flowing in the capacitor connected across AB using Norton's theorem. [$0.902\angle88.01°$ A]

4.54 Determine the value of the load impedance Z_L in Fig. 4.93 when maximum power is transferred to it. Calculate the value of the maximum power. [$1.867 + j3.4$, 40.46 W]

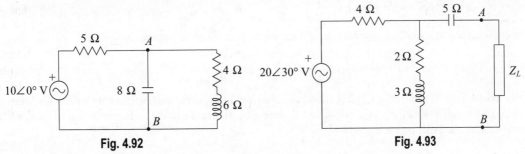

Fig. 4.92

Fig. 4.93

Additional Examples

AE 4.1   An inductor L of 200 mH, a capacitor C of 250 nF, and a resistor 150 Ω are connected in series across a voltage source of 6 V with frequency varying between 50 Hz to 10 kHz. Calculate (a) resonance frequency, (b) Q factor, (c) resonant frequency current, and (d) draw the phasor diagram at resonance.

Solution

$V_L = 35.78$ V

(a) Use of Eq. (4.146) gives $f_0 = \dfrac{1}{2\pi\sqrt{200\times10^{-3}\times250\times10^{-9}}} = 711.763$ Hz

$V_R = 6$ V

$I_R = 40$ mA

(b) Equation (4.150) gives $Q = \dfrac{1}{150}\sqrt{\dfrac{200\times10^{-3}}{250\times10^{-9}}} = 5.963$

(c) Resonant frequency current from Eq. (4.147) is $I_0 = 6/150 = 40$ mA

(d) At resonance, inductor voltage = capacitor voltage. Thus,

$V_C = 35.78$ V

$$V_L = V_C = 0.04\times2\pi\times711.763\times200\times10^{-3} = 35.777 \text{ V and } V_R = 0.04\times150 = 6 \text{ V}$$

The phasor diagram is shown here.

AE 4.2   A series *RLC* circuit produces a bandwidth of 100 rads./sec when in resonance and has a Q factor of 100. If the resistance R of the circuit is 5 Ω determine (a) inductance and capacitance, (b) resonance frequency, and (c) maximum current of the circuit. What are the lower and upper level frequencies of the bandwidth? Assume a supply voltage of 5 V.

Solution

(a) Use of Eq. (4.154) gives $L = 5/100 = 0.05$ H $= 50$ mH and use of Eq. (4.150) yields

$$C = \dfrac{0.05}{(100)^2\times5^2} = 0.2\ \text{μF}$$

(b) From Eq. (4.146) resonant frequency $\omega_0 = 1/(\sqrt{0.05\times0.2\times10^{-6}}) = 10$ kilo rads/sec

(c) Current at resonance $I_0 = 5/5 = 1$ A

   Frequency at lower level of bandwidth $\omega_1 = 10000 - 100/2 = 9950$ rads/sec
   Frequency at upper level of bandwidth $\omega_2 = 10000 + 100/2 = 10050$ rads/sec

AE 4.3   Develop equations for power consumed at (a) resonance and (b) at lower and upper levels of the bandwidth frequencies ω_1 and ω_2 respectively. There from prove that the power dissipated at the bandwidth frequencies ω_1 and ω_2 is half that of power at resonance.

Solution

Assume that the applied voltage has a maximum value of V_m volts and the resistance is R Ω.

Power consumed at resonance $P(\omega_0) = \left(\dfrac{V_m}{\sqrt{2}}\right)^2\times\dfrac{1}{R} = \dfrac{V_m^2}{2R}$ W   (AE 4.3.1)

Power consumed at lower and upper levels of band width frequencies is obtained as under

$$P(\omega_1) = P(\omega_2) = \left(\dfrac{V_m}{\sqrt{2}}\right)^2\times\dfrac{1}{2R} = \dfrac{V_m^2}{4R}\ \text{W} \qquad\qquad (AE\ 4.3.2)$$

From Eqs (AE 4.3.1) and (AE 4.3.2) it is seen that the power dissipated at the lower and upper levels of band width frequencies is half that of the power consumed at resonance. Hence the lower and upper levels of band width frequencies are called half power frequencies.

AE 4.4 The current in the *RLC* series circuit in Fig. AE 4.1 is 10 $\angle 0°$ A when it is operating at a frequency of ω rads/sec. When the operating frequency is increased by 10.5%, the circuit current becomes $7.07 \angle -45°$ A. Compute (a) ω, (b) capacitance C, and (c) inductive and capacitive reactances at the increased frequency. All data is shown in the figure.

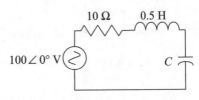

Fig. AE 4.1

Solution

Use of Eq. (4.146) gives $C = \dfrac{1}{\omega^2 \times 0.5} = \dfrac{2}{\omega^2}$ F

At the increased frequency of 1.105ω, the current in the circuit is written as

$$100\angle 0° = 7.07\angle -45° \times \left[10 + j\left(1.105\omega \times 0.5 - \frac{1}{1.105\omega \times \left\{2/\omega^2\right\}}\right)\right]$$

or $\quad \dfrac{100\angle 0°}{7.07\angle -45°} = [10 + j0.1\omega], \quad$ or, $\quad 10 + j10 = [10 + j0.1\omega]$ $\qquad$ (AE 4.4.1)

(a) From Eq. (AE 4.4.1) the operating frequency $\omega = 10/0.1 = 100$ rads/sec.

(b) Capacitance $C = 1/(100)^2 \times 0.5 = 200 \ \mu$F

(c) Inductive reactance $X_L = 110.5 \times 0.5 = 55.25 \ \Omega$

and capacitive reactance $X_C = 1/110.5 \times 200 \times 10^{-6} = 45.25 \ \Omega$

AE 4.5 Derive expressions for ω_1 and ω_2 in terms of R, L, and C for parallel resonance and show that (a) $BW = 1/RC$, and (b) $Q = \omega_0 RC = R/\omega_0 L$.

Solution Making use of Eq. (4.164) and knowing that current at half power frequencies is $I = \sqrt{2}VG$ it can be written that

$$V\sqrt{\frac{1}{R^2} + \left(\omega C - \frac{1}{\omega L}\right)^2} = V\sqrt{2}G = \frac{\sqrt{2}V}{R}$$

or $\qquad \left(\omega C - \dfrac{1}{\omega L}\right) = \pm\dfrac{1}{R}$ $\qquad$ (AE 4.5.1)

It may be noted that Eq. (AE 4.5.1) is a quadratic which would lead to two values of ω as follows

$$\omega_1 = -\frac{1}{2RC} + \sqrt{\left(\frac{1}{2RC}\right)^2 + \frac{1}{LC}} \qquad \text{(AE 4.5.2)}$$

And $\qquad \omega_2 = \dfrac{1}{2RC} + \sqrt{\left(\dfrac{1}{2RC}\right)^2 + \dfrac{1}{LC}}$ $\qquad$ (AE 4.5.3)

(a) Subtracting Eq. (AE 4.5.2) from Eq. (AE 4.5.3) gives

$\qquad \omega_{BW} = \omega_2 - \omega_1 = 1/RC$

(b) Q in parallel resonance is the ratio of current in inductor or capacitor to the circuit current. Thus,

$$|I_{L0}| = \frac{V}{\omega_0 L} = \frac{IR}{\omega_0 L}, \quad \text{or,} \quad \frac{|I_{L0}|}{I} = \frac{R}{\omega_0 L} = Q$$

Similarly, $|I_{C0}| = V\omega_0 C = IR\omega_0 C$,   or, $\dfrac{|I_{C0}|}{I} = R\omega_0 C = Q$

Hence $Q = \dfrac{R}{\omega_0 L} = \omega_0 RC$

AE 4.6   In the circuit in Fig. 4.72, $R = 10$ kΩ, $L = 0.5$ mH, and $C = 10$ µF. Compute (a) f_0, (b) Q, and f_{BW}, and (c) f_1, f_2. What is the power dissipated at resonance and the half frequency points? Assume the applied voltage equal to 20 sinωt V.
Solution

(a) From Eq. (4.164) $f_0 = \dfrac{1}{2\pi\sqrt{0.5\times10^{-3}\times10\times10^{-6}}} = 2.25$ kHz

(b) Using the results of example AE 4.5, $Q = 2\times\pi\times2250\times10\times10^3\times10\times10^{-6} = 1414$

Again using the results of example AE 4.5, $f_{BW} = 2\pi\times2250/1414 = 10$ rads/sec

(c) $f_1 = f_0 - f_{BW}/2 = 2250 - 10/2 = 2245$ kHz and $f_1 = f_0 + f_{BW}/2 = 2250 + 10/2 = 2255$ kHz

Assume the applied voltage as the reference phasor. At resonance $Z = R = 10$ kΩ and

$\quad I_0 = V/Z = 20\angle0°/10\times10^3 = 2\angle0°$ mA

Power dissipated at resonance $P = \dfrac{1}{2}|I_0|^2 R = \dfrac{1}{2}\times\left(2\times10^{-3}\right)^2\times10\times10^3 = 20$ mW

Power at half frequencies $\omega = \omega_1, \omega_2$ is given by $P = V_m^2/R = (20)^2/\left(4\times10\times10^3\right) = 10$ mW

AE 4.7   A voltage waveform is represented as $[v(t) = 4.5 + 16.5\sin(10\pi t)]$ V. Compute (a) frequency and time period, (b) maximum and minimum voltage, (c) peak to peak voltage, and (d) average value of the waveform.
Solution
(a) Using Eq. (4.8), frequency $f = 10\pi/2\pi = 5$ Hz, and time period $T = 1/5 = 02$ sec.
(b) Maximum voltage = 4.5 + 16.5 = 21 V, and minimum voltage = 4.5 − 16.5 = − 12 V.
(c) Peak to peak voltage = 21 − (− 12) = 33 V
(d) Average value of the waveform is the sum of the dc component and average of the varying component. Thus,

$\quad V_{avg} = 4.5 + \dfrac{2\times16.5}{\pi} = 4.5 + 10.5 = 15$ V

AE 4.8   Determine the effective value of the periodic voltage waveform shown in Fig. AE 4.2.
Solution   The signal in Fig. AE 4.2 has a time period $T = 8.0$ sec. and can be written as

$\quad v_1 = 2.0 \quad$ for $0 < t < 2$

$\quad v_2 = t \quad$ for $2 < t < 4$

$\quad v_3 = 4.0 \quad$ for $4 < t < 6$

$\quad v_4 = 0 \quad$ for $6 < t < 8$

Using Eq. (4.10) gives

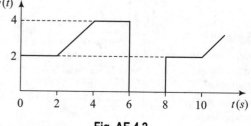

Fig. AE 4.2

$$V_{eff} = \sqrt{\dfrac{1}{8}\left\{\int_0^2 4\,dt + \int_2^4 t^2\,dt + \int_4^6 16\,dt\right\}} = \sqrt{\dfrac{1}{8}\left\{[4t]_0^2 + \left[\dfrac{t^3}{3}\right]_2^4 + (16t)_4^6\right\}}$$

$$= \sqrt{\dfrac{1}{8}\left\{8 + \dfrac{56}{3} + 32\right\}} = \sqrt{\dfrac{176}{24}} = 2.708 \text{ V}$$

Using Eq. (4.15) gives

$$V_{avg} = \frac{1}{8}\left[\int_0^2 2dt + \int_2^4 t\ dt + \int_4^6 4dt\right] = \frac{1}{8}\left[(2t)_0^2 + \left(\frac{t^2}{2}\right)_2^4 + (4\ t)_4^6\right]$$

$$V_{avg} = \frac{2 \times 2 + \frac{1}{2}(4^2 - 2^2) + 4(6-4)}{8} = \frac{18}{8} = 2.25 \text{ V}$$

AE 4.9   A current signal having the waveform shown in Fig. AE 4.3 is flowing through a resistor of R. Derive an expression for the (a) instantaneous power dissipated at any time t and (b) the average power dissipated by the resistor.

Solution   The signal represented by the waveform can be mathematically written as

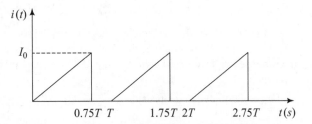

$$i(t) = \left(\frac{I_0}{0.75T}\right)t \quad \text{for } 0 \le t \le 0.75T$$

$$i(t) = 0 \quad \text{for } 0.75T \le t \le T$$

Fig. AE 4.3

(a) Power dissipated by the resistor at any time t is given by

$$p(t) = i^2(t)R = \left(\frac{I_0}{0.75T}\right)^2 t^2 R = 1.7778I_0^2 R \times t^2 \quad \text{for } 0 \le t \le 0.75T$$

$$p(t) = 0 \quad \text{for } 0.75\ T \le t \le T$$

(b) Expression for average power is derived by using Eq. (4.10) as under

$$I_{eff} = \sqrt{\frac{1}{T}\int_0^{0.75T} i^2(t)dt} = \sqrt{\frac{1}{T}\left(\frac{I_0}{0.75T}\right)^2 \int_0^{0.75T} t^2 dt} = \sqrt{\frac{I_0^2}{(0.75)^2 T^3}\left(\frac{t^3}{3}\right)_0^{0.75T}}$$

$$= \sqrt{\frac{I_0^2}{(0.75)^2 T^3} \times \frac{(0.75T)^3}{3}} = \sqrt{0.25I_0^2} = 0.5I_0$$

$$P_{avg} = I_{eff}^2 R = (0.5I_0)^2 R = 0.25I_0^2 R$$

AE 4.10   The triangular voltage wave shown in Fig. AE 4.4 has a maximum voltage of V_m volts and a time period of T. Derive expressions for (a) average and (b) effective values for the voltage.

Solution
(a) For waveforms which are symmetrical about the x-axis, the average value of the voltage waveform is computed over half a cycle only. Thus,

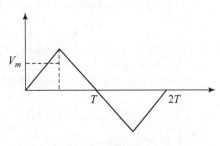

$$V_{avg} = \frac{\text{Area under the triangle}}{\text{Base}} = \frac{(1/2) \times V_m \times T}{T} = 0.5V_m \text{ V}$$

Fig. AE 4.4

(b) Mathematically the triangular signal is expressed as

$$v = 0 \quad \text{for } t < 0$$

$$v = \left(\frac{2V_m}{T}t\right) \qquad \text{for } 0 < t < T/2$$

$$v = \left(-\frac{2V_m}{T}t\right) \qquad \text{for } T/2 < t < T$$

Taking advantage of the symmetry, the effective value of the voltage signal is determined as under

$$V_{\text{eff}} = \sqrt{\frac{2}{T} \times \left[\int_0^{T/2} \frac{4V_m^2}{T^2}t^2 dt\right]} = \sqrt{\frac{8V_m^2}{T^3}\left[\frac{t^3}{3}\right]_0^{T/2}} = \frac{V_m}{\sqrt{3}} = 0.577V_m$$

AE 4.11 A 50 V, 50 Hz sinusoidal voltage is applied across a resistor of 25 Ω. (a) What is the power dissipated in the resistor? (b) What should be the effective and maximum voltages of a triangular waveform signal, to dissipate the same power, if it is applied across the 25 Ω resistor? What is the average value of such a waveform?
Solution

(a) Power dissipated in the resistor $= \dfrac{50^2}{25} = 100$ W.

(b) In order that the same power is dissipated, the effective value of the triangular wave has to be 50 V.
   Using the relationship for effective voltage of example AE 4.9, the maximum voltage

$$V_m = 50/0.577 = 86.603 \text{ V}$$

Average value of such a waveform $V_{\text{avg}} = 86.603/2 = 43.301$ V

Objective Type Questions

1. A coil of conductor length l m rotates in a magnetic field of flux density B Wb/m^2. If the position of the coil is θ radians with the respect to the vertical, the total voltage induced in the coil is
   (i) *Blu*
   (ii) *2Blu*
   (iii) *Blu* sin θ
   (iv) *2Blu* sin θ

2. If the width of the coil in Question 1 is b m and it is rotating at n revolutions per second, the maximum value of the induced voltage is
   (i) $2\pi BAn$
   (ii) $2Bl \times \pi bn$
   (iii) Blv
   (iv) all of these

3. The angular velocity of a sinusoidal wave is given by
   (i) θ/t
   (ii) $2\pi/T$
   (iii) $2\pi f$
   (iv) all of these

4. If the number of poles in an ac generator is doubled and its speed halved, the frequency of the induced emf
   (i) is halved
   (ii) remains unchanged
   (iii) is doubled
   (iv) is quadrupled

5. A direct current of I ampere flowing through a resistance produces a certain amount of heat. If an alternating current flows through the resistance, the same amount of heat will be produced by

   (i) maximum current
   (ii) effective current
   (iii) average current
   (iv) none of these

6. The average value of a half-wave-rectified sine wave is
   (i) $0.318 \times$ maximum value
   (ii) $0.637 \times$ maximum value
   (iii) $1.273 \times$ maximum value
   (iv) none of these

7. If the maximum value of a sinusoidal wave is 100 V, the effective value of the voltage is
   (i) 100 V
   (ii) 70.7 V
   (iii) 63.7 V
   (iv) none of these

8. A sinusoidal voltage wave whose maximum value is V_m can be represented as
   (i) $V_m \sin \omega t$
   (ii) $V_m \angle 0°$
   (iii) $\text{Im}\{V_m e^{j\omega t}\}$
   (iv) all of these

9. A phasor quantity **A** can be represented in the complex form as
   (i) $A \angle \theta$
   (ii) $A e^{j\theta}$
   (iii) $a + jb$
   (iv) all of these

10. When a sinusoidal voltage is applied to a pure resistance, the current flowing in the circuit
    (i) has the same maximum value as the voltage
    (ii) lags the applied voltage

(iii) leads the applied voltage

(iv) none of these

11. A capacitance of C farad is connected across a 50-Hz sinusoidal voltage source. The capacitive reactance is given by

(i) C

(ii) $50C$

(iii) $314C$

(iv) $1/314C$

12. A resistance of R ohm is connected in series with an inductance of L henry. The resistance of the inductance is r ohm. What is the heat loss in the circuit if a sinusoidal current of I ampere is flowing in the circuit?

(i) I^2L

(ii) I^2R

(iii) $I^2(R+r)$

(iv) $I^2(R+L)$

13. In Question 12, the power factor angle of the circuit is given by

(i) $\cos^{-1} \dfrac{(R+r)}{\sqrt{(R+r)^2+(\omega L)^2}}$

(ii) $\cos^{-1} \dfrac{R}{\sqrt{R^2+(\omega L)^2}}$

(iii) $\cos^{-1} \dfrac{\omega L}{\sqrt{(R+r)^2+(\omega L)^2}}$

(iv) none of these

14. A circuit has a resistance of R ohm, an inductance of L henry, and a capacitance of C farad connected in series. The power factor of the circuit will be unity when

(i) $X_L < X_C$

(ii) $X_C < X_L$

(iii) $X_L = X_C$

(iv) $R = 0$

15. A circuit consists of three complex impedances Z_1, Z_2, and Z_3 connected in series. The total impedance Z_{eq} of the circuit is given by

(i) $\dfrac{1}{Z_{eq}} = \dfrac{1}{Z_1} + \dfrac{1}{Z_2} + \dfrac{1}{Z_3}$

(ii) $Z_{eq} = Z_1 + Z_2 + Z_3$

(iii) $Z_{eq} = Z_1 + \dfrac{Z_2 \times Z_3}{Z_2 + Z_3}$

(iv) none of these

16. To draw the phasor diagram of a reactive circuit having a number of branches connected in parallel across a common voltage source, which of the following is more convenient to use as the reference phasor?

(i) circuit current

(ii) branch current

(iii) source voltage

(iv) none of these

17. In a parallel RC circuit, the equivalent admittance of the circuit is

(i) $\dfrac{1}{R} + j\omega C$

(ii) $\dfrac{1}{R} - j\omega C$

(iii) $R + \dfrac{1}{j\omega C}$

(iv) $R = j\omega C$

18. In a circuit, a resistance R, a pure inductance L, and a capacitance C are connected in parallel across a sinusoidal voltage source of V volt. The circuit current will lead the applied voltage if

(i) $I_C < I_L$

(ii) $I_C = I_L$

(iii) $I_C > I_L$

(iv) none of these

19. A pure inductance and a capacitance have reactances of $100\,\Omega$ each and are connected in parallel across an ac voltage supply of 220 V. What is the current drawn from the supply?

(i) 4.4 A

(ii) 1.1 A

(iii) 0 A

(iv) none of these

20. The average of a power waveform is given by

(i) $\dfrac{V_m I_m}{2}$

(ii) VI

(iii) I^2R

(iv) all of these

21. In a circuit having a resistance, reactance, and a power factor angle φ, the power absorbed by the circuit is a maximum when φ is equal to

(i) $90°$

(ii) $45°$

(iii) $0°$

(iv) none of these

22. Which of the following apply to power in a purely reactive circuit?

(i) $P = 0$ and $Q = 0$

(ii) P is maximum and $Q = 0$

(iii) $P = 0$ and Q is maximum

(iv) P and Q are both maximum

23. Complex power in an ac circuit is represented by

(i) VI

(ii) $V^* I^*$

(iii) $V^* I$

(iv) VI^*

24. Heat losses in any electrical device, such as a generator, motor, etc., are independent of

(i) voltage

(ii) current

(iii) power factor

(iv) all of these

25. In a series RLC circuit, resonance occurs when

(i) $L = C$

(ii) $R = C$

(iii) $R = L$

(iv) $X_L = X_C$

26. When a series circuit is in resonance, it allows a

large magnitude of current to flow on either side of the resonant frequency. In such a condition, the circuit behaviour is termed as

   (i) rejector   (ii) acceptor

   (iii) neutral   (iv) none of these

27. When a pure *LC* parallel circuit is in resonance, the circuit condition can be represented by

   (i) a short circuit

   (ii) an open circuit

   (iii) a normal parallel circuit

   (iv) none of these

28. In the circuit of Question 27, under the resonance condition, which of the following is correct?

   (i) real power is zero

   (ii) reactive power in *L* and *C* shows itself in the form of electric and magnetic energy

   (iii) power is continuously exchanged between *L* and *C*

   (iv) all of these

29. In order to employ the mesh and nodal methods of network analysis, which of the following conditions have to be adopted?

   (i) resistance replaced by complex impedance

   (ii) KCL and KVL applied to phasor quantities

   (iii) dc voltages and currents replaced by phasor voltages and currents

   (iv) all of these

30. The condition for maximum power to be transferred is represented by

   (i) $Z_L = Z_{TH}$ (ii) $Z_L = Z_{TH}^*$

   (iii) $Z_L < Z_{TH}$ (iv) $Z_L > Z_{TH}$

Answers

1. (iv)	2. (iv)	3. (iv)	4. (ii)	5. (ii)	6. (i)	7. (ii)
8. (iv)	9. (iv)	10. (iv)	11. (iv)	12. (iii)	13. (i)	14. (iii)
15. (ii)	16. (iii)	17. (i)	18. (iii)	19. (iii)	20. (iv)	21. (iii)
22. (iii)	23. (iv)	24. (iii)	25. (iv)	26. (ii)	27. (ii)	28. (iv)
29. (iv)	30. (ii)					

THREE-PHASE SYSTEMS

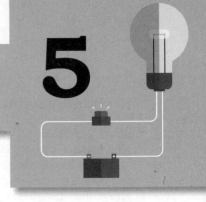

5

Learning Objectives

This chapter will enable the reader to
- Understand the popularity of the three-phase system over the single-phase system for power generation, transmission, and distribution.
- Visualize the generation of three-phase voltages, their representation and specifications.
- Understand how the three-phase voltages can be connected in star or delta form to obtain different levels of voltages and currents.
- Apply the single-phase circuit analysis techniques to obtain currents and voltages in three-phase circuits supplying balanced and unbalanced loads.
- Understand how to measure the three-phase power and compute the power factor from the measurements.

The topics presented in Chapter 4 have exclusively dealt with single-phase ac circuits constituted of single sinusoidal sources. The transmission and distribution of electric power to residences, commercial establishments, and industries are accomplished mostly by means of three-phase networks consisting of transformers, transmission and distribution lines, and associated protective systems, such as relays, circuit breakers, fuses, etc. In fact, most of the electricity generating systems universally use three-phase systems. In a balanced three-phase system, there are three sinusoidal voltage sources of equal amplitude and equal frequency of variation but out of phase with each other by $2\pi/3$ or $120°$ electrical. Three-phase power systems are more efficient and robust and possess operational advantages over single-phase systems. Moreover, single-phase voltage sources originate from three-phase systems.

In this chapter, the principles associated with three-phase systems including the two methods of connection, delta and star, are introduced. The relationships between phase and line currents and voltages for both types of connection are developed. The methods of connection for measuring the power in three-phase systems are also discussed.

5.1 SINGLE-PHASE SYSTEMS AND THREE-PHASE SYSTEMS—COMPARISON

The earliest applications of alternating current were for resistive loads, such as electric lamps and electric heaters. For these purposes the electric load is resistive and a single-phase system works satisfactorily. However when ac motors were developed, it was found that the single-phase ac supply did not work satisfactorily as it was unable to produce the starting torque. To make a single-phase ac motor self-starting, it is fitted with an auxiliary winding. Further, the power delivered to a load from a single-phase ac source is pulsating as no matter what the power factor may be, there is always an alternating component with frequency that is double the supply frequency. For small motors and electric lamps, this is not a serious matter, but it may cause problems, such as vibration and noisy operation, in large motors.

The speed of the prime mover that drives a single-phase synchronous generator tends to fluctuate in response to the cyclic variation of the electrical output power of the alternator. The rotor of the alternator gains speed at instants when the electrical output is less than average and slows down at instants when the output is more than the mechanical input from the prime mover. Further, mmf set up by the armature currents (armature reaction) is pulsating and this causes eddy currents in the field winding and field structure, thereby resulting in heating.

An important advantage of three-phase power supply is that three-phase motors have a non-zero starting torque, unlike their single-phase counterpart. For a given size of a frame, the output of a three-phase machine is greater than that of a single-phase machine.

A balanced poly-phase circuit has the characteristic that the currents and voltages everywhere in the circuit are balanced. The most important property of such a circuit is that though the power in an individual phase is pulsating and varies at twice the supply frequency, the sum of powers delivered by all the phases is constant at every instant of time.

The generation, transmission, and distribution of large blocks of power are usually accomplished by means of balanced three-phase circuits. For a fixed voltage between conductors, a three-phase system requires 75% of the weight of the conductor as compared to a single-phase system. Thus, a three-phase system is more efficient than a single-phase system.

Although theoretically any number of phases might be present, such as two phases, six phases, or twelve phases, it is common practice to employ three-phase circuits. However, in some applications, such as supply of power to rectifiers, six-phase or twelve-phase systems are preferred.

5.2 THREE-PHASE SUPPLY VOLTAGE

Generation and analysis of a three-phase voltage supply are discussed in this section.

5.2.1 Generation of Three-phase Voltage

Figure 5.1 shows the schematic diagram of a three-phase generator in which three identical stationary coils AA_1, BB_1, and CC_1, with their axes 120° apart from each other, are placed over the periphery of the stator of the machine. Here A, B, and C are the finish terminals and A_1, B_1, and C_1 are the start terminals of the three coils. Each of these coils constitutes a phase of the machine. When the magnetic field produced by NS magnet, placed on the rotor of the machine, is driven at constant speed in the anticlockwise direction, a sinusoidally varying voltage is generated

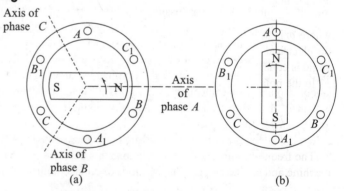

Fig. 5.1 Generation of three-phase voltages

in each phase. As the three coils are similar and symmetrically placed around the stator, the generated voltages have the same magnitude but differ in phase by 120°. It is assumed that the flux density on the face of the pole is sinusoidally distributed with the maximum occurring at the centre of the face of the pole.

For the position of coil AA_1 with respect to the poles as shown in Fig. 5.1(a), the voltage generated or the emf induced in the coil is zero. When the poles move in an anticlockwise direction through 90°, as shown in Fig. 5.1(b), the emf induced in the coil is maximum and its direction around the loop is from start A_1 to finish A. Let this direction be regarded as positive. The emf induced in loop AA_1 for one complete rotation of the poles is represented by the curve e_A in Fig. 5.2(a).

Since the poles rotate in the anticlockwise direction, as is evident from Fig. 5.1, the emf induced in loop BB_1 has exactly the same amplitude as that generated in coil AA_1, but lags by 120°, and its direction around the loop is from start B_1 to finish B. Similarly, the emf induced in loop CC_1 has the same amplitude as that generated in coil BB_1 but lags the phase in coil B by 120°, and its direction around the loop is from start C_1 to finish C. Hence the emfs generated in loops AA_1, BB_1, and CC_1 are represented by the three equally spaced curves e_A, e_B, and e_C, respectively, in Fig. 5.2(a). The induced emfs are assumed to be positive when their directions around the loops are from the start terminal to the finish terminal of their respective loops. Thus the three coils, generating voltages equal in magnitude but differing in phase by 120° from each other, can be viewed as three independent voltage sources.

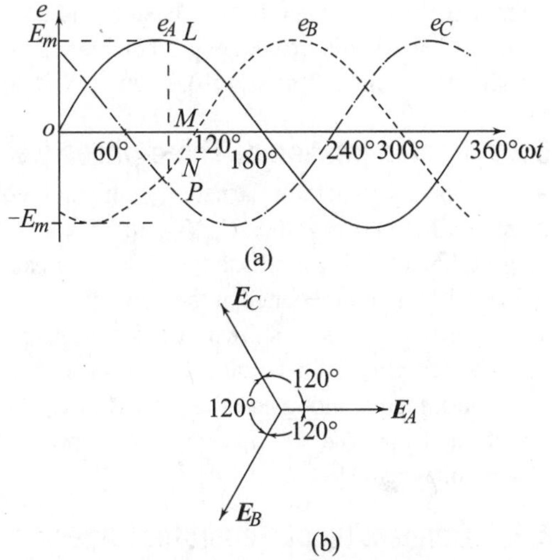

The rms values of the induced emfs may be represented by the phasors E_A, E_B, and E_C, having the same lengths but differing in phase by 120° [see Fig. 5.2(b)]. Thus, the rms value of the three phasors is given by $E_m/\sqrt{2}$, where E_m is the maximum or the peak value of the induced emf. If the instantaneous value of the generated voltage in phase AA_1 at time t is represented by

$$e_A = \sqrt{2}\,E \sin \omega t \qquad (5.1)$$

then the instantaneous value of the generated voltage in BB_1 is given by

$$e_B = \sqrt{2}\,E \sin (\omega t - 120°) \qquad (5.2)$$

and the instantaneous value of the generated voltage in CC_1 is given by

$$e_C = \sqrt{2}\,E \sin (\omega t - 240°) \qquad (5.3)$$

where ω is the angular speed of the rotor expressed in electrical angle per second. For a two-pole machine, the electrical angle and the mechanical angle are the same; while for a P-pole machine one unit of mechanical angle is equal to $P/2$ units of electrical angle.

In the polar form, the three phase rms voltages can be written as

$$E_A = E \angle 0° \qquad (5.4a)$$
$$E_B = E \angle -120° \qquad (5.4b)$$
$$E_C = E \angle -240° \qquad (5.4c)$$

The frequency f of the induced emf is related to the speed of the rotor N and the number of poles P of the machine and may be expressed as

$$f = \frac{NP}{120} \qquad (5.5)$$

5.2.2 The Phase Sequence

The phase sequence is defined as the sequence or the order in which the phase voltages attain their maximum values. The phase sequence is referred to as A-B-C or positive sequence when the voltage of phase A leads that of phase B and the voltage of phase B leads that of phase C. The waveforms and the phasor diagrams for the positive sequence voltages are shown in Figs 5.2(a) and (b). Similarly, when the

Fig. 5.2 (a) Positive sequence waveforms of three-phase voltages and (b) Phasor diagram of positive sequence voltages

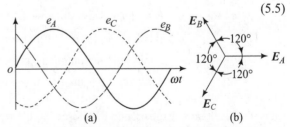

Fig. 5.3 (a) Negative sequence waveforms of three-phase voltages and (b) phasor diagram of negative sequence voltages

voltage of phase *A* leads that of phase *C* and the voltage of phase *C* leads that of phase *B*, the sequence is said to be *A-C-B* or negative sequence. The waveforms and the phasor diagram for the negative sequence voltages are shown in Figs 5.3(a) and (b).

5.2.3 Representation of Three-phase Generator

A three-phase generator is usually represented in the form of three emfs with rms values E_A, E_B, and E_C, which have the same magnitude but differ in phase by 120° from each other, in series with the internal impedance of the generator Z_S, assumed to be the same for each phase, as shown in Fig. 5.4. As per the convention for representation of a voltage source in Section 2.2.1, the positive mark placed alongside the source of emf indicates that the direction of positive current is out of the positive terminal mark of the voltage source.

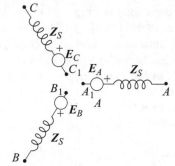

Fig. 5.4 Representation of a three-phase generator

5.2.4 Connection of Generator Phases

Each phase of the generator is isolated from the other two phases [see Fig. 5.4] and has two terminals that can be connected to a single-phase load. The rms values of three phase emfs E_A, E_B, and E_C, in series with internal impedance Z_S, can be connected to loads L_1, L_2, and L_3 as shown in Fig. 5.5. The rms currents I_A, I_B, and I_C flow from the source to the load. This arrangement requires six conductors for connecting the load to the source of emfs and is therefore complicated and expensive. For this reason the phases are always interconnected so that the number of conductors needed for connection to the loads is reduced. There are two general methods of interconnection of the phases, called the delta connection and the star connection. These are discussed in the following sections.

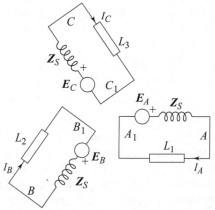

Fig. 5.5 Three-phase windings with six conductors

5.2.4.1 Delta-connected Generator

In a delta connection, the three phases are connected in series by joining the start of one coil to the finish of the other coil, that is, joining A_1 to B, B_1 to C, and C_1 to A, so as to form a closed circuit as shown in Fig. 5.6.

The instantaneous value of the total voltage in the closed loop is given by

$$e_A + e_B + e_C = \sqrt{2}\, E \sin \omega t + \sqrt{2}\, E \sin(\omega t - 120°) + \sqrt{2}\, E \sin(\omega t - 240°)$$
$$= \sqrt{2}\, E\, [\sin \omega t + \sin (\omega t - 120°) + \sin (\omega t - 240°)]$$
$$= \sqrt{2}\, E\, (\sin \omega t + \sin \omega t \cos 120° - \cos \omega t \sin 120° + \sin \omega t \cos 240° - \cos \omega t \sin 240°)$$
$$= \sqrt{2}\, E(\sin \omega t - 0.5 \sin \omega t - 0.866 \cos \omega t - 0.5 \sin \omega t + 0.866 \cos \omega t)$$
$$= 0$$

Thus, the resultant voltage around the closed circuit is zero at every instant, and no circulating current is set up around the closed loop. The three line conductors are joined to the three junctions thus formed. This circuit arrangement is referred to as a *delta* (from the Greek capital letter Δ) connection, also known as a mesh connection.

It might be useful at this stage to consider the actual values and directions of the induced voltages at a particular instant. At instant M [see Fig. 5.2], the voltage generated in phase A is positive and is represented by ML acting from A_1 to A. The induced voltage in phase B is negative and is represented by MN acting from B to B_1, and that in phase C is also negative and is represented by MP acting from C to C_1. But the sum of MN and MP is exactly equal numerically to ML. Consequently, the algebraic sum of the induced voltages around the closed circuit formed by the three windings is zero.

5.2.4.2 Star-connected Generator

The three start terminals A_1, B_1, and C_1 in Fig. 5.4 are joined together at N as shown in Fig. 5.7, so that the three conductors of Fig. 5.5 can be replaced by a single conductor MN as shown in Fig. 5.7. This common conductor is called the neutral wire. Since the generated emf has been assumed positive when acting from the start terminal to the finish terminal, the current in each phase must also be regarded as positive when flowing in that direction. The positive mark placed alongside the source of emf indicates that the direction of positive current is out of the positive terminal mark of the voltage source as shown in Fig. 5.7. If i_A, i_B, and i_C are the instantaneous values of the currents in the three phases, the instantaneous value of the current in the neutral wire MN is $i_N = (i_A + i_B + i_C)$, having its positive direction from M to N. This arrangement is referred to as a *four-wire star-connected* system and the junction N is referred to as the star point, also known as the neutral point.

Three-phase loads, such as three-phase induction motors and synchronous motors, are connected to the three line conductors A, B, and C whereas single-phase loads, such as lamps, fans, heaters, and other electrical household appliances, are usually connected between the line and the neutral conductors, as indicated by L_1, L_2, and L_3. However, the total single-phase load is distributed as equally as possible between the three lines and the common neutral.

If the three single-phase loads L_1, L_2, and L_3 are exactly the same, the instantaneous values of phase currents i_A, i_B, and i_C have the same peak value $I_m = \sqrt{2}I$, where I is the rms current phasor and the three phase currents differ in phase by $120°$. Hence, if the instantaneous value of the current in load L_1 is represented by

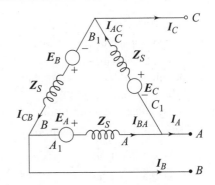
Fig. 5.6 Delta-connected three-phase winding

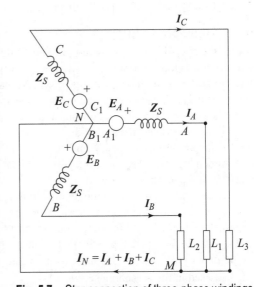
Fig. 5.7 Star connection of three-phase windings

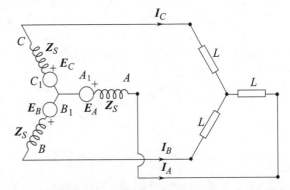
Fig. 5.8 Three-wire star-connected system with balanced load

$$i_A = \sqrt{2}\, I \sin \omega t \tag{5.6a}$$

the instantaneous value of current in L_2 is

$$i_B = \sqrt{2}\, I \sin (\omega t - 120°) \tag{5.6b}$$

and the instantaneous value of current in L_3 is

$$i_C = \sqrt{2}\, I \sin (\omega t - 240°) \tag{5.6c}$$

Hence, the instantaneous value of the resultant current in the neutral conductor *MN* of Fig. 5.7 is

$$i_N = i_A + i_B + i_C = \sqrt{2}\, I \sin \omega t + \sqrt{2}\, I \sin(\omega t - 120°) + \sqrt{2}\, I \sin (\omega t - 240°)$$

$$= \sqrt{2}\,\ I[\sin \omega t + I_m \sin (\omega t - 120°) + I_m \sin (\omega t - 240°)]$$

$$= \sqrt{2}\, I \times 0 = 0$$

Thus with a balanced load the resultant current in the neutral conductor is zero at every instant. Hence this conductor can be dispensed with, thereby, giving the *three-wire star-connected* system shown in Fig. 5.8.

5.2.5 Three-phase Supply

So far the simple case of a three-phase generator feeding a three-phase load has been discussed. In actual practice this is not a normal mode of operation. Typically, the generator would be 'synchronized' onto a large interconnected grid, thus operating in synchronism or in parallel with all other generators of the integrated system supplying energy to all the loads connected to it. It suffices to say that the generator will supply to the network a symmetrical set of currents at a rated voltage.

The distribution systems provide power at various voltage levels. Large industrial consumers can accept power directly at 66 to 220 kV and step down the voltage further to lower voltage levels. Small industrial or commercial consumers accept power at 11 to 33 kV. Residential consumers accept power at 415 V, three phase and 240 V, single phase.

A balanced three-phase supply available to the users of electric energy may be closely approximated by three sinusoidal ideal voltage sources, which are equal in magnitude and 120° out of phase with each other. The three voltage sources forming a three-phase source are V_A, V_B, and V_C, with $V_A = V\angle 0°$, $V_B = V\angle - 120°$, and $V_C = V\angle - 240°$ [see Fig. 5.9(a)]. The phasor diagram for these voltages is shown in Fig. 5.9(b). The three single-phase voltages may be arranged in a star or delta configuration to form a three-phase supply system (as discussed in the preceding section).

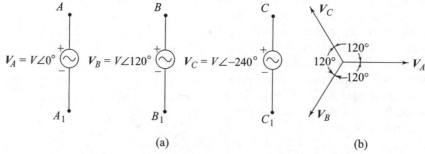

(a)   (b)

Fig. 5.9   (a) Three-phase source and (b) Phasor diagram

5.2.5.1 Star-connected Supply—Voltages and Currents

In a star connection the negative polarity terminals A_1, B_1, and C_1 of the voltage sources [see Fig. 5.9(a)] are joined together to form the neutral terminal N; and A, B, and C, the positive polarity terminals, are brought out to form a three-phase star-connected supply as shown in Fig. 5.10.

In Fig. 5.10, the voltages across phases V_A, V_B, and V_C, at the terminals A, B, and C, each with respect to the neutral point N, are called phase voltages. The rms phase voltage phasors of the three-phase supply are then given as $V_A = V\angle\, 0°$, $V_B = V\angle - 120°$, and $V_C = V\angle - 240°$. The three phase voltages appearing in Fig. 5.10 may be represented by a phasor diagram as shown in Fig. 5.11 and the direction in which the voltage or the current is assumed to be positive may be indicated on it.

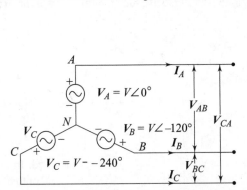

Fig. 5.10 Three-phase star-connected supply

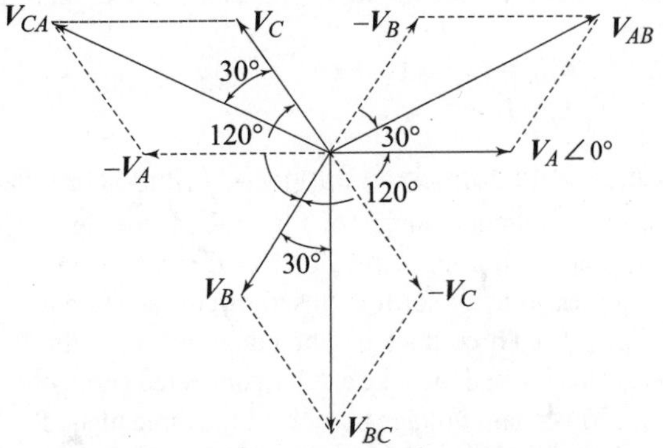

Fig. 5.11 Phasor diagram for star-connected supply

The three-phase star-connected supply has three terminals A, B, and C, called the line terminals, and may or may not have the fourth terminal N, called the neutral terminal. The voltages across the terminals, V_{AB}, V_{BC}, and V_{CA}, are called line voltages. The relationships between the line and the phase quantities for a star-connected system may be derived from the circuit diagram and the phasor diagram for the same supply system.

The value of voltage acting between the terminals A and B is the phasor difference of V_A and V_B. Hence $-V_B$ is drawn equal and opposite to V_B and added to V_A giving V_{AB} as the voltage acting from B to A via N. It may be noted that the positive direction of the voltage V_{AB} is from B to A. Thus, in terms of phasor voltages

$$V_{AB} = V_A - V_B = V\angle\, 0° - V\angle - 120°$$

$$= V(1 + j0) - V(-0.5 - j0.866) = V\,(1.5 + j0.866)$$

$$= V\sqrt{1.5^2 + 0.866^2}\, \angle \tan^{-1}\frac{0.866}{1.5} = \sqrt{3}\ V \angle 30° \tag{5.7}$$

Similarly, $\quad V_{BC} = V_B - V_C = V\angle - 120° - V\angle - 240°$

$$= V(-0.5 - j0.866) - V(-0.5 + j\,0.866) = -j1.732V$$

$$= \sqrt{3}\ V \angle - 90° \tag{5.8}$$

and $\qquad V_{CA} = V_C - V_A = V\angle - 240° - V\angle\, 0°$

$$= V(-0.5 + j\,0.866) - V(1 + j0) = V(-1.5 + j\,0.866)$$

$$= V\sqrt{(-1.5)^2 + (0.866)^2}\, \angle \tan^{-1}\frac{0.866}{-1.5} = \sqrt{3}\ V \angle 150° \tag{5.9}$$

It can be seen from Eqs (5.7)—(5.9) that the system of line voltages constitutes a balanced three-phase voltage system. The magnitude of the line voltages is $\sqrt{3}$ times the phase voltage. Thus, in general, if V_L represents the line voltage, that is, the potential difference between two line conductors and V_P represents the phase voltage, or the potential difference between a line conductor and the neutral point, then

$$V_L = 1.732V_P \tag{5.10}$$

Further, it is obvious from Fig. 5.10 that in a star-connected three-phase system connected to a balanced load, the current in a line conductor is identical to that of the phase to which that line conductor is connected. If I_L and I_P represent the line and phase currents, respectively, then

$$I_L = I_P \tag{5.11}$$

If KVL and KCL are applied to the star connection or the node N in Fig. 5.10, then the following expressions result:

$$V_{AB} + V_{BC} + V_{CA} = 0 \tag{5.12}$$
$$I_A + I_B + I_C = 0 \tag{5.13}$$

5.2.5.2 Delta-connected Supply—Voltages and Currents

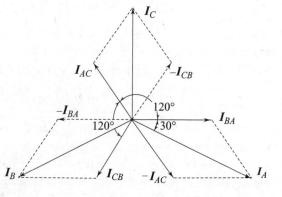

The three voltage sources of Fig. 5.9 can be delta connected by joining A_1 to B, B_1 to C, and C_1 and A together so as to form a closed circuit as shown in Fig. 5.12. Three lines can be taken out from the nodes thus formed to make a delta-connected supply.

The rms phase voltage phasors of the three phases are given by $V_A = V\angle 0°$, $V_B = V\angle -120°$, and $V_C = V\angle -240°$. The line voltages across the line terminals are V_{AB}, V_{BC}, and V_{CA}. It is obvious from Fig. 5.12 that the line voltages in a delta-connected supply are equal to the phase voltage and are given by

$$V_{AB} = V_A = V\angle 0° \tag{5.14a}$$
$$V_{BC} = V_B = V\angle -120° \tag{5.14b}$$
$$V_{CA} = V_C = V\angle -240° \tag{5.14c}$$

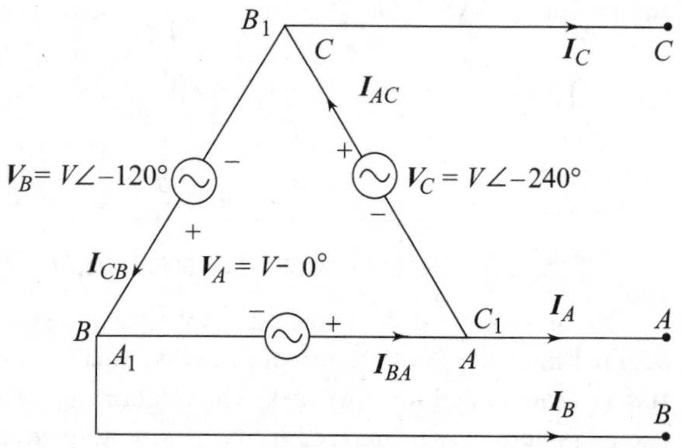

Fig. 5.12 Three-phase delta-connected supply

Thus, in general, if V_L represents the line voltage and V_P represents the phase voltage, then

$$V_L = V_P \tag{5.15}$$

The currents flowing in the phases of a delta-connected supply are denoted by I_{BA}, I_{CB}, and I_{AC}; I_A, I_B, and I_C denote the line currents. The value of line currents may be found out by applying KCL at the nodes A, B, and C.

$$I_A = I_{BA} - I_{AC} \tag{5.16}$$
$$I_B = I_{CB} - I_{BA} \tag{5.17}$$
$$I_C = I_{AC} - I_{CB} \tag{5.18}$$

The phasor diagram of the balanced phase currents is shown in Fig. 5.13. When the phase currents are balanced, the line currents can be obtained from the phasor diagram as well.

From the phasor diagram, taking I_{BA} as the reference phasor, that is, $I_{BA} = I_{BA}\angle 0° = I_P \angle 0°$, then the line currents are given by

$$I_A = \sqrt{3}\, I_P \angle -30° \tag{5.19}$$
$$I_B = \sqrt{3}\, I_P \angle -150° \tag{5.20}$$
$$I_C = \sqrt{3}\, I_P \angle 90° \tag{5.21}$$

where I_P is the magnitude of the phase current.

If I_P and I_L represent the magnitudes of the phase and the line current, respectively, in a balanced delta-connected system, then

Fig. 5.13 Phasor diagram of the phase and line currents for a balanced delta-connected supply system

$$I_L = \sqrt{3}\,I_P \qquad (5.22)$$

The co-relations between *line and phase voltages and line and phase currents* are summarized in Table 5.1

Table 5.1 Summary of co-relations between line and phase voltages and currents

Quantity	Type of connection	
	Star	Delta
Voltage	$V_L = V_P/\sqrt{3}$	$V_L = V_P$
Current	$I_L = I_P$	$I_L = \sqrt{3}\,I_P$

5.2.5.3 Specifications of a Three-phase Supply

For a three-phase system, there are two voltages, the phase voltage and the line voltage, that may be specified. It is a normal practice to specify a three-phase supply in terms of its line voltage, number of phases, and frequency. For example, the specification 415 V, 3 phase, 50 Hz implies a three-phase, 50-Hz system whose line-to-line voltage is 415 V and pertains to low-tension distribution lines in India.

5.3 POWER IN THREE-PHASE AC SYSTEMS WITH BALANCED LOAD

Power in a three-phase circuit is defined as the summation of the power delivered to all the three phases. In a balanced three-phase circuit, though the power in the individual phases varies sinusoidally at twice the supply frequency, it is observed that the total instantaneous power in a three-phase circuit is constant and independent of time. The concepts of instantaneous, active, reactive, apparent, and complex power and power factor, as applied to the three-phase circuits, are discussed in this section.

5.3.1 Instantaneous Power in Three-phase Circuits

Power in a three-phase balanced network, having the phase sequence *A-B-C*, is the sum of the products of instantaneous voltages and currents. If p is the total instantaneous power, then

$$p = v_A i_A + v_B i_B + v_C i_C \qquad (5.23)$$

The instantaneous values of phase voltages v_A, v_B, and v_C and phase currents i_A, i_B, i_C can be written as

$$v_A = \sqrt{2}\,V_P \sin \omega t$$
$$v_B = \sqrt{2}\,V_P \sin(\omega t - 120°)$$
$$v_C = \sqrt{2}\,V_P \sin \omega(t - 240°)$$
$$i_A = \sqrt{2}\,I_P \sin(\omega t - \varphi)$$
$$i_B = \sqrt{2}\,I_P \sin(\omega t - \varphi - 120°)$$
$$i_C = \sqrt{2}\,I_P \sin(\omega t - \varphi - 240°)$$

where V_P and I_P are the rms values of the phase voltage and phase current, respectively, while φ is the phase angle between the phase voltage and the corresponding phase current. The instantaneous total power in the three phases then becomes

$$p = v_A i_A + v_B i_B + v_C i_C$$
$$= \sqrt{2}\,V_P \sin \omega t \sqrt{2}\,I_P \sin(\omega t - \varphi) + \sqrt{2}\,V_P \sin(\omega t - 120°)\sqrt{2}\,I_P \sin(\omega t - \varphi - 120°)$$
$$+ \sqrt{2}\,V_P \sin(\omega t - 240°)\sqrt{2}\,I_P \sin(\omega t - \varphi - 240°) \qquad (5.24)$$

On simplifying Eq. (5.24) the instantaneous power becomes

$$p = 3V_P I_P \cos \varphi \qquad (5.25)$$

Hence, the average three-phase power P is given by

$$P = \frac{1}{T}\int_0^T p\,dt = 3V_P I_P \cos \varphi \qquad (5.26)$$

In a star-connected system, the line voltage $V_L = \sqrt{3}\,V_P$ and $I_L = I_P$. Substituting for V_P and I_P in Eq. (5.26), the power in a star-connected network becomes

$$P = \sqrt{3}\,V_L I_L \cos \varphi$$

Similarly, for a delta-connected system, $V_L = V_P$ and $I_L = \sqrt{3}\,I_P$. Substituting for V_P and I_P in Eq. (5.26), the power in a delta-connected network becomes

$$P = \sqrt{3}\,V_L I_L \cos \varphi$$

Hence, it can be concluded that power in a balanced three-phase network, star or delta, is given by

$$P = \sqrt{3}\,V_L I_L \cos \varphi \tag{5.27}$$

It may be noted that unlike in single-phase systems, where the instantaneous power is pulsating, the total instantaneous power for a three-phase system is constant. Thus a three-phase supply system has a distinct advantage over a single-phase system, especially in the running of large industrial drives.

5.3.2 Reactive and Apparent Power

The reactive power in a three-phase system is equal to the sum of the reactive powers of the individual phases. For a balanced three-phase system, the reactive power Q is

$$Q = 3V_P I_P \sin \varphi \tag{5.28a}$$
$$= \sqrt{3}\,V_L I_L \sin \varphi \tag{5.28b}$$

The apparent power or the volt-ampere for a three-phase circuit is the sum of the volt-amperes of the three individual phases. For a balanced system, the apparent power S is

$$S = 3V_P I_P \tag{5.29a}$$
$$= \sqrt{3}V_L I_L \tag{5.29b}$$

It may be observed that in a balanced three-phase system

$$P = 3V_P I_P \cos \varphi = S \cos \varphi$$
$$Q = 3V_P I_P \sin \varphi = S \sin \varphi$$
$$S = \sqrt{P^2 + Q^2} \tag{5.30}$$

5.3.3 Complex Power

For convenience in calculations, the concept of complex power is useful. It is a complex number; the real part of this number is the total active power and the imaginary part is the total reactive power in a circuit. Thus,

$$S = P + jQ$$
$$= S(\cos \varphi + j \sin \varphi) = S \angle \varphi$$

Complex power in terms of V_P and I_P can be expressed as

$$S = 3V_P I_P^* \tag{5.31}$$

Let $V_P = Ve^{j0}$ and $I_P = I\,e^{j(0-\varphi)}$

where I_P lags V_P by φ. Then, for balanced load,

$$S = 3V_P I_P^* = 3V_P e^{j0}\,I_P e^{-j(0-\varphi)}$$
$$= 3V_P I_P e^{j\varphi} = 3VI(\cos\varphi + j \sin \varphi)$$
$$= P + jQ = S \angle \varphi = S(\cos\varphi + j \sin \varphi) \tag{5.32}$$

5.4 ANALYSIS OF THREE-PHASE CIRCUITS

The analysis of a three-phase circuit may be carried out in exactly the same way as that of a single-phase circuit. The main objective of the analysis of a three-phase circuit is to determine the currents and the voltages in all the phases in the various elements/components of the network. The source and the load in a three-phase system may be treated as three independent single-phase sources and loads. However, in a balanced three-phase system, the computations for only one phase need to be carried out as the response of the other two phases follows from that of the first phase. A three-phase load or source may either be star- or delta-connected. The analysis of some combinations of sources and loads has been included in this section.

5.4.1 Star-connected Supply and Star-connected Balanced Load

Figure 5.14 shows a star-connected balanced load, with each limb having an impedance $Z = Z \angle \theta$, fed from a three-phase star-connected supply with balanced phase voltages V_A, V_B, and V_C and the balanced line voltages V_{AB}, V_{BC}, and V_{CA}. The three-phase three-wire supply connection is shown in Fig. 5.14. This type of supply is normally employed in high- and medium-voltage distribution systems. For this connection, the currents from the phases of the source and the load are the same.

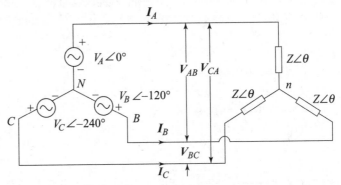

Fig. 5.14   Star-connected supply connected to star-connected load

Applying Ohm's law along the three phases, the potential difference between the source neutral N and the load neutral n may be written as

$$V_N - V_n = -V_A + Z \cdot I_A \tag{5.33}$$
$$= -V_B + ZI_B \tag{5.34}$$
$$= -V_C + ZI_C \tag{5.35}$$

Adding Eqs (5.33), (5.34), and (5.35)

$$3(V_N - V_n) = -(V_A + V_B + V_C) + Z(I_A + I_B + I_C) \tag{5.36}$$

Also, applying KCL to node N

$$I_A + I_B + I_C = 0$$

and as the three-phase supply voltage is balanced

$$V_A + V_B + V_C = 0$$

Applying these two conditions, Eq. (5.36) gives

$$V_N - V_n = 0$$
$$V_N = V_n \tag{5.37}$$

Then from Eq. (5.33) current I_A may be obtained as

$$I_A = \frac{V_A}{Z} = \frac{V \angle 0°}{Z \angle \theta} = \frac{V}{Z} \angle -\theta = I_A \angle -\theta \tag{5.38}$$

Figure 5.15 gives the phasor diagram showing all the currents and the voltages. It may be noted that joining the tips of phase voltage phasors in the phasor diagram gives the line voltages.

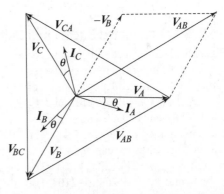

Fig. 5.15   Phasor diagram of balanced star-connected three-phase load connected to three-phase star-connected supply

For a three-phase four-wire star-connected supply system, the three phase voltages and the neutral conductor are available at the supply terminals. The neutral conductor connects the neutral point of the source with the neutral point of the star-connected load. This connection is generally employed for low-voltage distribution systems. If the three-phase load is unbalanced, the neutral conductor carries current I_N given by

$$I_N = I_A + I_B + I_C$$

For a balanced load, I_N is zero.

Example 5.1   In a three-phase four-wire system the line voltage is 400 V. Non-inductive loads of 12 kW, 10 kW, and 8 kW are connected between the three line conductors and the neutral point as shown in Fig. 5.16. Calculate (a) the current in each line, and (b) the current in the neutral conductor.

Solution   The phase voltages are given by

$$V_A = \frac{400}{\sqrt{3}} \angle 0° = 230.94 \angle 0°$$

$$V_B = 230.94 \angle -120°$$

$$V_C = 230.94 \angle -240°$$

(a) The line currents are given by

$$I_A = \frac{12 \times 10^3}{230.94 \angle 0°} = 51.96 \angle 0° \text{ A}$$

$$I_B = \frac{10 \times 10^3}{230.94 \angle -120°} = 43.3 \angle 120° \text{ A}$$

$$I_C = \frac{8 \times 10^3}{230.94 \angle -240°} = 34.64 \angle 240° \text{ A}$$

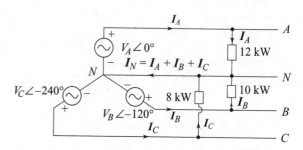

Fig. 5.16   Circuit diagram for Example 5.1

(b) The current in neutral

$$I_N = I_A + I_B + I_C$$
$$= 51.96 \angle 0° + 43.3 \angle 120° + 34.64 \angle 240°$$
$$= 51.96(1 + j0) + 43.3(-0.5 + j0.866) + 34.64(-0.5 - j0.866)$$
$$= 51.96 + (-21.65 + j37.5) + (-17.32 - j30.0)$$
$$= 12.99 + j7.5 = 15 \angle 30° \text{ A}$$

5.4.2 Star-connected Supply and Delta-connected Balanced Load

Figure 5.17 shows a delta-connected balanced load, with an impedance $Z = Z \angle \theta$ inserted in each pair of lines, fed from a three-phase star-connected supply with balanced phase voltages $V_A = V_P \angle 0°$, $V_B = V_P \angle -120°$, and $V_C = V_P \angle -240°$ and balanced line voltages, each having magnitude $V_L = \sqrt{3} V_P$. The line voltage phasors can then be expressed as $V_{AB} = \sqrt{3} V_P \angle 30°$, $V_{BC} = \sqrt{3} V_P \angle -90°$, and $V_{CA} = \sqrt{3} V_P \angle -210°$. For this connection, the phase currents from the source and the line currents drawn by the delta-connected loads are the same.

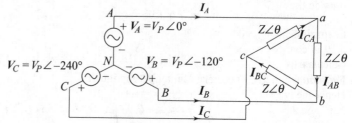

Fig. 5.17   Delta-connected load across star-connected supply

The load currents may be obtained as

$$I_{AB} = \frac{V_{AB}}{Z} = \frac{\sqrt{3}V_P \angle 30°}{Z\angle\theta} = \frac{\sqrt{3}V_P}{Z} \angle (30 - \theta)° \quad (5.39)$$

$$I_{BC} = \frac{V_{BC}}{Z}$$

$$= \frac{\sqrt{3}V_P \angle -90°}{Z\angle\theta} = \frac{\sqrt{3}V_P}{Z} \angle(-90 - \theta)° \quad (5.40)$$

$$I_{CA} = \frac{V_{CA}}{Z}$$

$$= \frac{\sqrt{3}V_P \angle -210°}{Z\angle\theta} = \frac{\sqrt{3}V_P}{Z} \angle(-210 - \theta)° \quad (5.41)$$

The line currents can now be calculated as the difference of appropriate load phase currents and are given by

$$I_A = I_{AB} - I_{CA} \qquad (5.42)$$
$$I_B = I_{BC} - I_{AB} \qquad (5.43)$$
$$I_C = I_{CA} - I_{BC} \qquad (5.44)$$

It may be seen from Eqs (5.39)–(5.41) that the phase currents in the load are equal in magnitude and displaced from each other by 120°. The magnitude of the phase current $I_P = \sqrt{3}\, V_P/Z$. Similarly, the line currents are also a balanced set of currents. Figure 5.18 gives the phasor diagram of voltages and currents of a delta-connected load across a star-connected supply.

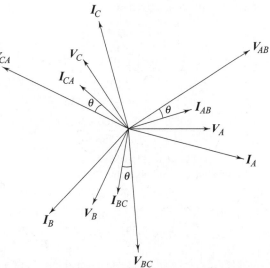

Fig. 5.18 Phasor diagram of balanced delta-connected three-phase load connected to star-connected three-phase supply

5.4.3 Unbalanced Three-phase Circuits

An unbalanced three-phase system is one in which the phases are not balanced. This may happen due to unequal magnitudes of the phase quantities, unequal phase differences between the phase quantities, or both. This may also be caused by unbalanced loading, unbalanced supply voltage, or both.

The mesh analysis or nodal analysis, as discussed in the case of single-phase circuits in Chapter 4, may be used for the analysis of unbalanced three-phase circuits. Some simplifications are possible under specific conditions. For instance, the simplification for an unbalanced star-connected load, connected to a balanced three-phase four-wire supply, can be carried out on per-phase basis as the neutral connection is available. The method of solution is more involved in the case of unbalanced loads. The application of symmetrical components simplifies the analysis for unbalanced networks. However, this will not be discussed here as it is outside the scope of this book.

Example 5.2 A delta-connected load as shown in Fig. 5.19 is connected to a 400-V, 3-phase, 50-Hz supply. Calculate (a) the phase currents and (b) the line currents.
Solution   (a) The supply phase sequence is *A-B-C*. Hence the line voltage V_{AB} leads the line voltage V_{BC} by 120° and the line voltage V_{BC} leads the line voltage V_{CA} by 120°. The line voltage V_{AB} is chosen as the reference phasor. Thus, the line voltages are given by

$$V_{AB} = 400 \angle 0° \text{ V}$$
$$V_{BC} = 400 \angle -120° \text{ V}$$
$$V_{CA} = 400 \angle -240° \text{ V}$$

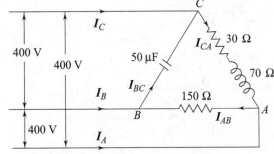

Fig. 5.19 Circuit diagram for Example 5.2

The phase currents are given by

$$I_{AB} = \frac{400\angle 0°}{150} = 2.667 \text{ A}$$

$$I_{BC} = \frac{400\angle -120°}{-j/(314 \times 50 \times 10^{-6})} = \frac{400 \times 314 \times 50 \angle -120°}{10^6 \angle -90°} = 6.2832\angle -30°$$

$$= 6.2832 \times \{\cos(-30) + j\sin(-30°)\} = (5.4414 - j3.1416) \text{ A}$$

$$I_{CA} = \frac{400\angle -240°}{30 + j70} = \frac{400\angle -240°}{76.16\angle 66.8} = 5.252\angle 53.1986°$$

$$= (3.1463 + j4.2056) \text{ A}$$

(b) The line currents are given by

$$I_A = I_{AB} - I_{CA} = 2.667 - (3.1463 + j4.2056) = -0.4797 - j4.2056 = 4.2328\angle -96.51° \text{ A}$$

$$I_B = I_{BC} - I_{AB} = 2.7747 - j3.1416 = 4.1915 \angle -48.55° \text{ A}$$

$$I_C = I_{CA} - I_{BC} = 2.2951 + j7.3472 = 7.6973 \angle 107.35° \text{ A}$$

Example 5.3   A balanced star-connected load of $(4 + j\,3)$ Ω per phase is connected to a 400-V, 3-phase, 50-Hz supply. Find the (a) line current, (b) power factor, (c) power, (d) reactive volt-ampere, and (e) total volt-ampere.

Solution   (a) Load impedance, $Z_L = 4 + j3 = 5 \angle 36.87°$

Assuming the phase voltage $V_A = \angle 0°$ as the reference phasor and using Eq. (5.38) the line current I_A may be calculated as

$$I_A = \frac{V_A}{Z_L} = \frac{(400/\sqrt{3}) \angle 0°}{5 \angle 36.87°} = 46.19 \angle -36.87° \text{ A}$$

As the load is balanced, the phase currents are also balanced. They have equal magnitude but are displaced from each other by 120°. Therefore,

$$I_B = 46.19\angle -36.87° - 120° = 46.19\angle -156.87° \text{ A}$$

$$I_C = 46.19\angle -36.87° - 240° = 46.19\angle -276.87° = 46.19\angle 83.13° \text{ A}$$

(b) The power factor, $\cos \varphi = \cos(-36.87°) = 0.8$ laging

(c) The three-phase power, $P = \sqrt{3} \ V_L I_L \cos\varphi = \sqrt{3} \times 400 \times 46.19 \times 08 = 25.6 \text{ k}$

(d) The reactive volt-ampere, $Q = \sqrt{3} \ V_L I_L \sin\varphi = \sqrt{3} \times 400 \times 46.1 \times 0.6 = 19.2 \text{ kvar}$

(e) The total volt-ampere, $\sqrt{3} \ V_L I_L = \sqrt{3} \times 400 \times 46.19 = 32 \text{ kVA}$

Example 5.4   A symmetrical 440-V, 3-phase system supplies a star-connected load with the following impedances: $Z_A = 50$ Ω, $Z_B = j15$ Ω, and $Z_C = -j15$ Ω. Calculate the voltage drop across each branch and the potential of the neutral point with reference to earth. Also, sketch the phasor diagram of the voltages, the phase sequence being *A-B-C*.

Solution   Figure 5.20 shows a star-connected load connected to a 440-V supply. I_A and I_B are the assumed loop currents flowing in meshes *ANBA* and *BNCB*, respectively. If the line voltage V_{AB} is taken as the reference phasor, then

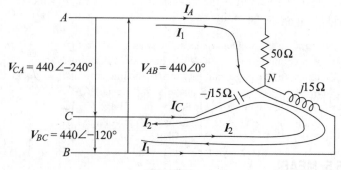

Fig. 5.20

$$V_{AB} = 440 \angle 0°$$
$$V_{BC} = 440 \angle -120°$$
$$V_{CA} = 440 \angle -240°$$

Applying KVL to meshes *ANBA* and *BNCB*,

$$50 \, I_1 + j15(I_1 - I_2) - 440 \angle 0° = 0 \qquad (I)$$
$$j15(I_2 - I_1) + (-j15) I_2 - 440 \angle -120° = 0 \quad (II)$$

Solving Eqs (I) and (II)

$$I_1 = I_A = 29.33 \angle -30° = (25.4 - j14.665) \text{ A}$$
$$I_2 = (-23.48 - j70) = 73.83 \angle -108.54° \text{ A}$$
$$I_B = I_2 - I_1 = (-23.48 - j70) - (25.4 - j14.665)$$
$$\quad = -48.88 - j55.335 = 73.83 \angle -131.45° \text{ A}$$
$$I_C = -I_2 = 23.48 + j70 = 73.82 \angle 71.5° \text{ A}$$

Hence, the voltage drops V_R, V_L, V_C, across the resistance, inductance and capacitance branches, respectively, are given by

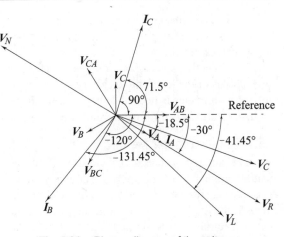

Fig. 5.21   Phasor diagram of the voltages

$$V_R = 29.33 \angle -30° \times 50 = 1466.5 \angle -30° \text{ V}$$
$$V_L = 73.83 \angle -131.45° \times j15 = 1107.45 \angle -41.45° \text{ V}$$
$$V_C = 73.83 \angle 71.5° \times -j15 = 1107.45 \angle -18.5° \text{ V}$$

The potential of the neutral point is given by

$$V_N = V_{AN} - V_R = (440/\sqrt{3}) \angle -30° - 1466.5 \angle -30°$$
$$\quad = (220 + j127) - (1270 - j733.25)$$
$$\quad = (-1050 + j606.25) = 1212.45 \angle 150° \text{ V}$$

The sketch of the phasor diagram of the voltages and the currents, not to scale, is shown in Fig. 5.21.

It can be seen from the phasor diagram that the current I_A is in phase with the voltage drop V_R, while the currents I_B and I_C, respectively, lag and lead the voltage drops V_L and V_C.

Example 5.5   A three-phase motor operating on a 440-V, 50-Hz supply system is developing 25 kW at an efficiency of 90% and a power factor of 0.85. Calculate (a) the line current and (b) the phase current if the windings are delta-connected.

Solution

(a) Efficiency $\eta = \dfrac{\text{Output power}}{\text{Input power}}$

Now,   Input power $= \sqrt{3} \, V_L I \cos \varphi = \sqrt{3} \times 440 \times I_L \times 0.85$

Then, $0.90 = \dfrac{25 \times 10^3}{\sqrt{3} \times 440 \times 0.85 \times I_L}$

$\therefore \qquad I_L = \dfrac{25 \times 10^3}{\sqrt{3} \times 440 \times 0.85 \times 0.9} = 42.88 \text{ A}$

(b) For a delta-connected winding, the phase current is given as

$$I_P = \dfrac{25 \times 10^3}{3 \times 440 \times 0.85 \times 0.9} = 24.76 \text{ A}$$

5.5 MEASUREMENT OF ACTIVE POWER IN THREE-PHASE NETWORKS

A wattmeter is an instrument used for the measurement of power. Power consumed in a dc circuit is the product of the voltage and the current in the circuit. Power consumed in an ac circuit is the product of the rms voltage, the rms current, and the power factor of the circuit. Therefore, ac circuits require the use of wattmeters to measure power. The principle of operation and the construction details of a wattmeter are discussed in Chapter 10.

The total active power in a three-phase network is the summation of the active power delivered to all the three phases of the load. In the case of a star-connected balanced load, with the neutral terminal accessible, only one wattmeter is needed to measure the three-phase power; for an unbalanced three-phase load, two wattmeters are needed for measurement of active power. Both the methods are described in this section.

5.5.1 One-wattmeter Method

In a three-phase star-connected balanced load, with the neutral terminal accessible, power can be measured by connecting the current coil of the wattmeter in one phase and its voltage coil between the same phase and the neutral as shown in Fig. 5.22.

The reading shown by the wattmeter is the active power per phase. The total active power of the star-connected balanced load is equal to three times the wattmeter reading.

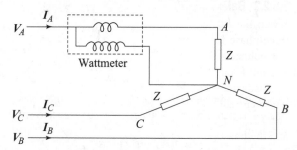

Fig. 5.22 Measurement of active power in a star-connected balanced load

5.5.2 Two-wattmeter Method

The active power and the power factor measurement of unbalanced and balanced three-phase loads are discussed in this section.

5.5.2.1 Unbalanced Three-phase Load

Figure 5.23 shows the circuit diagram for connecting two wattmeters in a three-phase star-connected unbalanced load for the measurement of the three-phase active power. The current coils of the two wattmeters W_1 and W_2 are connected in any two phases, in this case A and B phases, respectively, and the voltage or pressure coils are connected between these two phases and the third phase, in this case between AC and BC, respectively.

If v_A, v_B, and v_C are the instantaneous values of voltages appearing across the loads L_1, L_2, and L_3, respectively, and i_A, i_B, and i_C are the corresponding instantaneous values of currents, then the instantaneous power indicated by p_1 and p_2 can be written as

$$p_1 = i_A v_{AC} = i_A(v_A - v_C) \tag{5.45}$$
$$p_2 = i_B v_{BC} = i_B(v_B - v_C) \tag{5.46}$$

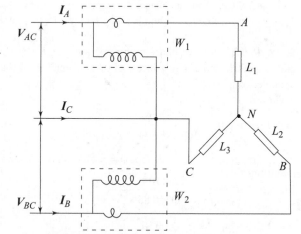

Fig. 5.23 Measurement of power by two wattmeters

where v_{AC} is the pd between the phases A and C and v_{BC} is the pd between the phases B and C.

The total instantaneous power is given by

$$p_1 + p_2 = i_A v_A + i_B v_B - (i_A + i_B)v_C \tag{5.47}$$

But, $i_A + i_B + i_C = 0$

Therefore, $p_1 + p_2 = i_A v_A + i_B v_B + i_C v_C \tag{5.48}$

The power measured by the wattmeters varies from instant to instant, but the inertia of the moving parts of the wattmeters make the pointers of the wattmeters read the average values. Thus, the sum of the wattmeter readings provides the total power in a three-phase network. The same result is obtained for a delta-connected load also. It is left as an exercise for the readers to prove that the total power for a delta-connected load is also given by $p_1 + p_2$. Thus, irrespective of the type of connection (star or delta) and the nature of load (balanced or unbalanced), three-phase power can be obtained using two wattmeters and adding the readings of the wattmeters.

5.5.2.2 Balanced Three-phase Load

When the loads in the circuit shown in Fig. 5.23 are equal, the three-phase load is balanced. The phasor diagram for the balanced star-connected load is shown in Fig. 5.24. In the phasor diagram, V_A, V_B, and V_C are the rms values of the phase voltages appearing across the loads and I_A, I_B, and I_C are the corresponding values of the rms phase currents. The phasor V_{AB} is the phasor difference between V_A and V_B, the phasor V_{BC} is the phasor difference between V_B and V_C, and the phasor V_{CA} is the phasor difference between V_C and V_A [see Fig. 5.24].

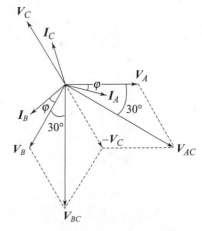

Fig. 5.24 Phasor diagram for a balanced star-connected load

Since the line current I_A flows through the current coil of wattmeter W_1, its potential coil measures the voltage between phase A and phase C. The power P_1 is then given by

$$P_1 = V_{AC}I_A \cos(30° - \varphi) \tag{5.49}$$

Similarly, the current coil and the potential coil of wattmeter W_2 measure I_B and V_{BC}. Hence, power P_2 is given by

$$P_2 = V_{BC}I_B \cos(30° + \varphi) \tag{5.50}$$

The total power is given by

$$P_1 + P_2 = V_{AC}I_A \cos(30° - \varphi) + V_{BC}I_B \cos(30° + \varphi) \tag{5.51}$$

Since the load is balanced, substituting $V_{BC} = V_{AC} = V$ and $I_A = I_B = I$ in Eq. (5.51),

$$P_1 + P_2 = VI\{\cos(30° - \varphi) + \cos(30° + \varphi)\}$$
$$= \sqrt{3}\ VI \cos \varphi \tag{5.52}$$

Subtracting Eq. (5.50) from Eq. (5.49) and simplifying

$$P_1 - P_2 = VI\{\cos(30° - \varphi) - \cos(30° + \varphi)\}$$
$$= VI \sin \varphi \tag{5.53}$$

Dividing Eq. (5.53) by Eq. (5.52) gives

$$\tan \varphi = \sqrt{3}\ \frac{P_1 - P_2}{P_1 + P_2} \tag{5.54}$$

$$\varphi = \tan^{-1} \sqrt{3}\ \frac{P_1 - P_2}{P_1 + P_2} \tag{5.55}$$

Equation (5.55) provides an expression for the determination of the power factor angle φ of the load by the two-wattmeter method.

It is important to note that, in general, the two wattmeters work under quite different phase angle conditions. Thus, the phase angle of one is $(30 - \varphi)$ and that of the other is $(30 + \varphi)$. Hence, only in the special case of $\varphi = 0$ (resistive load), that is, unity power factor, can the two readings be equal. For all other power factors, even under balanced conditions, the two readings are not equal.

Figure 5.25 shows the variation in wattmeter readings with a change in the power factor angle. As stated earlier, each wattmeter shows identical reading when the power factor is unity ($\varphi = 0$). It may also be noted that if one of the phase angles becomes greater than 90°, that is $\varphi > 60°$, the wattmeter will give a negative reading, which must be corrected by reversing the connections to the terminals of its pressure coil. Under these circumstances its reading is reckoned as negative, and the total power is then the difference of the readings. Thus, in general, the total power is given by the algebraic sum of the two readings.

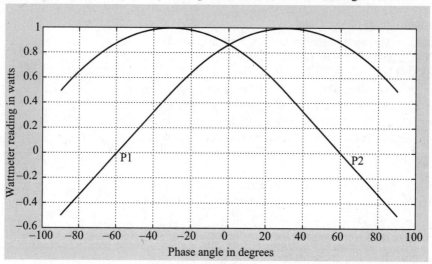

Fig. 5.25 Variation in wattmeter readings with a change in power factor angle

Table 5.2 provides a quick reference to how the wattmeter readings vary within certain range of the power factor angle.

Table 5.2 Nature of wttmeter readings for certain ranges of power factor angle

pf angle φ pf = cos φ	Wattmeter readings P_1/P_2
$\varphi = 0°$ pf = 1.0	$P_1 = P_2$ Both P_1 and P_2 are positive Total power = $2P_1$
φ lies between 0° and 60° pf ranges between unity and 0.5	$P_1 > P_2$ Both P_1 and P_2 are positive Total power = $P_1 + P_2$
$\varphi = 60°$ pf = 0.5	P_1 = positive $P_2 = 0$ Total power = P_1
φ lies between 60° and 90° pf ranges between 0.5 and 0	P_1 = positive P_2 = negative Total powr = $P_1 - P_2$
$\varphi = 90°$ pf = 0	$P_1 = -P_2$ Total power = 0

Example 5.6 In order to measure the power supplied to a star-connected balanced load, the current coil of the wattmeter is connected in line C and the potential coil is connected betweeen lines A and B as shown in Fig. 5.26.

Draw the phasor diagram and derive an expression for the power measured. Confirm whether the wattmeter reading relates to real power. If not what is the type of power measured?

Solution The phasor diagram is shown in Fig. 5.27.

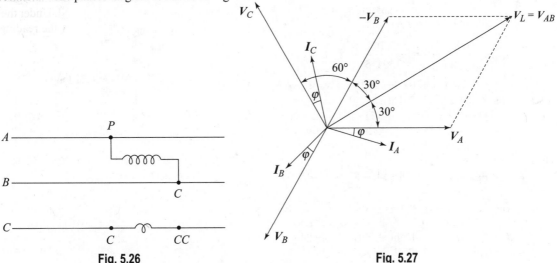

Fig. 5.26   **Fig. 5.27**

From the phasor diagram, it is seen that the line voltage between phases A and B is

$$V_L = V_{AN} - V_{BN} = \sqrt{3}\, V_{AN}$$

Further, it is seen from the phasor diagram that the phase angle between the line voltage V_L and the current in line C is $(90° - \varphi)$. Hence, if the wattmeter reads W, the power per phase is given by

$$W = V_L \times I_C \times \cos(90° - \varphi) = \sqrt{3}\, V_{AN} I_C \sin\varphi \tag{I}$$

From Eq. (5.28) it may be observed that the wattmeter reads reactive power and the total reactive power in the circuit is

$$Q = 3 V_L I_L \sin\varphi$$

Substituting from Eq. (I) yields the total reactive power in the circuit as

$$Q = \sqrt{3}\, W$$

Example 5.7 A 3-phase, 415-V, mesh-connected system shown in Fig. 5.28 has the following loads: 25 kW at power factor 1.0 for branch AB, 40 kVA at power factor 0.85 lagging for branch BC, 30 kVA at power factor 0.6 leading for branch CA. Find the line currents and the readings on wattmeters whose current coils are in phases A and C. Also, sketch the phasor diagram.

Solution Let the line voltage V_{AB} be the reference phasor. Then the line voltages are

$$V_{AB} = 415\angle 0°$$
$$V_{BC} = 415\angle{-120°}$$
$$V_{CA} = 415\angle{-240°}$$

The phase currents are

$$I_{AB} = \frac{\text{kW} \times 10^3}{\sqrt{3} \times V_{AB} \times \text{pf}} = \frac{25 \times 10^3}{\sqrt{3} \times 415 \times 1}$$

$$= 34.78 \text{ A in phase with respect to } V_{AB}$$

or $I_{AB} = 34.78 \angle 0° = 34.78 + j0$

$$I_{BC} = \frac{\text{kVA} \times 10^3}{\sqrt{3} \times V_{BC}} = \frac{40 \times 10^3}{\sqrt{3} \times 415}$$

$$= 55.65 \text{ A at pf } 0.85 \text{ lagging with respect to } V_{BC}$$
$$\cos^{-1} 0.85 = \angle 31.8°$$

Then $I_{BC} = 55.65\angle(-120 - 31.8)° = 55.65\angle - 151.8°$

lagging with respect to V_{AB}

$$= -49.04 - j26.3 \text{ A}$$

$$I_{CA} = \frac{\text{kVA} \times 10^3}{\sqrt{3} \times V_{AB}} = \frac{30 \times 10^3}{\sqrt{3} \times 415}$$

$$= 41.74 \text{ A at pf } 0.6 \text{ leading with respect to } V_{CA}$$

Also, $\cos^{-1} 0.6 = \angle 53.13°$. Then

$$I_{CA} = 41.74\angle(-240 + 53.13)° = 41.74\angle - 186.87°$$

lagging with respect to V_{AB}

$$= -41.44 + j5.0 \text{ A}$$

The line currents are

$$I_A = I_{AB} - I_{CA} = (34.78 + j0) - (-41.44 + j5.0)$$
$$= 76.22 - j\,5.0 = 76.38\angle - 3.75° \text{ A}$$
$$I_B = I_{BC} - I_{AB} = (-49.04 - j26.3) - (34.78 + j0)$$
$$= -83.82 - j26.3 = 87.85 \angle - 162.6° \text{ A}$$
$$I_C = I_{CA} - I_{BC} = (-41.44 + j5.0) - (-49.04 - j26.3)$$
$$= 7.6 + j\,31.3 = 32.21\angle 76.35° \text{ A}$$

The wattmeter readings in Phase A,

$$W_1 = V_{AB} \times I_A \times \cos \varphi_A$$

where φ_A is the phase angle between the phasors V_{AB} and I_A. Then

$$W_1 = 415 \times 76.38 \times \cos(-3.75°) = 31.63 \text{ kW}$$

The wattmeter readings in Phase C,

$$W_2 = V_{CB} \times I_C \times \cos \varphi_C$$

where φ_C is the phase angle between the phasors V_{CB} and I_C

$$V_{CB} = -V_{BC} = -415 \angle -120° = 415\angle 60°$$
$$\varphi_C = 76.35 - 60 = 16.35°$$

Then, $W_2 = 415 \times 32.21 \times \cos(16.35°) = 12.827 \text{ kW}$

The sketch of the phasors is shown in Fig. 5.29.

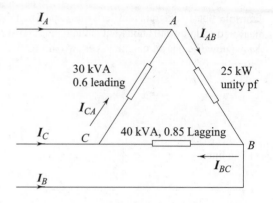

Fig. 5.28 Circuit diagram for a three-phase delta connected loads

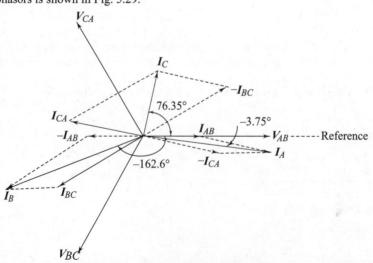

Fig. 5.29 Phasor diagram for Example 5.7

Example 5.8 The power input to a 2200-V, 50-Hz, 3-phase motor, running on full load at an efficiency of 90%, is measured by two wattmeters, which indicate 500 kW and 200 kW, respectively. Calculate (a) the total input power, (b) the power factor, (c) the line current, and (d) the horse power output.

Solution (a) The total input power $= W_1 + W_2 = 500 + 200 = 700$ kW

(b) $\tan \varphi = \sqrt{3} \dfrac{W_1 - W_2}{W_1 + W_2} = \sqrt{3} \dfrac{500 - 200}{500 + 200} = 0.7423$

$\qquad \varphi = \tan^{-1}(0.7423) = 36.58°$

The power factor, $\cos \varphi = \cos 36.58 = 0.803$.

(c) The line current can be determined from

$\qquad P = \sqrt{3} V_L I_L \cos \varphi = 3 \times 2200 \times I_L \times 0.803$

$\therefore \qquad I_L = \dfrac{700 \times 10^3}{\sqrt{3} \times 2200 \times 0.803} = 228.77$ A

(d) Efficiency $\eta = $ output/input

Then,   output $= \eta \times$ input $= 0.9 \times 700 = 630$ kW

Now,   1 hp $= 746$ W

$\therefore \qquad$ Output $= \dfrac{630 \times 10^3}{746}$ 844.5 hp

Recapitulation

For a star-connected system: $V_L = \sqrt{3} V_P$ and $I_L = I_P$

For a delta-connected system: $V_L = V_P$ and $I_L = \sqrt{3} I_P$

Active power in a three-phase network,

$\qquad P = 3 V_P I_P \cos \varphi = \sqrt{3} V_L I_L \cos \varphi$

Reactive power in a three-phase network,

$\qquad Q = 3 V_P I_P \sin \varphi = \sqrt{3} V_L I_L \sin \varphi$

Volt-amperes in a three-phase network,

$\qquad VA = 3 V_P I_P = \sqrt{3} V_L I_L$

The total power in a balanced three-phase network with the neutral available is three times the reading shown by the wattmeter.

The total power measured in a three-phase network (balanced or unbalanced), by the two-wattmeter method, is $P_1 + P_2$, where P_1 and P_2 are the readings on the wattmeters.

Power factor angle, $\varphi = \tan^{-1} \sqrt{3} \dfrac{P_1 - P_2}{P_1 + P_2}$

Assessment Questions

1. Enumerate with reasons, the advantages of using three-phase supply systems over single phase supply systems.
2. (a) Explain the method of generating three-phase voltages.
   (b) What is positive and negative phase sequence?
3. (a) Specify the number of ways in which a three-phase generator can be connected.
   (b) Draw phasor diagrams of each type of connections and derive expressions to co-relate phase and line voltages and currents.
4. (a) What is a balanced three-phase supply system? How is a three-phase supply system specified?
   (b) Distinguish between balanced and unbalanced loads.
5. Prove that the instantaneous three-phase power in a circuit is equal to the average power.
6. (a) Explain the two-wattmeter method of measuring power in a circuit feeding a three-phase unbalanced load.
   (b) Discuss how the reading on each wattmeter varies when the power factor angle changes from 0° to 90°.
7. With the help of suitable phasor diagrams, explain the significance of
   (a) equal wattmeter readings and (b) a zero reading on one of the wattmeters.

Problems

5.1 A balanced three-phase load consisting of three coils, each with a resistance of 3 Ω and inductance of 0.04 H, is connected to a three-phase 415-V, 50-Hz supply. Determine the total active power when the coils are (a) star-connected, (b) delta-connected.   [(a) 3.1 kW, (b) 9.302 kW]

5.2 A balanced star-connected load of $(4 + j\,3)$ Ω per phase is connected to a 3-phase, 415-V, 50-Hz supply. Find (a) the line current, (b) the power factor, (c) the power, (d) the reactive volt-amperes, and (e) the total volt amperes.   [(a) 47.92 A, (b) 0.8, (c) 27.556 kW, (d) 20.667 kVAR, (e) 34.445 kVA]

5.3 A balanced 3-phase, star-connected load of 6 kW takes a leading current of 10 A with a line voltage of 415 V, 50 Hz. Find the circuit constants of the load per phase. $[R = 20\ \Omega,\ C = 241.37\ \mu F]$

5.4 A balanced star-connected load is supplied from a symmetrical 3-phase, 440-V system. The current in each phase is 20 A and lags 30° behind the phase voltage. Find (a) the phase voltage and (b) the total power. Also, draw the phasor diagram showing the currents and voltages.   [(a) 254 V, (b) 8.8 kW]

5.5 A three-phase delta-connected load, with each phase having inductive reactance of 20 Ω and resistance of 10 Ω, is fed from the secondary of a three-phase star-connected transformer which has a phase voltage of 231 V. Draw the circuit diagram for the system and calculate (a) the current in each phase of the load, (b) the pd across each phase of the load, (c) the current in the transformer's secondary windings, (d) the total active power taken from the supply and its power factor.   [(a) 17.89 A, (b) 400 V, (c) 31 A, (d) 9607 W]

5.6 Three similar coils, connected in star, take a total power of 3 kW from a three-phase, three-wire, 415-V, 50-Hz supply at a power factor of 0.3. Calculate (a) the resistance and inductance of each coil and (b) the line currents if one of the coils is short-circuited.   [(a) 5.17 Ω, 52.3 mH (b) 24.1 A, 41.75 A]

5.7 Three non-reactive resistors of 15, 25, and 30 Ω are star-connected to the *A*, *B*, and *C* phases, respectively, of a 415-V, symmetrical system. Determine the current and the power in each resistor and the voltage between the star point and the neutral terminal, the phase sequence being *A-B-C*.   [12.57 A, 2.367 kW; 10.46 A, 2.735 kW; 9.22 A, 2.55 kW; 52.27 V]

5.8 Determine the line currents in an unbalanced star-connected load supplied from a symmetrical 3-phase, 400 V system. The branch impedances of the load are $Z_A = 10\angle 30°$ Ω, $Z_B = 10\angle 45°$ Ω, and $Z_C = 10\angle 60°$ Ω. The phase sequence is *A-B-C*.   [21.65 A, 26.4 A, 21.7 A]

5.9 A three-phase, 400-V, star-connected motor has an output of 25 kW, with an efficiency of 85% and a power factor of 0.8. Calculate the line current. Sketch a phasor diagram showing the voltages and the currents. If the motor windings were mesh-connected, what would be the correct voltage of a three-phase supply suitable for the motor?   [45.1 A, 230.9 V]

5.10 A balanced mesh-connected load of $(12 + j9)$ Ω per phase is connected to a 3-phase, 400-V supply. Find (a) the line current, (b) the power factor, (c) the power, (d) the reactive volt-amperes, and (e) the total volt-amperes.   [(a) 26.67 A, (b) 0.8, (c) 25.607 kW, (d) 19.205 kVAR, (e) 32 kVA]

5.11 A symmetrical 3-phase, 415-V system supplies a balanced mesh-connected load. The current in each branch circuit is 25 A and the phase angle is 25° (lagging). Find (a) the line current and (b) the total power. Draw the phasor diagram showing the voltages and currents in the lines and circuits for the three phases.   [(a) 43.3 A, (b) 28.2 kW]

5.12 The currents in the *AB*, *BC*, and *CA* branches of a mesh-connected system with symmetrical voltages are 30 A at power factor 0.85 lagging, 25 A at power factor 0.6 leading, and 30 A at unity power factor respectively. Determine the current in each line. The phase sequence is *A-B-C*. $[I_A = 16.43\ A,\ I_B = 37.31\ A,\ I_C = 25\ A]$

5.13 Three coils are connected in delta to a three-phase, three-wire, 400-V, 50-Hz supply and take a line current of 10 A, 0.85 power factor lagging. Calculate (a) the resistance and the inductance of the coils, (b) the line current and the total power if the coils are star-connected to the same supply, and (c) the line currents if one coil becomes open-circuited when the coils are star-connected.   [(a) 69.28 Ω, 0.1162 H; (b) 3.33 A, 1961 W; (c) 2.886 A,]

5.14 A three-phase, 400 V system has the following load connected in delta: a non-reactive resistor of 80 Ω between the lines A and B, a coil having a reactance of 100 Ω and negligible resistance between the lines B and C, a loss-free capacitor having a reactance of 160 Ω between the lines C and A. Calculate (a) the phase currents and (b) the line currents. Assume the phase sequence to be A-B-C. Also, draw the complete phasor diagram. [(a) 5 A, 4 A, 2.5 A; (b) 7.27 A, 8.7 A, 3.5 A]

5.15 The load taken by a three-phase induction motor is measured by the two-wattmeter method and the readings are 750 W and 200 W. What is the active power taken by the motor and at what power factor is it working? [0.95 kW, 0.706]

5.16 Two wattmeters connected to read the total power in a three-phase system supplying a balanced load read 12.5 kW and $-$ 2.5 kW, respectively. Calculate the total active power. With the help of suitable phasor diagrams, explain the significance of (a) equal wattmeter readings and (b) a zero reading on one of the wattmeters. [10 kW]

5.17 Two wattmeters connected to measure the input to a balanced three-phase circuit indicate 2000 W and 400 W, respectively. Find the power factor of the circuit (a) when both the readings are positive and (b) when the latter reading is obtained after reversing the connections to the current-coil of one of the instruments. Draw the phasor and connection diagrams. [(a) 0.6546, (b) 0.3592]

5.18 A 400-V, 50-Hz, three-phase motor takes a line current of 20.0 A when operating at a lagging power factor of 0.60. When a capacitor bank is connected across the motor terminals, the line current is reduced to 13.0 A. Calculate the rating (in kVA) and the capacitance per phase of the capacitor bank for (a) star connection, (b) delta connection. Also, find the new overall power factor. [(a) 7.621 kVAR, 151.7 μF; (b) 50.56 μF, 0.923 lagging]

Objective Type Questions

1. Modern-day power systems are three-phase systems because compared to single-phase systems they
   (i) are robust
   (ii) are efficient
   (iii) possess operational advantages
   (iv) are all of these

2. Which of the following do not characterize a three-phase system?
   (i) three single-phase sinusoidal voltage sources
   (ii) unequal amplitude
   (iii) out of phase by 120°
   (iv) same frequency

3. Three-phase systems are used for power
   (i) distribution   (ii) transmission
   (iii) generation   (iv) all of these

4. Which of the following is not a disadvantage of a single-phase power system?
   (i) resistive load
   (ii) not suitable to produce a starting torque in ac motors
   (iii) pulsating power
   (iv) presence of an alternating component of double the frequency

5. In a three-phase system, the phase sequence is used to indicate the
   (i) amplitude of the phase voltages
   (ii) frequency of the phase voltages
   (iii) order in which the phase voltages attain their maximum values
   (iv) phase angle between the phase voltages

6. When the phase sequence of a three-phase sinusoidal system is stated as a-b-c, it implies that
   (i) phase voltage a leads the phase voltage b by 120°
   (ii) phase voltage a lags the phase voltage c by 120°
   (iii) phase voltage b leads the phase voltage c by 120°
   (iv) all of these

7. Which of the following represents a negative phase sequence?
   (i) a-b-c (ii) a-c-b
   (iii) b-a-c (iv) None of these

8. The delta connection is formed when
   (i) the starting point of one coil is connected to the finish of the other
   (ii) all the starting points of the coils are connected together

(iii) the starting and the finish points of each coil are left unconnected

(iv) none of these

9. In a delta-connected generator, the sum of the instantaneous voltages around the delta is equal to
   (i) the maximum phase voltage
   (ii) twice the maximum phase voltage
   (iii) $\sqrt{2}$ times the rms value of the phase voltage
   (iv) zero

10. At what level of voltage is power supplied to domestic consumers?
    (i) 415-V three-phase, 240-V single-phase
    (ii) 1.1-kV three-phase, 635-V single-phase
    (iii) 2.2-kV three-phase, 1270-V single-phase
    (iv) 3.3-kV three-phase, 1905-V single-phase

11. In terms of phasor representation, V_{CA} implies
    (i) $V_C + V_A$ (ii) $V_C - V_A$
    (iii) $V_A - V_C$ (iv) none of these

12. Which of the following are applicable to a three-phase star-connected system supplying a star-connected balanced load?
    (i) $I_L = I_P$ (ii) $V_L = 1.732 \, V_P$
    (iii) $V_{AB} + V_{BC} + V_{CA} = 0$ (iv) all of these

13. A star-connected balanced load draws a line current of 10 A from a balanced three-phase four-wire star-connected system. The current flowing in the neutral wire is equal to
    (i) 30 A   (ii) $\sqrt{3} \times 10$ A
    (iii) $\sqrt{2} \times 10$ A   (iv) 0 A

14. Which of the following cannot be applied to a three-phase delta-connected system?
    (i) $V_A + V_B + V_C = 0$ (ii) $V_L = V_P$
    (iii) $I_L = I_P$ (iv) $I_L = \sqrt{3} \times I_P$

15. In order to fully specify a three-phase system, which of the following is essential?
    (i) effective value of the magnitude of the line voltage
    (ii) frequency
    (iii) number of phases
    (iv) all of these

16. A 415 V, three-phase, star-connected system is connected to a delta-connected balanced load of 83 Ω each. How much is the line current?
    (i) 0 A   (ii) 5 A
    (iii) 8.66 A   (iv) 10 A

17. Which of the following formulae is used to express true power in a balanced three-phase circuit?
    (i) $V_P I_P \cos\varphi$ (ii) $V_L I_L \cos\varphi$
    (iii) $3 V_P I_P \cos\varphi$ (iv) $\sqrt{3} \, V_P I_P \cos\varphi$

18. Which of the following represents complex power in a three-phase circuit?
    (i) $3 V_P I_P^*$ (ii) $P + jQ$
    (iii) $3 V_P I_P e^{j\varphi}$ (iv) All of these

19. Reactive power in balanced three-phase system is denoted by
    (i) $3 V_L I_L$ (ii) $\sqrt{3} \, V_L I_L$
    (iii) $\sqrt{3} \, V_L I_L \sin\varphi$ (iv) all of these

20. When an unbalanced load is supplied by a three-phase four-wire star-connected voltage system, with the neutral points of the voltage system connected to the neutral point of the load, the current I_N through the neutral wire is expressed by
    (i) $I_N = 0$ (ii) $I_N = I_A + I_B + I_C$
    (iii) $I_N = 3I_P$ (iv) none of these

21. Which of the following conditions are associated with an unbalanced system?
    (i) unequal magnitude of the phase quantities
    (ii) unequal angular displacement between the phase quantities
    (iii) unequal magnitude and angular displacement between the phase quantities
    (iv) all of these

22. One-wattmeter method can be used to measure power in a three-phase circuit when the load is
    (i) balanced
    (ii) balanced and star-connected
    (iii) balanced and star-connected, with the neutral accessible
    (iv) balanced and delta-connected

23. To which of the following circuits, the two watt-meter method is not suitable to measure power in a three-phase circuit?
    (i) unbalanced star-connected loads with the neutral inaccessible
    (ii) balanced star- or delta-connected loads
    (iii) unbalanced delta-connected loads
    (iv) none of these

24. The power in a three-phase circuit is measured by the two-wattmeter method. The readings indicated on the two wattmeters are W_1 and W_2. The total power in the circuit is not given by
    (i) $W_1 + W_2$ (ii) $W_1 - W_2$

(iii) $\sqrt{3}\ V_L I_L \cos\varphi$ (iv) $3V_P I_P \cos\varphi$

25. If W_1 and W_2 are the readings on the two wattmeters in a three-phase circuit, which of the following is the correct expression for the power factor angle of the circuit?

(i) $\varphi = \tan^{-1}\sqrt{3}\ \dfrac{P_1 + P_2}{P_1 - P_2}$

(ii) $\varphi = \tan^{-1}\dfrac{P_1 + P_2}{P_1 - P_2}$

(iii) $\varphi = \tan^{-1}\sqrt{3}\ \dfrac{P_1 - P_2}{P_1 + P_2}$

(iv) $\varphi = \tan^{-1}\dfrac{P_1 - P_2}{P_1 + P_2}$

Answers

1. (iv)	2. (ii)	3. (iv)	4. (i)	5. (iii)	6. (iv)	7. (ii)
8. (i)	9. (iv)	10. (i)	11. (ii)	12. (iv)	13. (iv)	14. (iii)
15. (iv)	16. (iii)	17. (iii)	18. (iv)	19. (iii)	20. (ii)	21. (iv)
22. (iii)	23. (iv)	24. (ii)	25. (iii)			

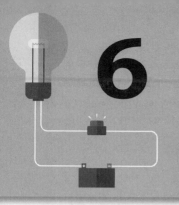

6 TRANSFORMER PRINCIPLES

―――― **Learning Objectives** ――――

This chapter will enable the reader to
- Realize transformer as a high-efficiency static device, in changing the voltage levels of power for transmission, distribution, and utilization
- Understand the principles of working of a transformer based on an understanding of the laws of electromagnetic induction
- Obtain an idea about losses in the core of a transformer and visualize the physical construction of a transformer and the functions of various components
- Develop the equivalent circuit of a practical transformer and draw phasor diagrams of a transformer operating on no-load as well as on load
- Understand the importance of transformer regulation and compute regulation for varying load conditions and power factors
- Obtain various three-phase transformer connections from single-phase transformers
- Differentiate between an auto-transformer and a single-phase transformer

6.1 INTRODUCTION

Electric energy is generated at generating stations where resources for electric power generation are available. This energy is transmitted over long distances to distant places where it is utilized in industries, houses, offices, etc. Normally, power generation is at a voltage level that ranges between 6.6 kV and 33 kV. It is uneconomical to transmit large amounts of power over long distances, at this voltage level due to large voltage drops and excessive power losses, because of the flow of very high currents. Reduction in voltage drops and power losses can be achieved by increasing the transmission voltage level. The ac transmission voltage used in India is usually 110 kV, 132 kV, 220 kV, or 400 kV. However, the utilization voltage for the domestic, commercial, and industrial users is 230/415 V. Large industrial motors operate at 3.3 kV, 6.6 kV, or 11 kV, three-phase or single-phase as the case may be. Thus, at generating stations, voltage has to be stepped up to the transmission voltage level. At the consumer end, the transmission voltage has to be stepped down to the value suitable for utilization. A transformer is an electromagnetic device used for transfer of energy between two or more electrical systems with a magnetic field as the coupling device. It can be used to step up or to step down voltage levels. Thus, a transformer is an important and essential component of an electric energy system as it makes it possible to generate electric energy at the most economic voltage level, transmit electric energy at the most economic transmission voltage, and distribute electric power at the voltage suitable for electric drives, lighting loads, heaters, etc. at the consumer end.

Transformers find applications in fields other than electrical power systems. They are used in low-power applications, as instrument transformers in protection systems, as impedance-matching devices in electronic and control circuits, and as devices to isolate a portion of a network from another, or isolate dc while maintaining ac continuity between two circuits. In communication systems, transformers can be used as filament transformers to supply power to heat filaments of devices, such as vacuum tubes, cathode ray tubes in television receivers, computer monitors, and other display devices. Also, input transformers, inter-stage transformers, and output transformers are used in audio and video circuits. Based on the application, transformers are known by different names, such as power transformers, distribution transformers, impedance-matching transformers, isolating transformers, instrument transformers, pulse transformers, filament transformers, input/inter-stage/output transformers, IF transformers, etc. In this chapter, however, discussions are mainly focussed on applications of transformers in electric power systems.

The common form of transformer consists of two coils mutually coupled through a ferromagnetic core. The ferromagnetic core ensures a high value of magnetic flux linkage between the coils. Such a transformer is called an iron-core transformer. One of the coils, called the primary winding, is connected to the power supply to receive power. The other coil, called the secondary winding, is connected to the load circuit and delivers power to it.

As a magnetic circuit is an integral part of a transformer as well as other electrical machines, the response of a magnetic circuit to ac excitation will be discussed first. Some of the properties of the ferromagnetic core affect the construction of a transformer; these factors are partly responsible for the power losses in a transformer and require a brief explanation before considering the construction details and the principle of operation of a transformer.

6.2 RESPONSE OF MAGNETIC CIRCUITS TO AC VOLTAGES

An ideal zero-resistance coil of N turns wound around one of the limbs of a ferromagnetic core is shown in Fig. 6.1. When a sinusoidal ac voltage v of frequency f is connected across the coil, it causes a small value of steady-state sinusoidal alternating current i_ϕ, called the exciting or magnetizing current, to flow through the coil. Assuming that even the maximum flux produced by this current is not saturating the core, the flow of sinusoidal current through the coil produces an mmf that sets up a sinusoidally varying flux in time phase with the current.

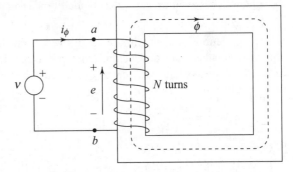

Fig. 6.1 Schematic arrangement of one coil on an iron core

Applying Kirchhoff's voltage law to the circuit shown in Fig. 6.1, the following equation results:

$$v - e = 0 \tag{6.1}$$

where e is the voltage drop caused by the flow of current i_ϕ through the coil. The voltage drop e can be expressed in terms of i_ϕ as

$$e_{ab} = e = L \frac{di_\phi}{dt} \tag{6.2}$$

where L is the inductance of the coil. Substituting Eq. (6.2) in to Eq. (6.1) and rearranging gives

$$v = e = L \frac{di_\phi}{dt} = e_{ab} \tag{6.3}$$

Since the applied voltage is sinusoidal, assuming linear magnetization of the ferromagnetic core, the magnetizing current i_ϕ is also sinusoidal and is represented by

$$i_\phi = \sqrt{2}\, I_\phi \sin \omega t \tag{6.4}$$

where I_ϕ is the rms value of the magnetizing current and ω is the angular frequency given as $2\pi f$ rad/sec (electrical).

Substituting Eq. (6.4) in Eq. (6.3) and simplifying gives

$$v = e = \sqrt{2}\, \omega L I_\phi \cos \omega t$$

$$= \sqrt{2}\, \omega L I_\phi \sin(\omega t + \pi/2) \tag{6.5a}$$

$$= \sqrt{2}\, E \sin(\omega t + \pi/2) = E_m \sin(\omega t + \pi/2) \tag{6.5b}$$

where

$$E = \omega L I_\phi = \text{the rms value of inductive reactance drop} \tag{6.5c}$$

$$E_m = \sqrt{2}\, E = \text{the maximum value of the induced emf} \tag{6.5d}$$

From Eqs (6.4) and (6.5b), it may be noted that the magnetizing current lags the applied voltage by 90° electrical. The phasor diagram for this condition is shown in Fig. 6.2(a). In the phasor diagram, the rms magnetizing current phasor I_ϕ lags the supply voltage phasor V by 90° electrical and the voltage drop phasor $E = V$. This phasor diagram for an ideal iron-core reactor is arrived at from the point of view of circuit theory.

An alternate approach for arriving at the phasor diagram for an ideal iron- core reactor using Faraday's law of electromagnetic induction is now considered. In the circuit in Fig. 6.1, sinusoidal magnetizing current i_ϕ, expressed by Eq. (6.4), produces flux ϕ in the core, which is assumed to vary sinusoidally and is represented by

$$\phi = \phi_m \sin \omega t \tag{6.6}$$

where ϕ_m denotes the maximum amplitude of the core flux. Applying Faraday's law to the circuit shown in Fig. 6.1, the emf e induced in N turns of the coil may be expressed either as a voltage drop e_{ab} given by

$$e_{ab} = e = +N\,\frac{d\phi}{dt} \tag{6.7}$$

or as a voltage rise e_{ba} given by

$$e_{ba} = -e = -N\,\frac{d\phi}{dt} \tag{6.8}$$

In Eqs (6.7) and (6.8), the sign is assigned by using Lenz's law. The negative sign states that the emf induced due to the changing flux is always in the direction in which current would have to flow to oppose the changing flux. The positive sign in Eq. (6.7) denotes that the polarity of a is positive whenever the time rate of change of flux is positive as determined by the direction of current flow.

Substitution of Eq. (6.6) into Eq. (6.8) yields

$$e_{ba} = -e = -N\,\frac{d\phi}{dt} = -N\,\frac{d}{dt}(\phi_m \sin \omega t)$$

$$= -N\omega\phi_m \cos \omega t = E_m \sin(\omega t - \pi/2) \tag{6.9}$$

where

$$E_m = N\omega\phi_m = 2\pi f N \phi_m \text{ volts} \tag{6.10}$$

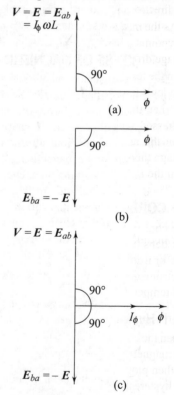

Fig. 6.2 Phasor diagrams of iron-core reactor: (a) circuit viewpoint, (b) field viewpoint, (c) combined approach

The rms value of the induced voltage, E, is obtained from Eq. (6.10) as

$$E = \frac{E_m}{\sqrt{2}} = \frac{2\pi fN\phi_m}{\sqrt{2}}$$

$$= 4.44fN\phi_m \text{ volts} \tag{6.11}$$

Equation (6.11) is the emf equation of the coil as this gives the emf induced in the coil when a flux with a maximum value ϕ_m varies sinusoidally with frequency f.

If A m^2 is the area of cross section of the core, l metres is the mean length of the flux path, and B_m Wb/m^2 is the maximum flux density of the core, Eq. (6.10) modifies to

$$E = 4.44fNB_m \times A \tag{6.12}$$

and the rms value of the emf per turn E_t may be obtained as

$$E_t = 4.44fB_m \times A = 4.44f\phi_m \tag{6.13}$$

Since the current produces the flux, the current and flux may be considered to be in phase.

On the basis of field viewpoint, comparing Eq. (6.6) with Eq. (6.9), it is observed that the induced emf $-e$ lags the changing flux ϕ by 90° electrical, and the corresponding phasor diagram is shown in Fig. 6.2(b). Thus the emf $-e$ is a voltage rise which has a direction that, if it were free to act, would cause a current to flow opposing the direction of I_ϕ. This leads to the polarity mark of inducted emf e shown in Fig. 6.1.

As the magnetizing current phasor I_ϕ and the flux ϕ are in time phase, it is quite normal to combine the two viewpoints into a single phasor diagram as shown in Fig. 6.2(c). In the phasor diagram $E_{ab} = E$ represents the voltage drop, with $E_{ba} = -E$ being the voltage rise from the field point of view.

In the text that follows in this chapter, the induced emf is treated as voltage drop. In other words, in applying Faraday's law, the principle described by Eq. (6.7) will be used.

When the resistance of the coil is considered, the resistance drop does not exceed a few per cent of the impressed voltage, and as a close approximation resistance drop may be neglected at least to begin with. Then the applied voltage and the induced voltage may be considered equal and opposite. The applied voltage then determines the flux ϕ_m in accordance to Eq. (6.11), and the excitation current can be determined from the B-H curve of the material of the core.

6.3 CORE LOSSES

The core in Fig. 6.1 is subjected to cyclic flux density variations due to the sinusoidal current. The cyclic flux density variations give rise to losses, which appear in the form of heat. These losses are referred to as core or iron losses and are constituted of two parts, namely, hysteresis losses and eddy-current losses. In all electromagnetic equipment, core losses play a significant role in determining equipment rating, efficiency, and temperature rise.

6.3.1 Hysteresis Losses

When the core is energized from an ac source, the magnetizing current varies sinusoidally with time and hence the magnetizing force H applied to the magnetic circuit is also alternating. The variation of the flux density B when plotted against the magnetizing force H for one cycle of ac excitation is in the form of a loop, called the hysteresis loop, as discussed in Section 3.11.1. In each cycle of magnetization the hysteresis loss appears as heat, which can be determined from Eq. (3.45) of Chapter 3. The larger the loop area, the greater is the hysteresis loss. In Section 3.11.2, it has been proved that the area of the loop represents the energy loss due to hysteresis per unit volume of the magnetic core. This is in fact the extra work done to reverse the magnetization direction. The power loss due to hysteresis may be obtained using the empirical relation given by Eq. (3.48) and is given below for ready reference.

$$P_h = K_h \mathcal{V} f B_m^n \text{ watts}$$

where K_h is a constant whose value depends on the material of the core and range of flux density, f is the frequency of alternating magnetization in hertz, $\mathcal{V}$ is the volume of the magnetic core in m^3, B_m is the maximum value of flux density in Wb/m^2, and n is the Steinmetz constant. The value of n varies between 1.5 and 2.0. The value of n is usually taken as 1.6.

Hysteresis losses can be reduced by using core materials that have a narrow hysteresis loop, for example, nickel-iron alloys, such as mumetal and permalloy, or a special steel known as silicon steel.

6.3.2 Eddy-current Losses

The time-varying flux in the ferromagnetic core induces emfs in it and causes circulating currents in closed paths within the body of the ferromagnetic material. These currents are called eddy currents and give rise to i^2r losses. These losses are called eddy-current losses. The losses in the core would depend on the effective resistance and the length of the eddy-current paths. If the magnetic circuit is made of solid iron, then the circulating current is higher as the resistance of the core is relatively small, and hence the power loss is appreciable. An increase in the effective resistance of the eddy-current paths can be achieved by using silicon steel (4% silicon) as core material. Laminated sheets parallel to the direction of the flux, having small thickness, are used for the construction of the core. The laminations significantly increase the resistance of the core. An empirical equation for determining eddy-current loss P_e is

$$P_e = k_e f^2 B_m^2 \tau^2 \mathcal{V} \quad \text{watt} \tag{6.14}$$

where k_e is the characteristic constant of the core material, τ is the thickness of lamination in metre, f is the frequency of alternating magnetization in hertz, B_m is the maximum value of flux ensity in Wb/m^2, and $\mathcal{V}$ is the volume of the magnetic core in m^3.

Example 6.1   A single-phase transformer has 500 primary and 1000 secondary turns. The net cross-sectional area of the core is 50 cm^2. If the primary winding is connected to a 50-Hz supply at 400 V, calculate (a) the peak value of the flux density in the core and (b) the voltage induced in the secondary winding.

Solution   (a) Let B_m Wb/m^2 be the peak value of the flux density in the core. Then the peak flux ϕ_m is given by

$$\phi_m = B_m \times A = B_m \times 50 \times 10^{-4} \text{ Wb}$$

From Eq. (6.11), the equation for induced voltage gives

$$400 = 4.44 f N_1 \phi_m = 4.44 \times 50 \times 500 \times B_m \times 50 \times 10^{-4}$$

$$\therefore \qquad B_m = \frac{400}{4.44 \times 50 \times 500 \times 50 \times 10^{-4}} = 0.7207 \text{ Wb/m}^2$$

(b) Voltage induced in the secondary is obtained using Eq. (6.12) as under:

$$E_2 = 4.44 \times 50 \times 1000 \times 0.7207 \times 50 \times 10^{-4}$$
$$= 799.977 \approx 800 \text{ V}$$

Example 6.2   A single-phase transformer has 600 primary and 80 secondary turns. The mean length of the flux path in the ferromagnetic core is 1.6 m, the value of flux in the core for a magnetic field strength of 1.2 T is 425 AT/m, and the corresponding core loss is 1.5 W/kg at 50 Hz. The density of the core is 7400 kg/m^3. If the maximum value of flux density is 1.2 T when the primary is connected to a 3300-V, 50-Hz supply, calculate (a) the cross-sectional area of the core, (b) the secondary voltage on no load, (c) the primary magnetizing current, and (d) the core loss.

Solution   (a) Using Eq. (6.12),

$$3300 = 4.44 \times 600 \times 5 \times \phi_m$$

$$\therefore \qquad \phi_m = 0.0248 \text{ Wb}$$

The cross-sectional area of the core $\dfrac{\phi_m}{B_m} = \dfrac{0.0248}{1.2} = 0.02067 \text{ m}^2$

(b) The secondary voltage on no load $= 3300 \times \dfrac{80}{600} = 440 \text{ V}$

(c) The primary magnetizing current $= \dfrac{\text{mmf}}{\text{primary turns}} = \dfrac{H \times l}{600} = \dfrac{425 \times 1.6}{600} = 1.133 \text{ A}$

Assuming sinusoidal current, the rms value of the magnetizing current $= 1.133/\sqrt{2} = 0.80 \text{ A}$

(d) Volume of the core $= 1.6 \times 0.02067 = 0.03307 \text{ m}^3$

Mass of the core $= 0.3307 \times 7400 = 244.73 \text{ kg}$

Therefore, the core loss $= 244.73 \times 1.5 = 367 \text{ W}$

6.4 CONSTRUCTION OF TRANSFORMERS

The construction of transformers is simple as there are no rotating parts. Figure 6.3 shows a cross-sectional view of magnetic circuits of core and shell type transformers. The main parts of a transformer are: a magnetic core, windings and insulations, and a transformer tank.

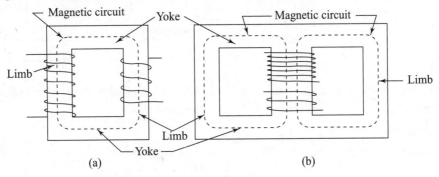

Fig. 6.3　(a) Core type, (b) Shell type

6.4.1 Magnetic Core

For single-phase transformers the two designated types are core type and shell type. The core type has a core in which there is a single magnetic circuit as shown in Fig. 6.3(a). The two vertical members, called the limbs, each carry one-half of the primary and one-half of the secondary windings. The two horizontal members are the yokes. The shell type can be regarded as two core-type cores placed side by side giving a central limb of twice the cross section of each outer limb, as shown in Fig. 6.3(b). In the shell type, the windings are wound around the centre leg of the three-legged core.

The various forms of sheet steel punching employed for assembling transformer cores are shown in Fig. 6.4.

L-shaped punching are employed to assemble cores whose overall dimensions are not very large and E or F-shaped punching are used to build small shell type transformers.

In a transformer, steel sheet laminations are used as core materials. The magnetic sheet materials are normally alloy steels containing silicon and carbon in small quantities, which increase the permeability at low flux densities, reduce hysteresis and also reduce eddy-current losses by increasing resistivity.

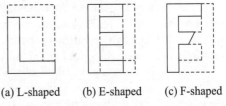

(a) L-shaped　　(b) E-shaped　　(c) F-shaped

Fig. 6.4　Different forms of sheet steel punching

The silicon content increases tensile strength and impairs ductility. The present trend is to use cold-rolled grain-oriented steel (CRGOS) which shows excellent magnetic properties in the direction of rolling, and a flux density of 1.6–1.85 T at 50 Hz is easily obtained. A considerable reduction of core dimensions and weight is obtained by the use of CRGOS.

In order to keep the reluctance of the magnetic circuit to a minimum, the strips of alternate layers are staggered so as to give an interleaved joint as shown in Fig. 6.5. After the core has been completed, the top yoke is removed, the windings are placed in position, and the yoke is replaced. The cross-section of small transformers is rectangular. But, in large transformers, an approximately circular cross section is adopted since this has the smallest perimeter for a given area, thus requiring less copper than the rectangular section. In very large transformers the strips are arranged in packets with ducts between them. This helps in providing ventilation and cooling of transformers. Some typical cross sections are shown in Fig. 6.6.

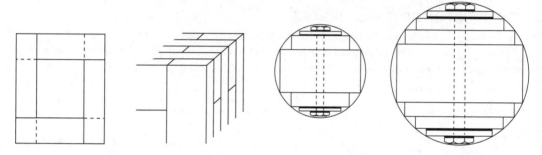

Fig. 6.5 Interleaved joints   **Fig. 6.6** Typical cross sections for large transformers

6.4.2 Windings and Insulation

Transformer windings may be either concentric or sandwiched. Concentric windings are used in core type transformers, as shown in Fig. 6.7(a). Sandwiched windings are almost exclusively used in shell type transformers, as shown in Fig. 6.7(b). From the point of view of insulation requirement, the positioning of the low-voltage (LV) and high-voltage (HV) windings with respect to the core is quite important. The low-voltage (LV) winding is placed nearer to the core in the case of concentric windings and in outside positions in the case of sandwiched windings.

The conductor used for winding is a copper wire of circular section for small currents and straps of rectangular section for large currents. It is undesirable that the current in any one conductor becomes 70 to 100 A. Consequently, windings for large currents consist of several conductors wound in parallel.

The insulation of individual conductors, called the minor insulation, can be paper, cotton, or glass tape. The insulation is wrapped around the conductor without any overlap of adjacent turns. The windings of the transformers are generally impregnated with varnish in vacuum-drying and pressure-process cycle. Now-a-days oil impregnation is used in place of varnish.

In the core type transformer the LV and HV windings are in the form of concentric cylinders, the LV cylinder being nearer to the core. Each cylinder is wound on an insulating cylinder made of either bakelized paper or pressboard. These two cylinders insulate not only the primary and secondary windings from one another but also from the core as shown in Fig. 6.7(d). The insulation at the top and bottom ends of the coil from the core is carried out by the end pieces or blocks, which may be bakelized paper or impregnated wood. The bushings insulate the leads going to the transformer terminals from the tank. The two insulating cylinders along with the end pieces and bushings are known as the major insulation. For low voltages, each coil can be in the form of a continuous cylinder while HV winding is subdivided into a number of coils of short axial lengths. These coil sections are separated from one another by radial tabs of insulating material [see Fig. 6.7(c)], each tab being fitted to a longitudinal wedge-shaped strip placed parallel along the length of the insulating cylinder as shown in Fig. 6.7(d).

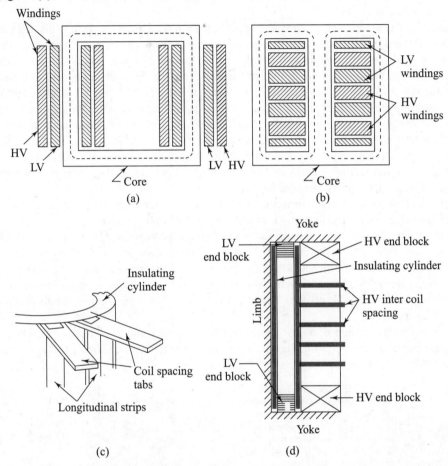

Fig. 6.7 (a) Concentric windings, (b) sandwich winding, (c) coil spacing tabs, and (d) Insulation details of a small/medium-size distribution transformer

There is sufficient space between the core and the LV cylinder which provides efficient ducts for cooling. Similarly, the inside diameter of the HV cylinder is greater than the outside diameter of the LV coil so that there is also a duct between LV and HV. These are longitudinal ducts. When a winding is split up into section coils, there are radial ducts, in addition. As such the cooling facilities provided by such a construction are very good.

The shell type transformer, unlike the core type, has thin coils. The conductor is a rectangular strap wound on the flat, and the conductor insulation is a strip of pressboard wound with the strap. As the core section is rectangular, each coil is rectangular in form and is commonly called a pancake coil because of its thinness. The HV coils are interleaved with the LV and the coil groups are so arranged that there is a half group of LV coils at each end so as to facilitate insulation from the core. The major insulation is in the form of sheets and spacing strips.

6.4.3 Transformer Tank

The function of a transformer tank is to provide a protective cover to the core, windings, and other internal parts including transformer oil. The tank provides a rigid support to the fittings and accessories. Transformer tanks are made of thin sheet steel.

The oil provides a medium for insulation and heat dissipation. The oil in the ducts and at the surface of the coils and cores absorbs by conduction the heat generated due to the core loss and I^2R losses of the windings and rises up. Cool oil from the bottom of the tank rises up to take its place. A continuous circulation of oil takes place with the heated oil flowing to the tank sides, where the heat gets dissipated to the surrounding atmosphere, thereby cooling the oil to an ambient temperature that makes the oil fall to the bottom of the tank. Thus, the transformer tank provides an external surface for dissipating heat to the surrounding atmosphere. To increase the heat-dissipating area, fins or corrugations are welded between the cover flange and the tank base. This type of a tank is used for distribution transformers of up to 50 kVA. Transformers of all sizes above 50 kVA use tanks with external cooling tubes welded vertically to the tank sides so as to provide an additional cooling surface and a path for the circulation of oil. In power transformers of large ratings, tanks with separate cooling systems (heat exchangers) are provided.

The transformer tank is filled with oil when cold. Heat generated inside the transformer expands the oil. A conservator is an airtight drum mounted on the top of the tank or on a neighbouring wall and is connected to the tank. The conservator maintains a fixed head of oil in the tank for all operating conditions of the transformer. When oil gets heated it expands and enters the conservator from the tank and forces some air out of the conservator. On the other hand, when the oil contracts, some oil enters the tank from the conservator and air enters the conservator from the atmosphere. To prevent the transformer oil from absorbing moisture, the conservator is connected to the atmosphere through a narrow passage called the breather. Inside the passage dry silica gel is placed for drying the air that the transformer breathes in.

An external photographic view of a three-phase distribution transformer is shown in Fig. 6.8(a) where different fittings/accessories are indicated. Figure 6.8(b) is a photograph which shows the active parts such as HV bushing, LV bushing, core, windings, etc. of the transformer.

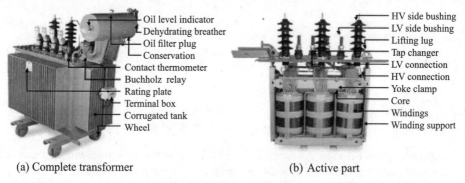

(a) Complete transformer   (b) Active part

Fig. 6.8   Distribution transformer

6.5 WORKING PRINCIPLE OF A TRANSFORMER

A core type transformer consisting of a primary winding with N_1 turns and a secondary winding with N_2 turns wound around two limbs of a ferromagnetic core is shown in Fig. 6.9. The two windings are insulated from each other as well as from the core. The primary winding P is connected across an ac voltage of frequency f hertz, which causes an alternating current to circulate through the winding. The ac current I_1 flowing through the winding produces an mmf and sets up an alternating flux ϕ in the ferromagnetic core. This, in turn, induces in the coil an emf whose frequency is the same as excitation frequency f and the magnitude is proportional to the number of turns N_1 of the winding. The emf induced is given by Eq. (6.11). As secondary winding is wound on the same core, the mutual alternating flux ϕ_m linking the winding S would also induce anemfin h coil, whose frequency is f and magnitude is proportional to the number of turns N_2 of the winding. Thus,

$$\frac{\text{emf induced in } S}{\text{emf induced in } P} = \frac{N_2 \times \text{emf per turn}}{N_1 \times \text{emf per turn}} = \frac{N_2}{N_1}$$

When the secondary winding terminals are open, the terminal voltage V_2 is the same as the emf induced in it. The current drawn by the primary winding from the suppl is very small and the induced emf in the primary winding can be taken as almost equal to the applied voltage V_1. Then

$$\frac{V_2}{V_1} \approx \frac{N_2}{N_1} \tag{6.15}$$

If a load is connected to the secondary winding, power flows from the primary to the secondary winding load circuit. The transformer in this process will incur some losses. Usually transformers have an efficiency exceeding 95%. If the losses are neglected, then the total power taken from the source by the transformer is transferred to the load side. Therefore,

$$V_1 I_1 \times \text{primary power factor} = V_2\, I_2 \times \text{secondary power factor}$$

At full-load operating condition, the primary power factor is approximately equal to the secondary power factor. Then,

$$V_1 I_1 = V_2 I_2 \tag{6.16}$$

or

$$\frac{I_2}{I_2} = \frac{V_2}{V_2} \tag{6.17}$$

The leakage fluxes ϕ_{L1} and ϕ_{L2} of the primary and secondary windings, respectively, are also shown in Fig. 6.9. The leakage flux of one winding is shown to complete the path through the air space around the winding, and therefore does not link the other winding.

The relationship between the primary and secondary currents obtained in Eq. (6.16) can alternatively be obtained from the comparison of the primary and secondary ampere-turns as follows. When the secondary is on open circuit, the primary winding will draw a very small current I_1 and the mmf in the primary

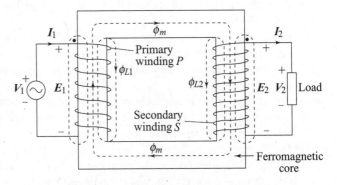

Fig. 6.9 A transformer connected to a load

winding $\mathcal{F}_1 = N_1 I_1$ AT is very small and just sufficient to produce the flux in the ferromagnetic core. The alternating flux induces emf E_1 in the primary winding that is practically equal and opposite to the applied voltage V_1. This primary current drawn is called the magnetizing current or no-load current of the transformer and is usually about 3% to 5% of the full-load primary current.

When a load is connected across the secondary terminals, the secondary current I_2 flows from the instantaneous positive polarity to the load, as shown in Fig. 6.9, and this will cause mmf $\mathcal{F}_2 = N_2 I_2$ AT in the magnetic circuit. According to Lenz's law, the current I_2 in the secondary winding will produce flux in a direction opposite to the mutual flux ϕ_m. Thus, the secondary ampere-turns have a demagnetizing effect on the main flux.

Consequently, the flux and the emf induced in the primary E_1 both are slightly reduced. The difference between the fixed applied voltage V_1 and the reduced value of emf E_1 causes an increase in the primary current and a counter mmf is generated to oppose the mmf produced by the secondary coil. The demagnetizing ampere-turns of the secondary are thus nearly neutralized by the increase in the primary ampere-turns. Since the primary ampere-turns on no load are very small compared with the full-load ampere-turns, the full-load primary ampere-turns are approximately equal to full-load secondary ampere-turns. Thus,

$$I_1 N_1 \cong I_2 N_2 \tag{6.18}$$

and hence

$$\frac{I_1}{I_2} \cong \frac{N_2}{N_1} \cong \frac{V_2}{V_1} \tag{6.19}$$

It will be seen that the connecting link between the primary and secondary circuits is the mutual magnetic flux. Any variation in the secondary current is accompanied by a small variation in the flux and also the emf induced in the primary, thereby enabling the primary current to vary approximately proportionally to the secondary current. This balance of primary and secondary ampere-turns is an important relationship for transfer of power from the primary side to the secondary side of a transformer.

In Fig. 6.9, the dots marked at the top of the primary and secondary windings indicate the instantaneous positive polarity of the ac voltage induced in both the primary and secondary windings. Further, the currents flowing into the dot in one winding and flowing out of the dot in the other winding will tend to produce magnetic fluxes in the core that are in opposite directions, and the corresponding current directions are taken as instantaneously positive.

6.6 IDEAL TRANSFORMER

The theory of operation of a transformer and its applications are best understood by viewing the transformer as an ideal device. The following simplifying assumptions are made to enable a transformer to be treated as an ideal transformer.

(i) All magnetic flux created by the primary winding links with the secondary winding. There is no leakage flux, that is, the coefficient of coupling is unity.

(ii) The primary and secondary windings have zero resistance, hence there is no resistive voltage drop and no resistive heat loss.

(iii) The leakage of the windings being zero, there is no reactive voltage drop in the windings.

(iv) The power transfer efficiency is 100%, that is, there are no hysteresis losses, eddy-currents losses, and heat losses due to resistance.

(v) The permeability of the core is infinite so that it requires zero mmf to create flux in the core.

(vi) The transformation ratio or the turns ratio is equal to the ratio of its primary to secondary voltages and also the ratio of its secondary to primary current.

6.6.1 Ideal Transformer on No Load

An ideal transformer with a ferromagnetic core is shown in Fig. 6.10. It is assumed that the secondary winding of the transformer is open-circuited, that is, the transformer is on no load and the secondary rms current $I_2 = 0$. The primary winding is connected to an ac source $v_1 = V_m \sin \omega t$. The primary winding will draw negligible steady-state sinusoidal magnetizing current i_1, which flows through N_1 turns of primary winding, that in turn establishes a sinusoidally varying flux $\phi = \phi_m \sin \omega t$ in the phase with the current. An emf e_1 is induced in the primary winding according to Eq. (6.7) and is given by

$$e_1 = -N_1 \frac{d\phi}{dt} = -\omega N_1 \phi_m \cos \omega t = \omega N_1 \phi_m \sin(\omega t - \pi/2) \tag{6.20}$$

In view of the assumptions (ii) and (iii), applying KVL to the primary circuit gives

$$v_1 = e_1 = N_1 \frac{d\phi}{dt} \qquad (6.21)$$

As per assumption (i), the whole of the flux ϕ links the secondary winding as well. Hence, the emf induced in the secondary winding is given by

$$e_2 = -N_2 \frac{d\phi}{dt} = \omega N_2 \phi_m \cos \omega t$$
$$= \omega N_1 \phi_m \sin(\omega t - \pi/2) \qquad (6.22)$$

It may be seen from Eqs (6.21) and (6.22) that e_1 and e_2 are in phase. As the secondary is open-circuited, the following equation holds good:

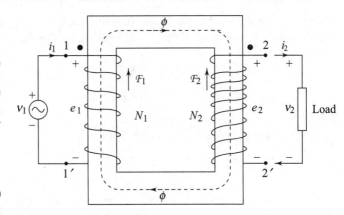

Fig. 6.10   Ideal transformer

$$v_2 = e_2 = N_2 \frac{d\phi}{dt} \qquad (6.23)$$

Dividing Eq. (6.21) by Eq. (6.23), the ratio $v_1 : v_2$ is obtained as

$$\frac{v_1}{v_2} = \frac{e_1}{e_2} = \frac{N_1}{N_2} \qquad (6.24)$$

The relationship between the instantaneous voltages of Eq. (6.24) can be expressed alternatively in terms of the effective or rms values as

$$\frac{V_1}{V_2} = \frac{E_1}{E_2} = \frac{N_1}{N_2} \qquad (6.25)$$

It can be seen from Eq. (6.25) that by appropriately varying the number of turns in the secondary winding, the voltage induced in it can be made greater or smaller than the voltage in the primary winding. When $E_2 > E_1$ (or $V_2 > V_1$), the voltage is stepped up and the transformer is called a step-up transformer; when $E_2 < E_1$ (or $V_2 < V_1$), the voltage is stepped down and the transformer is called a step-down transformer. Therefore, Eq. (6.25) provides a relation for transforming voltages and is called the transformation ratio or turns ratio a. Thus,

$$a = \frac{V_1}{V_2} = \frac{N_1}{N_2} \qquad (6.26)$$

Equations (6.6) and (6.9) provide the relationship between the flux phasor ϕ and the induced voltage phasor E. From Eqs (6.21) and (6.23), it is obvious that the voltage phasors $E_1 = V_1$ and $E_2 = V_2$ are themselves in phase. Hence, the voltages V_1, V_2 and the emfs E_1, E_2 lead the flux phasor ϕ by an angle $\pi/2$. The phasor diagram for the ideal transformer on no load is shown in Fig. 6.11.

If the instantaneous flux is

$$\phi = \phi_m \sin \omega t \qquad (6.27)$$

then from Eq. (6.21) the induced voltage is

$$e_1 = N_1 \frac{d\phi}{dt} = \omega N_1 \phi_m \cos \omega t \qquad (6.28)$$

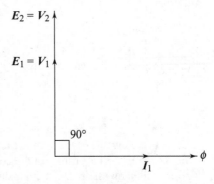

Fig. 6.11   Phasor diagram of an ideal transformer under no-load condition

where ϕ_m is the maximum value of the flux and $\omega = 2\pi f$, the frequency being f hertz. For the positive directions shown in Fig. 6.11, the induced emf leads the flux by 90°. The rms value of the induced emf is given by

$$E_1 = \frac{2\pi}{\sqrt{2}} fN_1\phi_m = \sqrt{2}\,\pi f N_1 \phi_m$$

$$= 4.44 f N_1 \phi_m \qquad (6.29)$$

As the voltage drop in the primary circuit is zero, the counter emf E_1 equals the applied voltage V_1. Thus, the maximum value of the flux may be expressed as

$$\phi_m = \frac{V_1}{\sqrt{2}\pi f N_1} = \frac{V_1}{4.44\,f N_1} \qquad (6.30)$$

From Eq. (6.30) it may be observed that the flux is determined solely by the applied voltage, its frequency, and the number of turns of the winding. For a transformer N_1 is constant, hence the ratio V_1/f decides the maximum core flux. If V_1/f is constant the maximum core flux is also constant.

6.6.2 Ideal Transformer Under Load

Assume that the terminals $2 - 2'$ of the secondary winding of the transformer in Fig. 6.10 are connected across a load and a current i_2 flows from instantaneous positive polarity of the load. This current, which flows through N_2 turns of the winding, sets up an mmf $\mathcal{F}_2 = i_2 N_2$ AT, which by Lenz's law will oppose the core flux ϕ. Since the core flux is determined by the magnetizing current in the primary winding, it cannot change. Therefore, a current i_1 must flow in the primary winding such that the mmf in the secondary winding is balanced by the primary winding. In other words,

$$\mathcal{F}_1 = \mathcal{F}_2$$

or $\qquad i_1 N_1 = i_2 N_2 \qquad (6.31)$

Equation (6.31) is called the *mmf balance* equation and it shows that the mmfs in the primary and secondary windings are always balanced under all load conditions. The voltage ratio in Eq. (6.25) and the mmf balance in Eq. (6.31) show that the volt-ampere rating of a transformer is the same for both the primary and secondary windings. Hence, a transformer may be viewed as a high-efficiency static device that can transform the voltage levels of energy but cannot vary the energy levels.

Through the constancy of the mmf, the primary knows of the presence of the secondary current. The net mmf acting on the core therefore is zero, since for an ideal transformer it is assumed that the excitation current is zero. From Eq. (6.19)

$$\frac{i_1}{i_2} = \frac{N_2}{N_1} = \frac{1}{a} \qquad (6.32)$$

If the rms values of currents are used, Eq. (6.32) modifies to

$$\frac{I_1}{I_2} = \frac{N_2}{N_1} = \frac{1}{a} \qquad (6.33)$$

Equation (6.33) shows that, in an ideal transformer, the transformation of current is the inverse of the ratio of transformation and that the phasors of the currents in the primary and secondary windings are in phase.

In order to show the phase relationship between the primary and secondary voltages and the currents, let it be assumed that the load (secondary) current lags the load (secondary) voltage by a power factor angle φ. Figure 6.12 shows the phasor diagram of an ideal transformer under load condition.

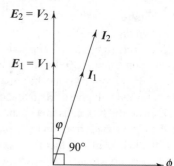

Fig. 6.12   Phasor diagram of an ideal transformer under load condition

Complex power S_1 and S_2 in the primary and secondary windings, respectively, can be written as

$$S_1 = V_1 \times I_1^* = V_2 \times I_2^* = S_2 \qquad (6.34)$$

From Eq. (6.34) it can be concluded that, in an ideal, loss-less transformer both the active and reactive powers in the primary and secondary windings are completely balanced.

6.6.3 Equivalent Circuit of an Ideal Transformer

With the help of the transformation ratio, using Eqs (6.26) and (6.33), secondary voltage and secondary current can be referred to the primary side, that is,

$$V_1 = V_2' = a V_2 \qquad (6.35)$$

$$I_1 = I_2' = \frac{1}{a} I_2 \qquad (6.36)$$

where V_2' and I_2', respectively, denote the secondary voltage and current referrd to the primary. Similarly, the primary voltage and current can be referred to the secondary side as follows:

$$V_2 = V_1' = \frac{1}{a} V_1 \qquad (6.37)$$

$$I_2 = I_1' = a I_1 \qquad (6.38)$$

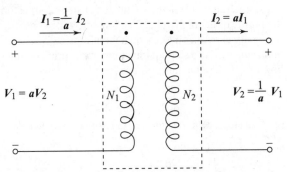

Fig. 6.13 Equivalent circuit of an ideal transformer

The equivalent circuit of an ideal transformer is shown in Fig. 6.13.

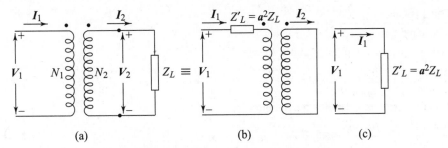

Fig. 6.14 Transformation of load impedance

Figure 6.14(a) shows an ideal transformer, the secondary of which is connected across a load of impedance Z_L and draws a current of I_2 amperes.

$$Z_L = \frac{V_2}{I_2} \qquad (6.39)$$

Substituting the values of V_2 and I_2 from Eqs (6.37) and (6.38), Eq. (6.39) becomes

$$Z_L = \frac{V_1/a}{a I_1} \qquad (6.40)$$

or $\quad \dfrac{V_1}{I_1} = a^2 Z_L = Z_L' \qquad (6.41)$

Z_L' is called the equivalent impedance referred to the primary side or simply the reflected impedance. Figure 6.14(b) shows the load impedance connected after transformation to the primary of an ideal transformer. Figure 6.14(c) is the final reduced network as looked at from the source of supply. It may be noted that so far as the performance of an ideal transformer is concerned the three circuits of Fig. 6.14 are the same. It is also clear that, if needed, the voltages and the currents may also be referred to the secondary side.

Equation (6.41) shows impedance-modifying property of a transformer. This property can be used for matching fixed impedance to the source for maximum power transfer.

Example 6.3   A 10-kVA, 6600/220-V, 50-Hz transformer is rated at 2.5 V/turn of the winding coils. Assume the transformer to be ideal and calculate (a) step-up transformation ratio, (b) step-down transformation ratio, (c) the total turns of the high-voltage and low-voltage coils, (d) the primary current as a step-down transformer, (e) the secondary current as a step-down transformer.

Solution   (a) As a step-up transformer, the primary voltage $V_1 = 220$ V.
From Eq. (6.26), the turns ratio is found as

$$a = \frac{V_1}{V_2} = \frac{220}{6600} = \frac{1}{30}$$

(b) As a step-down transformer, the primary voltage $V_1 = 6600$ V.

Then, $a = \dfrac{V_1}{V_2} = \dfrac{6600}{220} = 30$

(c) The number of turns of the high-voltage coil is

$$\frac{\text{rated voltage}}{\text{voltage per turn}} = \frac{6600}{2.5} = 2640$$

The number of turns of the low-voltage coil is

$$\frac{\text{rated voltage}}{\text{voltage per turn}} = \frac{220}{2.5} = 8$$

(d) The primary current as a step-down transformer is

$$\frac{\text{kVA} \times 10^3}{6600} = \frac{10 \times 10^3}{6600} = 1.515 \text{ A}$$

(e) The secondary current as a step-down transformer

$$\frac{\text{kVA} \times 10^3}{220} = \frac{10 \times 10^3}{220} = 45.45 \text{ A}$$

Example 6.4   A load of 32 Ω is connected to a source having an internal resistance of 2 Ω through an ideal transformer so that maximum power can be obtained from the source. Determine the turns ratio of the transformer for impedance matching.

Solution   Let the turns ratio be a.
The load impedance referred to the primary side $= a^2 R_L = a^2 \times 32$
Under maximum power transfer condition, $R_S = a^2 R_L$
Then, $2 = a^2 \times 32$

$\therefore$ $a = \dfrac{1}{4}$

Thus, the turns ratio for impedance matching is 4, i.e., $N_1/N_2 = 1/4$.

6.7 PRACTICAL TRANSFORMER

Practical transformers are far different from ideal transformers discussed in Section 6.6. The assumptions made in the case of an ideal transformer are relaxed one by one so as to understand and analyse the performance of a practical transformer.

6.7.1 Adding Core Losses to an Ideal Transformer

Idealism is against nature. Therefore, it is impossible to have a magnetic core that has infinite permeability and does not have hysteresis and eddy-current losses. The core loss causes heating of the transformer core.

Let it be assumed that the magnetic core has a reluctance of $\mathcal{R}$ AT/Wb and is operated on the linear portion of the magnetizing curve. Using Eqs (3.5) and (3.25) the mmf $\mathcal{F}$ AT may be expressed as

$$\mathcal{F} = N_1 \times i_\phi = \mathcal{R} \times \phi$$

or

$$i_\phi = \frac{\mathcal{R} \times \phi}{N_1} \tag{6.42}$$

where i_ϕ is the magnetizing current. Substituting Eq. (6.6) in Eq. (6.42)

$$i_\phi = \frac{\mathcal{R} \times \phi_m \sin \omega t}{N_1} \tag{6.43}$$

It can be seen from Eq. (6.43) that the magnetizing current is in phase with the core flux. The rms value of the magnetizing current I_ϕ can be written as

$$I_\phi = \frac{\mathcal{R} \times \phi_m}{\sqrt{2} \times N_1} \tag{6.44}$$

Substituting for ϕ_m from Eq. (6.30) and using it in Eq. (6.44) leads to

$$I_\phi = \frac{\mathcal{R}}{\sqrt{2} \times N_1} \times \frac{V_1}{\sqrt{2}\pi f N_1} \tag{6.45}$$

$$I_\phi = \frac{\mathcal{R} \times V_1}{2\pi f \times N_1^2} = \frac{\mathcal{R} \times V_1}{\omega \times N_1^2}$$

$$= \frac{V_1}{X_m} \tag{6.46}$$

where the magnetizing reactance X_m in ohms is given by

$$X_m = \frac{\omega N_1^2}{\mathcal{R}} \tag{6.47}$$

Equation (6.46) in the phasor form becomes

$$\boldsymbol{I_\phi} = \frac{V_1}{jX_m} = -jB_m\boldsymbol{V}_1 \tag{6.48}$$

where $B_m = 1/X_m$ is the susceptance.

Equation (6.48) shows that the magnetizing current lags the applied voltage V_1 by $\pi/2$. Thus, the circuit model of a transformer shows the magnetizing current as the current drawn by the magnetizing reactance X_m from the voltage source connected to the primary winding. The magnetizing reactance is a function of the frequency of the supply voltage, reluctance (or permeability) of the magnetic core, and the number of turns of the primary winding.

It was shown in Section 6.3 that due to the alternating flux produced by the current drawn by the transformer primary winding the core of the transformer experiences core or iron loss. From Eq. (6.30) it is observed that the applied voltage V_1 and the frequency f determine the flux ϕ_m produced in the core of a transformer when the supply frequency is constant. The core loss is constituted of hysteresis loss proportional to $V_1^{1.6}$ and eddy-current loss proportional to V_1^2 since ϕ_m is proportional to V_1. Both the components of power loss can be represented by an equivalent resistance R_C (or conductance G_C) conected across the supply voltage V_1. Thus core loss

$$P_C = \frac{V_1^2}{R_C} = G_C V_1^2 \tag{6.49}$$

and the component of current causing core loss

$$I_C = \frac{V_1}{R_C} = G_C V_1 \tag{6.50}$$

It may be mentioned that assuming the hysteresis loss component proportional to the square of the supply voltage in Eq. (6.49) does not introduce any significant inaccuracy. Both G_C and B_m are usually determined at the rated voltage and frequency. They are then assumed to remain constant for the small departures from the rated values associated with normal operation.

Equations (6.48) and (6.50) show that the exciting (or no-load or magnetizing) current I_0, required to set up the flux in the core and to account for the core loss, is drawn from the supply source and consists of two components, namely, the magnetizing current I_ϕ and the core loss current I_C. In the phasor form, the relation between the currents can be expressed mathematically as

$$I_0 = I_\varphi + I_C \tag{6.51}$$

The phasor diagram under no-load condition is shown in Fig. 6.15(a), with the core flux as the reference phasor. Figure 6.15(b) shows the phasor diagram for the transformer under loaded condition, assuming resistances and leakage reactances of the windings to be negligible, but taking into account the core losses. The primary current I_1 is the phasor sum of the no-load current I_0 and the current I_2', which is the secondary load current I_2 referred to the primary side.

From the phasor diagram of Fig 6.15(a), the total exciting current I_0 drawn from the supply is given as

$$I_0 = I_\varphi + I_C \tag{6.52}$$

where $I_0 = \sqrt{I_\varphi^2 + I_C^2}$ (6.53a)

and $\cos \varphi_0 = \dfrac{I_C}{I_0}$ (6.53b)

$\cos \varphi_0$ is the power factor of the primary current on no-load. Since $I_\phi < I_0$, the power factor angle is close to $90°$.

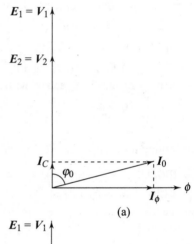

(a)

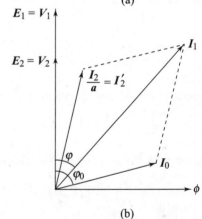

(b)

Fig. 6.15 (a) No-load phasor diagram and (b) on-load phasor diagram after accounting for core losses

The equivalent circuit of the transformer, after accounting for only the no-load losses and neglecting the resistances and leakage reactances of the primary and secondary windings, is drawn in Fig. 6.16. It may be noted from Fig. 6.16 that since the magnitude of the flux ϕ_m is independent of the extent of the load on a transformer, the core loss has been modelled by a fixed resistance R_C (or G_C) to account for the core loss. Further, since both the magnetizing current I_ϕ and the core loss current I_C are drawn from the supply, G_C and B_m have been shown connected in parallel with the supply voltage. From the equivalent circuit,

$$I_1 = I_0 + \frac{I_2}{a} = I_0 + I'_2 \tag{6.54}$$

where (I_2/a) is the secondary load current referred to the primary. The no-load current is predominantly re-active. As such transformer cores are made up of lminated sheet steel of high permeability. The magnitude of the no-load current in transformers is of the order of 2% to 5%.

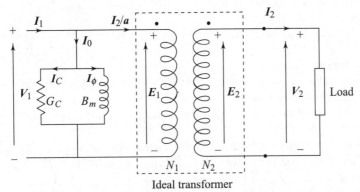

Ideal transformer

Fig. 6.16 Equivalent circuit of a transformer after accounting for core losses only

6.7.2 Incorporating Resistances and Leakage Reactances in the Ideal Transformer and Equivalent Circuits

When the transformer is on no load, the entire flux set up by the primary current links the secondary winding as the amount of the leakage flux through the air is negligible compared to the flux passing through the core.

When the transformer is on load, rms current I_2 flows from the instantaneous positive polarity to the load. This current, flowing through N_2 turns of the secondary winding, sets up an mmf $\mathcal{F}_2 = I_2 N_2$ AT, which by Lenz's law will oppose the core flux ϕ. Since the core flux is constant and determined by the supply voltage V_1 connected to the primary winding, it cannot change. Therefore, the primary current I_1 must increase by an amount such that the increased mmf of the primary opposes or balances the mmf of the secondary winding.

The total flux produced by the primary winding can be divided into two components—the resultant mu-tual flux restricted to the magnetic core and produced by the combined effect of the primary and secondary currents, and the primary leakage flux that links the primary winding only. The leakage path is mainly in air and the voltage induced by it varies linearly with the primary current I_1. The effect of this leakage flux can be taken into account by attaching to the primary a leakage inductance equal to the leakage flux linkage of the primary per unit primary current, or a leakage reactance x_1 equal to $2\pi f$ times the leakage inductance. In addition, there is a voltage drop due to the resistance of the primary winding.

The impressed terminal voltage phasor V_1 is then opposed by the phasor voltage drops $I_1 r_1, I_1 x_1$, the leakage reactance x_1 of the primary, and the phasor emf E_1 induced in the primary by the resultant mutual flux. All these components are shown in Fig. 6.17, which is the equivalent circuit of a transformer.

The resultant mutual flux links both the primary and secondary windings and is created by their combined mmfs. The primary current has two components—an exciting component I_0 required to produce the resultant mutual flux and a load component I'_2 which would exactly balance the mmf of secondary current I_2, that is, $I'_2 N_1 = I_2 N_2$.

The exciting component I_0 has a core-loss component I_C in phase with the emf E_1 and a magnetizing com-ponent I_ϕ lagging E_1 by 90°. In the equivalent circuit shown in Fig. 6.17, this is represented by a conductance G_C in parallel with a susceptance B_m connected in shunt across E_1 (which is the same as in Fig. 6.16).

Just as in an ideal transformer, the resultant mutual flux ϕ induces an emf E_2 in the secondary and since this flux links both the windings, the induced-voltage ratio $E_1/E_2 = N_1/N_2$. The voltage and the current transformation ratios in Eq. (6.33) can be accounted for by introducing an ideal transformer in the equivalent circuit of Fig. 6.17.

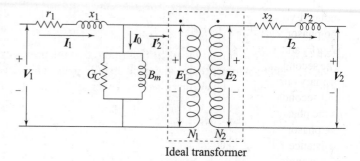

Fig. 6.17 Complete equivalent circuit of a transformer

The secondary current I_2 creates secondary leakage flux. The secondary terminal voltage V_2 differs from the induced voltage E_2 by the voltage drops in the secondary resistance r_2 and the secondary leakage reactance x_2 shown in Fig. 6.17 to the right of E_2.

The actual transformer is therefore equivalent to an ideal transformer plus external impedances. By referring all quantities to the primary or to the secondary, the ideal transformer in Fig. 6.17 can be moved either to the right or to the left of the shunt circuit.

It may be noted that the leakage flux is proportional to the load current, while the core flux is largely independent of the load current. Also, the energy transfer takes place via the core flux and the leakage flux plays no part in it.

As discussed in Section 6.6.3, the resistance r_2 and the leakage reactance x_2 of the secondary windings can be referred to the primary side as r_2' and x_2', respectively. Using Eq. (6.41)

$$r_2' = a^2 r_2 \tag{6.55a}$$
$$x_2' = a^2 x_2 \tag{6.55b}$$

Similarly, the secondary current and voltage can be referred to the primary side as

$$I_2' = \frac{I_2}{a} \tag{6.56a}$$
$$V_2' = aV_2 \tag{6.56b}$$

The equivalent circuit is usually drawn as in Fig. 6.18, with the ideal transformer not shown and all voltages, currents, and impedances referred to the same side. In Fig. 6.18 the referred values are indicated with primes to distinguish them from the actual values of Fig. 6.17.

The equivalent circuit can be used to draw the phasor diagram of the transformer. Assume that the load connected on the secondary side draws a current of I_2 amperes at a lagging power angle of φ. Using the equivalent circuit of

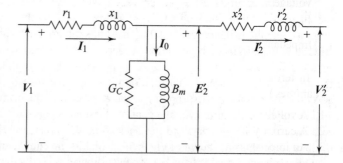

Fig. 6.18 Equivalent circuit of a transformer with the ideal transformer left out

Fig. 6.18 the phasor diagram has been drawn with the flux ϕ as the reference phasor shown in Fig. 6.19. The phasor aV_2 represents the secondary load voltage and (I_2/a) is the secondary current phasor referred to the primary side. The secondary current lags the secondary voltage by the power factor angle φ_2. The secondary induced voltage E_2 is obtained by the phasor addition of the phasor voltage aV_2 and the secondary resistance drop $(I_2/a)r_2$ and the secondary leakage reactance drop $j(I_2/a)x_2$. The secondary induced voltage E_2 leads the flux by $\pi/2$. The primary current is the phasor sum of the no-load current I_0 and secondary current (I_2/a). Thus, the primary current I_1 is the phasor sum of (I_2/a) and I_0. The supply voltage V_1 is the phasor sum of E_1, the voltage drops due to the primary resistance I_1r_1, and the leakage reactance drop jI_1x_1. The phase angle φ_1 between V_1 and I_1 is the power factor angle of the primary side.

6.7.3 Usefulness of the Equivalent Circuit

An equivalent circuit of a transformer forms the basis of determining its performance in respect of its efficiency and regulation. (Transformer efficiency and regulation are discussed later in this chapter.) Since, in a transformer electrical quantities are present both on the primary and secondary sides, it would be useful to work with quantities by referring them either to the primary or the secondary side. Tables 6.1(a) and (b) summarize the rules for referring the various quantities to a particular side.

Transformation ratio, $a = \dfrac{V_1}{V_2} = \dfrac{I_2}{I_1} = \dfrac{N_1}{N_2}$

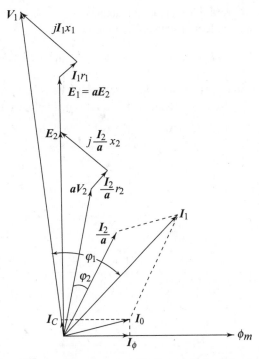

Fig. 6.19 Phasor diagram for the equivalent circuit shown in Fig. 6.17

Table 6.1(a) Transfer of primary quantities to secondary quantities

Quantity	From primary	To secondary
Voltage	V_1	$V_1' = V_1/a$
Current	I_1	$I_1' = a I_1$
Impedance	Z_1	$Z_1' = Z_1/a^2$

Table 6.1(b) Transfer of secondary quantities to primary quantities

Quantity	From secondary	To primary
Voltage	V_2	$V_2' = aV_2$
Current	I_2	$I_2' = I_2/a$
Impedance	Z_2	$Z_2' = a^2/Z_2$

In terms of the high-voltage (HV) and low-voltage (LV) sides, the following correlation between the quantities may also be noted:

1. A voltage when referred to LV side will have a lower value than its HV side value.
2. A current when referred to LV side will have a higher value than its HV side value.
3. An impedance when referred to LV side will have a lower value than its HV side value.

6.7.4 Approximate Equivalent Circuits

The equivalent circuit shown in Fig. 6.18 may be modified to form approximate equivalent circuits whereby the computations can be appreciably reduced. The shunt branch representing the exciting current may be moved from the middle of the circuit of Fig. 6.18 to either the primary or the secondary terminal as shown in Figs 6.20(a) and (b), respectively. All parameters in these circuits are referred to either the primary or the secondary. The series branch, thus, reduces to the equivalent impedance consisting of the equivalent resistance

R_{eq} and the equivalent reactance X_{eq} as shown in Figs 6.20(a) and (b). The equivalent resistance R_{eq} and the equivalent reactance X_{eq} are given by

$$R_{eq} = r_1 + r_2' = r_1 + \frac{r_2}{a^2} \quad \text{and} \quad X_{eq} = x_1 + x_2' = x_1 + \frac{x_2}{a^2} \tag{6.57}$$

An error is introduced by neglecting the voltage drop in the primary or the seconry leakage impedance caused by the exciting current, but this error is insignificant in power transformers of large ratings.

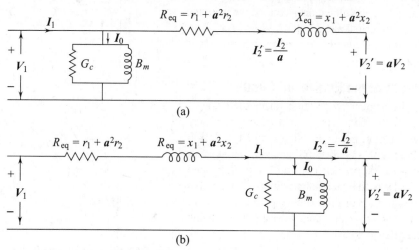

(a)

(b)

Fig. 6.20 Approximate equivalent circuits

The equivalent circuit is further simplified by neglecting the exciting current as in Fig. 6.21(a), in which case the transformer is represented by equivalent series impedance $Z_{eq} = R_{eq} + jX_{eq}$. For large power transformers, the equivalent resistance R_{eq} is small as compared to the equivalent reactance X_{eq} and can be neglected. The equivalent circuit reduces to that shown in Fig. 6.21(b).

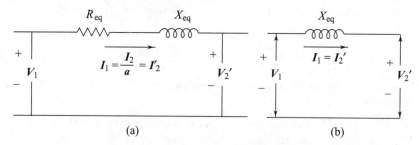

(a)   (b)

Fig. 6.21 Approximate equivalent circuits

Example 6.5 A single-phase 100-kVA, 2200/220-V, 50-Hz transformer has a resistance of 0.75 Ω and a leakage reactance of 1.15 Ω for its HV winding. The corresponding values of the resistance and the leakage reactance of the LV winding are 0.007 Ω and 0.009 Ω. The shunt excitation branch parameters G_C and B_m determined at the rated voltage and frequency, as viewed from the LV side are 0.0035 S and 0.025 S, respectively. Draw (a) the equivalent circuit referred to (i) the LV side, (ii) the HV side; (b) an approximate equivalent circuit referred to (i) the LV side, (ii) the HV side.

Solution The turns ratio,

$$a = \frac{N_1}{N_2} = \frac{2200}{220} = 10$$

(a) (i) The LV side leakage impedance $= r_2 + jx_2 = (0.007 + j0.009)\,\Omega$
The HV side leakage impedance $= r_1 + jx_1 = (0.75 + j1.15)\,\Omega$
The HV side impedance referred to the LV side

$$r_1' + jx_1' = \frac{1}{a^2}\,(r_1 + jx_1) = (1/10^2) \times (0.75 + j1.15) = (0.0075 + j0.0115)\,\Omega$$

The shunt branch admittance referred to the LV side

$$G_C - jB_m = (0.0035 - j0.025)\,\text{S}$$

Then the equivalent circuit referred to the LV side is shown in Fig. 6.22(a-i).
(ii) The LV side impedance referred to the HV side

$$r_2' + jx_2' = a^2\,(r_2 + jx_2) = 10^2\,(0.007 + j0.009) = (0.70 + j0.90)\,\Omega$$

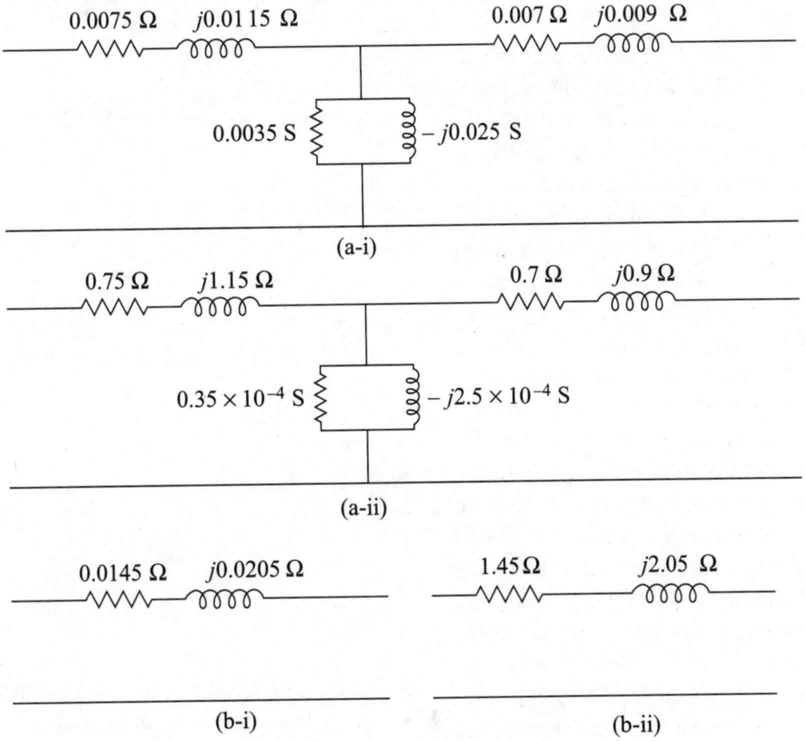

(a-i)

(a-ii)

(b-i)   (b-ii)

Fig. 6.22

The magnetizing admittance referred to the HV side

$$\frac{1}{a^2}\,(G_C - jB_m) = 10^{-2} \times (0.0035 - j0.025) = (0.000035 - j0.00025)\,\text{S}$$

Then the equivalent circuit referred to the LV side is shown in Fig. 6.22(a-ii).
(b) For an approximate equivalent circuit, the magnetizing admittance is neglected from the exact equivalent circuit of Figs 6.22(a-i) and (a-ii).

(i) The equivalent impedance referred to the LV side

$$(r_2 + r_1') + j(x_2 + x_1') = (0.007 + 0.0075) + j(0.009 + 0.0115) = (0.0145 + j0.0205)\ \Omega$$

Then the equivalent circuit referred to the LV side is shown in Fig. 6.22(b-i).

(ii) The equivalent impedance referred to the HV side

$$(r_1 + r_2') + j(x_1 + x_2') = (0.75 + 0.7) + j(1.15 + 0.9) = (1.45 + j2.05)\ \Omega$$

Then the equivalent circuit referred to the LV side is shown in Fig. 6.22(b-ii).

6.8 TRANSFORMER TESTING

Testing of transformers is undertaken to determine the parameters of the equivalent circuit of a transformer and then compute its efficiency and regulation without actually loading it. Transformer testing by actual loading is not practical due to (i) large amount of energy required for testing, which is simply a loss; (ii) accuracy required in measuring the input and output power since the two are close to each other, and (iii) difficulty in arranging the load for testing purposes.

The two tests employed to determine the parameters are: (a) open-circuit test and (b) short-circuit test.

6.8.1 Open-circuit (OC) Test

The connection diagram for performing the oc test is shown in Fig. 6.23. Normally, the oc test is conducted on the LV side of the transformer as the normal rated voltage on the LV side is more likely to be in the range of usually available instruments, and also because of less hazard of working with the LV side. Further, in this test a low-power-factor wattmeter is used as only low-power-factor magnetizing current is drawn from the supply in this test.

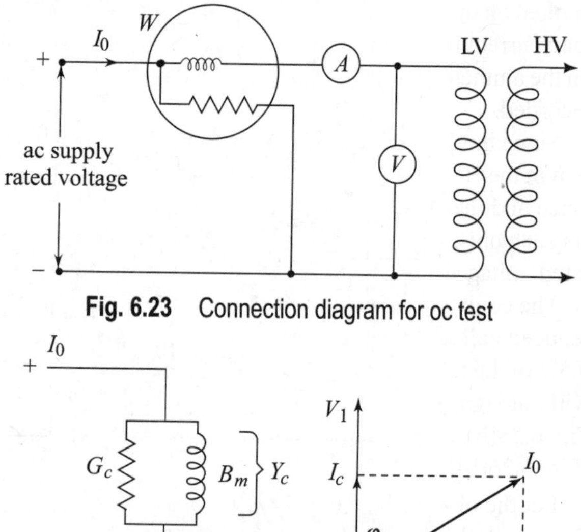

In this test, the LV winding is subjected to the rated primary voltage and frequency, and the other side is open-circuited. The readings on the voltmeter, the ammeter, and the wattmeter are recorded.

Figure 6.24 shows the equivalent circuit of the transformer under no-load condition along with the resultant phasor diagram. Since the HV side is open-circuited, the ammeter reading shows the no-load current. As already stated, the no-load current is 2% to 5% of the full-load current of the transformer and, therefore, the I^2R loss in the primary winding is small and can be neglected. The wattmeter reading indicates the core loss of the transformer.

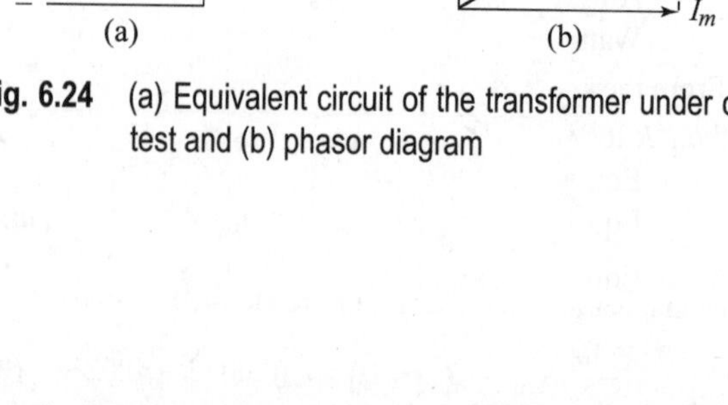

Fig. 6.23   Connection diagram for oc test

Fig. 6.24   (a) Equivalent circuit of the transformer under oc test and (b) phasor diagram

Let the observations recorded be as follows:

Primary voltmeter reading, V_1

Ammeter reading, I_0

Wattmeter reading, P_C

From these observations, the parameters for the shunt branch of the equivalent circuit can be computed as under:

Admittance of the core, $Y_C = \dfrac{I_0}{V_1}$

$$(6.58)$$

Using Eq. (6.49), the conductance of the core, $G_C = P_C/V_1^2$ (6.59)

Susceptance of the core, $B_m = \sqrt{Y_C^2 - G_C^2}$ (6.60)

No-load power factor, $\cos \varphi_0 = \dfrac{P_C}{V_1 I_0}$ (6.61)

Employing the method of transformation, if needed, the parameters of the shunt branch can be referred to the HV side also, as given below.

$$G_{C(hv)} = a^2 G_{C(lv)} \quad \text{and} \quad B_{m(hv)} = a^2 B_{m(lv)}$$

where $a = N_{lv}/N_{hv}$

The equivalent circuit referred to the HV side will show the core conductance and magnetizing susceptance values of $G_{C(hv)}$ and $B_{m(hv)}$, respectively

6.8.2 Short-circuit (sc) Test

The connection diagram for performing the sc test is shown in Fig. 6.25. Usually the LV winding of the transformer is short-circuited and the reduced voltage applied on the HV side is adjusted such that the full-load current flows in the HV winding. The readings on the ammeter, the voltmeter, and the wattmeter are recorded.

Since the LV winding is short-circuited, the current in the HV winding is limited by the series resistance and the leakage reactance of the transformer. As such only a small magnitude (2% to 10%) of the

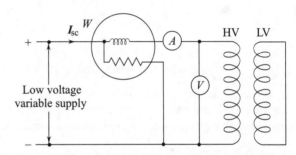

Fig. 6.25   Connection diagram for sc test

rated voltage of the HV winding is sufficient to circulate the full-load current in the HV side.

The equivalent circuit of the transformer, under the short-circuit condition, is drawn in Fig. 6.26(a). Since reduced voltage is applied to the HV winding, the no-load component of the short-circuit current $I_{0(sc)}$ is 0.1 to 0.5% of the full-load current. Hence, the effect of the magnetizing shunt branch is negligible and can be ignored without significantly affecting the accuracy of the computations. The simplified equivalent circuit drawn in Fig. 6.26(b) and the corresponding phasor diagram is shown in Fig. 6.26(c). The simplified equivalent circuit [Fig. 6.26(b)] can be adopted for computing the series parameters of the transformer from the sc test results.

Let the observations recorded be as follows:

Voltmeter reading, V_{sc}
Ammeter reading, I_{sc}
Wattmeter reading, P_{sc}

From the simplified equivalent circuit of Fig. 6.26(b), it can be seen that the wattmeter reading is indicative of the $I^2 R$ loss in the transformer windings at full load. Hence,

Equivalent resistance of the transormer, $R_{eq} = P_{sc}/I_{sc}^2$ (6.62)
Equivalent impedance of the transformer, $Z_{eq} = V_{sc}/I_{sc}$ (6.63)

Equivalent reactance of the transformer,

$$X_{eq} = \sqrt{Z_{eq}^2 - R_{eq}^2}$$ (6.64)

It is clear that the sc test provides data to determine the series parameters of the equivalent circuit of the transformer.

As in the oc test, the equivalent resistance and reactance values obtained from the sc test data can be transferred to the LV side as follows:

$$R_{eq(lv)} = a^2 R_{eq(hv)}$$

and $\quad X_{eq(lv)} = a^2 X_{eq(hv)}$

Again it may be remembered that while drawing the equivalent circuit referred to the LV side, the values of R_{eq} and X_{eq} referred to the LV side are to be employed.

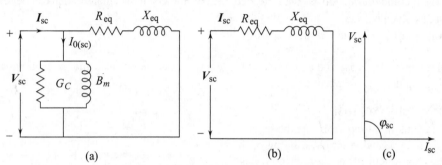

Fig. 6.26   Equivalent circuit of the transformer and its phasor diagram under sc test

In the sc test, the HV side is chosen as primary because the rated current on the HV side is lower than the LV side and instruments of lower current ranges can be used. Further, a high-power-factor wattmeter is used in this test as the magnetizing current drawn is negligible and unity-power-factor current flows during this test.

Example 6.6   The open-circuit and short-circuit tests on a 4-kVA, 200/400-V, 50-Hz, single-phase transformer gave the following results:

    oc test on the LV side: 200 V, 1 A, 100 W

    sc test with the LV side shorted: 15 V, 10 A, 85 W

(a) Determine the parameters of the equivalent circuit. (b) Draw the equivalent circuit referred to the LV side.
Solution   (a) From the oc test data, the parameters of shunt admittances are determined as follows:
Admittance of the core from Eq. (6.58),

$$Y_C = \frac{I_0}{V_1} = \frac{1}{200} = 0.005 \text{ S (or mho)}$$

Conductance of the core from Eq. (6.59),

$$G_C = \frac{P_C}{V_1^2} = \frac{100}{200^2} = \frac{1}{400} = 0.0025 \text{ S}$$

Susceptance of the core from Eq. (6.60),

$$B_m = \sqrt{0.005^2 - 0.0025^2} = 0.0433 \text{ S}$$

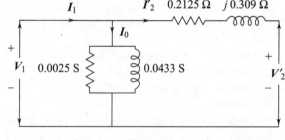

Fig. 6.27

From the sc test data, the equivalent impedance parameters referred to the HV side are determined as follows:

Equivalent resistance from Eq. (6.62), $R_{eq} = \dfrac{P}{I_{sc}^2} = \dfrac{85}{10^2} = 0.85 \ \Omega$

Equivalent impedance from Eq. (6.63), $Z_{eq} = \dfrac{V_{sc}}{I_{sc}} = \dfrac{15}{10} = 1.5 \ \Omega$

Equivalent leakage reactance from Eq. (6.64), $X_{eq} = \sqrt{1.5^2 - 0.85^2} = 1.236\ \Omega$

(b) The turns ratio, $\quad a = \dfrac{200}{400} = \dfrac{1}{2} = 0.5$

The equivalent impedance parameters referred to the LV side are determined as follows:

Equivalent resistance $R'_{eq} = 0.5^2 \times 0.85 = 0.2125\ \Omega$

Equivalent leakage reactance $X'_{eq} = 0.5 \times 1.236 = 0.309\ \Omega$

Equivalent impedance $Z'_{eq} = 0.5^2 \times 1.5 = 0.375\ \Omega$

The equivalent circuit referred to the LV side is shown in Fig. 6.27.

6.9 TRANSFORMER REGULATION

The voltage regulation of a transformer is defined as the ratio of the change in the magnitude of the secondary voltage when full load at a specified power factor is thrown off, with the primary voltage and the frequency held constant. It is usually expressed in per cent as follows:

$$\% \text{ Regulation} = \frac{\{\text{change in voltage magnitude with load}\}}{\{\text{rated full-load secondary voltage magnitude}\}} \times 100 \tag{6.65}$$

$$= \frac{V_{2,\text{no-load}} - V_{2,\text{full-load}}}{V_{2,\text{full-load}}} \times 100 \tag{6.66}$$

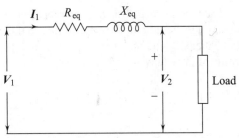

Fig. 6.28　Approximate equivalent circuit of a transformer

Practicing engineers use regulation to maintain voltage within specified limits at the consumer terminals by regulating the primary voltage.

Figure 6.28 shows an approximate equivalent circuit in which the shunt branch has been left out since it does not effect the voltage computation.

From the approximate equivalent circuit of Fig. 6.28, with full load at the secondary terminals, the supply voltage V_1 at the primary terminals can be expressed as

$$V_1 = V_2 + (R_{eq} + j\, X_{eq})I_1 \tag{6.67}$$

$$V_1 - V_2 = (R_{eq} + j\, X_{eq})I_1 \tag{6.68}$$

where I_1 is the full-load current. When the load is thrown off, then $I_1 = 0$ and $V_1 = V_{2,\text{no-load}}$.

The phasor diagram (not to scale) for the equivalent circuit of Fig. 6.28 is drawn and is shown in Fig. 6.29. In a transformer, the voltage drops due to the winding resistance R_{eq} and the leakage reactance X_{eq} are small in magnitude as compared to the voltages V_1 and V_2. Also, the angle δ between V_1 and V_2 is small (only a few degrees). Therefore, for the phasor diagram of Fig. 6.29(a), which is for a lagging power factor, it may be assumed that $OC = OB$ without introducing any significant error in the calculation of the regulation of a transformer.

Thus, $\quad V_1 = OB$

$$V_1 - V_2 = OB - OA = AB = I_1(R_{eq}\cos\theta + X_{eq}\sin\theta) \tag{6.69}$$

Therefore,

$$\% \text{ Voltage regulation} = \frac{V_1 - V_2}{V_2}$$

$$= \frac{I_1\,(R_{eq}\cos\theta + X_{eq}\sin\theta)}{V_2} \tag{6.70}$$

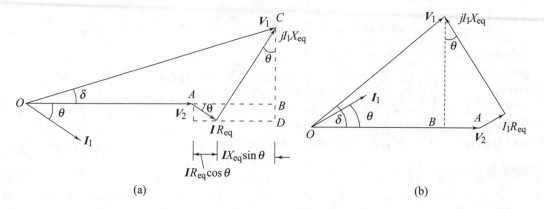

Fig. 6.29 Phasor diagram for the circuit of Fig. 6.28 (a) Lagging power factor and (b) leading power factor

For a leading power factor, the phasor diagram is shown in Fig. 6.29(b). In this case, the + sign in Eq. (6.70) would be replaced by the − sign. The voltage regulation for the leading power factor is given by

$$\% \text{ Voltage regulation} = \frac{I_1(R_{eq}\cos\theta - X_{eq}\sin\theta)}{V_2} \tag{6.71}$$

The voltage regulation is often represented in per unit (pu) rather than in per cent. The per unit value of a quantity is the ratio of the actual value to the base or reference value of a quantity. In the per unit system, if the voltage V_2 and the current I_1 are aken as the base voltage and the base current, respectively, then the base value of impedance is given by

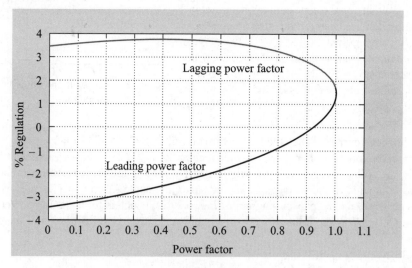

Fig. 6.30

$$Z_{base} = \frac{V_2}{I_1} \tag{6.72}$$

then

$$R_{pu} = \frac{R_{actual}}{Z_{base}} = \frac{R_{actual} \times I_1}{V_2}; \quad X_{pu} = \frac{X_{actual}}{Z_{base}} = \frac{X_{actual} \times I_1}{V_2} \tag{6.73}$$

Substituting from Eq. (6.73) in Eq. (6.70) and (6.71) and combining,

$$\text{pu regulation} = R_{\text{pu}} \cos\theta \pm X_{\text{pu}} \sin\theta \qquad (6.74)$$

For a lagging power factor, pu regulation would be calculated with the $+$ sign and for a leading power factor, pu regulation would be calculated with the $-$ sign.

For typical values of R_{pu} and X_{pu}, percentage regulation is plotted in Fig. 6.30 for varying power factors. From the plot, it is seen that transformer regulation is higher for a lagging power factor than a leading power factor of the same magnitude. For example, at a power factor of 0.8 lagging, the transformer regulation is of the order of 3.25% while it is approximately -0.9% when the power factor is 0.8 leading.

Example 6.7 A 4-kVA, 200/400-V, 50-cycles/sec, single-phase transformer gave the following test figures:

Low-voltage data for no load: 200 V, 0.7 A, 60 W

High-voltage data for short circuit: 9 V, 6 A, 21.6 W

Calculate (a) the magnetizing current and the component corresponding to iron loss at normal voltage and frequency; (b) the efficiency on full-load and unity power factor; (c) the secondary terminal voltage on full load at power factors of unity, 0.8 lagging, and 0.8 leading.

Solution No-load test data for the LV side: $V_1 = 200$ V, $P_c = 60$ W, and $I_0 = 0.7$ A

The parameters of the shunt branch of the equivalent circuit can be computed as under:

$$Y_C = \frac{I_0}{V_1} = \frac{0.7}{200} = 0.0035 \text{ mho}$$

$$G_C = \frac{P_C}{V_1^2} = \frac{60}{(200)^2} = 0.0015 \text{ mho or S}$$

$$B_m = \sqrt{(0.0035)^2 - (0.0015)^2} = 0.00316 \text{ mho or S}$$

$$I_m = B_m \times V_1 = 0.00316 \times 200 = 0.632 \text{ A}$$

sc test data for the HV side: $V_{sc} = 9$ V, $I_{sc} = 6$ A, and $P_{sc} = 21.6$ W

The parameters of the series branch of the equivalent circuit can be determined as under:

$$R = \frac{P_{sc}}{I_{sc}^2} = \frac{21.6}{6^2} = 0.6 \ \Omega$$

$$Z_{eq} = \frac{E_{sc}}{I_{sc}} = \frac{9}{6} = 1.5 \ \Omega$$

$$X_{eq} = \sqrt{Z_{eq}^2 - R_{eq}^2} = \sqrt{1.5^2 - 0.6^2} = 1.375 \ \Omega$$

Since the sc test data are recorded on the HV side, the parameters of the series branch are required to be referred to the LV side. This is done as follows:

$$a = \frac{200}{400} = 0.5$$

Resistance referred to the LV side,
$$R'_{eq} = a^2 \times R = 0.25 \times 0.6 = 0.15 \ \Omega$$
Reactance referred to the LV side,
$$X'_{eq} = a^2 \times X = 0.25 \times 1.375 = 0.344 \ \Omega$$

Figure 6.31 shows the equivalent circuit of the transformer with all parameters referred to the LV side.

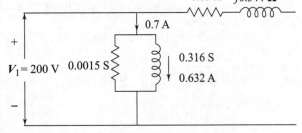

Fig. 6.31 Equivalent circuit of the transformer

Output on full-load at unity power factor = 4000 W

Full load current, $I_{FL} = \dfrac{4 \times 10^3}{200} = 20$ A

Copper loss at full load $= I_{FL}^2 \times R = (20^2) \times 0.15 = 60$ W

Iron or core loss = 60 W

% Efficiency, $\eta = \dfrac{\text{output}}{\text{output} + \text{iron loss} + \text{copper loss}} \times 100 = \dfrac{4000}{4000 + 60 + 60} \times 100 = 97.08\,\%$

Equation (6.69), which is reproduced below, is convenient to compute the secondary voltage.

$$V_1 - V_2 = I_1(R_{eq}\cos\theta + X_{eq}\sin\theta)$$

$$V_2 = \frac{1}{a}\,[V_1 - I_1(R_{eq}\cos\theta + X_{eq}\sin\theta)]$$

It may be noted that the above equation has been divided by the transformation ratio *a* to obtain the voltage on the secondary side.

The terminal voltage on full load at unity power factor,

$$V_2 = \frac{200 - 20 \times 0.15}{0.5} = 394 \text{ V}$$

The terminal voltage on full load at 0.8 power factor lagging,

$$V_2 = \frac{200 - 20\,(0.15 \times 0.8 + 0.344 \times 0.6)}{0.5} = 386.94 \text{ V}$$

The terminal voltage on full load at 0.8 power factor leading,

$$V_2 = \frac{200 - 20\,(0.15 \times 0.8 - 0.344 \times 0.6)}{0.5} = 403.45 \text{ V}$$

6.10 TRANSFORMER EFFICIENCY

Power losses in a practical transformer are mainly of two types:

(i) The iron or core losses which comprise hysteresis and eddy-current losses. As already discussed, both these losses occur in the iron core and are proportional to the maximum value of the flux in the core. In a normal transformer, the change in flux from no load to full load is of the order of 2%. Therefore, core losses are taken as constant and independent of the load on the transformer.

(ii) I^2r losses take place in the transformer windings. These losses, being proportional to the square of the load current, are dependent on the loading condition of the transformer.

Besides these, there may be dielectric losses, which are due to the leakage of current in the insulating material. The dielectric losses are so small that they do not effect the efficiency computations and, therefore, can be neglected. In addition, there may be stray-load losses due to the leakage of flux, which induces eddy currents in conductors and walls of the transformer tank. These losses being small, are also neglected in the efficiency computations.

The efficiency of a transformer, η, is given as

$$\eta = \frac{\text{output power}}{\text{input power}} = \frac{\text{input power} - \text{losses}}{\text{input power}}$$

$$= 1 - \frac{\text{losses}}{\text{input power}} \tag{6.75a}$$

$$= 1 - \frac{\text{losses}}{\text{output power} + \text{losses}} \tag{6.75b}$$

6.10.1 Maximum Efficiency Condition

Let it be assumed that a transformer has the primary and secondary winding resistances r_1 and r_2, respectively and the transformation ratio is a. The equivalent resistance of the transformer referred to the secondary side can be written as

$$R_{eq} = \frac{r_1}{a^2} + r_2$$

If the load current is I_2, then the heat loss in the transformer windings is given by $I_2^2 R_{eq}$.

Let the load voltage be V_2 and the power factor be $\cos\theta$ lagging. If the core loss in the transformer is represented by P_C, then

$$\eta = \frac{V_2 I_2 \cos\theta}{V_2 I_2 \cos\theta + P_C + I_2^2 R_{eq}} \tag{6.76a}$$

$$= \frac{V_2 \cos\theta}{V_2 \cos\theta + (P_C/I_2) + I_2 R_{eq}} \tag{6.76b}$$

For a transformer, the load voltage V_2 is approximately constant. Thus, for a load of given power factor, the efficiency is maximum when the denominator in Eq. (6.76b) is minimum. Therefore,

$$\frac{d}{dI_2}[V_2 \cos\theta + (P_C/I_2) + I_2 R_{eq}] = 0$$

or $$-\frac{P_C}{I_2^2} + R_{eq} = 0$$

$$\therefore \qquad P_C = I_2^2 R_{eq} \tag{6.77}$$

To check that the condition stated by Eq. (6.77) is the condition for the minimum value of the denominator in Eq. (6.76b), the second derivative of the denominator in Eq. (6.76b) with respect to I_2 is obtained. Thus,

$$\frac{d}{dI_2}\left(-\frac{P_c}{I_2^2} + R_{eq}\right) = \frac{2P_c}{I_2^3} \tag{6.78}$$

Since the quantity $(2P_c)/I_2^3$ is positive, it proves that for maximum transformer efficiency, the constant core loss must be equal to the variable $I^2 R_{eq}$ loss.

The load for which the maximum efficiency occurs can also be determined. From Eq. (6.77),

$$I_2^2 = \frac{P_C}{R_{eq}} \tag{6.79}$$

Dividing both sides of Eq. (6.79) by the square of the full-load current $I_{2,fl}$ gives

$$\left(\frac{I_2}{I_{2,fl}}\right)^2 = \frac{P_C}{I_{2,fl}^2 R_{eq}} = \frac{P_C}{\text{full-load } I^2 R \text{ loss}}$$

or $$\frac{I_2}{I_{2,fl}} = \sqrt{\frac{P_C}{\text{full-load } I^2 R \text{ loss}}} = m, \text{ a constant} \tag{6.80}$$

or $$I_2 = m I_{2,fl} \tag{6.81}$$

Thus, the efficiency is maximum when the load current is $m I_{2,fl}$, where m is given by Eq. (6.80).

Example 6.8   (a) A single-phase transformer has a core loss of P_C W and a full load copper loss of P_{cu}W. If the kVA rating of the transformer is S, derive a generalised expression for the transformer efficiency when it is supplying a load $x\%$ of the rated load at a power factor of $\cos\theta$. (b) Draw the transformer efficiency curve when the load is varied from 50% to 125% of the full load for $\cos\theta = 1.0$, $\cos\theta = 0.8$, and $\cos\theta = 0.65$.

Solution   Transformer output = $0.01xS$ kVA

or   transformer output = $0.01xS \times 1000 \cos\theta$ W

Copper loss at $x\%$ of the full load = $0.0001x^2P_{cu}$

Transformer efficiency, $\eta = \dfrac{10\,x\,S\cos\theta}{10\,x\,S\cos\theta + P_C + 0.0001\,x^2P_{cu}}$

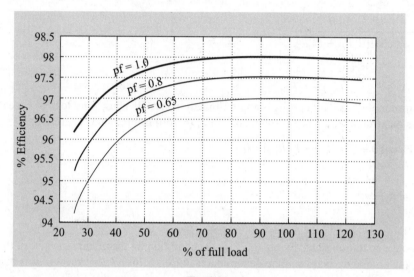

Fig. 6.32

Based on Eq. (I), the transformer efficiency versus percentage of full load is plotted in Fig. 6.32 for varying power factors. It may be observed from the plot that maximum efficiency is independent of the power factor and is entirely governed by the load at which the core loss is equal to the copper loss.

Example 6.9   Calculate the efficiencies a 1/2 load, full load, and 1¼ load of a 100-kVA transformer for power factors of (a) unity, (b) 0.8. The copper loss is 1200 W at full load and the iron loss is 1200 W.

Solution   Iron loss is 1.2 kW and independent of the load on the transformer. I^2R loss is dependent on the load of the transformer. For power factors of unity and 0.8, copper loss for varying loads and transformer efficiencies have been computed and are given in Table 6.2.

Table 6.2

Power factor	Unity			0.8		
Load	Half full load = 50 kW	Full load = 100 kW	1¼ full load = 125 kW	Half full load = 40 kW	Full load = 80 kW	1¼ full load = 100 kW
I^2R loss	0.3 kW	1.2 kW	1.875 kW	0.3 kW	1.2 kW	1.875 kW
% Efficiency	97.087	97.656	97.6	96.38	97.087	97.016

6.10.2 All-day Efficiency of a Transformer

A transformer, particularly in an electrical distribution system, is energized round the clock for 24 hours a day and supplies varying loads. Therefore, it is subjected to core losses throughout the day independent of the load. On the other hand I^2R loss being proportional to the square of the load current, varies according to the load.

In order to assess the performance of such a transformer, all-day efficiency (also known as operational efficiency) is computed. All-day efficiency is define as

$$\eta_{\text{all-day}} = \frac{\text{energy output in kWh for 24 hours}}{\text{energy input in kWh for the corresponding 24 hours}} \qquad (6.82)$$

Higher $\eta_{\text{all-day}}$ can be achieved by minimizing core losses. The minimization of core losses can be achieved by designing transformers with low flux densites.

Example 6.10 A transformer has its maximum efficiency of 0.98 at 15 kVA at unity power factor. During the day it is loaded as follows:

      2 kW at power factor 0.5 for 10 hours
      12 kW at power factor 0.8 for 8 hours
      18 kW at power factor 0.9 for 6 hours

Find the all-day efficiency.

Solution   Output at unity power factor = 15 kW

$$\eta_{\text{max}} = 0.98$$

At η_{max} and unity power factor

$$\text{Input} = \frac{\text{Output}}{\eta_{\text{max}}} = \frac{15}{0.98} = 15.306 \text{ kW}$$

Thus,   losses = input − output = 15.306 − 15.0 = 0.306 kW
At maximum efficiency, iron loss

$$P_C = I^2R \text{ loss}$$

Hence, $P_C = I^2R \text{ loss} = \dfrac{0.306}{2} = 0.153 \text{ kW}$

Since the iron loss is independent of the load, the energy loss due to iron loss in 24 hours is
      $0.153 \times 24 = 3.672$ kWh

The energy output for the given load cycle along with the energy loss due to copper loss is computed and given in Table 6.3.

Table 6.3

Duration, hours	10	8	6
Load, kW	2 at 0.5 pf	12 at 0.8 pf	18 at 0.9 pf
Rated output at given pf, kW	$15 \times 0.5 = 7.5$	$15 \times 0.8 = 12$	$18 \times 0.9 = 13.5$
I^2R loss at given pf, kW	$(2/7.5)^2 \times 0.153$ $= 0.0109$	$(12/12)^2 \times 0.153$ $= 0.153$	$(18/13.5)^2 \times 0.153$ $= 0.272$
Energy loss due to I^2R loss, kWh	0.0109×10 $= 0.109$	0.153×8 $= 1.224$	0.272×6 $= 1.632$
Energy output, kWh	$2 \times 10 = 20$	$12 \times 8 = 96$	$18 \times 6 = 108$

Total energy output during the day = 20 + 96 + 108 = 224 kWh
Energy loss due to I^2R loss = 0.109 + 1.224 + 1.632 = 2.965 kWh

$$\eta_{\text{all-day}} = \frac{224}{224 + 3.672 + 2.965} = 0.9712$$

6.11 TYPES OF TRANSFORMERS

The cores of transformers, out of necessity, have to be made of iron or its alloys since they are required to operate at frequencies that vary in the range 25 to 400 Hz. Air or a non-magnetic core is required for transformers operating at very high frequencies.

Transformers are classified on the following basis: (i) type of construction, (ii) type of connections, and (iii) special types.

6.11.1 Type of Construction

Based on the type of core construction, normal transformers are mainly divided into two categories: (i) core type and (ii) shell type. These are already discussed in Section 6.4.1.

6.11.2 Type of Connections

Based on the number of windings and the type of connections, transformers can be classified as (i) auto-transformers and (ii) three-phase transformers.

6.11.2.1 Auto-transformers

An auto-transformer is a transformer having a part of its winding common to the primary and secondary circuits. The input and the output are electrically connected and the input and output currents are superimposed in the common part of the winding.

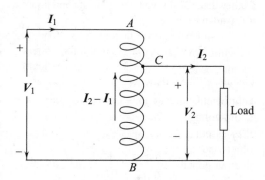

Fig. 6.33   An auto-transformer

Figure 6.33 shows an auto-transformer with winding AB and total primary turns N_1. The winding AB is tapped at any convenient point C and the number of secondary turns between C and B is N_2. The supply voltage V_1 is applied across AB and the load, being connected across CB, has a voltage V_2. The primary input current is I_1 and the load current is I_2. Neglecting any voltage drop due to leakage impedances,

$$\frac{V_1}{V_2} = \frac{N_1}{N_2} = \frac{E_1}{E_2} = a \tag{6.83}$$

where E_1 and E_2 are the induced emfs and a is the turns ratio. The mmfs in the two windings balance each other. Neglecting the exciting current, the following holds good:

$$N_2 (I_2 - I_1) = (N_1 - N_2) I_1$$

or $$N_1 I_1 = N_2 I_2$$

or $$\frac{I_2}{I_1} = \frac{N_1}{N_2} = a \tag{6.84}$$

Power transferred inductively or by transformer action is given by

$$V_2 (I_2 - I_1) = \frac{a-1}{a} V_2 I_2$$

The remainder of the power, $1/a$ times the output, is transferred conductively through electrical connection.

With respect to a two-winding transformer and the same input–output specification, copper is saved in the case of an auto-transformer. Assuming the same core flux and current density in both the cases,

$$\frac{W_{AT}}{W_T} = \frac{I_1(N_1 - N_2) + (I_2 - I_1)N_2}{I_1 N_1 + I_2 N_2}$$

$$= \frac{2(I_2 - I_1)N_2}{2I_2 N_2} = 1 - \frac{I_1}{I_2} = 1 - \frac{V_2}{V_1} = \frac{a-1}{a} \qquad (6.85)$$

where W_T and W_{AT} are the conductor materials in a normal transformer and an auto-transformer, respectively.

Equation (6.85) shows that as nearer is V_2 to V_1, the greater is the saving. In addition, the I^2R loss in the auto-transformer is lower and the efficiency higher than in the two-winding transformer.

Auto-transformers are mainly used for interconnecting systems that are operating at roughly the same voltage, for starting cage-type induction motors, as boosters to raise the voltage in ac feeders, and as a regulating transformer for illumination control, etc. If an auto-transformer is used to supply a low-voltage system from a high-voltage system, it is essential to earth the common connection; otherwise there is a risk of serious shock. However, in such cases, it is preferable to use ordinary transformers to ensure electrical circuit isolation.

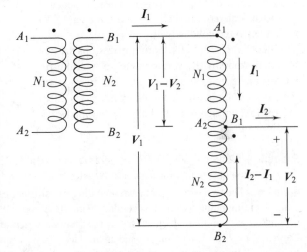

Fig. 6.34 Auto-transformer connection of a two-winding transformer

An auto-transformer can also be obtained by connecting a normal two-winding transformer in a special manner in which the two windings are electrically connected as shown in Fig.6.34.

Example 6.11 A 200/400 V, 10 kVA, 50 Hz single-phase transformer has a core loss of 75 W and full load copper loss of 150 W. (a) Calculate the efficiency of the transformer at full load 0.8 pf. (b) The transformer is now connected as an auto-transformer to give 200/600 V. Calculate the kVA rating and efficiency at full load 0.8 pf.

Solution Two-winding transformer:

Power output $= 10 \times 0.8 = 8$ kW

Losses $= 75 + 150 = 225$ W

Efficiency $= \dfrac{8}{8.225} \times 100 = 97.26\%$

Auto-transformer: With reference to the auto-transformer connection shown in Fig. 6.35,

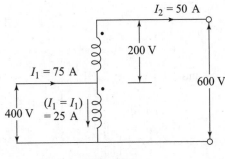

Fig. 6.35

$$I_2 = \frac{10 \times 1000}{200} = 50 \text{ A}$$

$$I_1 - I_2 = \frac{10 \times 1000}{400} = 25 \text{ A}$$

$$\therefore \qquad I_1 = 75 \text{ A}$$

$$\text{kVA rating} = \frac{600 \times 50}{1000} = 30$$

Power output $= 30 \times 0.8 = 24$ kW

$$\text{Efficiency} = 1 - \frac{0.225}{24} = 99.06\%$$

It may be noted that when a two-winding transformer is connected as an auto-transformer, the rating goes up from 10 kVA to 30 kVA and its efficiency increases from 97.26% to 99.06%. This is possible because 20 kVA is delivered inductively and the balance 10 kVA is delivered conductively.

One of the most common applications of auto-transformers is the variac having a winding on the toroidal core as shown in Fig. 6.36. In a variac, the output voltage can be varied from 0% to 120% of the input voltage with the help of a carbon brush moving contact.

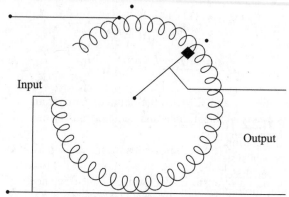

Input

Output

Fig. 6.36 A single-phase variac

Example 6.12 An 11,500/2300 V transformer is rated at 150 kVA as a two-winding transformer. If the two windings are connected in series to form an auto-transformer, what will be the voltage ratio and the output?

Solution The two windings of a two-winding transformer can be connected in series to form an auto-transformer, either winding being used as a secondary. Therefore, the voltage ratio and the output of the auto-transformer will depend on the winding which is used as the secondary winding.

Case-I: The 2300 V winding is used as the secondary.

Rating of the two-winding transformer, $S_T = 150$ kVA

Primary voltage of the auto-transformer, $V_1 = 11500 + 2300 = 13.8$ kV

Secondary voltage of the auto-transformer, $V_2 = 2.3$ kV

With reference to Fig. 6.34,

$$\text{The two-winding transformer voltage ratio, } a = \frac{V_1 - V_2}{V_2} = \frac{N_1}{N_2} = \frac{11.5}{2.3} = 5$$

$$\text{The auto-transformer voltage ratio, } a' = \frac{V_1}{V_2} = \frac{V_1 - V_2 + V_2}{V_2} = a + 1 = 6$$

$$\text{The turns ratio, } a = \frac{13.8}{2.3} = 6$$

$$\text{Rating of the transformer, } S_T = (V_1 - V_2)\, I_1 = (I_2 - I_1)\, V_2 \tag{I}$$

$$\text{Rating of the auto-transformer, } S_{AT} = V_1 I_1 = V_2 I_2 \tag{II}$$

$$\text{But,} \quad \frac{I_2 - I_1}{I_1} = \frac{N_1}{N_2} = a \tag{III}$$

$$\text{Then,} \quad I_1 = \left(\frac{1}{1 + a}\right) I_2 \tag{IV}$$

substituting Eq. (IV) in Eq. (I),

$$S_T = V_2 \left(\frac{V_1}{V_2} - 1\right)\left(\frac{1}{1 + a}\right) I_2 = (1 + a - 1)\left(\frac{1}{1 + a}\right) V_2 I_2 = \left(\frac{a}{1 + a}\right) S_{AT} \tag{V}$$

Using Eq. (V), $S_{AT} = \left(\dfrac{1 + a}{a}\right) \times 150 = 180$ kVA

Case-II: The 11500 V winding is used as the secondary.

$V_1 = 13.8$ kV

$V_2 = 11.5$ kV

Voltage ratio, $a = \dfrac{13.8}{11.5} = 1.2$

Voltage ratio, $a = \dfrac{13.8 - 11.5}{11.5} = \dfrac{2.3}{11.5} = 0.2$

Using Eq. (V), $S_{AT} = \left(\dfrac{1+a}{a}\right) \times 150 = 900$ kVA

6.11.2.2 Three-phase Transformers

Generation, transmission, and distribution in ac power systems are usually accomplished by balanced three-phase circuits because of the advantages the three-phase systems have over single-phase systems (see Section 5.1). Three-phase transformers are employed to step up or step down the voltage levels in power systems. A three-phase transformer can be a single three-phase unit or can be made up of three single-phase trasformers.

A single-unit three-phase transformer has the high- and low-voltage three-phase windings placed on the limbs of a common magnetic core, with the low-voltage windings located closest to the limbs, as shown in Fig. 6.37. The advantage of using a

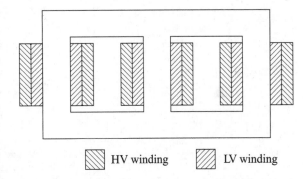

HV winding ▧   LV winding ▨

Fig. 6.37   Cross-sectional view of a single-unit three-phase transformer

single-unit three-phase transformer is that it requires less space, weighs less, has a higher efficiency, and costs approximately 15% less than a bank of single-phase transformers. However, its disadvantage is that in case of a fault in one phase, the transformer is required to be replaced.

6.11.2.3 Three-phase Transformer Connections

To obtain three-phase connections, the primary and secondary windings can be connected in star or delta, which leads to the following four types of connections:

(a) Primary in star (Y) and secondary in delta (Δ)
(b) Primary in Y and secondary in Y
(c) Primary in Δ and secondary in Δ
(d) Primary in Δ and secondary in Y

While making the winding connections, it is important to ensure that the winding connections are made by taking care of the polarities of the voltages, as provided by the dot markings. In order to study the above-stated four types of connections, consider three identical single-phase transformers A, B, and C with $a = V_1/V_2 = I_2/I_1$. For the sake of distinction, the primary wind-

A_1 ⚊⚊⚊ A_2 a_2 ⚊⚊⚊ a_1 Transformer *A*

B_1 ⚊⚊⚊ B_2 b_2 ⚊⚊⚊ b_1 Transformer *B*

C_1 ⚊⚊⚊ C_2 c_2 ⚊⚊⚊ c_1 Transformer *C*

Primary winding   Secondary winding

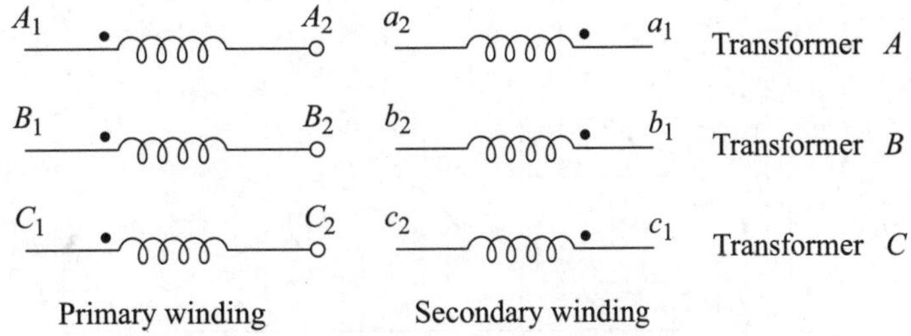

Fig. 6.38   Three single-phase two-winding transformers

ings of the transformers A, B, and C in Fig. 6.38 have been, respectively, marked as $A_1 A_2$, $B_1 B_2$, and $C_1 C_2$. Correspondingly, the secondary windings have been shown with lower case letters as $a_1 a_2$, $b_1 b_2$, and $c_1 c_2$. The instantaneous polarity of each winding is indicated by the dot markings (see Section 3.12.2), which indicates that at the instant the primary terminal A_1 has a positive polarity, the polarity of the secondary terminal a_1 will also be positive.

(a) Primary in star and secondary in delta   In this case, the primary windings of the transformers are connected in star by joining terminals A_2 B_2 C_2 as shown in Fig. 6.39(a). If the primary terminals A_1 B_1 C_1 are now connected across a three-phase voltage supply of V volts (line to line), the primary phase voltages $V_{AN} = V_{BN} = V_{CN} = V/\sqrt{3}$. If the voltage phase sequence is taken as ABC, the phase voltages, though equal in magnitude, will have a phase displacement of 120° between themselves as shown in the phasor diagram in Fig. 6.39(d).

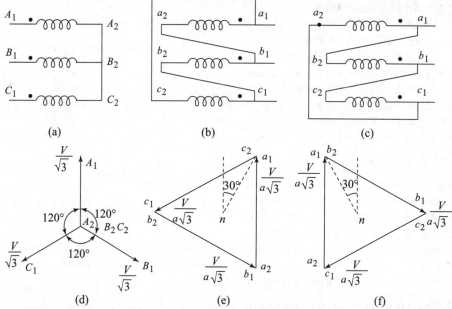

(a)   (b)   (c)

(d)   (e)   (f)

Fig. 6.39   Star-delta connection with phasor diagrams: (a) primary winding star connection, (b) secondary winding delta connection, (c) secondary winding alternate delta connection, (d) phasor diagram for primary winding, (e) phasor diagram for secondary delta connection, –30° phase shift, and (f) phasor diagram for secondary alternate delta connection, +30° phase shift

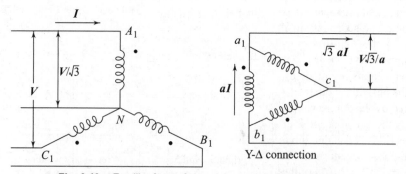

Fig. 6.40   Familiar form of star–delta transformer connection

The secondary windings of the three transformers are connected in delta. The delta connection is basically a series connection and one form of this connection is obtained by connecting a_2 to b_1, b_2 to c_1, and c_2 to a_1 and the junctions of these terminals are connected to feed a three-phase load, as shown in Fig. 6.39(b). The phasor diagram for the delta-connected secondary windings is obtained by drawing secondary voltage phasors parallel to the primary voltage phasors, since the induced voltages in secondary windings are in phase with the corresponding primary voltages. The phasor diagram for the delta-connected secondary is shown in Fig. 6.39(e). It may be noted from the phasor diagrams in Figs 6.39(d) and (e) that there is a phase shift of −30° (or, phase lag of 30°) between the voltage of the secondary phase a to neutral n (which is fictitious in this case) and the voltage of the primary phase A to neutral N. On the other hand, if the delta connections of the secondary are reversed [Fig. 6.39(c)], i.e., a_1 is connected to b_2, b_1 is connected to c_2, and c_1 is connected to a_2, the phase shift between the voltage of the secondary phase a to neutral n and the voltage of the primary phase A to neutral N changes to +30° (or, phase lead of 30°) as shown in Fig. 6.39(f). The more familiar form of the three-phase star–delta connection is shown in Fig. 6.40.

The star–delta connection is used to step down high voltages to medium and low voltages. A desirable feature of this connection is that the neutral on the high-voltage side (star side) can be grounded.

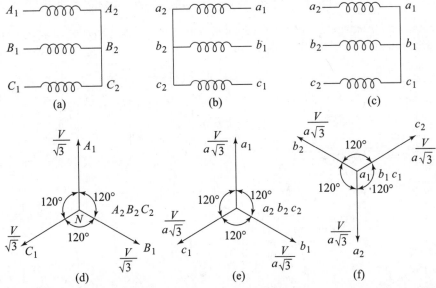

Fig. 6.41 Star–star connection with phasor diagrams: (a) primary winding star connection, (b) secondary winding star connection, (c) secondary winding alternate star connection, (d) phasor diagram for primary winding, (e) phasor diagram for secondary winding star connection, and (f) phasor diagram for secondary winding alternate star connection

(b) Primary in star and secondary in star   The star connection, obtained by connecting primary terminals A_2, B_2, and C_2 to a common point N is shown in Fig. 6.41(a). As in the previous case, it is seen that the primary phase to neutral voltages have a magnitude of $V/\sqrt{3}$ volts and possess a phase displacement of 120° between themselves. The phasor diagram is shown in Fig. 6.41(d).

The two methods of connecting the secondary are shown in Figs 6.41(b) and (c). According to the polarity markings, the secondary phase voltages are in phase with the primary phase voltages and each has a magnitude equal to $V/(a\sqrt{3})$ volt. Reference to the corresponding secondary voltages in the phasor diagram shown in Fig. 6.41(e) shows that there is no phase shift between the primary and secondary phase voltages. On the other hand, if the star connection on the secondary side is obtained by joining terminals a_1, b_1, and c_1 [see Fig. 6.41(c)], there occurs 180° phase shift between the primary and secondary phase voltages. Figure 6.41(f) shows the phasor diagram for the alternate star connection on the secondary side, while Fig. 6.42 shows the commonly used form of star–star connection.

This form of connection is rarely employed due to the problem of magnetizing current inrush.

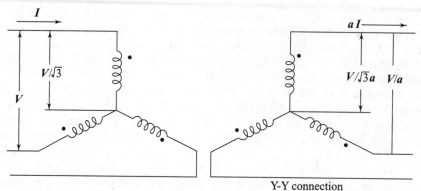

Fig. 6.42 Familiar form of star–star connection

(c) Primary in delta and secondary in delta Figure 6.43 shows the primary connected in delta along with the two types of delta connections for the secondary.

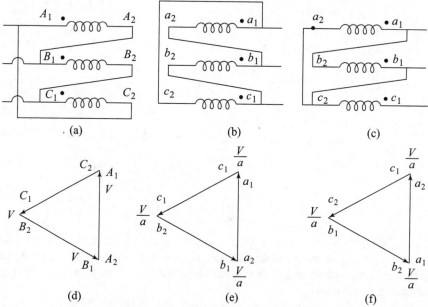

Fig. 6.43 Delta–delta connection with phasor diagrams: (a) primary winding delta connection, (b) secondary winding delta connection, (c) secondary winding alternate delta connection, (d) phasor diagram for primary winding, (e) phasor diagram for secondary winding delta connection, 0° phase shift, and (f) phasor diagram for secondary winding alternate delta connection, 180° phase shift

A comparison of the phasor diagrams in Figs 6.43(d) and (e) shows that there is 0° phase shift between the voltages on the secondary and primary sides and from Fig. 6.43(f) it is seen that there is 180° phase shift between the secondary and primary voltages. Figure 6.44 shows the frequently employed delta–delta connection.

The advantage of the delta–delta connection is that even in case of a fault, repair, or maintenance in one of the transformers, the other two transformers continue to function and supply power at a reduced rating of 58% of the original rated capacity.

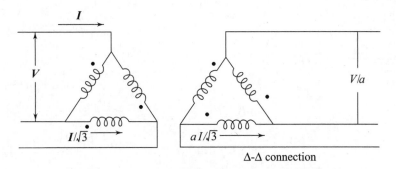

Fig. 6.44 Commonly employed delta–delta connection

(d) Primary in delta and secondary in star The circuit diagrams for the primary connected in delta and two types of star connections of the secondary winding are shown in Fig. 6.45(a) to (c). The corresponding phasor diagrams are drawn in Figs (d) to (f).

In the delta-star connection, it may be noted that the secondary star side phase voltage leads primary delta side fictitious phase voltage by 30° for the secondary connection of Fig. 6.45(b); while for the secondary connection of Fig. 6.45(c), the star side phase voltage lags primary delta side phase voltage by 150°. This type of connection is used for stepping up the voltages. Figure 6.46 shows the familiar form of the delta–star connection.

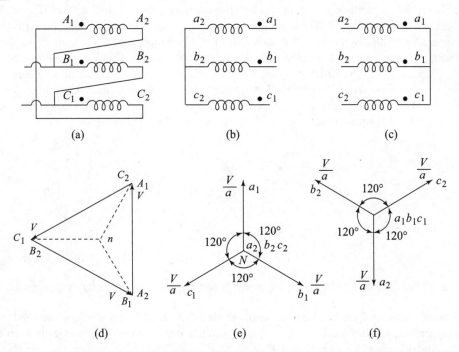

Fig. 6.45 Delta–star connection with phasor diagrams: (a) primary winding delta connection, (b) secondary winding star connection, (c) secondary winding alternate star connection, (d) phasor diagram for primary winding, (e) phasor diagram for secondary winding star connection, 0° phase shift, and (f) phasor diagram for secondary winding alternate star connection, 180° phase shift

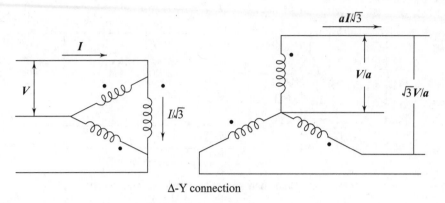

Δ-Y connection

Fig. 6.46 Familiar form of delta–star connection

6.11.2.4 Representation of Phase Shift in Transformer Specifications

The above discussion shows that a phase shift occurs in the secondary voltages of a three-phase transformer. Depending upon the type of connection, this phase shift could be ±30° or 0° (no phase shift) or 180° (phase reversal) with respective line to neutral or phase to neutral voltages. Transformer specifications, therefore, in addition to specifying frequency, primary and secondary voltages, also include phase shift in the secondary voltage on the name plate. To represent the phase shift or the vector group (phasor group), a combination of upper/lower case alphabets and numerals is adopted. The first alphabet is Y or D (upper case), which represents the primary winding connection. Similarly, the second alphabet is y or d, which represents the secondary winding connection. The third, which is a numeral, is employed to signify the phase shift. Clock convention is used to show phase shift. The minute hand is customarily used to show the primary phase to neutral voltage and always occupies the 12 o'clock position. The position of the hour hand occupies a position depending upon the phase shift in the phase to neutral voltage on the secondary side. Figure 6.47 indicates the clock convention used to symbolize the phase shift between the voltages on the primary and secondary sides.

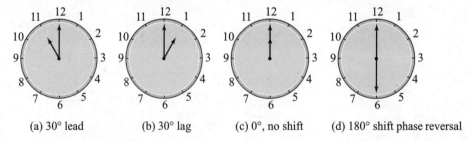

(a) 30° lead   (b) 30° lag   (c) 0°, no shift   (d) 180° shift phase reversal

Fig. 6.47 Clock convention to represent phase shift between voltages on the primary and secondary sides

As an example, the transformer name plate would show $Yd1$ or $Yd11$ for the primary (star) and the two secondary (delta) connections shown in Fig. 6.39. Table 6.4 summarizes the transformer name plate representation of the vector groups shown in Figs 6.39 to 6.46.

Table 6.4 Summary of transformer vetor (phasor) groups

Primary winding connection	Secondary winding connection	Vector (phasor) group
Star	Delta	$Yd1$
	Alternate delta	$Yd11$
Star	Star	$Yy0$
	Alternate star	$Yy6$
Delta	Delta	$Dd0$
	Alternate delta	$Dd6$
Delta	Star	$Dy11$
	Alternate star	$Dy1$

Example 6.13 A three-phase step-down transformer is connected to 6600 V mains and draws a current of 10 A. The ratio of the turns per phase is 15. Calculate the secondary line voltage, line current, and output for the following connections: (a) mesh–mesh, (b) star–star, (c) mesh–star, (d) star–mesh.

Solution (a) Mesh–mesh

$$V_{1P} = 6600 \text{ V}, \ V_{2P} = V_{2L} = \frac{6600}{15} = 440 \text{ V}$$

$$I_{1P} = \frac{10}{\sqrt{3}} \text{ A}, \ I_{2P} = 15 \times \frac{10}{\sqrt{3}} \text{ A}, \ I_{2L} = 15 \times \frac{10}{\sqrt{3}} \times \sqrt{3} = 150 \text{ A}$$

(b) Star–star

$$V_{1P} = 6600 \text{ V}, \ V_{2P} = \frac{6600}{15 \times \sqrt{3}} \text{ V}, \ V_{2L} = \frac{6600}{15 \times \sqrt{3}} \times \sqrt{3} \text{ V} = 440 \text{ V}$$

$$I_{1P} = 10 \text{ A}, \ I_{2P} = 15 \times 10 = 150 \text{ A}, \ I_{2L} = 15 \times 10 = 150 \text{ A}$$

(c) Mesh–star

$$V_{1P} = 6600 \text{ V}, \ V_{2P} = \frac{6600}{15} = 440 \text{ V}, \ V_{2L} = \frac{6600}{15} \times \sqrt{3} = 762.1 \text{ V}$$

$$I_{1P} = \frac{10}{\sqrt{3}} \text{ A}, \ I_{2P} = 15 \times \frac{10}{\sqrt{3}} \text{ A}, \ I_{2L} = 15 \times \frac{10}{\sqrt{3}} = 86.6 \text{ A}$$

(d) Star–mesh

$$V_{1P} = \frac{6600}{\sqrt{3}} \text{ V}, \ V_{2P} = \frac{6600}{15 \times \sqrt{3}} \text{ V}, \ V_{2L} = \frac{6600}{15 \times \sqrt{3}} = 254 \text{ V}$$

$$I_{1P} = 10 \text{ A}, \ I_{2P} = 15 \times 10 = 150 \text{ A}, \ I_{2L} = 15 \times 10 \times \sqrt{3} = 259.8 \text{ A}$$

The output of the transformer will be the same for all types of connections:

$$\text{Output} = 3 \times V_{1P} \times I_{1P} = 3 \times 6600 \times \frac{10}{\sqrt{3}} = 114.3 \text{ kVA}$$

Example 6.14 A 75-hp, 415-V, three-phase induction motor with an efficiency of 0.9 and a power factor of 0.85 on full load is supplied from a 6600/415 V, mesh–star connected transformer. Ignoring the magnetizing current, calculate the currents in the high- and low-voltage transformer phases when the motor is running at full load.

Solution Motor output = $75 \times 746 = 55{,}950$ W

$$\text{Motor input} = \frac{55{,}950}{0.9} = 62{,}166.7 \text{ W}$$

Now, $\sqrt{3} \times V_L \times I_L \times \cos\varphi = 62{,}166.7$

where V_L and I_L are the line voltage and the line current, respectively.

The line current on the low voltage star side of the transformer

$$I_{LV} = \frac{62166.7}{\sqrt{3} \times 415 \times 0.85} = 101.75 \text{ A}$$

The transformation ratio,

$$a = \frac{6600}{415/\sqrt{3}} = 27.546$$

The ratio of phase currents

$$a = \frac{I_{LV}}{I_{HV}} = \frac{101.75}{I_{HV}}$$

$$\therefore \qquad I_{HV} = \frac{101.75}{27.546} = 3.6938 \text{ A}$$

$$I_L = \frac{41444.44}{\sqrt{3} \times 440 \times 0.85} = 63.98 \text{ A}$$

$$a = \frac{6600}{440 \times \sqrt{3}} = 15\sqrt{3}$$

$$\frac{V_{1P}}{V_{1P}} = \frac{I_{2P}}{I_{1P}} = 15\sqrt{3}$$

$$I_{1P} = \frac{63.98}{15\sqrt{3}} = 2.46 \text{ A}$$

6.11.3 Special Types of Transformers

Transformers can be used in fields other than electrical power transmission and distribution. They find use in low-power applications, as instrument transformers, in protection systems, and in electronic circuits. The two low-power applications of transformers are as (i) current transformer in protection systems and (ii) audio-frequency transformer in electronic amplifiers.

6.11.3.1 Current Transformer

For measurement, control, and protection systems involving alternating currents greater than about 100 A, it is difficult to construct ammeters, current coils of wattmeters, energy meters, relays, etc. designed to carry the full magnitude of current. To overcome such difficulties, current transformers are used. A current transformer (CT) steps down high currents to low currents, of the order 1–5 A. Current transformers have lower VA ratings of the order of 10 VA.

A current transformer usually consists of a single-turn primary (normally the line itself) and a secondary winding, which is short-circuited through an ammeter or connected to the current coil of a relay.

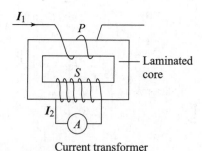

Current transformer

Bar primary current transformer

Fig. 6.48 Schematic representation of current transformers

The circuit connected to the secondary winding is termed as the burden of the current transformer. Current transformers are designed for low VA or burden. Figure 6.48 shows schematic representations of current transformers.

The secondary winding of a current transformer should never be open-circuited while the primary winding is carrying current, since all the primary ampere turns would then be available for producing flux in the absence of the secondary ampere turns. The core losses due to the high flux density would cause excessive heat and damage the core and windings. Under such conditions a very high voltage may be induced in the secondary winding. If at all the ammeter is to be removed, the secondary winding must be short-circuited.

6.11.3.2 Audio-frequency Transformer

An audio-frequency transformer is used in the output stage of an audio-frequency electronic amplifier. It is used to match the output impedance of the power amplifier stage with the load. The load on the transformer in this case is constant but the frequency is variable over the audio-frequency range of 20 Hz to 20 kHz.

The characteristics of an iron-core transformer depart considerably from that of an ideal transformer as far as its frequency response is concerned. Figure 6.49(a) shows the equivalent circuit of a transformer that is valid for a wide range of frequencies. This circuit is a modification of the equivalent circuit of a power transformer. To account for the response at higher frequencies, the capacitors C_1 and C_2, which represent the equivalent interwinding distributed capacitance of the input and the output transformer windings, are added. A desirable characteristic of an audio transformer is that the ratio V_2/V_1, called the response, should remain constant over the frequency range 0 Hz to 20 kHz, called the band. The phase response of an audio-frequency transformer is defined as the phase angle of V_2 with respect to V_1. The frequency response of a transformr is shown in Fig. 6.49(b).

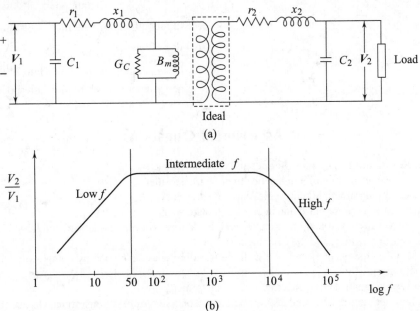

Fig. 6.49 Equivalent circuit and logarithmic frequency response of an iron-core transformer

Since the frequency is varying over a wide range, the leakage reactances of the windings are no longer constant and have been shown to depend on frequency. At low frequencies, inductances behave as a short circuit. The shunt branch X_m, produced by the mutual flux, increases as the frequency is increased up to 100Hz. At 100 Hz, X_m is sufficiently high to restrict the magnetizing current to a relatively small value, causing the response to flatten out from 100 Hz to about 5 kHz. As the frequency increases, the voltage drop across the leakage reactances increases and seriously limits the output. This accounts for the drooping characteristic at high frequencies.

Well-designed audio transformers are relatively expensive since their size must be large and they must possess an extremely high mutual inductive reactance X_m or susceptance B_m.

Recapitulation

Induced voltage, $e = E_m \cos \omega t$

Rms value of the induced voltage,
$$E = 4.44 f N \phi_m = 4.44 f N B_m \text{ A}$$

Hysteresis loss, $P_h = k_h f V B_m^n$

Eddy-current loss, $P_e = k_e f^2 B_m^2 \tau^2 V$

Transformation ratio, $a = \dfrac{E_1}{E_2} = \dfrac{V_1}{V_2} = \dfrac{N_1}{N_2} = \dfrac{I_2}{I_1}$

Complex power, $S_1 = S_2 = V_1 I_1^* = V_2 I_2^*$

Magnetizing current, $I_\phi = \dfrac{V_1}{jX_m} = -j B_m V_1$

Core-loss current, $I_C = \dfrac{V_1}{R_C} = G_c V_1$

No-load current, $I_0 = I_\phi + I_C$

Magnitude of no-load current, $I_0 = \sqrt{(I_\phi^2 + I_c^2)}$

Power factor on no-load, $\cos \varphi_0 = \dfrac{I_\phi}{I_0}$

Equivalent circuit parameters, $R_{eq} = r_1 + r_2'$
$$X_{eq} = x_1 + x_2'$$
$$Z_{eq} = R_{eq} + X_{eq}$$

$\% \text{ Regulation} = \dfrac{V_{2,\text{ no-load}} - V_{2,\text{ full-load}}}{V_{2,\text{ full-load}}} \times 100$

$= \dfrac{V_1 - a V_2}{V_1} = \dfrac{I_1 \left(R_{eq} \cos\theta \pm X_{eq} \sin\theta \right)}{V_1}$

pu regulation $= R_{pu} \cos\theta \pm X_{pu} \sin\theta$

Efficiency, $\eta = 1 - \dfrac{\text{losses}}{\text{output power } + \text{ losses}}$

Maximum efficiency condition, $P_C = I_2^2 R_{eq}$

All-day efficiency, $\eta_{\text{all-day}}$

$= \dfrac{\text{Energy output in kWh in 24 hours}}{\text{Energy input in kWh for the corresponding 24 hours}}$

Assessment Questions

1. (a) State the various parts of a transformer and their functions.
   (b) Explain the construction of a magnetic core of a transformer.
2. Explain the working principle of a single phase transformer.
3. (a) Enumerate the assumptions made in an ideal transformer.
   (b) Starting with an ideal transformer on no-load, explain its working under load condition. Draw the phasor diagrams and the equivalent circuit.
4. Explain how the equivalent circuit of a working transformer is developed by doing away with each of the ideal conditions one by one.
5. What is an approximate equivalent circuit of a transformer?
6. Describe how open and short circuit tests are performed on a single phase transformer. Discuss the significance of the tests.
7. Define regulation of a transformer. Derive an expression for the regulation of a transformer when it is delivering load at a leading power factor.
8. Define transformer efficiency and derive the condition under which it will have maximum efficiency.
9. Define all-day efficiency of a transformer and explain its significance in selecting a transformer for servicing variable loads on a daily basis.

10. (a) What is an auto-transformer? Explain its working and derive an expression for the turns ratio.

    (b) Explain how power is transferred in an auto-transformer.

11. Prove that for an equivalent input-output specification there is a saving in copper in an auto-transformer when compared with a conventional two winding transformer.

12. Write a note on the uses of an auto-transformer.

13. (a) State the relative merits of using a bank of three-phase transformers and a single unit transformer in a three phase system.

    (b) Explain with circuit and phasor diagrams the various three phase systems that can be obtained from a bank of three single phase identical transformers.

14. Explain the clock convention to represent phase shift between the primary and secondary voltages in three phase transformers.

15. Enumerate the various types of three phase transformer connections. What is the advantage of using delta–delta connections?

Problems

6.1 (a) A coil of N turns is wound around a rectangular magnetic core. From the first principle, derive an expression for the voltage induced in the coil when it is connected across an ac voltage source of maximum voltage V_m and frequency f Hz. Show that the magnetizing current lags the induced voltage by $\pi/2$ radians. State all the assumptions made.

    (b) Prove that the rms value of the voltage induced per turn is $4.44\, f\phi_m$ volt.

6.2 The primary winding of a transformer has 300 turns. When connected across a 220-V, 50-cycles/sec supply, it draws a current of 5 A at a power factor of 0.25 on no-load. Determine (a) the maximum value of the core flux; (b) magnetizing current, and (c) the loss in the core. How many turns are required in the secondary to step-up the voltage to 415 V?   [(a) 3.03 mWb, (b) 4.84 A, (c) 275 W, 566]

6.3 The secondary of a single-phase transformer is supplying a current of 300 A at a power factor of 0.8 lagging. The no-load current of the transformer is 5 A at a power factor of 0.2 lagging. Calculate the primary current and its power factor. Assume the ratio of transformation to be 3 and neglect voltage drops due to the impedance of the windings.   [103.79 A, 0.78]

6.4 A load impedance of $(40 + j30)$ Ω is connected to the secondary of an ideal single-phase transformer. If the primary and secondary voltages are 400 V and 200 V, respectively, determine (a) secondary current, (b) primary current, (c) real and reactive power, and (d) the equivalent load impedance, when it is connected across the primary, for the transformer to supply the same load.

$\qquad$ [(a) $4\angle -36.87°$, (b) $2\angle -36.87°$, (c) 640, 480, (d) $200\angle 36.87°$ Ω]

6.5 A voltage source supplies a current of 50 A at a lagging power factor of 0.75 to the primary of a single-phase transformer. If the no-load current is 1 A at a power factor of 0.25 lagging, determine the secondary current and its power factor. Assume that there is no drop in the windings and the transformation ratio is 3.

$\qquad$ [147.39 A, 0.76 lagging]

6.6 For the single-phase transformer shown in Fig. 6.50, calculate (a) the no-load current, (b) primary current and voltage, (c) secondary current and voltage, (d) voltage at the load terminals, and (e) real and reactive power. The primary of the transformer is connected to a 1100-V voltage supply and the latter supplies a current of 50 A at a power factor of 0.8 lagging. Take the transformation ratio of the transformer as 5.

$\qquad$ [(a) $0.49 \angle -63.43°$ A; (b) $49.56 \angle -36.62°$ A, $950,85 \angle -7.83°$ V;

$\qquad$ (c) $247.8 \angle -36.62°$ A, $190.17 \angle -7.83°$; (d) $163.61 \angle -17.18°$ V;

$\qquad$ (e) 38.23 kW, 13.49 kVAR]

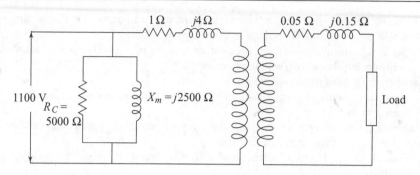

Fig. 6.50

6.7 For Problem 6.6, determine the equivalent impedance and the equivalent load referred to the primary circuit. $[(16.80 + j9.23)\,\Omega, (2.25 + j7.75)\,\Omega]$

6.8 The secondary of a single-phase transformer feeds a load of 50 kW at a power factor of 0.8 lagging. The voltage at the load terminals is 415 V. Determine the supply voltage and the current if the ratio of transformation is 5. Compute the regulation of the transformer for the given load. The parameters of the transformer are shown in Fig. 6.51. $[2133.18 \angle 1.6°\text{ V}, 33.22 \angle 39.0°, 2.80\%]$

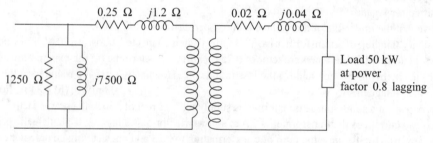

Fig. 6.51

6.9 The following is the test data for a 50-kVA, 1100/110-V transformer:
oc test performed on the 110 V side: 110 V, 1.0 A, 100 W
sc test performed on the 1100 V side: 80 V, full-load current, 800 W
Draw the equivalent circuit of the transformer referred to (a) the high-voltage side and (b) the low-voltage side. $[(a)\ G_C = 0.83\text{ S}, B_m = 0.37\text{ S}, R_{eq} = 0.387\,\Omega, X_{eq} = j1.72\,\Omega;$
$(b)\ G_C = 0.00835\text{ S}, B_m = 0.0037\text{ S}, R_{eq} = 0.00387\,\Omega, X_{eq} = j0.0172\,\Omega]$

6.10 Determine the regulation of the transformer in Problem 6.9 when it supplies a load of 50 kW at (a) 0.8 lagging power factor, (b) unity power factor, and (c) 0.8 leading power factor. Also, draw the vector diagram in each case. $[(a)\ 7.09\%, (b)\ 1.85\%, (c)\ -3.37\%]$

6.11 Compute the regulation of the transformer in Problem 6.9 when it is supplying a load current at (a) 50%, (b) 75%, and (d) 125% of its rated capacity, at a lagging power factor of 0.8.
$[(a)\ 2.80\%, (b)\ 4.22\%, (c)\ 7.09\%]$

6.12 The primary and secondary resistances of a 75 kVA transformer are 0.25 Ω and 0.01 Ω, respectively. The corresponding leakage reactances are 1.0 Ω and 0.025 Ω. If the primary voltage is 1100 V and the ratio of transformation is 2.5, calculate the per unit values of the resistances and reactances and show that they are independent of the side to which they are referred. Also, determine the per unit regulation of the transformer for 0.8 power factor (a) lagging and (b) leading. What will be the secondary terminal voltage on full load at unity power factor? $[R = 0.019\text{ pu}, X = 0.072\text{ pu}, (a)\ 0.058\text{ pu}, (b)\ -0.028\text{ pu}]$

6.13 A 20-kVA, 1100/110-V transformer has primary and secondary winding resistances of 4.0 Ω and 0.04 Ω, respectively. If the core loss is 2.5 kW, determine the efficiency of the transformer when it is delivering (a) full load and (b) half load at 0.8 power factor. Calculate the load at which the transformer has maximum efficiency. What is the maximum efficiency of the transformer?

$$[(a)\ \eta = 75.67\%;\ (b)\ 71.68\%;\ \text{Load at}\ \eta_{max} = 15.56\ \text{kW},\ \eta_{max} = 75.68\%]$$

6.14 A single-phase 100-kVA distribution transformer is loaded as under during the day:

No load for 4 hours
50% load at unity power factor for 8 hours
75% load at 0.8 power factor for 6 hours
Full load at 0.9 power factor for 6 hours
If the full load copper loss is 5 kW and the core loss is 2 kW, calculate the all-day efficiency of the transformer. [92.53%]

6.15 A 50-kVA, 2200/220-V two-winding transformer has a copper loss of 400 W at full load and an iron loss of 100 W. The transformer windings are now connected in series to form an auto-transformer. Compute (a) the efficiency of the two-winding transformer at full load, (b) the transformation ratio and the output when operating as an auto-transformer, either winding is used as the secondary winding, and (c) the efficiency at full load with a power factor of 0.8.   [(a) 98.77%; (b) 11, 55 kVA, 1.1, 550 kVA; (c) 98.88%, 99.89%]

6.16 A two-winding transformer has primary and secondary voltages of V_1 and V_2, respectively. The corresponding primary and secondary turns are N_1 and N_2. If the primary and secondary currents are I_1 and I_2, respectively, determine the saving in copper if the transformer is wound as an auto-transformer to have the same voltage and current ratings. $[1 - (1/a)]$

6.17 The secondary of an auto-transformer supplies 200 V to a load of 10 kW at 0.8 power factor. If the primary voltage is 500 V, determine (a) the transformation ratio, (b) the primary and secondary currents, (c) the power transformed in secondary, and (d) the power from the supply. What is the ratio of weight of copper of the auto-transformer to the weight of copper of a similar two-winding transformer?

$$[a]\ 2.5;\ (b)\ 25\ \text{A},\ 62.5\ \text{A};\ (c)\ 6000\ \text{W};\ (d)\ 4000\ \text{W};\ 3:5]$$

6.18 A balanced star-connected load of 25 kW at 0.8 power factor is connected to the secondary of a three-phase transformer. The primary winding is delta-connected. If the secondary winding has a line-to-line voltage of 415 V and the transformation ratio is 5, calculate the phase and line currents and the voltage in the primary and secondary windings.   [Secondary winding: $I_L = I_P = 43.48$ A, $V_L = 415$ V, $V_P = 239.6$ V

Primary winding: $I_L = 15.06$ A, $I_P = 8.70$ A, $V_L = V_P = 1198$ V]

6.19 The primary windings of three identical single-phase transformers having a transformation ratio of 10 are connected in star across a three-phase 2200 V supply. (a) If the secondary windings are delta-connected, determine the secondary phase and line voltages. (b) If the primary windings are then delta-connected in delta across the supply, what will be the secondary phase and line voltages?

$$[(a)\ V_L = V_P = 127\ \text{V},\ (b)\ V_L = V_P = 220\ \text{V}]$$

Additional Examples

AE 6.1   A single phase 300 kVA, 3300/400 V, single phase transformer has the following winding data: $R_1 = 0.1\ \Omega, X_1 = 2\ \Omega, R_2 = 0.001\ \Omega, X_2 = 0.02\ \Omega$. If the transformer is operating at full load and 0.8 power factor (lagging), compute (a) primary voltage, (b) transformer efficiency.

Solution   Transformation ratio $a = N_1/N_2 = 3300/400$

Referring primary side parameters to the secondary side yields using relationships of Table 6

$$r_1' = 0.1 \Big/ \left(\frac{3300}{400}\right)^2 = 0.1 \times \left(\frac{400}{3300}\right)^2 = 0.0015\ \Omega \text{ and } x_1' = 2\ \times \left(\frac{400}{3300}\right)^2 = 0.0294\ \Omega$$

Then equivalent impedance on the secondary side

$$Z'_{eq} = (0.0015 + 0.001) + j(0.0294 + 0.02) = (0.0025 + j0.0494) \ \Omega$$

Full load secondary current $I_2 = \dfrac{300 \times 1000}{400} = 750 \ \text{A}$

Secondary voltage $E_2 = 400 + 750(0.8 - j0.6) \times (0.0025 + j0.0494) = (423.73 + j2.852) \ \text{V}$

Magnitude of secondary voltage $E_2 = \sqrt{(423.73)^2 + (2.852)^2} = 424.688 \ \text{V}$

(a) Primary voltage $E_1 = 424.688 \times \left(\dfrac{3300}{400}\right) = 3503.7 \ \text{V}$

(b) Copper loss at full load $= (750)^2 \times 0.0025 = 1406.3 \ \text{W}$

Transformer efficiency $\eta = \dfrac{\text{output}}{\text{output} + \text{losses}} \times 100 = \dfrac{300 \times 1000 \times 0.8}{300 \times 1000 \times 0.8 + 1406.3 + 500} \times 100 = 99.21\%$

(c) Voltage drop across the equivalent secondary impedance

$$= 750(0.8 - j0.6) \times (0.0025 + j0.0494) = (23.73 + j28.52) = 37.097 \angle 50.23° \ \text{V}$$

Thus, % voltage regulation $= \dfrac{37.097}{400} \times 100 = 9.27$

AE 6.2   A 50 kVA, 6.6 kV/220 V single phase power transformer has primary winding resistance and reactance of 12 Ω and 20 Ω respectively. Corresponding resistance and reactance for the secondary winding is 0.02 Ω and 0.0222 Ω. (a) What is the magnitude of the primary voltage necessary to circulate full load current when the secondary is short circuited? (b) What is the copper loss under short circuit condition? (b) Compute (i) percentage regulation and (ii) secondary load voltage, when the transformer is supplying a load 40 kVA at 0.7 leading power factor.
Solution   Using Eq. (6.55), the secondary data is referred to the primary side as follows:

$$r'_2 = 0.02 \times \left(\dfrac{6600}{220}\right)^2 = 18 \ \Omega \quad \text{and} \quad x'_2 = 0.0222 \times \left(\dfrac{6600}{220}\right)^2 = 19.98 \ \Omega$$

Equivalent primary impedance $Z_{eq} = (12 + 18) + j(20 + 19.98) = 30 + j39.98 = 49.98 \angle 53.17° \ \Omega$

Full load primary current $I_{FL} = 50/6.6 = 7.58 \ \text{A}$

(a) Magnitude of primary voltage under short circuit $E_1 = I_{FL} \times Z_{eq} = 7.58 \times 49.98 = 378.64 \ \text{V}$

(b) Copper loss under short circuit $P_C = I_{FL}^2 \times R_{eq} = (7.58)^2 \times 30 = 1.72 \ \text{kW}$

Power factor angle $\phi = \cos^{-1}(0.7) = 45.57°$

Primary current when the connected load of 40 kVA, $I_1 = 40/6.6 = 6.06 \ \text{A}$

(i) Use of Eq. (6.71) leads to the calculation of voltage regulation as under

$$\% \text{ Voltage regulation} = \dfrac{6.06 \times \left[30 \times \cos(45.57°) - 39.98 \times \sin(45.57°)\right]}{6600} \times 100 = -0.7\%$$

Primary voltage $E_1 = 6600 - 6.06 \times \left[30 \times \cos(45.57°) - 39.98 \times \sin(45.57°)\right] = 6645.8 \ \text{V}$

(ii) Secondary voltage $E_2 = 6645.8 \times \left(\dfrac{220}{6600}\right) = 221.53 \ \text{V}$

AE 6.3   It is required to obtain a 400 V output voltage from a 250 V supply. Show how a 7.5 kVA, 250 V/150 V single phase transformer can be connected as an auto-transformer for the purpose. Draw the connection diagram and determine the kVA rating of the auto-transformer. What is the output kVA rating of the auto-transformer? Is the HV winding overloaded? Compute the amount of kVA transferred (a) magnetically and (b) conductively.

Solution   In order to obtain an output voltage of 400 V, it is necessary to connect the HV and LV windings in series. Assuming that the dot markings of the windings are known, the HV and LV windings are connected as shown in Fig. AE 6.1.

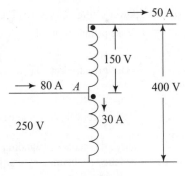

Current rating of HV winding $= 7500/250 = 30$ A

Current rating of LV winding $= 7500/150 = 50$ A

According to the dot convention, the windings are connected in series yielding an output voltage of 400 V. Since the current rating of either the HV or the LV cannot be exceeded, the output rating of the auto-transformer $= 50 \times 400 = 20$ kVA.

Fig. AE 6.1

Supply current $= 20000/250 = 80$ A

Application of KCL at A leads to the current in the HV winding $= 80 - 50 = 30$ A

Thus, the HV winding is not overloaded.

(a) kVA transferred magnetically = kVA rating of HV or LV winding $= 30 \times 250 = 7.5$ kVA

(b) kVA transferred electrically = total kVA – kVA transferred magnetically $= 20 - 7.5 = 12.5$ kVA

AE 6.4   An auto-transformer is connected to two loads of 55 Ω and 88 Ω as shown in Fig. AE 6.2. If the total number of transformer turns is 220, and if 55 Ω load is supplied by tapping the winding at point B such that the number of turns between B to D is 165, calculate (a) voltage across each load, (b) respective load currents, (c) kVA rating and supply current, and (d) magnitude and direction of current in the CD part of the winding. Assume the number of turns between C and D to be 110.

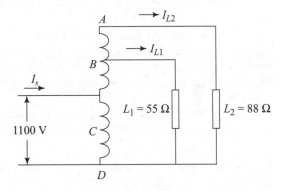

Solution   The assumed load currents and supply current is shown in Fig. AE 6.2.

Voltage/turn $= 1100/110 = 10$

Using the principle that voltage/turn remains constant,

Fig. AE 6.2

(a) Voltage across each load is given by

$$V_{L1} = 165 \times 10 = 1650 \text{ V} \quad \text{and} \quad V_{L2} = 220 \times 10 = 2200 \text{ V}$$

(b) Current through each load is calculated as under

$$I_{L1} = 1650/55 = 30 \text{ A and } I_{L2} = 2200/88 = 25 \text{ A}$$

(c) Rating of the transformer $= 1650 \times 30 + 2200 \times 25 = 104.5$ kVA

Supply current $I_S = 104500/1100 = 95$ A

(d) Application of KCL at node C gives the magnitude of the current flowing in the winding CD.
    Thus the current from point C to D, $I_{CD} = 95 - 30 - 25 = 40$ A.

AE 6.5   A single phase transformer has a power rating of S kVA and its primary and secondary voltages are E_1 kV and E_2 kV respectively. If the corresponding primary and secondary winding impedances are Z_1 and Z_2 ohms respectively, derive expressions for per unit primary and secondary impedances and there from prove that impedance is independent of the primary or secondary side.

Solution   Commence by making the following assumptions:

(i) Base power rating: S kVA, (ii) Base primary voltage: E_{1b} kV, and (iii) Base secondary voltage: E_{2b} kV.

Base primary current $I_{1b} = S/E_{1b}$ A and base secondary current $I_{2b} = S/E_{2b}$ A

$$\text{Base primary impedance } Z_{1b} = \frac{E_{1b} \times 1000}{I_{1b}} = \frac{\left(E_{1b}\right)^2 1000}{S} \ \Omega \tag{6.4.1}$$

$$\text{Base secondary impedance } Z_{2b} = \frac{E_{2b} \times 1000}{I_{1b}} = \frac{\left(E_{2b}\right)^2 1000}{S} \ \Omega \tag{6.4.2}$$

If the primary and secondary winding impedance is assumed to be Z_1 and Z_2 ohms, the corresponding per unit values of the impedance is obtained by using Eqs. (6.4.1) and (6.4.2) in turn as under:

$$Z_{1pu} = \frac{Z_1}{Z_{1b}} = \frac{Z_1 \times S}{\left(E_{1b}\right)^2 1000} \tag{6.4.3}$$

$$Z_{2pu} = \frac{Z_2}{Z_{2b}} = \frac{Z_2 \times S}{\left(E_{2b}\right)^2 1000} \tag{6.4.4}$$

Division of Eq. (6.4.3) by (6.4.4) results in

$$\frac{Z_{1pu}}{Z_{2pu}} = \frac{Z_1}{Z_2 \left(E_{1b}/E_{2b}\right)^2} \tag{6.4.5}$$

It may be recalled that the denominator in Eq. (6.4.5) represents the relationship to refer secondary side impedance to the primary side. Thus, Eq. (6.4.5) takes the form $Z_{1pu} = Z_{2pu}$ which proves that the per unit impedance of a transformer is independent of primary or secondary side.

AE 6.6   Show that per unit resistance of a transformer represents per unit value of the transformer copper loss at full load.

Solution   If the primary and secondary winding full load currents and equivalent resistances are respectively I_1, I_2, R_{1eq}, and R_{2eq}, the total copper loss in the transformer is given by

$$P_C = I_1^2 R_{1eq} = I_2^2 R_{2eq}$$

If the power rating of the transformer is assumed to be S, the per unit value of transformer copper loss can be expressed as

$$P_{Cpu} = \frac{I_1^2 R_{1eq}}{S} = \frac{I_2^2 R_{2eq}}{S} \tag{6.5.1}$$

where $S = E_{1b} \times I_{1b} = E_{2b} \times I_{2b}$ $\hspace{2cm}$ (6.5.2)

At full load, the primary and secondary voltages and currents are rated voltages and currents. Hence, $E_{1b} = E_1$, $E_{2b} = E_2$, $I_{1b} = I_1$, and $I_{2b} = I_2$. Substituting, these values in Eq. (6.5.2) and replacing S with the resultant expression in Eq. (6.5.1) leads to

$$P_{C\,pu} = \frac{I_1^2 R_{1eq}}{E_1 I_1} = \frac{I_2^2 R_{2eq}}{E_2 I_2} = \frac{R_{1eq}}{E_1/I_1} = \frac{R_{2eq}}{E_2/I_2} = \frac{R_{1eq}}{Z_{1b}} = \frac{R_{2eq}}{Z_{2b}} = R_{eq-pu} \tag{6.5.3}$$

Eq. (6.5.3) shows that per unit resistance of a transformer represents its copper loss at full load.

AE 6.7   A 15 kVA, 1100/220 V transformer is supplying 80% load at 0.75 lagging power factor. If the per unit impedance of the transformer is $(0.02 + j0.05)$, compute its voltage regulation.

Solution   Power factor angle $\varphi = \cos^{-1}(0.75) = 41.41°$

Per unit load current $I_2 = 0.8 \left[\cos(41.41°) - j\sin(41.41°) \right] = 0.6 - j0.529$

Assume per unit secondary voltage $V_2 = 1.0\angle 0°$

Per unit primary voltage is computed as follows

$$V_1 = 1 + (0.6 - j0.529)(0.02 + j0.05) = (1.0385 + j0.0195) = 1.0386\angle 1.07°$$

$$\text{Voltage regulation} = \frac{|V_1| - |V_2|}{|V_2|} \times 100 = \frac{1.0386 - 1}{1} \times 100 = 3.86$$

AE 6.7　Determine the efficiency of the transformer in AE 6.6 at the given load. Assume a core loss of 150 W.

Solution　Per unit core loss = 150/15000 = 0.01

Using Eq. (6.5.3), per unit copper loss at full load = 0.02

Per unit copper loss at 0.8 × full load $= (0.8)^2 \times 0.02 = 0.0128$

Per unit power output $P = (\text{complex power output}) \times$ power factor

$$= (V_2 \times I_2) \times 0.6 = (1.0 \times 0.8) \times 0.6 = 0.48$$

Thus, transformer efficiency $\eta = \dfrac{0.48}{0.48 + 0.0128 + 0.01} \times 100 = 94.47\%$

Objective Type Questions

1. A transformer is a device used to
   (i) convert energy
   (ii) generate energy
   (iii) change the level of energy utilization
   (iv) all of these

2. When a sinusoidal voltage of frequency f hertz is applied to a coil of N turns wound around a magnetic core, a sinusoidally varying flux is set up due to
   (i) production of an mmf
   (ii) frequency of the applied voltage
   (iii) N turns
   (iv) none of these

3. In a transformer which of the following cannot be classified as core loss?
   (i) hysteresis loss
   (ii) eddy-current loss
   (iii) copper loss due to current in the windings
   (iv) all of these

4. Which of the following constitute the basic components of a single-phase transformer?
   (i) tank　　　　　　(ii) windings
   (iii) magnetic core　(iv) all of these

5. It is desired to have the current in the transformer windings to not exceed which of the following range?
   (i) 50–70 A　　　　(ii) 70–100 A
   (iii) 100–120 A　　(iv) 120–150 A

6. The mutual flux set up in the core of a transformer links
   (i) the primary winding only
   (ii) the secondary winding only
   (iii) both the windings
   (iv) none of these

7. The magnitude of the emf induced in the secondary depends upon
   (i) N_1　　　　　　　(ii) N_2
   (iii) $\dfrac{N_1}{N_2}$　　　　　　(iv) none of these

8. Which effect does the secondary load current have on the main flux?
   (i) magnetization　　(ii) demagnetization
   (iii) no effect　　　　(iv) none of these

9. The maximum flux in a transformer is ϕ_m when the primary voltage is V and the frequency is f cps. If the primary voltage is doubled and the frequency halved, the new value of flux will be
   (i) $0.5\phi_m$　　　　　(ii) ϕ_m
   (iii) $2\phi_m$　　　　　　(iv) $4\phi_m$

10. The secondary load impedance Z_L can be represented on the primary side as
    (i) Z_L/a　　　　　　(ii) Z_L
    (iii) $a\,Z_L$　　　　　(iv) $a^2\,Z_L$

11. The no-load component of current is constituted of

    (i) magnetizing current I_ϕ
    (ii) core-loss current I_C
    (iii) both I_ϕ and I_C
    (iv) none of these

12. When a transformer having a turns ratio of a supplies a secondary load current of I_2, the primary current consists of
    (i) the phasor sum of I_0 and I_2
    (ii) the phasor sum of I_0 and $a\,I_2$
    (iii) the phasor sum of I_0 and $a^2 I_2$
    (iv) the phasor sum of $\dfrac{I_2}{a}$ and I_0

13. The no-load current of a transformer normally varies from
    (i) 0% to 2%   (ii) 2% to 5%
    (iii) 5% to 10%   (iv) 10% to 15%

14. Which of the following cannot be determined by the no-load test?
    (i) winding resistance and reactance
    (ii) shunt branch parameters
    (iii) no-load power factor
    (iv) no-load losses

15. The short-circuit test is conducted to determine
    (i) power factor at short circuit
    (ii) series parameters
    (iii) efficiency
    (iv) regulation

16. Which of the following are not considered in computing the efficiency of a transformer?
    (i) $I^2 r$ loss   (ii) core loss
    (iii) dielectric loss   (iv) all of these

17. Maximum efficiency occurs in a transformer when
    (i) constant loss > variable loss
    (ii) constant loss < variable loss
    (iii) constant loss = variable loss
    (iv) none of these

18. The load remaining constant, which of the following is applicable to the efficiency of a transformer when the power factor changes from 0.8 lagging to 0.8 leading?

    (i) no change   (ii) increases
    (iii) decreases   (iv) none of these

19. Which of the following is used to classify transformers?
    (i) type of construction (ii) number of phases
    (iii) special applications(iv) all of these

20. Which of the following is a power transformer?
    (i) auto-transformer
    (ii) three-phase transformer
    (iii) distribution transformer
    (iv) all of these

21. Which of the following apply to an auto-transformer when compared with an equivalent two-winding transformer?
    (i) copper loss is lower
    (ii) higher efficiency
    (iii) as V_2 approaches V_1 saving in copper is more
    (iv) all of these

22. Which of the following can be used to describe a variac?
    (i) an electrically connected two-winding transformer
    (ii) an auto-transformer with a toroidal core
    (iii) an auto-transformer capable of supplying a uniformly varying output voltage
    (iv) all of these

23. In a delta–star-connected transformer, the primary voltage is V, the primary current is I, and the turns ratio is $N_1/N_2 = a$. Which of the following does not represent the correct transformation?
    (i) on the delta side: $V_L = V_P = V$
    (ii) on the star side: $V_P = V/a$
    (iii) $I_L = I_P = I/\sqrt{3}$
    (iv) on he star side: $I_L = I_P = a\,I/\sqrt{3}$

24. In which of the following fields of applications can transformers be used?
    (i) low power   (ii) protection systems
    (iii) electronic circuits   (iv) all of these

25. The rating of the secondary of a current transformer is normally in the range of
    (i) 1–5 A   (ii) 5–15 A
    (iii) 15–30 A   (iv) 30–50 A

Answers

1. (iii)	2. (i)	3. (iii)	4. (iv)	5. (ii)	6. (iii)	7. (ii)
8. (ii)	9. (iv)	10. (iv)	11. (iii)	12. (iv)	13. (ii)	14. (i)
15. (ii)	16. (iii)	17. (iii)	18. (i)	19. (iv)	20. (iv)	21. (iv)
22. (iv)	23. (iii)	24. (iv)	25. (i)			

SYNCHRONOUS MACHINES

7

Learning Objectives

This chapter will enable the reader to
- Obtain a general overview of the principle of electro-mechanical energy conversion in rotating machines
- Enumerate the constructional features of synchronous machines
- Derive an expression for the generated emf and modify the same to include the effects of distributed and short-pitched windings
- Understand the process of production of constant-amplitude synchronously rotating magnetic field by three-phase currents
- Analyse and study the effect of armature reaction on a synchronous generator when it is supplying resistive, inductive, and capacitive loads
- Develop phasor diagram and equivalent circuit of a synchronous generator under no-load and load conditions
- Comprehend the significance of generator regulation and the effect of power factor on it
- Perform oc and sc tests to determine the synchronous impedance of a generator
- Analyse the effect of varying torque and field current on synchronous generators operating in parallel
- Grasp the principle of torque production and operation of a synchronous motor
- Develop equivalent circuit and phasor diagrams of a synchronous motor and therefrom analyse its behaviour (i) on load and (ii) under varying excitations at constant load
- Understand the operation of a synchronous motor as a synchronous condenser and itemize the merits and demerits of synchronous motors

7.1 INTRODUCTION

Electric energy is one form of energy that is most flexible and pollution free, and can be transmitted efficiently and reliably over long distances. Moreover, it can be easily controlled and converted to other forms of energy and converted back from other energy forms to electric energy. Thus, energy-conversion devices are required both at the generating end and at the receiving end of the electrical power systems. The most important conversion of energy, from mechanical to electrical form (as in generators) or from electrical to mechanical form (as in motors), is termed as electromechanical energy conversion (EMEC).

The electromechanical energy conversion process is essentially reversible. Thus, generators can run as motors and motors as generators. The coupling between the electric and mechanical systems is through the medium of electromagnetic field.

The principles of generation of induced emf in a conductor and the production of force on a conductor carrying current, both in the presence of magnetic fields, are discussed in Chapter 1. Three basic important phenomena are as follows.

(a) *Generation of voltage*   It may be seen from Section 1.10.3 that voltage is induced in a coil when there is a change in the flux linking the coil. The induced emfs are of two types: when the conductor is stationary and changing magnetic flux links the conductor, the induced voltage is called statically induced emf or transformer emf; while the voltage induced in the coil by motion of either the coil or the magnetic field relative to each other is called motional voltage, speed voltage, or rotational voltage. The change in flux is associated with mechanical motion and the energy conversion that takes place is called electromechanical energy conversion.

(b) *Force on conductor*   In Section 1.10.2 it was observed that a mechanical force is exerted on a current-carrying conductor when it is placed in a magnetic field. The energy conversion process is reversible as a voltage is induced in the conductor moving in a magnetic field. This process, in association with the one mentioned before, is found to be the dominant one in the study of behaviour of rotating machines.

(c) *Force on iron*   A mechanical force is exerted on ferromagnetic materials when they are brought near a magnetic field, tending to align them with or bring them into the position of the densest part of the magnetic field. The force acting on a plane iron face parallel to an electromagnet may be determined using Eq. (3.87) and is given by $F = B^2A/2\mu_0$, where B is the air-gap flux density in Wb/m^2, A is the cross-sectional area in m^2, and μ_0 is the permeability for free space and is equal to $4\pi \times 10^{-7}$ Wb/ATm. This force is used in many electromagnetic devices such as lifting magnets, magnetic chucks, magnetically operated contactors or switches, relays, etc.

For continuous conversion of energy, rotary motion is employed. Rotating electro-mechanical conversion devices are called rotating electrical machines. Broadly, the rotating electrical machines can be classified into three categories as under:

(a) Synchronous machines
(b) Asynchronous (induction) machines
(c) Direct current (dc) machines

The first two categories of machines operate on alternating current.

All the three types of machines have some similarities. For example, each has a stationary member called the stator and a rotating member called the rotor. Between the stator and the rotor is an annular air gap through which the magnetic flux is set up in the rotor.

In order to allow for rotation in the rotor, an air gap is an essential requirement of any rotating machine. On the other hand, since flux is established in the rotor through the air gap, an additional mmf is required to overcome the reluctance of the air gap. To keep the magnitude of the mmf expended in the air gap at a minimum value, the length of the air gap is kept as small and uniform as is mechanically possible.

A synchronous generator is the most commonly used machine for the generation of electrical power for commercial purposes. A synchronous generator is also referred to as an alternator as it generates alternating (ac) voltages. A striking feature of synchronous generators is that they must run on a definite fixed speed called the synchronous speed. The relationship between synchronous speed N_S in rpm, frequency f in hertz, and the number of poles, P, as given by Eq. (4.9) is $f = N_SP/120$ Hz. The construction of a synchronous generator depends upon the type of prime mover used to rotate the rotor. The prime mover for thermal, nuclear, gas-turbine-based power stations is the high-speed steam turbine (3000 rpm). In these, the turbine generator system is mounted horizontally with the synchronous generator having two poles. For hydro-power generating stations, hydro turbines are used as the prime mover, where the speed of the prime mover varies from 1000 to 500 rpm. Due to low speeds, the rotor of the synchronous generator has a large number of poles and the turbine generator combination is mounted vertically. Diesel engines are used as prime movers for synchronous generators that have low power ratings. The speed of the prime mover is usually 1500 rpm. The generators have four poles and are mounted horizontally.

In this chapter, the construction, operating principle, and characteristics of synchronous machines are discussed.

7.2 CONSTRUCTION FEATURES OF SYNCHRONOUS MACHINES

The various construction features of synchronous machines are discussed in this section.

7.2.1 Advantages of Stationary Armature and Rotating Field

Synchronous machines are those in which the generator or the motor is usually constructed with three-phase armature windings uniformly distributed in slots around the periphery of the stator with field poles on the rotor. A synchronous machine can also be constructed with a rotating armature and stationary field poles. But the former type, with a stationary armature and rotating field pole, is preferred due to the significant advantages discussed in the following paragraph.

7.2.1.1 Higher Operating Voltage Due to Ease in Insulating Armature Windings

It is easy to insulate a static winding, thus making it possible to design synchronous machines that can operate on high voltages. This arrangement makes it possible to have a synchronous machine with operating voltages up to 33 kV.

7.2.1.2 Robust Rotor and Less Slip Rings

Due to the simple construction of the rotor, it can be made more robust, thereby, making it possible to attain higher surface velocities, which in turn leads to a higher output. If the armature was rotating, then it would require three slip rings. In addition, the transfer of high currents at high voltages from the armature slip rings to stationary brushes is quite difficult. Moreover, insulating the slip rings from the shaft would also be difficult and wear and tear of brushes would be very fast due to high currents. A stationary armature eliminates these problems. Only two slip rings are required to feed the current to field winding at a low dc voltage around 400 V.

7.2.1.3 Lighter Rotor and Low Inertia

A low-voltage field winding uses many turns of thin copper wire to produce the field mmf, and the total weight of copper and insulation is very less. Thus, the rotor has a reduced weight and inertia.

7.2.1.4 Efficient Heat Dissipation

The heat produced in the armature due to internal losses can be easily removed using a suitable arrangement. The cooling can be efficiently done when the armature is stationary as provision for radial air ducts and ventilation holes for cooling can be easily incorporated.

7.2.2 Construction of Stator

The stator consists of an armature made of laminations of silicon steel having slots on its inner periphery to accommodate armature windings. Figure 7.1 shows a cross-sectional view of the stator of a three-phase two-pole synchronous machine. Double layer armature windings of three phases *a*, *b*, and *c* are placed in the slots. Since an alternating flux is produced in the stator due to the flow of alternating current in the armature winding, the stator is made of high permeability laminated steel stampings in order to reduce hysteresis and eddy-current losses. The provision of radial and axial ventilating spaces in the stampings assists in

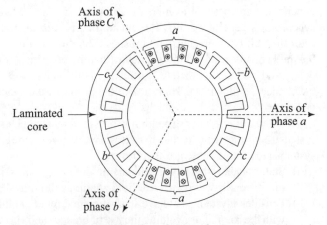

Fig. 7.1   Cross-sectional view of stator of a three-phase two-pole synchronous machine

cooling the machine. The laminations are stamped out in complete rings for smaller machines and stamped in segments for large machines and are insulated from each other with varnish. The whole structure is held in a frame made of cast steel or welded-steel plates.

Slots provided on the stator core are mainly of two types: open slots and semi-closed slots. Open slots are commonly used for commercial generators because the coils can be form-wound and insulated prior to placing them in the slots and the removal and replacement of defective coils can be easily carried out. However, non-uniform air gaps due to open slots may produce ripples in the emf waveform. The use of semi-closed type slots can minimize ripples by distributing the flux as uniformly as possible.

Coils of armature winding are made from insulated copper conductors. For this, hard drawn annealed copper having high conductivity is used. For machines of small ratings, double-cotton-

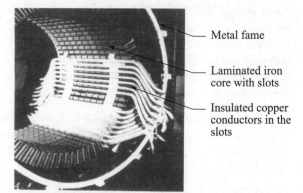

Fig. 7.2 Stator of synchronous machine

covered copper wires are used. For large machines, conductor with rectangular cross section is used and each conductor has impregnated cotton tape insulation.

For slot insulation, leatheroid, mica folium, or manila paper of proper thickness is used. The overhang on the back side of the coils not lying in slots is insulated using varnished and impregnated cotton tape. Figure 7.2 gives an isometric view of a stator of synchronous machine showing component details of the stator.

7.2.3 Construction of Rotor

The rotor of a synchronous machine can be designed to have either salient/projecting field poles or non-salient/cylindrical poles as shown in Fig. 7.3. The rotor carries dc field winding that produces the required flux. The rotor is constructed from laminated high permeability steel in order to reduce eddy-current losses. The field structure is designed to produce an even number of north-south magnetic poles. The excitation is usually provided from a small dc shunt or compound generator, known as an exciter, mounted on the same shaft of the synchronous generator itself. The exciting current is supplied to the rotor through two slip rings and brushes. The direction of current in the field coils is arranged in such a way that the polarities of the field poles produced are alternately North and South. The power rating of the exciter is ordinarily 0.3% to 1% of the power of the synchronous generator.

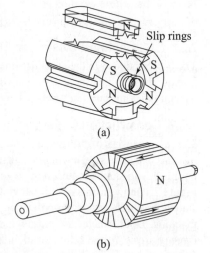

Fig. 7.3 Rotor of synchronous machine: (a) salient pole, and (b) round rotor

An alternative form of excitation is provided by a static excitation system using silicon diodes and thyristors. Two types of static excitation systems are in use. These are as follows.

(a) Static systems that have stationary diodes or thyristors, in which the current is fed to the rotor field windings through slip rings.

(b) Brushless systems that have shaft-mounted rectifiers that rotate with the rotor, thus avoiding the use of brushes and slip rings. The problems of maintenance and cooling of brushes, slip rings, and commutators associated with conventional dc generating systems are minimized.

7.2.3.1 Salient-pole Machines

Figure 7.4 shows a cross-sectional view of a salient-pole machine with four poles of small diameter. For a frequency of 50 Hz, the synchronous speed for this machine will be 1500 rpm. Since the peripheral speed of the machine would be very high, it would be difficult to obtain the needed mechanical strength to withstand the high centrifugal force. On the other hand, for a 12- pole rotor, the synchronous speed of the machine is 500 rpm. In order to accommodate the 12 poles and the associated field winding, the diameter of the machine would have to be large. Since the output is approximately proportional to the volume, the axial length of such a machine will be small. It is shown in design treatises that for a synchronous machine the following relation holds good:

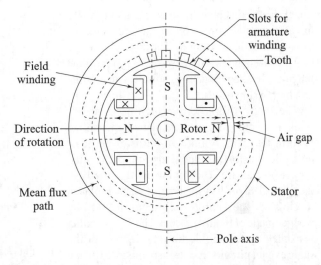

Fig. 7.4 Cross-sectional view of a salient pole machine with four poles

$$\text{Output in kVA} = KD^2LN$$

where D is the diameter of stator bore, L is the core length, N is the machine speed, and K is the design factor. For machines of the same type and not too widely varying output range, the variation of K will not be too large, so that the following relation can be written:

$$D^2L \propto \frac{\text{output}}{N} \propto \frac{1}{N} \text{ for a given output}$$

Salient-pole construction, therefore, is employed for synchronous machines having a comparatively low output and slow/moderate speeds. Such types of synchronous machines are used with hydraulic turbines and their operating speeds range between 100 and 375 rpm. Salient-pole cnstruction is also employed for synchronous motors.

The salient poles are made of thick steel laminations rivetted together and are fixed to the rotor by a dove-tail joint. The pole faces are usually provided with damper windings that are embedded in the slots parallel to the axis. These dampers are useful in preventing the hunting phenomenon. The pole faces are so shaped that the radial air gap length increases from the pole centre to the pole tips so that the flux distribution over the armature is sinusoidal and the waveform of the generated emf is sinusoidal. The field coils are placed on the pole pieces and connected in series. The field windings are connected to the exciter through slip rings and are mounted on the shaft.

Example 7.1 A 50 Hz, two-pole alternator has a length of 1.0 m and a maximum peripheral velocity of 120 m. Calculate (i) the diameter of the stator bore and (ii) the output in kVA. Assume a design factor of 2.0 for the alternator.

Solution (i) Assume that the diameter of the stator bore is D metre.

Rotor speed, $N = \dfrac{120f}{P} = \dfrac{120 \times 50}{2} = 3000$ rpm

Peripheral velocity can be written as

$$\frac{\pi \times D \times 3000}{60} = 120 \quad \text{or} \quad D = \frac{120 \times 60}{\pi \times 3000} = 0.76 \, \text{m}$$

(ii) Output of the alternator in kVA = $KD^2LN = 2 \times (0.76)^2 \times 1 \times 3000 = 3501.66$

7.2.3.2 Non-salient Pole Machines

Synchronous machines used with steam turbines have non-salient pole rotors and are designed to have two or four poles. These machines are characterized by a small diameter, very long axial length, robust construction, noiseless operation, and high rotor speeds.

The rotor is forged from high permeability steel. The rotor forgings are planed and milled to form the teeth. About two-thirds of the rotor pole pitch is slotted, leaving one-third unslotted for the pole centre. The field winding, usually in the form of silver-bearing copper strips, is placed in the slots and appropriately held in place by wedges made of bronze or steel. The field winding is excited by feeding dc from the exciter or the static exciter circuits. Exciters are mounted on the same shaft as the synchronous machine.

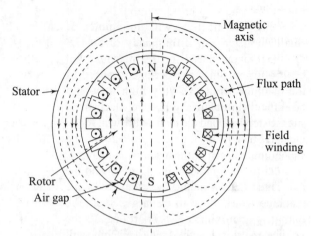

Fig. 7.5 Cross-sectional view of non-salient pole rotor

Figure 7.5 provides a cross-sectional view of a two-pole cylindrical rotor machine with 16 slots and 1 conductor per slot. In an actual rotor, there are more slots and more conductors per slot.

When the rotor current has the direction represented by dots and crosses shown in Fig. 7.5, the flux distribution is as shown by the dotted lines. It may be noted that as one moves away from the pole axis, the flux paths link decreasing field mmf and produce a flux distribution around the periphery that approximates more closely to a sine wave than in a salient-pole rotor. Hence, the emf generated has a better sine wave form.

7.2.3.3 Flux Density Distribution in Synchronous Machines

The flux density distribution for a four-pole-salient synchronous machine with concentrated field coils is shown in Fig. 7.6(a). Since the air gap is uniform over the pole faces, the flux density is constant in this region. Due to fringing at the pole tips, the flux density begins to reduce and becomes zero in the inter-polar region. Since such a flat-topped flux density distribution is not conducive to generating a sinusoidal emf, chamfering the pole faces increases the air gap outward. This ploy helps to make the flux density distribution almost sinusoidal. In further discussions of synchronous machines, the air-gap flux density distribution will be assumed to be a sine wave as shown in Fig. 7.6(b).

In a non-salient pole machine, since the rotor is cylindrical, the air gap is uniform. The armature windings are uniformly distributed. These two factors reduce the requirement of the number of ampere-turns progressively due to a decrease in the flux linkages and also because

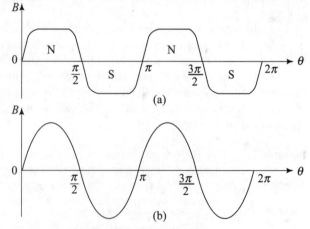

Fig. 7.6 Air-gap flux density distribution in a salient pole synchronous machine

the movement is away from the magnetic axis of the field. In this arrangement, the flux density distribution has a stepped waveform corresponding to the rotor slots, but it can be approximated to a sine wave for the purpose of analysis since the effect of high-frequency harmonics can be ignored without introducing significant errors in the results.

Figure 7.7 shows the air gap flux density distribution for a two-pole non-salient synchronous machine. It may be noted that the stepped air gap flux density distribution, which corresponds to stator and rotor slots, approximates to a sinusoidal distribution.

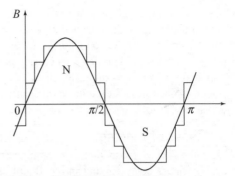

Fig. 7.7 Air gap flux density distribution in a non-salient pole machine

7.3 THREE-PHASE ARMATURE WINDINGS

For the production of a set of three-phase voltages displaced by 120 electrical degrees in time, a minimum of three coils phase displaced by 120 electrical degrees in space must be used, as is discussed in Section 5.2.1. Figure 7.8(a) shows the distribution of stator coils *a*, *b*, and *c* in a two-pole three-phase synchronous machine. Each individual coil spans one pole or one-half wavelength of flux. The voltage in any one coil passes through one cycle of values in one complete revolution of the two-pole machine. For a four-pole machine, a minimum of two such sets of coils must be used as shown in Fig. 7.8(b). Each

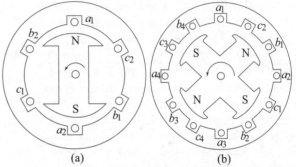

(a)   (b)

Fig. 7.8 (a) Three-phase two-pole synchronous generator and (b) three-phase four-pole synchronous generator

individual coil spans one-half wavelength of flux. The voltage in any one coil passes through two complete cycles of values in one revolution of the two-pole machine. The frequency *f* in hertz is twice the speed of rotation in revolutions per second. The two coils of each phase of Fig. 7.8(b) are connected either in series so that their voltages add or in parallel so that their currents add but voltage is the same as coil voltage. The three phases may be connected either in star or in delta as shown in Fig. 7.9.

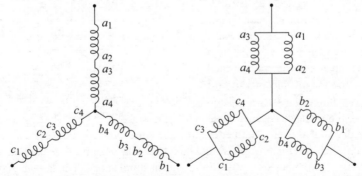

Fig. 7.9 Schematic diagrams for star connection of windings in Fig. 7.8(b)

The armature winding is usually made in the form of diamond-shaped coils (single or multi-turn) and is uniformly distributed in slots around the periphery of the stator. Figure 7.10 shows a diamond-shaped full-pitch coil. When the span or width of the coil is 180 electrical degrees, that is, coil side a_2 occupies a slot in the stator, which is π electrical radians apart from side a_1, it is a full-pitch coil.

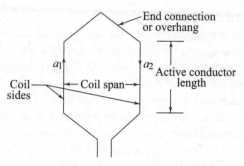

Fig. 7.10   Diamond-shaped coil

7.3.1 Types of Windings

Windings of three-phase synchronous machines are of two types: (a) single-layer winding and (b) double-layer winding.

7.3.1.1 Single-layer Winding

In a single-layer winding, each slot contains only one coil side. Each coil should have a large number of turns to produce the mmf and flux to generate considerable emf per phase. Such a winding is called concentrated winding. The size of the slot and the number of conductors per slot are large. Dissipation of heat, particularly from the conductors in the centre of the slot, causes a problem due to the large size of the slot, which has to accommodate the requisite number of conductors also. Further, the flux density distribution around the inner periphery is flat-topped as shown in Fig. 7.6(a). Since the flux density is not sinusoidal, the waveform of the generated emf would also not be sinusoidal. With this arrangement it is difficult to arrange the end connections without causing obstruction to each other. In actual practice, distributed winding is used where the winding of a phase is distributed over a number of slots under each pole. One of the advantages of a distributed winding is better dissipation of heat that is produced by the I^2R losses in the windings. Also, the waveform of the emf generated is nearer to a sinusoidal curve as compared to the concentrated windings. However, the magnitude of the emf developed per phase is marginally less as compared to the concentrated winding. This is because when

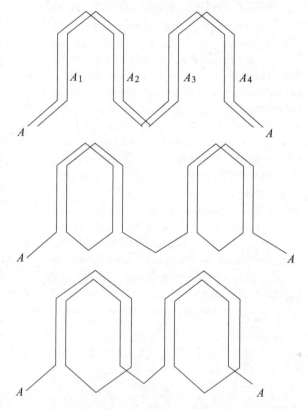

Fig. 7.11   Single-layer distributed winding for a three-phase, four-pole machine

the winding coils are distributed over a number of slots, the emfs induced in the coil sides placed in adjacent slots are displaced from each other in time phase and their total phasor sum is less than the algebraic sum. The windings of a phase may be connected in wave, in lap, or in spiral forms. The layout of phase *A* distributed winding connected in wave, in lap, and in spiral for a four-pole machine with two slots per pole per phase is shown in Fig. 7.11.

7.3.1.2 Double-layer Winding

In a double-layer winding, two coil sides occupy one slot. In such an arrangement, one side of the coil forms the top layer in a slot and the other side of the same coil constitutes the bottom layer and is placed π electrical radians apart for a full-pitch winding. Such type of winding connections ensure that coils can be made similar and the coil overhang can be properly organized. Figure 7.12 illustrates the layout of phase A of a double-layer winding for a four-pole machine with one slot per pole per phase. The four coils of phase A are A_1, A_2, A_3, and A_4. The solid lines indicate the coil sides at the top of a slot and the dotted lines indicate the coil sides at the bottom of the slot.

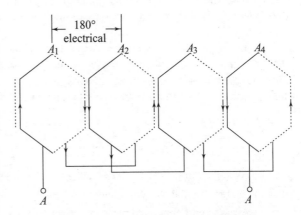

Fig. 7.12   Three-phase double layer winding

7.4 GENERATED EMF IN A SYNCHRONOUS MACHINE

An elementary two-pole, three-phase ac machine as shown in Fig. 7.8(a) is considered. The armature has a single-turn full-pitch coil per phase. It is assumed that the field winding of the rotor produces a sinusoidal flux distribution at its surface along the air gap of the machine. The flux density may be expressed as

$$B = B_m \cos \theta \tag{7.1}$$

where B_m is the maximum flux density at the centre of the pole and θ is the angle in electrical radians measured from the rotor pole axis.

Let l be the length of the conductor or the axial stator length and r be the mean radius at the air gap. Then the air gap flux per pole ϕ is the integral of the flux density over the pole area. Thus, for a two-pole machine,

$$\phi = \int_{-\pi/2}^{\pi/2} B_m \cos \theta \, l r d\theta = 2 B_m l r \tag{7.2}$$

For a P-pole machine,

$$\phi = \frac{2}{P} \times 2 B_m l r = \frac{4}{P} B_m l r \tag{7.3}$$

since the pole area is $2/P$ times that of a two-pole machine of the same length and diameter.

When the rotor magnetic axis and the coil axis coincide, the stator coil links flux ϕ. As the rotor is rotated at a constant angular velocity ω, the flux linking T turns of the stator coil at any time t is

$$\psi = T\phi \cos \omega t \tag{7.4}$$

where $t = 0$ when the rotor magnetic axis and the coil axis coincide. By Faraday's law, the voltage induced in the stator coil is

$$e = -\frac{d\psi}{dt} = \omega T \phi \sin \omega t \tag{7.5}$$

The voltage induced is sinusoidally varying with time, with a peak value

$$E_m = \omega T \phi$$

and the rms value of the induced phase voltage is given by

$$E = \frac{\omega T \phi}{\sqrt{2}} = \frac{2\pi f T \phi}{\sqrt{2}} = 4.44 f T \phi \tag{7.6}$$

The emf induced in a synchronous machine may also be determined as follows.

Let a synchronous machine be assumed to operate as a generator (alternator) and be running at a synchronous speed which is governed by the relation given in Eq. (4.9). Further, let the armature coils be assumed to be of full pitch.

Let Z be the number of conductors in series per phase, ϕ be the useful flux per pole in weber, P be the number of poles of the generator, and N_S be the synchronous speed of the generator in rpm.

As the coil moves from one inter-polar axis to the other, the flux cut by the conductors of the coil is ϕ weber. The rate at which the coil conductors cut the flux is $1/(2f)$ sec. Hence, the rate of change of flux

$$\frac{d\phi}{dt} = \frac{\phi}{1/(2f)} = \phi \times 2f$$

By definition,

Average emf generated per conductor $= \phi \times 2f$ V

Since there are Z numbers of conductors per phase,

Average emf generated per phase $= \phi \times 2f \times Z$ V

Since the flux density distribution has been assumed to be sinusoidal, the generated phase emf will also be sinusoidal. Therefore, the rms value of the generated emf per phase, E, is given by

$$E = 1.11 \times \phi \times 2f \times Z = 2.22 \times \phi \times f \times Z \text{ V} \tag{7.7}$$

In terms of the number of turns, Eq. (7.7) changes to

$$E = 4.44\phi fT \text{ V} \tag{7.8}$$

and $Z = 2T$.

7.4.1 Distributed Winding

In electrical engineering practice, the generation system voltages are as high as 11 kV and possibly 33 kV. In order to build synchronous machines to generate such high voltages, it is necessary to fully utilize the armature periphery and increase the number of slots per pole per phase. This leads to the concept of distributed armature winding.

For a three-phase synchronous machine having the number of poles, P, total number of slots S uniformly distributed a round the stator armature, and number of slots per pole per phase m, then

$$m = \frac{S}{3P} \tag{7.9}$$

In general, if q is the number of phases, Eq. (7.9) may be written as

$$m = \frac{S}{qP} \tag{7.10}$$

Figure 7.13(a) shows a two-pole synchronous machine with the coil sides of the three-phase full-pitch armature windings concentrated in one slot. In Fig. 7.13(b), the two-pole synchronous machine has three slots per pole per phase and the armature windings are distributed in three adjacent armature slots.

In distributed winding, the angle γ between two adjacent armature slots is given by

$$\gamma = \frac{\pi P}{S} \tag{7.11}$$

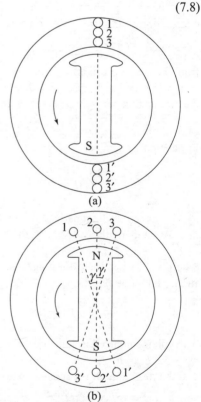

Fig. 7.13 Two-pole synchronous machine with (a) concentrated winding and (b) distributed winding ($m = 3$)

When the coils are concentrated in one slot, the emfs generated in the coil sides are in phase as shown in Fig. 7.14(a). From the phasor diagram, it is seen that if the coils are connected in series, the total per phase generated emf E_p is the arithmetic sum of the individual coils.

When the armature winding is uniformly distributed, unlike the concentrated winding, the emfs generated in the coils 11', 22', and 33' have the same magnitude but progressively differ in phase by γ electrical radians. Thus, the emf generated per phase would be the phasor sum of the coil emfs as shown in the phasor diagram of Fig. 7.14(b).

It can be seen from the phasor diagram shown in Fig. 7.14(b) that the coil emf phasors E_{C1}, E_{C2}, and E_{C3} in each of the coils 11', 22', and 33', respectively, form sides of a regular polygon. The rms value of emf induced in each coil is the same. Let it be represented by E_C, then $E_C = E_{C1} = E_{C2} = E_{C3}$. Let E_P be the rms value of phase voltage. Then E_P is the phasor sum of the emfs generated in the coils of the phase. In Fig. 7.14(b) the radius of the circle circumscribing the coil voltages E_{C1}, E_{C2}, and E_{C3} is denoted by OA and the following may be written from the phasor diagram:

$$E_C = 2 \times OA \times \sin(\gamma/2) \qquad (7.12)$$

and $\qquad E_P = 2 \times OA \times \sin(3\gamma/2) \qquad (7.13)$

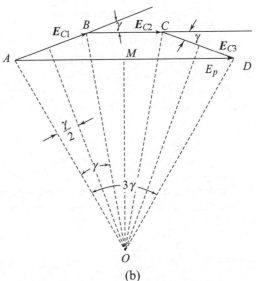

Fig. 7.14 Phasor sum of emfs generated in the phase coils when the windings are (a) concentrated and (b) distributed ($m = 3$).

The ratio of the resultant phase emf to the algebraic sum of the individual coil emfs constituting the phase is called the breadth factor or the distribution factor and is denoted by k_d. For $m = 3$, that is, three slots per phase per pole, the distribution factor is

$$k_d = \frac{E_P}{3 \times E_C} = \frac{\sin[(3\gamma)/2]}{3\sin(\gamma/2)} \qquad (7.14)$$

It may be noted that the coil emfs for a general case of m slots per pole per phase would form the sides of a polygon with m sides and the distribution (or breadth) factor k_d is then given as

$$k_d = \frac{E_P}{mE_C} = \frac{\sin[(m\gamma)/2]}{m\sin(\gamma/2)} \qquad (7.15)$$

Since $E_P < mE_C$, the distribution factor k_d is less than unity.

The expressions for generated voltages given by Eqs (7.7) and (7.8), therefore, get modified as under

$$E_P = 2.22k_d Zf\phi \text{ V} \qquad (7.16)$$

$$E_p = 4.44k_d Tf\phi \text{ V} \qquad (7.17)$$

7.4.2 Short-pitched Coils

When the coil span is less than full pitch, that is, less than π radians (electrical), the coil is said to be a short-pitched coil. Short-pitched coils are employed to save conductor material, reduce high-frequency harmonic content in the generator armature voltage, and improve the waveform of the generated phase voltage. Short-pitched coils reduce the overhang.

Figure 7.15(a) shows a short-pitched coil with a span of $(\pi - \beta)$ radians (electrical). The voltages generated by the two coil sides a and a' are not in phase and therefore cannot be added algebraically. Let the magnitudes of the generated voltage in the coil sides a and a' be E_a and E_a', respectively. These voltages are equal in magnitude but differ in phase by an angle β as shown in Fig. 7.15(b). The generated coil emf E_C is the phasor addition of the emfs E_a and E_a' and can be written as

$$E_C = E_a + E_{a'}$$

$$= 2E_a \cos \frac{\beta}{2} \qquad (7.18)$$

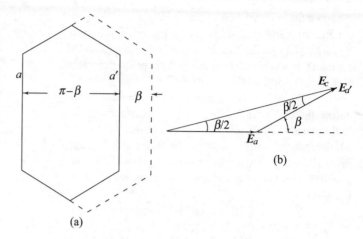

(a)

(b)

Fig. 7.15 Short-pitched coil

Pitch factor k_p is defined as the ratio of the voltage generated in a short-pitched coil to the voltage generated in a full pitch coil. The voltage generated in a full pitch coil is twice the coil side voltage, that is, $2E_a$. Hence, the pitch factor k_p is given by

$$k_p = \frac{E_C}{2E_a} = \cos \frac{\beta}{2} \qquad (7.19)$$

The value of the pitch factor K_p is always less than unity.

The expression for generated voltage given by Eqs (7.16) and (7.17) are further modified as

$$E_P = 2.22 k_d k_p Z f \phi \text{ V} \qquad (7.20)$$

$$E_p = 4.44 k_d k_p T f \phi \text{ V} \qquad (7.21)$$

7.4.3 Winding Factor

Winding factor represents the combined effects of distributing the windings and short pitching the coils. It is denoted by k_w. The winding factor is less than both k_d and k_p.

Substituting the winding factor $k_w = k_d k_p$ in Eqs (7.20) and (7.21) gives

$$E_P = 2.22 k_w Z f \phi \text{ V} \qquad (7.22)$$

$$E_p = 4.44 k_w T f \phi \text{ V} \qquad (7.23)$$

Example 7.2 A 50-Hz synchronous clock shows correct time at 8 am IST. The average frequency is 49.95 Hz from 8 am to 2 pm and 49.90 Hz from 2 pm to 7 pm. What must be the average frequency for the remainder of the 24 hours in order that the clock be correct again at 8 am? By how much is the clock incorrect at 7 pm?

Solution The total number of cycles the clock should perform in 24 hours for correct time is

$$24 \times 60 \times 60 \times 50 = 4320 \times 10^3$$

The number of cycles the clock performs from 8 am to 7 pm is

$$(6 \times 49.95 + 5 \times 49.90) \times 60 \times 60 = 1977.12 \times 10^3$$

For correct time, the number of cycles required in the remaining 13 hours is

$$4320 \times 10^3 - 1977.12 \times 10^3 = 234288 \times 10^3$$

The desired average frequency for correct time for the remaining 13 hours is

$$\frac{2.34288 \times 10^6}{13 \times 60 \times 60} = 50.061538$$

The shortfall in the number of cycles from 8 am to 7 pm,

$$0.05 \times 6 + 0.10 \times 5 = 0.8$$

The time by which the clock is incorrect at 7 pm

$$\frac{0.8 \times 60 \times 60}{50} = 57.6 \text{ sec slow}$$

Example 7.3 A three-phase, 12-pole synchronous machine has a star-connected full-pitch winding with 108 slots and 12 conductors per slot. The flux per pole is 50 mWb and sine distributed. The speed of rotation is 500 rpm. Find (a) the frequency, (b) phase emf, and (c) the line emf.

Solution (a) Frequency $f = \dfrac{NP}{120} = \dfrac{500 \times 12}{120} = 50$ Hz

(b) As the winding is full pitch, $k_p = 1$.

From Eq. (7.15), the distribution factor,

$$k_d = \frac{\sin[(m\gamma)/2]}{m\sin(\gamma/2)}$$

Now, $m = \dfrac{108}{12 \times 3} = 3$ slots/pole/phase

$$\gamma = \frac{180°}{\text{slots per pole}} = \frac{180°}{9} = 20°$$

$\therefore$ $k_d = \dfrac{\sin[(3 \times 20)/2]}{3\sin(20/2)} = \dfrac{\sin 30°}{3\sin 10°} = \dfrac{0.5}{0.521} = 0.96$

$$Z = \frac{108 \times 12}{3} = 432$$

Then using Eq. (7.20),

$$E_P = 2.22 \times 1.0 \times 0.96 \times 432 \times 50 \times 50 \times 10^{-3} = 2301.7 \text{ V}$$

(c) The line voltage, $E_L = \sqrt{3}E_P = 3986.65$ V

Example 7.4 A three-phase, 10-pole, 50-Hz synchronous machine has a star-connected winding with a coil span of 150° electrical. It has the following data: slots per pole per phase are 2, conductors per slot (double-layer) are 6, the flux per pole is 0.15 Wb. All the coils of a phase are connected in series. Determine (a) the speed of rotation of the machine, (b) phase emf, and (c) the line emf.

Solution (a) Speed of rotation, $N = \dfrac{120f}{P} = \dfrac{120 \times 50}{10} = 600$ rpm

(b) From Eq. (7.19) the pitch factor

$$k_P = \cos(\beta/2) = \cos(30°/2) = 0.966$$

Now, $m = 2$.

$$\gamma = \frac{180° \times P}{S} = \frac{180° \times 10}{2 \times 3 \times 10} = 30°$$

From Eq. (7.14),

$$k_d = \frac{\sin[(m\gamma)/2]}{m\sin(\gamma/2)} = \frac{\sin 30°}{2\sin 15°} = 0.966$$

$$Z = 6 \times 2 \times 10 = 120$$

Then using Eq. (7.20),

$$E = 2.22 \times 0.966 \times 0.966 \times 120 \times 50 \times 0.15 = 1864.44 \text{ V}$$

(c) The line voltage, $E_L = \sqrt{3} E_P = 3229.3$ V.

7.5 ROTATING MAGNETIC FIELD DUE TO THREE-PHASE CURRENTS

The ac windings used in three-phase synchronous and induction machines produce magnetic fields of constant amplitude rotating at a uniform speed around the air-gap circumference when the windings carry three-phase currents. This fact may be demonstrated physically by considering a three-phase, two-pole synchronous machine having one slot per pole per phase shown in Fig. 7.16(a). The coils of the three-phase armature windings are displaced in space from each other by 120°. The starts of the coils are at A, B, and C and the finish of the coils are at A_1, B_1, and C_1. The currents in the three phases A, B, and C as functions of time are shown in Fig. 7.16(b).

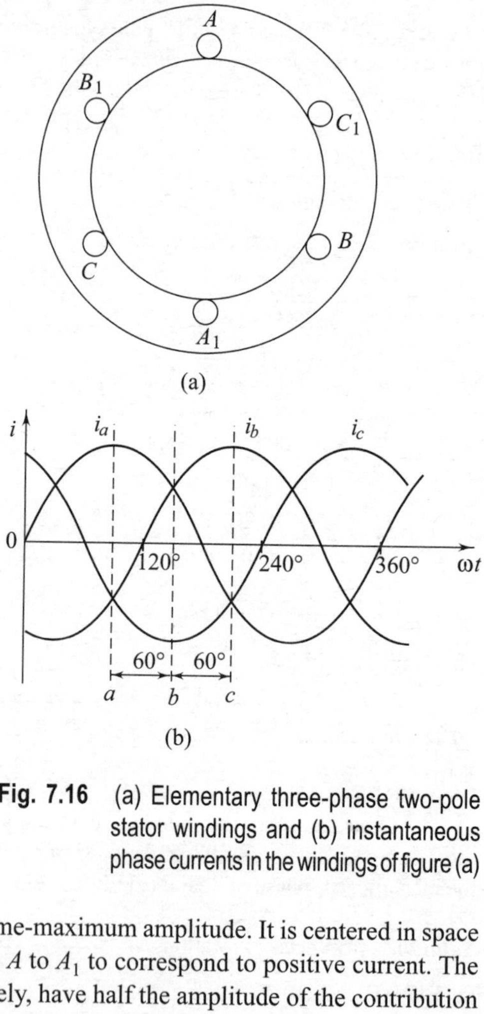

To illustrate the magnetic field produced by this winding, the developed diagram of the cylinder forming the stator shown in Fig. 7.16(a) is drawn with the conductor A being shown twice in order to identify the ends clearly. The components and the resultant field distributions produced by the currents flowing in the three windings are plotted in Fig. 7.17 for several successive instants of time a, b, and c as shown in Fig. 7.16(b). The flux distribution of any one phase is assumed to be sinusoidal, although this is not true in case the phase consists of a concentrated coil. Practical windings, however, consist of a group of distributed coils for each phase and the flux distribution is very nearly sinusoidal.

Consider first the instant a [see Fig. 7.16(b)], when the current in phase A is positive and at its maximum value whereas the currents in phases B and C are negative and each is at half the maximum value. The flux distribution

(a)

(b)

Fig. 7.16 (a) Elementary three-phase two-pole stator windings and (b) instantaneous phase currents in the windings of figure (a)

produced by phase A alone, as shown in Fig. 7.17(a), has its time-maximum amplitude. It is centered in space about the axis of coil A and is arbitrarily drawn positive from A to A_1 to correspond to positive current. The flux distributions produced by phases B and C each, respectively, have half the amplitude of the contribution from phase A, are centered in space about their own axes, and are drawn negative from B to B_1 or C to C_1 to correspond to negative currents. The resultant flux distribution, obtained by adding the individual contributions of the three phases, is a sinusoid centered on phase a and has an amplitude 1.5 times the maximum contribution of phase A, as shown in Fig. 7.17(a).

The phasor diagram of flux produced by instantaneous currents in the three phases is also shown in Fig. 7.17(a), where Φ_a, Φ_b, Φ_c represent, respectively the flux produced by currents i_a, i_b, i_c; Φ_m is the maximum magnitude of flux; and Φ_{res} is the resultant flux. The values of flux produced by i_a, i_b, i_c, at the instant a of Fig. 7.16(b) are

$$\Phi_a = \Phi_m \sin 90° = \Phi_m$$
$$\Phi_b = \Phi_m \sin (90° - 120°) = -0.5\Phi_m$$
$$\Phi_c = \Phi_m \sin (90° - 240°) = -0.5\Phi_m$$

Then $\quad \Phi_{res} = \Phi_m + 2 \times (0.5\ \Phi_m) \times \cos 60° = 1.5\ \Phi_m$

The resultant flux Φ_{res} acts along the same direction as that of flux Φ_a.

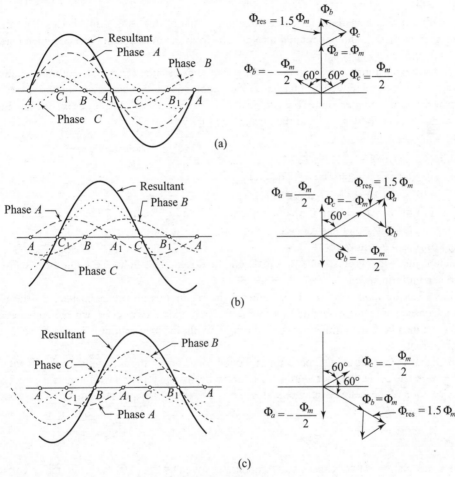

Fig. 7.17 Distribution of component and resultant flux caused by three-phase armature currents and the corresponding phasor diagrams.

For the instant b [see Fig. 7.16(b)], the current in phase C is a negative maximum and currents in phases A and B are positive, each having a value half the maximum. Thus, the phase A flux contribution has half its previous amplitude, as shown in Fig. 7.17(b). The phase B contribution has the same amplitude but of reverse polarity to that at the instant a because its current is reversed. The phase C contribution has the same polarity as that at the instant a, but twice the amplitude. It may be noted that the same resultant flux distribution is obtained, but the wave has moved to the right by 60°.

The phasor diagram of flux produced by instantaneous currents in the three phases is shown in Fig. 7.17(b). The values of flux produced by i_a, i_b, i_c, at the instant b of Fig. 7.16(b) are

$$\Phi_a = \Phi_m \sin 150° = 0.5\Phi_m$$
$$\Phi_b = \Phi_m \sin (150° - 120°) = 0.5\Phi_m$$
$$\Phi_c = \Phi_m \sin (150° - 240°) = -\Phi_m$$

Then $\quad \Phi_{res} = \Phi_m + 2 \times (0.5\Phi_m) \times \cos 60° = 1.5\Phi_m$

The resultant flux Φ_{res} acts along the same direction as that of flux $-\Phi_c$. Thus the resultant flux Φ_{res} has the same magnitude $1.5\Phi_m$, and has rotated through 60° from the initial position it occupied corresponding to the instant of time a of Fig. 7.16(a).

Similarly, for the third instant c [see Fig. 7.16(b)], the same resultant flux distribution is obtained, but as shown in Fig. 7.17(c), it has moved still farther to the right by 60° and is centered on phase B.

The phasor diagram of flux produced by instantaneous currents in the three phases is shown in Fig. 7.17(c). The value of flux produced by i_a, i_b, i_c, at the instant c of Fig. 7.16(b) are

$$\Phi_a = \Phi_m \sin 210° = -0.5\Phi_m$$
$$\Phi_b = \Phi_m \sin (210° - 120°) = \Phi_m$$
$$\Phi_c = \Phi_m \sin (210° - 240°) = -0.5\Phi_m$$

Then $\quad \Phi_{res} = \Phi_m + 2 \times (0.5\Phi_m) \times \cos 60° = 1.5\Phi_m$

The resultant flux Φ_{res} acts along the same direction as that of flux $-\Phi_b$. Thus the resultant flux Φ_{res} has the same magnitude $1.5\Phi_m$, and has rotated through 60° from the initial position it occupied corresponding to the instant of time b of Fig. 7.16(a).

The resultant flux wave corresponds to a magnetic field rotating around the cylindrical stator shown in Fig. 7.16(a) at a uniform speed.

One cycle after the instant a [see Fig. 7.16(b)] the resultant field must be back in the position of Fig. 7.17(a). Consequently, the magnetic flux rotates through one revolution or two pole pitches. The speed of the field must then be f revolutions per second or $60f$ revolutions per minute, f being the stator frequency of the stator current.

Generalizing, if the synchronous generator is wound for P poles, the magnetic field will rotate through $2/P$ revolutions in one cycle. In other words, if the frequency of the three phase currents is f cycles/sec, then the speed of rotation of the stator magnetic field will be $2f/P$ revolutions per second. The synchronous speed, N_S, is then given by

$$N_S = \frac{120f}{P} \tag{7.24}$$

which is the same relation between speed and frequency as given by Eq. (4.9) for generated ac voltages. Thus, it can be concluded that in a synchronous generator the axis of the armature magnetic field mmf rotates at synchronous speed given by Eq. (7.24). In other words, in a synchronous generator supplying a balanced load, the three-phase armature currents set up a magnetic field in the air gap. The axes rotate at a synchronous speed, which is also the speed of rotation of the field poles placed on the rotor. This is the manifestation of Lenz's law.

Further, if the sequence of supply to the three-phase windings is changed by interchanging any two of the three phases, keeping the third unchanged, the resultant magnetic field will rotate in the opposite direction.

7.5.1 Mathematical Analysis of the Rotating Magnetic Field

Let the three phases of an ac machine winding carry balanced alternating currents. If I_m is the peak value of the currents, then the current in the three phases may be expressed as

$$i_A = I_m \sin \omega t$$

$$i_B = I_m \sin \left(\omega t - \frac{2\pi}{3} \right)$$

$$i_C = I_m \sin \left(\omega t - \frac{4\pi}{3} \right)$$

If the magnetic circuit is unsaturated and the maximum flux density due to maximum current in any phase is B_m, then the flux densities for the three phases may be expressed as

$$B_A = B_m \sin \omega t$$

$$B_B = B_m \sin \left(\omega t - \frac{2\pi}{3} \right)$$

$$B_C = B_m \sin \left(\omega t - \frac{4\pi}{3} \right)$$

At any point θ radians (electrical) in space (θ being measured from the point of zero flux density for phase A), the flux density distribution due to the three phases A, B, and C can be mathematically represented as

$$B_1 = B_m \sin \omega t \sin \theta$$

$$B_2 = B_m \sin \left(\omega t - \frac{2\pi}{3} \right) \sin \left(\theta - \frac{2\pi}{3} \right)$$

$$B_3 = B_m \sin \left(\omega t - \frac{4\pi}{3} \right) \sin \left(\theta - \frac{4\pi}{3} \right)$$

The resultant flux density distribution at θ radians (electrical) is the sum of the flux density distributions due to currents in the three phases. Therefore,

$$B_{\text{resultant}} = B_1 + B_2 + B_3$$

$$= B_m \left[\sin \omega t \sin \theta + \sin \left(\omega t - \frac{2\pi}{3} \right) \sin \left(\theta - \frac{2\pi}{3} \right) + \sin \left(\omega t - \frac{4\pi}{3} \right) \sin \left(\theta - \frac{4\pi}{3} \right) \right]$$

$$= 1.5 B_m (\sin \omega t \, \theta + \cos \omega t \cos \theta)$$

$$= 1.5 B_m \cos(\omega t - \theta) \tag{7.25}$$

From Eq. (7.25) it can be seen that when $\theta = \omega t = 2\pi f t$, the resultant flux density distribution has a maximum value of $1.5 B_m$. For a value of $t = 1/f$, the time duration of one cycle, $\theta = 2\pi$ electrical radians, which means the position of the peak value of the resultant flux rotates through two pole pitches in one cycle.

7.6 CHARACTERISTICS OF A THREE-PHASE SYNCHRONOUS GENERATOR

The characteristics of a three-phase synchronous generator are discussed in this section.

7.6.1 Armature Reaction

Section 7.5.1 shows that balanced three-phase currents in the three-phase stator windings set up a resultant magnetic field, which is constant in magnitude and rotates at the synchronous speed. The effect of the armature flux on the flux produced by the rotor field ampere-turns is called armature reaction. The armature flux will distort, oppose, or aid the field flux causing the air gap flux to increase or decrease depending on the power factor of the load.

7.6.1.1 Synchronous Generator Loaded with Resistive Load

A two-pole, three-phase synchronous generator with one slot per pole per phase, as shown in Fig. 7.18(a), is considered. When the generator terminals are on open circuit, the armature carries no current. The magnetic flux set up in the air gap is due to the rotor field ampere-turns and is shown in Fig. 7.18(a). If the direction of rotation of the poles is clockwise, the direction of emf induced in the armature conductors is shown by crosses and dots in Fig. 7.18(b) with emf generated in phase *A* at its maximum.

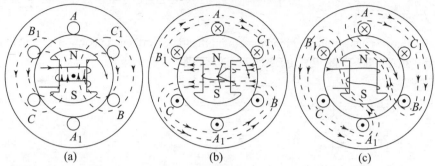

Fig. 7.18   (a) Magnetic flux due to rotor mmf alone and (b) magnetic flux due to stator currents alone and (c) Resultant magnetic flux due to armature reaction

If the generator is loaded with a resistive load, the instantaneous direction of current in the armature conductors will be the same as the direction of emfs induced in them and they are in time phase. Flux produced by the armature ampere-turns alone is shown in Fig. 7.18(b). This armature flux will be rotating at synchronous speed. The flux produced by the rotor ampere-turns is also rotating at synchronous speed. Thus, these two fluxes will be stationary with respect to each other. The two fluxes will give rise to a resultant air gap flux distribution [see Fig. 7.18(a) and (b)] that is distorted, as shown in Fig. 7.18(c). The flux lines along the air gap have been lengthened. The flux lines, in trying to shorten their path through the air gap, will exert a backward pull on the rotor. The prime mover driving the generator should, therefore, develop more torque to enable the rotor to continue to rotate at the synchronous speed. The extent of flux distortion and hence the magnitude of backward pull will depend upon the magnitude of armature current which in turn will depend on the load.

7.6.1.2 Synchronous Generator with Inductive Load

With a purely inductive load connected across the armature terminals, the armature current will lag the induced emf by 90°. When the emf of phase *A* is maximum, then the position of the rotor pole is as shown in Fig.7.18(a). For a purely inductive load, current in phase *A* will be maximum only after lapse of time by 90° when the rotor would have moved forward half a pole pitch as shown in Fig. 7.19.

It is seen that the armature flux is in direct opposition to the rotor field flux thereby reducing the air gap flux created by the rotor field poles. The distribution of flux will remain symmetrical over the two halves of the pole faces. Hence, no additional torque has to be developed by the prime mover to enable the rotor to rotate at the synchronous speed. The reduction in the air gap flux will cause a reduction in the induced emf and therefore a drop in the terminal voltage.

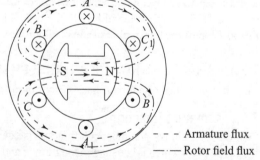

Fig. 7.19   Magnetic flux distribution for a purely inductive load and the corresponding rotor pole position

7.6.1.3 Synchronous Generator with Capacitive Load

When a purely capacitive load is connected across the armature terminals, the current flowing through the windings will lead the induced emf by 90°. Maximum current in phase *A*, in this case, will occur 90° before the occurrence of a maximum induced emf in that phase. Therefore, when a maximum current in phase A occurs, the position of the rotor would remain 90° behind as compared to its position under resistive load. The direction of the armature flux and the corresponding position of the rotor are shown in Fig. 7.20. It is seen from the figure that the armature flux is in the same direction as the rotor field flux. The two fluxes will help each other and therefore strengthen the air gap flux, thereby increasing the emf induced in the armature. Thus, the terminal voltage will increase. As the distribution of flux will remain symmetrical over the two halves of the pole faces, no additional torque has to be developed by the prime mover to enable the rotor to rotate at the synchronous speed.

7.6.1.4 Effect of Armature Reaction on Terminal Voltage

The effect of armature reaction on the terminal voltage at various power factor loads with the rotor field current held constant at a value giving an emf OV on open circuit is shown in Fig. 7.21.

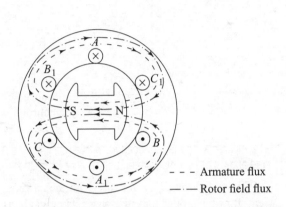

Fig. 7.20 Magnetic flux distribution for a purely capacitive load and the corresponding rotor pole position

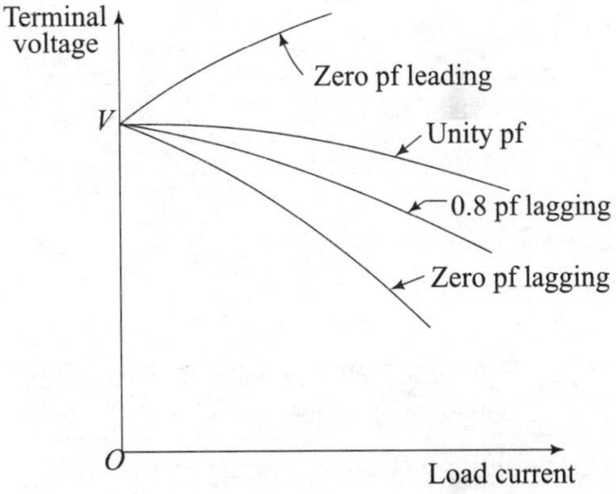

Fig. 7.21 Variation of terminal voltage with load at constant excitation

At unity power factor load, the change in the terminal voltage with load is somewhat less as compared to inductive and capacitive loads. At zero power factor lagging and leading load, the change in the terminal voltage with load is large due to the demagnetizing and magnetizing effects, respectively, of the armature reaction.

7.6.2 Phasor Diagram and Equivalent Circuit

In order to comprehend the operation and characteristics of a synchronous generator, it is important to understand its behaviour under no-load and loaded conditions. The behaviour of synchronous generators is explained and, therefrom, the phasor diagram and the equivalent circuit are developed in the following sections.

7.6.2.1 Synchronous Generator at No Load

Under a no-load operation the stator winding current is zero. The field winding flux ϕ_f induces an emf E volts per phase, the rms value of which is given by Eq. (7.6). It follows directly from Eqs (7.4) and (7.6) that the emf E lags the flux ϕ_f by 90° as shown in Fig. 7.22. E is called the excitation voltage, as this voltage always appears at the armature terminals at no load with the field winding excited.

7.6.2.2 Synchronous Generator on Load

If a balanced three-phase unity-power-factor load is connected across the armature terminals as shown in Fig. 7.23(a), the load draws an rms current I_a amperes at a terminal voltage V_t volts per phase. Let the stator armature winding resistance per phase be denoted by R_a. In addition, some of the flux produced by the armature ampere-turns, which do not cross the air gap, is called leakage flux and is denoted by X_L. If the synchronous generator is loaded, then the terminal voltage changes due to the voltage drops in the armature resistance and the armature leakage reactance. The terminal voltage of the synchronous generator also changes due to the armature reaction. This will be explained in the text that follows.

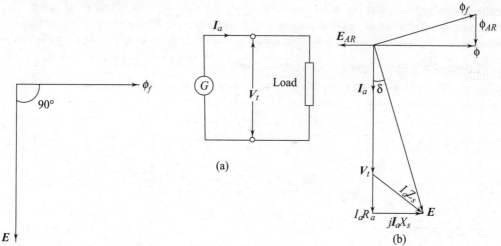

Fig. 7.22 Phasor diagram of synchronous generator on no load

Fig. 7.23 (a) Synchronous generator connected to a resistive load and (b) phasor diagram for unity power factor load

The resulting phasor diagram is shown in Fig. 7.23(b). The load current I_a is in phase with the terminal voltage V_t because the load is resistive. The three-phase currents flowing through the three-phase stator windings give rise to a rotating field flux ϕ_{AR} in phase with the armature current I_a, as the armature flux and the current that produce the rotating field flux are in time phase. The armature flux ϕ_{AR} combines with the field flux ϕ_f of the field winding to give the resultant flux per pole ϕ. If the stator-winding resistance and leakage reactance are neglected, then this resultant flux ϕ must lead V_t by 90°.

The armature flux ϕ_{AR} rotates at the synchronous speed and cuts the stator armature winding, thereby inducing an armature-reaction voltage per phase, E_{AR}. The voltage E_{AR} lags the flux ϕ_{AR} or the stator current I_a that creates it by 90°. This effect is exactly similar to that of inductive reactance. Thus, the armature-reaction voltage E_{AR} can be considered as a voltage drop in a fictitious reactance X_a.

The relation between the excitation voltage E and the terminal voltage V_t may be expressed as follows:

$$E = V_t + I_a R_a + j I_a (X_L + X_a)$$
$$= V_t + I_a R_a + j I_a X_S = V_t + I_a (R_a + j X_S)$$
$$= V_t + I_a Z_S \tag{7.26}$$

The parameter X_S is called the synchronous reactance per phase of the stator winding. It consists of the leakage reactance X_L plus a fictitious reactance X_a, which replaces the effect of the armature winding mmf on the field flux. The complex impedance $Z_S = R_a + j X_S$ is called the sychronous impedance of the generator. The magnitude of the synchronous impedance Z_S and its angle θ is given by

$$Z_S = \sqrt{R_a^2 + X_S^2} \tag{7.27a}$$

$$\theta = \tan^{-1} \frac{X_S}{R_a} \tag{7.27b}$$

7.6.2.3 Equivalent Circuit

The equivalent circuit of a synchronous machine follows directly from Eq. (7.26). E is considered the source voltage and Z_S is treated as an internal source impedance. Figure 7.24 shows the resulting equivalent circuit.

The phasor diagrams for a synchronous generator, with the terminal voltage V_t as the reference, for lagging and leading power-factor (pf) loads are shown in Fig. 7.25(a) and Fig. 7.25(b), respectively. The angle φ is the phase angle (lag or lead) of the armature current I_a with respect to the terminal voltage V_t. In the diagram, the phasor $I_a R_a$, the voltage drop in the stator resistance, is in phase with the load current phasor I_a. The phasor $I_a X_S$, the voltage drop in the synchronous reactance X_S of the generator, leads the current I_a by 90°. The voltage drop in the synchronous impedance Z_S is $I_a Z_S$ and is the phasor sum of the voltage drops due to the stator winding resistance and synchronous reactance. The angle δ is the phase displacement between the generated voltage E and the terminal voltage V_t. The generated voltage E leads the terminal voltage, a condition necessary for the generator action to take place. The flux ϕ_f, the armature flux ϕ_{AR}, and the air gap flux ϕ are also shown on the phasor diagrams.

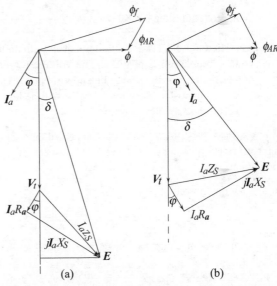

(a)   (b)

Fig. 7.25 Phasor diagram for (a) lagging pf and (b) leading pf synchronous generator

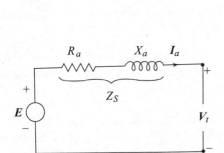

Fig. 7.24 Equivalent circuit of a

If the values of the load current, power factor, resistance, and synchronous reactance per phase of the stator winding and the terminal voltage are known, the value of the generated emf E can be computed from the phasor diagram. The voltage drops $I_a R_a$ and $I_a X_S$ may be resolved along and perpendicular to the voltage phasor V_t using trigonometrical relations. For lagging pf, the excitation voltage may be found from the following:

$$E^2 = (V_t + I_a R_a \cos\varphi + I_a X_S \sin\varphi)^2 + (I_a X_S \cos\varphi - I_a R_a \sin\varphi)^2 \qquad (7.28)$$

For leading pf, the excitation voltage may be found from the following:

$$E^2 = (V_t + I_a R_a \cos\varphi - I_a X_S \sin\varphi)^2 + (I_a X_S \cos\varphi + I_a R_a \sin\varphi)^2 \qquad (7.29)$$

7.7 VOLTAGE REGULATION

A synchronous generator is required to maintain a specified voltage at its terminals for the rated full load and power factor. When the terminal voltage varies due to a change in the load current or a change in the power factor, the generator induced emf E is adjusted by varying the field current so as to maintain the rated voltage at the generator terminals. The voltage regulation of a synchronous generator is the voltage rise at the terminals when a given load is thrown off, the excitation and speed remaining constant.

Per unit voltage regulation, at a given power factor, may be defined as the ratio of change in the terminal voltage from full load to no load to the full-load voltage, i.e.,

$$\text{Voltage regulation} = \frac{\text{No-load voltage} - \text{full-load voltage}}{\text{full-load voltage}} \text{ pu} \quad (7.30)$$

Figure 7.26 shows the variation of the terminal voltage with varying load current at 0.8 lagging pf and unity pf. In the figure, OA represents the rated terminal voltage and OL represents the rated current at full load. From the figure,

$$\text{Voltage regulation at unity pf} = \frac{OB - OA}{OA} = \frac{AB}{OA} \text{ pu}$$

where OB represents the no-load terminal voltage. Similarly,

$$\text{Voltage regulation at 0.8 lagging pf} = \frac{OC - OA}{OA} = \frac{AC}{OA} \text{ pu}$$

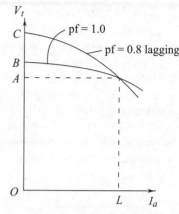

Fig. 7.26 Variation of terminal voltage with load current

It can be seen that there is always a voltage drop with an increase of load for unity and lagging power factors and the voltage regulation is positive. The voltage regulation for lagging power factors is higher than that at unity power factor. Further, it can be shown that for leading power factors, the voltage regulation will be negative, that is, the no-load terminal voltage will be less than the full-load terminal voltage due to the magnetizing effect of the armature reaction.

Example 7.5 A 250-kVA, 3300-V, star-connected three-phase synchronous generator has resistance and synchronous reactance per phase of 0.25 Ω and 3.5 Ω, respectively. Calculate the voltage regulation at full load 0.8 power factor lagging.

Solution Output power in VA = $\sqrt{3}\, V_L I_L$

$$I_L = \frac{250 \times 10^3}{\sqrt{3} \times 3300} = 43.74 \text{ A}$$

For a star-connected machine, the line current is equal to the phase current. Hence, the phase current, $I_a = 43.74$ A

Now, phase voltage, $V_t = \dfrac{V_L}{\sqrt{3}} = \dfrac{3300}{\sqrt{3}} = 1905.26$ V

If $\cos \varphi = 0.8$, then $\sin \varphi = 0.6$.

$R_a = 0.25$ Ω and $X_S = 3.5$ Ω

Using Eq. (7.28), $E = \sqrt{(V_t + I_a R_a \cos\varphi + I_a X_S \sin\varphi)^2 + (I_a X_S \cos\varphi - I_a R_a \sin\varphi)^2}$

$$= \sqrt{(1905.26 + 43.74 \times 0.25 \times 0.8 + 43.74 \times 3.5 \times 0.6) + (43.74 \times 3.5 \times 0.8 - 43.74 \times 0.25 \times 0.6)^2}$$

$$= 2009.21 \text{ V}$$

∴ Percentage voltage regulation = $\dfrac{2009.21 - 1905.26}{1905.26} \times 100 = 5.455$

Alternate method

Let $\quad V_t = 1905.26 \angle 0°$

Now $\quad \varphi = \cos^{-1} 0.8 = 36.87°$

Then $\quad I_a = 43.74 \angle -36.87°$

$\qquad Z_S = 0.25 + j3.5 = 3.51 \angle 85.91°$

Using Eq. (7.26),

$$E = V_t + I_a Z_S$$
$$= (1905.26 + j0) + 43.74 \times 3.51 \angle (85.91 - 36.87)°$$
$$= 1905.26 + 153.53 \angle 49.04°$$
$$= 1905.26 + 153.53 \times (0.6558 + j0.7551)$$
$$= 1905.26 + 100.68 + j115.93$$
$$= 2009.03 \angle 3.31$$

$\therefore$ Percentage voltage regulation $= \dfrac{2009.03 - 1905.26}{1905.26} \times 100 = 5.46$

7.8 OPEN-CIRCUIT (OC) AND SHORT-CIRCUIT (SC) TESTS ON A THREE-PHASE SYNCHRONOUS GENERATOR

As in the case of a transformer, open-circuit (oc) and short-circuit (sc) tests are performed on a three-phase generator to determine its synchronous reactance. Figure 7.27 shows the connection diagram for performing the oc and sc tests.

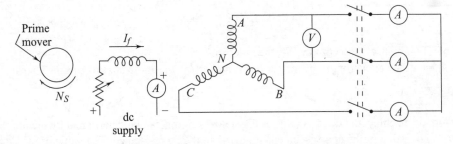

Fig. 7.27 Connection diagram for oc and sc tests on a three-phase synchronous generator

In the connection diagram, the rotor field winding is connected to a dc voltage source through an ammeter and a rheostat. The three-phase stator winding is connected to a three-pole switch for short-circuiting the winding. The rotor is driven by a prime mover at the synchronous speed. The open-circuit voltage is measured by the voltmeter connected between two lines and the three ammeters measure the short-circuit current.

For the oc test, the switch is kept open and the open-circuit line voltage V_{oc} is read from the voltmeter for different values of the field excitation current I_f. The excitation current can be varied using the rheostat in the field circuit. The field current is increased in steps up to an open-circuit voltage, which is approximately 20–25% above the rated voltage. The open-circuit characteristic (OCC) is plotted with the phase voltage $V_{oc}/\sqrt{3}$ on the Y-axis against I_f on the X-axis as shown in Fig. 7.28. The OCC is the magnetization characteristic of the generator.

For the sc test, the field current is adjusted to a low value such that when the switch is closed the rated full-load current flows in the stator armature winding and the terminal voltage is zero. Since I_f is adjusted to a low value, the magnetic circuit is unsaturated. Therefore, the variation of the armature current under short circuit, I_{sc}, with the variation in the field current is linear. Only one reading is sufficient to obtain the short-circuit characteristic (SCC). The SCC is also plotted with the current I_{sc} on the Y-axis and the field current I_f on the X-axis as shown in Fig. 7.28.

Under the sc condition of the synchronous generator $V_t = 0$, and it follows from Eq. (7.26) that

$$Z_S = \frac{V_{oc}/\sqrt{3}}{I_{sc,\ full\text{-}load}} \quad \text{at } I_f \text{ corresponding to } I_{sc,\ full\text{-}load} \tag{7.31}$$

As shown in Fig. 7.28, at a field current OA, the short-circuit current is AC, which is equal to the full-load current, and AB is the induced line voltage of the synchronous generator. Then

$$Z_S = \frac{AB}{AC}$$

Because of the non-linear nature of OCC shown in Fig. 7.28, the ratio of the open-circuit voltage to the short-circuit current at different values of excitation currents is different. Hence, Z_S when plotted at different excitation currents, as shown in Fig. 7.28, indicates a higher value at lower excitations as compared to higher excitations. When the machine is loaded, it operates under the condition of magnetic saturation and the value of Z_S will be lower than that obtained by using Eq. (7.31). A more realistic value of Z_S may be computed as

$$Z_S = \frac{V_{oc,\ rated}/\sqrt{3}}{I_{sc}} \quad \text{at } I_f \text{ corresponding to } V_{oc,\ rated\ value}$$
$$(7.32)$$

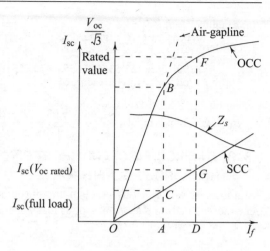

Fig. 7.28   Open-circuit and short-circuit characteristics of a three-phase generator

As shown in Fig. 7.28, at field current OD, the induced rated line voltage of the synchronous generator is DF and DG is the short-circuit current. Then

$$Z_S = \frac{DF}{DG}$$

For a synchronous machine, the dc resistance of the stator winding can be determined by the ammeter–voltmeter method. The value of the ac resistance is higher than the dc resistance. The value of the dc resistance may be multiplied by a factor 1.4 to get the ac resistance value R_a. The synchronous reactance then can be calculated using Eq. (7.27a) as

$$X_S = \sqrt{Z_S^2 - R_a^2} \tag{7.33}$$

Example 7.6   Determine the voltage regulation of a 2200-V, single-phase alternator giving a current of 120 A at (a) unity power factor, (b) power factor 0.8 leading, and (c) power factor 0.707 lagging. From the oc and sc test results, full-load current of 120 A is produced on short circuit by a field excitation of 2.5 A and an electromotive force of 480 V is produced on open circuit by the same excitation. The armature resistance is 0.4 Ω.

Solution   From the oc and sc test results, the synchronous impedance Z_S can be computed.
Using Eq. (7.3),

$$Z_S = \frac{480}{120} = 4\ \Omega$$

Using Eq. (7.27a),

$$X_S = \sqrt{4^2 - 0.4^2} = 3.98\ \Omega$$

The impedance angle,

$$\theta = \tan^{-1} \frac{3.98}{0.4} = 84.26°$$

Then in polar form,

$$Z_S = 4\angle 84.26°$$

The equivalent circuit of the alternator is shown in Fig. 7.29.

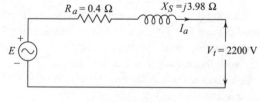

Fig. 7.29

(a) At unity pf, $\cos \varphi = 1$, then $\varphi = 0°$.

Let the terminal voltage V_t be the reference phasor. Then

$$V_t = 2200 + j0$$
$$I_a = 120 \angle 0°$$

Using Eq. (7.26),

$$
\begin{aligned}
E &= V_t + I_a Z_S \\
&= 2200 + 120\angle 0° \times 4\angle 84.26° \\
&= 2200 + 480\angle 84.26° \\
&= 2200 + 48 + j477.6 = 2248 + j\,477.6 \\
&= 2298.17\angle 12°
\end{aligned}
$$

$\therefore$ Percentage voltage regulation $= \dfrac{2298.17 - 2200}{2200} \times 100 = 4.46$

(b) At 0.8 leading pf, $\cos \varphi = 0.8$, then $\varphi = \cos^{-1} 0.8 = \angle 36.87°$. Then

$$
\begin{aligned}
I_a &= 120 \angle 36.87° \\
E &= V_t + I_a Z_S \\
&= 2200 + 120\angle 36.87° \times 4\angle 84.26° \\
&= 2200 + 480\angle 121.13° \\
&= 2200 - 248.15 + j410.88 = 1951.85 + j\,410.88 \\
&= 1994.63 \angle 12°
\end{aligned}
$$

$\therefore$ Percentage voltage regulation $= \dfrac{1994.63 - 2200}{2200} \times 100 = -9.335$

(c) At 0.707 lagging pf, $\cos \varphi = 0.707$, then $\varphi = \cos^{-1} 0.707 = \angle 45°$.

Then
$$
\begin{aligned}
I_a &= 120 \angle -45° \\
E &= V_t + I_a Z_S \\
&= 2200 + 120\angle -45° \times 4\angle 84.26° \\
&= 2200 + 480\angle 39.26° \\
&= 2200 + 371.65 + j303.76 = 2571.65 + j303.76 \\
&= 2589.53\angle 6.74°
\end{aligned}
$$

$\therefore$ Percentage voltage regulation $= \dfrac{2589.53 - 2200}{2200} \times 100 = 17.706$

7.9 SYNCHRONOUS GENERATOR CONNECTED TO AN INFINITE BUS BAR

Synchronous generators are invariably connected to a very large electric power system. An important characteristic of the power system network is that the system voltage at the point of connection is constant in magnitude and phase angle. The system frequency is also constant. Such a point in a power system is termed as *infinite bus*. Thus, when a generator is connected to an infinite bus, the terminal voltage V_t of the generator will not be altered by any changes in the excitation of the generator and no frequency change will occur regardless of the change in prime mover input of the synchronous machine. Hence, an infinite bus is an ideal voltage source and has infinite inertia.

A synchronous machine, therefore, when connected to an infinite bus has its speed and terminal voltage fixed. However, the field current and the mechanical torque on the shaft are the two controllable variables of the synchronous machine. The variation of the field excitation current I_f, when applied to a synchronous machine, operating as a generator or motor, supplies or absorbs a variable amount of reactive power. Because the synchronous machine runs at constant speed, the real power can only be controlled through the variation of torque imposed on the shaft of the machine by either the prime mover in the case of a generator or the mechanical load in the case of a motor.

An infinite bus bar may be viewed as a system from which infinite (very large) power can be withdrawn or added without any change in the bus bar voltage or frequency. Theoretically, the bus bar has infinite inertia.

Figure 7.30(a) shows a round rotor synchronous generator connected to an infinite bus bar whose per phase voltage is V volts. The machine has a negligible armature resistance and a reactance of X_s Ω/phase. It is supplying power to the bus bar at a lagging power factor angle of φ degrees. Figure 7.30(b) shows the phasor diagram with the bus bar voltage as the reference phasor. The generated armature voltage is E volts/phase.

If I ampere/phase is the armature current, the power supplied by the generator is

$$P_0 = VI \cos\varphi \text{ W} \tag{7.34}$$

Application of the sine rule to the triangle ABC in Fig. 7.30(b) leads to

$$\frac{E}{\sin(90+\varphi)} = \frac{IX_s}{\sin\delta}$$

or $\qquad I\cos\varphi = \dfrac{E}{X_s}\sin\delta$

Substitution for $I\cos\varphi$ in Eq. (7.34) leads to

$$P_0 = \frac{EV}{X_s}\sin\delta = P_{max}\sin\delta \text{ W} \tag{7.35}$$

where δ, called the power angle, is the phase displacement between the bus bar and the armature voltages and $P_{max} = (EV)/X_s$ watt is the maximum power.

If the armature resistance loss, hysteresis, and eddy current losses in the generator are neglected, then the generator output is equal to its input. Thus, if P_i denotes the generator input power, it can be expressed as

$$P_i = \frac{EV}{X_s}\sin\delta \text{ W} \tag{7.36}$$

The expression for input power P_i represented by Eq. (7.36) is of extreme importance since it represents the power supplied by a generator to an infinite bus bar. It may be observed that the power output of a generator is sinusoidal and it delivers maximum power at $\delta = 90°$. Equation (7.36) is plotted in Fig. 7.31.

It may be noted from the plot in Fig. 7.31 that if δ exceeds 90°, the power output becomes greater than the input and the generator is said to lose *snchronism*.

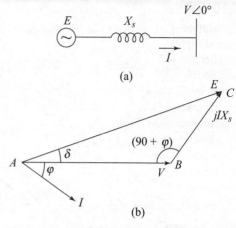

(a)

(b)

Fig. 7.30 Synchronous generator feeding power to an infinite bus bar: (a) synchronous generator connected to an infinite bus bar and (b) phasor diagram

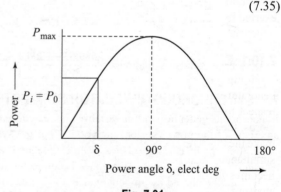

Fig. 7.31

7.10 SYNCHRONOUS GENERATORS IN PARALLEL

Synchronous generators are invariably connected in parallel at the power-generating stations to the infinite bus bars to supply a common load. In power-generating stations, instead of one large capacity generator, a number of smaller generating units are installed and all of these units are connected in parallel. Nowadays all the generating stations as well as loads are interconnected through a network of transmission and distribution lines to form a power system. Thus, all synchronous generators connected to the power system operate in parallel and feed electrical power to the system at the point of their connections. The power fed to the system is used by the electrical load of the consumers connected to the system at different points. An electric supply system with a large number of synchronous generators connected in parallel is called an *infinite bus*. The infinite bus bar is assumed to be a large fictitious synchronous machine with zero synchronous impedance and a very large inertia constant; with the result that both the terminal voltage and frequency are constant irrespective of the load. In this section, the behaviour of synchronous generators connected in parallel to the infinite bus bars is discussed.

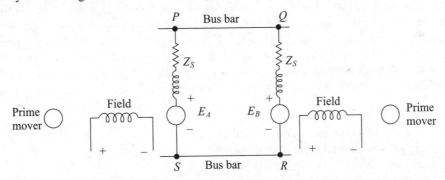

Fig. 7.32   Synchronous generators connected in parallel

For easier comprehension of the process of parallel operation, two similar single-phase generators A and B connected in parallel to the bus bars, as shown in Fig. 7.32, are considered. The emf generated in series with synchronous impedance represents the equivalent circuit of each generator. It is assumed that (i) the generators are on no load; (ii) the generated emf of each machine is equal, that is, $E_A = E_B$; (iii) the torque exerted by each independent prime mover on the respective generator is the same; and (iv) the synchronous impedance Z_S of each machine is the same.

7.10.1 Effect of Varying the Prime Mover Torque

From Fig. 7.32, it can be seen that the emf generated by each generator has the same direction at a particular moment with respect to the bus bars to which they are connected. In other words, the directions of the generated emfs are such that both make the top bus bar positive with respect to the bottom bus bar simultaneously. However, if the closed loop $PQRS$ is traced, it becomes clear that E_A and E_B are in opposition. Since the generated emfs are equal and are in phase opposition within the loop, no current circulates in the closed loop circuit. Figure 7.33(a) shows the phasor diagram that depicts this condition.

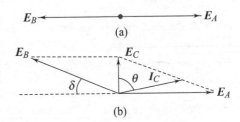

Fig. 7.33   Effect of varying prime mover torque

Let the input to the prime mover of generator B be reduced. This results in a reduction in the driving torque of B. The effect of the reduced torque is to cause the rotor of generator B to fall back by an angle δ with respect to that of A. Figure 7.33(b) depicts the phase relation between E_A and E_B when the rotor of B has fallen back by angle δ. Since the phasors E_A and E_B are no longer in phase opposition, a resultant emf E_C is set up in the closed circuit, which circulates a lagging current I_C. The current lags E_C by an angle θ, as shown in Fig. 7.33(b). The magnitude of the circulating current and angle θ are obtained as under:

$$I_C = \frac{E_C}{2Z_S} \text{ and } \theta = \tan^{-1}\left(\frac{X_S}{R_a}\right) \tag{7.37}$$

where R_a and X_S are the resistance and synchronous reactance, respectively, of the stator winding of each generator.

Since R_a is very small compared to X_S, $\theta \approx 90°$. Therefore, the circulating current I_C is nearly in phase with E_A and in phase opposition with E_B, that is, the machine A is acting like a generator since it is supplying power to machine B to make up for the reduced power being supplied by the prime mover of B.

From the phasor diagram shown in Fig. 7.33(b), it is seen that the larger the magnitude of δ, the greater is the circulating current and more is the power supplied by generator A. In order to maintain the supply frequency, the prime mover of generator A must increase its driving torque by an amount equal to the decrease in the driving torque of the prime mover of B. If the machine B further falls back in relation to A, the prime mover of machine A will have to increase its torque to compensate for the new loss in power in machine B. As long as generator A is able to compensate for the loss in the driving power of the prime mover of B and maintain the balance, machines A and B will continue to run in synchronism.

7.10.2 Effect of Varying the Field Current

The machines A and B are connected to the bus bars and are operating in parallel with equal prime mover torques and generated emfs. Let the field excitation of generator B be increased. The effect of increasing the field current is that the generated emf E_B increases but remains in phase opposition with the generated emf E_A.

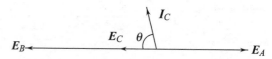

Fig. 7.34   Effect of varying the field current

A resultant emf E_C is established due to the difference in the magnitudes of E_A and E_B. This causes a current I_C to circulate through the stator windings. The circulating current causes a voltage drop in the synchronous impedances of the stator windings, thereby affecting the terminal voltage. Figure 7.34 shows the phasor diagram which depicts the generated emfs, the resultant voltage, and the circulating current phasor positions.

From the phasor diagram, we can write the relationship between various quantities as under

$$E_C = E_B - E_A \tag{7.38}$$

Terminal voltage $= E_B - I_C \times Z_S = E_A + I_C \times Z_S$

Since the machines are similar, $\dfrac{E_C}{2} = I_C \times Z_S$

Hence, the terminal voltage $= E_B - \dfrac{E_C}{2} = E_A + \dfrac{E_C}{2}$ $\tag{7.39}$

The effect of varying the field excitation is to affect the terminal voltage. Further, since $\theta \approx 90°$, the power factor of the circulating current is zero. Consequently, there is a negligible circulation of the active power from one generator to the other, while there is a significant change in the reactive power resulting in the variation of the power factor of the machine.

In light of the above discussions with respect to the operation of synchronous generators in parallel, the following conclusions can be drawn.

(a) The active power load sharing between generators is governed by the driving torques of their prime movers. Varying the field excitation of the generators has a negligible effect on load sharing.

(b) Varying the field excitation will vary the terminal voltage and the power factor of the machines. The reactive power distribution between the machines is changed.

7.10.3 Advantages of Operating Synchronous Generators in Parallel

Operating synchronous generators in parallel offer the following advantages:

(i) A generator operates at its maximum efficiency when it is supplying load close to its full load capacity. Therefore, several small units can be made to operate at their near-full load capacities to supply a varying load, making it possible to add or remove units when the load increases or decreases.

(ii) With several generators operating in parallel, reliability of power supply is assured in the event of a fault or when a unit has to be removed for repair or maintenance.

(iii) A single generator of high volt-ampere capacity would be uneconomical to build due to technical limitations.

(iv) To start with, small-capacity generators may be used to supply the demand, thereby saving on the initial capital investment, and more generators can be added depending on the increase in demand.

Example 7.7 Two single-phase alternators, each having generated emfs of E_1, E_2 volts and synchronous impedances of Z_1 and Z_2 Ω, respectively, are connected in parallel to supply a load of Z_L Ω at V_L volts. Derive expressions for the load voltage in terms of the generated voltages and the synchronous impedances.

Given $Z_1 = (0.25 + j\, 1.5)$ Ω, $Z_2 = (0.2 + j\, 1.25)$ Ω, $Z_L = 10\angle 36°$ Ω, and $E_1 = E_2 = 220\angle 0°$ V, calculate (i) load voltage, (ii) load current, and (iii) load supplied by each generator in VA.

Solution Figure 7.35 shows the two generators connected in parallel and supplying the given load.
Assume that the load voltage and current are V_L and I_L, respectively. From the circuit diagram, the following equations are easily obtained:

$$I_L = I_1 + I_2$$

The load voltage is given by

$$V_L = I_L Z_L = (I_1 + I_2) Z_L \qquad \text{(I)}$$

Since the generators are connected in parallel, their terminal voltages are equal and are also equal to the load voltage. Thus,

$$E_1 - I_1 Z_1 = E_2 - I_2 Z_2 = V_L \qquad \text{(II)}$$

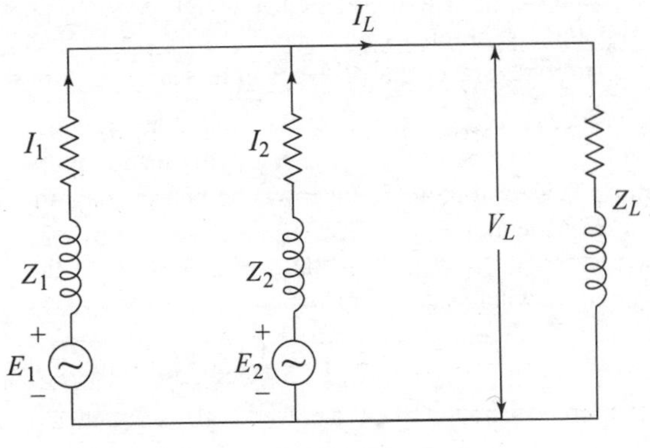

From Eq. (II), the generator currents are obtained as follows:

Fig. 7.35

$$I_1 = \frac{E_1 - V_L}{Z_1} = (E_1 - V_L)Y_1 \quad \text{and} \quad I_2 = \frac{E_2 - V_L}{Z_2} = (E_2 - V_L)Y_2 \qquad \text{(III)}$$

Substitution of Eq. (III) in Eq. (I) and simplification of the resultant equation produces

$$V_L = \frac{E_1/Z_1 + E_2/Z_2}{1/Z_1 + 1/Z_2 + 1/Z_L} = \frac{E_1 Y_1 + E_2 Y_2}{Y_1 + Y_2 + Y_L} \qquad \text{(IV)}$$

where Y_1, Y_2, and Y_L correspond to the respective synchronous admittances of the generators and the load admittance. From the data:

$$Y_1 = 1/(0.25 + j1.5) = (0.11 - j0.65)\text{S}, \quad Y_2 = 1/(0.2 + j1.25) = (0.13 - j0.78)\text{S}.$$

$$Y_L = 1/10\angle 36° = 0.1\angle -36° = (0.08 - j0.06)\text{S}$$

$$E_1 = E_2 = 220\ \angle 0°\ \text{V}$$

(i) Substitution of these values in Eq. (IV) results in

$$V_L = 209.26 - j9.7 = 209.48\angle -2.65°\,V$$

Use of Eq. (III) leads to the following gen r tor currents:

$$I_1 = (220 - [209.26 - j9.7])(0.11 - j0.65) = 7.45 - j5.92 = 9.52\angle -38.45°\,A$$

$$I_2 = (220 - [209.26 - j9.7])(0.13 - j0.78) = 8.91 - j7.17 = 11.43\angle -38.83°\,A$$

(ii) The load current is given by

$$I_L = I_1 + I_2 = 16.36 - j13.08 = 20.95\angle -38.65°$$

(iii) Output of generator 1 = $E_1 \times \text{abs}(I_1) = 220 \times 9.52 = 2093.65\,VA$

Output of generator 2 = $E_2 \times \text{abs}(I_2) = 220 \times 11.43 = 2515.04\,VA$

Example 7.8 A single-phase 6.6 kV synchronous generator is supplying power to an infinite bus at unity power factor. (a) If the armature current is 200 A, calculate (i) armature voltage and (ii) power angle. (b) With the prime mover input remaining unchanged, the field excitation of the generator is varied and the armature voltage is increased by 25%. Determine (iii) armature current and (iv) power factor. Assume an armature synchronous reactance of 8 Ω and neglect armature resistance and all types of losses.

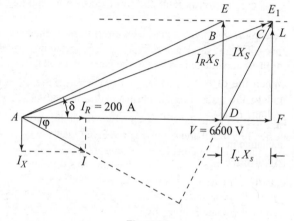

Fig. 7.36

Solution (a) First step is to determine the armature voltage when the generator is supplying load at unity power factor. Figure 7.36 shows the phasor diagram of the generator.

(i) Reference to triangle ABD in Fig. 7.36 gives the armature voltage as

$$E = \sqrt{V^2 + (I_R X_s)^2} = \sqrt{6600^2 + (200 \times 8)^2} = 6791.17\,V$$

(ii) From triangle ABD, the power angle is obtained as

$$\delta = \tan^{-1}(I_R X_s / V) = \tan^{-1}(200 \times 8 / 6600) = 13.63°$$

Alternate method

The power output of the generator, $P_O = VI_R = 6600 \times 200 = 1320\,kW$
Making use of eq. (7.36) gives

$$\delta = \sin^{-1}\left(\frac{P_o X_s}{EV}\right) = \sin^{-1}\left(\frac{1320 \times 10^3 \times 8}{6791.17 \times 6600}\right) = 13.63°$$

Thus, the computations are verified.

Due to an increase in the excitation, the armature voltage changes to $E_1 = 1.25E = 1.25 \times 6791.17 = 8488.96\,V$. Since the prime mover output has been kept constant, the power output remains constant. The effect of this is twofold, namely, the locus of the armature voltage is a horizontal line parallel to the bus voltage (shown by dotted line in Fig. 7.36) and the armature current begins to lag to neutralize the effect of the increase in excitation voltage. From the triangle ACF, it is seen that

$$AC^2 = (AD + DF) + CF^2$$

or $\quad (6600 + I_x \times 8) = \sqrt{(8488.96)^2 - (200 \times 8)^2} = 8336.82$

or $\quad I_x = \dfrac{8336.82 - 6600}{8} = 217.10$ A

(iii) From the phasor diagram, it is easily seen that armature current can be written as

$$I = \sqrt{I_R^2 + I_x^2} = \sqrt{200^2 + 217.10^2} = 295.18 \text{ A}$$

(iv) Power factor, $\cos\varphi = \dfrac{I_R}{I} = \dfrac{200}{295.18} = 0.68$

Example 7.9 With the magnitude of the generator voltage fixed at E_1 volts in Example 7.8, the input to the generator is raised by 50% by increasing the prime mover output. Calculate the (i) armature current and (ii) power factor.

Solution The generator output P_o now becomes $1.5 \times 6600 \times 200 = 1980$ kW

(i) Use of Eq. (7.36) leads to

$$\delta = \sin^{-1}\left(\frac{P_0 X_s}{E_1 V}\right) = \sin^{-1}\left(\frac{1980 \times 10^3 \times 8}{8488.96 \times 6600}\right) = 16.42°$$

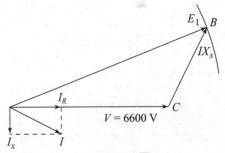

Fig. 7.37

When the excitation is kept constant and the prime mover input to the generator is increased, the locus of the generated voltage E_1 is the arc of a circle as shown in the phasor diagram in Fig. 7.37. The power supplied by the generator is brought about by a change in the power angle δ.

Application of the cosine rule to triangle ABC results in

$$IX_s = \sqrt{E_1^2 + V^2 - 2EV\cos\delta}$$

$$= \sqrt{8488.96^2 + 6600^2 - 2 \times 8488.96 \times 6600 \times \cos(16.42°)}$$

$$= 2853.02 \text{ V}$$

(i) Hence, armature current, $I = 2853.02/8 = 356.63$ A

(ii) Power factor $= \cos\varphi = P_o/(VI) = 1,980,000/(6600 \times 356.63) = 0.84$

7.11 PRINCIPLE OF OPERATION OF THREE-PHASE SYNCHRONOUS MOTORS

A synchronous motor is a device where electrical energy is supplied to both the rotor field and the armature winding. Alternating current is supplied to the stator armature winding and dc excitation is supplied to the rotor field winding. The magnetic field of the armature current produces a field rotating at a synchronous speed as discussed in Section 7.5. To produce a steady electromagnetic torque, the magnitude of the magnetic fields of the stator and the rotor must be constant and these two fields must be stationary with respect to each other.

A simple two-pole machine may be considered to substantiate this point. A three-phase supply when connected to the stator windings of a synchronous motor will result in a synchronously rotating magnetic field of speed N_S given by Eq. (4.9). This rotating field may be considered to be equivalent to a magnet rotating clockwise at a synchronous speed as shown in Fig. 7.38. The dc excitation in the rotor field winding sets up a two-pole field which is stationary. Thus, there exists a pair of revolving armature poles and a pair of stationary rotor field poles. At some instant of time let the position of the N and S poles of the rotating field be as shown in Fig. 7.38(a). The N and S poles of the stationary rotor may be at any position, say at X and Y as indicated in Fig. 7.38(a). From Fig. 7.38(a) it is seen that in one half-cycle, the N pole of the stator rotating field repels the N pole of the rotor pole and the S pole of the stator rotating field repels the S pole of the rotor pole. As a result of this, there will be an anticlockwise torque on the rotor in this half-cycle. In the next half-cycle of the stator supply the orientation of the stator rotating field reverses as shown in Fig. 7.38(b). Then, the S pole of the stator rotating field attracts the N pole of the rotor field and vice versa. As a result of this, there will be a clockwise torque on the rotor in this half cycle. Hence, over one complete cycle of the stator supply, the average torque on the rotor is zero and the rotor cannot start rotating.

A synchronous motor is not a self-starting machine. However, by the use of some auxiliary drive, if the rotor be rotated in the same direction and at the same speed as that of the rotating field, then the relative positions of the poles as shown in Fig. 7.38(b) remain unaltered with time. Thus, the torque on the rotor remains unidirectional (clockwise here) in all the half cycles and the rotor starts rotating in the same direction and speed as that of the rotating field. This also reveals the fact that a synchronous motor can run only at the synchronous speed.

Sometimes a small dc motor mounted on the rotor shaft is used to rotate the unexcited rotor with a speed close to the synchronous speed. The dc excitation is then given to the field winding at the

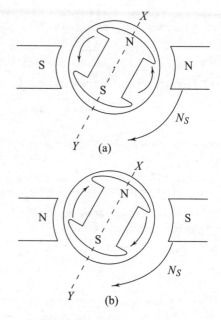

Fig. 7.38   Torque production in synchronous motor

right moment to produce a positive torque for a sufficiently long period so that the rotor and stator poles get locked into synchronism. Alternatively, a squirrel cage winding, similar to an induction motor embedded in the pole faces, is utilized for the purpose. With the help of this winding (also known as amortisseur winding), the synchronous motor is brought close to the synchronous speed.

7.11.1   Phasor Diagram and Equivalent Circuit

The equivalent circuit of a motor is shown in Fig. 7.39. It may be noted that the circuit is the same as that of a generator, except that the direction of the armature current is reversed.

The synchronous motor phasor diagrams may be derived from that of the synchronous generator by making two changes in Eq. (7.26). The terms E and V_t are interchanged in that V_t is now the source voltage applied to the motor and E is the counter emf generated in the machine. The voltage equation for the motor becomes

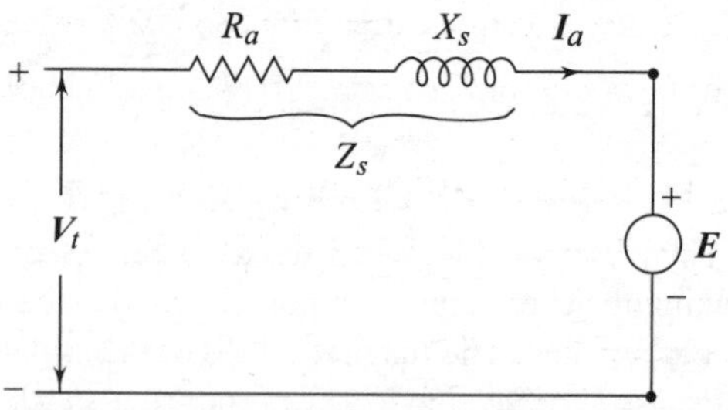

$$V_t = E + I_a Z_S \qquad (7.40)$$

Fig. 7.39   Equivalent circuit of a synchronous motor

The angle δ between E and V_t which was positive in the case of a generator because of the driving action of the prime mover is now negative in the case of a motor. The angle δ is known as the torque angle. It may be noted that E is associated with the rotor field and V_t is associated with the resultant field. At no load, the angle δ is zero implying that the axes of the field poles and the resultant field are in time phase. However, as the load on the motor shaft is applied, the rotor falls slightly behind its no-load position and this is opposite to that in the case of a generator. Based on these two changes, the phasor diagrams of the synchronous motor for unity pf, lagging pf, and leading pf currents are shown in Figs 7.40(a), 7.40(b), and 7.40(c), respectively.

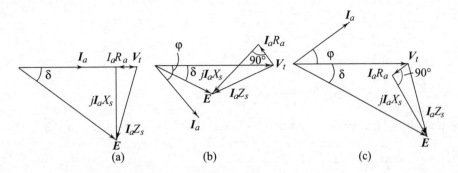

Fig. 7.40 Phasor diagram for a synchronous motor: (a) unity pf, (b) lagging pf, and (c) leading pf

7.11.2 Electrical and Mechanical Power

From any of the phasor diagrams in Fig. 7.40, it may be seen that the power input P_E, in watts, is given by

$$P_E = 3V_t I_a \cos \varphi \tag{7.41}$$

By subtracting the armature resistance loss $(3I_a^2 R_a \text{ W})$ from the power input P_E, the mechanical power or the electromagnetic power P_M is obtained as shown in Eq. (7.42).

$$P_M = 3V_t I_a \cos \varphi - 3I_a^2 R_a \tag{7.42}$$

Again reference to the phasor diagrams in Fig. 7.40 shows that the mechanical power P_M developed by the motor can be written as

$$P_M = 3EI_a \cos(\varphi \pm \delta) \text{ W} \tag{7.43}$$

It may noted that a $+\delta$ is to be used for leading power factor and $-\delta$ is to be employed for lagging power factor. Equating Eqs (7.42) and (7.43), yields

$$V_t \cos \varphi - I_a R_a = E \cos(\varphi \pm \delta) \text{ W} \tag{7.44}$$

Taking V_t as the reference phasor, $(V_t = V_t \angle 0°)$ and $E = E \angle -\delta°$, then the armatue current is

$$I_a = \frac{V_t \angle 0° - E \angle -\delta°}{Z_S \angle \theta°}$$

$$= \frac{V_t \angle -\theta°}{Z_S} - \frac{E \angle -(\delta + \theta)°}{Z_S} \tag{7.45}$$

Taking the component of I_a along V_t,

$$I_a \cos \varphi = \frac{V_t}{Z_S} \cos \theta - \frac{E}{Z_S} \cos(\delta + \theta) \tag{7.46}$$

Substituting the value of $I_a \cos \varphi$ from Eq. (7.46) into Eq. (7.41),

$$P_E = \frac{3V_t^2}{Z_S} \cos \theta - \frac{3V_t E}{Z_S} \cos(\delta + \theta) \tag{7.47}$$

To determine the electromagnetic power P_M, if emf E is chosen as reference, then $E = E \angle 0°$ and $V_t = V_t \angle \delta°$. The armature current is

$$I_a = \frac{V_t \angle \delta° - E \angle 0°}{Z_S \angle \theta°}$$

$$= \frac{V_t \angle \delta - \theta^\circ}{Z_S} - \frac{E \angle - \theta^\circ}{Z_S} \tag{7.48}$$

Taking the component of I_a along E,

$$I_a \cos(\delta \pm \varphi) = \frac{V_t}{Z_S} \cos(\delta - \theta) - \frac{E}{Z_S} \cos\theta \tag{7.49}$$

Substituting the value of $I_a \cos(\delta \pm \varphi)$ from Eq. (7.49) in to Eq. (7.44) yields the magnitude of P_M as under

$$P_M = \frac{3EV_t}{Z_S} \cos(\delta - \theta) - \frac{3E^2}{Z_S} \cos\theta \tag{7.50}$$

Usually the resistance of the armature R_a is small in comparison with the synchronous reactance of the machine. If R_a is neglected, then $Z_S = j X_S$, and $\theta = 90^\circ$. The expressions for the electrical power input and electromagnetic power become

$$P_E = \frac{3V_t E}{X_S} \sin\delta \tag{7.51}$$

and

$$P_M = \frac{3EV_t}{X_S} \sin\delta \tag{7.52}$$

The synchronous motor while in operation incurs rotational losses and core losses. If these losses are subtracted from the electromagnetic power, the mechanical output at the rotor shaft of the machine can be determined.

7.11.3 Synchronous Motor Operation at Constant Load and Variable Excitation

If a synchronous motor operates at a constant load, then neglecting the armature resistance, Eq. (7.42) may be written as

$$I_a \cos\varphi = \frac{P_M}{3V_t} = \text{constant, as } V_t \text{ is constant} \tag{7.53}$$

and Eq. (7.52) becomes

$$E \sin\delta = \frac{P_M X_S}{3V_t} = \text{constant, as } V_t \text{ and } X_S \text{ are constants} \tag{7.54}$$

If the excitation is changed by varying the rotor field current, then the induced emf E changes. From Eq. (7.54) it is seen that for constant load $E \sin\delta$ remains constant. Thus, with the change in the magnitude of E, the term $\sin\delta$ will also change to maintain the product $E \sin\delta$ constant. Therefore, it may be said that when excitation is increased, E increases and $\sin\delta$ decreases so also torque angle δ decreases and vice versa.

Figure 7.41 shows the phasor diagram of a synchronous motor driving a constant load. The effect of the variation of excitation is shown in the phasor diagram. To satisfy the condition developed in Eq. (7.54) for constant load, $E \sin\delta$ is constant, hence the tip of the phasor E must move along a line parallel to the phasor V_t. From Eq. (7.40), $V_t = E + j I_a X_S$, as $Z_S \approx j X_S$. Thus, when the phasor E changes due to a change in excitation current, the armature current phasor I_a has to change as the terminal voltage V_t and synchronous reactance X_S are assumed to be constant. But from Eq. (7.53), $I_a \cos\varphi$ has to be maintained constant. This is only possible if the tip of the current phasor I_a moves along a vertical line perpendicular to the direction of V_t. It may be further noted that the synchronous reactance drop is perpendicular to the current phasor I_a as shown in the phasor diagram. At unity pf, I_a and V_t are in phase and the excitation corresponding to unity pf is called normal or 100% excitation and the emf corresponding to this excitation is called as normal excitation voltage.

The emf phasor at normal excitation is denoted by **E**. Excitation greater than the normal excitation is called over excitation and excitation smaller than the normal excitation is called under excitation. From the phasor diagram it may further be noted that the synchronous motor draws lagging pf current when underexcited, while it draws leading pf current when overexcited.

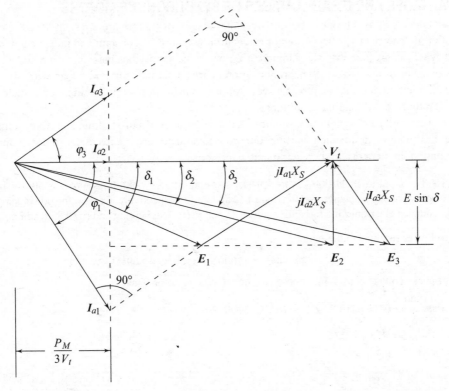

Fig. 7.41   Phasor diagram of a synchronous motor at constant load and variable excitation

The ability of the synchronous motor to draw leading current when overexcited can be used to improve the power factor at the input lines to an industrial establishment that makes heavy use of induction motors and other equipment drawing power at a lagging power factor. When a synchronous motor is used solely for power factor improvement, it is called a *synchronous condenser*.

It has been noticed that variation of excitation current I_f of the rotor field winding changes the magnitude of armature current I_a and pf (cos φ) such that I_a cos φ is constant. As I_f increases, from a low value to a nominal excitation value, the armature current I_a decreases in magnitude and pf lags till at unity pf I_a becomes zero. If I_f is increased beyond the nominal excitation value,

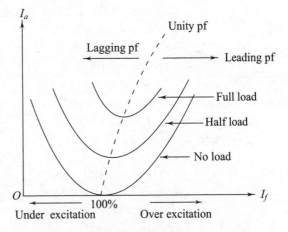

Fig. 7.42   *V* curves of a synchronous motor

the magnitude of armature current I_a increases and the power factor changes from lagging to leading. The graph showing the variation of armature current I_a with the variation of the excitation current I_f is called V curves because of its shape. A set of V curves for different constant loading conditions are shown in Fig. 7.42.

7.12 ADVANTAGES AND DISADVANTAGES OF SYNCHRONOUS MOTORS

The main advantages of synchronous motors over other types of motors are constant speed operation and easy control of the power factor. The speed of a synchronous motor is constant and independent of load. An overexcited synchronous motor draws leading current from the supply bus bars and can be used to raise the overall power factor of the system. Synchronous motors are sometimes run on no load with over excitation and under excitation modes, that is, run as a synchronous condenser, for improvement of voltage regulation of transmission lines for all loading conditions.

The main disadvantage of synchronous motors is that these are not self-starting, therefore, some auxiliary arrangement has to be made for their starting. Other disadvantages are the necessity of a dc voltage source for field excitation, higher cost, and greater maintenance cost.

Example 7.10 The input to an 11,000-V, three-phase, star-connected synchronous motor is 50 A. The effective resistance and synchronous reactance per phase are 1 Ω and 25 Ω, respectively. Find (a) the power supplied to the motor, (b) the induced electromotive force for a power factor of (i) 0.85 lagging and (ii) 0.85 leading.

Solution   (a) Power supplied to the motor = $\sqrt{3}\ V_L I_L \cos \varphi$

$$= \sqrt{3} \times 11000 \times 50 \times 0.85 = 809.73\ \text{kW}$$

(b) The motor phase voltage phasor V_t is chosen as the reference phasor. Then

$$V_t = \frac{11000}{\sqrt{3}} \angle 0° = 6350.85 \angle 0° = 6350.85 + j0$$

$$Z_S = 1 + j25 = 25.02 \angle 87.71°$$

(i) $\cos \varphi = 0.85$ lagging

∴ $\varphi = \cos^{-1} 0.85$ lagging $= \angle -31.79°$

The phase current phasor, $I_a = 50 \angle -31.79°$

The emf induced, $E = V_t - I_a Z_S$

$$= 6350.85 - 50 \angle -31.79° \times 25.02 \angle 87.71°$$
$$= 6350.85 - 1251 \angle 55.92°$$
$$= 6350.85 - 701.0 - j1036.15$$
$$= 5649.85 - j1036.15$$
$$= 5744.08 \angle -10.39°$$

(ii) $\cos \varphi = 0.85$ leading, then $\varphi = \angle 31.79°$

The phase current phasor, $I_a = 50 \angle 31.79°$

The emf induced, $E = V_t - I_a Z_S$

$$= 6350.85 - 50 \angle 31.79° \times 25.02 \angle 87.71°$$
$$= 6350.85 - 1251 \angle 119.5°$$
$$= 6350.85 + 616.02 - j1088.81$$
$$= 6966.87 - j1088.81$$
$$= 7051.44 \angle -8.88°$$

Example 7.11 A synchronous motor, which is driving a mechanical load of 12 horse power at an efficiency η of 85%, is used to improve the power factor of a three-phase 415 V, 50 Hz circuit. If the circuit load current is 25 A and the power factor is 0.8 lagging, determine (i) the input kVA to the motor and (ii) its power factor for improving the power factor to unity.

Solution Three-phase circuit load, $P = \sqrt{3}V_L I_L \cos\varphi = \sqrt{3} \times 415 \times 25 \times 0.8 = 14.38 \text{ kW}$
Power factor angle, $\varphi = \cos^{-1} 0.8 = 36.87°$

Figure 7.43(a) shows the power triangle of the three-phase circuit load.

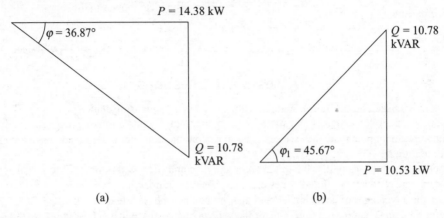

(a)   (b)

Fig. 7.43

From the power triangle, the reactive power component Q of the circuit load is given by
$$Q = P \tan\varphi = 14.38 \times \tan(36.87°) = 10.78 \text{ kVAR}$$
Mechanical load driven by the motor, $P_m = (12 \times 746)/1000 = 8.95 \text{ kW}$
In order that the power factor of the circuit load improves to unity, the synchronous motor must supply to the circuit leading reactive power equal to Q. Hence, the total input power S to the motor may be written as
$$S = \frac{P_m}{\eta} + jQ = \frac{8.95}{0.85} + j10.78 = (10.53 + j10.78)\text{kVA}$$
The power triangle for the input to the synchronous motor is shown in Fig. 7.43(b). Thus, the input kVA to the motor = $\sqrt{10.53^2 + 10.78^2} = 15.07 \text{ kVA}$.
From the power triangle, the power factor angle φ_1 is given by
$$\varphi_1 = \tan^{-1}\left(\frac{10.78}{10.53}\right) = 45.67°$$
and the power factor, $\cos\varphi = \cos 45.67° = 0.70$

Recapitulation

Frequency, $f = (NP/120)$ Hz

Slots per pole per phase, $m = S/(qP)$

Angle between two armature slots,
$\gamma = (\pi P)/S$ electrical radians

Distribution (or breadth) factor,
$$k_d = \frac{\sin[(m\gamma)/2]}{m\sin(\gamma/2)}$$

Pitch factor, $k_p = \cos(\beta/2)$

Winding factor, $k_w = k_d\, k_p$

Voltage induced, $E = 2.22 \times k_w \times Z \times f \times \phi$, V

$$= 4.44 \times k_w \times T \times f \times \phi, \text{ V}$$

Voltage regulation

$$= \frac{\text{No-load voltage } - \text{ Full-load voltage}}{\text{Full-load voltage}} \text{ pu}$$

Generated voltage, $E = V_t + I_a Z_S$

Synchronous impedance,

$$Z_S = \frac{V_{oc}/\sqrt{3}}{I_{sc,\ \text{full-load}}} \text{ at } I_f \text{ corresponding to } I_{sc,\ \text{full-load}}$$

Realistic synchronous impedance

$$Z_S = \frac{V_{oc,\ \text{rated}}/\sqrt{3}}{I_{sc}} \text{ at } I_f \text{ corresponding to } V_{oc,\ \text{rated value}}$$

Synchronous motor

$$V_t = E + I_a Z_S$$

Input power to the motor, $P_E = 3\, V_t I_a \cos\varphi$

Electromagnetic power in the motor,

$$P_M = 3 V_t I_a \cos\varphi - 3I_a^2 R_a = 3EI_a \cos(\varphi \pm \delta)$$

If $Z_S \approx jX_S$, then $P_E = P_M = [(3V_t E)/X_S]\sin\delta$

Assessment Questions

1. (a) Explain with sketches the constructional features of a synchronous machine.
    (b) Enumerate the relative merits of a stationary armature winding and a rotating magnetic field.
2. Explain why salient pole machines are designed to run on moderate or slow rpm and have a comparatively low output rating.
3. (a) Describe the flux density distribution in (i) salient pole and (ii) round rotor synchronous machines.
    (b) Classify and sketch the different types of three-phase armature windings.
    Compare the advantages of concentrated and distributed three-phase armature windings.
4. (a) Derive an expression for the rms value of the generated emf/phase in a synchronous machine.
    (b) Define and derive an expression for the distribution factor. How is the expression for the generated emf changed due to the distribution factor?
5. Define (i) pitch factor and (ii) width factor. How is the generated emf modified by the width factor?
6. (a) A stator armature is made up of three-phase windings displaced from each other by 120° in space. Show by sketching the waveforms and phasor diagrams that the resultant field produced by the flux, due to the currents in the three phases, has a constant amplitude and rotates at the synchronous speed. Specify the generalized expression for synchronous speed of the rotating field.
    (b) Derive a mathematical expression for the rotating magnetic field set up in the armature of a synchronous machine.
7. Discuss armature reaction and there from describe its effect on terminal voltage when the synchronous generator is supplying (i) a resistive load, (ii) an inductive load, and (iii) a capacitive load. Sketch the terminal voltage versus load current for varying power factors.
8. Describe the behaviour of a synchronous generator under (i) no load and (ii) loaded conditions and there from develop its equivalent circuit.
9. Describe the open-circuit and short-circuit tests on a synchronous generator and explain how the data is employed to determine its synchronous reactance.
10. Define regulation of a synchronous alternator. Under what condition, the terminal voltage is greater than the generated armature voltage?
11. Define an infinite bus. Derive an expression for the power delivered by a round rotor synchronous generator connected to an infinite bus. What is the magnitude and the angle at which the generator delivers maximum power?
12. Enumerate the advantages of operating generators in parallel.
13. (a) Explain the working principle of a synchronous motor.
    (b) Derive a relationship for the power input to a motor.
14. Enumerate the merits and demerits of synchronous motors.
15. Explain the behaviour of a synchronous motor under constant load and variable excitation. Thus, show how a synchronous motor can be made to function as a synchronous condenser.

Problems

7.1   A two-pole 2500 kVA turbo alternator runs at 3600 rpm. Calculate the (i) frequency of the generated voltage, (ii) diameter of the stator bore, and (iii) stator length. Assume a maximum peripheral rotor speed of 140 m/sec and a design factor of 3.0.   [(i) 60 Hz, (ii) 0.74 m, (iii) 0.42 m]

7.2   A stator of an 8-pole, three-phase synchronous alternator has 4 slots per pole per phase. Calculate the breadth factor and pitch factor of the coils. Assume a coil pitch of 10 slots.   [0.958, 0.966]

7.3   The alternator in Problem 7.1 is star-connected and is wound for three phases. It has a double-layer winding with each coil having 16 conductors per coil. If the flux per pole is 0.03 Wb determine (a) the speed of the alternator, (b) per phase voltage generated at a frequency of 50 cycles per second. What is the line current if the alternator supplies a balanced three-phase load of 55 kW at 0.8 power factor lagging?   [(a) 750 rpm; (b) 5465.57 V; 7.26 A]

7.4   The stator winding of a three-phase synchronous machine has 5 slots per pole per phase. If the span of the coils is 10 slot pitches, calculate the breadth and pitch factors of the coils.   [0.985, 0.866]

7.5   A salient-pole synchronous alternator generates a voltage at 50 cycles per second when its rotor is run at 300 rpm. Its stator has 300 slots with 10 conductors per slot. If the maximum value of the distributed flux density is 1.5 T, the stator conductor length is 0.75 m, and the mean air gap radius is 1.5 m, determine the rms value of the alternator line voltage when (a) the stator is connected as a single-phase machine, (b) the stator is connected as a two-phase machine, and (c) the stator is connected as a three-phase machine. In each case, calculate the kVA rating of the alternator if the winding is rated to carry a current of 15 A. Assume full pitch coils and a sinusoidal distribution of flux.
    [(a) 71.68 kV, 1075.19 kVA; (b) 71.67 kV, 1520.4 kVA; (c) 62.09 kV, 1613.25 kVA]

7.6   A star-connected three-phase synchronous alternator is required to generate a no-load line to line voltage of 415 V at 50 cycles/sec when running at 600 rpm. The number of slots per pole per phase is 2 and each slot has a double-layer winding with 6 conductors per slot. Determine (a) the number of poles and (b) the flux per pole required to generate the voltage. Assume that the coil span is short pitched by 1 slot pitch.
    [(a) 10, (b) 0.0145 Wb]

7.7   In Problem 7.6, what should be the coil span if the fifth harmonic is zero?   [4.8 slot pitches]

7.8   A three-phase, star-connected, 1.1-kV, 25-kVA synchronous alternator gave the following data when tested on open circuit.

Field current I (amperes)	2	4	6	8	10	12	14	16	18	20
Open-circuit line voltage (volts)	110	330	500	665	840	990	1100	1195	1280	1340

The short-circuit test gave the full-load current when the field current was 10 A. (a) Draw the open-circuit and short-circuit characteristics and determine the synchronous reactance of the alternator. (b) If the alternator is supplying a full load at 0.8 power factor lagging, calculate the regulation of the alternator.
    [(a) 34.52 Ω (b) 53.76%]

7.9   A single-phase synchronous alternator has a rating of 200 kVA at 2200 V. If the stator resistance and reactance are 0.5 Ω and 5 Ω, respectively, determine the voltage regulation at (a) 50%, (b) 100%, and (c) 120% full load at 0.8 power factor lagging.   [(a) 7.3% (b) 15.07% (c) 18.29%]

7.10 A star-connected three-phase alternator supplies a three-phase load of 500 MW at 33 kV. The synchronous reactance of the alternator is 0.15 Ω/phase and the resistance is negligible. Calculate the magnitude of the generated voltage, voltage regulation, and efficiency for (a) 0.8 lagging power factor, (b) unity power factor, and (c) 0.9 leading power factor. Draw the phasor diagram in each case.
    [(a) 34.78 kV, 5.41%, 99.98%; (b) 33.08 kV, 0.26%, 99.99%;
    (c) 31.981 kV, −3.07%, 99.9%]

7.11 In Problem 7.10(a) determine the phase displacement between the generated and terminal voltage. Calculate the terminal voltage and phase displacement if the power factor of the alternator is to be improved to unity. Assume that the load current remains unaltered. [3.74°, 34.93 kV, 6.54°]

7.12 A 3000 kVA synchronous alternator is connected to a 1.1 kV infinite bus. If the alternator is excited to an open circuit voltage of 1.0 kV, calculate the maximum load delivered. If the synchronous reactance of the alternator is 4 Ω, determine the armature current and power factor at the maximum load. [2229.29 A, 0.67]

7.13 A three-phase, star-connected, 50-Hz alternator supplies a three-phase star-connected purely resistive load of 20 Ω per phase. The stator of the alternator has a flux of 0.2 Wb and 90 conductors per phase. If the stator winding reactance is 2.5 Ω per phase, calculate (a) the generated line voltage, (b) the terminal line voltage, (c) the line current, and (d) the phase difference between the generated and terminal voltage. Draw the phasor diagram. Assume a winding distribution factor of 0.955 and negligible resistance. [(a) 3.3 kV (b) 3.28 kV (c) 94.67 A (d) 7.13°]

7.14 A three-phase star-connected load of magnitude $2.9\angle36.87°$ Ω/phase is serviced by two three-phase synchronous alternators each of capacity 7.5 MVA. The line-to-line open-circuit emf and synchronous reactance of each alternator is $6.6\angle0°$ kV, $j1.25$ Ω/phase and $7.3\angle5°$ kV, $j1.30$ Ω/phase, respectively. Compute the (i) load voltage, (ii) total load in MW, (iii) load supplied by each alternator. [$6.06\angle-6.22°$kV, (ii) 10.11 MW, (iii) 3.45 MW, 6.6 MW]

7.15 Two identical single-phase synchronous alternators are operating in parallel with their generated emfs having a magnitude of 1100 V and a phase displacement of 45° electrical. Both the machines have negligible resistance and a synchronous reactance of 5 Ω. Draw the phasor diagram showing the generated voltages, resultant emf, and circulating current. Determine (a) the magnitude of circulating current, (b) the terminal voltage, and (c) the active power supplied from one alternator to the other. [(a) 84.19 A (b) 1016.27 V (c) 92.61 kW]

7.16 In Problem 7.15, the excitation of one alternator is adjusted to give an open-circuit voltage of 1400 V and that of the other to give an open-circuit voltage of 820 V to ensure that the voltages are in exact opposition relative to each other. Calculate (a) the circulating current and (b) the terminal voltage. Draw the phasor diagram. [(a) 58 A (b) 1110 V]

7.17 A three-phase, star-connected, 1200-metric hp, 50-Hz, 20-pole synchronous motor is connected to a 3.3 kV infinite bus. The motor draws power at a 0.85 leading power factor. If the excitation of the motor is held constant, calculate (a) the maximum power and torque the motor can develop, (b) the current, (c) the power factor, and (d) the reactive power. Neglect resistance and assume a synchronous reactance of 3 Ω per phase. Draw the phasor diagram. [Take 1 hp = 735.5 W.] [(a) 4268.26 kW, 135.93 kN m; (b) 980.93 A; (c) 0.762 lagging; (d) 3629.98 kVAR]

7.18 A three-phase, star-connected, 750-kVA, 11-kV synchronous motor is supplying a load at the rated leading current and a power factor angle of 20°. If the synchronous reactance is 100 Ω per phase, determine (a) the generated emf and (b) the power factor. Draw the phasor diagram. [(a) 16.01 kV (b) 0.80 leading]

7.19 The machine in Problem 7.18 is delivering 600 kW. Determine the minimum value of the generated line emf at which it can deliver the above load without the motor being pulled out of synchronism. Also, calculate the line current and power factor of the motor under the above condition. Draw the phasor diagram indicating the positions of the generated emf and line current phasors. [5.46 kV, 70.88 A, 0.444 lagging]

7.20 A three-phase, star-connected, 1100-V synchronous motor takes 500 kW from the mains at the rated voltage and 0.8 leading power factor. The motor has a synchronous reactance of 2.5 Ω per phase. Calculate the power factor of the motor when it is supplying a load of 750 kW with the same excitation. [0.55 leading]

7.21 A three-phase, star-connected synchronous motor is connected across a 415-V three-phase supply. The motor draws 50 kW of power at 0.8 power factor leading. The stator resistance and synchronous reactance are 0.1 Ω/phase and 2.0 Ω/phase, respectively, and the field winding resistance is 40 Ω. Determine (a) the output power and (b) efficiency of the motor. Assume rotational losses of 1200 W. [(a) 46.53 kW (b) 87.2%]

7.22 The composite load of an industrial unit is made up of the following:

Lighting load: 120 kW

Induction motor load at 0.85 power factor (lagging), with an efficiency of 90% 420 hp

Synchronous condenser output: 450 kW

Efficiency of the condenser: 95%

Determine the power factor of the condenser to ensure a unity power factor for the industrial unit.   [0.97]

7.23 A load of 1000 kW at 0.8 lagging power factor is supplied by an alternator. The power factor of the alternator is improved to unity by a synchronous motor. If the armature current of the alternator is kept constant, determine the power that can be supplied by the alternator to the motor and the horse power developed by the latter at an efficiency of 95%. What is the power factor of the synchronous motor? Will the power factor be leading or lagging?   [250 kW, 318.36 hp, 0.32 (leading)]

Additional Examples

AE 7.1 Two synchronous generators, with different number of poles, are mounted on the same prime mover shaft. Compute the number of poles in each machine so that the prime mover is operated at its maximum speed to ensure that the two machines generate voltages at 50 Hz and 20 Hz? What is the speed of the prime mover?

Solution Assume that the number poles are P_1 and P_2 in the machines generating voltages at 50 Hz and 20 Hz respectively. With help of Eq. (4.9) the speed of the primer shaft can be written as

$$\frac{120 \times 50}{P_1} = \frac{120 \times 20}{P_2}$$

or $P_1 = 2.5 P_2$ (7.1.1)

In order that the prime mover operates at maximum speed, the number of poles in Eq. (7.1.1) must have lowest integer values.

Thus, $P_1 = 10$, and $P_2 = 4$.

Speed of the prime mover $N = \dfrac{120 \times 50}{10} = \dfrac{120 \times 20}{4} = 600$ rpm

AE 7.2 A 50 MVA, 11 kV round rotor synchronous generator has an armature resistance and synchronous reactance of 0.25 Ω and 3.0 Ω. If the generator is connected to an 11 kV infinite bus and is supplying a load current of 1800 A at 0.8 leading power factor, calculate (a) internal voltage, and (b) power angle. Also, determine (c) open circuit voltage, and (d) steady state short circuit current at the level of excitation at which the machine is operating. Ignore saturation effect.

Solution *Data*: Bus terminal voltage $V_t = 11000 \angle 0°$, $Z_S = (0.25 + j3.0)\,\Omega$

Load current $I_L = 1800(0.8 + j0.6) = (1440 + j1080)$ A

(a) Using Eq. (7.26), the internal voltage is written as

$$E = 11000 + (1440 + j1080) \times (0.25 + j3.0) = (8120 + j4590) = 9327.5 \angle 29.48° \text{ V}$$

(b) Power angle $\delta = 29.48°$

(c) Open circuit voltage = 9327.5 V

(d) Short circuit current $= E/Z_S = 9327.5 / \sqrt{0.25^2 + 3.0^2} = 3098.4$ A

AE 7.3 A three phase round rotor is rated at 100 MVA and 11 kV and has a synchronous reactance of 1.66 per unit. When connected to 11 kV infinite bus, its three phase internal voltage is 15 $\angle 25.84°$ kV. Compute (a) line current, (b) three phase load, (c) real and reactive power supplied by the generator.

Solution Computations are facilitated by working in per unit quantities.

Selecting 11 kV and 100 MVA as voltage and power base, the other quantities are converted into per unit as per the equations given below.

$$\text{Base current } I_{Base} = \frac{\text{MVA}_{Base}}{\text{kV}_{Base}} \text{, kA}$$

$$\text{Base impedance } Z_{Base} = \frac{\text{kV}_{Base}}{I_{Base}} = \frac{\text{kV}_{Base}}{\dfrac{\text{MVA}_{Base}}{\text{kV}_{Base}}} = \frac{(\text{kV}_{Base})^2}{\text{MVA}_{Base}} \text{, } \Omega$$

where MVA$_{Base}$, and kV$_{Base}$ are respectively the base MVA, and the base kV chosen for the system.

In this problem MVA$_{Base}$ = 100 MVA, kV$_{Base}$ = 11 kV

Then, bus terminal voltage = 11/11 = 1.0 per unit

Internal generator voltage $E = (15/11)\angle 25.84° = 1.364\angle 25.84°$ per unit

$$\text{Base current } I = \frac{100 \times 10^3}{\sqrt{3} \times 11} = 5.25 \text{ kA}$$

Since the internal voltage is higher than the bus voltage, the generator is supplying a lagging load. If the armature current is assumed to be $I\angle -\varphi$ per unit, the expression for voltage is written as

$$I \times 1.66 \angle (90° - \varphi) = 1.364 \big[\cos(25.84°) + j\sin(25.84°) \big] - 1.0\angle 0°$$

$$= 0.2276 + j0.5945 = 0.6366\angle 69.05°$$

or $\quad I\angle(90° - \varphi) = \dfrac{0.6366\angle 69.05°}{1.66} = 0.3835\angle 69.05°$ per unit

or $\quad I\angle(-\varphi) = \dfrac{0.6366\angle(69.05° - 90)}{1.66} = 0.3835\angle -20.95°$ per unit

(a) Generator armature current $I = 0.3835 \times 5.25 = 2.0134$ kA

(b) Three phase power $S = 1.0 \times 0.3835 = 0.3835$ per unit $= 0.3835 \times 100 = 38.35$ MVA

(c) Real power $P = 38.35 \times \cos(-20.95°) = 35.815$ MW

Reactive power $Q = 38.35 \times \sin(-20.95°) = -13.712$ MVAR

AE 7.4   Calculate (a) the internal voltage and (b) the power angle when the mechanical input and excitation of the generator in AE 7.3 is changed to reduce the armature current by 20%. The power factor of the generator remains at the same value as in AE 7.3.

Solution   Reduced armature current $I = 0.80 \times 0.3835\angle -20.95° = 0.3068\angle -20.95°$ per unit

Internal generator voltage $\quad E\angle\delta = 1.0\angle 0° + 0.3068\angle -20.95° \times 1.66\angle 90°$ per unit

$$= 1.0 + 0.5093\angle 69.05° = 1.0 + 0.5093 \times (0.358 + j0.934) \text{ per unit}$$

$$= 1.1821 + j0.4756 = 1.274\angle 21.92° \text{ per unit} \tag{AE 7.4.1}$$

(a) From Eq. (AE 7.4.1), line to line generator voltage $E = 14.016$ kV

(b) From Eq. (AE. 7.4.1) it is seen that power angle $\delta = 21.92°$

AE 7.5   In AE 7.4, the mechanical input and the field excitation of the generator are again adjusted to operate it at unity power factor at the bus terminals while supplying reduced line current. Compute the internal voltage and power angle of the generator under these conditions.

Solution   In the present conditions the armature current $I = 0.3068\angle 0°$ per unit.

The internal generator equation is given by

$$E\angle\delta = 1.0\angle 0° + 0.3068\angle 0° \times 1.66\angle 90° = 1.0 + j0.5093 = 1.122\angle 27° \text{ per unit} \tag{AE 7.5.1}$$

(a) From Eq. (AE 7.5.1), line to line generator voltage $E = 1.122 \times 11 = 12.344$ kV

(b) From Eq. (AE. 7.5.1) it is seen that power angle $\delta = 27°$

AE 7.6 A 2200 V, 50 Hz star connected three phase synchronous machine whose mutual inductance is 250 mH, consumes 100 kW at 0.9 leading power factor when it is operated as a motor. Determine the field current and load angle of the motor. Assume a rotor DC field current to equivalent AC stator current ratio $(i_f / I_a) = 0.5$.

Solution Voltage per phase $V_t = 2200 / \sqrt{3} = 1270.20$ V

Mutual reactance $X_m = 2\pi \times 50 \times 0.25 = 78.54 \, \Omega$

Magnetizing current $\boldsymbol{I_m} = 1270.20 / j78.54 = -j16.172$ A (AE 7.6.1)

Selecting the motor phase voltage as the reference voltage, the supply phase current is written as

$$|I_S| = \frac{(100/3) \times 10^3}{1270.2 \times 0.9} = 29.158 \, A$$

Since the motor is drawing power at a leading power factor, the current leads the applied voltage by $\delta = \cos^{-1}(0.9) = 25.84°$. Hence, the motor current $\boldsymbol{I_S} = 29.158 \angle 25.84°$ A

AC equivalent stator current

$$\boldsymbol{I_a} = \boldsymbol{I_m} - \boldsymbol{I_S} = -j16.172 - 29.158 \left[\cos(25.84°) + j\sin(25.84°) \right]$$

$$= -(26.24 + j28.88) = 39.02 \angle -132.26° \text{ A}$$

Thus, DC field current $i_f = 0.5 \times |I_a| = 0.5 \times 39.02 = 19.51$ A

AE 7.7 A 33-kV, three-phase, 45-MVA synchronous generator is connected to a 50-Hz infinite bus. The generator is supplying rated power at 0.8 power factor lagging. If the synchronous armature has negligible resistance and reactance per pahse of the generator is 10 Ω, determine (i) the excitation voltage per-phase and the load angle, (ii) the armature current, load angle, and power factor when the generator output is reduced to 30 MW by reducing the turbine input, and (iii) the maximum power that the generator can deliver without losing synchronism, armature current, and the power factor. In parts (i) and (ii) assume constant excitation. Determine the minimum excitation required for the generator to deliver rated output without losing synchronism.

Solution Rated power output $= 45 \times 0.8 = 36$ MW

Armature current $I_a = \dfrac{36 \times 1000}{\sqrt{3} \times 33 \times 0.8} = 787.2958$ A

Terminal voltage per-phase $V_1 = \dfrac{33}{\sqrt{3}} = 19.0526$ kV

Put $V_t = 19.0526 \angle 0°$ kV. Thus, $I_a = 787.2958 \, (0.8 - j0.6)$ A

From Eq. (5.12) with $R_a = 0$,

(i) $\boldsymbol{E} = 19052.6 + j10 \times 787.2958 \, (0.8 - j0.6) = 23776 + j6298.4$

$\quad\quad = 24.596 - 14.8369°$ kV per phase

(ii) From Eq. (5.22a)

$$\delta = \sin^{-1}\left(\frac{(30/3) \times 10}{24.596 \times 19.0526} \right) = 12.3213°$$

$$\boldsymbol{I_a} = \frac{(24.596 \angle 12.3213° - 19.0526 \angle 0°) \times 1000}{j10} = 524.86 - j497.69$$

$$= 723.3068 \angle -43.477° \text{ A}$$

Power factor of the load = cos (−43.4776°) = 0.7256 lagging
(iii) For maximum power $\delta = 90°$ and Eq. (5.22a) takes the form

$$P_{max} = \frac{3 \times 24.596 \times 19.0526}{10} = 140.5853 \text{ MW}$$

Armature current $I_a = \frac{(24.596 \angle 90° - 19.0526 \angle 0°) \times 1000}{j10} = 2459.6 + j1905.3$

$$= 3111.2 \angle 37.7621° \text{ A}$$

Power factor = cos (37.7621°) = 0.7903 leading

The upper limit of the load angle at which power can be supplied without losing synchronism is 90°. Since, at minimum excitation, power factor is equal to unity, the power output = (36/3) = 12 MW/phase

AE 7.8   A star connected, three phase synchronous generator is connected to an infinite bus. If the generator is supplying 8 MVA, 0.8 power factor lagging load at 11 kVA, calculate (a) magnitude of internal voltage, (b) torque angle, (c) maximum power output and the reserve power available at full load. Assume per phase armature resistance and synchronous reactance of 1.2 Ω and 10 Ω respectively.
Solution

Voltage per phase $V_t = 11000 / \sqrt{3} = 6351$ V

Armature current per phase $I_a = \frac{8 \times 10^6}{\sqrt{3} \times 11000} = 419.89$ A

With V_t, assumed as the reference phasor, then generator internal voltage E is computed as

$$E = 6351 + 419.89 \times (0.8 - j0.6) \times (1.2 + j10) = 9273.3 + j3056.8 = 9764 \angle 18.24° \text{ V}$$

(a) Magnitude of internal voltage = 9764 V/phase
(b) Power angle $\delta = 18.24°$
(c) Equation (7.51) is used by neglecting armature resistance, for maximum power as follows

$$P_{max} = \frac{3 \times 9764 \times 6351}{10} = 18.60 \text{ MW}$$

Reserve power = Maximum power − Full load output = 18.60 − 8 × 0.8 = 12.20 MW

AE 7.9   When the generator in AE 7.8 is supplying maximum possible power calculate (a) the armature current, and (b) the reactive power generated. Is the generator supplying or consuming reactive power? Draw the phasor diagram.
Solution   When the generator is delivering maximum power the power angle $\delta = 90°$.
(a) The armature current is calculated as shown below

$$I_a = \frac{jE - V_t}{(1.2 + j10)} = \frac{j9764 - 6351}{1.2 + j10} = 887.42 + j741.58 = 1156.5 \angle 39.88° \text{ A}$$

(b) The three phase reactive power is given by

$$Q = 3 \times V_t \angle 0° \times I_a \sin\varphi$$

$$= 3 \times 6351 \times 1156.5 \sin(0 - 39.88°) = -14.13 \text{ MVAR}$$

Negative reactive power means that the generator is absorbing power. The phasor diagram is drawn here.

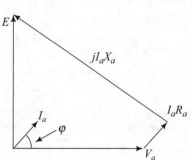

AE 7.10   Two synchronous generators A and B are connected in parallel to supply a motor load of 6000 kW at 0.75 power factor and a lighting

load of 4000 kW. Machine A is loaded to 5000 kW at 0.85 lagging power factor. Compute (a) the output in kW and (b) the power factor of machine B.

Solution Lighting load = 4000 kW

Power factor angle $\varphi = \cos^{-1}(0.75) = 41.41°$

Real power motor load = 6000 kW, and

Reactive power motor load = $6000 \times \tan(41.41°) = 5219.5$ kVAR.

Thus, total real power system load = (4000 + 6000) = 10000 kW

Total reactive load = 5219.5 kVAR

Load on machine A:

Power factor angle = $\varphi_1 = \cos(0.85) = 31.79°$

Real load = 5000 kW

Reactive load = $5000 \times \tan(31.79°) = 3098.7$ kVAR

Load on machine B:

Real load = 10000 − 5000 = 5000 kW

Reactive load = 5219.5 − 3098.7 = 2120.8 kVAR

Power factor angle of the machine $\varphi = \tan^{-1}(2120.8/5000) = 22.98°$

   (a) Thus, output of machine B = 5000 kW

   (b) Power factor of machine B = $\cos(22.98°) = 0.92$ lagging

AE 7.11 A V volts synchronous generator is supplying P W at a lagging power factor $(\cos\varphi_1)$. It is desired to improve the generator power factor to $(\cos\varphi_2)$ without varying the real power P by injecting reactive power. Show that the magnitude of the capacitor to be connected in parallel is given

by $\dfrac{P(\tan\varphi_1 - \tan\varphi_2)}{\omega V^2}$ F where f Hz is the frequency of the supply voltage.

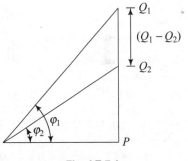

Fig. AE 7.1

Solution Power factor improvement uses the principle of cancelling a part of the reactive power by connecting a capacitor in parallel. Assume that a capacitor of C F is connected in parallel to improve the power factor to $(\cos\varphi_2)$. The power triangle with the capacitor connected is shown in Fig. AE 7.1.

From the power triangle reactive power at power factor $(\cos\varphi_1) = P\tan\varphi_1$ VA

Similarly reactive power at power factor $\cos\varphi = P\tan\varphi_2$ VA

Reactive power to be cancelled = $P(\tan\varphi_1 - \tan\varphi_2)$ VA   (AE 7.11.1)

Leading reactive power supplied by the capacitor = $2\pi f\, CV^2 = \omega\, CV^2$ VA   (AE 7.11.2)

Equating Eqs (AE 7.11.1) and (AE 7.11.2) and simplifying yields

$$C = \frac{P(\tan\varphi_1 - \tan\varphi_2)}{\omega V^2}\ \text{F}$$

AE 7.12 A synchronous generator is servicing a 600 kW load at 0.8 lagging power factor. Calculate the additional kW the generator can supply, at the same kVA loading, if its power factor is improved to unity. Comment on the results.

Solution Apparent power generated by the machine = 600/0.8 = 750 kVA

Hence, real power generated by the machine at unity power factor $= 750 \times 1.0 = 750$ kW

Additional real power generated at unity power factor $= 750 - 600 = 150$ kW

From the foregoing, it is seen that the generator output is 600 kW at 0.8 power factor even when it is loaded to its rated capacity of 750 kVA. On the other hand, at unity power factor, the generator is supplying 750 kW at its full capacity, hence, the significance of power factor improvement.

AE 7.13   The breakup of an industrial load is as under

(i) Lighting load $L_1 = 25$ kW at unity power factor

(ii) Induction motor load $L_2 = 120$ kW at 0.8 lagging power factor

(iii) Synchronous motor load $L_3 = 40$ kW at 0.95 leading power factor.

If the above load is supplied by a single phase synchronous generator, compute (a) individual apparent power of each load, (b) total kW and (c) total kVA generated by the machine. What is the power factor at which it is generating power? Is the generator power factor leading or lagging?

Solution   Power factor angle of (i) lighting load $\varphi_1 = \cos^{-1}(1.0) = 0°$, (ii) induction motor load $\varphi_2 = -\cos^{-1}(0.8) = -36.87°$, and synchronous motor load $\varphi_3 = \cos^{-1}(0.95) = 18.19°$,

(a) Apparent power of lighting load $= 25/\cos\varphi_1 = 25/1.0 = 25$ kVA

Apparent power of induction motor load $= 120/\cos\varphi_2 = 120/0.8 = 150$ kVA

Apparent power of synchronous motor load $= 40/\cos\varphi_3 = 40/0.95 = 42.11$ kVA

(b) Total industrial real power load $= 25 + 120 + 40 = 185$ kW

(c) Total reactive power load $= \left[25 \times \tan\varphi_1 + 120 \times \tan\varphi_2 + 40 \times \tan\varphi_3 \right]$ kVAR

$$= [0 - 90 + 13.15] = -76.85 \text{ kVAR} \qquad (\text{AE } 7.13.1)$$

Total kVA developed by the generator $= \sqrt{185^2 + 76.85^2} = 200.33$ kVA

Power factor of the generator $= 185/200.33 = 0.92$

From Eq. (AE 7.13.1) it is seen that the total reactive power is negative. Therefore, the power factor of the generator is lagging.

AE 7.14   A 440 V, 50 Hz, single phase synchronous motor is connected as shown in Fig. AE 7.2. The power factor of the motor is 0.75 lagging when it supplies a load of 15 HP. Draw the phasor diagram and calculate the magnitude of the capacitor to be connected to improve the power factor of the motor to 0.95. Assume the motor efficiency at 95%.

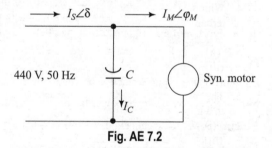

Fig. AE 7.2

Solution $\phi_M = \cos^{-1}(0.75) = 41.41°$

and $\delta = \cos^{-1}(0.95) = 71.81°$

Based on the assumed currents, the phasor diagram is shown in Fig. AE 7.3

$$\text{Motor current } I_M = \frac{15 \times 746}{0.95 \times 440 \times 0.75} = 35.69 \text{ A}$$

From the phasor diagram it is seen that

$$I_S \cos\delta = I_M \cos\phi_M$$

That is, $I_S = \dfrac{35.69 \times 0.75}{0.95} = 28.18 \text{ A}$

Reactive component of $I_S = 28.18\sin(18.19°) = 8.80$ A

Reactive component of $I_M = 35.69\sin(41.41°) = 23.607$ A

From the phasor diagram it is seen that

I_C = Reactive component of I_M − Reactive component of I_S

$= 23.607 - 8.80 = 14.807$ A

Assume that the capacitor has capacitance C farads. Thus,

$$C = \frac{14.807}{440 \times 2 \times \pi \times 50} = 107.12\ \mu\text{F}$$

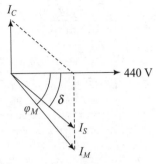

Fig. AE 7.3

Objective Type Questions

1. Which of the following forms of energy conversion takes place in an electrical energy system?
   (i) mechanical to electrical
   (ii) electrical to mechanical
   (iii) mechanical to electrical to mechanical
   (iv) all of these
2. Which type of coupling occurs between a mechanical and electrical system?
   (i) electrostatic   (ii) electromechanical
   (iii) mechanical   (iv) electrical
3. In a rotary machine the purpose of an air gap is to
   (i) provide a path for the flux
   (ii) reduce the mmf required
   (iii) allow for rotation of the rotor
   (iv) improve the reluctance of the magnetic circuit
4. Which of the following is not a feature of an alternator?
   (i) It runs on fixed speed.
   (ii) It generates alternating voltage.
   (iii) It generates alternating current.
   (iv) It is used commercially.
5. The number of poles of a synchronous generator in a hydro-power station are
   (i) 2   (ii) 4
   (iii) 6 to 12   (iv) none of these
6. The number of slip rings on an alternator is
   (i) 2   (iii) 3
   (iii) 4   (iv) 6
7. Which of the following is a disadvantage of placing the field winding on the rotor?
   (i) low dc voltage is required for excitation
   (ii) low inertia of the rotor
   (iii) only two slip rings are used
   (iv) none of these

8. Which of the following is not a feature of a salient-pole machine?
   (i) speeds of 1500 rpm and above
   (ii) low output
   (iii) moderate speeds between 100 and 375 rpm
   (iv) small axial length
9. Which of the following is not a characteristic of a cylindrical rotor synchronous machine used with a steam turbine?
   (i) 2 to 4 number of poles
   (ii) large diameter and small axial length
   (iii) robust construction
   (iv) noiseless operation
10. Which of the following is not applicable to a coil in a four-pole machine?
    (i) Coil span is one pole pitch.
    (ii) Frequency of the generated voltage is equal to the speed of rotation in rps.
    (iii) Frequency of the generated voltage is equal to twice the speed of rotation in rps.
    (iv) None of these.
11. In a wave winding, the number of paths is equal to
    (i) 2   (ii) 4
    (iii) number of poles   (iv) none of these
12. Which of the following represents the effective value of the generated voltage in a three-phase ac machine?
    (i) $4.44\phi fT$ V   (ii) $2.22\phi fTZ$ V
    (iii) $4.44\phi \left(\dfrac{PN_s}{120}\right) T$ V (iv) all of these
13. A three-phase alternator has P poles and S slots in the stator. Which of the following gives the distribution factor?

(i) $\dfrac{\sin[(3\gamma)/2]}{3\sin(\gamma/2)}$ (ii) $\dfrac{\pi P}{S}$

(iii) $\dfrac{S}{3P}$ (iv) none of these

14. The coil span of a short-pitched coil is
   (i) less than π radians mechanical
   (ii) π radians mechanical
   (iii) less than π radians electrical
   (iv) π radians electrical

15. If k_d and k_p represent the distribution and pitch factor, respectively, the winding factor k_w is given by
   (i) k_d/k_p (ii) $k_d k_p$
   (iii) k_p/k_d (iv) none of these

16. A three-phase alternator has 8 poles and 96 slots. If the pitch of the coils is 10 slots, the short-pitch angle is equal to
   (i) 30 degree electrical (ii) 15 degree electrical
   (iii) 12 degree electrical (iv) 10 degree electrical

17. When a three-phase synchronous generator is supplying a lagging load, the effect of the armature flux on the rotor field flux is to
   (i) increase the air gap flux
   (ii) maintain the air gap flux constant
   (iii) decrease the air gap flux
   (iv) none of these

18. The synchronous impedance is determined from the short-circuit test data. Which of the following formula determines the synchronous impedance for a star-connected armature of a three-phase alternator?
   (i) $\dfrac{V_{oc,\,rated}}{I_{sc}}$ (ii) $\dfrac{V_{oc,\,rated}}{I_{sc}/\sqrt{3}}$

   (iii) $\dfrac{V_{oc,\,rated}/\sqrt{3}}{I_{sc}}$ (iv) None of these

19. Which of the following is a characteristic of an infinite bus in a power system?
   (i) voltage and frequency independent of load
   (ii) voltage and frequency vary with load
   (iii) capacity to supply limited power
   (iv) none of these

20. Which of the following is applicable to a synchronous motor?
   (i) It is not self-starting.
   (ii) It is self-starting.
   (iii) It can run at any speed.
   (iv) It can run at a speed close to the synchronous speed.

21. Which of the following is the appropriate expression for the terminal voltage of a synchronous motor?
   (i) $V_t = E + I_a Z_S$ (ii) $V_t = E - I_a Z_S$
   (iii) $E = V_t + I_a Z_S$ (iv) $E = V_t - I_a Z_S$

22. Which of the following does not occur when the excitation of the rotor field is increased at constant load?
   (i) The generated emf increases.
   (ii) The load angle δ decreases.
   (iii) Sin δ decreases.
   (iv) The terminal voltage increases.

23. Which of the following function is performed when a synchronous motor is used as a synchronous condenser?
   (i) a normal motor supplying a mechanical load
   (ii) acts as an induction motor
   (iii) power factor improvement
   (iv) none of these

24. Which of the following is not an advantage of a synchronous motor?
   (i) requires a separate dc source
   (ii) constant speed
   (iii) speed independent of load
   (iv) can function as a synchronous generator

25. How does the armature reaction affect the terminal voltage when the load on an alternator is increased?
   (i) no change
   (ii) effect depends on power factor
   (iii) decreases
   (iv) increases

Answers

1. (iv)	2. (ii)	3. (iii)	4. (iii)	5. (iii)	6. (i)	7. (iv)
8. (i)	9. (ii)	10. (ii)	11. (i)	12. (iv)	13. (i)	14. (iii)
15. (ii)	16. (i)	17. (iii)	18. (iii)	19. (i)	20. (i)	21. (i)
22. (iv)	23. (iii)	24. (i)	25. (ii)			

INDUCTION MOTORS

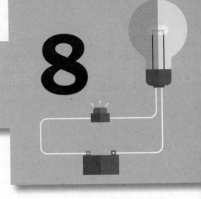

8

Learning Objectives

This chapter will enable the reader to

- Familiarize with the constructional features of three-phase induction motors
- Understand how torque is produced in an induction motor based on the principle of a rotating field produced by three-phase currents
- Develop expressions for induced rotor voltage and current based on transformer action and therefrom draw the equivalent circuit
- Enumerate the different types of losses in an induction motor and determine the mechanical power developed in its rotor
- Draw the torque–slip characteristics and derive conditions for a motor to develop maximum torque
- Perform no-load and blocked rotor tests and from the data, compute the parameters of the equivalent circuit of an induction motor
- Understand the need for using starters to start induction motors and develop knowledge about the various methods of starting and their relative merits
- Enumerate, understand, and apply the different methods of speed control of induction motors

8.1 INTRODUCTION

Induction motor is the most commonly used ac motor. As a rough estimate, nearly 80% of the world's ac motors are poly-phase induction motors, while most of the rest are single-phase induction motors. It is considered to be the workhorse of the industry. Induction motor has a simple and rugged construction, is self-starting, runs at almost a constant speed from no load to full load, operates at a reasonably good efficiency, requires less maintenance, and is less expensive than other motors of the same kilowatt and speed ratings. Induction motors are available for both single-phase and three-phase operations. Three-phase induction motors are widely used for industrial applications, such as in lifts, cranes, pumps, exhaust fans, lathes, etc. Single-phase induction motors are used mainly for domestic electrical appliances such as fans, refrigerators, washing machines, exhaust pumps, etc.

In 1891 Nikola Tesla presented for the first time a crude type of a poly-phase induction motor. Since then great improvements have been made in the design and performance of induction motors and numerous types of poly-phase and single-phase induction motors have been developed. Induction motors are also called *asynchronous motors* as they run at a speed other than the synchronous speed of their flux. This chapter discusses the construction features, principle of operation, and operating characteristics of three-phase induction motors.

8.2 CONSTRUCTION FEATURES OF THREE-PHASE INDUCTION MOTORS

A three-phase induction motor primarily consists of two parts: an outer part, called *stator*, and an inner part, called *rotor*. In the induction motor a rotating magnetic flux is produced by connecting a three-phase supply to the three-phase winding housed in the stator. The rotor is called the *armature*, and carries short-circuited windings in which the current is induced, because of which the starting torque is produced.

8.2.1 Stator

The stator of an induction motor is practically identical with the stator of an alternator. It is designed to carry a uniformly distributed three-phase winding, called stator or primary winding, which can be connected in star or delta across the main source of supply. In such a case, all six terminals are usually brought out to the terminal box to provide flexibility to change the connection from star to delta if ever there is a need to do so. The stator winding draws a current, from the three-phase ac voltage source, which sets up a flux in the air gap. Since the stator is subjected to an alternating flux, it is made of laminations of high permeability steel, such as silicon steel, so as to minimize hysteresis and eddy-current losses.

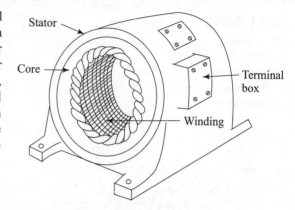

Fig. 8.1 Stator of a three-phase induction motor

In small motors, where low core loss is not so important, thicker laminations are used. Ventilating ducts are provided along the length of the core by providing spacers between batches of laminations. In general, windings, which are suitable for the stator of alternators, can also be used for the stator of induction motors. A double-layer winding is always used because of its greater ease of manufacture, assembly, and repair. Moreover, stator windings are almost always short-pitched as this reduces the copper weight and winding resistance and also reduces the leakage reactance and harmonic torque disturbances. Stator winding is always completely insulated corresponding to the voltage rating of the induction motor. A perspective view of the stator of a three-phase induction motor is shown in Fig. 8.1.

8.2.2 Rotor

The rotor of an induction motor is cylindrical and is assembled by using sheet steel laminations of the same material as the stator. However, because of the lower frequencies of the rotor flux, thicker laminations can be used without excessive loss. In small rotors, the lamination is of one piece; in large motors, segmented laminations are used, which are held together with the help of a central spider. Ventilating ducts are often provided in the rotor core. Fan blades are generally used at the ends of the rotor core to force circulating air through the machine.

The rotor surface is slotted, and rotor conductors are placed in these slots. Usually semi-closed slots are used in rotors as the open slots increase the effective length of the air gap considerably and the closed slots increase the leakage reactance, which in turn decreases the maximum torque of the motor. The number of teeth in the rotor is always made different from the number of stator teeth in such a way that a magnetic locking of the rotor at starting does not occur. Magnetic wedges are often used to hold coils in these slots. An induction motor is provided with a narrow air gap. Hence, the rotor is supported on both sides by two end shields which house the bearings. Ball-and-roller bearings are used in induction motors as this makes centering of the rotor simpler and accurate. The rotor must be properly balanced to avoid the danger of the rotor striking the stator.

Based on the type of rotor windings, rotors of induction motors may be classified into two types, namely, (i) the squirrel cage rotor and (ii) the wound rotor. The squirrel cage rotors are simple, more rugged, and cheaper than the coil wound rotors. Because of the robust structure of this type of rotor, about 90% of the induction motors employed in industry are of this type. However, the main advantage of the coil wound rotor is the flexibility to vary the rotor resistance by inserting suitable external resistances between the slip rings, which is normally used to provide a high starting torque.

8.2.2.1 Squirrel Cage Rotor

The windings of squirrel cage rotors consist of solid, bare conductors of aluminium, copper, or brass, which are in the form of bars placed in the rotor slots close to the rotor surface and almost parallel to the shaft. Un-insulated conductors are used since they do not short-circuit the core; the latter has a high resistance compared to the conductors, and the current takes a path of low resistance through the conductors. The conductor bars are welded, brazed, or bolted to two end rings, thus short-circuiting them; thereby providing a complete path for the flow of current in the rotor. Many squirrel cage rotors are built with aluminium bars and end rings cast directly in the rotor. A squirrel cage rotor is shown in Fig. 8.2. Some squirrel cage induction motors have rotor slots which are not parallel to the shaft but are skewed at a desired angle, as shown in Fig. 8.2, which result in a more uniform torque, reduce humming noise during running, and avoid magnetic locking.

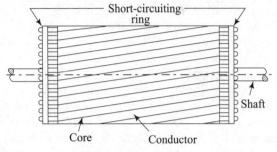

Fig. 8.2　Squirrel cage rotor

8.2.2.2 Wound Rotor

The windings of wound rotors are similar to stator windings. The coil wound rotor must be wound for the same number of poles as that of the stator. The conductors used are bar type or strap type when fewer turns per phase are required, or wire type when more turns per phase are to be used. Either a delta or star connection may be used, but the star connection is generally preferred. The terminals of the winding are brought out to three slip rings, which are tapped by brushes. During starting, external resistances are connected in the rotor circuit, which are cut out by shorting of slip rings during running. Figure 8.3 shows a wound rotor with slip rings.

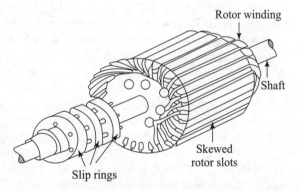

Fig. 8.3　A slip-ring or wound rotor of an induction motor

8.3 PRINCIPLE OF OPERATION OF THREE-PHASE INDUCTION MOTOR

When the stator of an induction motor is connected across a balanced three-phase ac supply, it draws a balanced three-phase current, which sets up a rotating magnetic field of constant magnitude (already discussed in Section 7.5). The magnetic flux rotating at the synchronous speed sweeps past the rotor conductors, which are stationary at the start, and induces emfs in them. The frequency of the induced emfs in the rotor conductors is same as the supply frequency. As the rotor conductors are short-circuited the induced emfs produce three-phase rotor currents, which in turn produce a magnetic field that revolves at the same speed as the stator field. A starting torque is produced due to the interaction of these two fields and tends to turn the rotor in the direction of rotation of the stator field. Now, according to Lenz's law, the developed torque must oppose the cause to which it is due, that is, cutting of flux lines by rotor conductors. Therefore, the developed torque makes the rotor move in the direction of the flux waves so as to reduce the relative speed between the stator flux wave and the rotor conductors, thereby reducing the cutting of flux lines by the rotor conductors. If the starting torque is sufficient to overcome the load torque on the motor shaft, the motor will start rotating and reach its operating speed. However, the operating speed can never be equal to the synchronous speed N_S as at that speed the rotor conductors would be stationary with respect to the stator field. As a result, there will be no flux cutting by the rotor conductors and, hence, the developed torque will be zero. As a matter of fact, the rotor speed, N, of an induction motor is always slightly less than the synchronous speed.

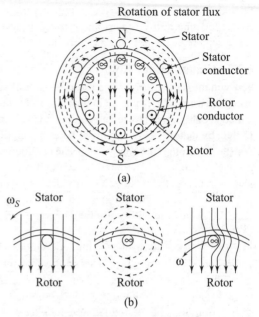

Figure 8.4(a) shows the distribution of flux, at a particular instant in a two-pole stator, of an induction motor when the former is connected across a balanced three-phase ac supply of frequency f hertz. If the stator rotating flux rotates anticlockwise at an angular velocity $\omega_S = 2\pi f$, the direction of the induced emf generated in the stationary rotor conductors can be determined by Fleming's right hand rule. The emf generated in one rotor conductor is shown separately in Fig. 8.4(b). In Fig. 8.4(b), the first

Fig. 8.4   (a) Induction motor with cage rotor; (b) force on rotor

part shows the stator flux rotating at an angular speed ω_S and links the rotor conductor; the second part shows that the induced current circulating in the rotor conductor produces a flux around it in the clockwise direction, determined by Maxwell's corkscrew rule; and the third part shows the effect of this flux, which strengthens the flux density on the right hand side and weakens that on the left hand side of the conductor. This causes the conductor to be pushed toward the left. Thus the rotor also begins to rotate in the anticlockwise direction, that is, in the direction of the rotating magnetic field. Thus, induction motors are self-starting.

Just before the rotor starts rotating, the frequency of the induced emf in the rotor is equal to that of the supply frequency since the relative speed between the rotor and the rotating magnetic field is the synchronous speed. As the rotor picks up speed, the relative speed between the rotor and the synchronously rotating magnetic field reduces, thereby reducing the frequency of the induced emf. In other words, closer is the speed of the rotor to the synchronous speed of the rotating magnetic field, smaller is the frequency and magnitude of the induced emf in the rotor.

When the motor is on no load, its speed is very nearly equal to the synchronous speed. Therefore, a current of a very low value circulates in the rotor conductors due to the small magnitude of the induced emf. The torque generated on no load, therefore, is just enough to supply the rotor friction and windage losses and maintain it in motion. As the mechanical load on the motor shaft increases, the rotor speed falls and the relative speed between the rotor and the synchronously rotating magnetic field increases. The values of the rotor emf and current and, therefore, the torque increase to meet the demand of the increased load and the rotor losses. The rotor comes to a dynamic equilibrium when the generated torque is equal to the load torque plus losses.

8.3.1 Slip and Rotor Frequency

The rotating flux wave produced by the stator rotates at the synchronous speed N_S rpm with respect to the stator. N_S is given by

$$N_S = \frac{120 f}{P} \text{ rpm} \qquad (8.1)$$

where f is the supply frequency in hertz and P is the number of poles of the stator winding.

As already discussed, the rotor can never attain the synchronous speed, N_S, because then the relative speed between the flux wave and the rotor conductors will be zero. Now, in the case of induction motors, the relative speed between the rotating flux and the rotor is usually represented as a fraction of the synchronous speed, which is called *slip*. The term is descriptive of the manner in which the rotor slips back from the rotating flux. Hence, by definition, the slip s may be expressed as

$$s = \frac{N_S - N}{N_S} \qquad (8.2)$$

Thus slip, in per unit, is defined as the difference between the synchronous speed N_S of the stator rotating magnetic field and the actual speed N of an induction motor expressed as a ratio of the synchronous speed. Slip is denoted by s. Slip s, at full load, in induction motors varies from 2%, in small motors, to nearly 6%, in large machines. Thus, induction motors are usually regarded as constant speed machines.

From Eq. (8.2) it may be found that the relative speed between the stator flux and rotor conductors, $(N_S - N)$, is equal to sN_S. Now the frequency f_r of the rotor induced emf is given by

$$f_r = \frac{(\text{relative speed between rotating flux and the rotor in rpm}) \times P}{120}$$

$$= \frac{(N_S - N) P}{120} \qquad (8.3)$$

$$= \frac{s N_S P}{120} = sf \qquad (8.4)$$

The induced currents in the rotor windings set up a resultant magnetic field, which rotates at a speed $(N_S - N)$ relative to the rotor surface and is in the same direction, while the speed of the rotor itself is N. Thus the speed of the resultant magnetic field set up by the rotor currents relative to the stator is given by

$$(N_S - N) + N = N_S$$

In other words, the resultant magnetic field due to the rotor currents revolves at the same speed as the resultant stator magnetic field and hence the two fields appear stationary with respect to each other. As both the stator and the rotor fluxes rotate at synchronous speed N_S with respect to stator, the actual air gap flux will now be the resultant effect of these two fluxes. From this discussion, it is evident that the stator and rotor mmfs are stationary with respect to each other at all possible rotor speeds. This enables the three-phase induction motor to produce a uniform torque at all possible rotor speeds. Furthermore, the induction motor can be regarded as equivalent to a transformer with the stator winding acting as the primary and the rotor winding acting as the secondary. Unlike a transformer, the magnetic circuit carrying the primary and secondary windings in the induction motor is separated by an air gap. Because of this gap, the magnetizing current and the magnetic leakage for an induction motor are large compared with the corresponding values for a transformer of the same apparent power rating. Also, the friction and windage losses contribute towards making the efficiency of the induction motor less than that of the corresponding transformer. In an induction motor the stator field mmf has to provide the magnetizing and no-load loss components of the stator current and also balance the rotor field mmf, as in a transformer. Hence, an increase in slip due to an increase in load is accompanied by an increase in rotor currents and, hereby, a corresponding increase in stator currents.

Example 8.1 A three-phase induction motor is wound for four poles and is supplied from a 50-Hz supply. Calculate (a) the synchronous speed, (b) the speed of the rotor when the slip is 3%, and (c) the rotor frequency when the speed of the rotor is 900 rpm.

Solution   (a) From Eq. (8.1),

$$\text{Synchronous speed} = \frac{120f}{P} = \frac{120 \times 50}{4} = 1500 \text{ rpm}$$

(b) From Eq. (8.2),

$$0.03 = \frac{1500 - \text{rotor speed}}{1500}$$

∴   Rotor speed $= 1500 - 1500 \times 0.03 = 1455$ rpm

(c) From Eq. (8.2) per unit slip $= \dfrac{1500 - 900}{1500} = 0.4$

From Eq. (8.4), rotor frequency $= 0.4 \times 50 = 20$ Hz

Example 8.2   A three-phase induction motor is supplying a mechanical load at a full load speed of 1170 rpm. The voltage supply to the motor is from a 10-pole synchronous generator, which is operating at a synchronous speed of 720 rpm. Determine (i) the number of poles and (ii) per unit slip of the induction motor.

Solution   The frequency of the voltage which is fed to the motor is obtained from the synchronous generator conditions as follows:

$$f = \frac{PN_S}{120} = \frac{10 \times 720}{120} = 60 \text{ Hz}$$

It has been shown that an induction motor runs at a speed that is lower than the synchronous speed. In the present case, the nearest synchronous speed possible is 1200 rpm.

(i) Hence, the number of poles of the induction motor is given by

$$P = \frac{120f}{N_S} = \frac{120 \times 60}{1200} = 6$$

(ii) Slip is then obtained as

$$s = \frac{N_S - N}{N_S} = \frac{1200 - 1170}{1200} = 0.025 \text{ pu}$$

8.3.2   Voltage and Current Equations and Equivalent Circuit of an Induction Motor

As shown in the preceding section, induction motors operate on the same principle as transformers. The stator and rotor voltage equations of an induction motor are similar to those for the primary and secondary windings of a transformer.

Let V_P be phase voltage applied to the stator windings, E_1 the emf per phase induced in the stator windings, I_1 the stator phase current, and $Z_1 = (r_1 + jx_1)$ be the impedance consisting of per phase resistance r_1 and leakage reactance per phase x_1 of the stator. Then

$$V_P = E_1 + I_1(r_1 + jx_1) = E_1 + I_1 Z_1 \tag{8.5}$$

If the voltage drop in the stator impedance is ignored, then $V_P = E_1$. The emf induced per phase may be obtained from Eq. (7.23) developed in the case of a synchronous machine as

$$V_1 = E_1 = 4.44 k_{w1} f T_1 \phi \text{ volt} \tag{8.6}$$

where k_{w1} is the winding factor of the stator winding, f is the supply frequency, T_1 is the stator number of turns in series per phase, and ϕ is the air gap flux per pole.

When the rotor is at standstill, the flux ϕ per pole in the air gap induces a voltage E_2 per phase in the rotor, which is given by

$$E_2 = 4.44 k_{w2} f T_2 \phi \text{ volt} \tag{8.7}$$

where k_{w2} is the winding factor of the rotor winding and T_2 is the rotor number of turns in series per phase. When the induction motor is running at a slip s, since the rotor frequency $f_r = sf$, the magnitude of the induced emf, E_r, in the rotor is given by

$$E_r = sE_2 \tag{8.8}$$

The ratio of transformation a is obtained by dividing Eq. (8.6) by Eq. (8.7) as

$$a = \frac{E_1}{E_2} = \frac{4.44 k_{w1} \, fT_1 \phi}{4.44 k_{w2} \, fT_2 \phi} = \frac{k_{w1} T_1}{k_{w2} T_2} \tag{8.9}$$

At no load, the primary (stator) winding of the induction motor draws a current I_0 which consists of two components: I_ϕ, the magnetizing component to set up the air gap flux ϕ, and I_C, the core loss component to supply the losses. The no-load current, therefore, is

$$I_0 = I_\phi + I_C$$

Compared to a transformer of a similar rating, the magnitude of the no-load current in an induction motor can go up to 40% of the full-load current, as against 2%–5% of full-load current in the case of the former because of the presence of the air gap in the magnetic circuit.

The rotor speed falls as the mechanical load on the motor shaft increases, and the slip s increases, the values of the rotor (secondary) emf and the rotor (secondary) current increase, and therefore, the torque increases to meet the demand of the increased load and the rotor losses. If the secondary winding current is I_2, the primary (stator) winding must draw an additional current I_2' to balance the ampere turns of the rotor. The primary (stator) current, drawn from the supply, therefore, is

$$I_1 = I_0 + I_2' \tag{8.10}$$

and

$$\frac{I_2'}{I_2} = \frac{1}{a} \tag{8.11}$$

If r_2 is the per phase resistance and x_2 is the leakage reactance of the rotor (secondary) winding at standstill, the rotor circuit impedance Z_2 at slip s is given by

$$Z_2 = r_2 + jsx_2 \tag{8.12}$$

As the rotor is shorted, the rotor current I_2 at slip s can be written as

$$I_2 = \frac{sE_2}{r_2 + jsx_2} \tag{8.13}$$

The electrical power crossing the air gap from the stator to the rotor is divided into two components—the mechanical power output, P_m, and the loss in the rotor resistance, $3I_2^2 r_2$.

Based on the above equations of the induction motor, its equivalent circuit can be drawn, on an analogy with the principle of the operation of a transformer, and is shown in Fig. 8.5.

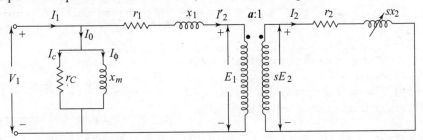

Fig. 8.5 Equivalent circuit of an induction motor

In the equivalent circuit shown in Fig. 8.5, the shunt branch represents the magnetizing reactance x_m and the core loss component r_C. The transformation ratio is a. The magnitude of the per phase rotor impedance and the rotor current at slip s are given by

$$Z_2 = \sqrt{r_2^2 + (sx_2)^2}$$

$$I_2 = \frac{sE_2}{\sqrt{r_2^2 + (sx_2)^2}} \tag{8.14}$$

The impedance angle θ_2 of the secondary (rotor) winding is given by

$$\theta_2 = \tan^{-1}\frac{sx_2}{r_2} = \cos^{-1}\frac{r_2}{\sqrt{r_2^2 + (sx_2)^2}} \tag{8.15}$$

Power is made invariant by reducing the effective turns ratio to 1:1, which is easily done by referring the rotor quantities to the stator side. By making use of Eqs (8.9) and (8.11), Eq. (8.13) is rewritten as

$$I_2' = \frac{I_2}{a} = \frac{s(E_1/a)}{a(r_2 + jsx_2)} = \frac{sE_1}{a^2 r_2 + jsa^2 x_2} = \frac{sE_1}{Z_2'} \tag{8.16}$$

Equation (8.16) provides a basis for referring rotor quantities to the stator as shown below.

$$Z_2' = r_2' + jx_2' = a^2 r_2 + jsa^2 x_2 \tag{8.17}$$

and the equivalent circuit gets modified to that shown in Fig. 8.6.

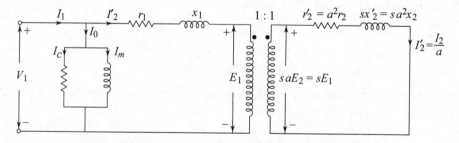

Fig. 8.6 Modified equivalent circuit with effective turns ratio 1:1

In the equivalent circuit shown in Fig. 8.6,

$$I_2' = \frac{saE_2}{r_2' + jsx_2'} = \frac{sE_1}{r_2' + jsx_2'} \tag{8.18}$$

Dividing the numerator and denominator on the right-hand side of Eq. (8.18) by s,

$$I_2' = \frac{E_1}{\dfrac{r_2'}{s} + jx_2'} \tag{8.19}$$

Based on Eq. (8.19), the equivalent circuit of the rotor is modified and is shown in Fig. 8.7.

From Eq. (8.19), it can be seen that the rotor resistance r_2' is a function of the slip. It can be modified as shown below:

$$\frac{r_2'}{s} = r_2' + r_2'\left(\frac{1}{s} - 1\right) \tag{8.20}$$

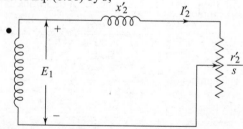

Fig. 8.7 Equivalent circuit of a rotor of an induction motor

Thus, the variable resistance is split into two components: r'_2 the rotor resistance component and an imaginary component, $r'_2[(1/s) - 1]$, which represents the electrical equivalent of the mechanical load on the shaft. Figure 8.8 shows the modified equivalent circuit of the rotor after incorporating the above transformation. The effect of the transformation is to refer the rotor impedance to the stator frequency.

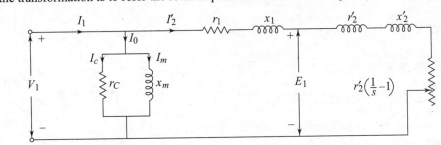

Fig. 8.8 Modified equivalent circuit of an induction motor

Example 8.3 The stator of a 50 Hz three-phase induction motor with a synchronous speed of 1000 rpm has 90 slots with 4 conductors per slot. Determine (i) the winding factor and (ii) the maximum flux in the air gap when the winding is connected to a three-phase 415-volt supply. Assume that the winding is star connected and the coil span of the winding is 160°. If the three-phase star-connected rotor has 120 slots with 2 conductors per slot, calculate (iii) the ratio of transformation and (iv) the voltage induced in the rotor.

Solution The number of poles of the motor, $P = \dfrac{120f}{N_s} = \dfrac{120 \times 50}{1000} = 6$

Slot per pole per phase, $m = \dfrac{90}{6 \times 3} = 5$

Angle between two adjacent stator slots,

$$\gamma = \frac{180°}{\text{slots per pole}} = \frac{180°}{90/6} = 12°$$

Using Eq. (7.15) gives the distribution factor $k_d = \dfrac{\sin\left[(m\gamma)/2\right]}{m\sin(\gamma/2)} = \dfrac{\sin 30°}{5 \times \sin 6°} = 0.96$

From Eq. (7.19) the pitch factor, $k_p = \cos(20/2) = 0.98$

(i) Therefore, the winding factor, $k_{w1} = k_d k_p = 0.96 \times 0.98 = 0.94$

The number of turns in the stator per phase, $T_1 = \dfrac{90 \times 4}{3 \times 2} = 60$

(ii) Employing Eq. (8.6) determines the maximum flux in the air gap as follows:

$$\phi = \frac{E_1/\sqrt{3}}{4.44 \times k_{w1} \times f \times T_1} = \frac{415/\sqrt{3}}{4.44 \times 0.94 \times 50 \times 60} = 19.1 \text{ mWb}$$

The number of rotor turns per phase, $T_2 = \dfrac{120 \times 2}{3 \times 2} = 40$

(iii) Equation (8.9) provides the transformation ratio as $a = T_1/T_2 = 60/40 = 1.5$

(iv) Thus, the induced rotor voltage per phase is $E_2 = E_1/a = \left(415/\sqrt{3}\right)/1.5 = 159.73$ V

Example 8.4 With rated voltage applied to the stator of a three-phase induction motor, the open-circuit voltage developed across the slip rings of a star-connected rotor is 75 V. If the rotor resistance and reactance per phase are 0.75 Ω and 5 Ω, respectively, determine the rotor current when (a) at standstill with an external resistance and reactance of 5.25 Ω and 7 Ω, respectively, connected between each rotor phase and (b) running at a slip of 4% with the rotor short circuited.

Solution   At standstill, slip $s = 1.0$. Therefore, the rotor circuit impedance, $Z_2 = \{(0.75 + 5.25) + j(5 + 7) = (6.0 + j12)\}\,\Omega$.

From Eq. (8.13), the rotor current is written as

$$I_2 = \frac{E_2}{Z_2} = \frac{75/\sqrt{3}}{6 + j12} = (1.44 - j2.89) = 3.23\angle -63.43°\ \text{A}$$

At $s = 0.04$ and with the rotor short circuited, $Z_2 = (0.75 + 0.04 \times j5) = (0.75 + j\,2.0)\,\Omega$
Again use of Eq. (8.13) results in

$$I_2 = \frac{sE_2}{Z_2} = \frac{0.04 \times \left(75/\sqrt{3}\right)}{0.75 + j2.0} = (0.28 - j0.76) = 0.81\angle -69.44°\ \text{A}$$

8.3.3 Power Balance in an Induction Motor

The power input to a three-phase induction motor is given by

$$P_i = 3V_1I_1\cos\varphi_1 \tag{8.21}$$

where $\cos\varphi_1$ is the power factor of the input phase current I_1 drawn from the supply phase voltage V_1. Part of this power is consumed by the stator resistance loss $3I_1^2r_1$ and the core loss P_C. The balance power is transferred across the air gap to the rotor. This input power to the rotor P_{ag} is usually called the air gap power. Thus

$$P_{ag} = P_i - 3\,I_1^2r_1 - P_C \tag{8.22}$$

From the equivalent circuit of the rotor in Fig. 8.7, it is seen that the air gap power can also be expressed as

$$P_{ag} = 3I_2'^2\frac{r_2'}{s} \tag{8.23}$$

$$= \frac{3I_2'^2 r_2'}{s} = \frac{\text{total rotor resistance loss}}{\text{slip}} \tag{8.24}$$

Therefore,

$$\text{Total rotor resistance loss} = sP_{ag}$$

$$= s \times \text{input power to rotor} \tag{8.25}$$

The rotor resistance loss when subtracted from the air gap power gives the gross mechanical power output P_m. Thus

$$P_m = P_{ag} - sP_{ag} = (1-s)P_{ag} \tag{8.26}$$

Substitution of Eq. (8.23) into Eq. (8.26) gives

$$P_m = (1 - s)\,\frac{3I_2'^2 r_2'}{s}$$

$$= 3\,I_2'^2 r_2'\left(\frac{1}{s} - 1\right) \tag{8.27}$$

This means that the mechanical power output is equal to the power absorbed by the load resistance $r_2'\,[(1/s) - 1]$ as shown in Fig. 8.8.

The power distribution and the losses occurring in various parts of an induction motor are depicted in Fig. 8.9. In an induction motor the core losses occur mainly in the stator core. The rotor core loss is negligible, because, in running condition, the frequency of the rotor current is small. Thus, in the diagram of Fig. 8.9 only the stator core loss is shown.

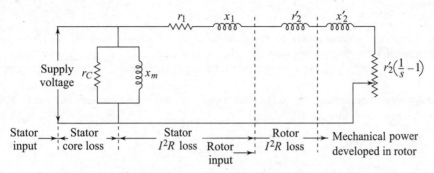

Fig. 8.9 Power distribution of a three-phase induction motor

The mechanical power delivered to the shaft is less than the developed mechanical power at the rotor by an amount equal to the sum of the mechanical losses of the rotor due to friction and windage, the rotor iron losses in the surface of and in the teeth of the stator and rotor caused by the rotation of the rotor, and the losses in the slip rings in case the induction motor is of wound rotor type.

If T is the torque in newton-metre (Nm) exerted on the rotor by the rotating flux and N_S is the synchronous speed in rpm, the power transferred from the stator to the rotor is

$$P_{ag} = 2\pi T \left(\frac{N_S}{60} \right) \text{ watt} \tag{8.28}$$

The torque in terms of the rotor current and rotor parameters can be obtained by first rewriting Eq. (8.28) and then substituting Eq. (8.24) as shown below:

$$T = \frac{1}{2\pi \left(N_S / 60 \right)} P_{ag} = \frac{1}{2\pi \left(N_S / 60 \right)} \times 3 \left(I_2' \right)^2 \frac{r_2'}{s} \tag{8.28a}$$

or $\quad T = K \left(I_2' \right)^2 \dfrac{r_2'}{s} \tag{8.28b}$

where $\quad K = \dfrac{3}{2\pi \left(N_S / 60 \right)}$

Substituting for I_2' from Eq. (8.13) in the above expression, we get

$$T = \frac{1}{2\pi \left(N_S / 60 \right)} \times 3 \frac{s^2 E^2}{r_2^2 + \left(s x_2 \right)^2} \times \frac{r_2'}{s} = \frac{3}{\omega_S^2} \times \frac{s E_2^2 r_2}{r_2^2 + \left(s x_2 \right)^2} \tag{8.28c}$$

where $\omega_S = 2\pi \left(N_S / 60 \right)$ is the synchronous speed of the motor in radians/sec.

If N is the rotor sped in rpm, the total mechanical power developed in the rotor is

$$P_m = 2\pi T \left(\frac{N}{60} \right) \text{ watt} \tag{8.29}$$

But, from Fig. 8.9, it is seen that the total $I^2 R$ loss in the rotor is rotor input P_{ag} – total mechanical power developed $P_m = \dfrac{2\pi T(N_S - N)}{60}$

$\therefore \qquad \dfrac{\text{Total } I^2 R \text{ loss in rotor}}{\text{Input power to rotor}} = \dfrac{2\pi T(N_S - N)}{2\pi T N_S} = s$ $\qquad\qquad$ (8.30)

Example 8.5 The stator impedance and rotor standstill impedance of a three-phase 415-V induction motor is $(0.05 + j\,0.15)\Omega$/phase each. If the magnetizing current of the motor is 32 A, calculate the (i) power factor of the supply, (ii) horse power output, and (iii) efficiency of the motor when it is running at 975 rpm. Assume the following data for the motor:

Number of poles = 6, frequency = 50 Hz, core loss $P_C = 1000$ W, mechanical loss = 600 W, and turns rtio between the stator and rotor $a = 1.0$

Solution Synchronous speed, $N_S = \dfrac{120 f}{P} = \dfrac{120 \times 50}{6} = 1000$ rpm

Slip, $s = \dfrac{N_S - N}{N_S} = \dfrac{1000 - 975}{1000} = 0.025$

Figure 8.10 shows the equivalent circuit of the induction on the stator side.

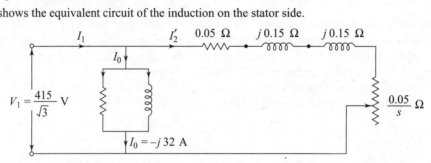

Fig. 8.10

The rotor impedance referred to the stator side,

$\qquad Z'_2 = (0.5/0.025 + j0.15) = (2.0 + j0.15)\,\Omega$

Assuming the per phase supply voltage as the reference phasor, from the circut diagram, it is seen that the stator load current can be written as

$\qquad I'_1 = \dfrac{415/\sqrt{3}}{(0.05 + j0.15) + (2.0 + j0.15)} = \dfrac{239.60}{2.05 + j0.3} = (114.43 - j16.75)\ \text{A} = 115.65\angle - 8.33°\text{A}$

Since the magnetizing current lags the voltage, $I_0 = -j32\text{A}$ $\qquad\qquad$ (I)
The current drawn from the supply is given by

$\qquad I_1 = I'_1 + I_0 = (114.43 - j16.75) - j32 = (114.43 - j48.75) = 124.38\angle - 23.07°\text{A}$

(i) Power factor of the supply (motor) $= \cos(-23.07°) = 0.92$

Power input to the motor, $P_i = 3 \times \left(415/\sqrt{3}\right) \times 124.38 \times 0.92 = 82.25$ kW

From Eq. (8.22), the input power to the rotor is given by

$\qquad P_{ag} = P_i - 3I_1^2 \times 0.050 - P_C = 82250 - 3 \times (124.38)^2 \times 0.05 - 1000 = 78.93$ kW

From Eq. (8.25), the gross mechanical power output is give by

$$P_m = (1-s)P_{ag} = (1-0.025) \times 78.93 = 76.96 \text{ kW}$$

(ii) Output of the motor $= \dfrac{P_m - \text{mechanical loss}}{746} = \dfrac{76960 - 600}{746} = 102.36 \text{ HP}$

(iii) Efficiency of the motor, $\eta = \dfrac{P_m - \text{mechanical loss}}{P_i} = \dfrac{76960 - 600}{82250} \times 100 = 92.83\%$

Example 8.6 The power input to the rotor of a 415-V, 50-Hz, 6-pole, three-phase induction motor is 75 kW. The rotor electromotive force is observed to make 120 complete cycles per minute. Calculate (a) the slip, (b) the rotor speed, (c) the mechanical power developed, (d) the rotor resistance loss per phase, and (e) the rotor resistance per phase if the rotor current is 60 A.

Solution (a) Synchronous speed $N_S = \dfrac{120f}{P} = \dfrac{120 \times 50}{6} = 1000 \text{ rpm}$

Given rotor frequency, $f_r = \dfrac{120}{60} = 2 \text{ Hz}$

From Eq. (8.4), $f_r = sf$

$\therefore \qquad s = \dfrac{f_r}{f} = \dfrac{2}{50} = 0.04 \text{ pu or } 4\%$

(b) Rotor speed $N = (1-s)N_S = (1-0.04) \times 1000 = 960 \text{ rpm}$

(c) Air gap power per phase, $P_{ag} = \dfrac{75}{3} = 25 \text{ kW}$

From Eq. (8.26), mechanical power developed,

$$P_m = 3P_{ag}(1-s) = 3 \times 25 \times (1-0.04) = 72 \text{ kW}$$

(d) From Eq. (8.25),

Rotor resistance loss per phase $= sP_{ag} = 0.04 \times 25 = 1.0 \text{ kW} = 1000 \text{ W}$

(e) If r_2 and I_2 per phase are the rotor resistance and current, respectively,

$$I_2^2 r_2 = \text{rotor resistance loss} = 1000 \text{ W}$$

$$r_2 = \dfrac{1000}{60^2} = 0.278 \ \Omega$$

Example 8.7 An induction motor has an efficiency of 0.88 when the load is 60 kW. At this load, the stator resistance and rotor resistance loss each equals the iron loss. The mechanical losses are one-fourth of the no-load loss. Calculate the slip.

Solution Power input per phase $= \dfrac{\text{power output per phase}}{\text{efficiency}}$

$$= \dfrac{60}{3 \times 0.88} = 22.7272 \text{ kW}$$

Let the iron loss per phase be X kilowatt. Then,

Mechanical loss $= 0.25X \text{ kW}$

Stator resistance loss per phase = Rotor resistance loss per phase $= X \text{ kW}$
Air gap power per phase

P_{ag} = Input – (iron loss + stator resistance loss + rotor resistance loss) = $22.727 - 3X$

But, P_{ag} = power output per phase + mechanical losses = $20 + 0.25X$

Then $22.727 - 3X = 20 + 0.25X$

∴ $X = \dfrac{2.7272}{3.25} = 0.839$ kW

$P_{ag} = 20 + 0.25 \times 0.839 = 20.21$

Now, Rotor resistance loss = sP_{ag}

Hence $s = \dfrac{0.839}{20.21} = 0.0415$

Example 8.8 A four-pole, three-phase induction motor develops 25 kW, including mechanical losses of 1 kW, at a speed of 1440 rpm on 440 V, 50 Hz mains. The power factor is 0.85. For this load calculate (a) slip, (b) the rotor resistance loss, (c) the total input if the stator losses are 1.75 kW, (d) the efficiency, (e) the line current, and (f) the number of complete cycles of the rotor elctromotive force per minute.

Solution   (a) Synchronous speed

$$N_S = \frac{120f}{P} = \frac{120 \times 50}{4} = 1500 \text{ rpm}$$

$$\text{Slip } s = \frac{1500 - 1440}{1500} = 0.04 \text{ pu}$$

(b) The total mechanical power developed

P_m = power developed + mechanical losses

     = 25 + 1 = 26 kW

From Eq. (8.26),

$P_m = (1 - s)\, P_{ag}$

∴ $P_{ag} = \dfrac{P_m}{1-s} = \dfrac{26}{1-0.04} = 27.083$ kW

Then,   rotor resistance loss = $sP_{ag} = 0.04 \times 27.083 = 1.083$ kW

(c) Total input   = P_{ag} + stator loss

         = 27.083 + 1.75 = 28.833 kW

(d) Efficiency = $\dfrac{\text{output}}{\text{input}} \times 100 = \dfrac{25}{28.833} \times 100 = 86.7\%$

(e) Input power = $\sqrt{3}\, E_l I_l \cos\varphi$

Therefore, the line current

$$I_l = \frac{28.833 \times 10^3}{\sqrt{3} \times 440 \times 0.85} = 44.51 \text{ A}$$

(f) Rotor frequency

$f_r = sf = 0.04 \times 50 = 2$ Hz

The number of complete cycles of the rotor emf per minute is $2 \times 60 = 120$.

8.3.4 Torque–slip Characteristics

The electrical power P_{ag} generated in the rotor circuit of a three-phase induction motor in terms of the rotor circuit shown in Fig. 8.5 is given by

$$P_{ag} = 3sE_2I_2 \cos\theta_2 \text{ watt} \tag{8.31}$$

Substituting the values of I_2 and $\cos\theta_2$ from Eq. (8.14) and Eq. (8.15), respectively, into Eq. (8.31),

$$P_{ag} = 3sE_2 \times \frac{sE_2}{\sqrt{r_2^2 + (sx_2)^2}} \times \frac{r_2}{\sqrt{r_2^2 + (sx_2)^2}}$$

$$= 3 \times \frac{s^2 E_2^2 r_2}{r_2^2 + (sx_2)^2} \tag{8.32}$$

All this power is dissipated as I^2R in the rotor circuits. As per Eq. (8.28), the input power to the rotor is also equal to $2\pi T(N_S/60)$. Thus, from Eqs (8.30) and (8.32),

$$s\frac{2\pi T N_S}{60} = 3 \times \frac{s^2 E_2^2 r_2}{r_2^2 + (sx_2)^2} \tag{8.33}$$

Consequently, for a given synchronous speed,

$$T \propto \frac{sE_2^2 r_2}{r_2^2 + (sx_2)^2} \tag{8.34}$$

Since E_2 is proportional to ϕ, it can be said that

$$T \propto \frac{s\phi^2 r_2}{r_2^2 + (sx_2)^2} \tag{8.34a}$$

$$T \propto \frac{\phi^2 (r_2/s)}{(r_2/s)^2 + x_2^2} \tag{8.34b}$$

Assuming that the voltage drop due to the stator resistance and leakage reactance is negligible and, therefore, the rotor induced emf E_2 is constant, Eqs (8.34a) and (8.34b) are modified as

$$T \propto \frac{sr_2}{r_2^2 + (sx_2)^2} \tag{8.35}$$

$$T \propto \frac{(r_2/s)}{(r_2/s)^2 + x_2^2} \tag{8.35a}$$

At a low slip, $s \le 0.1$ pu, $(sx_2)^2$ is very small compared to r_2^2 and can be neglected. Equation (8.35) simplifies to

$$T \propto \frac{sr_2}{r_2^2} \propto \frac{s}{r_2} \tag{8.36}$$

From Eq. (8.36) it can be seen that torque T varies directly as the slip, for low values of s, for a given value of r_2. Lower the value of r_2, greater is the slope of the straight line, as is seen from Fig. 8.11, which shows a plot of torque–slip characteristics for different values of the rotor resistance, $r_2^1, r_2^2, r_2^3, r_2^4$ and the given value of the rotor leakage reactance x_2. The rotor resistance values are such that $r_2^1 < r_2^2 < r_2^3 < r_2^4$.

For a slip $s > 0.1$ pu, $(sx_2)^2$ is very large compared to r_2^2, hence the latter can be neglected. In this case, Eq. (8.35) for a given value of the rotor leakage reactance x_2 gets changed to

$$T \propto \frac{sr_2}{(sx_2)^2} \propto \frac{r_2}{s} \tag{8.37}$$

From Eq. (8.37), it is clear that for a slip greater than 0.1 pu, torque T varies inversely as the slip for a given leakage reactance. A study of the torque–slip characteristics drawn in Fig. 8.11 shows the inverse variation of the torque with slip for different combinations of the rotor resistance and leakage reactance.

As the value of slip increases from a low value to a high value, the variation of torque T, varying directly with s for low values, changes to the inverse variation with s for high values. Thus, it can be concluded that torque is maximum for a particular value of s. Figure 8.11 shows that the maximum value of torque occurs at a particular slip for given values of r_2 and x_2. Also, it is seen that the maximum torque developed is the same for all combinations of the rotor resistance and leakage reactance.

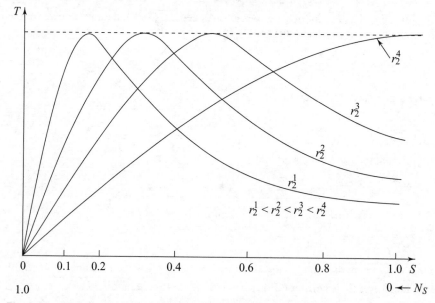

Fig. 8.11 Torque–slip characteristics of an induction motor for different values of r_2 and x_2

The condition for the maximum torque can be derived by differentiating Eq. (8.34) with respect to slip as follows:

$$\frac{d}{ds}\left\{ \frac{sr_2}{r_2^2 + (sx_2)^2} \right\} = 0$$

or

$$\frac{\{r_2^2 + (sx_2)^2\}\, r_2 - sr_2\,(2sx_2^2)}{\{r_2^2 + (sx_2)^2\}^2} = 0$$

Simplifying gives

$$r_2^2 - s^2 x_2^2 = 0$$

or

$$s = \frac{r_2}{x_2} \tag{8.38}$$

As can be seen, the condition for the maximum developed torque occurs when the slip is equal to the ratio of the rotor resistance and leakage reactance. Substituting Eq. (8.38) in Eq. (8.35) and Eq. (8.35a), the maximum value of the torque can be obtained. Thus,

$$T_{max} \propto \frac{sr_2}{2r_2^2} \qquad (8.39)$$

$$\propto \frac{1}{2x_2} \qquad (8.39a)$$

It can be seen from Eq. (8.39a) that since the rotor leakage reactance x_2 is approximately a constant for an induction motor, the value of the maximum torque is independent of the value of the rotor resistance r_2.

Example 8.9 A six-pole 50-Hz, star-connected, 220-V three-phase induction motor has a stator to rotor turns ratio of 1.5. If the rotor resistance and reactance per phase is 0.15 Ω and 1.0 Ω, respectively, and the full load slip of the motor is 4%, calculate (i) full load torque, (ii) metric horse power developed at full load, (iii) maximum torque, and (iv) speed at maximum torque.

Solution Synchronous speed, $N_s = \dfrac{120f}{P} = \dfrac{120 \times 50}{6} = 1000$ rpm

Speed of the motor at full load, $N = (1-s)N_s = (1-0.04) \times 1000 = 960$ rpm

Rotor resistance referred to the stator side, $r_2' = r_2 \times a^2 = 0.15 \times (1.5)^2 = 0.34 \ \Omega$

Rotor reactance referred to the stator side, $x_2' = x_2 \times a^2 = 1 \times (1.5)^2 = 2.25 \ \Omega$

(i) Torque at full load is given by

$$T_{fl} = \frac{3}{(2\pi \times N_S/60)} \times \frac{sE_2^2 r_2'}{(r_2')^2 + (sx_2')^2}$$

$$= \frac{3}{(2\pi \times 1000/60)} \times \frac{0.04 \times \left(220/\sqrt{3}\right)^2 \times 0.34}{0.34^2 + (0.04 \times 2.25)^2} = 51.14 \ \text{Nm}$$

(ii) 1 metric hp = 735.5 W

So, metric horse power at full load $= \dfrac{2\pi N T_{fl}}{60 \times 735.5} = \dfrac{2\pi \times 960 \times 51.14}{60 \times 735.5} = 6.99$ hp

(iii) Condition for maximum torque to develop is $s = r_2/x_2 = 0.15/1.0 = 0.15$

Maxium torque developed is expressed as

$$T_{max} = \frac{3}{(2\pi \times 1000/60)} \times \frac{0.15 \times \left(220/\sqrt{3}\right)^2 \times 0.34}{0.34^2 + (0.15 \times 2.25)^2} = 102.71 \ \text{Nm}$$

(iv) Speed at maximum torque $= (1-0.15) \times 1000 = 850$ rpm

8.3.5 Induction Motor Testing

The efficiency of an induction motor can be determined by directly loading the motor and by measuring its input and output power. However, it is not convenient to perform a load test on a large induction motor, as it may be difficult to arrange load for the motor. Even if load is provided, there will be a heavy power loss in the process. The wastage of power can be avoided by determining various losses of an induction motor from the no-load test and blocked-rotor test. Through these tests parameters of the equivalent circuit of an induction motor can be determined and performance characteristics can be obtained.

8.3.5.1 No-load Test

The no-load test on an induction motor is performed by connecting its stator to a variable three-phase voltage supply at rated frequency. The supply voltage is now gradually increased to the rated value and the motor is allowed to rotate freely. Before proceeding to record the data from this test, it is important to understand that a freely rotating motor does indeed represent an open-circuit or no-load condition.

When an induction motor is connected across its rated supply voltage and the rotor is revolving at a speed very close to the synchronous speed, slip s is very small and can be assumed to be zero. Hence, the electrical equivalent of the mechanical load on the shaft has a value, given by

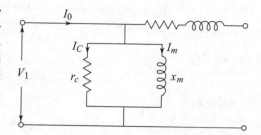

Fig. 8.12   Equivalent circuit of an induction molar under no-load condition

$$r'_2 \left(\frac{1}{s} - 1 \right) = \infty$$

The above reflects an open-circuit condition and the equivalent circuit of the induction motor shown in Fig. 8.9 takes the form shown in Fig. 8.12.

The observations noted down from the test are rated line voltage (V_0), no-load current (I_0), and no-load power input (P_0). The power P_0 is equal to the sum of the stator core loss and friction and windage loss of the induction motor on no load. Neglecting the friction and windage loss, the power P_0 gives the stator core loss. The parameters r_c and x_m of the shunt branch of the equivalent circuit shown in Fig. 8.12 are determined in the same manner as in the case of no-load test on a transformer (see Section 6.8.1).

8.3.5.2 Blocked Rotor Test

The stator is connected to a reduced supply voltage (15%–20%), with the rotor blocked, that is, prevented from rotating, so as to circulate full-load rated current. When the rotor is kept at standstill by blocking its rotation, the slip s is equal to unity and the electrical equivalent of the mechanical load on the shaft has a value given by

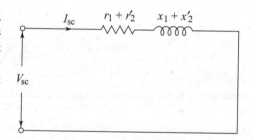

Fig. 8.13   Equivalent circuit of an induction motor under short circuit

$$r'_2 \left(\frac{1}{s} - 1 \right) = 0$$

This condition is equivalent to the short-circuit test on a transformer. As such, the equivalent circuit shown in Fig. 8.9 is redrawn to reflect the short-circuit condition and is shown in Fig. 8.13. The observations noted down from the test are rated line voltage (V_{sc}), short circuit current (I_{sc}), and short circuit power input (P_{sc}). Since the full-load rated current is circulated under the short-circuit condition, P_{sc} reflects the full-load copper loss. From the recorded data, the equivalent resistance and reactance are first computed in much the same way as in the case of the short-circuit test on a transformer (see Section 6.8.2). The equivalent resistance so determined represents the combined stator resistance r_1 and the transformed rotor resistance r'_2. The value of the stator resistance r_1 can be measured by a dc test and then converted to an ac value. Then the values of r_1 and r'_2 can be separated. The equivalent impedance Z_{eq} computed from V_{sc} and I_{sc} represents the total impedance of the stator and reflected impedance of the rotor. Since $(r_1 + r'_2)$ have already been determined, the value of the combined reactance $(x_1 + x'_2)$ is given by

$$\left(x_1 + x_1' \right) = \sqrt{Z_{eq}^2 - r_{eq}^2}$$

For separating the stator and rotor leakage reactances, the following approximation is employed

$$\frac{x_1}{x_2'} = \frac{r_1}{r_2'}$$

Example 8.10 No load and blocked rotor tests are performed on a three-phase delta-connected 415 V induction motor. To measure the core loss (no-load power) P_C and the short-circuit power P_{sc}, two wattmeter method is employed. The test data (line values), along with the readings of the wattmeters are given below:

No-load test: 230 V   15 A   2200 W   −1200 W

Blocked rotor test: 95 V   50 A   3500 W   −1050 W

Determine the parameters and draw the equivalent circuit of the induction motor. Assume that both the rotor resistance and leakage reactance per phase, referred to the stator side, are respectively equal to per phase stator winding resistance and the leakage reactance. Compute the (i) line current, (ii) power input, (iii) power output, and (iv) efficiency when the motor is running at 4% slip.

Solution *No-load test:*

Wattmeter reading, $P_C = 2200 - 1200 = 1000$ W

For the mesh-connected stator, the no-load current per phase $= 15.0/\sqrt{3}$ A

Referring to Fig. 8.12 and using Eq. (6.58), the admittance of the core,

$$Y_C = \frac{15.0/\sqrt{3}}{230} = 0.038 \; \mho$$

No-load power per phase $= 1000/3$ W

From Eq. (6.59), the conductance of the core, $G_C = \dfrac{1000/3}{(230)^2} = 0.0063 \; \mho$

Employing Eq. (6.60), the core susceptance, $B_m = \sqrt{(0.038)^2 - (0.0063)^2} = 0.037 \; \mho$

Thus, core resistance, $r_c = 1/G_C = 1/0.0063 = 158.70 \; \Omega$, and magnetizing reactance $x_m = 1/B_m = 1/0.037 = 26.94 \; \Omega$

Blocked rotor test:

Short-circuit power, $P_{sc} = 3500 - 1050 = 2450$ W

Short-circuit power per phase $= 2450/3$ W

Short-circuit current per phase $= 50/\sqrt{3}$ A

Referring to Fig. 8.13 and using Eq. (6.62), equivalent resistance is written as

$$R_{eq} = \left(r_1 + r_2' \right) = \frac{(2450/3)}{\left(50/\sqrt{3} \right)^2} = 0.98 \; \Omega$$

Thus, $r_1 = r_2' = 0.98/2 = 0.49 \; \Omega$

From Eq. (6.63), equivalent short-circuit impedance, $Z_{eq} = 95/\left(50/\sqrt{3} \right) = 3.29 \; \Omega$

Equivalent reactance, from Eq. (6.64) is obtained as

$$X_{eq} = (x_1 + x_2') = \sqrt{3.29^2 - 0.98^2} = 3.14 \; \Omega$$

Hence, $x_1 = x_2' = 3.14/2 = 1.57 \; \Omega$

The equivalent circuit, along with the parameters of the induction motor, is drawn in Fig. 8.14. The circuit for determining the line current and power factor at 4% slip is shown in Fig. 8.14(c).

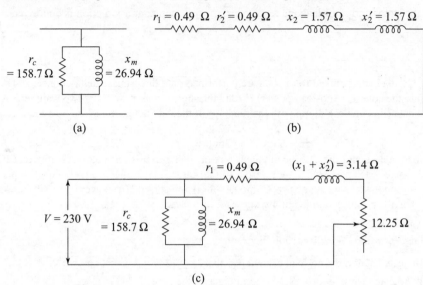

(a)   (b)

(c)

Fig. 8.14   (a) Open-circuit parameters, (b) short-circuit parameters, and (c) equivalent circuit

Equivalent load resistance at 4% slip = 0.49/0.04 = 12.25 Ω
From the circuit diagram in Fig. 8.14(c),

$$I_0 = \frac{230}{158.7} + \frac{230}{j26.74} = 1.4492 - j8.6013 \text{ A}$$

$$I_L = \frac{230}{12.74 + j3.14} = \frac{230}{13.1212\angle 13.85°} = 17.5288\angle -13.85° = 17.0191 - 4.1960 \text{ A}$$

$$I_1 = I_0 + I_L = 18.5183 - j12.7973 = 22.51\angle -34.65° \text{ A}$$

(i) Line current = $\sqrt{3}I_1 = \sqrt{3} \times 22.51 = 38.9884$ A

Phase current, $I_1 = \dfrac{230}{(17.19 + j29.33)} = (3.42 - j5.84) = 6.77\angle -59.63°$ A

(ii) Power input, $P_{in} = \sqrt{3} \times 230 \times 38.9884 \times \cos(34.65°) = 12777.1494$ A

(iii) Current through the load, $I_L = 17.5288\angle -13.85°$ A

(iv) Three-phase power output, $P_o = 3 \times (17.5288)^2 \times 12.25 = 11291.762$ W

(v) Efficiency of the motor, $\eta = \dfrac{P_o}{P_{in}} \times 100 = \dfrac{11291.762}{12777,1494} \times 100 = 88.374\%$

Example 8.11 A three-phase induction motor with a star-connected rotor has an induced electromotive force of 75 V between slip rings at standstill on open-circuit with a normal voltage applied to the stator. The resistance and standstill reactances of each rotor phase are 0.6 Ω and 6 Ω, respectively. Calculate the current per phase in the rotor

(a) when at standstill and connected to a star-connected rheostat of resistance 8 Ω and reactance 2 Ω per phase, (b) when running short-circuited with 5% slip.

Solution Impedance of the rotor, $Z_r = (0.6 + j6)\Omega$

Impedance of rheostat, $Z_{RH} = (8 + j2)\Omega$

At standstill $s = 1$,

Total impedance of rotor circuit $= Z_r + Z_{RH} = 8.6 + j8 = 11.75 \angle 42.93°\Omega$

Voltage per phase in the rotor circuit $= \dfrac{75}{\sqrt{3}}$ V

Rotor current per phase $= \dfrac{75/\sqrt{3}}{11.75\angle 42.93} = 3.685\angle -42.93°$ A

When slip = 5%, the rotor impedance

$Z_r = 0.6 + j(0.05 \times 6) = 0.6 + j0.3 = 0.671\angle 26.56°\Omega$

Rotor current at $s = 0.05$,

$$I_r = \frac{0.05 \times 75/\sqrt{3}}{0.671\angle 26.56°} = 3.22\angle -26.56°A$$

Example 8.12 A three-phase induction motor has a four-pole, star-connected, stator winding and runs on a 400 V, 50 Hz supply. The rotor resistance is 0.1 Ω and reactance 1.2 Ω. The ratio of the stator to rotor turns is 2.25. The full-load slip is 4%. For this load, calculate (a) the total torque, (b) the metric horse power (hp), (c) the maximum torque, and (d) the speed at maximum torque. (1 metric hp = 735.5 W)

Solution (a) The rotor phase voltage at standsill,

$$\frac{400}{\sqrt{3} \times 2.25} = 102.64 \text{ V}$$

Synchronous speed, $N_S = \dfrac{120f}{P} = \dfrac{120 \times 50}{4} = 1500$ rpm

From Eq. (8.32), $s \dfrac{2\pi T N_S}{60} = 3 \times \dfrac{s^2 E_2^2 r_2}{r_2^2 + (sx_2)^2}$

Then, $T = \dfrac{3 \times 60}{2\pi N_S} \times \dfrac{E_2^2(r_2/s)}{(r_2/s)^2 + x_2^2}$ (I)

$$= \frac{180}{2\pi \times 1500} \times \frac{(102.64)^2 \times (0.1/0.04)}{(0.1/0.04)^2 + (1.2)^2} = 65.41 \text{ Nm}$$

(b) Speed of the rotor

$N = (1 - s) \times N_S = (1 - 0.04) \times 1500 = 1440$ rpm

Output $= \dfrac{2\pi NT}{60} = \dfrac{2\pi \times 1440 \times 65.41}{60} = 9863.6$ W

Now, 1 metric hp = 735.5 W

Output $= \dfrac{9863.6}{735.5} = 13.41$ hp

(c) The condition for the maximum torque is given by $r_2 = sx_2$, or, $x_2 = r_2/s$.
Substituting this condition in Eq. (I) of this problem, the maximum torque is

$$T_{max} = \frac{3}{50\pi} \times \frac{E_2^2 x_2}{2x_2^2} = \frac{3E_2^2}{50\pi \times 2x_2}$$

$$= \frac{3 \times (102.64)^2}{50 \times \pi \times 2 \times 1.2} = 83.83 \text{ N m}$$

(d) The maximum torque will occur at

$$s = \frac{r_2}{x_2} = \frac{0.1}{1.2} = 0.08333$$

Speed at maximum torque

$$(1 - s)N_S = (1 - 0.08333) \times 1500 = 1375 \text{ rpm}$$

8.3.6 Starting of Three-phase Induction Motors

To start a three-phase induction motor directly, in case of a slip ring rotor, the terminals are short circuited and the rated three-phase voltage is applied to its stator terminals. Since at the time of starting, the rotor is at standstill ($N=0$ and $s = 1$), the starting current (neglecting the stator impedance), is obtained from Eq. (8.14) as

$$I_{st} = \frac{E_2}{\sqrt{r_2^2 + x_2^2}} \text{ A/phase} \tag{8.40}$$

When a three-phase induction motor is directly connected across a three-phase voltage supply, its rotor behaves like a short-circuited secondary winding and, therefore, the starting current drawn from the voltage supply is very high, about four to seven times the full-load current. The high current rapidly decreases as the motor starts up, but it is at a very low power factor. From Eq. (8.28b), it is seen that at starting the torque T_{st} $\propto I_{st}^2$ Therefore, it would be desirable to have a high starting current since it would result in higher torque. However, the main difficulty in starting induction motors by directly connecting them to the supply voltage arises due to the disturbance the high starting current produces in the voltage at the consumer terminals in the neighbourhood. Consider that a large-or medium-sized induction motor is being supplied from a fixed voltage distribution sub-station as shown in Fig. 8.15. When this induction motor is directly connected to the voltage supply, the large starting current flowing through the feeder is way above the magnitude of the value for which the feeder has been designed. Consequently, the resultant voltage dip along the feeder affects the consumers, both connected before and after this consumer. As such, the electric supply authorities limit the size of the motor that can be started by directly connecting it to the voltage supply to 3 kW. For induction motors of larger ratings, it is usual to start induction motors with a reduced voltage. In this section the various methods of starting three-phase induction motors are discussed.

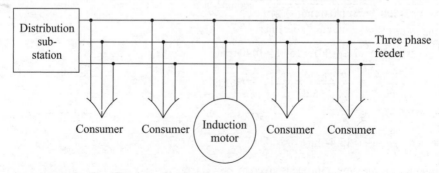

Fig. 8.15　Effect of directly connecting an induction motor to the voltage supply

8.3.6.1 Direct-on-line (DOL) Starting

The three-phase induction motor, up to 3 kW rating, can be started by directly connecting the stator terminals to the supply line. If T_{st} is the starting torque, then from Eq. (8.37), as slip $s = 1$ at starting, $T_{st} \propto r_2$. Thus in an induction motor, T_{st} is very low as the value of the rotor resistance r_2 is small. Hence the DOL method of starting of the induction motor may give rise to a high rate of rise in the temperature of the motor and if the start-up period is increased due to either an excessive load torque, insufficient rotor resistance, or excessive voltage drop in the supply, the motor may get damaged. Hence it is advisable to start only small capacity motors using this method.

The relationship between the starting torque T_{st} and full-load torque T_{fl} can be obtained. If I_{st} and I_{fl} are, respectively, the starting currents drawn from the main supply at the instant of starting and the full-load, then, from Eq. (8.28b) it is seen that,

$$\frac{T_{st}}{T_{fl}} = \frac{I_{2st}'^2 \, (r_2'/1)}{I_{2fl}'^2 \, (r_2'/s_{fl})}$$

$$= \left(\frac{I_{2st}'}{I_{2fl}'} \right)^2 s_{fl} \tag{8.41}$$

where s_{fl} is the slip at full load. If the no-load current is neglected, then

$$\frac{I_{st}}{I_{fl}} = \frac{I_{2st}'}{I_{2fl}'}$$

Hence Eq. (8.40) becomes

$$\frac{T_{st}}{T_{fl}} = \left(\frac{I_{st}}{I_{fl}} \right)^2 s_{fl} \tag{8.42}$$

$$= \left(\frac{I_{sc}}{I_{fl}} \right)^2 s_{fl} \tag{8.43}$$

where I_{sc} is the short-circuit current at the rated voltage and at starting $I_{st} = I_{sc}$.

Direct-on-line starting of a three-phase induction motor is shown in Fig. 8.16. When the ON push button is pressed, the contactor coil is energized and its open contacts are closed. The motor gets connected to the supply mains through the main contacts of the contactor. The motor continues to get supply even when the push button is released, since the contactor coil gets supply through the hold-on contact '*a*' of the contactor. When the OFF push button is pressed, the contactor coil gets de-energized, the main contacts of the contactor open and the motor stops. If the motor is overloaded, the contact *D* of the overload elay opens and the motor stops. Fuses are provided for short-circuit protection.

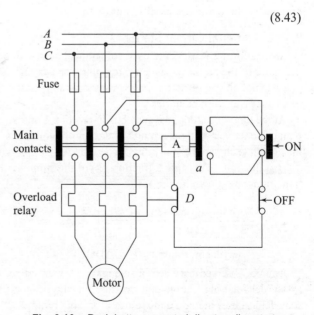

Fig. 8.16　Push button operated direct-on-line starter

8.3.6.2 Star–delta Starter

Star–delta starting is used for squirrel cage induction motors. In this method, the two ends of each phase of the stator windings of the motor are brought out and connected to a delta–star starter, which is a triple-pole double-throw switch. The connection diagram is shown in Fig. 8.17.

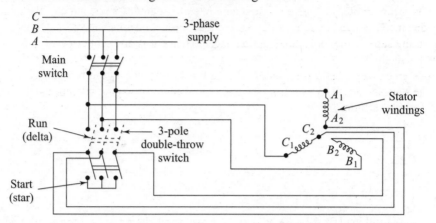

Fig. 8.17 Circuit diagram for star–delta starting of three-phase squirrel cage induction motors

At the time of starting the motor, the starter is in the 'start' position and the stator winding is connected in star. The voltage applied across the winding is, therefore, $1/\sqrt{3}$ times the line voltage and the starting torque is 1/3 times the full-load torque if the motor were to start DOL with the windings in delta. When the motor has accelerated, the switch position gets quickly changed to 'run' position, thus applying the full-line voltage to the stator winding.

This method of starting costs less and is effective and thus is used extensively. This method is suited for applications in which the starting torque is not required to exceed 50% of the full-load torque. It is used for pumps, machine tool drives, flour mills, etc.

8.3.6.3 Auto-transformer Starter

Auto-transformer starters are used for squirrel cage motors. In this method, the supply voltage is applied to the stator of the motor through a three-phase star-connected auto-transformer with two to three tappings per phase to obtain appropriate voltage ratios to be applied to a given motor. Figure 8.18 shows the circuit diagram of an auto-transformer starter.

At the time of starting, the switch is in position 1 and a voltage which is determined by the tap selected and which is only a certain percentage of the supply voltage is applied to the motor. As the motor picks up speed, the switch is changed to position 2 when the full-supply voltage is applied across the stator winding.

If at the time of starting a fraction kV_1 of the supply voltage is applied to the stator terminals of the auto-transformer as shown in Fig. 8.18,

$$\frac{T_{st}(\text{with autotransf.})}{T_{st}(\text{DOL})} = \left(\frac{kI_{sc}}{I_{sc}}\right)^2 = k^2$$

$$T_{st}(\text{with autotransf.}) = k^2 \, T_{st}(\text{DOL})$$

It may be observed that when the setting $k = 1/\sqrt{3}$ and is fixed at this value, the auto-transformer starting corresponds to star–delta starting. Starting squirrel cage induction motors using an auto-transformer are more expensive than the star–delta starter method and are used for large motors.

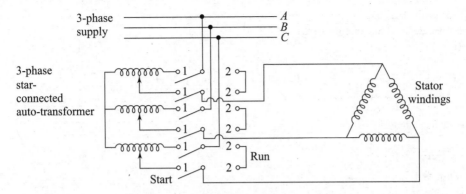

Fig. 8.18 Auto-transformer starting for three-phase squirrel cage induction motors

8.3.6.4 Rotor Resistance Starting

Rotor resistance starting is used for slip-ring induction motors. A star-connected external resistance is connected into the three-phase rotor winding through the slip rings on the shaft, as shown in the circuit diagram in Fig. 8.19. The arms *A* of the starting resistance are electrically and mechanically connected together.

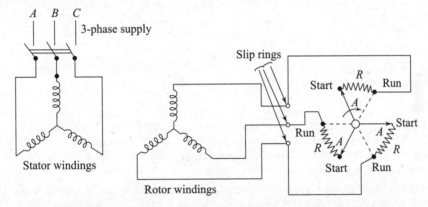

Fig. 8.19 Circuit diagram for rotor resistance starting of slip-ring induction motors

When the triple-pole switch is closed to connect the three-phase supply to the three-phase stator windings, at that instant the three arms *A* of the star-connected resistance are in the position as indicated in the circuit diagram shown in Fig. 8.19, and the maximum resistance is included in the rotor circuit so that the starting torque increases. Then the set of arms *A* is moved in the clockwise direction, so that the inserted resistance *R* in each phase of the rotor winding decreases as the induction motor speed picks up. Finally, as the speed reaches the rated value, the rotor resistance is completely cut out. The dotted lines show the position of the arms when the rotor resistance is completely cut out.

The rotor resistance starting has the advantage of a high starting torque with suitable external resistances. Thus it is ideal for on-load staring. It reduces the magnitude of starting current and improves the starting power factor.

Example 8.13 An induction motor has a short-circuit current equal to five times the full-load current. Find the starting torque as a percentage of the full-load torque if the motor is started by a (a) direct-on-line starter, (b) star–delta starter, (c) auto-transformer starter, or (d) rotor resistance starter. The starting current in (c) and (d) is limited to two times the full-load current and the full-load slip is 5%.

Solution (a) With DOL starting,

Starting current I_{st} = Short-circuit current $I_{sc} = 5 \times I_{fl}$

From Eq. (8.41),

$$\frac{T_{st}}{T_{fl}} = \left(\frac{I_{st}}{I_{fl}}\right)^2 s_{fl} = \left(\frac{5 \times I_{fl}}{I_{fl}}\right)^2 \times 0.05 = 1.25$$

$\therefore$ $\qquad T_{st} = 1.25 \times T_{fl}$

(b) For star–delta starter,

$$I_{st} = \frac{1}{\sqrt{3}} \times I_{sc} = \frac{1}{\sqrt{3}} \times 5 \times I_{fl}$$

Then, from Eq. (8.41),

$$\frac{T_{st}}{T_{fl}} = \left(\frac{5}{\sqrt{3}}\right)^2 \times 0.05 = \frac{1.25}{3} = 0.4166$$

$\therefore$ $\qquad T_{st} = 0.4166\, T_{fl}$

(c) With auto-transformer starter,

$$I_{st} = 2 \times I_{fl}$$

Then, from Eq. (8.41),

$$\frac{T_{st}}{T_{fl}} = \left(\frac{2}{1}\right)^2 \times 0.05 = 0.2$$

$\therefore$ $\qquad T_{st} = 0.2 \times T_{fl}$

(d) With a rotor resistance starter, the effect is the same as that with an auto-transformer starter since in both these cases the starting current is reduced to twice the full-load current.

$\therefore$ $\qquad T_{st} = 0.2 \times T_{fl}$

Example 8.14 A three-phase induction motor, with its stator connected in delta, has a full load rating of 15 metric horse power at 415 V. Determine the ratio of the starting current to full load current when a star–delta starter is used for starting the motor. The short-circuit current is 25 A at 150 V. Assume a full load efficiency and power factor of 90% and 0.8, respectively, and neglect magnetizing current.

Solution The given short-circuit current is experimentally determined at 150 V (line-to-line). Therefore, the short-circuit current per phase is $(25/\sqrt{3})$ A. Since at the time of starting, the rotor behaves like a short-circuited secondary, the given short-circuit current may be taken as the starting current. Further, at the time of starting, the stator is connected in star and the per phase voltage is $(415/\sqrt{3})$ V. In order to obtain the starting current, the given short-circuit current is proportionally increase as follows:

$$I_{st} = \frac{25}{\sqrt{3}} \times \frac{415}{\sqrt{3}} \times \frac{1}{150} = 23.06 \text{ A}$$

Full load current, $I_{fl} = \dfrac{15 \times 735.5}{\sqrt{3} \times 415 \times 0.9 \times 0.8} = 21.32$ A

Ratio of starting to full load current is given by

$$\frac{I_{st}}{I_{fl}} = \frac{23.06}{21.32} = 1.08$$

Example 8.15 A three-phase induction motor is started by employing an auto-transformer. The motor has a ratio of the starting current at normal voltage to the full load current at a slip of 5% of 4.5. (i) Compute the auto-transformer tapping so that the supply current at the time of starting is limited to 2.25 times the full load current. (ii) What is the ratio of the starting to full load torque given the above condition?

Solution   Assume that the voltage applied to the motor is reduced by a magnitude of a.
From the given condition of operation, it is required that the starting current

$$I_{st} = 4.5 I_{fl} \text{ A} \tag{I}$$

With the reduced voltage applied to the stator, the starting current is limited to I_{st}/a A
This reduced starting current when transformed to the primary (supply) side is further reduced to I_{st}/a^2 A
(i) The given condition that the starting current should not increase beyond 2.25 I_{fl} A leads to

$$\frac{I_{st}}{a^2} = 2.25 I_{fl} \tag{II}$$

Substituting Eq. (I) in Eq. (II) results in

$$\frac{I_{st}}{a^2} = \frac{2.25}{4.5} I_{st} \quad \text{or} \quad a = \sqrt{2} = 1.41$$

The auto-transformer tapping should be such that the voltage applied to the motor is $1/\sqrt{2} = 0.71$ of the supply voltage. Motor input current at starting is given by

$$I'_{2st} = \frac{I_{st}}{a} = \frac{4.5 I_{fl}}{1.41} = 3.18 I_{fl}$$

Using Eq. (8.40),

$$\frac{T_{st}}{T_{fl}} = \left(\frac{3.18 I_{fl}}{I_{fl}} \right)^2 s_{fl} = 10.125 \times 0.05 = 0.5062$$

Hence, the starting torque is 50.62% of the full load torque.

8.3.7 Speed Control of Three-phase Induction Motors

An induction motor finds numerous applications because of its simplicity and ruggedness, and fulfills the requirements of substantially constant speed drives. However, it suffers from the drawback that its speed cannot be easily and efficiently varied in a continuous manner over a wide range of operating conditions. In many applications a variable-speed drive is needed and, economically, it is attractive to use induction motor.

The speed of the induction motor can be changed by varying (a) the synchronous speed of the rotating field and (b) the slip. If the supply frequency is constant, varying the speed by changing the synchronous speed, by changing the number of poles, results only in discrete or step changes in the speed of the motor. On the other hand, the mechanical power output of an induction motor is approximately proportional to $(1 - s)$, thus any method of speed control which depends on the variation of slip is inherently inefficient.

The synchronous speed, N_S, of the rotating field in an induction machine, as given by Eq. (8.1), is

$$N_S = \frac{120 f}{P}$$

where P is the number of poles and f is the supply frequency. The synchronous speed, N_S, can be varied by changing (i) the number of poles P or (ii) the frequency f.

The operating slip can be changed by varying (i) the line voltage, (ii) the rotor resistance, or (iii) by applying voltage of appropriate frequency to the rotor circuits. A brief review of the various possible methods by which the speed of the induction motor can be varied either continuously or in discrete steps is presented here.

8.3.7.1 Pole-changing Method

In this method the stator winding of the motor is so designed that the terminals of the various coils are brought out and by changing the coil connections the number of poles of the winding can be changed in the ratio 2:1. Accordingly, two synchronous speeds result and only two speeds of operation are possible. The rotor is usually of the squirrel cage type. If more independent windings are provided, each arranged for pole changing, more synchronous speeds (e.g., four) can be obtained. The method of pole changing involves pole-amplitude modulation. Single-winding squirrel cage motors that give three operating speeds are reported to have been developed. Another method, phase-modulated pole changing, has been developed that produces three or five speeds.

8.3.7.2 Variable-frequency Method

The synchronous speed is directly proportional to the supply frequency. If it is practicable to vary the supply frequency, the synchronous speed of the motor can also be varied. The variation in speed is continuous or discrete, according to continuous or discrete variation of the supply frequency. However, for a constant maximum torque, both the supply voltage and the supply frequency should be varied simultaneously to maintain a constant flux density. The inherent difficulty in the application of this method is that the supply frequency, which is commonly available, is fixed. Thus, the method is applicable only if a variable-frequency supply is available. With the development of solid-state devices of comparatively large power ratings, it is now possible to use static inverters to drive induction motors for a variable-speed operation.

8.3.7.3 Variable Stator Voltage Method

Since the internal electromagnetic torque developed by the machine is proportional to the square of the voltage applied to the stator windings, different torque–speed curves are obtained for different voltages applied to the motor, as shown in Fig. 8.20. For a given rotor resistance r_2^1, two such curves are shown in Fig. 8.20 for two applied voltages V_1 and $V_2 = 0.5\ V_1$. If the load has the characteristic shown by the dashed line, it may be seen that the motor speed reduces from N_1 to N_4. If the voltage can be varied continuously from V_1 to V_2, the speed of the motor can also be varied continuously between N_1 and N_4 for the given load. This method is applicable to the cage as well as to wound-rotor type induction motors.

8.3.7.4 Variable Rotor Resistance Method

This method is applicable only to the wound-rotor motor. The effect on the speed–torque curves of inserting external resistances in the rotor circuit is shown in Fig. 8.20 for three different rotor resistances r_2^1, r_2^2, and r_2^3. For the given load, three speeds of operation N_1, N_2, and N_3 are possible. A continuous variation of speed is possible by a continuous variation of the rotor resistance.

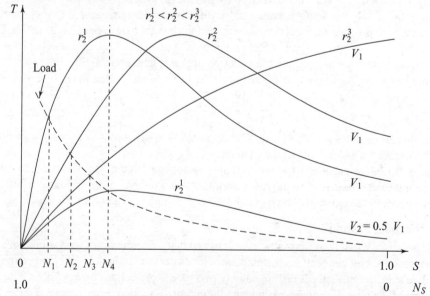

Fig. 8.20   Speed control by variable voltage and variable rotor resistance

8.3.7.5 Control by Solid-state Switching

The speed of the wound-rotor motor can be controlled either by inserting a solid-state inverter in the rotor circuit or by controlling the stator voltage by means of solid-state switching devices, such as silicon-controlled rectifiers (SCR) or thyristors. Adjusting the firing angle controls the output voltage from the thyristor bridge feeding the motor. The method is similar to the variable-voltage method outlined in Section 8.3.7.3. However, control by SCR gives a wider range of operation and is more efficient than other slip-control methods.

8.3.8 Squirrel Cage Motor Versus Wound-rotor Induction Motor

Squirrel cage motors are more robust, reliable, and inexpensive. Because of the robust structure of the rotor, about 90% of the induction motors employed in industry are of this type. Additionally, cage motors are slightly more efficient, have good overload capacity, operate at nearly constant speed, and need less maintenance than wound-rotor induction motors. In the absence of slip rings and brushes, there is no risk of sparking and consequent explosion. Hence cage motors are safer. The dissipation of rotor I^2R losses is better because of the bare conductors and end rings. As there is no winding in the rotor, more space for providing a cooling fan is available compared to a wound rotor motor. However, the cage rotor suffers from the disadvantages of a low starting torque, a high starting current, being sensitive to fluctuations in the supply voltage, and having a very low power factor at light loads. Squirrel cage induction motors, therefore, are suitable for industrial drives of low- and medium-power drives where speed control is not required, for example, blowers, fans, lathes, milling machines, grinding machines, flour mills, printing machinery, water pumps, compressors, etc.

Wound-rotor induction motors have the advantage of higher starting torques and lower starting currents because of the possibility of introduction of external resistances in the rotor circuit through slip rings. Such motors are used for drives having heavy loads and frequent starting and accelerating duty cycles. The external rotorresistances eliminate the problems of rotor cooling. However, the presence of slip rings and brush gear requires regular maintenance and thus additional cost. By employing solid-state switching, speed variation is possible in the slip-ring motors.

Slip-ring motors are used for industrial drives of medium and high power, such as driving shafts, lifts, elevators, conveyors, travelling cranes, punches, presses, compressors, etc.

Recapitulation

Three-phase induction motor

Slip $s = \dfrac{N_S - N}{N}$ pu

Synchronous speed, $N_S = \dfrac{120f}{P}$ rpm

Rotor frequency, $f_r = sf$

Induced voltage in stator, $V_1 = E_1 = 4.44k_{w1}fT_1\phi$ volt

Induced voltage in rotor, at standstill,

$\qquad E_2 = 4.44k_{w2}fT_2\phi$ volt

At slip s rotor voltage, $E_r = sE_2$

Ratio of transformation, $\boldsymbol{a} = \dfrac{E_1}{E_2} = \dfrac{k_{w1}T_1}{k_{w2}T_2}$

No-load current, $\boldsymbol{I}_0 = \boldsymbol{I}_m + \boldsymbol{I}_C$

Rotor current, $\boldsymbol{I}_2 = \dfrac{sE_2}{r_2 + jx_2}$

Rotor impedance, $\boldsymbol{Z}_2 = \sqrt{r_2^2 + s^2 x_2^2}$

Rotor copper loss, $sP_{ag} = s\,\dfrac{2\pi NT}{60}$

Rotor torque,

$$T \propto \frac{sE_2^2 r_2}{r_2^2 + s^2 x_2^2}$$

$$T \propto \frac{E_2^2 \times (r_2/s)}{(r_2/s)^2 + x_2^2}$$

Slip at maximum torque, $s = \dfrac{r_2}{x_2}$

Maximum rotor torque, $T_{max} \propto \dfrac{sr_2}{2r_2^2} \propto \dfrac{1}{2x_2}$

Ratio of starting torque to full-load torque:

Direct on line (DOL) $\dfrac{T_{st}}{T_{fl}} = \left(\dfrac{I_{sc}}{I_{fl}}\right)^2 S_{fl}$

Star–delta starter $T_{st} = \dfrac{1}{3}T_{fl}$

Auto-transformer starter, T_{st} (with autotransf.)

$\qquad = k^2\, T_{st}$ (DOL)

Assessment Questions

1. Enumerate the various parts of a three phase induction motor and describe the functions of each component.
2. Explain the principle of operation of a three phase induction motor and distinguish between slip and rotor frequency.
3. Using the transformer principle, develop the equivalent circuit of an induction motor.
4. Prove mathematically that if the sequence of the three-phase supply voltage connected to an induction motor is reversed, the direction of rotation of the motor is reversed, since the direction of the rotating field is also reversed.
5. Derive an expression for the torque developed by a three-phase induction motor.
   Show that the torque–slip characteristic can be estimated as a straight line when the motor is running at speeds close to the synchronous speed.
6. Discuss the power balance in a three-phase induction motor and show how it is distributed.
7. (a) Explain why an induction motor should not be tested by directly loading it.
   (b) Describe the (i) no-load and (ii) blocked rotor tests of an induction motor.
8. Explain how the data from the no-load and blocked rotor tests on an induction motor is used to develop its equivalent circuit and determine the performance characteristics.
9. Derive an expression for the ratio of starting to maximum torque developed by a three-phase induction motor.
10. Prove mathematically that (i) a free rotation of a three-phase induction motor at rated voltage simulates an open-circuit condition and (ii) a blocked rotor simulates a short-circuit condition.
11. Explain why it is not desirable to start a large three-phase induction motor by directly connecting it to the supply voltage.
12. Enumerate the various methods of starting slip ring motors and explain the operating functions of each.
13. Itemize and describe the various methods of speed control of three-phase induction motors.
14. Explain the relative merits of squirrel cage and wound rotor induction motors.

Problems

8.1 A three-phase induction motor has four poles and supplies a load at 3% slip. If the supply frequency is 50 Hz, calculate the (a) synchronous speed, (b) rotor speed in rev/min, and (c) rotor frequency.

   [(a) 1500 rpm, (b) 1455 rpm, (c) 1.5 Hz]

8.2 The rotor frequency of a three-phase induction motor when supplying a load is 2.5 cycle/sec and the slip is 5%. If the rotor speed is 950 rpm, determine the (a) supply frequency, (b) synchronous speed, and (c) number of pole pairs of the induction motor.

   [(a) 50 Hz, (b) 1000 rpm, (c) 3]

8.3 The no-load speed of an induction motor is 1500 rpm. When connected across a voltage source of frequency 50 cycle/sec, the rotor speed is 1200 rpm at full load. Determine: (a) the number of poles, (b) slip, (c) rotor frequency, and (d) speed of the rotor field with respect to the rotor. What is the rotor speed with respect to the stator? What is the speed of the rotor field in the air gap with respect to the stator field?

   [(a) 4; (b) 2%; (c) 1 Hz; (d) 30 rpm; 1480 rpm, zero rpm]

8.4 A three-phase, 50-Hz induction motor has eight poles. Its stator winding is constituted of 960 conductors distributed in 96 slots. Calculate (a) the distribution factor of the winding and (b) the maximum flux per pole in the air gap. Assume the stator winding to be connected across a three-phase, 230-V supply.

   [(a) 0.958, (b) 6.76 mWb]

8.5 A three-phase, 50-Hz, six-pole induction motor has delta-connected stator and star-connected rotor windings. The number of conductors per phase in the two windings is 300 and 50. If the rotor winding has a resistance of 0.02 Ω per phase and a standstill leakage reactance of 0.05 Ω per phase, calculate (a) the open-circuit rotor emf per phase, (b) rotor emf and current per phase when the motor is supplying a load at 950 rpm, (c) phase difference between the rotor current and the rotor voltage (i) for the slip in (b) and (ii) for 100% slip. Assume the stator winding is ideal and the supply voltage is 415 V.

[(a) 69.17 V; (b) 3.46 V; 173 A; (c) (i) 7.13°, (ii) 68.2°]

8.6 A three-phase, star-connected, six-pole, 220-V, 50-Hz induction motor has a stator impedance of $(1 + j\, 2.5)$ Ω/phase and an equivalent rotor impedance, referred to the stator side, of $(1.2 + j\, 0.8)\,\Omega$. Draw the equivalent circuit and calculate the (i) slip at maximum torque and (ii) maximum torque. Neglect magnetization current. [Hint: The condition for maximum torque will not apply since the stator impedance is specified.]

[(i) 0.21, (ii) 15.94 Nm]

8.7 The rotor of a three-phase, 50-Hz induction motor is star-connected and it gives 415 V between the slip rings on open circuit. What is the rotor current and power factor when a three-phase star-connected impedance, having a resistance of 12 Ω/phase and an inductance of 0.02 H/phase, is connected externally? What is the rotor current and power factor when the slip rings are short-circuited and the motor is running at a slip of 4%? Take the per phase resistance and inductance of the rotor to be 0.3 Ω and 0.02 H, respectively.

[13.63 A, 0.70; 15.31 A, 0.77]

8.8 The rotor resistance of a three-phase, 50-Hz, six-pole induction motor is 0.02 Ω/phase. The motor develops a maximum torque at a speed of 960 rpm. Calculate the (a) starting torque and (b) torque at a speed of 800 rpm as a percentage of the maximum torque. Take the flux in the air gap to remain constant at all loads.

[(a) 7.99%, (b) 38.46%]

8.9 The rotor of an eight pole, 50-Hz, three-phase induction motor develops a total torque of 80 N m when running at a speed of 720 rpm. Determine (a) the mechanical power developed by the rotor and (b) I^2R loss in the rotor. What is the output if the frictional loss is 300 W?   [(a) 6.03 kW; (b) 241.27 W, 5.73 kW]

8.10 A three-phase, four-pole, 50-Hz induction motor has a starting torque which is 20% of the maximum torque. If the rotor resistance is 0.3 Ω/phase, calculate (a) the rotor leakage reactance, (b) slip at maximum torque, and (c) speed at maximum torque. What is the value of the external resistance required to be connected in the rotor if the starting torque has to be increased to 50% of the maximum torque?

[(a) 2.97 Ω; (b) 3%: (c) 1455 rpm: 0.496 Ω or 10.78 Ω]

8.11 The equivalent circuit shown in Fig. 8.21 represents a three-phase, 415 V, star-connected, 50-Hz, four-pole induction motor. The speed of the motor when supplying a load is 1440 rpm. Determine (a) the supply current and power factor, (b) power input, (c) net power output, (d) net output torque, and (e) efficiency. Take the power loss due to windage, friction, and core losses (inclusive of the stator core loss) to be 400 W.

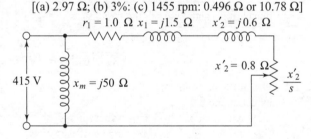

Fig. 8.21

[(a) 12.76 $\angle$−27.65°, 0.89 lagging; (b) 8.12 kW; (c) 7.03 kW; (d) 46.69 N m; (e) 86.58%]

8.12 A three-phase, star-connected, 415-V, six-pole induction motor gave the following results when tested in a laboratory:

No-load (oc) test: 415 V   10 A   1000 W
Blocked (sc) rotor test: 200 V   45 A   6000 W

(a) Develop the equivalent circuit of the motor. If the motor is running at a speed of 940 rpm when supplying a load, calculate (b) the supply current and power factor, (c) power output (d) efficiency, and (e) torque. Assume the stator resistance per phase to be 0.3 Ω.

[(b) 25.66 $\angle$−33.22°, 0.84 lagging; (c) 13.60 kW; (d) 87.80%; (e) 138.16 Nm]

8.13 What is the supply current and torque at starting for the induction motor in Problem 8.13? Calculate the slip and torque for the maximum power output.   [103.32 $\angle$−69.14°, 183.70 N m, 0.29, 306.19 Nm]

8.14 A three-phase induction motor has the following rating:

> Output 50 hp (metric)   Number of poles 4
> Supply voltage 415 V   Supply frequency 50 Hz

If the induced emf in the rotor makes 120 complete cycles per minute, determine the (a) slip, (b) rotor speed, (c) rotor resistance loss, and (d) power input to rotor. If the losses in the stator are neglected, what is the power input and efficiency? Ignore friction and windage losses in the rotor. Take 1 hp (metric) = 735.5 W.

[(a) 4%; (b) 1440 rpm; (c) 1.55 kW; (d) 38.31 kW, 38.31 kW, 96%]

8.15 A 415 V, 50 Hz, eight-poles, three-phase induction motor draws 50 kW from the supply when running at 720 rpm. The motor has a stator loss of 750 W and friction and windage losses are 1500 W. Determine the (a) slip, (b) gross power in the rotor, (c) metric horse power, and (d) efficiency.

[(a) 4%, (b) 49.25 kW, (c) 62.24 hp, (d) 91.56%]

8.16 A 30-hp (metric), four-poles, 415 V, 50 Hz, three-phase induction motor is supplying full load at 0.866 power factor and 1410 rpm. The friction and windage losses are 1200 W and the copper loss in the stator is 900 W. Calculate the (a) load current, (b) slip, (c) rotor copper loss, (d) power drawn from the supply, (e) current supplied, and (f) efficiency of the motor.

[(a) 35.45 A, (b) 6%, (c) 1.454 kW, (d) 25.62 kW, (e) 41.16 A, (f) 86.12%]

8.17 A three-phase 30 metric horse power, 415-V, 50-Hz, eight-pole induction motor has a stator and rotor impedance of $(0.3 + j0.8)$ Ω/phase and $(0.2 + j0.6)$ Ω/phase, respectively. Both the stator and rotor are star connected and the voltage trans formation ratio is 415/325. Determine the starting torque when an external resistor of 1.2 Ω/phase is inserted in each rotor phase.

[509.07 Nm]

8.18 A three-phase slip ring motor has a star-connect rotor with an impedance of $(0.6 + j\ 4.5)$ Ω/phase. Calculate the rotor current of the motor (i) at standstill when an impedance of $(3.4 + j\ 3.5)$ Ω/phase is connected in star across the slip rings of the motor and (ii) running normally at a slip of 4%. Assume that a voltage of 60 V is induced between the slip rings on open circuit.

[(i) 3.87 A, (ii) 2.21 A]

8.19 A three-phase induction motor is sequentially started (a) by directly connecting it to the supply, (b) through a delta–star starter, and (c) by connecting it to an autotransformer having a tapping of a%. Compare the starting line currents and toques in the three cases.

$$\left[1 : \frac{1}{\sqrt{3}} : a^2, 1\frac{1}{3} : a^2\right]$$

Objective Type Questions

1. Which of the following is another name for an induction motor?
   (i) synchronous   (ii) asynchronous
   (iii) linear   (iv) none of these

2. Which of the following types of construction is used for the rotor of an induction motor?
   (i) solid non-cylindrical
   (ii) laminated non-cylindrical
   (iii) solid cylindrical
   (iv) laminated cylindrical

3. The winding of a squirrel cage rotor is
   (i) similar to the stator winding
   (ii) solid uninsulated conductors
   (iii) solid insulated conductors
   (iv) laminated bare conductors

4. A starting torque in an induction motor is produced due to

   (i) the rotating magnetic field set up by the three-phase stator current
   (ii) the magnetic field induced in the rotor
   (iii) the interaction between the rotating stator and magnetic fields
   (iv) none of these

5. The frequency of the induced emf in the rotor of an induction motor about to start is
   (i) equal to the supply frequency
   (ii) less than the supply frequency
   (iii) greater than the supply frequency
   (iv) none of these

6. Which of the following is not a valid synchronous speed?
   (i) 500 rpm   (ii) 750 rpm
   (iii) 2000 rpm   (iv) 3000 rpm

7. For a 10 pole induction motor, its speed will be

(i) 700 rpm
(ii) 600 rpm
(iii) above but close to 600 rpm
(iv) below but close to 600 rpm

8. Which of the following represents the speed of a three-phase, 60 Hz, 12-pole induction motor running at a slip of 0.02?
(i) 490 rpm   (ii) 500 rpm
(iii) 588 rpm   (iv) 600 rpm

9. The range of variation of slip in induction motors is
(i) 2% – 6%   (ii) 6% – 8%
(iii) 8% –10%   (iv) none of these

10. The frequency f_r of the induced emf in the rotor, in terms of the stator frequency f and slip s, is given by
(i) $f_r = sf$ (ii) $f_r = (1-s)f$
(iii) $f_r = f/s$ (iv) $f_r = s/f$

11. The magnitude of the current drawn by an induction motor on no load, as a percentage of the full load current, is
(i) up to 5%   (ii) up to 15%
(iii) up to 30%   (iv) up to 40%

12. Which of the following is a valid representation of the mechanical load?
(i) $r_2(1-s)$ (ii) $[r_2(1-s)]/s$
(iii) $r_2(1+s)$ (iv) $[r_2(1+s)]/s$

13. The power input to the rotor is given by
(i) $\sqrt{3}\,V_L I_L \cos\varphi_1$ (ii) $3V_P I_P \cos\varphi_1$
(iii) $\sqrt{3}\,V_L I_L \cos\varphi_1 - 3I^2_1 r_1$
(iv) $3V_P I_P \cos\varphi_1 - 3I^2_1 r_1 -$ core loss

14. Which of the following does not represent the gross mechanical power output?
(i) $3I'^2_2 r'_2 [(1/s)-1]$ (ii) $(1-s)P_{ag}$
(iii) $P_i - 3I^2_1 r_1 - P_C$ (iv) all of these

15. Which of the following is a correct statement for the torque developed by a 50-Hz, two-pole, 3-phase induction motor when it is run at a speed equal to or more than 2700 rpm?
(i) torque varies inversely as the slip
(ii) torque varies linearly with slip
(iii) torque varies as the square of the slip
(iv) none of these

16. In Question 15 which of the following statements is correct when the motor is run at speeds below 2700 rpm?

(i) torque varies inversely as the slip
(ii) torque varies linearly with slip
(iii) torque varies as the square of the slip
(iv) none of these

17. The unit of mechanical loss is
(i) voltage   (ii) current
(iii) watt   (iv) var

18. When full-rated voltage is applied to the stator of a three-phase induction motor, the electrical equivalent of the mechanical load has a value which
(i) is zero
(ii) is ∞
(iii) lies between zero to ∞
(iv) none of these

19. The starting current in an induction motor is high because
(i) the motor is on no load
(ii) the rotor resistance is low
(iii) the stator resistance is low
(iv) the rotor behaves like a short-circuited secondary

20. In the star–delta starter method the voltage applied across the stator winding is
(i) equal to the supply voltage
(ii) $\sqrt{3}$ times the supply voltage
(iii) $1/\sqrt{3}$ times the supply voltage
(iv) none of these

21. Which of the following is an advantage of rotor resistance starting?
(i) improved power factor at starting
(ii) high starting torque
(iii) low starting current
(iv) all of these

22. Which of the following is a disadvantage of an induction motor?
(i) ruggedness
(ii) almost constant speed from no load to full load
(iii) speed variation restricted under different operating conditions
(iv) simple construction

23. Which of the following is not an advantage of a wound motor compared to a cage motor?
(i) high starting torque
(ii) low starting current
(iii) speed variation through solid-state switching
(iv) maintenance of slip rings and brushes

24. With an autotransformer tapping of 30% the starting torque is 60 Nm. What is the starting torque when the autotransformer tapping is 75%?
    (i) 24 Nm   (ii) 60 Nm
    (iii) 150 Nm   (iv) 375 N m

25. The power factor of a three-phase induction motor generally lies in the range
    (i) 0.4 – 0.45   (ii) 0.5 – 0.6
    (iii) 0.65 – 0.75   (iv) 0.8 – 0.85

Answers

1. (ii)	2. (iii)	3. (ii)	4. (iii)	5. (i)	6. (iii)	7. (iv)
8. (iii)	9. (i)	10. (i)	11. (iv)	12. (ii)	13. (iv)	14. (iv)
15. (ii)	16. (i)	17. (iii)	18. (ii)	19. (iv)	20. (iii)	21. (iv)
22. (iii)	23. (iv)	24. (iv)	25. (iv)			

DIRECT CURRENT MACHINES

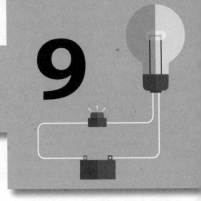

9

Learning Objectives

This chapter will enable the reader to

- Illustrate the constructional features of the various parts of dc machines
- Explain the generation of dc voltage
- Obtain an expression for the generated voltage and develop an equivalent circuit for the dc generator
- Classify generators based on the types of connections of field and armature windings and obtain their characteristics
- Appreciate armature reaction and its effect on magnetic field
- Understand the principle of torque development in a dc machine operating as a motor
- Derive mathematical expressions for speed and torque of a dc motor
- Develop characteristics of series, shunt, and compound motors
- Familiarize with the need for starting dc motors and the use of a three-point starter
- Enumerate the various methods of speed control of dc motors

9.1 INTRODUCTION

Direct current (dc) machines are electromechanical energy conversion devices which can operate as generators converting mechanical energy given to them from a prime mover to electrical energy. They can also operate as motors, taking electrical energy from a dc supply system and converting it into mechanical energy of rotation and driving a mechanical load.

Direct current machines were the first rotating electrical machines that were introduced to drive locomotives way back in 1839. They are very versatile, as they can be designed to have a wide range of performance characteristics based on various combinations of field windings such as shunt, series, and separately excited field. As generators, dc machines can have a wide range of output voltage-load ampere characteristics, while as motors, they can have a wide range of speed-torque characteristics. As the speed of dc motors can be controlled easily, these motors are widely used in industries. Industrial processes involving electroplating and electrolysis use dc generators as sources of supply.

With the tremendous development in solid-state technology in the recent past, dc power supply units using ac to dc converters of kW ratings varying from a few to hundreds of kilowatts are gaining wide acceptance. Further, the speed control of ac motors over a wide range of speeds is also possible because of the development of solid-state ac drives. Hence, drive systems previously associated exclusively with dc motor drives now have an alternative in ac drive systems. However, dc machines will continue to be used because they have relatively simpler and more flexible drive systems compared to ac machines. Presently, dc motors are being used in electric traction, cranes, etc.

9.2 CONSTRUCTION OF DC MACHINES

A dc machine consists of a stator and a rotor with a uniform air gap in between. The stator, in case of dc machines, invariably carries the field windings and the rotor carries the armature windings. Figure 9.1 illustrates the general arrangement of a four-pole dc machine, showing its main parts. The constructional features of the main parts are discussed in this section.

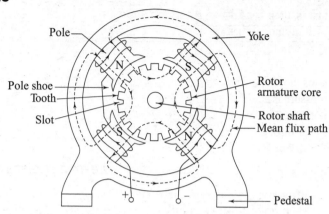

Fig. 9.1 General arrangement of a four-pole dc machine

9.2.1 Stator

The stator is mainly constituted of field poles, yoke, interpoles, etc. In the following section, various constituents are described.

9.2.1.1 Field Poles and Yoke

Pole cores are made of cast steel bolted to the yoke and usually not laminated. In some machines pole shoes are laminated and composed of electrical grade cast steel sheets of thickness 1 mm and insulated from one another. The pole core laminations are riveted together and attached to the yoke, secured in place either with the help of bolts or using a dovetail joint. The core supports the field windings, which produce the necessary magnetic field. The field windings are wound around the cores such that they set up alternate north and south poles. The pole shoe is so designed as to produce a uniform distribution of flux in the air gap. Pole shoes are bolted or dovetailed to the core in the case of solid poles, or secured by means of projections punched integrally with the laminations of the pole cores. The pole face or pole shoe is curved, and is wider than the pole core in order to spread the flux more uniformly in the air gap. The pole shoe is always laminated to avoid eddy current losses caused by the fluctuation in the flux distribution on the pole face due to the movement of the armature slots and teeth. In addition, the pole shoe provides support to the field windings.

The field windings consist of a few turns of wire for a series field and a large number of turns of fine wire for a shunt field and are supported on the field poles. The field coils act as electromagnets that produce the flux in the air gap needed to generate an emf.

The yoke is a part of the frame and provides the return path for the flux from one pole to another. It is made of cast steel or fabricated from rolled steel. For small machines, it is made of cast iron. In large and moderate-sized machines, the yoke is usually split along a horizontal diameter for convenience in assembling and repairing.

9.2.1.2 Interpoles

The interpoles and their windings (not shown in Fig. 9.1) are mounted on the yoke and are located in the interpolar region between the main poles. The interpole is generally smaller compared to the poles. The interpole winding is composed of a few turns of thick wire, since it is connected in series with the armature circuit, so that the magnetomotive force is proportional to the armature current.

9.2.1.3 Brushes, Brush Holders, and Rocker Ring

The rotating armature and the external circuit are connected through brushes. Brushes are employed to collect current from the commutator and deliver it to the load. These are rectangular and are made of carbon, carbon graphite, graphite, metal graphite, or copper, depending upon the hardness required to meet the needs of commutation. The graphite in the brush serves to lubricate the commutator, which takes on a polished surface dark brown in colour. Most motors have radially placed brushes, that is, their centre line is radial to the commutator.

Individual brushes are supported in metal brush holders, as shown in Fig. 9.2. Studs attached to the rocker ring support the holders. A brush is inserted into a holder, where a clock-type phosphor bronze spring presses it against the commutator. The pressure is adjusted by moving the adjusting lever into different notches. The connection between the brush and brush holder is made through a flexible lead of braided copper wire, called a *pig-tail* or *shunt*, as shown in Fig. 9.2.

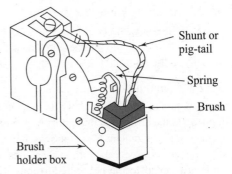

Fig. 9.2 Brush holder assembly

9.2.1.4 Compensating Windings

Compensating windings are optional (not shown in Fig. 9.1). They are connected in the same way as the interpole windings but are located in the axial slots of the field pole shoe.

9.2.1.5 End Bell Structure and Bearings

Mechanically connected to the yoke are end bells containing the bearings that support the armature shaft. The end bells are structural covers that enclose the machine. In some machines, the end bells support brush rigging, which in turn support the brushes.

9.2.2 Rotor

The rotor is essentially constituted of an armature core, armature windings, and a commutator. The following subsections describe the various parts of the rotor and their functions.

9.2.2.1 Armature Core

The armature core is the rotating part of a dc machine. It is assembled from steel laminations having a thickness of about 0.35–0.5 mm. In order to reduce eddy current losses, the laminations are insulated from each other.

Slots are punched at the periphery of the core. Core holes up to a diameter of 40 cm are generally made in one piece, which is usually keyed directly to the shaft and punched with holes near the shaft to form longitudinal ventilating passages as shown in Fig. 9.3. Cores of larger diameter are assembled on a cast steel spider. Radial ventilating ducts through the core are formed by means of spacers placed at intervals of 5–10 cm. The width of the ventilating ducts varies from 0.5 to 1 cm. Typical shapes of the teeth and slots are shown in Fig. 9.4. The slots are usually placed parallel to the axis of the armature, but are also sometimes skewed at a small angle to the axis to avoid the vibrations of the teeth.

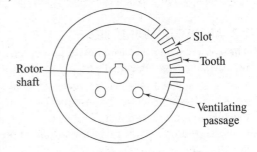

Fig. 9.3 Single armature stamping

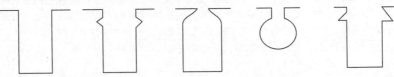

Fig. 9.4 Typical shapes of slots and teeth

9.2.2.2 Armature Windings

The armature winding coils (not shown in Fig. 9.1) are placed in the slots in various arrangements. The coils are insulated from one another and from the armature core. The armature winding coil ends are connected suitably to the commutator segments.

9.2.2.3 Commutator

The function of the commutator is to convert the ac voltages induced in the armature conductors to dc voltage in the external circuit in generator operation, whereas in dc motors it produces unidirectional torque. The ends of the armature coils are joined to the commutator consisting of a small number of wedge-shaped segments or bars of hard-drawn or drop-forged copper. Commutator segments are insulated from each other with the help of thin (0.5 – 1.0 mm thick) layers of mica strips, and from their supporting rings using mica cones and gaskets. The segments or bars are assembled side by side to form a ring. The commutator is pressed onto the shaft. The outer surface of the commutator is machined to provide a smooth surface so as to maintain a continuous contact with the stationary brushes, which press against it. The assembly of the commutator requires great care, since a little unevenness on its surface, or a slight eccentricity,

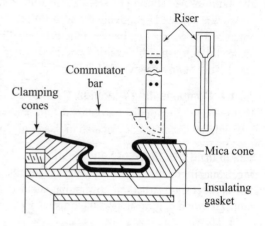

Fig. 9.5 Axial cross-section of a cylindrical commutator

can cause undue sparking between the commutator segments and brushes. Figure 9.5 shows the end view of a commutator.

The armature winding is joined to the commutator in various ways. If the armature and commutator diameters do not differ considerably, the winding ends are directly soldered to the commutator bars. Otherwise, they are soldered with risers as shown in Fig. 9.5.

9.3 ARMATURE WINDINGS

The rotor carries the armature winding in a dc machine. The winding consists of straight conductors arranged parallel to the shaft, housed in slots and grouped round the periphery of the armature core. The induced emf per conductor is small—a fraction of a volt for a small machine and only a few volts for larger machines; consequently many conductors are required. The conductors are connected together to form the winding. The armature conductors are diamond-shaped coils and usually have multiple turns. They are so shaped that if one side of a coil occupies the lower half of a slot, then the other side of the same coil will occupy the upper half of another slot. Thus, each slot accommodates two layers of coils. Such an arrangement is called a *double-layer armature winding*. The armature windings in dc machines are invariably double-layered.

When a coil is rotated inside a magnetic field, an emf is generated on either side of the coil, and it is necessary that the emf generated on both the sides has the same direction. This is possible when the two sides of the coil are one pole pitch apart, that is, when one side of the coil is under the north pole and the other side under the south pole. Mathematically, the coil span can be obtained as follows:

$$\text{Coil span} \approx \frac{\text{total number of slots}}{\text{total number of poles}}$$

The coil span must be a whole number. The ends of the coils are brought out and connected to the commutator segments. Each segment of a commutator has two ends of different coils connected to it. Thus,

Number of commutator segments = number of armature coils

Depending upon how the ends of the armature coils are joined to the commutator, armature windings can be classified into two categories: (i) lap winding and (ii) wave winding.

9.3.1 Lap Windings

In lap winding, the two ends of a coil, designated 'start' and 'finish', are connected to adjacent commutator segments. Figure 9.6 shows the connections of the coils to form a lap winding. Since the sides of successive coils overlap each coil, this winding is known as lap winding.

In Fig. 9.6(a), the 'finish' of each coil is connected to the commutator segment ahead of the 'start' of the coil; therefore, such a connection of coils is known as *progressive lap winding*. In Fig. 9.6(b), the connection of the coils is known as *retrogressive lap winding*, because here the 'finish' of the coil is connected to the commutator segment behind the 'start' of the coil. The winding finally closes onto itself.

A lap winding has as many paths in parallel be-

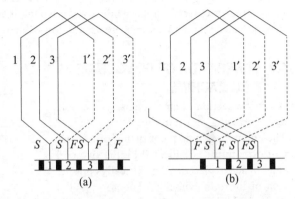

Fig. 9.6 Coils connected in lap winding: (a) Progressive (b) Retrogressive

tween the positive and negative brushes as the number of poles. Thus, for a lap-wound armature, the total current at the terminals of the dc machine is P times the current in one path of the winding, where P is the total number of poles of the dc machine.

9.3.2 Wave Windings

Figure 9.7 shows the coils connected in wave winding. From the figure, it can be seen that the 'finish' of one coil and the 'start' of another coil are connected to the same commutator segment. In this manner, the coils are progressively connected and the winding goes around all the poles till it returns to the coil side from where it started.

A wave winding has only two paths in parallel between the positive and negative brushes irrespective of the number of poles. Thus, for a wave-wound armature, the total current at the terminals of the dc machine is twice the current in one path of the winding. Table 9.1 gives a comparison of the characteristics of lap and wave winding connecions.

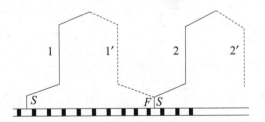

Fig. 9.7 Coils connected in wave winding

Table 9.1 Comparison of lap and wave winding

Quantity	Lap winding	Wave winding
Number of parallel paths a	Equal to the number of poles P	2 (independent of the number of poles)
Number of brushes	Equal to the number of poles P	Minimum 2, but actually P brushes are used.
Current in each coil	I_a/a, where I_a is the armature current	$I_a/2$, where I_a is the armature current

From Table 9.1, it can be seen that for a given current density and conductor cross section, the total armature current obtained from a machine connected in lap winding is $P/2$ times that obtained from the machine when it is connected in wave winding. On the other hand, for a given number of armature conductors, the numbers of conductors connected in series per path in a machine whose coils are connected in wave winding is $P/2$ times that when the machine is connected in lap winding. Thus, the output voltage obtained between the positive and negative brushes in a machine connected in wave winding is $P/2$ times that obtained when it is connected in lap winding. Generalizing, it can be stated that under similar conditions, a lap-wound machine gives a higher current output, while a wave-wound machine gives a higher voltage output. Alternatively, it may be stated that lap windings are used in low-voltage and heavy-current machines.

9.4 GENERATION OF DC VOLTAGE IN A DC MACHINE

A vertical section of an elementary two-pole dc generator is shown in Fig. 9.8.

The armature winding, consisting of a single coil of N turns, is indicated by the two coil sides a_1 and a_2 placed in diametrically opposite slots on the rotor, with the conductors parallel to the shaft. The rotor is revolved at a constant speed by a source of mechanical power (prime mover) connected to the shaft. The air-gap flux distribution usually approximates a flat-topped wave as shown in Fig. 9.9(a). The rotation of the coil induces a voltage, which varies as a function of time and has the same waveform as the spatial flux-density distribution.

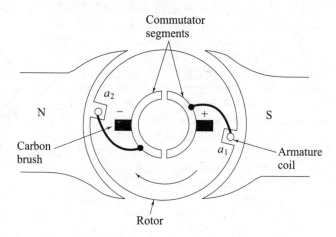

Fig. 9.8 Elementary dc generator with commutator

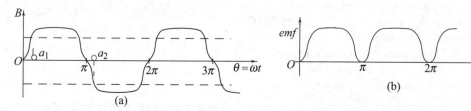

Fig. 9.9 (a) Flux-density distribution in space of an elementary dc generator and
(b) Waveform of voltage between the brushes

The voltage induced in an individual armature coil is an alternating voltage, which must therefore be rectified. In conventional dc machines, rectification is done mechanically using a commutator (shown in Fig. 9.8 for an elementary generator). The commutator is mounted on a rotor shaft and insulated from it. Stationary carbon brushes, which connect the winding to the external armature terminals, are held against the commutator segments.

9.4.1 Commutator Action

To understand commutator action in converting the generated ac voltage into dc voltage, the elementary two-pole generator shown in Fig. 9.8 is considered, which has been redrawn with the conductor slots in the vertical position as shown in Fig. 9.10(a). In this position, since the tangential velocity of the armature is parallel to the magnetic field, no voltage is generated and the brushes B_1 and B_2 are neutral.

When the generator rotates in the clockwise direction, conductor a_1 moves under the S pole. Application of Fleming's right hand rule in Fig. 9.10(b) shows that the direction of the induced emf in a_1 is such that if brushes B_2 and B_1 were connected to an external load resistor, current would flow from brush B_2 to brush B_1 via the load resistor. Thus, B_2 acquires a positive polarity and, correspondingly, B_1 has a negative polarity. The direction of the induced emf and, therefore, the polarities of both B_2 and B_1 remain unchanged as a_1 moves from 0° to 180° as can be seen from Figs 9.10(a) to (d). However, the magnitude of the induced emf in a_1 is maximum at $\theta = 90°$ and zero at $\theta = 180°$. At $\theta = 225°$, conductor a_1 is under the N pole [see Fig. 9.10(e)] and once again application of the Fleming's right hand rule will show that the direction of the induced emf will be such that the current will continue to flow from brush B_2 to the load resistor and then into the generator at B_1. It is easily seen that the direction of the induced emf in a_1 does not change as it moves under the N pole. As in the previous case, the magnitude of the induced emf will be a maximum, but negative, and zero at $\theta = 270°$ and at $\theta = 360°$, respectively.

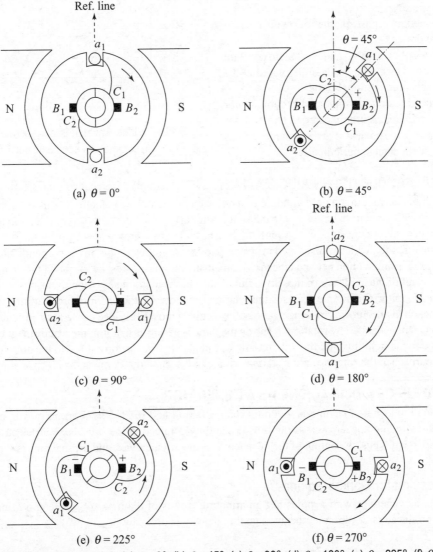

Fig. 9.10 Commutator action: (a) $\theta = 0°$, (b) $\theta = 45°$, (c) $\theta = 90°$, (d) $\theta = 180°$, (e) $\theta = 225°$, (f) $\theta = 270°$

From the foregoing, it is seen that the voltages at the brushes are unidirectional or dc. The waveform of the voltage obtained across the brushes is shown in Fig. 9.11(b) and has the same flat topped shape of the air gap flux distribution shown in Fig. 9.11(a). In practical machines, the waveforms of both the air gap flux distribution and the induced emf are made as nearly sinusoidal as feasible, by distributing the winding coils into a large number of slots distributed along the periphery of the armature and increasing the number of conductors per coil. Figure 9.11 shows waveforms of the induced emf, which appears across the commutators segments and the brushes.

Based on the above discussion, the following conclusions may be drawn:

(i) Polarities of brushes B_1 and B_2 remain constant with time. Hence, the voltage delivered at the brushes is unidirectional or dc.

(ii) Since conductor a_1 is connected to commutator segment C_1 and a_2 is connected to commutator segment C_2, the polarities of the voltages across the segment change periodically.

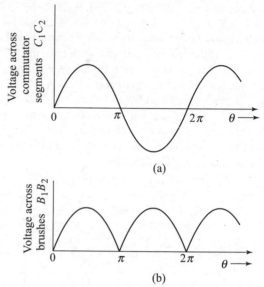

Fig. 9.11 Voltage waveforms of generated emf across (a) commutator segments and (b) brushes

9.5 TORQUE PRODUCTION IN A DC MACHINE

The flow of direct current in the field winding of a dc machine creates a magnetic flux distribution called the field flux, which is stationary with respect to the stator. Similarly, the effect of the commutator in a dc machine is such that when direct current flows through the brushes, the armature creates a magnetic flux distribution called the armature flux, which is also fixed in space. The axis of the armature flux is perpendicular to the axis of the field flux. The interaction between these two flux distributions creates the torque of the dc machine. The torque is the result of the tendency of these two flux distributions to align along the same axis. If the machine is acting as a generator, this torque opposes the rotation produced by the driving torque of the prime mover. This phenomenon also conforms to Lenz's law, as the torque opposes the very cause of its production, that is, the emf and current generated by rotation. If the dc machine is acting as a motor, the electromagnetic torque is developed due to the field flux and the armature flux produced by the dc current fed to the armature from the external dc source, and the rotor armature starts to rotate in the same direction as the electromagnetic torque.

9.6 OPERATION OF A DC MACHINE AS A GENERATOR

When the armature of a dc machine is given mechanical input and rotated by a prime mover, emf induced in the armature winding appears across the brushes, and by connecting the brushes to an electrical load, dc current can be drawn. Thus, the output of the dc machine is electrical power, and the machine is said to operate as a generator.

9.6.1 Expression for Generated emf

Assume that a dc generator with P poles has an armature (rotor) of diameter D and length L metres. If the average flux produced is Φ Wb/pole, the average flux density is given by

$$B_{av} = \frac{\Phi P}{\pi DL} \text{ Wb/m}^2$$

If the speed of rotation of the armature is N rpm, its tangential velocity is written as

$$v = \frac{\pi DN}{60} \text{ m/se}$$

Use of Eq. (1.51) leads to the induced voltage/conductor as

$$B_{av}Lv = \frac{\Phi PL}{\pi DL} \times \frac{\pi DN}{60} = \frac{\Phi PN}{60} \text{ V}$$

If Z is the total number of armature conductors and a the number of parallel paths, then,

$$\text{Number of conductors in series per parallel path} = \frac{Z}{a}$$

The average emf E generated is given by

$$E = \frac{\Phi NPZ}{60\,a} \tag{9.1}$$

From Eq. (9.1), it can be seen that for a given machine, P, Z, and a are constant.

Then, $\quad E \propto \Phi$ if the speed N of the machine is constant

and $\quad E \propto N$ if flux Φ per pole is kept constant

From Fleming's right-hand rule, it is evident that the direction of the generated emf is dependent on the direction of the magnetic field and the direction of rotation. Thus, when the direction of either the magnetic field or the rotation is changed, the direction of the generated emf is changed, thereby changing the polarity of the brushes.

9.6.2 Expression for Electromagnetic Torque in a DC Machine

As discussed earlier (see Section 1.10.2), a current-carrying conductor experiences a force when placed inside a magnetic field. In the present case, each of the armature conductors carries a current of magnitude I_a/a ampere and is rotating inside a magnetic field of flux density B_{av} Wb/m^2. The consequent torque produced in the dc machine is called the electromagnetic torque and an expression for it may be derived in the following manner.

$$\text{Force on a single conductor} = B_{av}\frac{I_a}{a} \times L = \frac{\Phi P}{\pi DL} \times \frac{I_a}{a} \times L = \frac{\Phi P}{\pi D} \times \frac{I_a}{a} \text{ V}$$

$$\text{Electromagnetic torque on single conductor} = \frac{\Phi P}{\pi D} \times \frac{I_a}{a} \times \frac{D}{2} \text{ Nm}$$

$$\text{Total electromagnetic torque on } z \text{ conductors } T = Z\frac{\Phi P}{\pi D} \times \frac{I_a}{a} \times \frac{D}{2} = \frac{\Phi PZ}{2\pi a}I_a \text{ Nm}$$

It is seen that for a given machine, Φ, P, Z, and a are constant and $T \propto \Phi I_a$

In terms of the generated voltage, the electromagnetic torque T may be expressed as

$$T = \frac{60E}{2\pi N}I_a \text{ Nm} \tag{9.2}$$

Equation (9.2) is often written in the more familiar form as

$$EI_a = \frac{2\pi NT}{60} \text{ W}$$

The expression for electromagnetic torque developed in Eq. (9.2) is applicable for both generator and motor action. However, in the case of a dc machine operating as a generator, the direction of the torque is to oppose the direction of rotation of the armature and EI_a represents the generator output in watts. On the other hand, when the dc machine is operating as a motor, the direction of the torque is in the direction of rotation of the armature and EI_a represents the input, in watts, to the motor.

Example 9.1

(a) A six-pole, wave-connected armature has 200 conductors and runs at 1500 rpm. The electromotive force generated in the open circuit is 600 V. Find the useful flux per pole.
(b) An eight-pole, lap-connected armature has 800 conductors, a flux of 0.04 Wb per pole, and a speed of 500 rpm. Calculate the emf generated in the open circuit.
(c) If the armature in (b) is wave-connected, at what speed must it be driven to generate 400 V?
(d) A four-pole generator has a flux of 0.05 Wb per pole and a lap-connected armature with 600 conductors. Find the emf generated in the open circuit at 800 rpm.

Solution   The induced emf E can be calculated using Eq. (9.1), which gives

$$E = \frac{\Phi NPZ}{60a}$$

(a) Data: $E = 600$ V, $P = 8$, $N = 1500$ rpm, $Z = 200$, and $a = 2$ for the wave winding. Then,

$$\Phi = \frac{600 \times 60 \times 2}{1500 \times 6 \times 200} = 0.04 \text{ Wb}$$

(b) Data: $\Phi = 0.05$ Wb, $P = 8$, $N = 500$ rpm, $Z = 800$, and $a = P = 8$ for the lap winding. Then,

$$E = \frac{0.05 \times 8 \times 500 \times 800}{60 \times 8} = 333.33 \text{ V}$$

(c) Data: $E = 400$, $a = 2$ for the wave winding, and the remaining data as per (b) is $\Phi = 0.05$ Wb, $P = 8$, and $Z = 800$. Then,

$$N = \frac{400 \times 60 \times 2}{0.05 \times 8 \times 800} = 150 \text{ rpm}$$

(d) Data: $\Phi = 0.05$ Wb, $P = 4$, $N = 800$ rpm, $Z = 600$, and $a = P = 4$ for the lap winding.

Then, $$E = \frac{0.05 \times 800 \times 4 \times 600}{60 \times 4} = 400 \text{ V}$$

Example 9.2   A six-pole dc generator has an armature of diameter 20 cm and length 25 cm. The armature is wave connected with 250 conductors and it runs at 800 rpm. If the average flux density is 0.90 Wb per m², calculate (i) the torque developed and (ii) the power output by the generator when a resistor of 5 Ω is connected across its brushes.

Solution   Data: $D = 0.2$ m, $L = 0.25$ m, $P = 6$, $Z = 250$, $B_{av} = 0.90$ Wb/m², $N = 800$ rpm, $a = 2$, and load = 50 Ω
Flux per pole, $\Phi = 0.90 \times 0.2 \times 0.25 = 0.045$ Wb

Using Eq. (9.1), the generated voltage, $E = \dfrac{\Phi PNZ}{60a} = \dfrac{0.045 \times 6 \times 800 \times 250}{60 \times 2} = 450$ V

The load current supplied to the resistor is the current flowing through the armature. Hence, the armature current, $I_a = 450/50 = 9.0$ A

(i) From Eq. (9.2), the generator torque, $T = \dfrac{60E}{2\pi N}I_a = \dfrac{60 \times 450}{2\pi \times 800} \times 9 = 48.34$ Nm

(ii) Power output of the generator $= EI_a = 450 \times 9 = 4050$ W

9.6.3 Equivalent Circuit of a DC Generator

A dc generator may be represented by a schematic circuit diagram as shown in Fig. 9.12(a). In the diagram, the field winding $F_1 F_2$ is shown to be excited from a dc voltage source, and produces a flux Φ required to generate the emf E in the armature $A_1 A_2$ of the generator. The rotor armature is shown enclosed in a rectangle of dashed lines as consisting of a source of emf E, generated in accordance with Eq. (9.1) in series with the armature resistance r_a. This armature is rotated by mechanical input from a prime mover (not shown in the diagram). From the schematic circuit diagram, the terminal voltage V can be obtained as

$$V = E - I_a r_a \qquad (9.3)$$

where I_a is and the armature current.

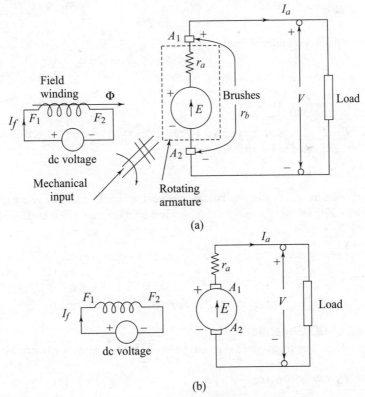

(a)

(b)

Fig. 9.12 (a) Complete schematic circuit diagram of a dc generator and (b) equivalent circuit of a dc generator

In computing the magnitude of the terminal voltage V, a voltage drop due to the brush contact resistance r_b at the brushes A_1, A_2, which is assumed constant and is of the order of 2 V, is usually accounted for and subtracted from the armature emf E. For the purpose of simplicity, the brush contact resistance r_b is included in the armature resistance r_a, and the equivalent circuit of the dc generator is then represented by a source of emf E in series with r_a, as shown in Fig. 9.12(b).

The power developed by the generator, P_g, is given y

$$P_g = E I_a \qquad (9.4)$$

The power supplied to the load, P_L, is

$$P_L = V I_a \qquad (9.5)$$

Example 9.3 A dc generator, driven at 500 rpm, is connected to a 200-V bus and is supplying a load of 25 kW. If the armature resistance, including the brush resistance, is 0.15 Ω, compute the (i) generated voltage, (ii) torque developed, (iii) power input, and (iv) power input and output at a speed of 420 rpm. Assume constant flux and neglect iron, friction, windage losses, etc.

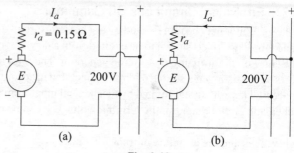

Fig. 9.13

Solution The equivalent circuit of the generator is shown in Fig. 9.13.

Armature current, $I_a = \dfrac{25 \times 1000}{200} = 125$ A

(i) From the equivalent circuit, generated voltage, $E = V + I_a r_a$

$$= 200 + 125 \times 0.15 = 218.75 \text{ V}$$

(ii) From Eq. (9.2), the developed generator torque is given by

$$T = \frac{60E}{2\pi N} I_a = \frac{60 \times 218.75}{2\pi \times 500} \times 125 = 522.23 \text{ Nm}$$

(iii) Power input $= E \times I_a + I_a^2 r_a$ (loss in armature) $= 218.75 \times 125 + (125)^2 \times 0.15 = 29.69$ kW

Since $E \propto N$, the voltage generated at 420 rpm is as follows:

$$E = \frac{420}{500} \times 218.75 = 183.75 \text{ V}$$

The generated voltage being lower than the bus voltage, the generator now draws current from the bus and functions as a motor. Figure 9.13(b) shows the flow of current when the generator is functioning as a motor. From the circuit diagram,

$$I_a = \frac{200 - 183.75}{0.15} = 108.33 \text{ A}$$

(iv) Power input $= 200 \times I_a = 200 \times 108.33 = 21.67$ kW

Power output $= 183.75 \times I_a - I_a^2 \times r_a = 183.75 \times 108.33 - (108.33)^2 \times 0.15 = 18.15$ kW

9.6.4 Classification of DC Generators

Direct current generators are classified based on the type of excitation employed to produce the required flux in the field circuit.

This classification is given below.

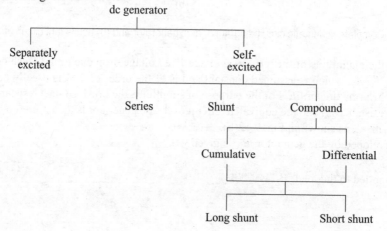

As seen from the classification of dc generators, there are two possible ways of connecting the field winding to a dc source. The field winding may be separately excited by connecting it to a dc source which is electrically independent of the machine, or self-excited by connecting it to the armature of the machine, as the armature itself is a dc source capable of supplying electric power to the load connected to it. In self-excited generators, the field winding is connected in a series, parallel, or series-parallel combination with the armature terminals. The initial flux is provided by the residual magnetism which is normally present in the field poles. When the machine is rotated, due to the residual magnetism, an emf of small magnitude is generated, which causes a small current to flow in the field winding. The flow of current in the field winding increases the magnetic flux, which in turn increases the magnitude of generated emf. The increased emf causes a current of higher magnitude, than the previous one, to flow in the field winding. In this manner the generator builds up to the rated voltage.

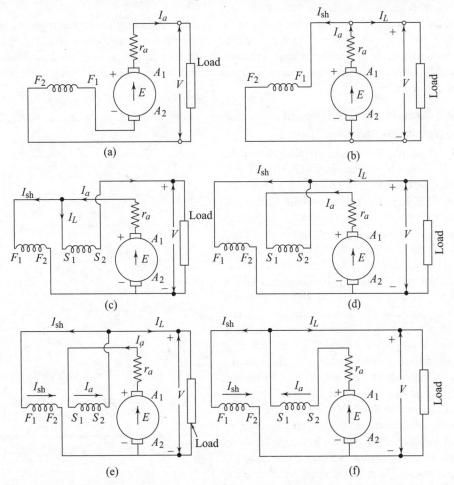

Fig. 9.14 Classification of dc generators: (a) series, (b) shunt, (c) short shunt, (d) long shunt, (e) cumulative compound, (f) differentially compound

The equivalent circuit diagram of a separately excited dc generator is shown in Fig. 9.12(b). By controlling the field current, both the magnitude of the flux-density wave and the generated voltage can be controlled.

When the control over the armature voltage and output is of special significance, separately excited dc generators are used.

Self-excited generators can be of the following three types.

(i) Series generator with field F_1F_2 connected in series with the armature A_1A_2 as shown in Fig. 9.14(a).

(ii) Shunt generator with field F_1F_2 connected in parallel with the armature A_1A_2 as shown in Fig. 9.14(b).

(iii) Compound generator with field having two sections, one field S_1S_2 in series and the other field F_1F_2 in parallel with the armature A_1A_2. Depending upon the electrical connection of the series field with respect to the armature and the shunt field, compound generators can be short shunt as shown in Fig. 9.14(c) or long shunt as shown in Fig. 9.14(d). Further, a compound generator can be cumulative as shown in Fig. 9.14(e) or differential as shown in Fig. 9.14(f) according to the relative magnetizing effects of series and shunt fields.

The permanent magnet generator is not included in the above classification. In this type of a generator, a permanent magnet is used to produce the necessary flux in the air gap of the dc machine. The flux produced has a constant magnitude. Permanent magnet generators, also known as *dynamos*, are low-power machines and, due to their compact construction, find applications in motorcycles, mopeds, etc. and as tachogenerators in motor control circuits.

9.6.5 Open Circuit Characteristics of a Separately Excited Generator

Figure 9.15(a) shows the connection diagram for determining the open circuit characteristics (OCC), also known as magnetization characteristic, of a separately excited generator. The generator is run at its rated speed, and by varying the field winding current I_f, the open circuit voltage E_{oc} is recorded. Figure 9.15(b) shows a plot of the field current I_f versus the open circuit voltage E_{oc}.

It was shown in Section 9.6.1 that if the speed of the armature remains constant, the generated emf E is proportional to the flux Φ. In the present situation, the flux Φ is produced by the field current I_f; therefore, Φ is proportional to I_f. Thus, E is proportional to I_f. When no current flows through the field winding, $I_f = 0$, a small emf is generated due to the residual magnetism in the magnetic circuit. As I_f is increased, the generated emf E increases. At lower values of the field current, the variation of E is linear along the air-gap line, represented by the dotted line tangential to the magnetization characteristic shown in Fig. 9.15(b). As I_f increases, due to magnetic saturation, the variation in E is no longer linear. Figure 9.15(b) shows the variation of E with the field current I_f for two different speeds. The dependence of E on speed N at constant excitation I_f is evident from the two curves shown in Fig. 9.15(b).

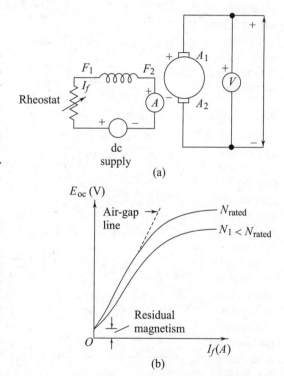

Fig. 9.15 OCC of a separately excited generator: (a) Connection diagram and (b) Magnetization characteristic

9.6.6 Open Circuit Characteristics of a Self-excited (Shunt) Generator

Figure 9.16(a) shows the connection diagram for determining the OCC of a shunt generator. The machine is run on no load, at its rated speed. By varying the resistance of the rheostat in the field circuit, the field current

I_f is varied and the voltage at the armature terminals is recorded for varying values of I_f. Since the generator is on no load, the voltage observed at the armature terminals is the induced generated voltage E. Therefore, I_f versus E is the magnetization curve and is shown in Fig. 9.16(b).

If R_f is the resistance of the field circuit, which includes the field winding resistance and the resistance of the rheostat, then $E = I_f R_f$. In Fig. 9.16(b), the R_f-line represents this relation between E and I_f with reference to the field circuit. The no-load voltage for a given field circuit resistance R_f is obtained by the intersection of the R_f-line and the magnetization curve. If R_f is varied by changing the resistance of the rheostat such that the R_f-line becomes tangential to the magnetization curve, the corresponding value of R_f is called the *critical resistance*. For values of R_f greater than the critical value, the generator will not excite because the value of the no-load voltage is very small, and consequently the magnitude of the field current is insufficient to produce the required flux.

Similarly, for a given R_f-line, the speed of the generator at which the R_f-line becomes tangential to the magnetization curve is called the *critical speed*. For a given value of R_f, the generator will not build up if it runs at a speed below the critical speed.

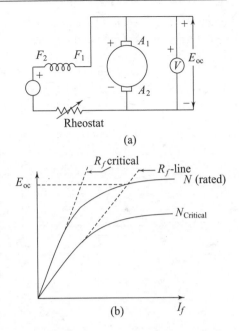

Fig. 9.16 OCC of a shunt generator: (a) Connection diagram and (b) No-load voltage characteristic

9.6.7 Armature Reaction

Armature reaction is the effect of the armature ampere-turns on the distribution of the magnetic flux in the air gap produced by the main field poles.

A two-pole dc machine with a uniformly distributed armature winding around the armature core is shown in Fig. 9.17(a). When the armature winding is not carrying current, it has no effect whatsoever on the main field magnetomotive force (mmf). F_f acts from the left to right, and the paths of the magnetic lines of force in the air gap are practically radial and uniformly distributed as shown in Fig. 9.17(a). A plane that is perpendicular to the stator field axis (or, alternately, a plane passing midway between any two magnetic poles) is defined as the *geometrical neutral plane* (GNP).

In Fig. 9.17(b), the field is unexcited, but the armature is shown as carrying a current. The brushes are placed along the GNP, so that all the conductors on the left-hand side of the brush axis carry currents into the paper, and all those on the right-hand side carry currents out of the paper. The armature clearly acts as an electromagnet, and the direction of its mmf, F_a, is vertically downwards. In the absence of the field mmf, the flux would take a vertical direction through the armature core, and the flux distribution is shown in Fig. 9.17(b). As the direction of this flux is at right angles to that due to the field winding, the flux due to the armature current is termed the *cross flux*. A plane passing through zero field is called the *magnetic neutral plane* (MNP). In Fig. 9.17(b), zero field lies along the two conductors that do not carry any armature current. Therefore, MNP lies along these two conductors. In the case of an armature carrying no current (or an unloaded machine), MNP coincides with GNP.

Figure 9.17(c) shows the actual condition when the field currents and armature currents act simultaneously, which occurs when the dc generator or motor is under load. The effect of the armature current is to distort the distribution of the air-gap flux density and displace the resultant field flux in the direction of rotation of the generator and against the direction of rotation of the motor. The resultant mmf F_t is shifted through an angle θ with respect to F_f, and therefore the magnetic neutral plane (MNP) is shifted through the same angle with respect to the GNP. Consequently, the armature mmf strengthens the symmetrical field flux, shown in Fig. 9.17(a), in the upper pole tip of the North Pole and in the lower pole tip of the South Pole. Similarly, the effect of the armature mmf is to weaken the field flux in the lower pole tip of the North Pole and in the upper pole tip of the South Pole.

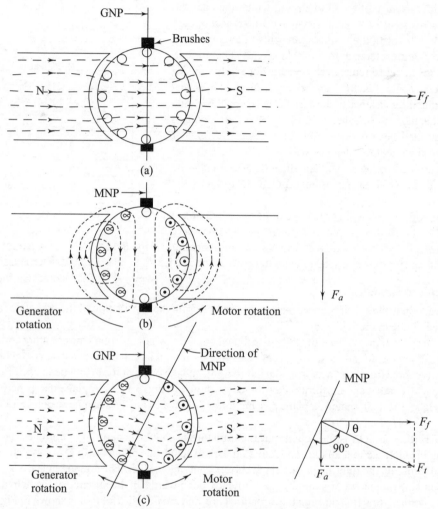

Fig. 9.17 Effect of armature current on the magnetic field of a dc machine: (a) Flux due to field mmf only (b) Flux due to armature mmf only and (c) Resultant flux due to field mmf and armature mmf

The degree of shift in the MNP is governed by the load carried by the machine. Another disadvantage of the shift in MNP is that an emf is generated in the coil that is undergoing commutation leading to sparking on the commutator surface. Such a situation can be avoided by shifting the brush axis so that it is aligned along the MNP as shown in Fig. 9.18(a).

As a result of the shift, the armature mmf F_a vector now acts along the brush axis and is shown in Fig. 9.18(b). The resolution of the vector F_a along and perpendicular to the GNP produces two components, namely, the cross magnetizing component F_{ac}, which is in quadrature with (or perpendicular to) the main field F_f and the demagnetizing component F_{ad}, which is in opposition to F_f. The effect of the demagnetizing component is to weaken the main field, which results in a reduction in both the generated emf and the terminal voltage of a generator with an increase in armature/load current. The effect of the cross magnetizing component, on the other hand, is to distort the flux distribution and reduce the field strength in the air gap, which again results in a reduction in the generated emf. In a dc motor, the effect of armature reaction is to twist or distort the flux against the direction of rotation.

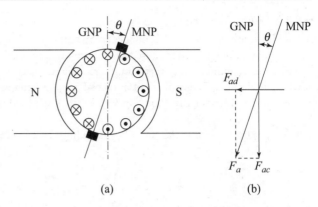

(a) (b)

Fig. 9.18 (a) Alignment of brush axis with MNP and (b) cross and demagnetizing effect of armature mmf

9.6.8 Characteristics of DC Generators

The basic characteristics of a dc generator are broadly classified into internal and external characteristics. The internal characteristic is a plot between the induced emf E and the load current I_L, with the dc machine being run at its rated speed N at no load, and the field circuit resistance held at a constant value.

The external characteristic is a plot between the terminal voltage and the load current, with constant excitation in the case of separately excited generators, and constant resistance in the field circuit for shunt and compound generators. These characteristics are plotted with the dc machine being run at its rated speed N.

9.6.8.1 Series Generator

In a series generator, the field winding is connected in series with the armature winding, as shown in Fig. 9.14(a). From the equivalent circuit it can be seen that the armature current I_a flows through both the field winding and the load. Since the series field winding carries the full armature current, it is designed to have a few turns and is made of thick wire or strips. The resistance of the series field winding is low and of the same order as the armature resistance. If R_s denotes the resistance of the series field winding, then

$$V = E - I_a(r_a + R_s) \tag{9.6}$$

The generated power P_g and the power P_L delivered to the load are given by

$$P_g = EI_a \tag{9.7}$$
$$P_L = VI_a \tag{9.8}$$

For series generators, the armature current, field current, and load current are the same. Therefore, the no-load and load characteristics can be determined only when the field is separately excited. The series generator running at constant speed, therefore, has only one characteristic, which is the external characteristic. The external characteristics of a dc series generator are shown in Fig. 9.19. In this figure, curve 1 is the magnetization characteristic (OCC). Owing to the effect of armature reaction, the flux with load for a given current is less than the flux with no load, and the induced emf E for a given current is less than the emf E_0 at no load. Curve 2 is the internal characteristic. Curve 3 is a straight line showing the internal voltage drop $I_a(r_a + R_s)$. The external characteristic is given by curve 4, which can be obtained by subtracting from the ordinate of curve E the voltage drop $I_a(r_a + R_s)$ for a given value of armature current I_a.

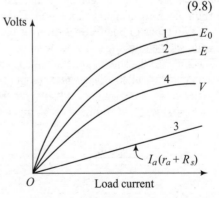

Fig. 9.19 Load characteristics of a series generator

9.6.8.2 Shunt Generator

The field winding in this case is connected in parallel with the armature winding. Thus, the voltage across the shunt field winding is the armature terminal voltage. In order to keep the shunt field current low, it is designed to have a large number of turns of thin wire. The resistance of a shunt field winding is high, of the order of a few hundred ohms. The equivalent circuit of a shunt generator is shown in Fig. 9.14(b).

From the equivalent circuit shown in Fig. 9.14(b), the load current I_L is given by

$$I_L = I_a - I_{sh} \tag{9.9}$$

where I_{sh} is the current in the shunt field winding. If R_{sh} is the resistance of the shunt field winding, then

$$V = E - I_a r_a \tag{9.10}$$

$$I_{sh} = \frac{V}{R_{sh}} \tag{9.11}$$

The expressions for P_g and P_L remain unchanged and are given by Eqs (9.7) and (9.8), respectively.

A shunt generator, on no load, generates an emf E whose value is decided by the field circuit resistance R_{sh}. The no-load terminal voltage $V_0 = E$ is shown in Fig. 9.20. If there were no armature reaction effect and no internal volt drop, the internal characteristic would be the horizontal, curve 1, through V_0. As the load current increases, the armature current also increases, and because of armature reaction, the flux per pole falls off, and so the internal characteristic droops somewhat, as shown by curve 2 of Fig. 9.20. Curve 3 is a straight line showing the internal voltage drop $I_a r_a$. The external characteristic is given by curve 4, which can be obtained by subtracting from the ordinate of curve E the voltage drop $I_a r_a$ for a given value of armature current I_a.

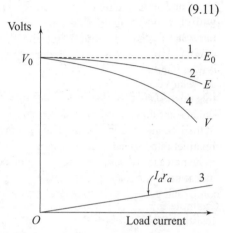

Fig. 9.20   Load characteristics of a shunt generator

9.6.8.3 Compound Generator

Both shunt and series field windings are employed to obtain a compound generator. The shunt field provides the main flux. However, the shunt field is designed to be somewhat weaker than the corresponding field in a shunt generator. The series field is considerably weaker than the corresponding field in a series generator. If the flux produced by the series field winding adds to the flux produced by the shunt field, the machine is called a *cumulative* compound generator, and if the flux produced by the series field opposes the flux produced by the shunt field, it is called a *differential* compound generator.

Two types of connections are feasible in a compound generator. In one type, the shunt field winding is connected across the armature terminal. Such a machine is called a *short shunt* compound generator. In the second type, the shunt field winding is connected in parallel with the armature and series field windings. Such a machine is called a *long shunt* compound generator. Equivalent circuits for short and long shunt compound generators are shown in Figs 9.14(c) and (d), respectively.

From Fig. 9.14(c) for a short shunt compound generator, the following expressions can be written:

$$I_a = I_{sh} + I_L \tag{9.12}$$

$$I_{sh} = \frac{E - I_a R_a}{R_{sh}} \tag{9.13}$$

$$V = E - I_a r_a - I_L R_s \tag{9.14}$$

Expressions for the developed power and the power delivered to the load are given by Eqs (9.7) and (9.8), respectively.

From Fig. 9.14(d) for a long shunt compound generator, the following expressions can be written:

$$I_a = I_{sh} + I_L \qquad (9.15)$$

$$V = E - I_a r_a - I_a R_s \qquad (9.16)$$

$$I_{sh} = \frac{V}{R_{sh}} \qquad (9.17)$$

Expressions for the developed power and the power delivered to the load are given by Eqs (9.7) and (9.8), respectively.

The nature of the external characteristic of a compound generator is dependent on the combined effect of the shunt and series fields (Fig. 9.21). The shunt field alone will give the drooping characteristic of curve 1, while the series field alone will give the rising characteristic of curve 2. The nature of the external characteristic depends on the relative strengths of the series and shunt fields. If the relative strengths of the two fields are adjusted such that the no-load terminal voltage is equal to the terminal voltage at full load as shown by the characteristic of curve 3, then the generator is said to be level compounded. With relatively strong series field windings, the terminal voltage may increase with increasing load current, and the generator is said to be over-compounded as shown by curve 4.

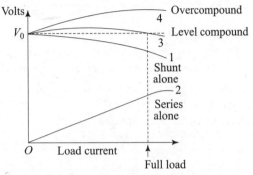

Fig. 9.21 Load characteristics of a compound generator

Example 9.4 A separately excited generator running at 1500 rpm supplies 250 A at 125 V to a circuit of constant resistance. What will the current be when the speed is dropped to 1200 rpm with the field current unaltered? The armature resistance is 0.05 Ω and the total drop at the brushes is 1.5 V. Ignore armature reaction.

Solution Load resistance $R_L = 125/250 = 0.5$ Ω
Generated emf $E = 125 + 200 \times 0.05 + 1.5 = 136.5$ V
As the field current remains constant, the flux Φ per pole is unchanged. Then from Eq. (9.1), $E \propto N$

The generated emf at 1200 rpm $E = 136.5 \times \dfrac{1200}{1500} = 109.2$ V

Let V be the terminal voltage at 1200 rpm. Then the armature current

$$I_a = \frac{V}{R_L} = \frac{V}{0.5} \text{ A}$$

But $\quad V = E - I_a r_a - \text{(voltage drop across brushes)}$

$$= 109.2 - \frac{V \times 0.05}{0.5} - 1.5 = 107.7 - 0.1 V = \frac{107.7}{1.1} = 97.91 \text{ V}$$

and $\quad I = \dfrac{97.91}{0.5} = 195.82$ A

Example 9.5 A series generator is connected as a booster between a station bus bar and a feeder of 0.40 Ω as shown in Fig. 9.22. The external characteristic of the generator is a straight line such that it has a terminal voltage of 60 V for a load current of 160 A. Determine the voltage between the far end of the feeder and the station bus bar when a current of (i) 150 A and (ii) 45 A flows through the feeder.

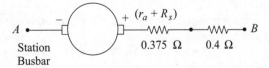

Fig. 9.22

Solution   From the external characteristic of the generator, the combined armature and series field resistance is given by

$$(r_a + R_s) = \frac{60}{160} = 0.375 \, \Omega$$

Determination of the voltage between the far end of the feeder B and the bus bar station A requires the calculation of the voltage drop, which for a current flow of *I* ampere can be expressed as $-0.375I + 0.4I = 0.025I$ V
(i) When $I = 150$ A, the voltage between points *A* and *B* is $0.025 \times 150 = 3.75$ V
(ii) When $I = 45$ A, the voltage between points *A* and *B* is $0.025 \times 45 = 1.125$ V

Example 9.6   A shunt generator has a no-load voltage of 250 V when running at a speed of 800 rpm. The terminal voltage drops by 8% when the generator is delivering full load. If the resistances of the armature and the field windings are 0.08 Ω and 92 Ω, respectively, compute the (i) output and (ii) input torque of the generator at full load.

Solution   Figure 9.23 shows the equivalent circuit of the shunt generator, with all the given data and the assumed current flows.

The terminal voltage on full load,
$$V = 250 \times (1 - 0.08) = 230 \text{ V}$$
The armature current,

$$I_a = \frac{E - V}{r_a} = \frac{250 - 230}{0.08} = 250 \text{ A}$$

Field current, $I_f = \dfrac{V}{R_f} = \dfrac{230}{92.0} = 2.5 \text{ A}$

Load current, $I_L = I_a - I_f = 250 - 2.5 = 247.5$ A

(i) Power output $= V \times I_L = 230 \times 247.5 = 56.93$ kW
(ii) The input torque is

$$T = \frac{60E}{2\pi N} I_a = \frac{60 \times 230}{2\pi \times 800} \times 247.5 = 679.49 \text{ Nm}$$

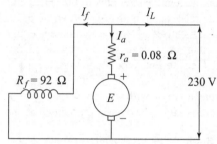

Fig. 9.23

Example 9.6   A long shunt compound generator delivers a load current of 50 A at 400 V and has armature, series field, and shunt field resistances of 0.04 Ω, 0.02 Ω, and 220 Ω, respectively. Calculate the generator's electromotive force and the armature current. Allow 1.5 V per brush for contact drop.

Solution   The equivalent circuit of the long shunt compound generator is shown in Fig. 9.24.

Shunt field current $I_{sh} = \dfrac{400}{220} = 1.818 \text{ A}$

Armature current $I_a = 50 + 1.818 = 51.818$ A

Armature voltage $E = V + I_a(r_a + R_s) +$ (voltage drop across brushes)

$$= 400 + 51.818 \times (0.04 + 0.02) + 2 \times 1.5$$
$$= 406.11 \text{ V}$$

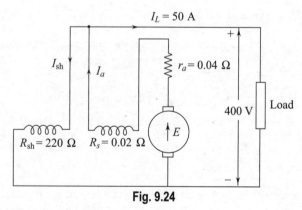

Fig. 9.24

9.7 OPERATION OF A DC MACHINE AS A MOTOR

If the armature terminals of a dc machine are connected to a dc voltage source, it begins to rotate and operate, like a motor, converting electrical energy into mechanical energy. Construction-wise, a dc motor is similar to a dc generator. Since the former has to operate in stringent environmental conditions, it has to be protected against moisture, fire hazards, chemical gases, and mechanical damages. Therefore, the frame of a dc motor is either fully or partially closed to provide sufficient protection, and is made flameproof.

9.7.1 Principle of Operation of a DC Motor

It was shown in Section 1.10.2 that when a current-carrying conductor is placed inside a magnetic field, it experiences a force [Eq. (1.47)], given by

$$F = BIl \text{ newton}$$

where B is the flux density in Wb/m^2, I is the current in amperes flowing in the conductor, and l is the effective length of the conductor in metres. The direction of the force on the conductor can be determined by applying Fleming's left-hand rule.

Figure 9.25(a) shows the armature of a two-pole dc machine connected to a dc voltage supply. It is assumed that the current in the conductors under the north pole is flowing into the plane of the paper and the current in the conductors under the south pole is flowing out of the plane of the paper, as shown in the figure. If Fleming's left-hand rule is now applied, it will be seen that the conductors, both under the north and the south poles, experience a force in the anticlockwise direction. Thus, the armature continues to rotate in the anticlockwise direction.

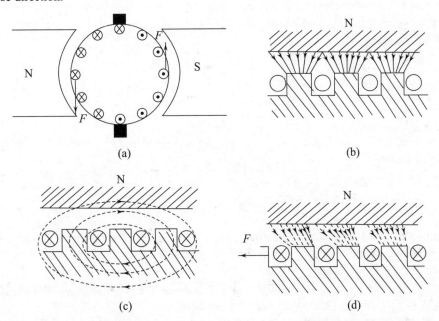

Fig. 9.25 Production of torque in a dc motor: (a) Force acting on the armature under the influence of the magnetic field, (b) Magnetic flux distribution without armature current, (c) Flux lines due to the armature current only (d) Resultant flux distribution

The physical process of torque production may be explained as follows. Figure 9.25(b) shows a few slots of a dc motor with its field and armature windings supplied by an outside source. It shows the paths taken by the lines of force crossing the air gap when there is no armature current. Figure 9.25(c) shows, in a very simplified form, the paths of the lines of force due to the armature currents alone. Figure 9.25(d) shows the combined effects of the main field and the armature field. It may be noted that the magnetic flux lines mostly concentrate through the teeth of the armature and not through the conductors. The resultant magnetic field in the air gap acts in an inclined direction, and consequently the lines of force entering the teeth are inclined in the same general direction. These tilted flux lines will tend to straighten themselves, as they behave like elastic fibres. Thus, the tops of the teeth are each pulled towards the pole face by an inclined force, which has a tangential component F acting in the direction shown in Fig. 9.25(d). Each force F multiplied by the radius gives a torque, and the sum of all these torques gives the total torque. Consequently, the armature will move in the direction of the torque.

9.7.2 Types of DC Motors

Based on the type of field connections, dc motors are also categorized in the same manner as dc generators. These are (a) permanent magnet motors, (b) separately excited motors, (c) series motors, (d) shunt motors, and (e) cumulative and differential compound motors.

9.7.3 Back emf in a DC Motor

When current-carrying armature conductors rotate within the magnetic field of a dc motor, the conductors cut the magnetic flux lines, thereby generating an emf. The direction of the generated emf, by Lenz's law, is opposite to that of the applied voltage V. This emf is called the back emf and is denoted by E_b. The equivalent circuit of a dc motor is shown in Fig. 9.26, in which the field circuit has purposely not been shown.

Since E_b is generated due to generator action, its magnitude is the same as that given by Eq. (9.1). Thus,

$$E_b = \frac{\Phi NPZ}{60a} \tag{9.18}$$

If the current supplied by the dc voltage source is I_a, by applying KVL to the equivalent circuit in Fig. 9.26, the following expression can be written:

$$V = E_b + I_a r_a \tag{9.19a}$$

$$\text{or} \quad I_a = \frac{V - E_b}{r_a} \tag{9.19b}$$

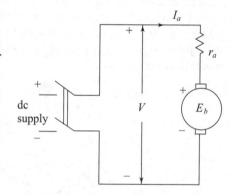

Fig. 9.26 Equivalent circuit of a dc motor

Equations (9.3) and (9.19a), which, respectively, provide expressions for the terminal voltage of a dc generator and a dc motor, can be combined and written in a generalized form as

$$V = E_a \pm I_a r_a \tag{9.20}$$

In Eq. (9.20), the positive sign will indicate motor action and the negative sign will indicate generator action. Since $E_b = E_a$, E_a in Eq. (9.20) is also used to designate E_b.

9.7.4 Speed of a DC Motor

For a given machine, P, Z, and a are constant. Thus, Eq. (9.18) can be rewritten as

$$E_b = k\,\Phi N \tag{9.21}$$

where k is a constant given by

$$k = \frac{PZ}{60\,a}$$

Equation (9.18) can be stated as

$E_b \propto \Phi$ for constant speed N

$E_b \propto N$ for constant flux Φ per pole

Substituting Eq. (9.21) in Eq. (9.19a) gives

$$V = k\Phi N + I_a r_a$$

or $$N = \frac{V - I_a R_a}{k\Phi}$$

$\therefore$ $$N \propto \frac{V - I_a R_a}{\Phi} \tag{9.22a}$$

or $$N \propto \frac{E_b}{\Phi} \tag{9.22b}$$

The armature voltage drop $I_a r_a$ is usually less than 5% of the applied voltage V. Neglecting the armature voltage drop, N may be expressed as

$$N \propto \frac{V}{\Phi} \tag{9.23}$$

Based on Eq. (9.23), it can be stated that the speed of a dc motor is approximately directly proportional to the applied voltage and inversely proportional to the flux.

9.7.5 Torque Developed in a DC Motor

If the torque developed by the armature of a motor to supply the mechanical load is T Nm and the motor is rotating at N rpm, then the mechanical power P_m developed is given by

$$P_m = \frac{2\pi NT}{60} \tag{9.24}$$

Multiplying Eq. (9.19a) by the armature current I_a, we get

$$VI_a = E_b I_a + I_a^2 r_a$$

or $$E_b I_a = VI_a - I_a^2 r_a \tag{9.25}$$

An inspection of Eq. (9.25) reveals that VI_a represents the electrical power input from the dc voltage supply and $I_a^2 r_a$ represents the heat loss in the armature winding. Therefore, $E_b I_a$ represents the power developed by the armature to supply the mechanical load on the shaft of the motor. However, the entire power represented by $E_b I_a$ is not available at the shaft, since a part of this power will be consumed in supplying frictional losses in the bearings, losses in the brushes, and eddy current and hysteresis losses in the magnetic core of the armature. Hence, if the losses are neglected, then

$$\frac{2\pi NT}{60} = E_b I_a$$

Substituting E_b from Eq. (9.18), the above expression becomes

$$\frac{2\pi NT}{60} = \frac{\Phi ZNP}{60a} I_a$$

$\therefore$ $$T = \frac{1}{2\pi} \Phi P \frac{Z}{a} I_a \tag{9.26}$$

For a given machine, P, Z, and a are constants. Hence Eq. (9.26) can be written as

$$T \propto I_a \Phi \tag{9.27}$$

It is seen from Eq. (9.27) that the torque developed by a dc motor is proportional to the product of the armature current I_a and the flux per pole Φ.

In respect of motor operation, the following conclusions may be drawn.

(i) The direction of the electromagnetic torque T developed by a motor is in the direction of rotation.

(ii) The direction of the load torque T_L is opposite to that of the electromagnetic torque T.

(iii) When $T > T_L$, the motor is *accelerating*, when $T = T_L$, the motor is operating at constant speed, and when $T < T_L$, it is *decelerating* or *braking*.

9.7.6 Characteristics of DC Motors

Based on the description of back emf developed and the speed relationship developed in the foregoing sections, the characteristics of various types of dc motors are explained in the following section.

9.7.6.1 Series Motors

In a series motor, the field winding is in series with the armature, and thus carries the same current I_a as shown in the equivalent circuit for a series motor in Fig. 9.27. It may be noted that I_a flows in opposition to the induced back emf E_b. If the supply voltage V is assumed to be constant, it may be represented by the dotted horizontal straight line shown by curve 1 of Fig. 9.28. The internal voltage drop is, for any current I_a, given by $I_a R_T$, where R_T is the total machine resistance, that is, the sum of the armature resistance r_a, series field resistance R_s, and the brush contact resistances, if any. This drop is represented by curve 2, which is a straight line. Subtracting the voltage drop $I_a R_T$ from curve 1, the graph of the back emf E may be obtained as shown by curve 3.

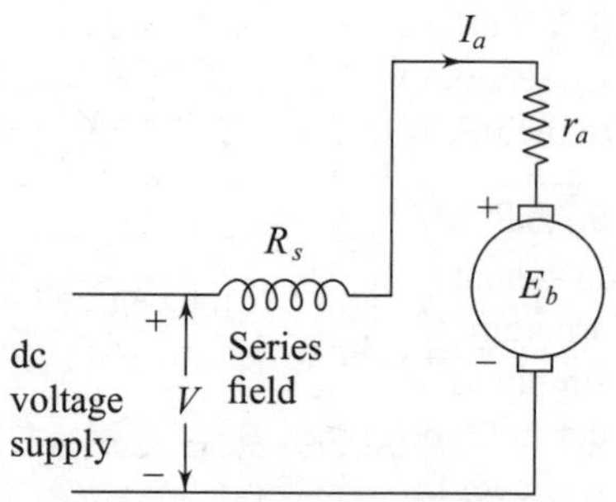

The magnetization characteristic of Φ against current I_a is illustrated by curve 4 in Fig. 9.28. For lower values of I_a, the variation of Φ is linear, whereas for higher values of I_a, due to magnetic saturation, the increase in Φ is very low with respect to the increase in I_a as may be noticed from the characteristic curve 4.

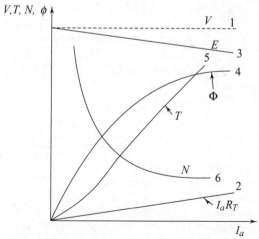

Fig. 9.27 Equivalent circuit of a dc series motor

The torque T developed in a dc motor is given by Eq. (9.27). In a series motor, the flux produced is directly proportional to the armature current. Thus, from Eq. (9.27),

$$T \propto \Phi I_a$$

or

$$T \propto I_a^2$$

Curve 5 of Fig. 9.28 shows the variation of torque with armature current. It can be seen that at low values of the armature current, since the magnetic circuit is unsaturated, the variation of torque is parabolic. As the current increases, due to magnetic saturation, there is no increase in flux Φ with increase in armature current. Therefore, $T \propto I_a$ and the characteristic becomes a straight line.

From Eq. (9.22), it can be seen that the speed is directly proportional to the voltage and inversely proportional to the flux. Thus,

$$N \propto \frac{V - I_a R_a}{\Phi} \propto \frac{E_b}{\Phi}$$

Fig. 9.28 Characteristics of a dc series motor

The armature current flows through both the series field and the armature winding, thereby producing a voltage drop. Since the series field windings and the armature windings have low resistances, the voltage drops due to these windings are negligible, and the applied voltage may be assumed to remain constant. Therefore, the speed N of the motor is inversely proportional to the flux Φ per pole. Expressed mathematically,

$$N \propto \frac{1}{\Phi}$$

A plot of speed versus armature current is represented by curve 6 of Fig. 9.28. At low values of the armature current, the increase in flux is proportional to the armature current; therefore, the decrease in speed is sharp, as it follows the rectangular hyperbolic relation. However, as the armature current increases, due to magnetic saturation, the flux tends to become constant and the speed characteristic becomes a straight line, with the speed reducing slightly because of the voltage drop in the series field and armature windings. It may be noted that at low armature currents, the flux is very small, and the speed is very high, being inversely proportional to the flux. Further, with a decrease in the load, the motor speed increases and is likely to develop to a high value. Under both conditions, very large centrifugal forces are set up, which can be dangerous and can cause damage to the rotating parts of the motor. Therefore, a series motor is never started with no load.

9.7.6.2 Shunt Motors

In a shunt motor, the field winding is connected directly across the armature terminals, which in turn are connected to a dc voltage supply as shown in Fig. 9.29. It may be seen that the armature current I_a flows in opposition to the back emf E_b. The line current I_L drawn from the supply is expressed as

$$I_L = I_a + I_{sh}$$

where I_{sh} is the current through the shunt field winding.

If the supply voltage V is assumed to be constant, it is represented by the dotted horizontal straight line shown by curve 1 of Fig. 9.30. The internal voltage drop is, for any current I_a, given by $I_a r_a$, where r_a is the armature resistance, which includes the

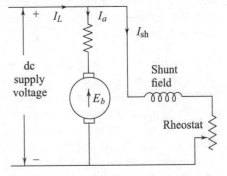

Fig. 9.29 Equivalent circuit of a shunt motor

brush contact resistances. This voltage drop is represented by curve 2, which is a straight line. Subtracting the voltage drop $I_a r_a$ from curve 1, the graph of back emf E_b may be obtained as shown by curve 3. This curve is a slightly drooping straight line.

For a given value of resistance in the shunt field circuit, the shunt current will be constant, and therefore the flux Φ will be constant. Now from Eq. (9.27),

$$T \propto \Phi I_a$$

But the flux Φ is a constant. Therefore,

$$T \propto I_a$$

Thus, the torque is proportional to the armature current. The graph of T versus I_a is a straight line passing through the origin if the flux is constant and the iron and friction losses are neglected, as shown by curve 4. When the friction loss, iron loss, and armature reaction are taken into account, the torque delivered to the shaft is slightly reduced; this is indicated by the dashed curve 4'. The speed

$$N \propto \frac{V - I_a R_a}{\Phi}$$

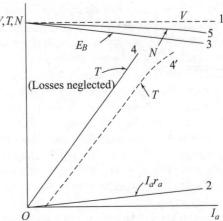

Fig. 9.30 Characteristics of a dc shunt motor

or
$$N \propto \frac{E_b}{\Phi}$$
$$\propto E_b$$

Thus, the speed is proportional to the back emf, as the flux is constant. The drop in speed from no load to full load is small, and the shunt motor may be regarded as a constant-speed motor. The graph of N versus I_a will be an almost horizontal line, drooping slightly at heavy loads as shown by curve 5.

Shunt motors due to their constant speed characteristic are most suited for applications where constant speed is desired, such as for driving lathes, milling machines, conveyors, fans, etc. Such machines are not convenient for use under varying load conditions, such as flywheels, or in parallel operation.

9.7.6.3 Compound Motors

Like a compound generator, a compound motor has both series and shunt field excitations. Figure 9.31 shows the equivalent circuit of a compound motor. There are two methods of operating this motor: (a) with shunt and series excitations helping one another, called *cumulatively compounded*; (b) with a series field excitation acting in opposition to the shunt field excitation, called *differentially compounded*.

Cumulatively compounded motors have characteristics which lie between those of series and shunt motors. As the load on the motor increases, the flux produced by the series field winding also increases, which increases the torque. On the other hand, an increase in the series flux when the load increases causes a drop in the speed of the motor. Thus with an increase in load, the increase in torque in a cumulatively

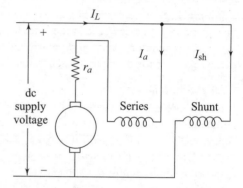

Fig. 9.31 Equivalent circuit of a compound motor

compounded motor is more than the normal straight-line increase in the torque in a shunt motor. On the other hand, the drop in speed in a cumulative compound motor is greater than that in a shunt motor when the load increases. The characteristics showing the variation of torque T and speed N with the variation of armature current are shown in Fig. 9.32.

It has been observed that a cumulatively compounded motor possesses the best characteristics of a series motor, in that it can be employed for starting heavy loads, and with light loads there is no risk of it acquiring dangerously high speeds. Simultaneously, the shunt field winding can be designed to provide a safe speed when the motor is run with no load. It can, therefore, be concluded that by adjusting the flux produced by the series and shunt field windings, a cumulatively compounded motor can be designed to provide the best characteristics of both the series motor and the shunt motor.

Cumulatively compounded motors are employed in applications that require a sudden application of heavy loads, as in lifts, mine hoists, rolling mills, and punching and shearing machines. These types of motors, however, are not useful in applications which require adjustable speeds by field control,

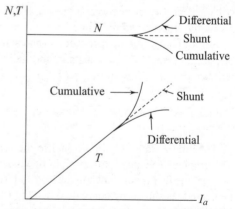

Fig. 9.32 Characteristics of a dc compound motor

since with a weakened shunt field, the series field becomes stronger, which may lead to unstable speeds with a change in load.

For differentially compounded motors, the flux produced by the series field is in opposition with the flux produced by the shunt field. Hence the speed of the compound motor is higher than that of the shunt motor for an increasing load. Similarly, with an increase in the load, the torque produced in a compound motor is less than that produced in a shunt motor. The characteristics showing the variation of torque T and speed N of a differential compound motor with the variation of armature current are given in Fig. 9.32. Due to problems in starting and difficulties during overloads, such motors are seldom used.

Example 9.8 A series motor draws a current of 35 A from a 220-V supply when operating at a speed of 1600 rpm. Compute the value of the resistor to be inserted to limit the speed (i) to 3200 rpm when the supply current is 15 A and the variation of flux between 15 A and 35 A is assumed linear and (ii) to 800 rpm when the supply current is 50 A and the flux at 50 A is 1.15 times the flux at 35 A. Determine the speed of the motor when it is directly connected to the voltage supply and it draws a current of 50 A. Assume the armature resistance of the motor to be 0.15 Ω.

Solution When the motor is running at 1600 rpm, the back emf E_{b1} from Eq. (9.19a) is given by

$$E_{b1} = 220 - 35 \times 0.15 = 214.75 \text{ V} \tag{I}$$

In a series motor, the flux Φ is proportional to the armature current I_a. Thus, when $I_{a1} = 35$ A, $\Phi_1 \propto 35$ and when $I_{a2} = 15$ A, $\Phi_2 \propto 15$. Therefore,

$$\frac{\Phi_1}{\Phi_2} = \frac{35}{15} \tag{II}$$

Also, in a series motor, $N \propto E_b/\Phi$. Hence, for the given two speeds, the correlation between E_b and flux Φ can be written as

$$1600 \propto \frac{E_{b1}}{\Phi_1} \tag{III}$$

and

$$3200 \propto \frac{E_{b2}}{\Phi_2} \tag{IV}$$

Dividing Eq. (IV) by Eq. (III) leads to

$$2 = \frac{E_{b2}}{\Phi_2} \times \frac{\Phi_1}{E_{b1}} \tag{V}$$

Substitution of Eqs (I) and (II) in Eq. (V) results in

$$E_{b2} = 2E_{b1} \times \frac{\Phi_2}{\Phi_1} = 2 \times 214.75 \times \frac{15}{35} = 184.07 \text{ V}$$

(i) Assume that the resistance to be connected in series is R_{se} Ω. Use of Eq. (2.18a) gives

$$(r_a + R_{se}) = \frac{V - E_{b2}}{I_{a2}} = \frac{220 - 184.07}{15} = 2.40 \text{ }\Omega$$

The resistance required to be inserted for a speed of 3200 rpm, $R_{se} = 2.40 - 0.15 = 2.25$ Ω
(ii) In this case, $\Phi_2 = 115 \Phi_1$ and Eq. (IV) takes the form of

$$800 \propto \frac{E_{b2}}{1.15\Phi_1} \tag{VI}$$

Division of Eq. (VI) by Eq. (III) lead to

$$\frac{1}{2} = \frac{E_{b2}}{1.15\Phi_1} \times \frac{\Phi_1}{E_{b1}} \tag{VII}$$

or

$$E_{b2} = 0.5 \times 1.15 \times 214.75 = 123.48 \text{ V}$$

Here also, assume that the resistance to be connected in series is R_{se} Ω. Use of Eq (9.19a) gives

$$(r_a + R_{se}) = \frac{V - E_{b2}}{I_{a2}} = \frac{220 - 123.48}{50} = 1.93 \text{ Ω}$$

The resistance required to be inserted for a speed of 800 rpm, $R_{se} = 1.93 - 0.15 = 1.78$ Ω

When the motor is drawing 50 A and is directly connected to the supply voltage, the back emf E_{b2}, from Eq. (9.1 a) is given by

$$E_{b2} = 220 - 50 \times 0.15 = 212.50 \text{ V}$$

The speed N_2 is then computed as follows:

$$N_2 = N_1 \times \frac{E_{b2}}{\Phi_2} \times \frac{\Phi_1}{E_{b1}} = 1600 \times \frac{212.50}{1.15\Phi_1} \times \frac{\Phi_1}{214.75} = 1377 \text{ rpm}$$

Example 9.9   The load torque of a dc series motor varies as the cube of its speed. If the motor has negligible armature resistance and it draws a current of 60 A at 450 V, determine the value of the resistance to be inserted to reduce its speed by (i) 60% and (ii) 25%.

Solution   Assume that the initial speed of the motor is N_1 and the load torque is T_1 Nm. As per the problem, $T_1 \propto N_1^3$. It is also known that torque $T \propto$ the square of the armature current.
(i) If the torque is T_2 at the reduced speed of $N_2 = 0.4N_1$, then

$$\frac{T_1}{T_2} = \frac{(60)^2}{I_a^2} = \left(\frac{N_1}{0.4N_1}\right)^3 = 15.63 \tag{I}$$

where I_a is the armature current at the reduced speed.

or $\qquad I_a = \sqrt{\dfrac{(60)^2}{15.63}} = 15.18$ A

Since the armature resistance is negligible, the back emf E_{b1}, when the motor is drawing a current of 60 A at 450 V, is given by

$$E_{b1} = 450 \text{ V}$$

Now making use of the relation that $N \propto E_b/\Phi$ and $\Phi \propto I_a$, it is easily seen that

$$N_1 \propto \frac{E_{b1}}{60} \tag{II}$$

and $\qquad N_2 = 0.4N_1 \propto \dfrac{E_{b2}}{15.18}$ $\tag{III}$

Dividing Eq. (III) by Eq. (II) and substituting from Eq. (I), yields

$$0.4 = \frac{E_{b2}}{450} \times \frac{60}{15.18}$$

or $\qquad E_{b2} = \dfrac{0.4 \times 450 \times 15.18}{60} = 45.54$ V

If the resistance to be inserted is assumed equal to R_{se} Ω, from Eq. (2.18a), it is seen that

$$R_{se} = \frac{450 - 45.54}{15.18} = 26.65 \text{ Ω}$$

(ii) In this case, the torque T_2 at the reduced speed of $N_2 = 0.75N_1$ is expressed as

$$\frac{T_1}{T_2} = \frac{(60)^2}{I_a^2} = \left(\frac{N_1}{0.75N_1}\right)^3 = 2.37 \qquad \text{(IV)}$$

where I_a is the armature current at the reduced speed.

or $\qquad I_a = \frac{60}{\sqrt{2.37}} = 38.97 \text{ A}$

Proceeding as in the previous case,

$$N_2 = 0.75N_1 \propto \frac{E_{b2}}{38.97} \qquad \text{(V)}$$

Division of Eq. (V) by Eq. (II) and substitution from Eq. (I) produces

$$0.75 = \frac{E_{b2}}{450} \times \frac{60}{38.97}$$

or $\qquad E_{b2} = \frac{0.75 \times 450 \times 38.97}{60} = 219.21 \text{ V}$

If the resistance to be inserted is assumed equal to R_{se} Ω, from Eq. (2.18a), it is seen that

$$R_{se} = \frac{450 - 219.21}{38.97} = 5.92 \; \Omega$$

Example 9.10 A shunt machine, connected to a 200 V mains, has an armature resistance (including brushes) of 0.15 Ω, and the resistance of the field circuit is 100 Ω. Find the ratio of its speed as a generator to its speed as a motor, the line current in each case being 75 A.

Solution

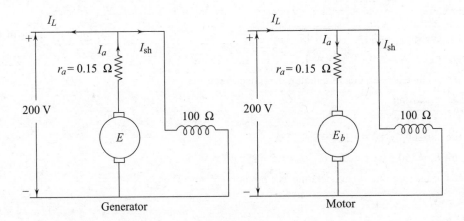

Fig. 9.33

The equivalent circuits of the machine when it is running as a generator as well as a motor are shown in Fig. 9.33. When the dc machine is operating as a generator, the field current

$$I_{sh} = \frac{200}{100} = 2 \text{ A}$$

The armature current

$$I_a = 75 + 2 = 77 \text{ A}$$

From Eq. (9.9),

$$V = E - I_d r_a$$

$$\therefore \quad E = V + I_d r_a = 200 + 77 \times 0.15 = 211.55 \text{ V}$$

When the dc machine is running as a motor, the armature current

$$I_a = 75 - 2 = 73 \text{ A}$$

The supply voltage

$$V = E_b + I_d r_a$$

$$\therefore \quad E_b = V - I_d r_a = 200 - 73 \times 0.15 = 189.05 \text{ V}$$

For a dc machine, the induced emf E is proportional to the speed N, that is,

$$E \propto N$$

Hence, $\quad E = k N_1$

and $\quad E_b = k N_2$

where N_1 and N_2 are the speeds at which the machine is operating as a generator and a motor, respectvely. Then,

$$\frac{N_1}{N_2} = \frac{211.55}{189.05} = 1.119$$

Example 9.11 A 120-kW, belt-driven shunt generator running at 500 rpm on 250 V bus bars continues to run as a motor when the belt breaks, then taking 20 kW. What will be its speed? Armature resistance $r_a = 0.02 \ \Omega$, field resistance $R_{sh} = 80 \ \Omega$, contact drop under each brush = 1.5 V. Ignore armature reaction.

Solution When the dc machine is running as a generator, the shunt field current

$$I_{sh} = \frac{250}{80} = 3.125 \text{ A}$$

The line current $I_L = \dfrac{120 \times 1000}{250} = 480 \text{ A}$

The armature current $I_a = 480 + 3.125 = 483.125 \text{ A}$

The induced voltage $E = 250 + 0.02 \times 483.125 + 2 \times 1.5 = 262.6625 \text{ V}$

When the dc machine is operating as a motor, the line current

$$I_L = \frac{20 \times 1000}{250} = 80 \text{ A}$$

The armature current $I_a = 80 - 3.125 = 76.875 \text{ A}$

The back emf $E_b = 250 - 0.02 \times 76.875 - 2 \times 1.5 = 245.4625 \text{ V}$

Let the speed of the machine when operating as a motor be N rpm. Then,

$$E \propto 500$$

and $\quad E_b \propto N$

Thus, $\quad E_b \propto \dfrac{E}{E_b} = \dfrac{500}{N}$

$$\therefore \quad N = \frac{500 \times 245.4625}{262.6625} = 467.26 \approx 467 \text{ rpm}$$

Example 9.12 A 200-V shunt motor has an armature resistance of 0.4 Ω and a field resistance of 200 Ω. When driving a load of constant torque at 500 rpm, the armature takes 25 A. If it is desired to raise the speed from 500 to 700 rpm, what resistance must be inserted into the shunt field circuit, assuming the magnetization curve to be linear?

Solution The shunt field current

$$I_{sh} = \frac{200}{200} = 1 \text{ A}$$

The line current $I_L = I_a + I_{sh} = 25 + 1 = 26 \text{ A}$

The motor is driving a constant torque load. However, the torque

$$T \propto \Phi I$$

Hence, $\Phi_1 I_1 = \Phi_2 I_2$ (I)

where Φ_1 and Φ_2 are the fluxes and I_1 and I_2 are the armature currents when the motor speed N is, respectively, 500 rpm and 700 rpm. As the magnetization curve is assumed to be linear,

$$\Phi_1 \propto I_{sh1} \text{ and } \Phi_2 \propto I_{sh2} \qquad\qquad\qquad \text{(II)}$$

where I_{sh1} and I_{sh2} are the currents in the shunt field winding when the speed of the motor is 500 rpm and 700 rpm, respectively. From Eqs (I) and (II), the following relation may be written:

$$I_{sh2}I_2 = I_{sh1}I_1$$

$$= \frac{V}{R_{sh}}I_1 = \frac{200}{200} \times 25 = 25$$

$$\therefore \qquad I_2 = \frac{25}{I_{sh2}}$$

When the motor is running at 500 rpm, the back emf

$$E_{b1} = 200 - 25 \times 0.4 = 190 \text{ V}$$

When the motor is running at 700 rpm, the back emf

$$E_{b2} = 200 - I_2 \times 0.4 = 200 - 0.4 I_2$$

Now, $E_{b1} \propto N_1 \Phi_1 \propto N_1 I_{sh1}$

$$E_{b2} \propto N_2 \Phi_2 \propto N_2 I_{sh2}$$

$$\therefore \qquad \frac{E_{b1}}{E_{b2}} = \frac{N_1 I_{sh1}}{N_2 I_{sh2}}$$

$$\frac{190}{200 - 0.4 \times (25/I_{sh2})} = \frac{500 \times 1}{700 I_{sh2}}$$

or $200 - \dfrac{10}{I_{sh2}} = 266 I_{sh2}$

or $266 I_{sh2}^2 - 200 I_{sh2} + 10 = 0$

Then, $I_{sh2} = \dfrac{200 \pm \sqrt{40000 - 40 \times 266}}{2 \times 266}$

$$= \frac{200 \pm 171.35}{532}$$

$$= 0.698 \text{ A (taking only the positive sign)}$$

Therefore, the total resistance in the field circuit = 200/0.698 = 286.53 Ω. The resistance to be inserted in the field circuit = 286.53 − 200 = **86.53 Ω**

Example 9.13 A 400-V, eight-pole shunt motor has a wave-connected armature winding with 1000 conductors. The useful flux per pole is 0.015 Wb and the armature and field resistances are 0.4 Ω and 200 Ω, respectively. Ignoring the effect of armature reaction, find the speed and the total developed torque when a current of 25 A is taken from the mains. If the iron, friction, and windage losses aggregate to 1000 W, find the useful torque, brake horsepower, and efficiency at this speed.

Solution Data: $\Phi = 0.015$ Wb, $P = 8$, $Z = 1000$, $a = 2$, $r_a = 0.4$ Ω, $R_{sh} = 200$ Ω. The shunt field current

$$I_{sh} = \frac{400}{200} = 2 \text{ A}$$

The armature current $I_a = 25 - 2 = 23$ A

The back emf

$$E_b = 400 - 25 \times 0.4 = 390 \text{ V}$$

From Eq. (9.18), the speed of the motor, N, is

$$N = \frac{E_b \times 60 \times a}{\Phi P Z} = \frac{390 \times 60 \times 2}{0.015 \times 8 \times 1000} = 390 \text{ rpm}$$

From Eq. (9.26), the torque is

$$T = \frac{1}{2\pi} \Phi P \frac{Z}{a} I_a$$

$$= \frac{1}{2\pi} \times 0.015 \times 8 \times \frac{1000}{2} \times 23 = 219.63 \text{ N m}$$

The total power developed by the motor $= E_b I_a = 390 \times 23 = 8970$ W

The aggregate of the iron, friction, and windage power losses = 1000 W. The net shaft power available = 8970 − 1000 = 7970 W. The useful shaft torque available

$$T_{shaft} = \frac{7970 \times 60}{2\pi \times 390} = 195.25$$

The brake horse power = 7970/746 = 10.683. The power input is given by

Supply voltage × line current = 400 × 25 = 10,000 W

$$\text{Efficiency } \eta = \frac{\text{power output}}{\text{power input}} = \frac{7970}{10,000} = 0.797 \text{ or } 79.7\%$$

9.7.7 Starting of DC Motors

While starting, the speed N of a dc motor is zero, and from Eq. (9.18) it is obvious that the back emf developed, E_b, is also zero. Thus, when a dc motor is connected to a dc supply voltage V, the armature current while starting will be given by

$$I_a = \frac{V}{r_a} \text{ for the shunt motor}$$

and

$$I_a = \frac{V}{r_a + R_S} \text{ for the series motor}$$

Now, both the resistance of the armature r_a and the resistance of the series field R_s being low, the armature current at the start will be damagingly high. To prevent this high current, some additional resistance is put in series with the armature to keep the starting current within a safe limit. As the machine picks up speed, induced emf E is gradually developed, reducing I_a in the process and thus permitting the gradual removal of additional armature circuit resistance. When the total additional armature circuit resistance is removed, the motor will be able to attain its full speed. Figure 9.34 shows a three-point starter for a dc shunt motor having resistance sections to reduce the starting current. When the handle of the starter is moved to the first stud, all the resistance sections are connected in the armature circuit, and a reduced voltage, is applied to the motor. Consequently, the motor starts to rotate at a reduced speed. As the handle is moved forward, the resistance sections are gradually cut out and the motor picks up speed, finally attaining the rated speed, as a result of the increased voltage applied across the armature. The handle of the starter is held in position by the no-volt release electromagnet.

Two protective devices, namely, a no-volt release and an overload release, are provided to protect the motor. The former prevents the motor receiving full voltage after a voltage failure. When the voltage is switched off, the no-volt release electromagnet gets demagnetized and the handle is pulled by the spring to the OFF position. Similarly, when the motor is overloaded, the electromagnet cuts off the supply to the no-volt release and the handle move to the OFF position.

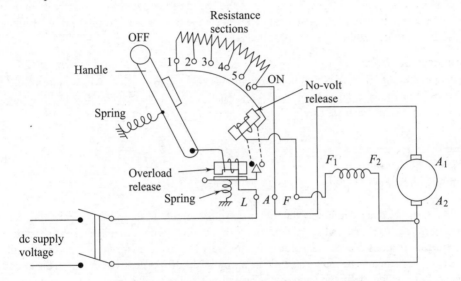

Fig. 9.34 Three-point starter for a dc shunt motor

9.7.8 Speed Control of DC Motors

It has been shown in Eq. (9.22) that the speed N of a dc motor is given by

$$N = \frac{V - I_a R_a}{k\Phi} = \frac{E_b}{k\Phi} \tag{9.28}$$

Hence the speed of a dc motor may be controlled by (a) varying the value of the flux or (b) varying the value of voltage applied to the motor armature.

9.7.8.1 Field Control

In this method, the armature voltage is kept fixed at the supply voltage level. The shunt field current I_{sh} can be reduced by increasing the resistance of the field circuit rheostat. As I_{sh} is reduced, the corresponding field flux Φ also reduces. If the motor operates from a fixed supply voltage, then the back emf E_b may be taken to be constant. As the resistance r_a of the armature is very small, the voltage drop $I_a r_a$ is also very small and thus can be neglected. Then from Eq. (9.28),

$$N \propto \frac{1}{\Phi} \tag{9.29a}$$

If magnetic saturation is neglected, then

$$\Phi \propto I_{sh} \tag{9.29b}$$

Equations (9.28a) and (9.28b) give

$$N \propto \frac{1}{I_{sh}} \tag{9.29c}$$

Equation (9.29c) shows that by reducing the shunt field current I_{sh}, the speed N of the motor can be increased and vice versa.

The main advantage of speed control of a dc motor by field control is that continuous control of speed can be maintained. In addition, low resistance loss in the field rheostat makes this method highly efficient. The disadvantage of the field control method is that the speed of the dc motor can only be increased above its normal value. Further, there is sluggish change in speed due to very slow response of the highly inductive shunt field circuit, and the field circuit rheostat needs to be of continuous rating.

9.7.8.2 Armature Control

In the armature control of speed, the shunt field excitation is kept constant. Reduced armature voltage is applied by dropping a portion of the supply voltage across an armature circuit rheostat employed for this purpose. Now, for a particular motor with a constant shunt field excitation, I_{sh} is constant, Eq. (9.28) gives

$$E_b \propto N \tag{9.30a}$$

At light loads,

$$E_b \approx V \tag{9.30b}$$

Hence from Eqs (9.30a) and (9.30b),

$$V \propto N \tag{9.30c}$$

Equation (9.30c) shows that by reducing the motor terminal voltage V, its speed N can be reduced.

The main merits of this method of speed control are (a) continuous control of speed and (b) quick change in speed due to the fast response furnished by the comparatively low inductance of the armature circuit. The main demerits are (a) speed can only be reduced below its normal value, (b) excessive I^2R loss in the armature circuit rheostat makes this method inefficient, and (c) the armature circuit rheostat needs to have a continuous rating.

Example 9.14 A 450-V series motor is started with the help of a starter, which has six studs. The resistance of the motor (armature + series field) is 0.25 Ω and it is desired to limit the currents between 125 A and 160 A during starting. Calculate the resistance of each step of the starter. Assume that the flux changes by 12% between the given current limits.

Solution   A schematic circuit diagram with the starter connected is shown in Fig. 9.35.

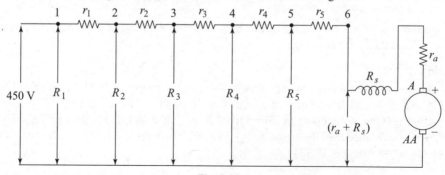

Fig. 9.35

It may be noted that at the stud positions 1, 2, ..., 5, the current is limited to 160 A, respectively, by resistance R_1, R_2, ..., R_5. At the time of starting, the motor is at standstill and, therefore, the back emf is zero. However, the current in the motor should be restricted to 160 A. The magnitude of R_1 is calculated as follows:

$$R_1 = \frac{450}{160} = 2.81 \, \Omega$$

When the starting handle is in position 1, the motor picks up speed to N_1 (say) and builds a back emf E_{b1} due to which the current drops to 125 A. E_{b1} is easily determined by employing Eq. (9.19) as follows:

$$E_{b1} = 450 - 125 \times R_1 = 450 - 125 \times 2.81 = 98.75 \text{ V}$$

When moving the handle from stud 1 to stud 2, the speed momentarily remains unchanged at N_1. However, since the resistance $r_1 \, \Omega$ has been disconnected, the current once again rises to 160 A. Consequently, the flux increases by 12% resulting in a proportionate increase in the back emf at N. The new back emf $E_{b2} = 1.12 \times E_{b1} = 1.12 \times 98.75 = 110.60$ V. Again, using Eq. (9.19), the resistance R_2 is given by

$$R_2 = \frac{450 - 110.60}{160} = 2.12 \, \Omega$$

When proceeding from stud 2 to 3, the speed has increased to a new value N_2 and the current has fallen to 125 A. Then the new back emf E_{b3} with resistance R_2 in the circuit is determined as shown below:

$$E_{b3} = 450 - 125 \times R_2 = 450 - 125 \times 2.12 = 185.0 \text{ V}$$

At stud 3, the current rises to 160 A and the new back emf $E_{b4} = 1.12 \times 185.0 = 207.20$ V at the speed of N_2 with resistance R_3 in the circuit. Hence,

$$R_3 = \frac{450 - 207.20}{160} = 1.52 \, \Omega$$

Following the above procedure, the resistance at the remaining studs is computed and is shown below.

Moving from stud 3 to 4, $E_{b5} = 450 - 125 \times R_3 = 450 - 125 \times 1.52 = 260.0$ V

At stud 4, $E_{b6} = 1.12 \times 260.0 = 291.20$ V

$$R_4 = \frac{450 - 291.20}{160} = 0.99 \, \Omega$$

Moving from stud 4 to 5, $E_{b7} = 450 - 125 \times R_4 = 450 - 125 \times 0.99 = 326.25$ V

At stud 4, $E_{b8} = 1.12 \times 326.25 = 365.40$ V

$$R_4 = \frac{450 - 365.40}{160} = 0.53 \, \Omega$$

Resistance of each section of the starter is determined as follows:

$$r_1 = R_1 - R_2 = 2.81 - 2.12 = 0.69 \, \Omega$$

$$r_2 = R_2 - R_3 = 2.12 - 1.52 = 0.60 \, \Omega$$

$$r_3 = R_3 - R_4 = 1.52 - 0.99 = 0.53 \, \Omega$$

$$r_4 = R_4 - R_5 = 0.99 - 0.53 = 0.46 \, \Omega$$

$$r_5 = R_5 - (\text{armature resistance + field resistance}) = 0.53 - 0.25 = 0.28 \, \Omega$$

9.8 LOSSES IN DC MACHINES

The various types of losses and their distribution in dc machines are pictorially depicted in Fig. 9.36.

  (i) The windage loss, which constitutes a part of the mechanical loss P_{mech}, is the friction loss due to the air resistance to armature rotation. P_{mech} is assumed constant for practically constant speeds.

 (ii) P_{cufl} symbolizes the loss in power in the field circuit and is equal to $VI_f = I_f^2 R_f$, where V, I_f, and R_f, respectively, represent applied voltage, current, and resistance of the field winding. P_{cufl} remains constant if the field current does not change during loading.

(iii) Core loss P_c, which occurs mainly in the rotating iron parts of the armature, may be assumed constant if the field does not change and the speed remains practically unchanged during loading.

 (iv) Copper loss P_{cuar} is the power loss in the armature winding and is represented by $I_a^2 r_a$, where I_a and r_a are the armature current and resistance, respectively. Since I_a is governed by the load on the machine, P_{cuar} is variable and is a function of the load.

  (v) Another type of loss, called the *stray loss*, also occurs in a machine apart from the four major losses detailed above. However, the stray loss is difficult to account for.

Efficiency of a dc generator is determined by commencing with the output and adding the various losses to obtain the power input. In the case of a motor, it is more useful to start with the electrical input and subtracting the losses to obtain the motor output.

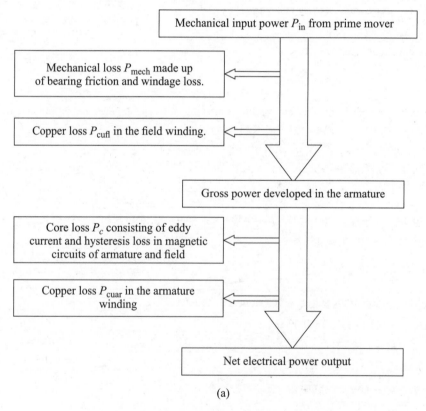

(a)

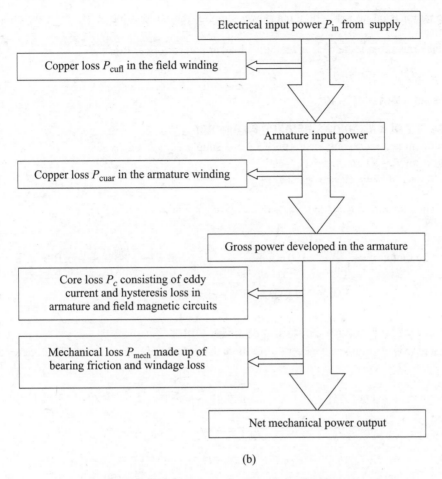

(b)

Fig. 9.36 Distribution of losses in a dc: (a) generator and (b) motor

9.9 TESTING OF A DC MACHINE

Testing of a dc machine is essentially required to confirm design data with performance characteristics. The Swinburne test (named after Sir James Swinburne) is an indirect method of determining the losses and efficiency of dc shunt and compound wound machines which have a constant flux. The circuit diagram for performing the Swinburne test is shown in Fig. 9.37.

Assume that the dc machine under test is a motor. A rated voltage V volt is applied to the terminals A-B and the speed of the motor is adjusted to the rated speed by varying the shunt resistance R. The ammeters 1, 2, and 3 in the circuit, respectively measure the no load current I_0, shunt field current I_{sh}, and armature current I_a.

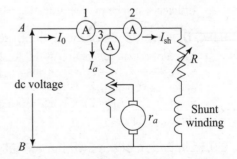

Fig. 9.37 Circuit diagram for performing Swinburne test

Since the motor is on no load, the input power supplies the no load losses W_C only, which are assumed constant.

Thus, if the armature resistance is r_a Ω, the no load armature copper loss

$$P_{cuar} = I_a^2 r_a = (I_0 - I_{sh})^2 r_a \ \text{W}$$

And the constant losses $\quad W_C = VI_0 - (I_0 - I_{sh})^2 r_a \ \text{W}$

9.9.1 Efficiency of a Machine Operating as a Motor

Assume that the motor is drawing a current of I A at a supply voltage of V volts.

$$\text{Power input} = VI$$

Copper loss in the armature $\quad P_{cu} = I^2 r_a = (I - I_{sh})^2 r_a \ \text{W}$

Total losses when the motor is loaded $\quad P_{cu} + W_C = (I - I_{sh})^2 r_a + VI_0 - (I_0 - I_{sh})^2 r_a \ \text{W}$

Efficiency of the motor $\quad \eta = \dfrac{\text{Output}}{\text{Input}} \times 100 = \dfrac{\text{Input} - \text{Losses}}{\text{Input}} \times 100 = \dfrac{VI - (P_{cu} + W_C)}{VI} \times 100$

$$= \dfrac{VI - [(I - I_{sh})^2 r_a + \{VI_0 - (I_0 - I_{sh})^2 r_a\}]}{VI} \times 100$$

9.9.2 Efficiency of a Machine Operating as a Generator

When the machine is operating as a generator, assume that it is supplying a current of I amperes at a voltage of V volts.

Current through the armature $= (I + I_{sh})$ A and armature copper loss $\quad P_{cu} = I^2 r_a = (I + I_{sh})^2 r_a \ \text{W}$

Total losses in the generator $= P_{cu} + W_C = (I + I_{sh})^2 r_a + VI_0 - (I_0 - I_{sh})^2 r_a \ \text{W}$

And generator efficiency $\quad \eta = \dfrac{VI - \left[(I + I_{sh})^2 r_a + \left\{VI_0 - (I_0 - I_{sh})^2 r_a\right\}\right]}{VI} \times 100$

9.9.3 Advantages of the Swinburne Test

- Since it is an indirect test large machines can be tested in a laboratory, therefore, economical and suitable to test dc machines.
- Efficiency can be predetermined at any load due to the constant losses being known.

9.9.4 Disadvantages of the Swinburne Test

- Effect of armature reaction at full load is not accounted for since iron loss is assumed constant.
- Since the test is performed on no load, it is not possible to establish whether commutation is satisfactory under load conditions.
- Similarly, it is also not feasible to establish whether the temperature rise is within limits when the machine is loaded.

9.9.5 Limitation of the Swinburne test

- Swinburne test cannot be used to test series dc machines since they cannot run on no-load, or light loads due to large variation in the flux and speed variations.
- The test can be applied to test shunt and compound (level) machines.

Example 9.15 The Swinburne test on a dc shunt produced the following readings: Terminal voltage = 400 V and input current = 5.6 A. If the field and armature resistances are 250 Ω and 0.5 Ω respectively, calculate (a) output and (b) efficiency of the motor on full load. Assume a full load current of 55 A.

Solution Shunt field current $= 400/250 = 1.6\,\text{A}$

No load armature current $I_a = 5.6 - 1.6 = 4\,\text{A}$

No load armature copper loss $P_{\text{cu}} = 4^2 \times 0.5 = 8\,\text{W}$ (I)

No load input power $= 400 \times 5.6 = 2240\,\text{W}$ (II)

Subtracting Eq. (I) from (II) yields the constant losses $= 2240 - 8 = 2232\,\text{W}$

When the motor is drawing a full load current of 55 A

Full load armature current $I_a = 55 - 4 = 51\,\text{A}$ and full load copper loss $= 51^2 \times 0.5 = 1300.5\,\text{W}$

Total loss on full load $= 1300.5 + 2232 = 3532.5\,\text{W}$

Power input at full load $= 400 \times 55 = 22000\,\text{W}$

(a) Full load output $= 22000 - 3532.5 = 18467.5\,\text{W}$

(b) Motor efficiency $= \dfrac{18467.5}{22000} \times 100 = 83.94\%\,\%$

9.10 CONDITION OF MAXIMUM EFFICIENCY OF A DC MACHINE

In order to determine the condition for maximum efficiency of a dc machine, the basic approach is to identify the variable and constant losses. The methodology for computing the condition for maximum efficiency is demonstrated for a shunt generator and motor separately.

9.10.1 Condition for Maximum Efficiency of a Generator

Refer to the shunt diagram shown in Fig. 9.14(b). The output of the generator is written as

$$P_o = VI_L$$

The mechanical loss P_{mech} and the field loss VI_f can be assumed constant since the speed remains practically constant and there is no change in the field current. Hence, the constant loss in written as

$$P_{\text{const}} = P_{\text{mech}} + VI_f$$

Power loss in armature, $P_{\text{cuar}} = I_a^2 r_a = I_L^2 r_a$ (since it can be approximated that $I_a \approx I_L$)

Mechanical input power, $P_{\text{in}} = P_o + P_{\text{const}} + I_L^2 r_a$

Efficiency of the generator, $\eta_{\text{gen}} = \dfrac{VI_L}{VI_L + P_{\text{const}} + I_L^2 r_a}$

For maximum efficiency, $\dfrac{d\left(\eta_{\text{gen}}\right)}{dI_L} = 0 = \dfrac{d}{dI_L}\left\{\dfrac{VI_L}{\left(VI_L + P_{\text{const}} + I_L^2 r_a\right)}\right\}$

$$= \dfrac{\left(VI_L + P_{\text{const}} + I_L^2 r_a\right) V - VI_L\left(V + 2I_L r_a\right)}{\left(VI_L + P_{\text{const}} + I_L^2 r_a\right)^2}$$

Upon simplificaion,the following result is obtained:

$$I_L^2 r_a \approx I_a^2 r_a = P_{const}$$

or $\quad I_a = \sqrt{\dfrac{P_{const}}{r_a}}$

9.10.2 Condition for Maximum Efficiency of a Shunt Motor

As in the case of a generator, the expression for the efficiency of a shunt motor supplying a load can be derived, commencing from the supply end. Upon differentiating the expression and equating the result to zero, it will be seen that the condition for maximum efficiency works to $I_a = \sqrt{P_{const}/r_a}$. The derivation, however, is left as a tutorial exercise for the reader.

From the foregoing discussion, it may be concluded that (i) the conditions for maximum efficiency of a shunt generator and a shunt motor are the same and (ii) the maximum efficiency occurs when the armature current variable loss is equal to the constant loss.

Example 9.16 A shunt generator is supplying a load of 120 kW at 450 V. The armature and shunt field resistances are 0.08 Ω and 250 Ω, respectively. Determine the efficiency of the generator at (i) full load and (ii) half load. Calculate the maximum efficiency of the generator and the percentage load at which it occurs. Assume a windage and friction loss of 1400 W and core loss of 2100 W.

Solution Figure 9.38 shows a schematic representation of a shunt generator.

Field current, $I_f = \dfrac{400}{250} = 1.6$ A

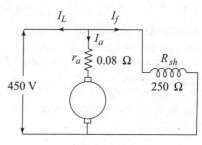

Fig. 9.38

Full-load current, $I_a = \dfrac{120 \times 1000}{400} = 300$ A

Loss in the field winding $= VI_f = 400 \times 1.6 = 640$ W constant loss,

$$P_{const} = \text{windage and friction loss} + \text{core loss} + \text{field loss}$$

$$= 1400 + 2100 + 640 = 4140 \text{ W}$$

Armature current, $I_a = I_L + I_f = 300 + 1.6 = 301.6$ A

Armature loss at full load, $P_{cuar} = I_a^2 r_a = (301.6)^2 \times 0.08 = 7277$ W

(i) Efficiency, $\eta_{gen} = \dfrac{P_o}{P_o + P_{cuar} + P_{const}} = \dfrac{120 \times 1000}{120 \times 1000 + 7277 + 4140} \times 100 = 91.31\%$

Armature current at half load, $I_a = (I_L/2) + I_f = 150 + 1.6 = 151.6$ A

Armature loss at full load, $P_{cuar} = I_a^2 r_a = (151.6)^2 \times 0.08 = 1838.60$ W

(ii) Efficiency, $\eta_{gen} = \dfrac{60 \times 1000}{60 \times 1000 + 1838.6 + 4140} \times 100 = 90.94\%$

For maximum efficiency, $I_a = \sqrt{P_{const}/r_a} = \sqrt{4140/0.08} = 227.49$ A

Maximum efficiency, $\eta_{max} = \dfrac{120 \times 1000}{120 \times 1000 + 2 \times 4140} = 93.55\%$

Since it can be assumed that $I_L \approx I_a$, the load at which maximum efficiency occurs is

$$227.49 \times 100/300 = 75.83\%$$

9.11 APPLICATIONS OF DC MACHINES

The advantages of dc machines are that they are easy to control and are adaptable. In spite of the requirement of high initial capital investment, about 25% motors manufactured are dc motors. Generally, however, the parameters that govern the selection of machines are applications and operating characteristics, technical feasibility, and economics.

9.11.1 DC Generators

A dc generator, despite the advent of SCR rectifiers and controlled power electronic devices, is employed as a component of an ac to dc generator motor set to transform ac voltage to dc voltage. The advantage of dc generators is that the generated dc voltage is superior to that obtained from SCR rectifiers and electronic devices. The latter suffer from the disadvantages of harmonics in generated voltage, poor power factor, poor braking, etc. Other applications of dc generators are in the fields of welding, as dynamometers, as tacho-generators, and cross field generators. Because a separately excited generator provides a wide range of control of output voltage, it is used for speed control in the Ward-Leonard system. The disadvantage of dc generators, which restricts their applications, is continuous maintenance and replacement of parts.

9.11.2 DC Motors

Since a dc compound motor possesses medium starting torque and speed regulation within 5% to 15%, it is employed to operate fans, blowers, centrifugal pumps, printing presses, etc. On the other hand, due to a high starting torque and drooping speed–load characteristics, a dc compound motor finds applications for supplying pulsating loads requiring flywheel action, sheers, crushers, conveyers, hoists, rolling mills, etc. Traction-type loads, battery-powered vehicles, cranes, etc. are driven by dc series motors due to their high starting torque, which is of the order of five times the full-load torque.

Recapitulation

Average emf generated, $E = \dfrac{\Phi PNZ}{60a}$

Terminal voltage (generator action), $V = E_a - I_a r_a$

Power generated, $P_g = E_a I_a$

Power supplied to the load, $P_L = VI_a$

Back emf generated $E_b = \dfrac{\Phi PNZ}{60a}$

Speed of a motor, $N = \dfrac{V - I_a r_a}{k\Phi}$

Torque developed, $T = \dfrac{1}{2\pi} \Phi P \dfrac{Z}{a} I_a$

The condition for maximum efficiency in a dc shunt generator/motor is

$$I_L^2 r_a \approx I_a^2 r_a = P_{const}$$

or $\quad I_a = \sqrt{\dfrac{P_{const}}{r_a}}$

Assessment Questions

1. Explain with the help of a sketch, the constructional features of a dc machine and briefly describe the functions of (i) armature core, (ii) commutator, and (iii) brushes.
2. (a) Write a note on armature windings.
   (b) Compare (i) lap, and (ii) wave windings.

3. Derive expressions for the (i) generated emf and (ii) electromagnetic torque of a dc machine. Explain what happens when the direction of rotation is reversed in each case.
4. Draw the schematic circuit diagram of a dc generator and sketch its equivalent circuit.
5. Classify generators according to their connections and draw the equivalent circuit for each category.
6. Draw the connection diagram for obtaining OCC of a separately excited generator and explain the nature of the characteristics.
7. Draw (i) the connection diagram, and (ii) no-voltage characteristic of a dc self excited generator.
8. (a) Define (i) GNP and (ii) MNP.
   (b) Explain commutator action with the help of sketches and show how dc voltage is obtained at the brushes.
9. Distinguish between internal and external characteristics of dc generators.
10. Sketch and explain the nature of load characteristics of (i) series, (ii) shunt, and (iii) compound generators.
11. Explain the principle of operation of a dc motor.
12. (a) Classify dc motors based on type of field connections.
    (b) Explain back emf in a dc motor
13. Discuss the characteristics of different types of motors.
14. Derive an expression for the speed of a dc motor and show that the speed of a dc motor is directly proportional to its back emf and inversely proportional to its flux per pole. There from derive an expression for the ratio of the speeds of a dc series motor when (a) the motor is operating on the linear portion of the magnetic characteristic and (b) the magnetic circuit is saturated.
15. Derive an expression for the torque developed by a dc motor. With the help of the torque expression explain what conclusions can be drawn in respect of motor operation.
16. Discuss the characteristics of (i) series, (ii) shunt, and (iii) compound motors.
17. (a) Explain the various methods of speed control of dc motors.
    (b) Explain with the help of a sketch the functioning of a three-point starter.
18. Derive the condition for maximum efficiency of a shunt motor.
19. Enumerate the different types of losses in a dc machine and with the help of a sketch show their distribution in a (i) generator, and (ii) motor.
20. Describe the Swinburne test for determining the efficiency of a dc machine.
21. Enumerate the advantages and disadvantages of the Swinburne test. What are its limitations?
22. Describe the various applications of dc generators and motors.

Problems

9.1 A four-pole, wave-connected shunt generator has 900 conductors. If the flux per pole is 0.03 Wb and the speed of the generator is 700 rpm, what is the magnitude of the armature voltage? If the armature current is 40 A, determine (a) the terminal voltage, (b) the field current, and (c) the load supplied. The armature and field resistances are 0.25 Ω and 100 Ω, respectively. If the generator is now lap-wound, what is the flux per pole required to supply the same load?   [630 V; (a) 610 V; (b) 6.1A; (c) 20.68 kW; 0.06 Wb]

9.2 Two four-pole generators, each having 600 conductors, are driven at 1000 rpm. If the flux in each generator is 15 mWb and the current in each conductor is 50 A, calculate (a) the total current, (b) the emf generated, and (c) the load supplied. Assume that one generator is wave-wound and the other is lap-wound.
           [Wave-wound: (a) 100 A; (b) 300 V; (c) 30 kW; Lap: (a) 200 A; (b) 150 V; (c) 30 kW]

9.3 A six-pole, lap-connected generator is driven at 600 rpm. It has 100 slots with 24 conductors per slot. What is the magnitude of the generated emf? If the number of conductors per slot is changed to 20, at what speed should the generator be run for the same voltage to be generated? The flux per pole is 0.02 Wb.
                                                                                        [480 V, 720 rpm]

9.4 A four-pole, wave-connected armature has 100 slots. If the flux per pole is 0.04 Wb, calculate the number of conductors required per slot to generate 220 V. Take the speed of the generator as 300 rpm. Calculate the new value of the flux due to the change in the number of conductors per slot, if any.   [6, 0.0367 Wb]

9.5 A shunt generator has an armature resistance of 0.1 Ω and a field winding resistance of 250 Ω. Determine (a) the load current, (b) the field current, (c) the armature current, and (d) the armature voltage when it supplies a load of 22 kW at 440 V.   [(a) 50 A, (b) 1.76 A, (c) 51.76 A, (d) 445.176 V]

9.6 A dc shunt generator gave the following data when tested with no load, at a speed of 1500 rpm.

I_f(A)	0	0.5	1.0	1.5	2.0	2.5	3.0	4.0	5.0	6.0
V_{oc} (V)	5	30	55	82	110	129	148	181	203	217

Draw the OCC of the generator and therefrom determine (a) the critical resistance, (b) the critical speed, (c) the field resistance and field current for a no-load voltage of 200 V, and (d) the value of the resistance to be added to the field circuit to reduce the voltage to 180 V.
                     [(a) 53.3 Ω; (b) 600 rpm; (c) 40.8 Ω, 4.9 A; (d) 4.2 Ω]

9.7 A short shunt compound wound generator is supplying a load of 25 kW at 400 V. If the armature, series field, and shunt field winding resistances are 0.1 Ω, 0.1 Ω, and 200 Ω, respectively, calculate (a) the load current, (b) the shunt and series field winding currents, (c) the armature current, (d) the generated voltage, (e) the power input, and (f) the efficiency. If the generator is run at 700 rpm, what is the input torque?
                     [(a) 62.5 A, (b) 62.5 A, (c) 64.53 A, (d) 412.703 V, (e) 26.63 kW, (f) 93.88%; 363.28 N m]

9.8 Repeat Problem 9.7 with the machine connected as a long shunt compound generator.
                     [(a) 62.5 A, (b) 64.5 A, (c) 64.5 A, (d) 412.90 V, (e) 26.63 kW, (f) 93.88%; 363.28 Nm]

9.9 A shunt generator having a lap-wound armature has six poles. If the armature and shunt field winding resistances are 0.2 Ω and 100 Ω, respectively, calculate (a) the armature current, (b) the field current, (c) the current per path, (d) the armature voltage, and (e) the input power, when the generator is supplying a load of 5 kW at 250 V. Assume a voltage drop of 1.5 V in each brush.
                     [(a) 22.5 A, (b) 2.5 A, (c) 3.75 A, (d) 257.44 V, (e) 5.792 kW]

9.10 A 300-V, four-pole dc motor draws a current of 50 A when supplying a certain load. The armature is wave-wound and has 600 conductors. If the flux per pole is 40 m Wb and the armature resistance is 0.4 Ω, determine (a) the back emf, (b) the speed, and (c) the output torque of the motor.
                     [(a) 280 V, (b) 350 rpm, (c) 381.97 N m]

9.11 The lap-wound armature of a 250-V, four-pole dc shunt motor draws a current of 60 A from the supply. The motor has 90 slots with four conductors per slot. Determine the speed of the motor if the flux per pole is 0.025 Wb and the armature resistance is 0.1 Ω. What is the power output and the torque developed?
                     [1627 rpm, 14.64 kW, 85.94 N m]

9.12 A dc shunt machine when operating as a generator supplies a load of 80 kW to an infinite bus at 250 V and 1000 rpm. The machine continues to run despite the failure of the prime mover and draws a current of 30 A from the infinite bus. If the armature and shunt field winding resistances are 0.1 Ω and 75 Ω, respectively, calculate the speed of the machine. Assume zero armature reaction and voltage drop in the brushes.
                     [876 rpm]

9.13 A six-pole, wave-wound dc motor has a useful flux of 0.03 Wb per pole and the number of conductors is 1000. The motor draws a current of 30 A at 500 V and has an armature resistance of 0.5 Ω. Calculate (a) the speed, (b) metric horsepower output, (c) the torque developed, and (d) the efficiency. Assume the windage, friction, etc. losses to be 2 kW.   [(a) 323.33 rpm, (b) 14.55 kW, (c) 17.06 metric hp, (d) 370.65 N m]

9.14 A dc machine when operating as a generator produces 10 kW at 300 V and 800 rpm. The resistances of the armature and field windings are 0.2 Ω and 150 Ω, respectively. (a) If the terminal voltage and armature current remain unchanged, what will the speed of the machine be if it is run as a motor with the flux per pole reduced by 20%? (b) Calculate the per cent change in the flux per pole if the machine has to run as a motor at the speed of the generator.   [(a) 953 rpm, (b) 4.6% reduction]

9.15 A separately excited dc motor has a full-load rated output of 300 hp (metric) at 500 rpm, at a rated voltage of 500 V. The armature has a lap-connected winding with 1000 conductors. If the I^2R loss in the armature winding is 10 kW, calculate the flux per pole. Assume that the motor has an efficiency of 90%. Take 1 hp (metric) = 735.5 W. [0.0575 Wb]

9.16 A separately excited dc motor delivers 50 hp (metric) at full load and 1000 rpm. Its no-load speed is 1100 and the armature voltage is 250 V. If the field current is held constant, calculate (a) the power at full load, and (b) the full-load current. What is the resistance of the armature? If the motor terminals are now connected to a supply source of magnitude 500 V, determine (c) the power output, (d) the speed, and (e) the torque. Assume that the field current does not change.

[(a) 36.775 kW; (b) 161.77 A, 0.14 Ω; (c) 77.22 kW; (d) 3000 rpm; (e) 351.18 N m]

9.17 In Problem 9.16, the field current is reduced by 25% while the armature voltage is held constant at 250 V. Calculate (a) the power developed, (b) the torque, and (c) the speed of the motor at full load.

[(a) 36.78 kW, (b) 263.52 N m, (c) 1333.5 rpm]

9.18 A dc series motor has a speed of 500 rpm when it draws a current of 50 A from a dc supply source of 300 V. Calculate the speed of the motor when it draws a current of 90 A. The series field and armature winding resistances are 0.2 Ω and 0.15 Ω, respectively. Assume a voltage drop of 1 V per brush.

$$\left[\text{(a)}\ \frac{N_1}{N_2} = \frac{E_{a1}I_{a2}}{E_{a2}I_{a1}},\ \text{(b)}\ \frac{N_1}{N_2} = \frac{E_{a1}}{E_{a2}};\ 263.91\ \text{rpm} \right]$$

9.19 A dc series motor is run at 250 rpm with the series winding separately excited. The voltage across its terminals is 1500 V when the field current is 300 A. Calculate the speed and torque of the motor if it is run by a 600 V supply and takes a current of 30 A. Take the resistance of the armature and field winding to be 0.05 Ω each. Neglect the voltage drop across the brushes and the armature reaction. Assume 20 Nm torque loss due to rotational losses. [995 rpm, 151.89 N m]

9.20 A 500-V dc series motor draws a current of 40 A when running at 500 rpm. It is desired to reduce the speed of the motor to 350 rpm. Calculate (a) the armature current and (b) the magnitude of the resistance to be connected in series with the motor for this purpose. Assume that the armature torque is half the torque when the motor is running at 500 rpm. Take the value of the motor resistance as 0.25 Ω. [(a) 28.28 A, (b) 8.91 Ω]

9.21 A 300-V dc series motor runs at 1000 rpm and draws a 45 A current when supplying a certain load. The armature and field winding resistances are 0.5 Ω each. Determine the value of the additional resistance to be connected in the motor circuit to reduce the speed to 750 rpm if the torque is kept (a) proportional to the speed, (b) proportional to the square of the speed, and (c) constant at its previous value.

[(a) 2.45 Ω, (b) 3.64 Ω, (c) 1.41 Ω]

9.22 A 300-V dc series motor has six poles and a wave-wound armature. The number of conductors per slot is 4 and the total number of slots is 300. If the armature and field winding resistances are 0.5 Ω and 0.25 Ω, respectively, and the flux per pole is 0.025 Wb, calculate (a) the speed of the motor, (b) the brake horsepower output, (c) the efficiency, and (d) the force on the pulley if the diameter is 0.5 m. The line current is 50 A. Assume that 1 kW is lost in windage and friction. [(a) 175 rpm, (b) 16.485 metric hp, (c) 80.83%, (d) 2646.52 N]

9.23 The armature winding of a six-pole dc shunt motor is wave-wound and has 666 conductors. It takes a full-load current of 60 A at 600 V. If the armature resistance is 0.3 Ω, the flux per pole is 0.025 Wb, and the voltage drop in each brush is 1.5 V, determine (a) the speed at full load, (b) the armature torque, (c) the output in kW, and (d) the efficiency. Assume windage and friction loss equal to 1 kW. Calculate also the above quantities for a motor whose armature is lap-wound.

[Wave-wound: (a) 695.5 rpm, (b) 476.98 N m, (c) 33.74 kW, (d) 93.72%;
Lap-wound: (a) 2086.5 rpm, (b) 158.99 N m, (c) 33.74 kW, (d) 93.72%]

9.24 A 15-kW dc shunt motor has armature and field winding resistances of 0.25 Ω and 200 Ω, respectively. Its no-load speed is 1200 rpm and it draws a current of 10 A when connected across a 300 V source. Calculate the speed of the motor on full load. Assume that there is no change in flux from no load to full load.

[1159.97 rpm]

9.25 A dc shunt motor has an armature resistance of 0.2 Ω and a field resistance of 250 Ω. When connected across a voltage supply of 300 V, it draws a current of 70 A and runs at 1200 rpm. Determine the resistance to be added to the field circuit to increase the speed of the motor to 1500 rpm at an armature current of 100 A. Assume that the motor operates on the linear portion of the magnetization curve. [49.46 Ω]

9.26 A 500-V dc shunt motor at full load draws a current of 52 A at a speed of 1000 rpm. The armature resistance is 0.3 Ω and the field resistance is 250 Ω. It is desired to reduce the speed of the motor to 800 rpm at 75% full load. Calculate (a) the resistance to be added to the armature circuit, (b) the efficiency, (c) the power loss in the armature circuit, and (d) the speed at half load with the additional resistance in the armature circuit. Assume that the flux remains constant. [(a) 2.69 Ω, (b) 73.67%, (c) 4.20 kW, (d) 876.80 rpm]

9.27 A dc shunt motor has an armature resistance of 0.4 Ω and is connected across a voltage supply of 300 V. An external resistance is connected in the armature circuit. Calculate the value of the resistance so that the armature current is limited to 60 A while starting. Calculate the magnitude of the generated emf when the armature takes a current of 30 A (with the additional resistance in the circuit) at constant speed. [4.6 Ω, 150 V]

9.28 A 300-V, six-pole dc shunt motor is wave-connected and has 1332 conductors. The flux per pole is 3 mWb when the no-load field current is 0.5 A. (a) What is the speed of the motor when the no-load current is 2 A? (b) What is the current drawn from the supply when the motor develops a torque of 35 Nm? (c) Determine the speed of the motor at the torque developed in (b). (d) Calculate the percentage change in speed from no load to the speed in (c). Assume an armature resistance of 0.6 Ω. [(a) 1497 rpm, (b) 18.84 A, (c) 1446 rpm, (d) 3.4%]

9.29 A 500-V dc shunt motor takes an armature current of 40 A when running at 1000 rpm. The armature resistance is 0.5 Ω. What is the speed of the motor when it is connected to a voltage supply of 250 V and the armature current is 25 A? Assume that the flux per pole reduces by 20% at 250 V in comparison with the flux per pole at 500 V. [618.49 rpm]

9.30 A 300-V, 60-kW dc shunt motor operates at 1200 rpm and draws a supply current of 15 A. The armature resistance is 0.2 Ω and the field resistance is 200 Ω. Determine the efficiency of the motor when it draws a current of 150 A. Also calculate the load current at maximum efficiency, and the maximum efficiency. Assume a windage and friction loss of 1 kW and a core loss of 0.5 kW. [85.87%, 100.24 A, 87.03%]

9.31 (a) A k-step starter is used to start a shunt motor shown in Fig. 9.39. Prove that

$$\frac{R_1}{R_k} = \gamma^{k-1}$$

where

$$\gamma = \frac{\text{Upper allowable current limit } I_1}{\text{Lower allowable current limit } I_2}$$

(b) A 400-V shunt motor has an armature resistance of 1.6 Ω. It is started by a four stud starter and it gives an initial starting current of 20 A. Calculate the resistance of each section of the starter given that the upper allowable current at each of the remaining three studs is limited to 42 A. What is the lower allowable current limit?

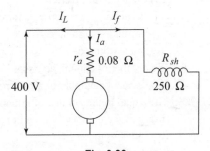

Fig. 9.39

[r_1= 11.38 Ω, r_2 = 4.90 Ω, r_3 = 2.11 Ω]

9.32 A shunt motor draws a full-load current of 400 A from a 400-V dc supply. The armature and field resistances of the motor are 0.3 Ω and 350 Ω, respectively. If the friction, windage, stray, eddy current, and hysteresis losses total to 4000 W, calculate the full-load efficiency of the motor. Plot the load supply current versus efficiency curve and estimate the (i) maximum efficiency and (ii) percentage load at which it occurs. Verify the result by applying the condition for maximum efficiency. [67.21%, (i) 81.72%, (ii) 30.47%]

Objective Type Questions

1. Which of the following is applicable to a dc machine with respect to the field winding?
   - (i) it is always placed on the stator
   - (ii) it is always placed on the rotor
   - (iii) it can also be placed on the rotor
   - (iv) none of these

2. Which of the following is a function of the yoke?
   - (i) to provide strength to the core
   - (ii) to support the field winding
   - (iii) to provide a return path for the flux
   - (iv) to distribute the flux uniformly

3. The armature core of a dc machine is made up of
   - (i) solid aluminum
   - (ii) laminated aluminum
   - (iii) solid steel
   - (iv) laminated steel

4. The armature laminations are insulated from each other
   - (i) to provide flexibility
   - (ii) to reduce noise
   - (iii) to reduce eddy current loss
   - (iv) to reduce hysteresis loss

5. Which of the following is a function of the commutator?
   - (i) to act as a rectifier
   - (ii) to act as an inverter
   - (iii) to act as a junction box for connecting the armature winding ends
   - (iv) none of these

6. The number of commutator segments must be
   - (i) twice the number of armature coils
   - (ii) equal to the number of armature coils
   - (iii) half the number of armature coils
   - (iv) none of these

7. Which of the following is not applicable to a lap-wound dc machine?
   - (i) the number of parallel paths is equal to the number of poles
   - (ii) the number of brushes is equal to the number of poles
   - (iii) the current in each coil is half the armature current
   - (iv) none of these

8. The flux produced by the field winding in a dc machine
   - (i) rotates at a synchronous speed with respect to the stator
   - (ii) rotates at a speed less than the synchronous speed with respect to the stator
   - (iii) rotates at a speed higher than the synchronous speed with respect to the stator
   - (iv) is stationary with respect to the stator

9. The interaction of the armature and field fluxes produces a torque. Which of the following is applicable when the machine is operating as a generator?
   - (i) the torque is non-directional
   - (ii) the torque is in the direction of the prime mover torque
   - (iii) the torque opposes the prime mover torque
   - (iv) none of these

10. A four-pole, lap-wound dc machine generates a voltage of 220 V. Which of the following gives the magnitude of the generated voltage when the machine is wave-wound?
    - (i) 440 V
    - (ii) 220 V
    - (iii) 110 V
    - (iv) none of these

11. Which of the following is not an expression for the power delivered to the load by a dc machine?
    - (i) VI_a
    - (ii) $(E - I_a r_a) I_a$
    - (iii) $V \dfrac{(E - V)}{R_L}$
    - (iv) EI_a

12. In a dc generator, the initial flux is produced due to
    - (i) the saturation of the core
    - (ii) residual magnetism
    - (iii) eddy currents
    - (iv) hysteresis

13. Which of the following gives the expression for the generated voltage in a dc generator?
    - (i) $4.44 \, fN\phi_m$
    - (ii) $4.44 \, fT\phi_m$
    - (iii) $\dfrac{\Phi NPZ}{60a}$
    - (iv) none of these

14. At which of the following will a given generator build up to generate voltage?
    - (i) a speed which is higher than the critical speed
    - (ii) a speed which is equal to the critical speed

(iii) a speed which is lower than the critical speed

(iv) any speed

15. With a dc generator running at rated speed on no load, which of the following will represent the internal characteristic?

   (i) E versus load current I_L

   (ii) E versus field current I_f

   (iii) E versus load voltage V

   (iv) none of these

16. Which of the following would represent the shunt field resistance of a dc compound generator?

   (i) $1.0\ \Omega$ (ii) $100\ \Omega$

   (iii) $1.5\ \Omega$ (iv) none of these

17. With which of the following dc generator connections is it possible to have the terminal voltage equal to the internal voltage?

   (i) shunt connection

   (ii) series connection

   (iii) compound connection

   (iv) all of these

18. Which of the following types of motors are possible?

   (i) permanent magnet   (ii) separately excited

   (iii) series   (iv) all of these

19. Which of the following would be applicable to a dc generator?

   (i) it is a prime mover

   (ii) it is a static converter

   (iii) it is a rotary converter

   (iv) it is an energy transformer

20. Which of the following gives the magnitude of the back emf E_b when the motor is about to start?

   (i) $E_b = \infty$ (ii) $E_b = V$

   (iii) $E_b = 0$ (iv) none of these

21. Which of the following represents the torque developed by a dc motor?

   (i) Torque $\propto I_a$ (ii) Torque $\propto \Phi$

   (iii) Torque $\propto I_a\Phi$ (iv) none of these

22. A dc shunt motor is not suitable for driving which of the following?

   (i) flywheels   (ii) milling machines

   (iii) conveyers   (iv) fans

23. The purpose of including an external resistance at the time of starting a dc motor is to

   (i) increase the starting torque

   (ii) increase the armature flux

   (iii) reduce the starting current

   (iv) none of these

24. Which of the following is applicable to the field control of the speed of a shunt motor?

   (i) speed can be increased above the normal speed

   (ii) low copper loss

   (iii) continuous control of speed

   (iv) all of these

25. How will the speed of a dc shunt-motor change when the applied voltage is half the normal voltage?

   (i) there will be no change

   (ii) the speed will become half the normal speed

   (iii) the speed will fall slightly below the normal speed

   (iv) the speed will increase slightly above the normal speed

Answers

1. (i)	2. (iii)	3. (iv)	4. (iii)	5. (i)	6. (ii)	7. (iii)
8. (iv)	9. (iii)	10. (i)	11. (iv)	12. (ii)	13. (iii)	14. (i)
15. (i)	16. (ii)	17. (iii)	18. (i)	19. (iii)	20. (iii)	21. (iii)
22. (i)	23. (iii)	24. (iv)	25. (iii)			

10 SINGLE-PHASE INDUCTION MOTORS AND SPECIAL MACHINES

Learning Objectives

This chapter will enable the reader to

- Realize why single-phase motors come in fractional kilowatt range and why their design is based on trial and error
- Understand the principle of forward and backward rotating magnetic fields and develop their torque–slip characteristics
- Catalogue the different types of single-phase motors and explain their advantages and disadvantages
- Grasp the main features of servo motors and categorize them according to their design principle
- Explain qualitatively the working of various types of dc (shunt and series) and ac servo motors and know their relative merits and demerits
- Specify the different types of stepper motors and explain their working, advantages, and disadvantages, along with their applications
- Figure out the principle of working of a hysteresis motor based on the torques produced due to hysteresis and eddy current loss
- Understand the working principle of a universal motor, a synchro system, and a tachometer generator

10.1 INTRODUCTION

Today, single-phase ac motors are manufactured in the fractional kilowatt range in varying assortments for applications in homes, shops, and offices because the individual load requirements, being low, can be met economically. Compared to three-phase ac motors, single-phase ac motors are simpler in construction. However, in the absence of established design practices, the analysis of fractional kilowatt motors is complex. Also, the development of such motors is based on developing a prototype on trial and error till the desired characteristic, or efficiency, or cost saving is achieved. Despite the lack of certain well-developed design concepts, the large and varied areas of applications of these motors motivate the engineers to continue to work to improve their performance. Of course, the advancements in computer applications have further aided in the development of these motors.

In this chapter, the following fractional kilowatt motors are qualitatively discussed to provide a basic understanding of their working.

- (i) Single-phase induction motors
- (ii) Servo motors
- (iii) Stepper motors
- (iv) Hysteresis motors

10.2 SINGLE-PHASE INDUCTION MOTORS

Single-phase induction motors are used in ceiling fans, refrigerators, food mixers, hair driers, portable drills, vacuum cleaners, washing machines, sewing machines, office machinery, blowers, centrifugal pumps, toys, etc. The cost of manufacture of a single-phase induction motor is low and is easy to operate, maintain, and repair. In this section, the starting and running characteristics of single-phase induction motors are described qualitatively in terms of rotating field theory.

Structurally, single-phase induction motor resembles a three-phase squirrel cage induction motor except that the stator winding is a single-phase winding. The motor is usually wound for an even number of poles, usually two, four, or six. Figure 10.1 shows a schematic representation of a single-phase squirrel cage induction motor.

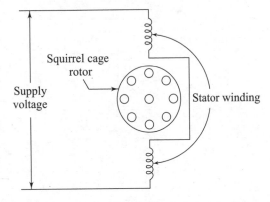

Fig. 10.1 Layout of a single-phase squirrel cage induction motor

10.2.1 Magnetic Field of a Single-phase Induction Motor

When the stator winding of a single-phase induction motor is connected to a single phase ac supply voltage, an alternating flux, but not a rotating flux, is produced. The alternating flux always acts along the axis of the stator winding but varies or pulsates sinusoidally with time. Due to transformer action, a voltage is induced in the stationary rotor circuit, which causes a current to flow in the closed rotor circuit. The direction of the induced current is such as to oppose the stator magnetic field. In other words, the axis of the rotor field coincides with the axis of the stator field and there is no relative motion between the two. Hence, no torque is produced. However, if the rotor is given a push in a particular direction, the motor continues to rotate in that direction. Single-phase induction motors are, therefore, not self-starting.

Figure 10.2 shows two phasors F_1 and F_2 representing two magnetic fields, each having a constant magnitude F and rotating in the opposite directions with the same angular speed of ω radians per second. Let it be assumed that both the fields start rotating from the horizontal axis OX at the time $t = 0$. At that instant of time, the resultant horizontal field is $2F$ as shown in Fig. 10.2(a). After a lapse of t seconds, each field will have rotated through an angle $\theta = \omega t$. Resolving, the two field phasors into horizontal and vertical components,

$$F_{1H} = F \cos \theta; \; F_{1V} = F \sin \theta$$
$$F_{2H} = F \cos (-\theta) = F \cos \theta; F_{2V} = F \sin (-\theta) = -F \sin \theta$$

Adding the horizontal and vertical components,

$$F_{1H} + F_{2H} = 2 F \cos \theta$$
$$F_{1V} + F_{2V} = 0$$

Figure 10.2(b) shows that the resultant mmf at $\theta = \omega t$ is $2F \cos \theta$, and the vertical components are equal and opposite and, therefore, cancel each other. At the instant of time when $\omega t = 90°$, two mmf phasors will be in direct opposition, as in Fig. 10.2(c), and the resultant mmf is zero. For angle ωt greater than $90°$, as in Fig. 10.2(d), the resultant mmf will be $-2F \cos \theta$. At $\omega t = 180°$, as in Fig. 10.2(e), the resultant will be $-2F$. When ωt is greater than $180°$ but less than $270°$, as in Fig. 10.2(f), the resultant mmf will be $-2F \cos \theta$. Thus, for a succession of values of θ from $0°$ up to $360°$, it is found that the resultant is an alternating mmf. Conversely, an alternating mmf can be resolved into two mmfs, each of one-half the amplitude of the original mmf, and rotating with equal angular velocities, but in opposite directions. This principle is known as Ferrari's principle.

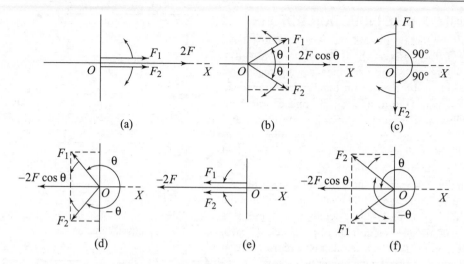

Fig. 10.2 Representation of single-phase field

From the above, it can be seen that a single-phase pulsating stationary magnetic field, set up by a single-phase ac voltage, can be resolved into two equal magnetic fields rotating at synchronous speed ($N_S = 120f/P$) but in opposite directions. The component of magnetic field which rotates in the same direction as the synchronous speed is named as *forward-revolving field F_f*, while the component of magnetic field that rotates in the direction opposite to the synchronous speed is named as *backward revolving field F_b*. The magnitudes of the two rotating magnetic fields is half of the pulsating stationary magnetic field.

10.2.2 Rotor Slip and Torque–Slip Characteristics

Let it be assumed that the rotor of a single-phase induction motor is rotating in the anti-clockwise direction at a speed of N rpm. The stationary pulsating magnetic field, set up by the stator mmf of the single-phase motor, can be resolved into two equal magnetic fields, as already discussed, the forward-revolving field F_f and the backward-revolving field F_b. If s_f and s_b, respectively, represent the rotor slip with respect to the forward-revolving field F_f and the backward-revolving field F_b, then

$$s_f = \frac{N_S - N}{N_S} = s \tag{10.1}$$

$$s_f = \frac{N_S - (-N)}{N_S} = \frac{2N_S - (N_S - N)}{N_S} = 2 - s \tag{10.2}$$

From Eqs (10.1) and (10.2), it is clear that the rotor slips s_f and s_b at standstill ($s = 1$) are unity. The torques produced by the forward and backward fields are equal and opposite and hence the motor cannot start by itself.

When the rotor starts moving, the forward field F_f induces low-magnitude and high-power factor currents in the rotor compared to that at standstill since the slip s_f is low ($s_f = s$), in the direction of motion of the rotor. The magnetic effect of the rotor currents is less than at standstill. As the speed increases, the field in the forward direction is strengthened. Similarly, the field F_b induces high-magnitude and low-power factor currents in the rotor since the slip s_b is high ($s_b = 2 - s$), in the direction opposite to the direction of rotation. The magnetic effect of the rotor currents is higher than at standstill. Their mmfs, which opposes the stator current, result in the weakening of the backward flux wave. Comparatively, since the field F_f is stronger than the field F_b, the torque produced in the forward direction is greater than the torque in the opposite direction and the motor continues to run.

As the speed increases, the forward flux wave increases while the backward flux wave decreases, their sum remaining roughly constant since it must induce the stator counter emf, which is approximately constant if the stator leakage-impedance voltage drop is small.

The torque versus speed/slip characteristic of a single-phase induction motor is shown in Fig. 10.3. The torque–speed characteristic for the forward and backward components of the field is shown by dashed curves and the characteristics are similar to those of a three-phase induction motor. The resultant torque–speed curve, which is the addition of the two component curves, shows that if the motor were started by auxiliary means, it would produce torque in whatever direction it is started.

In addition to the torques shown in Fig. 10.3, double-stator-frequency torque pulsations are produced by the interactions of the oppositely rotating flux and mmf waves, which glide past each other at twice the synchronous speed. These interactions produce no average torque, but they tend to make the motor noisier than a three-phase motor. Such torque pulsations are unavoidable in a single-phase

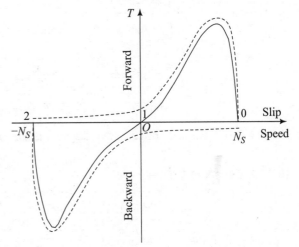

Fig. 10.3 Torque–speed/slip characteristic of a single-phase induction motor

motor because of the pulsations in instantaneous power input inherent in a single-phase circuit. The effects of the pulsating torque can be minimized by using elastic mountings for the motor.

The plot of the torque–slip characteristics shows that a single-phase induction motor has no initial torque. The challenge, in order to make a single-phase induction motor self-starting, is to strengthen the magnetic field in the direction of rotation relative to the backward field at the time of starting.

10.2.3 Types of Single-phase Induction Motors

Single-phase induction motors are classified in accordance with the methods of starting into two types—split-phase motors and shaded-pole motors. The starting methods and the torque–speed characteristics are discussed qualitatively in this section.

10.2.3.1 Split-phase Motors

Split-phase motors have two stator windings, a main winding and an auxiliary winding, with their axes displaced in space from each other by 90° electrical. Figure 10.4(a) shows the connection diagram. The auxiliary winding has higher impedance than the main winding. The auxiliary winding is made of thinner wire and fewer number of turns and thus has higher resistance to reactance ratio than the main winding. Therefore, the magnitude of the current in the auxiliary winding I_a is less than the current in the main winding I_m, and the current I_a leads I_m as shown in Fig. 10.4(b). At the time of starting, due to the phase difference in the two currents, the magnetic rotating field in the direction of the motion of the rotor is made stronger than the rotating magnetic field in the opposite direction, resulting in the production of starting torque. After the rotor has started rotating and reaches a speed of about 75 percent of synchronous speed, the auxiliary winding is usually cut out with the help of a centrifugal switch. A typical torque–speed characteristic is shown in Fig. 10.4(c).

Split-phase motors have low starting current and reasonably good starting torque. These motors have low cost and are used in fans, blowers, centrifugal pumps, grinders, and office equipment. These motors are available in the output range of 50 to 500 watts.

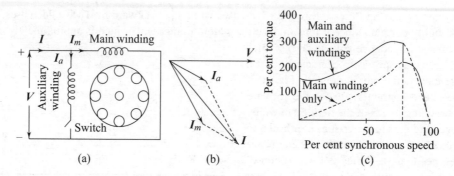

Fig. 10.4 Split-phase motor: (a) connection diagram, (b) phasor diagram at starting, and (c) torque–speed characteristics

10.2.3.2 Capacitor Motors

Capacitor motors are split-phase motors. Capacitors are used to improve the starting characteristic, running characteristic, or both of the single-phase motors.

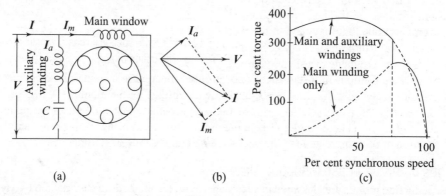

Fig. 10.5 Capacitor-start motor: (a) connection diagram, (b) phasor diagram at starting, and (c) torque–speed characteristics

In a capacitor start motor, the auxiliary winding has a capacitor connected in series with it as shown in Fig. 10.5(a), the current I_a in the auxiliary winding leads the supply voltage V. The current in the main winding I_m lags the supply voltage V. By selecting an appropriate capacitor, it is possible to make the auxiliary winding current I_a lead the main winding current I_m by 90° as shown in the phasor diagram of Fig. 10.5(b). In running condition, the auxiliary winding is cut out by the centrifugal switch. Figure 10.5(a) shows the connection diagram along with its phasor diagram. The torque–speed characteristics of such a motor are indicated in Fig. 10.5(c). It may be noted from the characteristics that the capacitor start motors have high starting torques. These motors are used for compressors, pumps, machine tools, and refrigeration and air-conditioning equipment. Capacitor split-phase induction motors are manufactured up to 1.5 kW rating.

In a permanent capacitor split-phase motor, the capacitor connected in series with the auxiliary winding is not cut out after starting but left in the auxiliary circuit permanently as shown in Fig. 10.6(a). Thus, the motor virtually is a two-phase motor running on single-phase supply. Due to the presence of a capacitor, the power factor, efficiency, and torque pulsations are improved. The capacitor helps in smoothing out pulsations in power input from the single-phase supply line. This makes the motor quieter. The starting torque of this motor is much less than that of capacitor start motor, as the selection of capacitor, a compromise between the best starting and best running value, is to be made. The torque–speed characteristic is shown in Fig. 10.6(b).

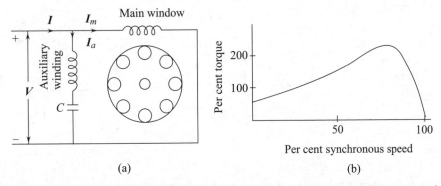

Fig. 10.6 Permanent capacitor split-phase motor: (a) connection diagram; and (b) torque–speed characteristic

In order to obtain the best starting as well as best running performance, two capacitors are used. In this way, a starting torque of value double the full-load torque can be obtained. The small value of capacitance required for optimum running condition is permanently connected in the auxiliary circuit, while a high value of capacitor at starting can be obtained by connecting a capacitor in parallel with the running capacitor as shown in Fig. 10.7(a). As the motor picks up speed and reaches about 75% of the rated speed, the starting capacitor is disconnected. Such a motor is called capacitor-start, capacitor-run motor. The torque–speed characteristic of such a motor is shown in Fig. 10.7(b).

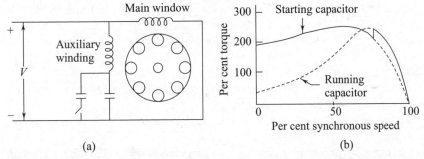

Fig. 10.7 Capacitor-start, capacitor-run split-phase motor: (a) connection diagram and (b) torque–speed characteristic

The cost of capacitor motors is related to the performance characteristics. Out of the three types of capacitor motors discussed, for motors with the same kilowatt rating, a permanent capacitor motor has the lowest cost, the cost of a capacitor-start is more, and the capacitor-start, capacitor-run motor has the highest cost.

10.2.3.3 Shaded-pole Motors

A shaded-pole motor consists of the main winding, which is wound around salient-poles on the stator. At about the centre of each pole, there is a shallow slot, which houses a short-circuited copper or aluminium ring, called the shading ring. The rotor usually has a squirrel cage construction. Figure 10.8 shows the basic construction of a shaded-pole motor. The single-phase current flowing in the exciting winding produces a two-pole alternating flux. Some of the flux through each pole links with the short-circuited shading ring, thereby inducing a current in the ring. This induced current sets up a magnetic flux in the shaded portion of the pole to lag the flux in the other portion. The phase differences between the two fields cause a torque to be produced. The field of the shaded-pole motor varies considerably in magnitude during rotation and, as a result, the starting torque is poor. The motor is suitable only for small powers and where the starting conditions are easy.

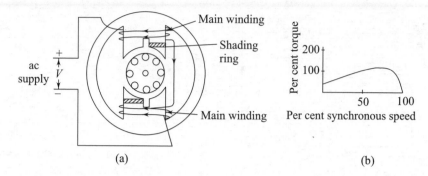

Fig. 10.8 Shaded-pole motor: (a) connection diagram and (b) torque–speed characteristic

Since these machines are simple in construction, they are the cheapest to manufacture. However, they are characterized by low efficiency, due to losses in the shading ring and the magnetic circuit, and low power factor. As such, they are available in ratings up to 1/20 kW and find applications in vending machines, convectors, small fans, advertising hoardings, etc.

10.3 SERVO MOTORS

A servo is a motor that has a rating of a few hundred watts and it finds widespread applications in aeroplanes, boats, radio-controlled cars, and robotics. A servo motor functions as an electromechanical device in which the position of the armature can be varied by an electrical input signal. Alternately, a servo motor transforms an electrical signal into a mechanical output such as torque and velocity. The electrical input is fed through an amplifier.

The features that a servo motor is expected to possess are as follows.

(i) The output torque of the motor should be approximately proportional to the error signal, which is amplified and fed as input voltage to the motor.

(ii) The output torque should be such that its direction is reversed with a reversal in the polarity of the input voltage.

Servo motors employed in present-day servo systems are classified according to their design principle into (i) dc servo motors and (ii) ac servo motors. The former are based on dc motor design and the latter are based on induction motor design.

10.3.1 DC Servo Motors

The various categories of dc servo motors are as follows:

(i) Shunt motors with field control

(ii) Shunt motors with armature control

(iii) Shunt motors with fixed field (permanent magnet)

(iv) Series motors

The shunt and series-field motors are operated as separately excited motors.

10.3.1.1 Shunt Motors with Field Control

The arrangement for a field-controlled dc servo motor is shown in Fig. 10.9. The field winding of the shunt motor is separately excited by a dc amplifier, which magnifies the error signal and excites the winding. The armature of the motor is supplied by a constant current source. In Section 9.7.6.2, it has been shown that the torque in a shunt motor is proportional to the product of the flux Φ and the armature current I_a. In this case, however, since I_a is kept constant, the torque produced will be directly proportional to flux Φ and the field current below the saturation level. The magnitude of Φ in turn depends upon the error signal. Thus, the torque produced by the motor will be zero when the error signal is zero since there will be no excitation. A reversal of the direction of the field current will reverse the direction of the flux, thereby reversing the direction of rotation of the motor.

A dc servo motor with field control suffers from the following weaknesses:

(i) A higher-rated servo motor would require a large constant current to be supplied to its armature, which may not be desirable.

(ii) Due to the large inductance of the field winding, it has a time constant of a magnitude, which results in a sluggish response to the changes in the error signal.

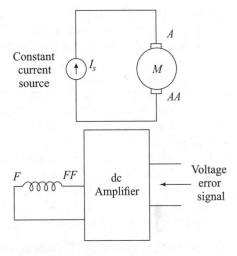

Fig. 10.9 Field-controlled dc servo motor

10.3.1.2 Shunt Motors with Armature Control

As is seen from the schematic diagram in Fig. 10.10, the armature of the motor is fed by an error signal through a dc amplifier and the field winding is excited by a constant current source. In order to minimize the sensitivity of the torque to minor changes in voltage fluctuations in the constant current source, the field winding is operated considerably above the knee of the magnetization curve. Further, due to the high magnitude of the field flux Φ, the torque of the motor ($T \propto \Phi I_a$) is more responsive to the changes in the armature current due to small variations in error signals from the amplifier. A reversal in the polarity of the error signal will reverse the direction of the armature voltage which in turn will reverse the direction of rotation of the motor.

The main advantage of armature-controlled servo motors is that due to the very low time constant of the armature circuit (which is largely non-inductive), the torque response is immediate due to large or small variations in the error signals. Motors of up to 1000 hp employ armature control.

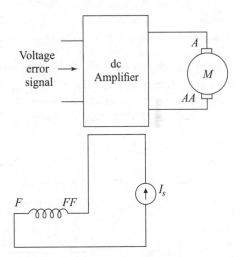

Fig. 10.10 Armature-controlled servo motor

10.3.1.3 Shunt Motors with Fixed Field (Permanent Magnet)

Figure 10.11 depicts the schematic arrangement of a permanent magnet armature controlled dc servo motor. Such types of motors operate in the same manner as armature voltage control servo motors discussed in the previous section. The permanent magnet provides a constant field excitation and is made from a high-quality magnetic material such as alnico or ceramic. The field magnet system consists of a cast circular ring made from alnico VI alloy. The ring completely surrounds the armature and provides a strong constant flux. The de-magnetizing effect of the field magnet due to reversal of dc armature voltage is compensated by the use of commutating windings. Hysteresis and eddy current effects are negligible and can be ignored.

These types of motors are very popular and are usually manufactured in the fractional horse power rating in 6 V and 28 V ranges. Up to 2 hp rating, the voltage rating is 150 V, in the integral horse power range.

Permanent magnet dc servo motors develop very high torques at standstill or low speeds and are, therefore, used for position control. When operating in the position control mode, these motors are also referred to as 'torque motors'. The second mode of operation of a permanent magnet dc servo motor is for velocity control. In this mode, since torque is inversely proportional to speed, the torques developed at high speeds are very small. The torque developed is also a function of the applied armature voltage.

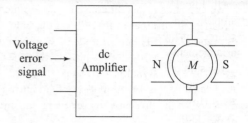

Fig. 10.11   Permanent magnet armature controlled dc servo motor

Permanent magnet dc servo motors have to run continuously to maintain a fixed velocity. Since the power developed by the motor is equal to $(2\pi NT)/60$ as per Eq. (9.24), as long as both torque and speed have low magnitudes, the power developed by it is small. However, since such motors are invariably closed type, a large frame size is used for efficient heat dissipation.

10.3.1.4 Series Motors with Split-field Windings

Series dc servo motors employ split-field windings with one called the main winding and the other the auxiliary winding. Both the windings are wound on the field poles in directions such that they set up fluxes that produce rotation in the direction opposite to each other. However, the mmf produced by each winding is generally equal in magnitude. The two circuit arrangements for exciting the armature are shown in Fig. 10.12. The advantages of employing two windings and splitting the field are as follows:

 (i) Since the windings are always excited and the time constant is negligible, due to a low inductance, the armature response to the error signal is enhanced.
 (ii) Since the direction of rotation is dependent upon the current difference between the auxiliary and main windings, a finer level of control is feasible.

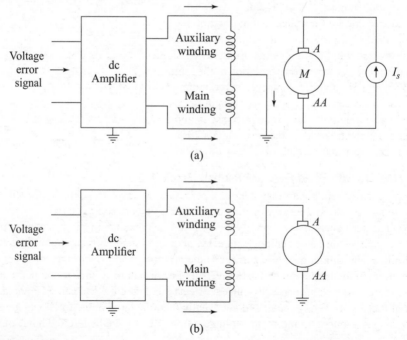

Fig. 10.12   Schematic circuit diagrams of dc series motors: (a) separately excited, and (b) directly excited

In Fig. 10.12(a), the main and auxiliary windings are excited by the error signal received via the dc amplifier and the armature is energized by a constant source. Since it is difficult to obtain constant current sources of high magnitudes, this arrangement is employed for small series motors. For larger motors, the arrangement shown in Fig. 10.12(b) is employed. In this configuration, the current flowing through the armature is the sum of the currents through the auxiliary and main windings.

The series servo motors generate a high starting torque and respond quickly to small changes in error signals. However, employing two windings in opposition results in a lower efficiency and their speed regulation is also poor. In small motors, low efficiency is of no significance and poor speed regulation can be overlooked since in servo systems the load is constant.

Generally, for the same horse power ratings, dc servo motors (both shunt and series types) have a higher rotor inertia than the ac servo motors due to heavier windings in the former. Also brush friction in dc motors discourages their use in small and sensitive servo systems.

10.3.2 AC Servo Motors

An ac servo motor is a two-phase induction motor with some modifications in its rotor for faster response. One of the phase windings, called the reference phase, is connected to a constant voltage ac source. The second phase, called the control phase, is driven by an ac amplifier. The control phase also includes a capacitor in series with the control winding to bring about a 90° phase displacement between the control and the reference voltages. The schematic arrangement of an ac servo motor is shown in Fig. 10.13.

The torque of the motor is controlled by changing the magnitude of the voltage applied to the control phase. The direction of rotation of the motor is reversed by bringing about a 180° phase displacement between the voltages.

A servo motor is used to control the motion of an object. Therefore, a desirable feature of such motors is that these should have a quick response. The construction of a two-phase ac servo motor is typically similar to that of any two-phase motor. However, in order to reduce the inertia, for a fast response, the iron cored rotor is removed and replaced by a shaft-mounted aluminium cup. Within the vacant space of the cup, a static iron core is reinserted to complete the magnetic circuit. The construction of an aluminium cup rotor is shown in Fig. 10.14. Such a rotor, having a low mass, accelerates more quickly than a squirrel cage rotor.

Compared with dc servo motors, ac servo motors offer the following advantages.

(i) Servo motors usually start from standstill or near-standstill positions. Heavy current flows, during slowly moving rotor conductors in dc servo motors, results in commutator pitting. Also due to arcing between the brushes and the

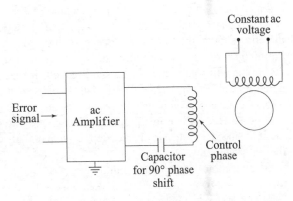

Fig. 10.13 Schematic arrangement of a two-phase ac servo motor

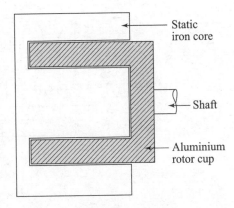

Fig. 10.14 Diagram showing the construction of a low-mass aluminium cup rotor of an ac servo motor

commutator, the former requires continuous replacement. On the other hand, rotors in ac servo motors are rugged and maintenance free since there are no brushes in contact with the commutator segments.

(ii) Commutator arcing in dc motors results in radiation and radio interference.

(iii) Low rotor inertia in an ac servo motor provides a quicker response in comparison with a dc servo motor, which has relatively higher rotor inertia.

(iv) Since rotors of ac servo motors do not involve any insulation, they can withstand higher temperatures.

10.4 STEPPER MOTORS

Like the dc and ac servo motors, a stepper motor is also an electromechanical device. It transforms electrical pulses into distinct mechanical movements. In other words, when an electrical pulse is applied to it, the shaft of the motor will rotate in distinct step increments. The principle of operation is explained with the help of Fig. 10.15, which shows a two-pole stepper motor.

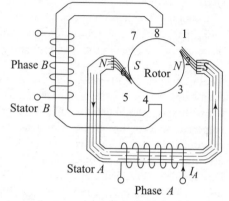

When phase A is excited by a dc current I_A flowing in the direction shown, the rotor moves so as to align itself to reduce the air gap reluctance and strengthen the flux. In this case, the motor will rotate in a clockwise direction. It may thus be noted that by exciting a single coil with a permanent magnet rotor, it is possible to obtain a 90° step. On the other hand, it is also possible to utilize this property of magnetic attraction to make the motor to rotate in distinct steps by sequentially exciting the stator windings.

Fig. 10.15 Principle of operation of a stepper motor

Stepper motors commonly employ two phases but motors with three and five phases are also available. A motor having two phases with a centre tap is called a unipolar machine and is shown in Fig. 10.16.

When a stepper motor has one winding per phase, it is called a bipolar motor. Figure 10.17 shows the schematic arrangement of a bipolar stepper motor.

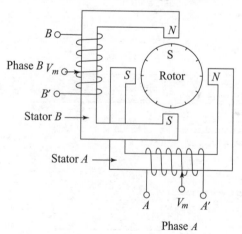

Fig. 10.16 Schematic diagram of a unipolar stepper motor

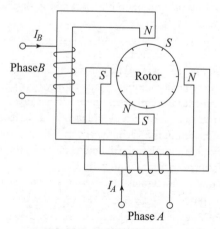

Fig. 10.17 Schematic diagram of a bipolar stepper motor

The full-step angle of a stepper motor is governed by the number of rotor poles P_R, which is equal to the number of stator poles per phase and the number of phases N_{ph}. Thus,

$$\text{Full-step angle, } \theta = \frac{360°}{P_R \times N_{ph}} \tag{10.3}$$

As an example, a three-phase motor with 16 rotor poles will give a full-step angle, $\theta = 360/(3 \times 16) = 7.5°$. On the other hand, the full-step size of a five-phase motor with 12 rotor poles is $\theta = 360/(5 \times 12) = 6°$.

10.4.1 TYPES OF STEPPER MOTORS

Broadly, steppers motors are classified into the following categories:
   (i)   Permanent magnet (PM)
   (ii) Variable reluctance (VR)
   (iii) Hybrid (HB)

10.4.1.1 PM Stepper Motors

As the name suggests, the rotor of these types of motors has a permanent magnet made from high-retentivity steel alloy, like alnico, and it always has an even number of poles. The stator phase windings whose terminals are brought out for dc excitation are similar to any conventional multiphase induction motor or synchronous motor windings. Figure 10.18(a) shows a typical two-pole two-phase stepper motor. PM stepper motors may use salient and/or non-salient poles stator and rotor construction.

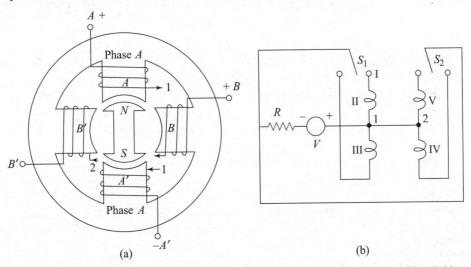

Fig. 10.18   Two-pole two-phase (bipolar) stepper motor: (a) connection of stator phases and (b) switching sequence

When stator phase AA′ alone is excited, the pole A acquires an N polarity while pole A′ gains an S polarity. Hence, the rotor remains locked in the position shown. Next, when phase BB′ alone is energized, the poles B and B′ are respectively S and N poles, causing the rotor to rotate through 90° in the clockwise direction. A continuous clockwise motion of the rotor is obtained by reversing the direction of current flow in the phases A and B consecutively and in turn. A static solid state switching device is employed to obtain the switching

Table 10.1   Full-step angles for a two-pole two-phase stepper motor

Switching step no.	Switch S_1	Switch S_2	Full-step angle
1	I	IV	90°
2	III	IV	180°
3	III	V	270°
4	I	V	360°
1	I	IV	90°

sequences for current reversal. Figure 10.18(b) shows the connections for the stator phases A and B. Table 10.1 shows the full-step angles as a result of the switching sequences.

PM stepper motors, also referred to as a 'tin can' or 'canstock' motors, are low-cost and low-resolution machines with typical step angles varying between 7.5° to 15°.

10.4.1.2 VR Stepper Motors

Construction-wise, a VR stepper motor has a wound stator and a multi-teethed rotor made of soft iron or a low-retentivity alloy. Figure 10.19 shows cross-section of a six-pole VR motor with four stator phases. It may be noted that for the VR motor shown, the full-stepper angle, $\theta = 360°/(4 \times 6) = 15°$

When the stator winding is energized, the resultant stator poles are attracted and the rotor begins to rotate. In other words, the rotor moves to a position to minimize the reluctance and maximize the air gap flux. The pull of the rotor to occupy a neutral position is the result of the stator excitation torque called the *reluctance torque*.

It may be mentioned that both the PM and VR stepper motors have inverse torque–speed characteristics, which are similar to induction motors.

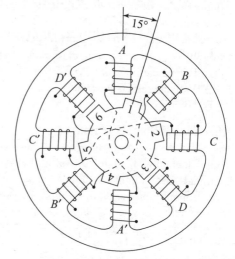

Fig. 10.19 Cross-sectional view of a six-pole and four-stator phases VR motor

10.4.1.3 HB Stepper Motors

HB stepper motors are assembled with a PM rotor and multi-teethed stator poles. To that extent, a HB motor utilizes the best characteristics of both the PM and the VR stepper motors. Typical HB stepper motor stators have eight poles with two to four teeth on each pole. The stator is wound for two phases with four poles per phase. The PM rotor has an even number (normally eighteen) of teeth. As against the PM and VR stepper motors, the full-step angle per change in stator excitation is a function of the number of rotor teeth only. Thus, if the number of rotor teeth in a HB motor is T, for each change in stator excitation, the full-step angle is given by

$$\theta = \frac{90°}{T} \tag{10.4}$$

Thus, full-stepping angle of a HB motor whose rotor has 40 variable reluctance teeth would be 2.25°. Typical full-step angles for a HB stepper vary between 0.9° to 3.6°. Figure 10.20 shows a cross section of a typical HB motor.

The advantage of using a teethed rotor is that the magnetic flux gets directed to the desired locations in the air gap, which provides it better dynamic torque properties than both the PM and VR types of motors.

Of the three types of stepper motors discussed above, PM and HB types of stepper motors are most commonly used. The two parameters that determine the selection of a stepper motor are cost and performance. PM motors are selected when cost is important and HB motors are chosen when performance is of significance.

A major advantage of HB motors is that small stepping angles are feasible with these types of motors. However, HB stepper motors are more expensive compared to PM stepper motors but provide better

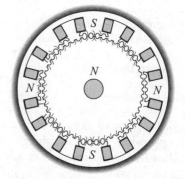

Fig. 10.20 Cross-section of a typical HB stepper motor

performance. On the other hand, use of PM stepper motors is restricted to smallest power ratings because of their inherent higher rotational speeds.

10.4.1.4 Ratings of Stepper Motors

Maximum power dissipation or heat dissipation limit determines the rating of a stepper motor. However, these are never clearly stated by the manufacturers. Generally, integrated circuit (IC)-driven motors are available for less than 1 W for small motors and in the range of 10 to 20 W for medium to large motors. The rating of a stepper motor is determined by using the well-known $V \times I$ formula. For example, a stepper motor rated at 6 V and 0.5 A/phase current would have a power rating of 3 W. However, if two phases of the motor are energized, it will have a power rating of 6 W. The normal power ratings of these motors are designed to allow a case or frame temperature rise of 65°C above the ambient.

10.4.2 Merits and Demerits of Stepper Motors

Following are the main merits of stepper motors:
   (i) Due to the absence of brushes, stepper motors have a high level of reliability.
  (ii) Accurate positioning since the step error in good designs of motors is of the order of 3 to 5% only.
 (iii) The angle of rotation is proportional to the input.
 (iv) These motors can be controlled precisely in an open loop system, that is, the position can be controlled by tracking the input pulses.
  (v) When the windings are excited, full torque is available at standstill.
The demerits of stepper motors are given below:
   (i) Operation is difficult at very high speeds.
  (ii) Good proper control is essential to overcome resonance.

10.4.3 Applications of Stepper Motors

Based upon the merits enumerated above, stepper motors find applications in areas where accurate positional control, such as rotational angle, speed, etc., is essential. Indicative areas of applications include fax machines, hard disc drives, printers and plotters, medical equipment, high-end office equipment, and numerous more.

10.5 HYSTERESIS MOTORS

Structurally, a hysteresis motor is a synchronous motor with a round rotor and without the dc field excitation. The smooth cylindrical rotor has no teeth and is made of magnetically hard steel. The stator, on the other hand, is slotted and carries distributed windings. The windings are designed to produce a flux, which has as near a sinusoidal distribution as practically feasible. As in the case of capacitor-type single-phase induction motors discussed earlier, the stator of a hysteresis motor is wound with a main and an auxiliary windings. The latter winding also includes a permanently connected capacitor, which creates a nearly two-phase conditions in the stator windings. Consequently, a magnetic field of constant waveform in space and rotating at the synchronous speed of ω_s rad/sec is produced by the stator windings. Figure 10.21(a) shows the magnetic field in the air gap and rotor of a hysteresis motor at any instant for a two-pole stator winding.

As a result of the magnetically hard steel, high levels of hysteresis and eddy current losses are induced due to the stator mmf. The interaction between the two losses is utilized to produce the torque. The stator magnetic field represented by the axis SS' is revolving in the anticlockwise direction at the synchronous speed of ω_s rad/sec. Due to the phenomenon of hysteresis in the rotor, the magnetization of the rotor will always lag behind the stator magnetic field. Thus, the rotor magnetization axis RR' in Fig. 10.21(a) will always lag the stator magnetic field axis SS'. The angle of lag is called the *hysteresis lag angle* and may be denoted by δ. At standstill, the torque produced is proportional to the product of the fundamental component of the stator mmf, rotor flux, and the sine of the hysteresis angle δ. If the torque due to the load on the motor is less than the torque developed, the rotor will start to rotate and continue to do so as long as the load torque is less than the developed torque. Thus, the speed of the rotor will be less than the synchronous speed of the stator mmf axis SS'. When the rotor is in motion at a speed less than ω_s rad/sec, the entire body of the rotor undergoes repeated hysteresis cycles at the slip frequency. Since the hysteresis angle δ is dependent only on the hysteresis loop of the rotor and is independent of the frequency at which it is crossed, the hysteresis angle δ remains constant even when the rotor is accelerating.

Thus, a hysteresis motor produces a constant torque, under ideal conditions, as shown in Fig.10.21(b).

The above statement may be proved mathematically as follows. The hysteresis losses P_h in watts (see Section 3.11.2) may be expressed as

$$P_h = K_h v f_r B^n \ \text{W} \tag{10.5}$$

All the symbols in Eq. (10.5) have the same meaning as in Eq. (3.46), except f_r, which is the rotor frequency and is given by $f_r = sf$. Thus, Eq. (10.5) is rewritten as

$$P_h = K_h v s f B^n \ \text{W} \tag{10.6}$$

From Eq. (8.24), the air gap power P_{ag} is given by

$$P_{ag} = \frac{P_h}{s} = \frac{K_h v s f B^n}{s} = K_h v f B^n = \text{constant} \tag{10.7}$$

If the unit for the torque developed is Nm, from the relation between the power and the torque, $T = P/\omega_s$, it is easily seen from Eq. (10.7) that the torque developed is also constant.

Along with the hysteresis loss, eddy current loss is also being produced in the rotor. With the help of Eq. (6.14), eddy current loss in the rotor may be written as

$$P_e = k_e f_r^2 B_m^2 \tau^2 v = k_e s^2 f^2 B_m^2 \tau^2 v \ \text{W}$$

From the foregoing discussion, the torque in Nm due to eddy current loss is given by

$$T = \frac{P_e}{\omega_r} = \frac{P_e}{s\omega_s} = \frac{k_e s^2 f^2 B_m^2 \tau^2 v}{s\omega_s} = \frac{k_e f^2 B_m^2 \tau^2 v}{\omega_s} s \tag{10.8}$$

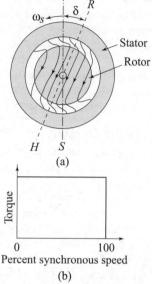

Fig. 10.21   (a) Magnetic field distribution in the air gap and rotor and (b) torque versus speed characteristics of a hysteresis motor

From Eq. (10.8), it is seen that the torque varies linearly with the slip, being the highest at starting when $s = 1$ and reducing to zero when $s = 0$, that is, at the synchronous speed. Since the torque due to the eddy current loss component adds to the hysteresis loss component, the hysteresis motor has an excellent, ripple-free, and nearly constant starting characteristic. As such, hysteresis motors find applications in areas where smooth starting is required, such as in clocks, other timing devices, and record player turntables. In the last application smooth starting torque is of particular advantage as it avoids record slippage.

10.5.1 Advantages of Hysteresis Motors
The advantages of hysteresis motors may be serialized as follows:
  (i) Because of constant starting torque, these motors are ideal to accelerate high-inertia loads.
 (ii) A hysteresis motor operates at a low noise level, in comparison to an induction motor, since the former operates at one synchronous speed and under undisturbed and nearly balanced two-phase conditions. The smooth rotor, without teeth, further helps in the motor's low noise performance.
(iii) Since the rotor of a hysteresis motor does not carry any windings, it quickly forms the same number of poles as the stator. This characteristic makes it possible to attain multispeed operation in hysteresis motors by setting up pole-changing stator windings.

10.6 UNIVERSAL MOTORS
A universal motor is a commutator motor and it is so named because it can be operated both with ac and dc voltages. It is a series wound dc motor which has been modified to operate on ac voltage. Due to its high speed, a universal motor develops the highest horse power per kg weight than any other ac motor.

10.6.1 Construction of a Universal Motor

The construction of a universal motor is very much like that of a dc machine. The field poles, which carry field windings, are mounted on the stator. The entire magnetic path consisting of the field and armature circuits is laminated to reduce eddy current loss when the motor is operating on ac voltage.

The wound type armature winding is placed on the rotor which has either straight or skewed slots. Stationary high resistance brushes press against the commutator and maintain a continuous contact. Due to induced currents in the armature coils, when the motor is operating on ac supply, commutation is poorer than under dc operation. Hence, high resistance brushes are used to reduce sparking.

10.6.2 Operation of a Universal Motor

The principle of operation of a universal motor when it is working on dc voltage is the same as that of a dc motor (see 9.7.1). Since the field and armature windings are connected in series, the same current flows in the two windings. The magnetic flux set up by this current exerts a force on the current carrying armature conductors causing them to rotate. The direction of rotation of the armature is determined by applying Fleming's left hand rule (see 1.10.2)

When connected to an ac supply, since both field and armature windings are connected in series, the directions of the field flux and the armature current change simultaneously every half cycle of the supply. Thus, the direction of the mechanical force on the armature conductors remains unchanged and the torque produced is unidirectional.

10.6.3 Characteristics of a Universal Motor

From the foregoing it becomes clear that a universal motor functions like a dc series motor irrespective of the fact whether it is operating on dc or ac supply voltage. Therefore, it is expected that its characteristics are similar to those of a dc series motor (see 9.7.6.1). Figure 10.22 shows the load versus speed characteristics.

From the characteristics it is seen that the speed of a universal motor is inversely proportional to the load. In order to obtain a desired speed at low loads, a gear train is used to reduce the speed of the motor. Further, it may also be noted that for the same torque, the motor speed is higher when it is operating on dc voltage.

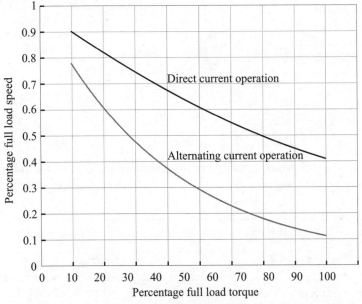

Fig. 10.22 Load versus speed characteristics of a universal motor

10.6.4 Applications of Universal Machines

Due to their high speeds and smaller size (per hp rating) than other ac motors, universal motors find wide applications in domestic appliances such as food mixers, vacuum cleaners, sewing machines and so on. Universal machines requiring higher ratings are used in blenders, hand held drills, and so on.

10.6.5 Drawbacks of Universal Machines

Universal motors are designed for high speeds, typically in the range of 3600 rpm to 20, 000 rpm, and since they employ a commutator and brushes, which tend to wear out, their life is limited.

10.7 TACHOMETER GENERATORS

A tachometer generator, or simply a tachometer, is a device used for measuring rotational speeds of motors and generators. It is usually mounted on the shaft of the motor/generator whose speed is required to be measured. It functions by generating a voltage signal which is proportional to the rotational speed of the shaft. Thus,

$$V = K_v v$$

where K_v is the tachometer gradient constant in volts per rad/sec and v is the shaft velocity in rad./sec.

10.7.1 Types of Tachometer Generators

Tachometers are classified into (i) dc tachometer generators and (ii) ac tachometer generators. In this sub-section the two types of tachometers are discussed.

10.7.1.1 DC Tachometer Generators

A present-day tachometer employs a permanent magnet with an armature mounted on a rotating shaft. The end of this shaft is attached to the machine shaft whose speed is required to be measured. Figure 10.23 shows the schematic layout of a dc generator.

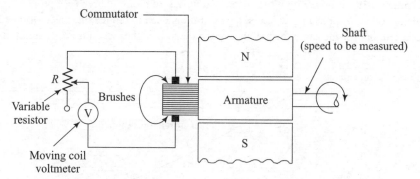

Fig. 10.23   Schematic of a dc tachometer generator

The moving coil voltmeter, which is connected across the brushes, measures the voltage generated by the tachometer generator. The variable resistance R in the voltmeter circuit maintains the current flow in the voltmeter to safe levels.

Since the voltage generated by the tachometer generator is proportional to its speed, the voltmeter is calibrated to read the speed in RPM directly.

10.7.1.2 AC Tachometer Generators

Figure 10.24 shows the layout of an ac tachometer generator. In this arrangement, the rotor has the field winding and the stator coil constitutes the stationary armature, as in the case of a synchronous generator. The alternating voltage induced in the armature (stator coil) is rectified and measured by the moving coil voltmeter

V. The capacitor C acts to eliminate any ripple in the generated ac voltage and provide a steady dc output voltage. Since both the amplitude of voltage and frequency are dependent on the speed of the shaft, any one of the parameters can be employed to read the speed in RPM directly.

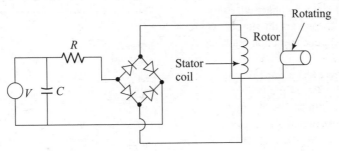

Fig. 10.24 Schematic layout of an ac tachometer generator

The advantage of an ac tachometer over its dc counterpart is that the commutation problems associated with dc tachometer are not present in ac tachometer.

10.8 SYNCHRO

In the expanded form a synchro stands for 'synchronous'. It represents a series of self synchronizing inductive devices to produce torques, which can force two mechanically independent shafts to rotate in synchronism or in the opposite direction, by electrically energizing and electrically connecting the windings of the devices. Depending upon the output, a synchro physically looks like a small cylindrical motor having a diameter of 1.5 cm to 10 cm. Due to their low torque producing capability, synchros find wide applications in control systems for conveying data on shaft positions. In the field, synchros are known by the tradename of *selsyns* or *autosysns*.

10.8.1 Construction

A synchro is essentially a transformer, in which the coupling between the windings may be varied by rotating one winding. A synchro transmitter consists of a single salient pole rotor which carries a single-phase winding wound in the iron slots of the laminated iron stack. Single-phase ac voltage is applied to the rotor through the slip rings. The stator, much like a synchronous generator, carries a three-phase winding which can be connected in star or delta. The three-phase stator winding is displaced in space by 120°.

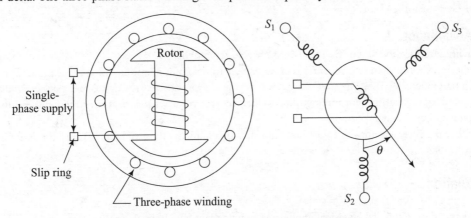

Fig. 10.25 Schematic layout of a control transmitter

10.8.2 Working Principle

When an ac voltage is applied to the rotor a magnetic field is set up which in turn induces an emf in each of the three stator phase windings. These induced voltages vary directly, as the sine of the angle between the actual rotor position and a zero reference position. Thus, a synchro may be viewed as a transformer wherein the coupling between the primary and the secondary winding can be changed by actually varying the orientation between the two windings. In the present case, since the stator is stationary and the rotor is free to rotate, the secondary winding (stator) tries to pull the rotor so that the primary and secondary magnetic fields are aligned. The

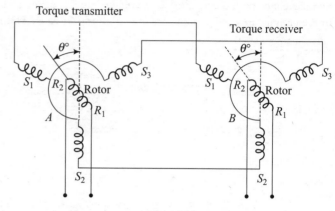

Fig. 10.26

torque so developed in the synchro, thus depends on the magnitude of the field produced in the stator winding, which in turn, depends on the amount of current flowing through it. Therefore, the magnitude of the torque developed in a synchro can be varied by altering the field strength.

The working of one of the commonly used transmitter and receiver synchros is explained below. When the three stator leads of transmitter 'A' are connected to the Y-connected stator of a second synchro 'B', as shown in Fig. 10.26, these voltages produce a resultant stator field in the second synchro, having the same angular orientation with respect to its zero reference as the transmitter rotor. When the second synchro, 'B' has its single-phase rotor winding connected to the same power supply that energizes the transmitter rotor, the rotor aligns itself to the same angle as the transmitter rotor. The synchro, 'B' is called a receiver. The transmitter and receiver are generally identical in construction.

10.8.3 Application

A torque synchro system for remote indication will meet the demand for a simple, robust, and accurate method of transmitting functional information to a remote observation point. The receiver is connected only electrically to the transmitter and can be remotely located. With proper cables, the distance between the transmitter and the receiver can be as far as 5 km. The pointer mounted on the receiver shaft will smoothly follow the motion of the transmitter shaft.

10.8.4 Advantages

The following are the advantages of a transmitted receiver synchro:
  (i) Sometimes more than one receiver can be driven from one transmitter for remote indication at more than one place.
 (ii) Transmission is self synchronizing, that is, the receiver rotor will align itself to the correct position as soon as power is switched on.
(iii) Neither the receiver nor the transmitter has any mechanical stops and can go through any number of revolutions.

Recapitulation

The load demand of individual electrical appliances, such as fans, food mixers, washing machines, refrigerators, fax, and photocopying machines, used in homes and offices lies within the fractional kilowatt range.

Motors in the fractional kilowatt range are

(i) Single-phase induction motors

(ii) Servo motors

(iii) Stepper motors

(iv) Hysteresis motors

Single-phase induction motors are classified as (i) split-phase motors, (ii) capacitor motors, and (iii) shaded-pole motors.

In single-phase induction motors, forward slip $s_f = s$ and backward slip $s_b = 2 - s$.

The main features of servo motors, which are used in aeroplanes, boats, radio-controlled cars, and robotics, are given below:

(i) Output torque is approximately proportional to the error signal.

(ii) Direction of the output torque reverses with a reversal in the polarity of the input voltage.

Servo motors based on dc motor and induction motor design principles are, respectively, classified as dc and ac servo motors.

The four categories of dc servo motors are shunt motors (i) with field control, (ii) with armature control, (iii) with fixed field (permanent magnet), and (iv) series motors.

For the same horse power rating, dc servo motors have higher rotor inertia than ac servo motors.

An ac servo motor is a two-phase induction motor with one phase acting as a reference phase and the second phase as a control phase, which also includes a capacitor for 90° phase displacement and is driven by an ac amplifier.

Magnitude of the control phase voltage controls the motor torque.

The inertia of an ac motor is reduced by employing a low-mass aluminium cup rotor.

In comparison with dc servo motors, ac servo motors (i) are rugged and require less maintenance, (ii) no radiation and radio interference, (iii) provide quick response, and (iv) can withstand higher temperatures.

Stepper motors are electromechanical devices, which transform electrical pulses into distinct mechanical movements.

A unipolar ac servo motor has two phases with a centre tap and a bipolar ac servo motor has one winding per phase.

Full-step angle, $\theta = \dfrac{360°}{P_R \times N_{\text{ph}}}$

where P_R and N_{ph} are the number of rotor poles and stator phases, respectively.

The rotor of a PM stepper motor has a permanent magnet with even number of poles and the stator phase windings, which are like any conventional multiphase induction motor or synchronous motor windings, receive dc excitation. The motion of the rotor is obtained by reversing the direction of current flow in the phase windings.

A VR stepper motor has a wound stator and a multi-teethed rotor made of soft iron or a low-retentivity alloy. Excitation of the stator generates a torque called the *reluctance torque*, which causes the rotor to rotate. Both the PM and VR stepper motors have inverse torque–speed characteristics, which are similar to induction motors.

HB stepper motors employ the rotor of a PM motor and multi-teethed stator poles of a VR motor and they possess the best characteristics of both these motors.

The full-step angle θ of a HB motor is a function of rotor teeth T only and is given by $\theta = 90°/T$.

The rating of a stepper motor, based on a case temperature rise of 65 °C, is determined by the formula $V \times I$, where I is the current/phase.

The main merits of stepper motors are (i) a high level of reliability, (ii) accurate positioning, (iii) an angle of rotation which is proportional to input, (iv) functioning in open loop system, and (v) full starting torque available when windings are excited.

The demerits of stepper motors are (i) the operation is difficult at very high speeds and (ii) to avoid resonance proper control is essential.

Areas of applications of stepper motors include fax machines, hard disc drives, printers and plotters, medical equipment, high-end office equipment, and numerous more.

A hysteresis motor is a synchronous motor with a round rotor of magnetically hard steel and without the dc field excitation.

The torque developed in the motor is due to the interaction between the synchronously revolving stator mmf and due to the lagging rotor magnetization axis because of the rotor hysteresis.

Under ideal conditions a HB motor produces a constant torque.

Advantages of HB motors are (i) ideal to accelerate high inertia loads, (ii) low noise level performance, and (iii) multispeed operation.

Hysteresis motors are used in areas where smooth starting is required, such as in clocks, other timing devices, and record player turntables.

A universal motor is a commutator motor. When operating on dc voltage it functions like a dc motor. On ac voltage, it produces a unidirectional torque since both the armature and field windings are connected in series and the field flux and armature current change directions with every half cycle. The universal motor has an inverse load-speed characteristic; however, for a given torque its speed is higher when it is operating on dc voltage.

Since universal motors operate at high speeds (3600 – 20,000 rpm), their life is limited due to wearing of the brushes and the commutator. Universal motors find applications in mixers, blenders, handheld drills, and so on.

Tachometer generators measure the rotational speed of shafts by generating a voltage signal which may be written as

$$V = K_v \, v$$

where K_v is the tachometer gradient constant in volts per rad/sec and v is the shaft velocity in rad./sec.

In a dc tachometer, a moving coil voltmeter is connected via a variable resistor, across the brushes of a permanent magnet generator. The variable resistor limits the current to safe values.

Constructionally, an ac tachometer is a synchronous generator, whose induced voltage is rectified and fed to a moving coil meter for measuring the generated voltage. Unlike its dc counterpart, an ac tachometer is free of commutation problems.

A synchro symbolizes a series of self synchronising inductive devices which produce torques, and force two mechanically independent shafts to rotate in synchronism or in the opposite direction, by electrically energizing and electrically connecting the windings of the devices.

Constructionally, a synchro is essentially a transformer, in which the coupling between the windings is varied by rotating one winding.

Advantages of synchros are

(i) Sometimes more than one receiver can be driven from one transmitter for remote indication at more than one place.

(ii) Transmission is self synchronizing.

(iii) Neither the receiver nor the transmitter have any mechanical stops and can go through any number of revolutions.

Assessment Questions

1. (a) Explain how a single-phase ac voltage can be resolved into two equal synchronously rotating magnetic fields in the opposite directions.
   (b) Discuss the torque–slip characteristics of a single-phase induction motor.
2. (a) Describe with the help of a connection diagram, the operation of a split-phase induction motor.
   (b) Draw the torque–slip characteristics of the motor.
3. (a) Describe the function of the series capacitor in the auxiliary winding of a capacitor start motor. Draw the phasor diagram at the time of starting of the motor.

(b) Discuss the working and torque–speed characteristics of a permanent capacitor split-phase motor.

4. (a) Describe the operation of a two-capacitor motor.
   (b) Enumerate the merits of split-phase and capacitor motors.

5. (a) Draw the connection diagram of a shaded-pole induction motor and explain how a torque is produced.
   (b) Discuss the merits and demerits of shaded-pole induction motors.

6. (a) State why single-phase motors are manufactured in fractional kilowatt range and enumerate a few of their applications.
   (b) Explain why trial-and-error approach is adopted in their development.

7. (a) What is a servo motor and what are the main features it is expected to possess?
   (b) Based on their design principles, categorize servo motors. Compare the dc and ac servo motors.

8. Classify dc servo motors and write notes to explain the working of each category.

9. (a) Write a detailed note on the working of an ac servo motor.
   (b) Explain with the help of a diagram, the construction of an aluminium cup rotor of an ac servo motor. Why is it necessary to use this type of a rotor?

10. (a) How are stepper motors different from servo motors?
    (b) Explain the principle of operation of stepper motors.

11. (a) Draw schematic diagrams of unipolar and bipolar stepper motors.
    (b) Classify stepper motors and explain the working of a permanent-magnet stepper motor.

12. Describe the working of a (a) VR motor and (b) HB motor. Explain why PM and HB motors are commonly used.

13. (a) Write a note on the rating of stepper motors.
    (b) Enumerate the merits and demerits of stepper motors.

14. (a) Describe with the help of a diagram, the structure of a hysteresis motor.
    (b) Explain qualitatively, how torque is produced in a hysteresis motor.

15. (a) Prove mathematically that the torque produced due to hysteresis loss is independent of the slip.
    (b) Derive an expression for the eddy current loss in the rotor of a hysteresis motor and show that the developed torque, in Nm, varies linearly with slip. Explain how it helps in obtaining a ripple free constant torque.

16. Discuss the advantages and disadvantages of hysteresis motors.

17. (a) Explain the principle of working of single-phase induction motors.
    (b) Describe slip in a single-phase induction motor and develop its torque–slip characteristic.

18. Write a detailed note on universal motors.

19. What is a tachometer? Describe the working of a dc tachometer generator.

20. Explain working of an ac tachometer. What are its advantages?

21. Write a detailed note on synchros.

Objective Type Questions

1. Which of the following is not a basis for developing fractional kilowatt motors?
   (i) established design practice
   (ii) prototype
   (iii) trial and error
   (iv) all of these

2. Which of the following statement is not associated with single-phase induction motors?
   (i) they are manufactured in kilowatt range.
   (ii) they have a cage rotor with a single-phase winding.
   (iii) they have a slip-ring type of a rotor with a single-phase winding.
   (iv) they are wound for even number of poles.

3. On which of the following is the principle of working of single-phase induction motors based?
   (i) three phase flux produced in the stator winding
   (ii) pulsating flux produced in the stator
   (iii) rotating field set up in the air gap
   (iv) none of these

4. A single-phase motor can be started if
   (i) it is suddenly switched on
   (ii) a starter is used
   (iii) there is no mechanical load
   (iv) the rotor is pushed in one direction
5. The speed of rotation of the two magnetic fields is
   (i) $\dfrac{120f}{P}$
   (ii) $\dfrac{240f}{P}$
   (iii) $\dfrac{60f}{P}$
   (iv) None of these
6. The backward rotor slip in an induction motor is equal to
   (i) $1-s$ (ii) s
   (iii) $2-s$ (iv) $s/2$
7. Which of the following is a characteristic of a capacitor start motor?
   (i) high starting torque
   (ii) can be manufactured up to 5 kW
   (iii) exhibits improved starting and running characteristics
   (iv) all of these
8. Which of the following is not a characteristic of a shaded-pole single-phase motor?
   (i) rotor has a slip ring construction.
   (ii) main winding is wound around salient poles on the stator.
   (iii) shading ring is placed inside a shallow slot at the centre of each pole.
   (iv) it has low efficiency.
9. Design of ac servo motors is based on
   (i) dc motor design
   (ii) synchronous motor design
   (iii) induction motor design
   (iv) none of these
10. Which of the following types of control can be used in a shunt motor to obtain a servo motor operation?
    (i) fixed field (ii) field control
    (iii) armature control (iv) all of these
11. In an armature-controlled servo motor, the error signal is fed to the
    (i) armature winding
    (ii) field winding
    (iii) armature and field windings
    (iv) none of these
12. At which of the following speeds, the torque developed by a permanent magnet dc servo motor is very small?
    (i) zero (ii) low

   (iii) very high (iv) all of these
13. In which of the following dc servo motors, a split field is employed?
    (i) shunt motors with field control
    (ii) shunt motors with armature control
    (iii) shunt motors with fixed field
    (iv) series motors
14. Which of the following is not a characteristic of dc series servo motor?
    (i) enhanced response to error signal
    (ii) possibility of finer level of control
    (iii) a high starting torque
    (iv) none of these
15. An ac servo motor employs a
    (i) squirrel cage rotor (ii) slip ring rotor
    (iii) salient pole rotor (iv) none of these
16. It is possible to have stepper motors with
    (i) two stator phases (ii) three stator phases
    (iii) five stator phases (iv) any one of these
17. A stepper motor has a full-step angle of 15°. The number of steps per revolution is
    (i) 48 (ii) 24
    (iii) 12 (iv) 6
18. Which of the following stepper motor is referred to as a 'tin can' or canstock'?
    (i) PM (ii) VR
    (iii) HB (iv) Hysteresis
19. Which of the following is not a feature of a HB stepper motor when compared to a PM stepper motor?
    (i) better performance
    (ii) more expensive
    (iii) higher rotational speed
    (iv) all of these
20. A HB motor has two stator phases and eight poles, with an eighteen-teethed rotor. Which of the following represents the full-step angle of the servo motor?
    (i) 22.5° (ii) 20°
    (iii) 5.625° (iv) 5.0°
21. Which of the following is a disadvantage of stepper motors?
    (i) no brushes (ii) open loop operation
    (iii) resonance (iv) all of these
22. The flux produced by the distributed stator windings in a hysteresis motor is
    (i) pulsating (ii) revolving
    (iii) sinusoidal (iv) none of these

23. Which of the following is responsible for low noise performance of a hysteresis motor?
    (i) rotor without teeth
    (ii) constant starting torque
    (iii) rotor carrying no winding
    (iv) constant starting torque

24. The operation of a universal motor is like that of a
    (i) dc series motor
    (ii) dc shunt motor
    (iii) dc compound motor
    (iv) none of these

25. Which of the following happens when a universal motor is operating on ac supply voltage?
    (i) direction of field flux changes
    (ii) direction of armature current changes
    (iii) direction of rotation does not change
    (iv) all of these

26. When compared with other ac motors, the horse power developed per kg by a universal motor is
    (i) less   (ii) more
    (iii) same   (iv) cannot say

27. Which of the following is not present in an ac tacho-generator?
    (i) Brushes   (ii) Stator
    (iii) Shaft   (iv) All of these

28. The torque produced in a synchro is due to
    (i) current in primary winding
    (ii) current in secondary winding
    (iii) magnetic fields
    (iv) all of them

Answers

1. (i)	2. (iii)	3. (ii)	4. (iv)	5. (i)	6. (iii)	7. (iv)
8. (i)	9. (iii)	10. (iv)	11. (i)	12. (iii)	13. (iv)	14. (iv)
15. (iv)	16. (iv)	17. (ii)	18. (i)	19. (iii)	20. (iv)	21. (iii)
22. (iii)	23. (i)	24. (i)	25. (iv)	26. (ii)	27. (i)	28. (iv)

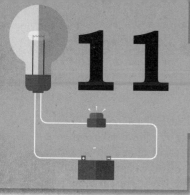

11 BASIC ANALOGUE AND ELECTRONIC INSTRUMENTS

Learning Objectives

This chapter will enable the reader to

- Categorize the different types of measuring instruments and identify them according to their principle of operation
- Enumerate the basic systems in measuring instruments
- Understand how different types of ammeters and voltmeters function and how multi-range instruments can be obtained
- Comprehend the working of various instruments used in the measurement of power
- Itemize the different types of instruments and describe their principles of operation for the measurement of energy in single- and three-phase circuits
- Describe the working principle of an insulation tester
- Acquire knowledge of the working of multi-meters and their applications
- Categorize the various types of errors likely to come about while taking measurements

11.1 INTRODUCTION

The quality of electric energy supplied is governed by the values of operating variables maintained at consumer terminals. The operating variables that define the condition of the supply system are voltage, frequency, current, power, and energy. As per the Indian Electricity Rules, the supplier is required to maintain declared values of voltage and frequency, within permissible variations, at consumers' terminals. For example, for domestic consumers, the voltage and frequency required to be maintained is $230 \pm 5\%$ V and $50 \pm 3\%$ Hz. Additionally, it is essential for the supplier to know whether the system is meeting the power demand of the consumers. For the purpose of billing the consumers, the supplier must also be able to measure the electric energy delivered to consumers.

In order to monitor the operating variables, it is essential to measure these quantities. Therefore, it is important to discuss the operating principles of different measuring instruments.

11.2 CLASSIFICATION OF INSTRUMENTS

Broadly, measuring instruments can be divided into the following categories:
- (a) Absolute instruments
- (b) Secondary instruments.
- (c) Type of input
    - (i) Analogue instruments
    - (ii) Digital instruments
- (d) Electrical instruments
- (e) Electronic instruments

11.2.1 Absolute Instruments

The value of an electrical quantity cannot be read directly from an absolute instrument. It is measured in terms of the constants of the instrument. The value of this quantity is then obtained by employing a mathematical expression that defines the relationship between the constants of the instrument. For example, in a tangent galvanometer, which is an absolute instrument, the current flowing through it is a function of the tangent of the angle of deflection, mean radius, number of turns of its coil, and the horizontal component of the earth's magnetic field. Absolute instruments, however, require no calibration.

11.2.2 Secondary Instruments

Secondary instruments, as opposed to absolute instruments, are direct reading type instruments. The value of an electrical quantity can be read directly from the deflection shown on the instrument. Before a secondary instrument can be made direct reading, it must be calibrated with an absolute instrument or an already calibrated secondary instrument. Deflections on an un-calibrated secondary instrument are not a true measure of the electrical quantities.

Due to the ease of measuring electrical quantities on secondary instruments, they are extensively used in practice. Absolute instruments find applications in laboratories for standardization, research, and calibration work.

11.2.3 Analogue Instruments

In analogue instruments a mechanical representation is used as the analogue of an electrical quantity to indicate the magnitude of the electrical quantity. For example, when a current is made to flow through a wire, a force is produced, which, in turn, produces a torque. If this wire is connected to a spring with a pointer, the latter will stretch in proportion to the magnitude of the current. The deflection of the pointer is indicative of the magnitude of the current.

Since the analogue instruments are relatively cheap, they find applications where high accuracy of measurements is not required.

11.2.4 Digital Instruments

In digital instruments, the electrical quantity or any other signal constitutes the input signal. This analogue input is converted to the binary coded decimal (BCD) input representation, which is then decoded and displayed as digits on a screen.

Digital instruments are used in applications in which accurate measurements are essential. For example, data transmission in communication circuits requires an extremely high accuracy in frequency setting.

11.2.5 Electrical Instruments

These instruments transform electricity into some other form of energy, such as heat, motion, etc. to measure electrical quantities like current, voltage, and so on.

11.2.6 Electronic Instruments

In these types of instruments, electronic devices are used for measurement. An electronic device on the other hand controls the current and adds significant information to it. For example a cell phone, as an electronic device, manipulates current to obtain sound information and makes conversation possible.

11.3 OPERATING PRINCIPLES

Electrical measuring instruments utilize one of the several effects of current or voltage for their operation. Table 11.1 shows these effects.

Table 11.1 Various effects utilized in different types of measuring instruments

Type of effect	Types of instruments utilizing the effect
Magnetic	Most electrical instruments, such as ammeters, voltmeters, wattmeters, etc.
Electromagnetic induction	ac ammeters, voltmeters, wattmeters, etc.
Heating	ac and dc ammeters and voltmeters
Chemical	dc ampere-hour meters
Electrostatic	Voltmeters, indirect application in ammeters and wattmeters

The first two effects of current and voltage are commonly applied in measuring instruments. Therefore, instruments operating on principles of magnetic and electromagnetic induction effects are described in this chapter.

11.4 ESSENTIAL FEATURES OF MEASURING INSTRUMENTS

Instruments working on the principle of magnetic or electromagnetic induction produce a deflection torque (also called operating torque), which is indicated by the measuring instrument. Hence, the first essential feature of a measuring instrument is a moving (deflecting) mechanism for measuring deflection. The second essential feature is a controlling system that controls the deflection of the moving mechanism. The third and final feature is a damping system to bring the deflection quickly to a steady state. These three essential systems of measuring instruments are discussed in the following sections.

11.4.1 Deflecting System

The deflection produced by the operating torque is proportional to the magnitude of the electrical quantity such as current, voltage, etc. being measured. The deflecting torque causes the moving mechanism to move from its initial zero position, that is, when there is no deflection torque.

11.4.2 Controlling System

The torque produced by the controlling system is in opposition to the deflecting torque. The pointer comes to rest at a position when the deflecting and controlling torques are equal. There are two types of controlling mechanisms: (i) spring control and (ii) gravity control.

11.4.2.1 Spring Control

A phosphor–bronze hair spring is attached to the spindle of the deflecting system of the instrument. When the pointer is subjected to a deflecting torque T_d the spring gets twisted in the opposite direction, thereby producing a controlling torque T_c.

Assuming T_d to be proportional to the current I passing through the instrument, and the controlling torque T_c proportional to the angle of twist, θ, then, when the pointer is at rest,

$$T_d = T_c$$

Therefore

$$\theta \propto I \tag{11.1}$$

From Eq. (11.1) it is clear that in instruments employing spring control, the scale is uniform or linear over the whole range. Figure 11.1 shows the spring control arrangement.

11.4.2.2 Gravity Control

In this arrangement, small adjustable weights are attached to the deflecting system. The weights are so arranged that they produce torques in the opposite direction. Figure 11.2 shows the arrangement of the weighs in a gravity control system.

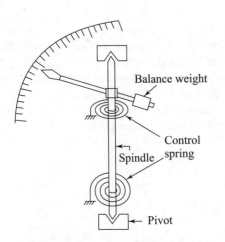

Fig. 11.1 Spring control arrangement

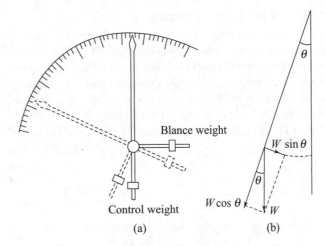

Fig. 11.2 Gravity control arrangement

In this case, the controlling torque

$$T_c = \sin \theta$$

Again, under the condition of equilibrium,

$$T_d = T_c$$

If it is assumed that the deflection torque $T_d \propto I$, then

$$I \propto \sin \theta \qquad (11.2)$$

It can be seen from Eq. (11.2) that in instruments employing gravity control the scale is not uniform. The scale is cramped for low values of currents while it is extended for higher values of currents.

The advantage of a gravity control system is that it is cheaper and sturdier than a spring control system. Its disadvantage is that it has to be kept in a vertical position for taking measurement.

11.4.3 Damping System

The deflecting torque and the controlling torque act in opposition. Hence, both deflecting and controlling systems, due to their inertias, have a tendency to cause the pointer to oscillate about a mean position. If an ammeter were to be suddenly connected in a circuit, the current I flowing through it will cause the pointer to oscillate under the action of the deflecting torque and controlling torque. In order to bring the pointer quickly to its steady-state position, it is necessary to damp the oscillations. For this purpose a damping system is essential. Curves in Fig. 11.3 show the effect of damping the oscillations of the pointer in

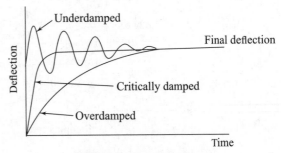

Fig. 11.3 Damping curves

bringing it to a steady final deflection. It may be noted that if the instrument is underdamped, oscillations around the final position will take place. If the instrument is overdamped, the pointer moves slowly from zero to final deflection. When the instrument is critically damped, the deflection of the pointer to the final position is achieved quickly. Instruments with such a type of damping (critically damped) are said to be dead-beat.

The two methods of damping commonly utilized in measuring instruments are (a) eddy current damping and (b) air friction damping.

11.4.3.1 Eddy Current Damping

Eddy current damping is considered to be the most efficient form of damping. Figures 11.4(a) and 11.5 show two arrangements of eddy current damping. In Fig. 11.4(a), a thin aluminium or copper (non-magnetic) disc is mounted on a spindle of the moving system. The edge of the disc is so adjusted that it moves between poles of a permanent magnet. Thus, when the disc rotates, it cuts the magnetic flux lines and an emf is induced, which causes currents, called *eddy currents*, to circulate in the disc. By applying Lenz's law, it can be seen that the direction of the eddy currents is such that they exert a force which opposes the direction of rotation of the disc. Figure 11.4(b) shows the distribution of the magnetic flux and direction of eddy curents.

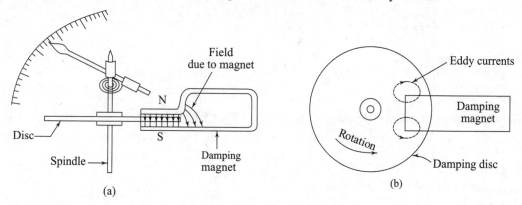

Fig. 11.4   Eddy current damping

Another type of eddy current damping is shown in Fig. 11.5. This type of damping is employed in permanent-magnet moving-coil instruments. In this arrangement the coil is wound on an aluminium frame (former) in which eddy currents are induced when the coil moves in the permanent magnet field. The direction of induced emf and eddy currents are shown when the coil rotates in the clockwise direction. Applying Lenz's law, it can be seen that the direction of the force exerted by eddy currents is opposite to the clockwise rotation of the coil.

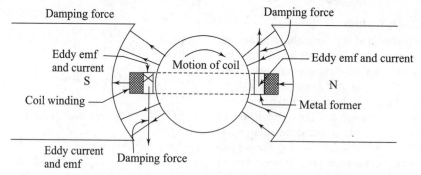

Fig. 11.5   Eddy current damping in a permanent-magnet moving-coil instrument

11.4.3.2 Air Friction Damping

Figure 11.6 shows a form of the air friction damping arrangement. In this arrangement, a small aluminium piston is attached to the spindle of the moving system. The piston itself moves in a circular or rectangular chamber with one side open to air. Damping of the moving system is brought about by compression and suction of the air in the chamber. When the piston moves into the chamber, the air inside it gets compressed. Thus, the pressure created due to the compressed air opposes the motion of the piston. Similarly, when the piston moves out of the chamber, the motion is again opposed due to the pressure being greater on the open side than on the closed side of the chamber.

This type of arrangement requires a precise adjustment of the angle of the arm carrying the piston so that the movement of the piston is not impaired. As such, this method is not preferred.

Figure 11.7(a) shows an alternate method of air friction damping. In this arrangement, one or two lightweight vanes are connected to the spindle of the moving system. The movement of the vanes is in a closed sector-shaped box which causes the motion of the deflecting system to be damped. Figure 11.7(b) shows the arrangement of the vanes.

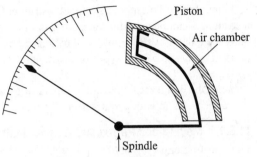

Fig. 11.6 Air friction damping

11.5 AMMETERS AND VOLTMETERS

Deflection of indicating pointers for both ammeters and voltmeters depends on the torque produced by the current flowing through them. In the case of an ammeter, the torque is produced by the current to be measured or a known fraction of it. The torque in a voltmeter is produced by the current proportional to the voltage to be measured. Thus, the principle of operation in both ammeters and voltmeters is the same.

A voltmeter is a high-resistance instrument since it is connected across the voltage to be measured. The high resistance permits only a small amount of current to flow through the voltmeter. On the other hand, an ammeter, which is connected in series, is required to have a low resistance so that the current to be measured is not altered.

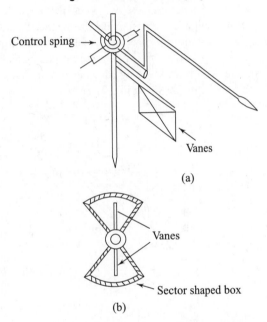

Fig. 11.7 Alternate method of air friction damping

Thus, it can be seen that ammeters and voltmeters can be used interchangeably. A voltmeter can be used as an ammeter by connecting a low-resistance shunt across it. Similarly, an ammeter can be converted into a voltmeter by connecting a high-resistance in series with the former.

11.5.1 Types of Ammeters and Voltmeters

The commonly employed ammeters and voltmeters are:
   (i) Moving-coil permanent-magnet type
  (ii) Moving-coil dynamometer type
 (iii) Moving iron type
  (iv) Hot wire (induction) type
   (v) Electrostatic type
  (vi) Induction type

In moving-coil permanent-magnet instruments the deflecting torque is produced by the current. Thus, such instruments can be used for measuring direct currents and voltages only. Induction type instruments depend upon the induction effect produced by alternating quantities. Therefore, such instruments measure alternating quantities only. All other types of instruments can be used for measuring direct and alternating currents and voltages.

Since the first three types of instruments are most commonly used in electrical/electronics laboratories and, in practice, it is intended to describe the operation of the first three types of instruments only.

11.5.1.1 Moving-coil Permanent-magnet Instruments

These instruments consist of a light rectangular coil of several turns wound on an aluminium or copper former. The aluminium/copper former provides mechanical strength to the coil. The coil is pivoted and its sides lie in the air gap between the two poles of a powerful U-shaped permanent magnet and a soft-iron cylinder. The soft-iron cylinder helps to make the magnetic field radial and reduce the reluctance of the air gap between the poles. The controlling system, which controls the movement of the coil, is constituted of two phosphor–bronze hair springs. These springs also serve as terminals for passing current through the coil. The damping of the moving coil is achieved by eddy currents induced in the aluminium former. Constructional details of such an instrument are shown in Fig. 11.8.

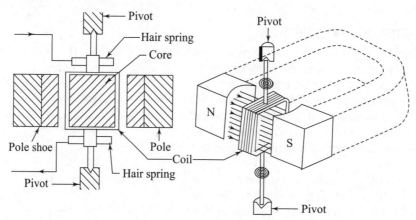

Fig. 11.8 Constructional details of a moving-coil permanent-magnet instrument

For discussions hereafter, the following parameters for the coil and the permanent magnet are assumed:

Number of turns in the coil = N
Length or depth of the coil = l metre
Breadth of the coil = b metre
Area of the coil = $A = l \times b$ m^2
Flux density = B Wb/m^2

Figure 11.9 shows the working of a moving-coil permanent-magnet instrument. When a current of I ampere flows through the coil, the two sides of the coil experience a force of F newton.

$$F = NBIl \text{ newton} \qquad (11.3)$$

Deflecting torque,

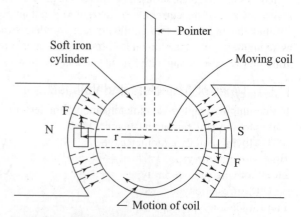

Fig. 11.9 Working of a moving-coil permanent-magnet instrument

$$T_d = F \times \text{perpendicular distance, N m}$$
$$= NBIl \times b \text{ N m}$$
$$= NBI \times A \text{ N m}$$

For a given instrument N, B, and A are constant, thus

$$T_d \propto I \qquad\qquad (11.4)$$

Since spring control has been used, T_c is proportional to θ, the deflection. Under the condition of equilibrium,

$$T_d = T_c$$
$$I \propto \theta \qquad\qquad (11.5)$$

It can be seen that such an instrument has a scale which is uniformly divided. The other advantages of such type of instruments are their high sensitivity and their not being affected by stray magnetic fields since they are well shielded.

If an alternating current is passed through a moving-coil permanent-magnet instrument, the deflecting torque produced on the moving system will be alternating. Therefore, the moving coil will oscillate around its zero position. Thus the main disadvantage of such type of instruments is that they cannot be employed for measuring ac measurements; they are suitable only for dc currents and voltages.

11.5.1.2 Moving-coil Dynamometer Instruments

Moving-coil dynamometer instruments are similar in construction to the moving-coil permanent-magnet instruments except that the permanent magnet is replaced by one or two fixed air-cored coils. Air-cored coils are used to avoid errors due to hysteresis and eddy currents when the instrument is used to measure alternating current quantities. The fixed coils are connected in series or in parallel with the moving coil and carry the current to be measured, when the instrument functions as an ammeter. When it operates as a voltmeter, the fixed coils carry a current proportional to the voltage to be measured. Figure 11.10 shows the general arrangement of a moving-coil dynamometer type of instrument.

The deflection torque T_d produced is proportional to the magnetic fields produced by the current flowing through the fixed and moving coils. If the instrument is working as an ammeter, $T_d \propto I^2$, where I is the current

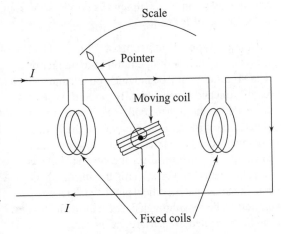

Fig. 11.10   General arrangement of a moving-coil dynamometer instrument

flowing through the fixed and moving coils. On the other hand, if the instrument is operating as a voltmeter, $T_d \propto I_1 I_2$, where I_1 and I_2 are the currents flowing in the fixed and moving coils, respectively. Since I_1 and I_2 are currents proportional to the voltage to be measured, hence $T_d \propto V^2$. Therefore, the deflection θ is proportional to the square of the current and the square of the voltage when the instrument is operating as an ammeter and voltmeter, respectively.

The dynamometer type of instruments, thus, can be used for measuring alternating quantities. Since the deflection $\theta \propto I^2 \propto V^2$, the scale of a dynamometer instrument is non-uniform, and it is cramped for low values of the quantities measured. These types of instruments can be used for both alternating and direct current measurements.

Due to air-cored coils, the magnetic field produced is weak. As such eddy current damping is insufficient. Fluid friction (pneumatic) damping, which is more effective, is employed.

The disadvantages of such type of instruments, when compared with moving-coil permanent-magnet instruments, are their low sensitivity and higher cost.

As such the dynamometer type of instruments as ammeters and voltmeters do not find application in the field. The dynamometer principle, however, is extensively employed in the wattmeter to measure alternating power.

11.5.1.3 Moving-iron Instruments

Moving-iron (MI) instruments are of two types, namely:
(a) Attraction type
(b) Repulsion type

Both the types of MI instruments require a magnetic field of minimum intensity for their operation. The field is set up by passing the current to be measured, or a current proportional to the voltage to be measured, through a coil having the requisite number of turns. In the attraction type of instruments, when current flows through the coil, a small soft-iron piece moves from the weaker magnetic field outside the coil towards the stronger field inside the coil.

The repulsion type of instruments have two pieces or rods, one fixed and the other movable, of soft iron inside the coil. When a current flows through the coil, both the soft-iron rods get magnetized similarly, thereby, causing them to repel each other. Thus the moving rod has a tendency to move away from the fixed rod.

Both forms of instruments are independent of the direction of the flow of current in the coil. In the attraction type, the moving iron always gets attracted towards the stronger field inside the coil. In the repulsion type, the two iron rods will always move away from each other. Moving-iron instruments are also known as un-polarized instruments.

Attraction type MI instrument Figure 11.11 shows the principle of operation of an attraction type MI instrument. The moving iron is a thin disc of soft iron and is eccentrically pivoted. When a current flows through the coil, the soft iron moves towards the inside of the coil where the field is stronger. Since the pointer is fixed to the spindle carrying the soft-iron disc, the former deflects when the latter moves towards the inside of the coil. Since the magnetic field produced is set up by the current flowing through the coil, the deflection can be calibrated to measure the current.

In Fig 11.11 gravity control has been shown to produce the controlling torque T_c. However, present-day moving-iron instruments invariably use spring control. Air friction damping is provided through a light piston moving in an air chamber.

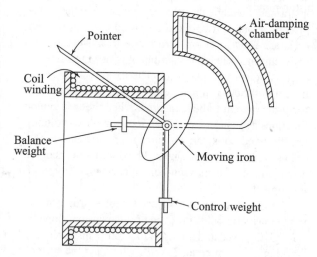

Fig. 11.11 Attraction type MI instrument

Figure 11.12 shows the magnetic field inside a coil, due to current I flowing through it. It is assumed that the axis of the disc has eccentricity β in a direction perpendicular to the direction of the magnetic field H produced by the coil when in zero position. Further, it is assumed that the disc deflects through an angle θ when current I flows through the coil.

The component of field H producing magnetization of the disc is proportional to $H \cos\{90 - (\beta + \theta)\}$, or $H \sin(\beta + \theta)$. The force pulling the disc into the coil is proportional to $H^2 \sin(\beta + \theta)$. Since field H is produced by current I, the force F on the disc is proportional to $I^2 \sin(\beta + \theta)$.

If it is assumed that this force is acting on the disc at a distance l from the pivot, the deflecting torque T_d is proportional to $l \cos(\beta + \theta)$. As the deflecting torque T_d is dependent on F,

$$T_d \propto I^2 \sin(\beta + \theta)\, l \cos(\beta + \theta) \tag{11.6}$$

Since l is a constant,

$$T_d = KI^2 \sin 2(\beta + \theta) \tag{11.7}$$

If the controlling torque is produced by a spring system, the controlling torque T_c is given by

$$T_c = K'\theta \tag{11.8}$$

Since in steady state $T_c = T_d$, Eqs (11.7) and (11.8) give

$$K'\theta = KI^2 \sin 2(\beta + \theta) \tag{11.9}$$

or

$$I = \frac{K}{K'}\sqrt{\frac{\theta}{\sin 2(\beta + \theta)}} \tag{11.10}$$

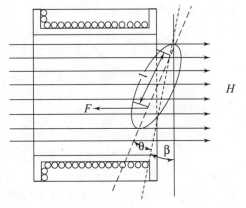

Fig. 11.12 Magnetic field inside a coil due to current I

Repulsion type of MI instrument   Figure 11.13 shows the sectional view of a repulsion type MI instrument. Two soft-iron rods, one fixed and the other movable, are placed inside the coil. The movable rod carries a pointer which slides over a calibrated scale. Damping of the pointer is through air friction provided by a light piston moving inside an air chamber. A spring system is used to produce the controlling torque T_c.

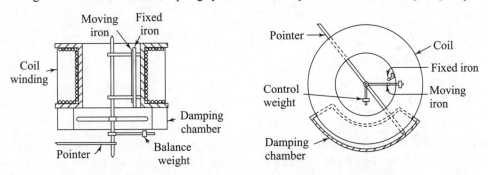

Fig. 11.13   Sectional view of a repulsion type MI instruction

When current I flows through the coil, the iron rods are similarly magnetized and a force of repulsion is experienced by them. Consequently, the moving soft-iron rod moves away from the fixed rod, thereby, causing the pointer to deflect.

Assume that due to current I magnetic field H is set up inside the coil; the pole strengths of the rods are m_1 and m_2; and the length of each of the rod is l and that they are at a distance D from each other, as shown in Fig. 11.14.

It is seen that $l \gg D$ and the initial angle of displacement between the rods is β. Let θ be the deflection of the pointer due to current I. Then, the force of repulsion, F, is given by

$$F = \frac{2m_1 m_2}{D^2} \tag{11.11}$$

If R is the radius of rotation of the moving rod from the centre of the coil, then

$$D = 2R \sin\left(\frac{\beta + \theta}{2}\right) \tag{11.12}$$

and the deflecting torque

$$T_d = FR \cos\left(\frac{\beta + \theta}{2}\right) \tag{11.13}$$

$$= \frac{2m_1 m_2}{D^2} R \cos\left(\frac{\beta + \theta}{2}\right) \tag{11.14}$$

On simplification

$$T_d = \frac{m_1 m_2 \cos\dfrac{\beta + \theta}{2}}{2R \sin^2 \dfrac{\beta + \theta}{2}} \tag{11.15}$$

Since the pole strengths of the rods are proportional to current I,

$$T_d = K' \frac{I^2 \cos\dfrac{\beta + \theta}{2}}{\sin^2 \dfrac{\beta + \theta}{2}} \tag{11.16}$$

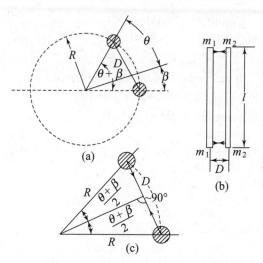

Fig. 11.14   Geometric arrangement of rods

In a spring controlled instrument,
$$T_c = K''\theta \tag{11.17}$$

In steady state,
$$T_c = T_d$$

Hence $$I = K \sin\frac{\beta + \theta}{2} \sqrt{\frac{\theta}{\cos\dfrac{\beta + \theta}{2}}} \tag{11.18}$$

where $$K = \frac{K''}{K'} \tag{11.19}$$

It can be seen that the relationship between current I and deflection θ in both types of MI instruments is non-linear. As such they can both be employed for direct current and alternating current measurements.

The advantages of MI instruments are that they are cheap and sturdy, and they possess a good degree of reliability. The disadvantage of such type of instruments is that the degree of precision of calibration is not high for direct current measurements due to hysteresis in iron parts.

11.5.2 Extension of Range

Ammeters and voltmeters are required to make measurements over a wide range of values. For example, if the resistance of an ammeter, which is connected in series, is not very much lower than the load resistance, the circuit current itself will get altered. Similarly, a voltmeter, which is connected across a load for voltage measurement, must have a high series resistance. In this manner, the voltmeter current is small and the voltmeter loading is avoided.

11.5.2.1 Permanent-magnet Moving-coil Types

In ammeters, it is difficult to design low inertia moving coil system for measuring currents in the micro-ampere range. The range of an ammeter, therefore, is extended by connecting a low resistance, called shunt, in parallel with it. The arrangement is shown in Fig. 11.15.

Figure 11.15(b) shows the schematic arrangement of a shunt having a low resistance R_{sh} connected in parallel with the ammeter. In moving-coil permanent-magnet instruments, the movement is sensitive to temperature. The moving-coil resistance increases and both the magnetic field and the spring tension decrease with an increase in temperature. The consequence of a reduction in the magnetic field and an increase in coil resistance is to make the ammeter read low for a given current. This disadvantage is easily overcome by connecting a manganin resistance r_m, called a *swamping resistance*, in series with the coil. A manganin resistor is used since it has a temperature coefficient of resistance that is almost zero, thereby resulting in a reduction in error which is of the order of 4%. For measuring currents in the higher range, shunts made of manganin are employed.

In order to obtain a multi-range ammeter, several shunts of variable values are employed as shown in Fig. 11.16. The range of the ammeter is selected by moving the rotary switch S to the appropriate shunt.

In using the above arrangement, it is important to note that while changing the position of S, when the ammeter is connected in a circuit, the shunt should not be open circuited even for a fraction of time. A shunt on open circuit will cause a current of high magnitude to flow through the ammeter, thus resulting in a possible damage to the instrument. Such a situation is avoided by using a *make-before-break* switch shown in Fig. 11.17.

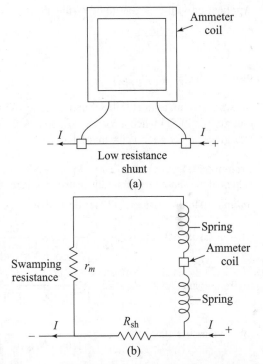

Fig. 11.15 Extension of ammeter range: (a) low resistance connected in parallel and (b) schematic arrangement

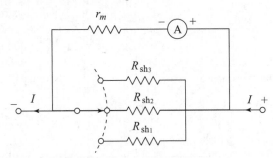

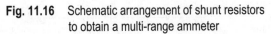

Fig. 11.16 Schematic arrangement of shunt resistors to obtain a multi-range ammeter

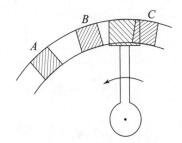

Fig. 11.17 Make-before-break switch

In the above arrangement, when the wide-ended contact is moved from position A to B, it connects two shunts in parallel with the ammeter and when the contact has fully moved to position B, only the appropriate shunt remains connected with the ammeter.

Suppose it is desired to measure a current of I ampere with an ammeter that has a resistance of $r_m \, \Omega$ and gives a full-scale deflection (FSD) with I_m ampere. The arrangement is shown in Fig. 11.18.

Application of the current divider rule gives

$$I_m = \frac{R_{sh}}{r_m + R_{sh}} I$$

or $\qquad R_{sh} = r_m \frac{I_m}{(I - I_m)}$

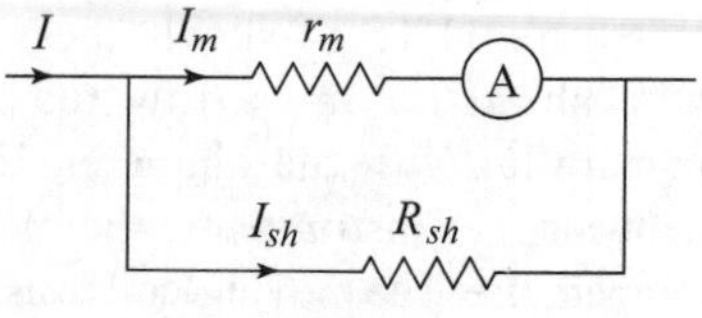

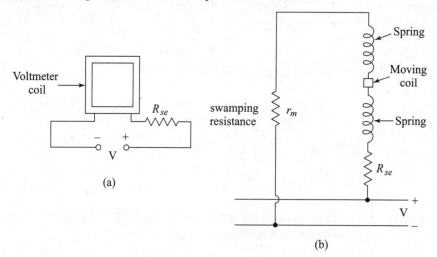

Fig. 11.18

Sensitivity of an ammeter is defined as the minimum current required to produce an FSD. Thus, an ammeter that gives an FSD when a current of 100 mA flows through it will have a higher sensitivity than an ammeter that gives an FSD for a current of 1 A.

Example 11.1 A current of 300 μA produces an FSD in an instrument whose coil resistance is 75 Ω. Calculate the values of the shunt resistors if the instrument is to measure currents of (i) 5 A, (ii) 7.5 A, and (iii) 10 A.

Solution In this case, $r_m = 75$ Ω and $I_m = 300$ μA

(i) To measure 5 A, $R_{sh} = r_m \dfrac{I_m}{(I - I_m)} = 75 \times \dfrac{300 \times 10^{-6}}{\left(5 - 300 \times 10^{-6}\right)} = 0.0045$ Ω

(ii) To measure 7.5 A, $R_{sh} = 75 \times \dfrac{300 \times 10^{-6}}{\left(7.5 - 300 \times 10^{-6}\right)} = 0.0030$ Ω

(iii) To measure 10.0 A, $R_{sh} = 75 \times \dfrac{300 \times 10^{-6}}{\left(10 - 300 \times 10^{-6}\right)} = 0.0023$ Ω

The range of a dc voltmeter can be extended by connecting a high-valued resistor R_{se} in series with the instrument as shown in Fig. 11.19. Since the series resistor R_{se}, which is also called a *multiplier resistor*, has a high resistance, the loading on the voltmeter is kept at a minimum level.

Fig. 11.19 Extension of voltmeter range: (a) high resistance connected in series and (b) schematic arrangement

A multi-range voltmeter is conveniently obtained by connecting a number of multiplier resistors in series through a rotary switch S as shown in Fig. 11.20.

In this case, the rotary switch S is a moving contact, which disconnects while moving from one position to the next position. In this case, the magnitude of the multiplier R_{se} is determined by referring to the arrangement shown in Fig. 11.21.

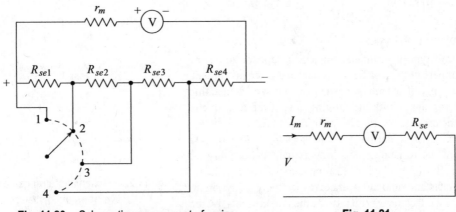

Fig. 11.20 Schematic arrangement of series resistors for a multi-range voltmeter

Fig. 11.21

If it is assumed that the range of the instrument is required to be extended to measure V volts and I_m ampere produces a full-scale deflection in it, application of the well-known Ohm's law yields

$$I_m = \frac{V}{r_m + R_{se}}$$

or

$$R_{se} = \frac{V}{I_m} - r_m$$

Both the ammeters and voltmeters are calibrated to directly read current and voltage, respectively. In case of multi-range instruments, multipliers are provided for each range.

Sensitivity of a voltmeter is defined as the ratio of the total resistance $(r_m + R_{se})$ in ohms to the voltage V, which produces an FSD in the instrument. Hence,

$$\text{Sensitivity} = \frac{(r_m + R_{se})}{V} = \frac{1}{V/(r_m + R_{se})} \ \Omega/V \tag{11.20}$$

In Eq. (11.20), the term $V/(r_m + R_{se})$ is easily recognized as the current I_m, which produces the FSD. Therefore, sensitivity of a voltmeter is alternately defined as the reciprocal of the instrument current, which produces FSD. Mathematically, sensitivity is stated as

$$\text{Sensitivity} = \frac{1}{I_m} /A \tag{11.21}$$

The sensitivity of an instrument, which gives an FSD when a current of 1 mA flows through it, will be $1/(1 \times 10^{-3})\Omega/V$ or 0.001 A.

Example 11.2 The instrument in Example 11.1 is now to be used to measure voltages of the magnitude (i) 50 V, (ii) 75 V, and (iii) 100 V. Determine the value of the multiplier resistors to be connected in series in each case.

Solution (i) To measure 50 V, $R_{se} = \dfrac{V}{I_m} - r_m = \dfrac{50}{300 \times 10^{-6}} - 75 = 166.59 \text{ k}\Omega$

(ii) To measure 75 V, $R_{se} = \dfrac{75}{300 \times 10^{-6}} - 75 = 249.92 \text{ k}\Omega$

(iii) To measure 100 V, $R_{se} = \dfrac{100}{300 \times 10^{-6}} - 75 = 333.26 \text{ k}\Omega$

11.5.2.2 Moving-iron Types

In moving-iron type instruments, the FSD depends upon the number of ampere turns, which is constant for a given instrument. However, the range of an ammeter can be extended by connecting a shunt coil in parallel with the moving coil of the instrument. The schematic arrangement is shown in Fig. 11.22.

As earlier, I_m is the current that produces the FSD and r_m Ω and L_m H are the resistance and inductance, respectively, of the ammeter. Similarly, I_{sh} is the shunt current and r_m Ω and L_m H are the resistance and inductance, respectively, of the shunt. If ω rad/sec is the frequency, the ratio of the two currents, by the application of the current division rule, is written as

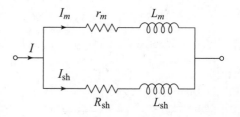

Fig. 11.22   Schematic arrangement for the extension of range of an MI ammeer

$$\frac{I_m}{I_{sh}} = \sqrt{\frac{R_{sh}^2 + (\omega L_{sh})^2}{r_m^2 + (\omega L_m)^2}} = \frac{R_{sh}}{r_m} \times \sqrt{\frac{1 + (\omega L_{sh}/R_{sh})^2}{1 + (\omega L_m/r_m)^2}} \tag{11.22}$$

If it is assumed that $\dfrac{L_{sh}}{R_{sh}} = \dfrac{L_m}{r_m}$, Eq. (11.22) becomes independent of the frequency and the ratio of the currents is given by

$$\frac{I_m}{I_{sh}} = \frac{R_{sh}}{r_m} \tag{11.23}$$

Application of the current law and subsequent substitution results in

$$I = I_m + I_{sh} = I_m\left(1 + \frac{r_m}{R_{sh}}\right) \tag{11.24}$$

Equation (11.24) provides the relationship for shunt multipliers. However, in practice, it is very difficult to design shunts having suitable resistance and inductance. As such, shunts are seldom used in moving iron ammeters. The range of these types of ammeters is extended by winding the instrument coils in sections, which are then connected in series, parallel, or series-parallel combinations to obtain the desired ampere turns.

For the extension of a moving-iron voltmeter, a non-inductive high resistance R_{se} is connected in series with the moving coil of the instrument as shown in Fig. 11.23.

The load voltage V and the voltage V_m across the voltmeter are, respectively, given by

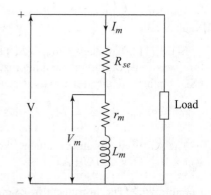

Fig. 11.23   Schematic arrangement for the extension of range of an MI voltmeter

$$V = I_m\sqrt{(R_{se} + r_m)^2 + (\omega L_m)^2} \tag{11.25}$$

and $V_m = I_m\sqrt{r_m^2 + (\omega L_m)^2} \tag{11.26}$

The value of the multiplier is obtained by dividing Eq. (11.25) by Eq. (11.26) as

$$m = \frac{V}{V_m} = \sqrt{\frac{\left(R_{se} + r_m\right)^2 + \left(\omega L_m\right)^2}{r_m^2 + \left(\omega L_m\right)^2}} \qquad (11.27)$$

It may be noted from Eq. (11.27) that this arrangement, that is, a non-inductive resistor in series with the instrument, will result in error in measurements, particularly when signal voltages are measured, since m is not independent of frequency. The error due to variations in frequency can be compensated by using an appropriate capacitor in parallel with the series resistor as shown in Fig. 11.24.

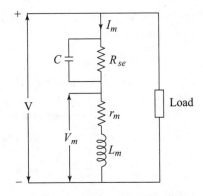

Fig. 11.24 Circuit for compensating for errors in voltage measurement due to variation in frequency

11.6 MEASUREMENT OF POWER

The power consumed in a direct current circuit is given by the product of voltage and current. In an alternating current circuit, power is the product of the rms value of the voltage, current, and power factor. Therefore, alternating current circuits require a wattmeter to measure power. The three types of instruments used for measurement of power are (a) dynamometer wattmeter, (b) induction wattmeter, and (c) electrostatic wattmeter.

Of these, dynamometer and induction type of instruments are commonly used and are described in the following sections.

11.6.1 Dynamometer Wattmeter

The fundamental principle of operation of this type of a wattmeter is the same as used for a moving-coil dynamometer ammeter and voltmeter. Figure 11.25 shows the connection diagram for a dynamometer type wattmeter.

The fixed air-cored coil is circular in shape and is divided into two halves. Both halves are parallel to each other and the distance between them is adjusted to obtain a uniform magnetic field. It carries the main circuit current and is called the current coil. The moving coil is centrally pivoted and carries a pointer which moves over a graduated scale. A current that is proportional to the applied voltage is fed to the coil through phosphor–bronze springs M. The moving coil is designated as the voltage coil. These

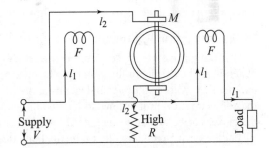

Fig. 11.25 Connection diagram for a dynamometer type wattmeter

springs also provide the controlling torque. Damping is provided by a piston moving inside an air chamber. R is a high resistance (called swamping resistance) connected in series with the moving coil to reduce the phase effect of the inductance of the voltage coil.

Let it be assumed that a direct current I flows in the fixed coils, a magnetic field of B Wb/m^2 is produced, and current I' (which is proportional to the dc voltage V) flows through the moving coil. Then, the deflecting torque T_d is proportional to BI'.

But $\qquad B \propto I$

and $\qquad I' \propto V$

Hence $\qquad T_d = KVI$ $\qquad\qquad\qquad\qquad\qquad\qquad$ (11.28) where the product VI is the dc power.

In the case of ac quantities,

$$T_d = Kvi \tag{11.29}$$

where v and i are instantaneous values of voltage and current, respectively.

Due to the inertia of moving parts, the reading shown is the average of the mean power. Therefore, the mean value of the deflecting torque $T_{d, \text{mean}}$ is given by

$$T_{d\,\text{mean}} = K \times (\text{average value of instantaneous power } vi) \tag{11.30}$$

If $v = V_m \sin \theta$ and $i = I_m \sin (\theta - \varphi)$, then

$$T_{d, \text{mean}} = K \frac{V_m I_m}{2\pi} \int_0^{2\pi} \sin \theta \, \sin (\theta - \varphi) \, d\theta \tag{11.31}$$

On simplification,

$$T_{d, \text{mean}} = VI \cos \varphi \tag{11.32}$$

where V and I are the rms values of voltage and current, respectively, and $\cos \varphi$ is the power factor of the circuit. Since a spring control system is used, T_c is proportional to θ, where θ is the deflection of the pointer. In steady state,

$$T_d = T_c$$

Hence, $\quad \theta \propto VI \cos \varphi \tag{11.33}$

This is the actual power in an alternating circuit.

Since the deflection in such instruments is proportional to the average power, they have almost uniform scales. The dynamometer type wattmeter, therefore, measures both direct and alternating current powers.

The advantage of a dynamometer wattmeter is the high degree of accuracy. Therefore, it is used as a standard for calibration work. Further, it can be used for the measurement of both dc and ac powers.

The instrument has the disadvantage that, due to the inductance of the voltage coil, the error is considerable when alternating power is measured on a low power factor. In addition, this type of wattmeters are expensive.

11.6.2 Induction Wattmeter

The working principle of induction instruments is based on the torque produced by a flux, whose magnitude depends on the current or the voltage to be measured, and eddy currents which are induced in a metal disc by another flux, whose value again depends upon the current or the voltage to be measured. As the magnitude of the eddy current is proportional to the flux inducing it, the torque at an instant is proportional to the square of the current or voltage to be measured.

Let the torque producing flux ϕ be mathematically represented by

$$\phi = \phi_m \sin \theta \tag{11.34}$$

and the induced eddy currents be given by

$$i = I_m \sin (\theta - \alpha) \tag{11.35}$$

where angle α is the angle by which the induced eddy currents lag the flux. Thus the average torque T may be expressed as

$$T \propto \frac{1}{\pi} \int_0^{\pi} \phi_m I_m \sin \theta \sin (\theta - \alpha) \tag{11.36}$$

Integrating and simplifying Eq. (11.36), the average torque becomes

$$T \propto \frac{\phi_m I_m}{2\pi} (\pi \cos \alpha) \tag{11.37}$$

or $\quad T \propto \phi I \cos \alpha \tag{11.38}$

ϕ and I are rms values of flux and current, respectively.

The working principle of an induction wattmeter is based on the induction effect. An induction wattmeter consists of two laminated electromagnets, between which a thin aluminium disc is mounted. The disc is so positioned that the flux from both the magnets cuts the disc. The winding on one of the magnets, called the pressure coil, is excited by a current proportional to the applied voltage. The other magnet, called the series magnet, carries the current coil and is excited by the load current or a definite proportion of it. The pressure coil electromagnet (called the shunt magnet) is fitted with one or more copper bands. The position of copper bands can be adjusted to make the resultant flux in it lag the applied voltage by precisely 90°. As a result of the flux produced in the two magnets, eddy currents are induced in the aluminium disc. The interaction between the eddy currents and the fluxes produces the deflection torque. Figure 11.26 shows two common forms of arrangements of series and shunt magnets and their windings employed in such type of wattmeters.

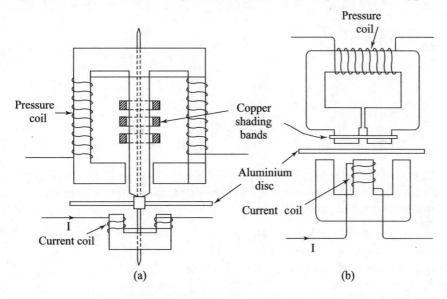

Fig. 11.26 Two commonly employed arrangements of the magnets and their windings in induction wattmeter

In Fig. 11.26(a) the pressure coil is in two parts and is connected in series in such a manner that the flux produced in both the limbs of the magnet sends the flux through the central limb. The current coil, which is of a few turns, is also in two parts. The coils are connected in series such that the current through them magnetizes the core in the same direction.

In the arrangement shown in Fig. 11.26(b), both pressure and current coils are in one part. Both forms of arrangement employ spring control, thereby providing a long uniform scale.

Let it be assumed that an ac voltage of magnitude V volts is applied to the pressure coil and an ac current of I amperes is flowing through the current coil. The current lags the voltage by an angle φ. Figure 11.27 shows the relationship between various quantities.

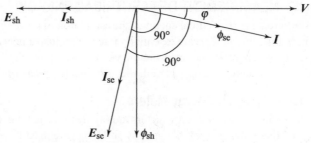

Fig. 11.27 Phasor diagram showing the relationship between various quantities

Since the copper bands have been adjusted, the flux ϕ_{sh} in the shunt magnet lags the voltage phasor V by 90°. Then, both the eddy emf induced e_{sh} and the eddy currents i_{sh} in the disc lag the flux ϕ_{sh} by 90°. Similarly the flux ϕ_{se} in the series magnet is proportional to and in phase with the line current phasor I in the current coil. The flux ϕ_{se} induces an emf e_{se}, which lags ϕ_{se} by 90°. The consequent eddy current i_{se} in the disc is in phase with e_{se}. In stating that the eddy currents in the disc are in phase with the induced emfs, the inductance of the disc has been neglected.

Le the applied instantaneous voltage v and instantaneous current i be expressed as

$$v = V_m \sin \omega t \quad \text{and} \quad i = I_m \sin (\omega t - \varphi)$$

Then, $\quad \phi_{sh} \propto -\int edt \propto -\int V_m \sin \omega t \propto \dfrac{V_m}{\omega} \cos \omega t$ \hfill (11.39)

since $\quad e_{sh} \propto \dfrac{d\phi_{sh}}{dt} \propto \dfrac{V_m}{\omega} \omega \sin \omega t \propto V_m \sin \omega t$ \hfill (11.40)

Therefore

$$i_{sh} \propto V_m \sin \omega t \tag{11.41}$$

Similarly,

$$\phi_{se} \propto i \propto I_m \sin (\omega t - \varphi) \tag{11.42}$$

Hence $\quad e_{se} \propto \dfrac{d\phi_{se}}{dt} \propto I_m \omega \cos (\omega t - \varphi)$ \hfill (11.43)

and $\quad i_{se} \propto I_m \omega \cos (\omega t - \varphi)$ \hfill (11.44)

From the phasor diagram, it can be seen that the average deflecting torque is given by

$$T \propto \Phi_{sh} I_{se} \cos \varphi - \Phi_{se} I_{sh} \cos (180 - \varphi) \tag{11.45}$$

where ϕ_{sh}, ϕ_{se}, I_{sh}, and I_{se} are the rms values of the various quantities and φ and $(180 - \varphi)$ are the phase angles between the interacting currents and fluxes.

Substituting the various values and simplifying gives

$$T \propto VI \cos \varphi \tag{11.46}$$

This is the power in the circuit.

The advantages of an induction wattmeter are their uniform and larger scale (up to 300°), greater torque, and not being affected by stray fields.

The disadvantages of an induction wattmeter are that it can be used for measuring ac power only, consumes more power, is less accurate and bulky due to a heavy moving system.

11.7 MEASUREMENT OF ELECTRICAL ENERGY

Measurement of electrical energy is the estimation of power consumed over a given period of time. Therefore, measuring instruments are essentially integrating instruments which keep track of the consumption of power. Such instruments do not measure the variations in power or energy consumed. Instead they measure the total energy consumed in a time period. Electrical energy measuring instruments are called energy or supply meters.

11.7.1 Types of Energy Meters

The various types of energy meters are (a) electrolytic type, (b) motor type, and (c) clock type.

Of these, the electrolytic type, by and large, are used for energy measurement in dc circuits. As such they find application to a limited extent. The clock type instruments on the other hand have lost their utility due to their complexity and high cost. However, owing to their high inbuilt accuracy, they find application as standard instruments. Motor type energy meters are most common in use.

Motor type energy meters are further classified into three categories as under:

  (a) Mercury motor   (b) Commutator motor   (c) Induction motor

    Mercury motor energy meters are generally used for energy measurement in dc circuits. Commutator motor instruments can be used for energy measurement, both in ac and dc circuits. The application of induction motor energy meters is limited to alternating current energy measurement only. The use of induction type instruments for measuring industrial and domestic electrical energy consumption has been accepted the world over. In the following section, the working of induction type energy meters is discussed.

11.7.2 Induction Type Energy Meters

Depending upon their application, induction type energy meters are classified as (a) single-phase energy meters and (b) three-phase energy meters.

    This type of energy meters operate basically on the same principle as the induction wattmeter. The prime difference between the two is that the indicating system is allowed to rotate continuously in the case of energy meters.

11.7.2.1 Single-phase Energy Meters

As already stated the principle of operation of this type of energy meters is the same as that of the induction type wattmeter. Also the construction of an energy meter is similar to the induction type wattmeter, except that the deflecting and spring control systems in the latter are replaced by a braking system. The braking system consists of a pair of permanent magnets, which are mounted diametrically opposite the shunt and series magnets. This pair of magnets induces eddy currents in the disc, which produces the braking effect. Figure 11.28 shows the construction of an induction energy meter.

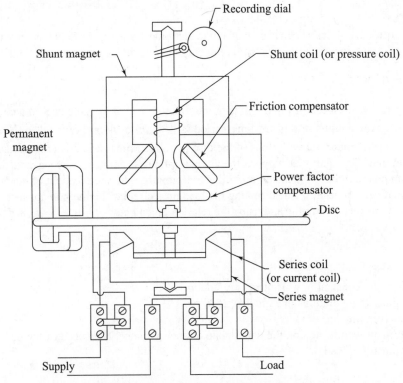

Fig. 11.28 Construction of a single-phase energy meter

Comparing the construction of the energy meter with that of the wattmeter, shown in Fig. 11.26(a), it is seen that the former has only one voltage coil and one shading ring on the central limb of the shunt magnet. In order to compensate for friction, two copper rings have been provided on each of the other two limbs of the shunt magnet additionally.

Let it be assumed that the flux of the braking magnet is ϕ and i is the magnitude of the eddy currents induced when the disc rotates in the magnetic field of the braking magnet. Let the speed of rotation of the disc be n and r be the resistance of the eddy current path.

The voltage induced is

$$e \propto \phi n \tag{11.47}$$

The braking torque is

$$T_B \propto \phi i \tag{11.48}$$

Substituting for i,

$$T_B \propto \phi \frac{e}{r} \propto \phi^2 \frac{n}{r} \tag{11.49}$$

In the case of energy meters, a revolving disc replaces the deflecting system. Therefore, a driving torque T_d in the latter case replaces the deflecting torque. In the steady state, when the disc is rotating at a constant speed,

$$T_d = T_B \tag{11.50}$$

Therefore

$$T_d \propto \frac{\phi^2 n}{r} \tag{11.51}$$

or

$$n \propto \frac{r}{\phi^2} T_d \tag{11.52}$$

In Section 11.6.1 it has already been shown that in induction wattmeters the deflecting torque T_d represents the real power, $EI \cos \varphi$, in a circuit. In the case of energy meters, the driving torque $T_d \propto EI \cos \varphi$, or

$$n \propto \frac{r}{\phi^2} EI \cos \varphi \tag{11.53}$$

It can thus be seen that the speed of rotation n is proportional to $EI \cos \varphi$, the power in a circuit. Therefore, the total number of revolutions in a given time period is $\int_0^t n\,dt$, or is proportional to $\int_0^t EI \cos \varphi \, dt$. In other words, the total number of revolutions of the disc in a given time period is proportional to the energy consumed in that time. The shaft that supports the aluminium disc is connected by a gear arrangement to the clock mechanism on the front of the meter, providing a decimally calibrated readout of the number of kWh.

An energy meter can be used for determining the load on a circuit. The method consists of counting the number of revolutions in a given time and using the relation given below:

$$P = \frac{3600 K_h R}{S} \tag{11.54}$$

where P is the power in watts,
K is the meter constant in watt-hr per revolution
R is number of revolutions in a given time in sec.
S is the given time in sec.

The meter constant is indicated on the name plate of energy meter. The speed of the disc in commercial meters is 1800 rpm at full load.

Example 11.3 The disc in an energy meter is observed to make 50 revolutions in one minute. Compute the load on the energy meter. The name plate on the energy meter shows the disc speed to be 1800 rpm/kWh.

Solution The constants defined in Eq. (11.54) are determined as follows

$$K_h = 1000/1800 = 0.56, R = 50, \text{ and } S = 60$$

Use of Eq. (11.54) gives $P = (3600 \times 0.56 \times 50)/60 = 1666.7$ W

Example 11.4 Calculate the disc speed of a 220 V, single phase energy meter when it is carrying a load of 15 A at 0.7 power factor. Assume the meter constant to be 1000 rpm/kWh.

Solution

Meter constant $K_h = 1000/1000 = 1.0$, $S = 60$, and $P = 220 \times 15 \times 0.7 = 2310$ W

Use of Eq. (11.54) leads to $R = (2310 \times 60)/(3600 \times 1) = 38.5$ rpm

11.7.2.2 Three-phase Energy Meters

This type of instruments operate on the same principle as used in the two wattmeter method for measuring power in a three-phase circuit. A three-phase energy meter is like two single-phase induction energy meters enclosed in one casing. It has two discs on the same spindle, which rotate between two shunt and series electromagnets independently, as shown in Fig. 11.29. Each unit has a separate shading ring, shading band, and a brake magnet. Revolutions of the discs are recorded by a single gear train attached to the spindle.

In order that the driving torques of the two instruments be exactly equal for the same watts, one of the units is provided with an adjustable magnetic field across its shunt magnet. The torques of the two instruments are adjusted to be exactly equal by connecting the current coils in series and the pressure coils in parallel. This ensures that polarization is such that the two driving torques are in opposition.

Under this arrangement if the torques are equal, none of the discs will rotate. However, if the meter tends to rotate, the shunt magnet is adjusted till it ceases to rotate with full-rated watts supplied to both the units.

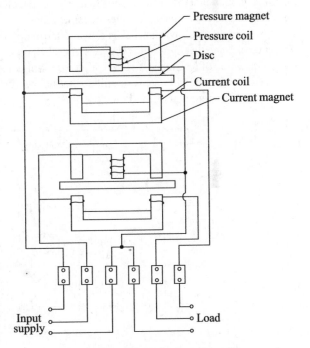

Fig. 11.29 Three-phase energy meter

11.8 MEASUREMENT OF INSULATION RESISTANCE

Before energizing any industrial or domestic installation or any network, it is important to check its insulation resistance. Since the magnitude of the insulation resistance is of the order of megaohms, the usual methods employed for the measurement of the resistance of conductors are unsuitable. Another requirement of an insulation resistance testing instrument is that it should be portable. The instrument used for insulation resistance testing is called a *megger*.

11.8.1 Megger

Figure 11.30 illustrates the principle of operation of a megger. It consists of two coils, a potential coil P and a current coil C, rigidly fixed at 90° with respect to each other. When the two coils are excited, each produces a torque which is in the opposite direction. A magnetic needle, to which is attached a pointer, tends to deflect under the influence of the opposing torques. X is the insulation resistance to be measured and is connected in series with the current coil.

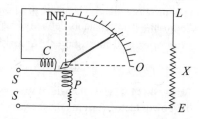

Fig. 11.30   Principle of operation of an insulation tester

When a voltage source, usually a hand-driven small generator giving a constant voltage, is connected to the terminals SS, a constant current independent of the value of X flows through the potential coil P. The direction of the torque produced in coil P is such that the pointer tends to move in the anticlockwise direction. On the other hand, the direction of the torque produced due to current in coil C is such that the pointer tends to deflect in the clockwise direction. The pointer occupies a position of rest, where the two opposing torques become equal.

When the value of resistance X is low, the torque produced by coil C exceeds the torque produced by coil P, hence, the needle aligns along the axis of coil C, that is, the position of zero or low resistance on the scale. When the value of resistance X is very high, or infinite, low or no current flows through coil C. Consequently, the needle aligns along the axis of coil P, that is, the position of infinite or high resistance. The scale is calibrated between zero and ∞.

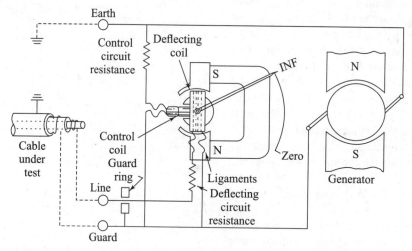

Fig. 11.31   Insulation tester—megger

Figure 11.31 shows the construction of a most commonly used insulation tester—megger. The potential coil P and current coil C herein are called the deflecting coil and control coil, respectively. The two are rigidly fixed at 90° and move within the air gap of a permanent magnet. The former is connected in series with a current-limiting resistance. Similarly, the control coil is also connected in series with a current-limiting resistance which comes into play when the resistance to be measured is low. The deflecting coil is supplied from a hand-driven direct current generator. The generator is driven through a clutch mechanism which slips when the generator speed exceeds its designed value. In this way, it is ensured that a constant voltage at a constant speed is applied to the deflecting coil.

11.9 ELECTRONIC INSTRUMENTS

In the previous sections various electrical instruments for measuring electrical quantities have been described. A very large number of electronic instruments are nowadays in use for measurements and testing, both in laboratories and practise.

11.9.1 Common Types of Instruments

The most common types of electronic instruments are:

  (a) Multimeter
  (b) Electronic voltmeter
  (c) Cathode ray oscilloscope

   In this section working, applications, advantages and disadvantages of a multimeter are described in detail. The other two types of instruments are not covered since they are out of the scope of the present work.

11.9.2 Multimeter

A multimeter is a three-in-one electronic instrument since it can measure (i) resistances, (ii) currents and (iii) voltages fairly accurately. Further, its ability to measure ac-dc voltages and currents makes it an essential instrument of any laboratory or a part of any field testing equipment.

11.9.2.1 Construction

The deflecting system of a multimeter is much like that of a moving coil instrument also called a galvanometer (see 11.5.1). Current flow through the coil produces a mechanical force which causes the pointer fixed to it to deflect. In order that a multimeter can be used to measure (a) voltages, (b) currents and (c) resistances, appropriate circuits are included with the galvanometer. In the following paragraphs, the working of a multimeter is described as (a) a voltmeter, (b) an ammeter, and (c) an ohmmeter.

Fig. 11.32 Circuit for voltage measurement

(a) *As a voltmeter*: To use a multimeter to measure voltages, a high magnitude resistor is connected in series with the galvanometer. Figure 11.32 shows the circuit for using a galvanometer to measure voltages.

   In Fig. 11.32, $R\ \Omega$ is a high valued resistor connected in series with the galvanometer G for voltage measurement. If current I_G, through the galvanometer, produces a FSD when a voltage V volts is applied, the magnitude of R is calculated as under

$$V = I_G(R + R_g), \quad \text{or,} \quad R = V/I_G - R_g \tag{11.55}$$

where $R_g\ \Omega$ is the galvanometer resistance.
Eq. (11.55) shows that by selecting the right value of R, the range of the multimeter can be set at $0 - V$ volts.

One of the objectives of a multimeter is to use it for measuring voltages of varying ranges. For this purpose several high valued resistors, each corresponding to the desired voltage range, with a range selector switch S, are connected in series with the galvanometer. Fig. 11.33 shows the circuit diagram.

By appropriate selection of resistors R_1, R_2, and R_3 the range of a multimeter can be set according to the voltage to be measured.

To use a multimeter, for measuring ac voltages, a full wave ac rectifier which converts ac into dc before feeding to the galvanometer is used. Fig. 11.34 shows the circuit for measuring ac voltages.

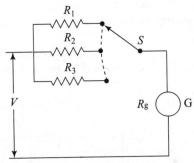

Fig. 11.33   Circuit diagram for multiple voltage ranges

For the measurement of ac voltages the selector switch on the multimeter is moved to ac position which activates the circuit in Fig. 11.34. Next, to measure the voltage the appropriate voltage range is selected by moving the switch to S. The multimeter scale is calibrated to read rms values of ac voltages.

(b) *As an ammeter*:   A galvanometer can be used to measure current by connecting a low resistance (shunt) in parallel with it. Figure 11.35 shows the arrangement.

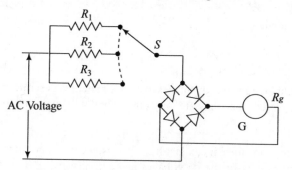

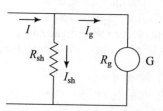

Fig. 11.34   Circuit diagram for multiple ac voltage ranges

Fig. 11.35   Circuit for alternating current measurement

In the arrangement in Fig. 11.35, I_g is the galvanometer current which produces a FSD. In order to limit the current to FSD, a low magnitude resistor of R_{sh} Ω is connected across the galvanometer. The current relations in the circuit are given by

$$I = I_{sh} + I_g \tag{11.56}$$

and $\quad I_{sh} R_{sh} = I_g R_g \tag{11.57}$

Eliminating *I*sh from Eqs (11.56) and (11.57) and simplifying yields

$$R_{sh} = \frac{I_g R_g}{(I - I_g)} \tag{11.58}$$

By using Eq. (11.58), magnitude of R_{sh} can be computed and the range of the multimeter can be set to read $0 - I$ amperes.

As in the case of a voltmeter, a multimeter can be used for measuring currents of varying ranges by connecting shunts of varying magnitudes in parallel with the galvanometer. Figure 11.36 shows the circuit arrangement.

Similar to the arrangement for using a multimeter for ac voltages, a full wave rectifier is included for reading alternating currents as shown in Fig. 11.37 below.

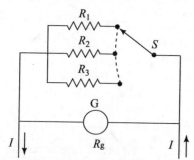

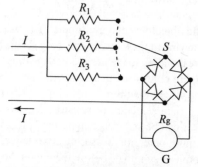

Fig. 11.36   Circuit arrangement for multiple direct current ranges

Fig. 11.37   Circuit arrangement for multiple alternating current ranges

In this case also, moving the selector switch to ac activates the rectifier circuit for ac measurement. RMS value of the current is measured by selecting the proper resistor with the help of the switch S.

(c) *As an ohmmeter*: The circuit to use a multimeter as an ohmmeter is shown in Fig. 11.38.

The ohmmeter circuit is constituted of an internal voltage source of V volts connected in series with a fixed resistor R, a variable resistor r and a galvanometer. Resistor R restricts the circuit current when terminal P and Q are short circuited and r is used for zero adjustment. Before any resistance measurement terminals P and Q are shorted and resistance r is adjusted so that the pointer shows zero reading for a FSD. In this manner the meter is calibrated and adjusted for battery voltage due to usage over time. After zero adjustment, the resistor to be measured is connected across the terminals $P - Q$ and the resistance is read off from the meter scale which is calibrated in ohms. Figure 11.39 shows the arrangement of a multi-range ohmmeter.

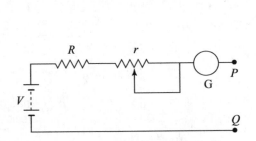

Fig. 11.38 Circuit arrangement to use a multimeter as an ohmmeter

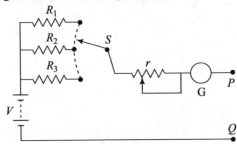

Fig. 11.39 Circuit arrangement for a multi-range ohmmeter

11.9.2.2 Sensitivity

Sensitivity of a multimeter is an important feature of the instrument. It is defined as resistance per volt of FSD and is indicative of the internal resistance of the multimeter. Assume that the current I_G produces a FSD when the meter reads V volts. The meter resistance $= V/I_G$ and by definition

$$\text{Meter sensitivity} = \frac{\left(\dfrac{V}{I_G}\right)}{V} = \frac{1}{I_G} \ \Omega/\text{volt} \tag{11.59}$$

It is always desirable to work with high sensitivity multi-meters since high sensitivity means high internal resistance. A high sensitivity multi-meter when used to measure voltages is connected in parallel and it will draw very little current. Therefore, the main circuit current is slightly affected and there is no distinguishable error in voltage measurement. Commercial multi-meters are available in sensitivity ranges varying between 5–20 kΩ/volt.

Example 11.5 A multimeter is using a 100 V scale to measure voltage across a 60 kΩ resistor. If the multimeter has a FSD of 2 mA, determine if the voltage across the resistor can be measured accurately.

Solution

The meter current $I_G = 2 \times 10^{-3}$ A. Using Eq. (11.59), sensitivity $= 1/I_G = (2 \times 10^{-3}) = 500 \ \Omega/\text{volt}$

Meter resistance $= 500 \times 100 = 50000 \ \Omega$

Figure 11.39 shows the circuit for measuring voltage across the 60 kΩ resistor.

Clearly, it can be seen that the current drawn by the meter will be of a magnitude which will disturb the circuit current. Thus, the voltage read from the meter will be incorrect.

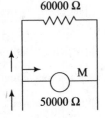

60000 Ω

M

50000 Ω

Fig. 11.39

Example 11.6 Will the multi-meter read correctly, if in example 11.5, the meter resistance is increased by 100 times? What will be the multi-meter FSD?

Solution   In this case, meter sensitivity $= 50000 \times 100/100 = 50000 \ \Omega$ /volt

Meter current $I_G = 1/(\text{sensitivity}) = 1/50000 = 20 \ \mu A$

Since the current through the meter is very small, the voltage reading will be correct. It may also be stated that a multi-meter will read accurately if its resistance is high (say about 100 times compared to the magnitude of the resistor across which the voltage is being measured.

Example 11.7 The circuit in Fig. 11.40 shows an arrangement for measuring currents of varying ranges. If the meter resistance is 100 Ω and FSD is 50 mA, calculate the shunt resistances for the meter to measure current ranges of (a) 0 – 100 mA, (b) 0 – 1 A, and (c) 0 – 5A.

Solution   Current divider rule is used for calculating the resistance.

(a) For 0 – 100 mA range:

$$50 \times 10^{-3} \times 100 = 50 \times 10^{-3}(R_{sh1} + R_{sh2} + R_{sh3})$$

or $\qquad R_{sh1} + R_{sh2} + R_{sh3} = 100 \qquad\qquad$ (I)

(b) For 0 – 1 A range:

$$50 \times 10^{-3}(100 + R_{sh1}) = (1 - 50 \times 10^{-3})(R_{sh2} + R_{sh3})$$

or $\qquad -R_{sh1} + 19R_{sh2} + 19R_{sh3} = 100 \qquad\qquad$ (II)

(c) For 0 – 5 A range:

$$50 \times 10^{-3}(100 + R_{sh1} + R_{sh2}) = (5 - 50 \times 10^{-3})(R_{sh3})$$

or $\qquad -R_{sh1} - R_{sh2} + 99R_{sh3} = 100 \qquad\qquad$ (III)

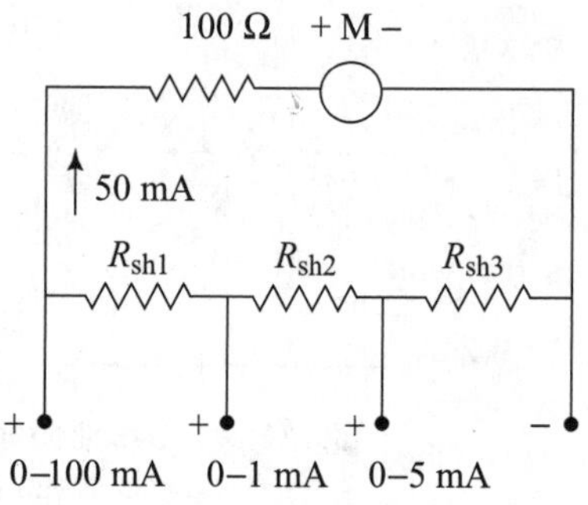

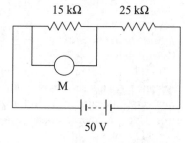

Fig. 11.40

Solving Eqs (I), (II), and (III) simultaneously yields the magnitudes of shunt resistors as under

$$R_{sh1} = 90 \ \Omega, \ R_{sh2} = 8 \ \Omega, \text{ and } R_{sh3} = 2 \ \Omega.$$

Example 11.8 Two resistors each of 15 kΩ and 25 kΩ are connected in series across a 50 V supply. (a) Determine the voltage drop across the 15 kΩ resistor. A multi-meter of sensitivity 2.5 kΩ/volt and a FSD of 20 V is used to measure the voltage across the 15 kΩ resistor. (b) What is the reading on the voltmeter? (c) Is the reading accurate?

Solution   Multi-meter resistance $= 2.5 \times 20 = 50 \ k\Omega$

(a) Voltage across the 15 kΩ resistor, when the multi-meter is not connected, is obtained by the application of the potential divider rule to the circuit in Fig. 11.41 as follows

$$V = \frac{15}{15 + 25} \times 50 = 18.75 \ V$$

Fig. 11.41

(b) When the multi-meter is connected across the 15Ω resistor,

$$\text{total circuit resistance} = \frac{15 \times 50}{65} + 25 = 36.539 \ k\Omega$$

Ohm's law is used to determine the circuit current $= 50/36.539 = 1.368 \ mA$

$$\text{Reading on the multi-meter} = 1.368 \times \frac{15 \times 50}{40} = 25.65 \ V$$

(c) Multi-meter reading is not accurate.

Example 11.9 Calculate the (a) resistance and (b) sensitivity of the multi-meter, in example 11.8, if it were to read the voltage more accurately, say 18.70 V, across the 15 kΩ resistor. Comment on the sensitivity of the meter.

Solution　Assume the resistance of the multi-meter to be R kΩ. Following the steps outlined in example 11.8, the resistance is computed as follows:

$$\text{Circuit resistance} = \frac{R \times 15}{R+15} + 25 = \frac{40R + 375}{R+15} \text{ k}\Omega$$

$$\text{Circuit current} = \frac{50(R+15)}{40R + 375}$$

$$\text{Voltage indicated by the multi-meter} = \frac{50(R+15)}{(40R+375)} \times \frac{15R}{(R+15)} = \frac{750R}{(40R+375)} = 18.70 \tag{I}$$

(a) From Eq. (I), $R = \dfrac{18.70 \times 375}{2} = 3.51 \text{ M}\Omega$

(b) Sensitivity of the multi-meter $= 3.51/20 = 175.32 \text{ k}\Omega/\text{volt}$

From the foregoing it can be stated that for more accurate measurements a multi-meter of higher sensitivity should be used. Alternately, it can be said that the resistance of the multi-meter should be at least 100 times the resistance across which the voltage is being measured.

11.9.2.3　Applications of a Multi-meter

A multi-meter finds extensive applications both in measurement laboratory and field testing. It is used for
  (a) Checking continuity of a circuit
  (b) Measurement of dc and ac voltages
  (c) Measurement of direct and alternating currents
  (d) Testing resistors for continuity.

11.9.2.4　Advantages and Disadvantages of a Multi-meter

Following are the advantages and disadvantages of a multi-meter:

Advantages: (a) Performs measurements with acceptable accuracy,(b) single instrument capable of undertaking several types of measurements, and (c) compact and portable.

Disadvantages: (a) Not capable of performing precise measurements, (b) High cost due to the need for high sensitivity (around Rs 1000 for a 20 kΩ/volt multi-meter), and (c) handling requires technical ability.

11.10　MEASUREMENT ERRORS

The utility of any measurement is the true representation of the value of the quantity being measured and the accuracy with which the measurement can be reproduced. The quality of a measuring instrument, therefore, depends on its accuracy and repeatability of the measured quantity.

　　As already stated, most instruments, in practice, are secondary type of instruments which require calibration before they can be used. However, due to the frequent use of an instrument under varying conditions, errors creep in, due to which both accuracy and repeatability get affected. The various types of common errors that affect the performance of measuring instruments are (a) human or operator errors, (b) instrument errors, (c) environmental errors, and (d) random errors.

11.10.1　Human or Operator Errors

In this category, errors can occur in reading from the instrument, recording the reading, and making computations. An error in reading the instrument can occur if the parallax between the pointer and its mirror image is not properly removed. Another type of operator error is the improper selection or incorrect connection of an instrument.

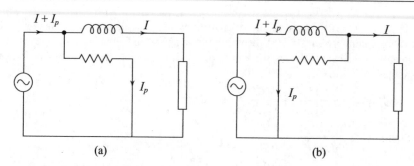

Fig. 11.42 Two types of connections of a potential coil of a wattmeter

Figure 11.42 shows how an error can occur with an improper connection of the potential coil of a wattmeter. When the current in the circuit is low and the voltage high, the potential coil should be connected as shown in Fig. 11.42(a). On the other hand, for high currents and low voltages, the potential coil should be connected as shown in Fig. 11.42(b). Errors are likely to occur if proper connections are not made. Errors in this category are of non-deterministic nature.

11.10.2 Instrument Errors

This type of errors occur due to fatigue and wear and tear of the mechanical parts of an instrument. For example, errors can occur due to friction in the bearings of the moving system, variation in tension of the spring in the control system, improper zero adjustment, etc. These errors can be eliminated by proper maintenance and periodic calibration of the instruments.

11.10.3 Environmental Errors

These are errors which occur due to the use of instruments in varying environmental conditions. Such errors cannot be predicted. The various factors due to which such errors occur are temperature, humidity, gravity, earth's magnetic field, stray electric and magnetic fields, etc. In installations where the accuracy of measurement can be crucial, these errors are eliminated by locating measurement instruments in places/enclosures whose temperature, pressure, and humidity conditions can be controlled.

11.10.4 Random Errors

As the very name suggests such type of errors are unpredictable and, therefore, cannot be determined. Their effect on measurements is minimized by taking several readings under the same conditions and taking the average value.

Recapitulation

Absolute instruments give readings in terms of constants of the instrument.

Secondary instruments are direct reading type, but require calibration.

An analogue instrument converts an electrical quantity (for example, current) into its mechanical analogue (for example, deflection).

A digital instrument converts an electrical quantity into binary coded decimals which are decoded and displayed as digits.

Electrical instruments transform electricity into other forms of energy to measure electrical quantities.

Electronic instruments utilize electronic devices for measurement.

Most analogue instruments utilize magnetic and electromagnetic effects of current for their operation.

Indicating instruments consist of deflecting, controlling, and damping systems.

The operating principle in both ammeters and voltmeters is the same.

The range of an ammeter can be extended by connecting a low-resistance (shunt) in parallel across its terminals.

The range of a voltmeter can be extended by connecting a high resistance in series with it.

Moving-coil instruments can be used for measuring direct current quantities only.

Dynamometer and moving iron type instruments can be used for measuring direct and alternating current quantities.

The principle of operation of dynamometer type wattmeters is the same as that of moving-coil dynamometer type ammeters and voltmeters.

Induction wattmeters utilize the induction effect of current for their operation.

Induction type energy meters operate on the same principle as wattmeters.

In an energy meter the deflecting system of a wattmeter is replaced by a continuously rotating disc.

A three-phase energy meter is like two single-phase energy meters inside a single casing.

Megger is used to measure the insulation resistance of an electrical insulation.

Common types of electronic instruments are (a) multimeter, (b) electronic voltmeter, and (c) cathode ray oscilloscope.

A multimeter is a three-in-one instrument used for measuring (i) resistances, (ii) currents, and (iii) voltages.

Sensitivity of a multimeter is defined as resistance per volt of FSD and is indicative of the magnitude of its internal resistance.

Operator, instrument, environmental, and random errors are the various types of errors in instruments.

Assessment Questions

1. (a) Classify the different types of measuring instruments.
   (b) State the various effects of voltage and current utilized in the operation of these instruments.
2. (a) Describe the important features of measuring instruments.
   (b) Why is it necessary to have a controlling system in a measuring instrument? Describe with sketches (i) spring control and (ii) gravity control.
3. (a) Why is it necessary to have a damping system? Describe the different types of oscillations that might result when the movement of a pointer is damped.
   (b) Describe (i) eddy current and (ii) air friction damping.
4. (a) Categorize the commonly used ammeters and voltmeters.
   (b) State with reasons which types of instruments are employed for ac and which types are used for dc measurements.
5. Describe the working of a moving-coil permanent-magnet instrument.
6. Explain how the range of an ammeter can be extended. State what is a make-before-break switch and what is its necessity.
7. Explain (i) ammeter sensitivity and (ii) voltmeter sensitivity.
8. Discuss the difficulties in developing multipliers for MI (i) ammeters and (ii) voltmeters.
9. (a) Explain the principle of operation of moving-coil dynamometer instruments.
   (b) What are the disadvantages of these instruments in comparison with the PMMC instruments? State why dynamometer-type instruments are not used as ammeters and voltmeters.
10. (a) State the two types of moving-iron instruments and discuss their working principle.
    (b) What are the advantages of MI instruments?
11. Derive expressions for deflection and controlling torques for (i) attraction- and (ii) repulsion-types of MI instruments.
12. (a) State why it is necessary to use a wattmeter to measure ac power.
    (b) Enumerate the different types of instruments used for measuring power. Of the various types of wattmeters, which types are commonly used and why?

13. (a) Derive an expression for the deflecting torque produced by a dynamometer-type of wattmeter. Does it really measure power? Why?

(b) List out the advantages and disadvantages of such types of wattmeters.

14. (a) Explain the principle of working of an induction-type wattmeter.

(b) Show with the help of sketches, the two types of magnet and winding arrangements employed in induction-type wattmeters.

15. (a) Prove that the torque produced in an induction-type wattmeter measures ac power.

(b) Itemize the advantages and disadvantages of induction-type wattmeters.

16. Distinguish between electrical and electronic instruments.

17. Explain with the help of circuit diagrams the working of a multi-meter when it is used to measure dc and ac voltages and currents.

18. Draw the circuit diagram of a multi-meter when it is used to measure resistance. What is the importance of zero adjustment?

19. Enumerate the applications of a multi-meter.

20. Enlist the advantages and disadvantages of a multi-meter.

21. Define sensitivity of a multi-meter and explain its importance.

22. (a) Explain why it is important to measure energy.

(b) Categorize the different types of energy meters and comment on their utilization in practice.

23. (a) Explain with the help of a diagram, the working of a single-phase induction-type energy meter.

(b) Prove mathematically that the driving torque produced in an induction-type energy meter is proportional to the speed of the disc.

24. How does a three-phase induction-type energy meter differs from a single-phase energy meter? Draw a sketch of a three-phase energy meter and indicate its various parts.

25. (a) Normal methods of resistance measurement cannot be employed for measurement of insulation resistance. Explain why.

(b) Explain the principle of operation of an insulation tester.

26. With the help of a sketch, show the construction of a most commonly employed megger. Explain the functions of the important parts. Why is it important that a megger should be portable?

27. Enlist and discuss the various types of errors in measurement.

Problems

11.1 A moving-coil permanent-magnet ammeter is utilizing several shunts to measure currents of different magnitudes. The arrangement is shown in Fig. 11.43. If a current of 500 μA produces an FSD in the ammeter, calculate the magnitude of the currents that can be measured for each of the settings. Assume the ammeter resistance to be 100 Ω.   [0.625 A, 0.833 A, 1.25 A, 2.5 A]

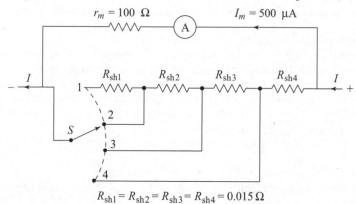

$R_{sh1} = R_{sh2} = R_{sh3} = R_{sh4} = 0.015\,\Omega$

Fig. 11.43

11.2 Two resistors of 350 kΩ and 250 kΩ are connected in series across a 240-V dc supply. Determine the voltage across the 350-kΩ resistor. What will be the voltage across the 350-Ω resistor if a voltmeter of sensitivity 8.33 kΩ/volt and FSD 350 V is connected across it? [140 V, 122.5 V]

11.3 Calculate the magnitudes of multiplier resistors required to measure voltages of (i) 250 V, (ii) 350 V, and (iii) 500 V with a PMMC instrument, which has a resistance of 100 Ω and an FSD current of 150 μA.

[(i) 1.67 MΩ, (ii) 2.33 MΩ, (iii) 3.33 MΩ]

11.4 A single phase 220 V, 20 A, 1200 revolutions/kWh energy meter was observed to make 60 revolutions in 72 sec when tested at half load. Compute the error at half load. [13.64%]

11.5 What is the reading shown by the voltmeter when it is connected across the 30 kΩ resistor shown in the circuit in Fig. 11.44? Assume the voltmeter has (a) infinite resistance, and (b) a sensitivity of 1500 Ω/volt. FSD of the voltmeter is 150 V. [(a) 72 V (b) 68.355 V]

11.6 Calculate the reading on the voltmeter of problem 11.5, for a voltmeter sensitivity of (a) 2000 Ω/volt and (b) 1000 Ω/volt.

[(a) 69.232 V (b) 66.667 V]

11.7 A meter has an internal resistance of 200 Ω. If the meter has a 50 μA movement, determine (a) the full scale voltage which can be displayed by the meter and (b) its sensitivity in ohm/volt.

[(a) 10 mV (b) 20000 Ω/volt]

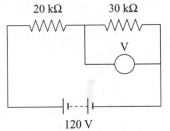

Fig. 11.44

11.8 A 6V battery is used in a meter of sensitivity 15000 Ω/volt for resistance measurement. If the internal resistance of the meter is 5 kΩ, what should be the magnitude of the internal resistor to be connected for proper zero adjustment. [90 kΩ]

11.9 Figure 11.45 shows a classic multi-meter circuit. Compute the magnitudes of the resistors when it is used to measure (a) currents, and (b) voltages. For the meter assume a resistance of 50 Ω and a FSD of 100 μA. The voltage and current ranges to be measured are shown in the figure.

[(a) $R_{sh1} = 40$ Ω, $R_{sh2} = 9$ Ω, and $R_{sh3} = 1$ Ω, (b) $R_1 = 900$ kΩ $R_2 = 90$ kΩ $R_3 = 9975$Ω]

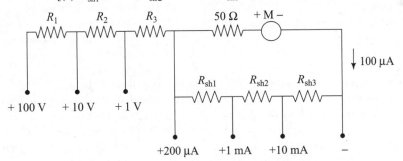

Fig .11.45

Objective Type Questions

1. Which of the following variable is not used to define the quality of energy supplied?
   (i) voltage   (ii) current
   (iii) frequency   (iv) power factor
2. Which of the following is an absolute measuring instrument?

   (i) moving coil
   (ii) tangent galvanometer
   (iii) permanent magnet
   (iv) moving iron
3. Which of the following effect of current or voltage is used in measuring instruments?

(i) chemical   (ii) electrostatic
(iii) heating   (iv) all of these

4. Which of the following is an essential system of a measuring instrument working on the principles of magnetic and electromagnetic induction?
   (i) moving mechanism (ii) controlling system
   (iii) damping system   (iv) all of these

5. Which of the following is responsible for producing a deflecting torque?
   (i) magnetic effect   (ii) damping effect
   (iii) controlling system (iv) all of these

6. Which of the following is not an advantage of a gravity control system?
   (i) it is not costly
   (ii) it is sturdy
   (iii) for measurements it has to be kept vertical
   (iv) none of these

7. Which of the following does not constitute a part of the eddy current damping?
   (i) coil   (ii) aluminium frame
   (iii) aluminium piston (iv) aluminium disc

8. Which of the following can be categorized as ammeter and voltmeter?
   (i) hot wire type
   (ii) moving iron type
   (iii) moving coil dynamometer type
   (iv) all of these

9. Which of the following is not an advantage of moving-coil instruments?
   (i) uniform scale
   (ii) high sensitivity
   (iii) immune to stray magnetic fields
   (iv) none of these

10. Power in an alternating circuit is a function of
    (i) power factor   (ii) voltage
    (iii) current   (iv) all of these

11. A thin aluminium disc is mounted between two laminated electromagnets. In which of the following types of wattmeter is this type of construction used?
    (i) electrostatic   (ii) induction
    (iii) dynamometer   (iv) none of these

12. Which of the following is an advantage of an induction type wattmeter?
    (i) can be used for measuring ac power only
    (ii) uniform and large scale
    (iii) heavy moving system
    (iv) consumes more power

13. Which of the following type of damping is employed in a dynamometer wattmeter?
    (i) air friction   (ii) pneumatic
    (iii) eddy current   (iv) none of these

14. Wattmeters are available with
    (i) single current and voltage ratings
    (ii) multiple current rating and single voltage rating
    (iii) single current rating and multiple voltage rating
    (iv) multiple current and multiple voltage ratings

15. Which of the following is not a function of a multi-meter?
    (i) measure ac voltage
    (ii) measure direct current
    (iii) measure resistance
    (iv) none of these

16. A multi-meter has a sensitivity of 2000 Ω/volt and FSD of 10 V. Which of the following represents its internal resistance?
    (i) 20 kΩ (ii) 2000 Ω
    (iii) 200 Ω (iv) none of these

17. The FSD of a current multi-meter is 20 μA. Its sensitivity is
    (i) 20 kΩ/volt   (ii) 50 kΩ/volt
    (iii) 40 kΩ/volt   (iv) 50 kΩ/volt

18. The measurement of energy is a measure of
    (i) power consumed over a period of time
    (ii) volt-amperes consumed over a period of time
    (iii) reactive volt-amperes consumed over a period of time
    (iv) none of these

19. Which of the following can be used to measure ac and dc energies?
    (i) electrolytic type   (ii) mercury motor
    (iii) commutator motor (iv) induction motor

20. Which type of a controlling system is adopted in an energy meter?
    (i) spring control   (ii) gravity control
    (iii) braking magnet   (iv) none of these

21. Which of the following is a function of a megger?
    (i) measure power
    (ii) measure energy
    (iii) measure insulation resistance
    (iv) measure conductor resistance

22. The scale of a megger is calibrated between
    (i) zero to ∞   (ii) zero to 1,000,000
    (iii) zero to 10,000   (iv) zero to 100

23. Which of the following errors are categorized as operator errors?
    - (i) improper connection
    - (ii) reading the instrument without removing parallax
    - (iii) wrong selection of the instrument
    - (iv) all of these

24. Error due to improper zero adjustment is classified as
    - (i) operator error
    - (ii) environment error
    - (iii) instrument error
    - (iv) random error

25. Which of the following controls the speed of an energy meter?
    - (i) shunt magnet
    - (ii) series magnet
    - (iii) shading band
    - (iv) braking magnet

Answers

1. (iv)	2. (ii)	3. (iv)	4. (iv)	5. (i)	6. (iii)	7. (iii)
8. (iv)	9. (iv)	10. (iv)	11. (ii)	12. (ii)	13. (i)	14. (iv)
15. (iv)	16. (i)	17. (iv)	18. (i)	19. (iii)	20. (iii)	21. (iii)
22. (i)	23. (iv)	24. (iii)	25. (iv)			

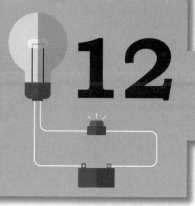

12 POWER SYSTEMS

Learning Objectives

This chapter will enable the reader to

- Perceive the growth of ac energy networks into complex power systems
- Delineate a modern power systems into various subsystems and their functions
- Identify the primary sources of energy
- Obtain a bird's eye view of the different types of generating stations, like coal-, gas-, and oil-fired, hydroelectric, nuclear, wind, geothermal, and tidal power stations and become familiar with the characteristics of each
- Obtain an idea about the operational voltage levels in the transmission subsystem and the sub-transmission system
- Demarcate a distribution subsystem into radial, loop, and network distribution systems, draw generalized layout plans, and understand the properties of each
- Comprehend the basic requirements of a good domestic wiring design based on the Bureau of Indian Standards (BIS) specifications
- Understand and draw layout plans of the distribution board system and tree system for distribution of energy from the main switch
- Familiarize with the loop-in and joint box/tee systems used in domestic wiring systems
- Understand the need for earthing and become familiar with the standardized codes and methods of earthing, such as pipe earthing and plate earthing, as per the BIS specifications

12.1 INTRODUCTION

Electric power systems is the branch of electrical engineering which deals with the technology of generation, transmission, and distribution of electrical energy. The first electric supply system was introduced by Thomas Edison in 1882 at the Pearl Street Station in New York City, USA. Power was generated in a steam engine driven dc dynamo (generator), and dc power was distributed through underground cables for lighting purposes only. The scope of distribution was limited to short distances because of the low voltage of the distribution circuits. The implementation of the transformer into the practical scheme in 1886 by William Stanley of Westinghouse Electric Corporation established the feasibility of ac systems, which revolutionized the electrical industry. It made possible stepping up/down ac voltages using transformers. This has made possible the transmission/distribution of electric power over long distances to reach every consumer. Nikola Tesla's invention of the induction motor in 1888 and the polyphase electric power system in 1894 pioneered the development of three-phase ac power systems. Present-day generation and transmission of electrical energy is mostly three-phase ac at 50 (or 60) Hz.

Since its inception in the 1880s, electric power systems have grown enormously in size and complexity. As power systems increased in size, so did the number of transmission and distribution lines, transformers, switchgear, protective devices, and so on, and their operation also became more complex and challenging. However, the basic objective of an electric power system is to generate electrical energy in sufficient quantities at the most suitable locations, transmit it in bulk quantities to the load centres, and then distribute it to the individual consumers in the proper form and quality at the most economic and ecological price.

12.2 COMPONENTS OF A POWER SYSTEM

An electric power system is constituted of several subsystems. A single-line diagram of such a system with four major subsystems is shown in Fig. 12.1. Broadly, an electric power system can be divided into the following subsystems.

Generation subsystem This system is constituted of groups of generating stations, where the conversion of energy from the primary energy source in one form (mechanical, thermal, nuclear, etc.) into electrical energy takes place in electric generators through the process of electromagnetic energy conversion.

Transmission subsystem The overhead transmission networks transfer electrical energy from generating stations located at various locations, usually over long distances, to the distribution systems, from where it is distributed. The transmission lines also interconnect neighbouring utilities for exchange of power during normal operation and emergencies.

Sub-transmission system The sub-transmission network is the portion of the transmission system that connects the high-voltage substations through step-down transformers to the distribution substations.

Distribution subsystem Distribution is the process by which energy is fed locally to various distribution substations in a given area from one or more main transmission substations. A distribution subsystem is constituted of overhead distribution lines and underground cables, and its function is to supply quality power to consumers.

Protection and control subsystem This subsystem is constituted of relays, switchgear, and other control devices that protect the various subsystems against faults and overloads, and ensures efficient, reliable, and economic operation of the electric power system.

It may be noted that transformers are used in all the subsystems. A transformer transfers power with very high efficiency from one voltage level to another voltage level. Insulation requirements limit the generated voltage to low values, up to 30 kV. Thus step-up transformers are used for the transmission of power at the sending end of the transmission lines. At the receiving end of the transmission lines, step-down transformers are used to reduce the voltage to suitable values for distribution to the consumers of electrical energy. Furthermore, depending on power-handling capacity, there are two types of transformers: power transformers and distribution transformers (Fig. 12.1). Usually, power transformers are rated between 250 kVA and 1000 MVA, and distribution transformers are rated between 20 kVA and 250 kVA.

In Fig. 12.1 the voltage levels at the various subsystems are indicated. EHV designates extra high voltage, usually above 220 kV and up to 800 kV. HV denotes high voltage, usually from above 66 kV to no more than 220 kV. MV means medium voltage, usually above 1 kV but less than 66 kV. LV stands for low voltages, which are 1 kV or less.

12.3 GENERATION SUBSYSTEM

The generation of electrical energy commenced with the setting up of individual power stations at pitheads to supply electric power to individual consumers. As the demand for electrical energy increased, power systems came into existence. Thus, the generation subsystem is constituted of groups of generating stations, which convert some form of primary energy into electrical energy.

The simplest form of a generating station is constituted of a prime mover coupled to an electric generator. A primary source of energy is employed as the input to the prime mover, which in turn rotates the generator to produce electrical energy.

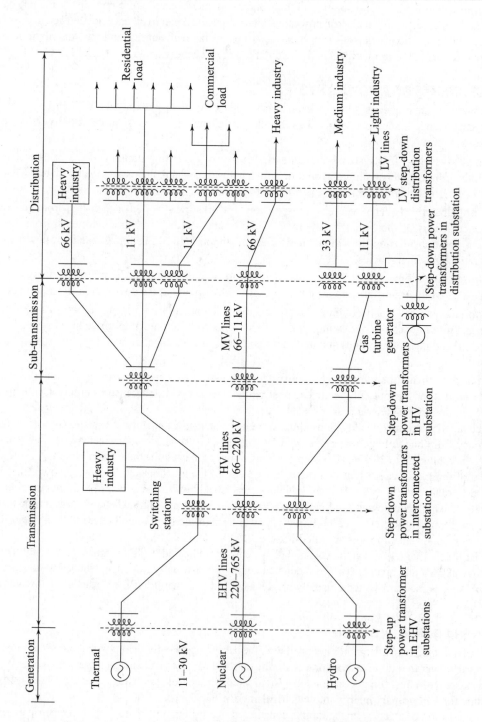

Fig. 12.1 Single-line diagram of an ac power system

12.3.1 Primary Sources of Energy

The important primary sources of energy employed for generation of electrical energy can be broadly classified into three categories.

(i) Fossil fuels; for example, coal (including lignite and peat), oil, and natural gas
(ii) Renewable energy generated from wind and water
(iii) Nuclear energy from uranium or plutonium

In modern-day electric power systems, most generating stations employ these three types of primary energy sources. The amount of electric power contributed by each type of generating station is governed primarily by the market costs of the primary energy sources. For example, water stored in dams and wind as primary input sources of energy have zero cost compared to the cost of fuel such as coal, oil, gas, uranium, etc. used in thermal and nuclear generating stations. In nuclear generating stations, energy costs are low compared to thermal generating stations. The economics of generating stations employing fossil fuels as the primary source of energy depends on the market prices of the fuels.

12.3.2 Types and Characteristics of Generating Stations

Generating stations can be classified into four categories, namely, (a) fossil-fuel, (b) hydroelectric, (c) nuclear, and (d) non-conventional.

In fossil-fuel generating stations, coal, oil, natural gas, etc. are employed as the primary source of energy, while the head and volume of water is employed as the primary source of energy in hydroelectric generating stations. Controlled nuclear fission is the source of energy in a nuclear power station. In unconventional generating stations, wind, geothermal (employing the heat present deep inside the earth) energy, tidal energy, etc. are used as sources of energy to generate electric power.

12.3.2.1 Fossil-fuel Generating Stations

Coal-fired A simple schematic diagram of a fossil-fuel generating station is shown in Fig. 12.2. The chemical energy in coal is utilized to generate electrical energy. Pulverized coal is burnt to produce steam, at high temperature and pressure, in a boiler. The steam so produced is passed through an axial flow steam turbine, where its internal heat energy is partially converted into mechanical energy. The steam turbine, which is the prime mover, is coupled to an electric generator. Thus, mechanical energy produced by the rotating turbine is converted into electrical energy.

The efficiency of the process of conversion of chemical energy into thermal energy and then into mechanical energy is poor. Due to heat losses in the combustion process, the rejection of a large quantity of heat in the condenser, and rotational losses, the maximum efficiency of the conversion process is limited to about 40%. In order to increase the thermal efficiency of the conversion of heat into mechanical energy, steam is generated at the highest possible temperature and pressure. To further increase the thermal efficiency, steam is reheated, after being partially expanded, by an external heater. This reheated steam is returned to the turbine, where it is expanded in the final stages of bleeding.

Modern practice is to design and build generating units having large megawatt generating capacity, since their capital cost per kilowatt decreases as the megawatt capacity increases. Increasing the unit capacity from 100 MW to 250 MW results in a saving of about 15% in the capital cost per kilowatt. It is also established that units of this magnitude result in a fuel saving of the order of 8% per kilowatt-hour. In addition, the cost of installation per kilowatt is considerably lower for large units. Presently, the maximum capacity of the turbo-generator sets being produced is nearly 1200 MW. In India, super-thermal units of capacity 500 MW are being commissioned by BHEL.

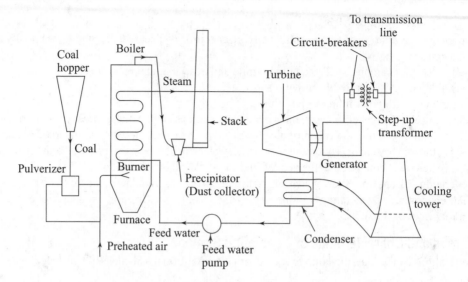

Fig. 12.2 Schematic diagram of a thermal generating station

Fossil-fuel generating stations also employ cogeneration in order to utilize the large amount of wasted heat. In cogeneration, electricity and steam or hot water are simultaneously made available for industrial use or space heating. It is claimed that cogeneration results in an overall increase in efficiency of up to 65%. Cogeneration has been found to be particularly advantageous for industries such as paper, chemicals, textiles, fertilizers, food, and petroleum refining.

The waste gases produced by coal-fired generating stations contain particulates and gases such as oxides of sulphur, NOX, etc. These gases are released into the atmosphere, resulting in air pollution. Thermal pollution also results due to the large amount of heat released via the condenser into the cooling water.

Oil-fired In an oil-fired steam station, oil is employed to produce the steam used to run the steam turbine. In oil-fired stations, two types of oil are used: crude oil, which is the oil pumped from oil wells, and residual oil, which is the oil left behind after the more valuable fractions have been extracted from crude oil. The cost of transporting oil through pipelines is less than that of shipping it by rail. However, residual-oil-fired stations have to be located close to oil refineries, since it is uneconomical to transport residual oil using pipelines because of its high viscosity.

Gas-fired The primary source of energy in gas-fired generating stations is natural gas. A gas turbine engine similar to a turboprop engine used in an aircraft is employed as a prime mover to run the generator. In order to achieve higher thermal efficiency, the combined-cycle method is used to generate electricity. In the first stage, gas turbine engines coupled to electric generators produce power. In the second stage, the hot gases exhausted from the gas turbines are passed through a heat exchanger to generate steam, which is used to run a conventional steam generator to produce electrical energy. Alternatively, the hot gases exhausted from the gas turbine can be used for producing steam for an industrial process. Figure 12.3 shows the schematic layout of a combined-cycle, gas-fired power station.

For the same amount of power generated, combined-cycle, gas-fired stations are more environment-friendly. The flue gases emitted by these stations contain almost zero sulphur dioxide, 50% carbon dioxide, and 25% NOX as compared to those produced in coal-fired power stations. In addition, the installation cost of gas-fired stations is lower and they can be started quickly. The operational cost of gas-fired stations is high, due to high fuel cost, when employed to supply power on their own. As such, they are used to supply peak load demand, for short periods.

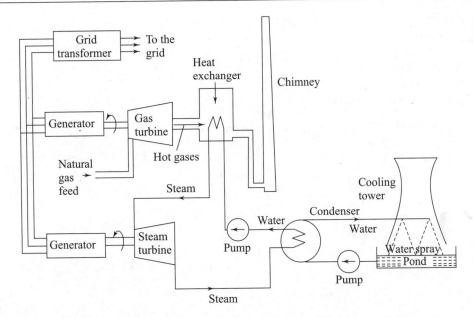

Fig. 12.3 Schematic layout of a combined-cycle, gas-fired power station

Diesel-oil-fired Diesel oil is used to run large internal combustion engines of the type employed in ships. Diesel-oil-fired stations exhibit characteristics similar to those of gas-fired stations. However, the speed of the former is considerably low and it has higher fuel efficiency than steam power stations. Due to the high cost of diesel compared with the cost of oil, in an oil-fired steam station, the use of diesel-oil-fired stations is limited to supplying standby power.

12.3.2.2 Hydroelectric Generating Stations

In a hydroelectric generating station, the potential energy and quantum of water is utilized to generate electric power. In other words, hydroelectric schemes function based on the flow of water and the difference in level known as *head*. Due to the difference in the head, considerable velocity is imparted to the water, which is used to drive a hydro-turbine. The hydro-turbine acts as a prime mover, and is coupled to an electric generator to produce electrical energy.

Hydroelectric stations depend on the availability of a head of water. As such, they are often sited in mountainous terrain and require long transmission lines to deliver power to the load centres. Hydroelectric schemes are classified on the basis of the head utilized to generate power, high-head storage type, medium head pondage type, and run-of-river. In low-head type of hydro-generators, both the velocity of the water and the difference in levels are used to rotate the turbine. In high-head generators, the difference in levels is used to impart high velocity to the water to run the turbine. As the name suggests, in run-of-river hydro-generators, the natural flow of river water is used to drive the turbines. Figure 12.4 shows the schematic diagram of a high-head storage type hydroelectric scheme.

The power P generated in a hydroelectric station is given by

$$P = 9.81 \rho Q h \, \eta \times 10^{-3} \text{ kW} \tag{12.1}$$

where Q is the discharge of water in m^3/sec through the turbine, ρ is the specific weight of water in 1000 kg/m^3, h is the head in metres, and η is the generation efficiency.

The merits of a hydroelectric station are the following.

(i) Operational costs are minimal since there is no fuel cost involved.
(ii) No air pollution.
(iii) No waste products.
(iv) Maintenance is minimal.
(v) Quick start-up time (within 5 min).
(vi) Long life (minimum 50 years).

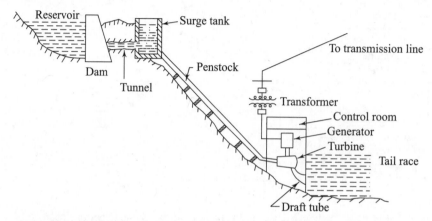

Fig. 12.4 Schematic diagram of a high-head storage type hydroelectric scheme

The demerits of a hydroelectric station are the following:
(i) Capital costs are high.
(ii) Long gestation period.
(iii) Ecological damage to the region.

12.3.2.3 Nuclear Power Stations

The fuel in a nuclear power station is uranium. Of the two isotopes of uranium, uranium-235 and uranium-238, found in natural uranium, only uranium-235 is capable of undergoing fission. Fission in U-235 is brought about by bombarding it with neutrons. In the fission reaction, heat energy and neutrons are released. The released neutrons further react with fresh U-235 atoms to generate more heat and more neutrons. Thus the fission reaction is a chain reaction and is required to be conducted under controlled conditions in a nuclear reactor.

In a nuclear power station, the nuclear reactor constitutes the heart of the station and replaces the boiler in a coal- or oil-fired station. Figure 12.5 shows the schematic layout of a nuclear power station. In the reactor pressure vessel, nuclear fuel rods are embedded in neutron speed reducing agents, such as heavy water, graphite, etc., called *moderators*. These moderators reduce the speed of neutrons to a *critical value*. The nuclear reaction is controlled by inserting boron steel rods, which have the property to absorb neutrons. Thus the rate of nuclear fission is controlled by controlling the neutron flux. A primary coolant, such as heavy water, carbon dioxide, etc., is used to transfer the heat generated due to the fission reaction to the heat exchanger. Steam is produced in the heat exchanger, which is used to run a conventional steam turbine.

The fuel requirements of a nuclear generating station are minimal compared to a coal-fired generating station. In addition, the cost of transporting nuclear fuel is negligible. Another advantage of nuclear power stations is that they do not cause any air pollution. Nuclear stations, therefore, can be sited close to load centres. However, since radioactive fuel waste is produced in a nuclear reactor, safety considerations demand that nuclear stations be sited away from populated areas. Nuclear stations require a high capital investment. The operational cost of such stations, however, is low.

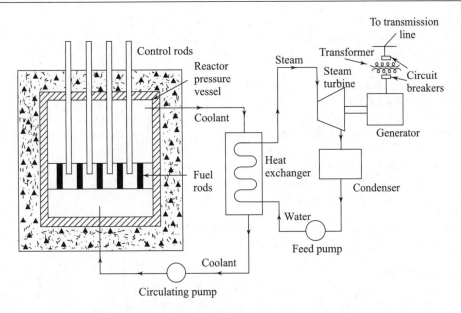

Fig. 12.5 Schematic layout of a nuclear power station

12.3.2.4 Non-conventional/Alternative Generating Stations

Wind power stations Wind as a source of energy has been used for centuries to grind grain and pump water. Wind as a source of primary energy is particularly attractive, since it is non-polluting. However, it is unpredictable and unsteady. The success of wind power generating stations is governed by the initial capital cost, maintenance cost, useful life, and power output. Wind power generating stations are useful for meeting low-power requirements in small isolated areas. In India, the gross potential of wind power has been assessed at approximately 45,000 MW, and the technical potential is estimated at 13,000 MW. Wind power stations have been set up in the states of Gujarat, Maharashtra, Orissa, Andhra Pradesh, and Tamil Nadu. The largest installation of wind turbines in the country so far is in the Muppandal and Perungudi area near Kanyakumari in Tamil Nadu, with an aggregate installed capacity of about 500 MW. State-of-the art technology is now available in India for manufacturing wind turbines of capacity up to 750 kW.

Geothermal power stations Geothermal power generation involves the conversion of the heat energy contained in hot rocks inside the core of the earth into electricity through steam. Water is used to absorb heat from the rocks and transport it to the earth's surface, where it is converted into electrical energy through a conventional steam-turbine generator. Geothermal energy has been employed to generate steam in a limited way in Italy, New Zealand, Mexico, USA, Japan, etc. In India, the use of geothermal energy is still at the developmental stage; the feasibility study for a 1-MW station in Ladakh has been undertaken. Though the efficiency of a geothermal station is less than that of a conventional fossil fuel plant, geothermal stations are attractive due to their low capital cost and zero fuel cost. The total available geothermal power globally has been estimated at 2000 MW, of which only about 500 MW has been tapped. In India, despite a number of hot springs, the availability of the exploitable geothermal energy potential appears to be unattractive.

Tidal power stations The gravitational effects of the sun and the moon and the centrifugal force of the earth's rotation on its axis cause sea tides. In about 25 hours, there are two high tides and two low tides. The minimum head required for power generation is about 5 m.

Tidal power stations use high-tide periods to fill reservoirs by opening sluice gates behind embankments along the seashore. During low-tide periods, when the tide is falling on the seaward side of the embankments, the sluice gates are closed and the stored water is made to flow through turbines coupled to generators. This is known as *ebb generation*.

A tidal power station is usually sited at the mouth of an estuary or a bay. A barrage or an embankment is constructed at the site to store water. Two-way generation can be achieved in tidal power stations. As the tidal waves come in, water flows through the reversible turbines to generate power and fill the estuary/bay. As the tide falls, water flows out of the estuary/bay and the turbines. Since the turbines are reversible, power is again generated.

A disadvantage of tidal power stations is that, due to variations in high- and low-tide timings, power stations may generate at peak demand on some days and remain idle on other days. Another disadvantage is the high cost of civil engineering works required for tidal power stations.

With hundreds of kilometres of coastline, a vast potential source of tidal energy is available in India. A 600-MW tidal power station has been planned at Kandala on the Gujarat coast by constructing a dam. Other sites under exploration are at Bhavnagar, Navalakhi (Kutch), Diamond Harbour, and Ganga Sagar.

Solar Power Stations

The earth receives radiation continually from the sun to the equivalent of 1.17×10^{17} W. It is this energy from the sun which is utilized to generate electricity. A solar cell is a thin silicon wafer of thickness 0.25 mm and can have a round or square form. It has the property of converting light energy of the sun into electric current. Figure 12.6 shows the one-dimensional geometric view of a *photovoltaic* cell.

Light photons from the sun penetrate into the junction diode (cell) and impart enough energy to the valence electrons to make them jump into the conduction band. Due to the unaccounted for numbers of photons penetrating the cell, an extremely large number of electrons enter the conduction band and are pushed out of the cell by the internal electric field which has been already produced during the manufacture of the junction diode. This flow of electrons leads to the flow of current. The process of direct conversion of solar light energy into electric current is called *photovoltaic (PV)* effect.

The electrons will continue to flow out of the cell as long as light photons from the sun continue to penetrate the cell. As such, a cell never loses power, like a battery, since cells do not 'consume' electrons. Therefore, a cell may be viewed as a converter, since it changes (sun) light energy into electric energy.

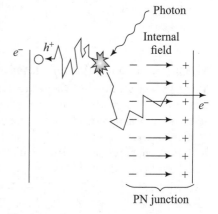

Fig. 12.6 Direct conversion of solar energy to electricity in a photovoltaic diode

Figure 12.7 shows a view of a basic circular PV device. The metal contacts placed in front and at the back draw and deliver the electrons to the cell. In this manner, the same electrons continue to travel the same path and in the process deliver light energy to the load.

A typical silicon PV cell produces only 0.5 V DC. Therefore, a PV cell forms the basic device to form modules or panels to obtain higher voltages. Since a minimum of 12 V is required to charge a storage battery, a typical panel 65 cm × 140 cm panel will be made of 36 PV solar cells connected in series to produce 18 V. When loaded, the output voltage of the panel drops to 14 V

Fig. 12.7 Basic PV device

which is the minimum voltage required to charge a storage battery. Thus, 36-solar cells panel has become the standard or basic module for the solar battery charger industry. Solar panels can be connected in series to obtain higher voltages of 24, 48, and more. Higher current capacity and therefore more power output can be obtained by connecting the basic modules of 36 cells each in parallel. Figure 12.8 provides a pictorial view of the basic modules of 36 cells.

Fig. 12.8 Pictorial view of a basic solar PV module

Under the National Action Plan on climate change, one of the eight missions that Government of India, has set up, in January 2010, is the Jawaharlal Nehru National Solar Mission (JNNSM). The aim of the Mission is to develop and promote the use of solar energy technology. The Mission aims to achieve, in three phases, a cumulative target of 20,000 MW and 2,000 MW respectively, in grid and off-grid solar power generation by 2022.

Table 12.1 shows the breakup of installed capacity in mega watt, at the all India level till November 2016, according to the data issued by Central Electricity Authority.

Table 12.1 All India installed capacity in MW of various types of power stations

Type of power station	Installed capacity in MW
Hydro	43133
Thermal	
Coal	187803
Gas	25282
Diesel	919
Sub-total	214004
Nuclear	5780
Renewable energy sources	
Small hydro power	4323
Wind power	28083
Biomass/Cogen	48822
Waste to energy	115
Solar power	8513
Sub-Total	45916
Grand Total	308833

12.4 TRANSMISSION SUBSYSTEM

A transformer and transmission line subsystem is designed to transmit bulk electric power for consumption at the load centres. In generating stations, power is generated at voltage levels varying between 11 and 30 kV. The transformers at the generating station end step up the voltage to the level suitable for the transmission of bulk power. Since these transformers step up the voltage, they are also known as step-up transformers.

The power transmitted over a transmission line is proportional to the square of the transmission voltage. Therefore, ideally it is desirable to have the highest possible transmission voltages. Hence, continuous efforts are undertaken to increase the transmission voltages. In western countries, power is transmitted at 765 kV. In India, the transmission voltage levels vary between 66 and 400 kV.

When bulk power is to be transmitted over distances greater than 600 km, high-voltage dc transmission is more economical than high-voltage ac transmission. Transmission takes place at dc voltages of 400 kV and above. At the generator end, the ac voltage generated is stepped up to the transmission voltage level by a step-up transformer, which is converted into high-voltage dc by a converter circuit. A converter is a three-phase, full-wave bridge circuit consisting of silicon-controlled rectifiers that can operate as rectifiers converting ac voltage into dc voltage. At the receiving or the load end of the high-voltage dc transmission, a converter operating as an inverter is used to change high-voltage dc to high-voltage ac. The ac voltage is then stepped down, using a step-down transformer, to a lower voltage level for distribution to consumers of electrical energy.

The level of voltage at which distribution of power is undertaken depends on the type of industries in the region. The first step down in voltage may be from the transmission or grid level to the sub-transmission level and may range between 33 kV and 132 kV.

For the purpose of supplying power to small industries and commercial and domestic consumers, the voltage is again stepped down at the distribution substation. The distribution of power is undertaken at two voltage levels: the primary or feeder voltage at 11 kV, and the secondary or consumer voltage at 415 V for a three-phase supply and 230 V for a single-phase supply.

12.5 SUB-TRANSMISSION SYSTEM

The portion of the transmission system that connects high-voltage substations through step-down transformers to the distribution substations is called the *sub-transmission network*. There is no clear demarcation between the transmission and sub-transmission voltage levels. The voltage level of the sub-transmission system ranges from 66 kV to 132 kV. Some heavy industrial consumers are connected to the sub-transmission system.

12.6 DISTRIBUTION SUBSYSTEM

A distribution subsystem is the part of an electric power system between the step-down distribution substation and the consumers' service switches. A distributed system is designed to supply continuous and reliable power at the consumers' terminals at minimum cost. A typical distribution system is shown in Fig. 12.9.

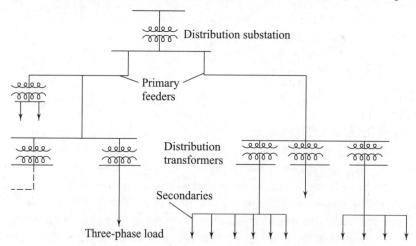

Fig. 12.9   Layout of a typical distribution system

At the distribution substation, the voltage level is brought down from 66 kV at the sub-transmission level to 11 kV at the distribution level. Each distribution substation normally serves its own area, which is a subdivision of the area served by the distribution system. Distribution transformers are ordinarily connected to each primary feeder and its sub-feeders and laterals. Each transformer or bank of transformers serves to step down the voltage to the utilization voltage 415 V (three-phase) or 230 V (single-phase), and supply a consumer or a group of consumers over its secondary circuit. Each consumer is connected to the secondary circuit through service leads and a meter.

12.6.1 Types of Distribution Systems

In order to supply power to consumers, different types of distribution systems are employed. A brief description of each type of distribution system is given below.

12.6.1.1 Radial Distribution System

The simplest and least expensive, and thus the most common form of, distribution system is the radial type of distribution system. Such a distribution system is employed to serve low- and medium-density load areas. Figure 12.10 provides the layout of a simple radial distribution system.

The main primary feeder branches out into various primary laterals, and the laterals in turn separate out into several sub-laterals, extending to all parts of the area served by the distribution system. Distribution transformers are connected to the primary feeders, laterals, and sub-laterals, usually through cut-out fuses. The reliability of the service continuity of radial distribution is low, as a fault at any location on the radial primary feeder causes power outage to every consumer on the feeder. Circuit-breaker

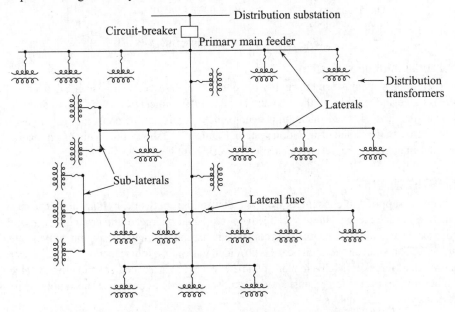

Fig. 12.10   Layout of a simple radial distribution system

12.6.1.2 Loop Distribution System

A loop distribution system has a single feeding point; the circuit forms a loop through the feeder load area and returns to the same feeding point bus. A loop distribution system is shown in Fig. 12.11. Each dot in the figure represents a balanced three-phase load at that location. The advantage of the loop distribution system over the radial system is that it is more reliable. In case of a fault, an alternative circuit/path is available. The loop distribution system is also known as a *parallel distribution system*.

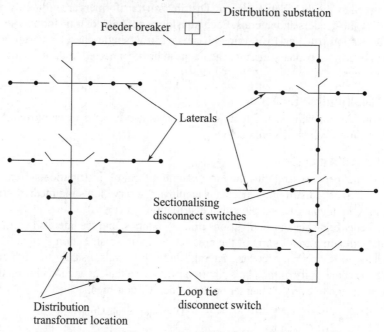

Fig. 12.11 A loop distribution system

12.6.1.3 Network Distribution System

A network distribution system is used to supply large loads spread over a large area. Such a distribution network is more reliable and exhibits all the advantages of an interconnected grid circuit. In addition, due to the small feeder lengths, it is easier to maintain proper voltage control at the consumer terminals in a network distribution system than in a radial distribution system. Figure 12.12 shows the layout of a network distribution system. Each dot in the figure represents a balanced three-phase load at that location.

12.7 DOMESTIC WIRING

Electrical energy is supplied to consumers from the distribution network, by the supplier (State Electricity Board), through a service feeder. The service feeder could be an overhead line or an underground cable. The service feeder terminates at the consumer's premises, via an energy meter, at a main switch called the *main distribution board*. The main distribution board is provided with a fuse for protection. The modern practice is to use a miniature circuit-breaker (MCB) in place of a fuse. An MCB is an accurate and efficient electromechanical device that provides protection against short circuits and overload currents. The supplier's responsibility is to deliver energy up to the main switch.

In order to utilize the electrical energy, the consumer has to appropriately design his wiring system. For example, in order to connect various electrical fixtures such as bulbs, fluorescent tube lights, water heaters, refrigerators, and other electrical appliances in a building, wiring circuits have to be designed for efficient distribution of electrical energy. The requirements of a good wiring circuit design are the following.

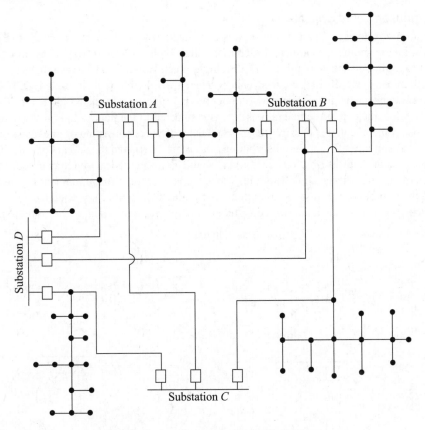

Fig. 12.12 Layout of a network distribution system

(i) Equal voltage distribution.
(ii) Same load current in each circuit.
(iii) Ease of repair and maintenance of the circuits.
(iv) Safety and minimum inconvenience to the consumer.

The Bureau of Indian Standards (BIS) has specified that the maximum load in each circuit shall be limited to 800 W and the maximum number of light and fan points, including socket outlets, shall not exceed 10. Table 12.2 provides the ratings assumed for different types of fixtures for calculating the load in each circuit. Therefore, if the load or the number of points to be connected is more than that specified, it becomes necessary to have more than one wiring circuit. In such a case, each wiring circuit is called a *sub-circuit*.

Table 12.2 Ratings for different types of fixtures

Type of fixture	Rating (W)
Fluorescent tube	40
Incandescent lamp	60
Ceiling fan	60
Socket outlet (5 A)	100
Power socket outlet (15 A)	1000
Exhaust fan	100

12.7.1 Electrical Energy Distribution Systems

From the main switch, electrical energy can be distributed to the various points using the following two systems: (a) the distribution board system and (b) the tree system.

12.7.1.1 Distribution Board Systems

In the distribution board system, each sub-circuit is connected to a sub-distribution board (SBD) via a fuse. In other words, all the sub-circuits are connected in parallel. Figure 12.13 shows the layout of a distribution board system. In the circuit shown in this figure, the single-phase service feeder of the supplier is connected to the main distribution board (MDB) through an energy meter and a double-pole, iron-clad (DPIC) main switch. The phase wire is connected through a fuse to a copper bus bar in the main distribution board. The neutral link in the main switch, indicated in the diagram as NL, is connected to the neutral link in the main distribution board. With this arrangement, it is possible to draw out pairs of wires consisting of one phase and the neutral to feed the various sub-circuits. The sub-distribution boards are connected to the pairs of wires from the main distribution board. Each sub-distribution board connects the various wired sub-circuits in order to connect the different types of electrical fixtures in a building. The advantage of the distribution board system is that there is no limit on the number of sub-distribution circuits, provided the supplier's service feeder is rated to carry the total load current. The disadvantage of the system is that it is expensive, since a lot of material is required.

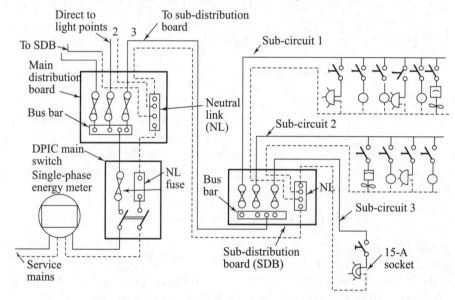

Fig. 12.13 Layout of a distribution board system Bus bar

12.7.1.2 The Tree System

In the tree system of distribution of electrical energy, the phase wire and the neutral wire are drawn from the main switch directly and are referred to as the *main branch*. From the main branch, which runs along the building, smaller branches are tapped off to supply the various sub-circuits. Each sub-circuit is provided with an independent fuse for its protection. Figure 12.14 shows the layout of a tree system. The disadvantages of the tree system are the following.

  (i) Unequal distribution of voltage across each sub-circuit.

 (ii) Large number of joints; fault location is difficult.

(iii) Fuses are scattered.

12.7.2 Domestic Wiring Systems

To connect or wire the various types of electrical fixtures in a building, two wiring methods are used, namely, (a) the loop-in system and (b) the joint box or tee system.

12.7.2.1 Loop-in System

In the loop-in system of wiring (Fig. 12.15), when an electrical fixture or an appliance is to be connected, both the phase and neutral wires are brought directly to the terminals of the appliance and then carried forward to the terminals of the next appliance to be connected. Thus the phase and neutral wires are looped-in. In this system of wiring, no joints are made along the run of the phase or neutral wire, and the fixtures and appliances are connected in parallel. This system of wiring is normally employed for wiring electrical fixtures and appliances.

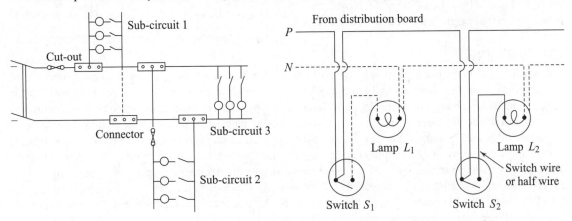

Fig. 12.14   Layout of a tree system   **Fig. 12.15**   Loop-in system of wiring

The advantages of the loop-in system of wiring are the following.
(i) Since all joints are made at the terminals of the electrical fixtures or appliances or in the distribution boxes, it is easy to locate and repair faults.
(ii) No joint boxes are required.

The disadvantage of the system is that longer lengths of wire are required. Consequently, the voltage drops and resistance losses are more. Due to the long lengths of wires, the system is more expensive.

12.7.2.2 Joint Box or Tee System

In the joint box system of wiring (Fig. 12.16), joint boxes are provided along the run of the wires. Electrical fixtures and appliances are connected either by making joints in the joint boxes or through suitable connectors.

Due to the large number of joints, the system is likely to suffer if the joints are not properly made. The savings made in the quantity of wire used are offset by the cost of the joint boxes and connectors. The joint box system is employed for temporary wiring, due to the low cost of wiring.

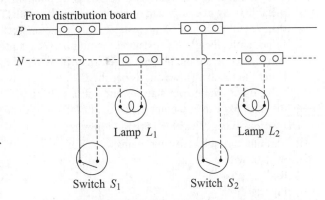

Fig. 12.16   Joint box or tee system of wiring

12.7.2.3 Staircase Lighting

In a staircase lighting a single light point is controlled from two different locations, one near the bottom of the stairs and the other near the top of the stairs. The wiring diagram of a staircase lighting system is shown in Fig. 12.17. A two-way switch S_1 is installed near the first step of the stairs, another two-way switch S_2 is installed at the upper floor where the stairs end, and the lamp L_1 is installed at a suitable height and location in between the first and last stairs. If the light is switched ON by the lower switch S_1, it can be switched OFF by the switch S_2 located near the top stair and vice versa.

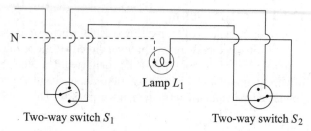

Lamp L_1

Two-way switch S_1 Two-way switch S_2

Fig. 12.17   Wiring diagram of a staircase lighting system

12.8 EARTHING

Whenever a non-current-carrying metallic part of an electrical appliance comes in contact with a live conductor, it accumulates a static charge and acquires a potential. If this charged metallic portion now comes in contact with a living body, the entire charge will flow through the body as current. Consequently the body will experience a shock. On the other hand, if the non-current-carrying metallic part of the electrical appliance is connected to the earth, no static charge can accumulate. In such a condition all static charge will be transferred to the earth when the body of the appliance comes in contact with a live wire. The process of connecting the non-current-carrying metallic parts of electrical appliances or the neutral of a supply system to the earth is called earthing. The earthing of electrical installations is required to be undertaken for the following reasons.

(i) To ensure that the potential, with respect to the earth, of any current-carrying conductor does not rise above its designed insulation level.

(ii) To avoid shocks to a living body.

(iii) To provide safety to operating and maintenance personnel.

(iv) To avoid fire hazards due to earth leakage current.

In view of the importance of earthing, the BIS [earlier called the Indian Standards Institution (ISI)] has standardized the specifications for earthing electrical installations, equipment, appliances, etc. The important and relevant specifications are the following.

(i) The minimum distance of the earthing electrode from the building whose installation is being earthed shall be 1.5 m.

(ii) The earth continuity conductor (ECC) is a conductor employed to connect the non-current-carrying metallic body of equipment or appliances to the earth. The cross section of the ECC should not be less than 2.9 mm^2, or half of the installation conductor size.

(iii) In principle, the earth's resistance should be sufficiently low to allow adequate flow of earth leakage current to operate the protective relays or blow out the fuse. Ideally, lower the resistance of the earth, better the results. However, as per the BIS, the permissible maximum resistance value for all installations, other than power stations and substations, is 8.0 W. The maximum value of the resistance from the earth plate to any point inside the installation (earth continuity) should be 1.0 W.

The various types of earthing are the following.

(i) Earthing through the water mains.

(ii) Strip or wire earthing.

(iii) Rod earthing

(iii) Pipe earthing

(iv) Plate earthing.

Using the water mains for earthing is not normally advised since it requires an iron water pipe, which should ensure electrical continuity. Wire or strip earthing is employed in rocky soils, where it is difficult to dig pits up to the required depths. Rod earthing, though cheap, is suitable for sandy soils. The pipe and plate methods are the two commonly employed methods for earthing. In the following sections, the pipe and plate methods of earthing have been described.

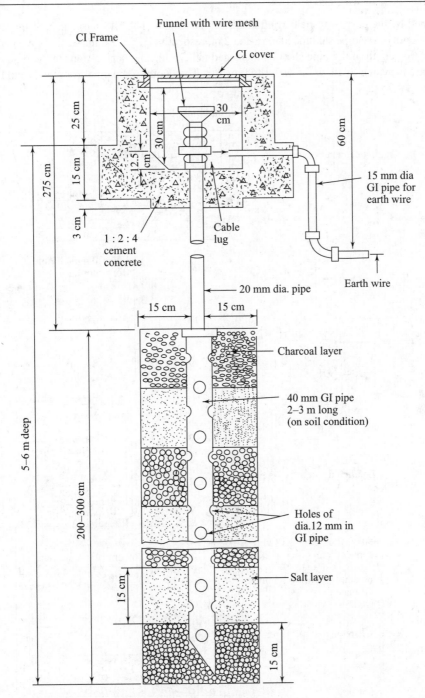

Fig. 12.18 Pipe earthing

12.8.1 Pipe Earthing

Figure 12.18 shows the details of pipe earthing. A galvanized perforated cast iron pipe (tapered at the bottom) of suitable diameter is vertically placed in a 40-cm² pit that has permanently wet soil. The diameter of the pipe is governed by the current to be carried and the type of soil. The BIS has specified that the minimum diameter and length of the pipe shall be 38 mm and 2 m, respectively, for ordinary soil. However, if the soil is rocky and dry, the length of the pipe shall be increased to 2.75 m. The perforations in the pipe are 12 mm in diameter and are provided to allow the water to flow out. The depth of the pipe from the ground level should be a minimum of 4.75 m.

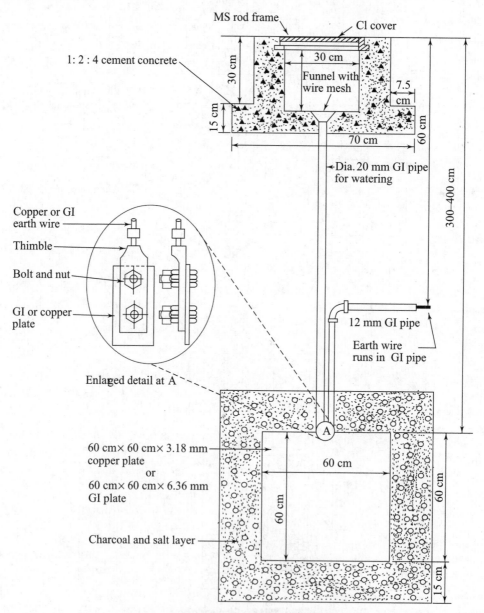

Fig. 12.19   Plate earthing

The pit is filled up to a distance of 15 cm around the pipe with alternate layers of charcoal and salt, up to a height of 2–3 m. The charcoal and salt layers are provided for retention of moisture, thereby helping to keep the earth resistance low. The height of each layer is 15 cm. The earthing pipe is connected to a galvanized iron (GI) pipe 20 mm in diameter. This GI pipe is connected to a funnel with a wire mesh at the top. Water is poured through the funnel from time to time to keep the soil wet. To protect the earth pipe from mechanical damage, a cement concrete work is provided with a cast iron (CI) cover at the ground level. The funnel is connected to another GI pipe, from the outside, to facilitate the connection to the earth wire.

12.8.2 Plate Earthing

In this method of earthing, a GI or copper plate of suitable dimensions is embedded in the earth at a sufficient depth. Due to the high cost of copper, GI plates are commonly employed. The minimum dimensions of the GI plate used are 60 cm × 60 cm × 6.36 mm.

The GI plate, also referred to as the *earth electrode*, is placed with its face vertical at a depth of 4 m in a pit. The space around the earth electrode is filled with alternate layers, each of 15 cm, of charcoal and salt. A GI pipe of diameter 12.7 mm is connected to the earth electrode or the plate. The GI pipe carries the earth wire, which is connected to the earth electrode with bolts, nuts, washers, and thimbles.

As in the case of pipe earthing, a GI pipe connects the charcoal and salt filled pit to the concrete work. The pipe is provided with a wire mesh covered funnel for pouring water in order to keep the earth resistance low. The concrete work is covered with a CI cover for protection against damage and inspection. Figure 12.19 shows the details of plate earthing.

Recapitulation

An electric power system is constituted of the following subsystems:
- Generation subsystem
- Transmission subsystem
- Sub-transmission system
- Distribution subsystem
- Protection and control subsystem

Transformers are used in all the subsystems to change the voltage levels of energy transfer.

The important primary sources of energy employed for generation of electrical energy are
- Fossil fuels; for example, coal (including lignite and peat), oil, and natural gas
- Renewable energy from wind and water
- Nuclear energy from uranium or plutonium
- The various categories of generators are
- Thermal
- Hydel
- Nuclear
- Unconventional

The various types of thermal stations are
- Coal-fired
- Oil-fired
- Gas-fired
- Diesel-oil-fired

Hydroelectric power stations utilize the head and quantum of water to generate electric power. The power P generated in a hydroelectric station is given by

$$P = 9.81\rho \, Qh\eta \times 10^{-3} \text{ kW}$$

In a nuclear power station, electric power is generated by the fission of uranium-235.

The various types of non-conventional/alternative generating stations are

- Wind power
- Geothermal
- Tidal
- Solar power

A transmission system functions to transmit bulk power over long distances.

In India, the ac transmission voltage levels vary between 66 and 400 kV, and the dc transmission voltages are 400 kV and above.

The voltage level of a sub-transmission system ranges from 66 kV to 132 kV.

A distribution subsystem is designed to supply continuous and reliable power at the consumers' terminals at minimum cost. The different types of distribution systems are

- Radial distribution
- Loop distribution
- Network distribution

The requirements of a good wiring circuit design are

- Equal voltage distribution
- Same load current in each circuit
- Ease of repair and maintenance of the circuits
- Safety and minimum inconvenience to the consumer

From the main switch, electrical energy can be distributed to the various points using the following two methods:

- Distribution board system
- The tree system

In order to connect the various types of electrical fixtures in a building, the following wiring methods are employed:

- Loop-in system
- Joint box or tee system

The earthing of electrical installations is undertaken for the following reasons:

- To ensure that the potential, with respect to the earth, of any current-carrying conductor does not rise above its designed insulation level.
- To avoid shocks to a living body.
- To provide safety to operating and maintenance personnel.
- To avoid fire hazards due to earth leakage current.

The various types of earthing are

- Earthing through the water mains
- Strip or wire earthing
- Rod earthing
- Pipe earthing
- Plate earthing

The two commonly employed methods for earthing are

- Pipe earthing
- Plate earthing

Assessment Questions

1. Draw a single line diagram of an ac power system and clearly show the various sub-systems and the range of voltages at which they operate.
2. (a) Enumerate the primary sources of energy and state how the price factor affects their utility in power generation.
   (b) Classify the four types of generating stations and the types of energy source used in each.

3. Draw the schematic diagram of a coal-fired thermal generating station and explain the process of power generation. Describe how the efficiency of the power station can be increased. Enumerate the disadvantages of a coal-fired thermal station.

4. Write detailed notes on (i) oil-fired and (ii) gas-fired thermal stations.

5. State the parameters on which hydroelectric power generation is categorized and name the different categories. Discuss the merits and demerits of hydroelectric stations.

6. Describe the working of a nuclear power station and draw its schematic layout. State the advantages and disadvantages of nuclear power stations.

7. Enumerate the different types of non-conventional generating stations and describe their characteristics and stages of development.

8. Describe a transmission subsystem and state how a sub-transmission system differs from it.

9. Draw layout diagrams of (i) distribution system, (ii) radial distribution system, and (iii) loop distribution system. State the relative merits of radial and loop distribution systems.

10. What is a network distribution system? Draw the layout plan of such a system.

11. State the essential requirements of a good wiring design. What is the permissible maximum load and the number of light points, sockets, etc. specified per circuit? Describe a sub-circuit and state when it becomes necessary.

12. Explain with layout plans (i) the distribution board system and (ii) the tree system. State the relative advantages of each.

13. Categorize the various domestic wiring systems. With the help of diagrams, explain the differences between them and state their relative merits.

14. (a) What is earthing and why is it necessary?
    (b) State the significant specifications for earthing installations, equipment, appliances, etc.

15. Show with the help of labelled diagrams the essential features of (i) pipe and (ii) plate earthing.

Objective Type Questions

1. Which of the following is a primary source of energy in a nuclear power station?
   (i) uranium   (ii) lignite
   (iii) peat   (iv) natural gas

2. The efficiency of the process of conversion of chemical energy into electrical energy is limited to a maximum of
   (i) 30%   (ii) 40%
   (iii) 50%   (iv) 60%

3. Which of the following is not a disadvantage of using super-thermal units?
   (i) low fuel cost
   (ii) low capital cost
   (iii) low installation cost
   (iv) all of these

4. Which of the following is not contained in the waste gases released by coal-fired generating stations?
   (i) carbon   (ii) sulphur dioxide
   (iii) NOX   (iv) carbon dioxide

5. Residual oil is
   (i) high-octane petrol
   (ii) oil pumped from an oil well
   (iii) oil extract
   (iv) none of these

6. Which of the following is the primary source of energy in a gas-fired power station?
   (i) LPG
   (ii) natural gas
   (iii) hot water spring gas
   (iv) none of the these

7. For which of the following reasons is a diesel-oil-fired power station used to supply standby power?
   (i) high cost of fuel
   (ii) environmental pollution
   (iii) noise pollution
   (iv) health hazard

8. Hydro-generating stations are characterized by which of the following?
   (i) high capital and high fuel cost
   (ii) high capital and zero fuel cost
   (iii) low capital and low fuel cost
   (iv) low capital and high fuel cost

9. The life of a hydro-generating station is taken as
    (i) minimum 20 years (ii) minimum 35 years
    (iii) minimum 50 years (iv) minimum 60 years
10. Which of the following is produced when fission takes place in uranium-235?
    (i) heat (ii) neutrons
    (iii) heat and neutrons (iv) heat and electrons
11. Boron steel rods have the property to
    (i) generate neutrons (ii) absorb neutrons
    (iii) generate electrons (iv) absorb electrons
12. Which of the following are classified as unconventional power stations?
    (i) geothermal (ii) wind
    (iii) tidal (iv) all of these
13. Which of the following is used to generate power in a geothermal station?
    (i) heat in the air
    (ii) heat in the ionosphere
    (iii) heat inside the earth
    (iv) heat of the sun
14. In a tidal power station, which of the following types of hydro-turbines is used?
    (i) Pelton wheel (ii) Kaplan turbine
    (iii) reversible turbine (iv) Francis turbine
15. The energy in a tidal wave is a function of
    (i) the volume of water in the wave
    (ii) the velocity of the wave
    (iii) the temperature of the wave
    (iv) the square of the height and period of motion of the wave
16. At which of the following voltage level ranges is power generated in a power station?
    (i) 1.1–6.6 kV (ii) 11–30 kV
    (iii) 33–66 kV (iv) none of these
17. In India, the transmission voltage level employed for transmitting ac power is
    (i) 1000 kV and above (ii) 765–1000 kV
    (iii) 400–765 kV (iv) 66–400 kV
18. Which of the following voltage levels is used for the distribution of three-phase power to domestic consumers?

(i) 230 V (ii) 415 V
(iii) 1100 V (iv) none of these
19. A radial distribution system is not used to supply
    (i) high-density load areas
    (ii) medium-density load areas
    (iii) low-density load areas
    (iv) none of these
20. Which of the following is a function of an MCB?
    (i) acting as a mains switch
    (ii) protection against short circuits
    (iii) protection against overload
    (iv) all of these
21. In deciding the load of a wiring circuit, which of the following ratings is to be taken for an exhaust fan?
    (i) as per specified rating
    (ii) 100 W
    (iii) 60 W
    (iv) 40 W
22. Which of the following is not a feature of the joint box system of wiring?
    (i) joints are made at the terminals of the fixtures only
    (ii) the system is employed for temporary wiring
    (iii) a large number of connectors is required
    (iv) all of these
23. Which of the following rating is to be assumed for a power socket outlet?
    (i) 1000 W (ii) 60 W
    (iii) 40 W (iv) none of these
24. As per specifications, which of the following is the minimum distance between the earthing electrode and the building whose installation is to be earthed?
    (i) 4.5 m (ii) 3.0 m
    (iii) 1.5 m (iv) 1.0 m
25. Which of the following types of plate is used as an earthing electrode?
    (i) aluminium (ii) galvanized iron
    (iii) steel (iv) brass

Answers

1. (i) 2. (ii) 3. (iv) 4. (i) 5. (iii) 6. (ii) 7. (i)
8. (ii) 9. (iii) 10. (iii) 11. (ii) 12. (iv) 13. (iii) 14. (iii)
15. (iv) 16. (ii) 17. (iv) 18. (ii) 19. (i) 20. (iv) 21. (i)
22. (i) 23. (i) 24. (iii) 25. (ii)

ILLUMINATION

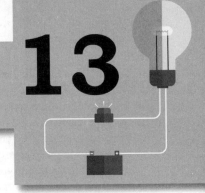

13

Learning Objectives

This chapter will enable the reader to
- Understand illumination and its importance in utility engineering
- Define and understand the various types of illumination related terminology
- Define and comprehend the different laws of illumination
- Apply the laws of illumination for computing luminosity

13.1 INTRODUCTION

In general the word 'illumination' has several meanings varying from spiritual or intellectual enlightenment to the simple act of using light for decorative illumination. A utility engineer is more interested in using light for decorative illumination purposes. This chapter explains the utilisation of light energy for illumination purposes along with the associated terms.

13.2 LIGHT RADIATIONS

Light energy is known to consist of electromagnetic waves. These light waves travel with a velocity of 3×10^8 m/s approximately but their wavelengths are different. The wavelength of red light is nearly 0.000078 cm and that of violet light 0.000039 cm. Since these wavelengths are very small, instead of using 1 cm as the unit for their measurement, a sub-multiple 10^{-8} cm is used. This sub-multiple is known as Angstrom Unit (Å).

Light energy can be perceived by the human eye in the wavelength range of 4000 Å to 7700 Å. Beyond the visual spectrum range of 4000 Å and 7700 Å wavelength is the ultraviolet and infrared spectrum, which are not visible to the naked eye.

Light, like heat, is radiant energy and can be produced artificially by heating a solid body or vapour till it begins to glow. Initially at low temperatures the energy radiated is in the form of heat waves. However, as the temperature rises the body begins to glow and both light and heat waves are emanated. As the temperature is further increased, the glowing body changes colour from red to orange and, finally at a very high temperature, to white. The wavelength of the visible radiation decreases with an increase in temperature. It may be mentioned that both heat and light waves are identical in nature, except that the former have longer wavelength and as such do not leave an impression on the retina. From the perspective of a utility engineer heat energy represents a loss since it cannot be employed for illumination.

13.3 DEFINITIONS

Commonly applied terminology employed in illumination is defined below.

13.3.1 Solid Angle

In order to define the luminosity (luminous intensity) of a light source, the concept of a solid angle is used. An ordinary angle, in radians, is defined as the angle subtended at the centre of a unit circle by the length of an arc. A *solid angle*, however, is the angle subtended at the centre of a sphere by a surface.

Figure 13.1 pictorially defines the solid angle. Hence as per the definition the solid angle, represented by ω, at the centre of the sphere is the angle enclosed by the surface area A. Thus,

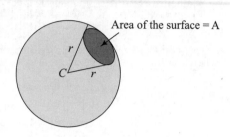

Fig. 13.1   Solid angle

$$\omega = \frac{A}{r^2} \text{ steradian} \qquad (13.1)$$

In Eq. (13.1), *steradian* is the unit of solid angle and is abbreviated 'sr'. If $A = r^2$, the solid angle $\omega = 1$. From the foregoing it follows that the solid angle of a complete sphere subtended at the centre will be given

by $\omega = 4\pi\ r^2/r^2 = 4\pi$ steradian

13.3.2 Luminous Flux

Luminous flux, represented by the Greek symbol Φ or F, is defined as the light energy radiated per second by a body as luminous light waves. The unit of measure of luminous flux is 'lumen' abbreviated as 'lm'. A *lumen* is defined as flux contained per unit solid angle of a source of one candela. Figure 13.2 pictorially defines a lumen.

Radiation of light waves represents a type of flow of energy; hence, it may be considered as a power unit. A lumen is defined as the flux contained per unit of solid angle of a source of one candela, that is, 1 lm = 0.0016 W (approx.) or 1 W = 625 lm.

Similarly, *lumen-hour* is defined as the quantity of light delivered by a flux of one lumen in one hour.

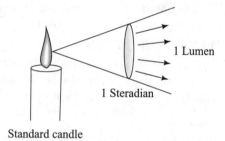

Fig. 13.2   Pictorial definition of a lumen

13.3.3 Candela

Candela is a unit for measuring light emitted by a source in a three-dimensional angular span. Since luminous intensity is depicted in terms of angle, it is independent of the distance at which it is measured. Figure 13.3 provides a two-dimensional view of a light source (burning candle).

It may be noted that since the angle covered by the screens A and B is the same, screen B will receive an equal amount of light as screen A from the source, when the obstruction due to screen A is not there.

As a unit of measure candela (cd) or lumens per steradian is defined as 1/60th of the luminosity per cm^2 of a black body radiator at 2045° K temperature, which is the temperature of solidification of platinum. A light source of 1 cd will emit 1 lm/sr. Therefore, total flux emitted by the source in a three-dimensional space will be $4\pi \times 1 = 4\pi$ lm.

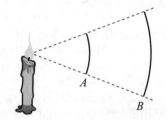

Fig. 13.3   Two-dimensional view of a light source

13.3.4 Luminous Intensity or Candle Power

Luminous intensity (I) may be seen as the solid angular flux density in a particular direction. The luminous intensity of a point source in a given direction is defined as luminous flux radiated out per unit solid angle in that direction. The unit of luminous intensity is 'cd' or 'lm/sr'.

Mathematically, luminous intensity is written as

$$I = \frac{d\Phi}{d\omega} \quad \text{cd or lm/sr} \tag{13.2}$$

where $d\Phi$ lm is the luminous flux radiated by a source within a solid angle of $d\omega$ sr.

Thus, if a light source has an average luminous intensity of I cd (or lm/sr), total flux radiated in the three-dimensional space is given by

$$\Phi = \omega I = 4\pi I \quad \text{lm} \tag{13.3}$$

13.3.5 Mean Spherical Candle Power (MSCP)

The distribution of light intensity or candle power (cp) from a light source is unequal in different directions. For the purpose of grading light sources average cp of a source is defined as the average value of its cp in all directions. The average cp of a source, designated as 'mean spherical candle power' (MSCP) is obtained by summing the total flux in lm in all directions divided by solid angle 4π sr as given below.

$$\text{MSCP} = \frac{\text{Total flux in a sphere in lm}}{4\pi} \tag{13.4}$$

13.3.6 Mean Hemispherical Candle Power (MHSCP)

MHSCP is defined as the ratio of the total flux released in a hemisphere and the solid angle subtended by the hemisphere at the point source. Thus,

$$\text{MHSCP} = \frac{\text{Total flux radiated in a hemisphere in lm}}{2\pi} \tag{13.5}$$

13.3.7 Reduction Factor

Reduction factor of a source is the ratio of mean spherical candle power (MSCP) and mean horizontal candle power (MHCP) and is written as

$$f = \frac{\text{MSCP}}{\text{MHCP}} \tag{13.6}$$

13.3.8 Illumination or Illuminance

A surface gets illuminated when luminous flux falls on it. Normal luminous flux per unit area received by a surface quantifies the illumination (E) of a surface. If $d\Phi$ is the luminous flux incident upon an area dA, illumination of the surface is given by

$$E = d\Phi/dA = \Phi/A \quad \text{lm/m}^2 \tag{13.7}$$

The alternative units of E are lux or metre-candle (cd).

Example 13.1 Show that the illumination at a distance d from the source is given by $E = I/d^2$.

Solution Assume that the distance between the area and the point where the solid angle is formed is d. The solid angle formed between the point and the area is written as

$$\omega = \frac{A}{d^2} \quad \text{sr} \tag{13.1.1}$$

Substituting Eq. (13.2) in Eq. (13.7) yields

$$E = \frac{\Phi}{A} = \frac{I \times \omega}{A} \tag{13.1.2}$$

Substituting Eq. (13.1.1) in Eq. (13.1.2) leads to

$$E = \frac{I \times A}{A \times d^2} = \frac{I}{d^2}$$

Table 13.1 summarises the various quantities related to illumination.

Table 13.1   Summary of illumination terminology

Term	Symbol	Unit
Luminous flux	Φ or F	Lumen
Luminous intensity	I	Candela
Illumination or Luminance	E	Lm/m^2 or lux

13.4 LAWS OF ILLUMINATION OR ILLUMINANCE

The laws of illumination which define the illumination E at a surface are based on the assumption that the light source is sufficiently distant from the surface for it to be considered as a point source.

13.4.1 Proportionality Law

This law states that illumination is directly proportional to the luminous intensity of a light source, that is, $E \propto I$.

13.4.2 Inverse Square Law

The illumination of a surface is inversely proportional to the square of the distance of the surface from a point source. Mathematically it is expressed as $E \propto I/d^2$, where d is the distance of the surface from the point source.

Figure 13.4 shows partial surfaces of three spheres of radii in the ratios of $r_1 : r_2 : r_3$ located in front of a point light source. It may be seen from the figure that they receive the same amount of total flux which is due to the portions of the spheres subtending the same solid angle at the source. On the other hand, the illumination of the partial sphere surfaces is in the ratio of $1/r_1^2 : 1/r_2^2 : 1/r_3^2$ since their radii are in the ratio of $r_1^2 : r_2^2 : r_3^2$, thereby proving the proportionality law.

13.4.3 Lambert's Cosine Law of Incidence

This law states that illumination E is directly proportional to the cosine of the angle made by the normal to the illuminated surface with the direction of the incident flux.

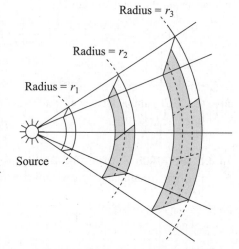

Fig. 13.4   Proof of proportionality law

Figure 13.5 shows two surfaces 1 and 2 each of area A. Assume that in position 1 the flux incident on the surface is Φ. In this position the illumination E_1 of the surface is given by

$$E_1 = \Phi/A \ \text{lm/m}^2 \tag{13.8}$$

In position 2, the surface is inclined at an angle θ to surface 1. In this position, the illumination of the surface is represented by

$$E_2 = \frac{\Phi}{A} \cos \theta \ \text{lm/m}^2 \tag{13.9}$$

Substituting Eq. (13.8) in (13.9) gives

$$E_2 = E_1 \cos \theta \ 1 \ m/m^2 \qquad (13.10)$$

With the help of the relation $E = I/d^2$ (see Example 13.1), Eq. (13.10) is written in a generalized form as

$$E = \frac{I \cos \theta}{d^2} \ 1 \ m/m^2 \qquad (13.11)$$

Equation (13.11) can be used for computing the illumination at any given point when the luminous intensity of a given light source(s) in the normal direction is known.

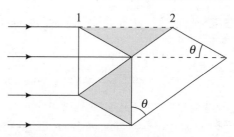

Fig. 13.5 Application of Lambert's cosine law of incidence

Example 13.2 Apply Lambert's cosine law and derive expressions for illumination at points 1, 2, 3, and 4 for the lighting arrangement shown in Fig. 13.6.

Solution Since point 1 is directly below the light point source, $\theta = 90°$. Use of Eq. (13.11) gives the illumination at the point as

$$E_1 = \frac{I}{d^2} \ lm/m^2$$

Illumination at point 2 is $E_2 = \dfrac{I \cos \theta_1}{(P2)^2} \ lm/m^2 \qquad (13.2.1)$

Since $\cos \theta_1 = d/P2$, its substitution in Eq. (13.2.1) results in

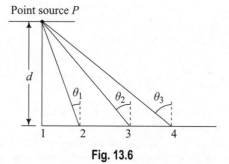

Point source P

Fig. 13.6

$$E_2 = \frac{I}{(P2)^2} \times \frac{d}{P2} = \frac{I}{d^2} \times \left(\frac{d}{P2} \right)^3 = E_1 \cos^3(\theta_1) \ lm/m^2 \qquad (13.2.2)$$

Proceeding on similar lines, illumination at points 2 and 3 can be written as $E_3 = E_1 \cos^3(\theta_2) \ lm/m^2$ and $E_4 = E_1 \cos^3(\theta_3) \ lm/m^2$. Verification of the expressions for illumination at points 3 and 4, however, is left as tutorial exercise for the reader.

Example 13.3 For the lighting arrangement shown in Fig. 13.7, compute the illumination (a) directly below the light source and (b) at a point in the horizontal plane 9 m away from the foot of the source. Assume that the light source is giving out 1500 l m all around.

Solution Equation (13.2) is used to determine the luminous intensity of the lamp as follows

$$I = 1500/4\pi = 119.37 \ cd$$

(a) Illumination directly below the light source is calculated as under

$$E = I/d^2 = 119.37/12^2 = 0.829 \ 1 \ m/m^2$$

(b) Distance between the light point source and point

$$B = \sqrt{12^2 + 9^2} = 15 \ m$$

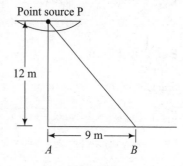

Point source P

12 m

9 m

A   B

Fig. 13.7

Hence $\cos \theta = 12/15 = 0.8$

Illumination at point B, is obtained by using Eq. (13.2.1) as under

$$E = (119.37 \times 0.8)/15^2 = 0.424 \ lm/m^2$$

Example 13.4 Calculate the distance of point B at which the illumination will 1/8th of the illumination at point A in Example 13.3.

Solution From Example 13.3, illumination at point B is obtained as

$$E_B = \frac{1}{8} \times E_A = \frac{1}{8} \times 0.829 = 0.104 \;\; \text{lm/m}^2$$

Assume that point B is located at x m from A.

Distance between the light point and the source $\sqrt{12^2 + x^2}$ m.

Thus, $\quad \cos\theta = 12 \big/ \sqrt{144 + x^2}$

Once again Eq. (13.2.1) is used to compute x as under

$$\frac{119.37}{\left(144 + x^2\right)} \times \frac{12}{\sqrt{144 + x^2}} = 0.104 \qquad\qquad (13.4.1)$$

Equation (13.4.1) is simplified for calculating x and is shown below

$$x = \sqrt{\left[\frac{119.37 \times 12}{0.104}\right]^{2/3} - 144} = 20.75 \;\; \text{m}$$

Example 13.5 A 60 m^2 drawing office is to be illuminated by 60 W incandescent bulbs which give out 12 l m/W. Calculate the number of bulbs required if 90 lux is needed on the drawing boards and only 50% of the light emitted is available for illuminating the drawing boards.

Solution Assume that the total number of bulbs required for the desired illumination is N.

$\qquad$ Output flux per lamp $= 60 \times 12 = 720 \;\; \text{lm}$

$\qquad$ Total output $= 720N$ l m

$\qquad$ Flux actually available for illumination $= 0.5 \times 720N = 360N \;\; \text{lm}$ $\qquad\qquad (13.5.1)$

$\qquad$ Required illumination for drawing office $= 90 \; \text{lux} = 90 \; \text{l m/m}^2$ $\qquad\qquad (13.5.2)$

Total flux required for drawing office is determined by using Eq. (13.5.2) as

$\qquad \Phi = 90 \times 60 = 5400 \;\; \text{lm}$ $\qquad\qquad (13.5.3)$

Equating Eqs (13.5.1) and (13.5.3) yields

$\qquad 360N = 5400$ or $N = 5400/360 = 15$

Recapitulation

Solid angle, $\omega = \dfrac{A}{r^2}$ steradian

For a complete sphere, $\omega = 4\pi r^2 / r^2 = 4\pi$ sr

Luminosity/candle power, $I = \dfrac{d\Phi}{d\omega}$ cd or lm/sr

Total flux emitted by a light source in all directions,

$\qquad \Phi = \omega I = 4\pi I$ lm

Mean spherical candle power, MSCPC

$\qquad = \dfrac{\text{Total flux in a sphere in lm}}{4\pi}$

Mean hemispherical candle power, MHSCP

$\qquad = \dfrac{\text{Total flux radiated in a hemisphere in lm}}{2\pi}$

Reduction factor,

$\qquad f = \dfrac{\text{MSCP}}{\text{MHCP}}$

Illumination or Illuminance, $E = d\Phi/dA = \Phi/A$ lm/m^2

Proportionality law, $E \propto I$

Inverse square law, $E \propto I/d^2$

Lambert's cosine law of incidence, $E = \dfrac{I \cos\theta}{d^2}$ l m/m^2

Assessment Questions

1. Write notes on (a) illumination, and (b) radiation.
2. Explain solid angle and how it differs from ordinary angle.
3. Write short notes on (a) luminous flux, (b) candela, and (c) candle power.
4. Define (a) MSCP, (b) MHCP. What is the significance of reduction factor?
5. Derive an expression for illumination at a point.
6. State and explain (a) proportionality law and (b) inverse square law.
7. Define Lambert's cosine law and derive an expression for determining illumination on an inclined surface.

Problems

13.1 Use Lambert's cosine law and show that illumination at point n is $E_n = E_1 \cos^3 (\theta_{n-1})$ lm/m^2 where θ_{n-1} is the vertical angle at point n.

13.2 Compute the distance of point B in Example 13.3 if the intensity at the point is 1/8th of the intensity at point A. [20.78 m]

13.3 A light source of 2000 lm is vertically suspended by a pole at a height of 5 m and it radiates light in all directions. Derive a generalized relationship for determining illumination at any point on the ground. Compute the illumination at the ground level at (a) directly below the lamp, (b) at 1 m, and (c) at 2 m away from the pole.
[(a) 6.366 lm/m^2, (b) 6.003 lm/m^2, (c) 5.096 lm/m^2]

13.4 For the lighting system shown in Fig. 13.8, compute the illumination at point P (a) due to each source, and (b) total illumination. All data is provided in the figure.
[(a) P1 0.053 lm/m^2, P2 0.027 lm/m^2 (b) 0.08 lm/m^2]

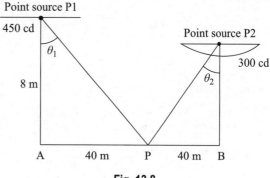

Fig. 13.8

Objective Type Questions

1. Which of the following is a unit of solid angle?
    (i) degree (ii) radian
    (iii) steradian (iv) all of these
2. Which of the following represents solid angle of a complete sphere?
    (i) π (ii) 2π
    (iii) 3π (iv) 4π
3. Which of the following is the wave length range of the visible spectrum?
    (i) less than 4000 Å
    (ii) 4000 A to 7700 Å
    (iii) more than 7700 Å
    (iv) none of these
4. Which of the following is used as a symbol to represent flux?

5. Identify the unit for luminous intensity.

    (i) Φ (ii) θ
    (iii) I (iv) E

    (i) sr (ii) cd
    (iii) lm (iv) lm/m^2
6. Which of the following is the full form of MSCP?
    (i) Minimum spherical candle power
    (ii) Maximum spherical candle power
    (iii) Mean special candle power
    (iv) Mean spherical candle power
7. Which of the following represents illumination at a point?
    (i) $\dfrac{d\Phi}{dA}$ (ii) Φ/A

    (iii) I/d^2 (iv) all of these

8. Which of the following is a statement of the proportionality law of illuminance?
   (i) $E \propto I$ (ii) $\Phi \propto I$
   (iii) $E \propto d^2$ (iv) $E \propto I/d^2$

9. Surfaces of three spheres of radii 1, r_1, and r_2 are partially exposed to a light point sources. Total flux received by the spheres is
   (i) same
   (ii) in the ratio of their radii

(iii) in inverse ratio of their radii
(iv) in inverse ratio of the square of their radii

10. Illumination of a surface in front of a light source inclined at an angle θ to the vertical is given by

   (i) Φ/A (ii) $\dfrac{I\cos\theta}{d}$

   (iii) $\dfrac{I\cos\theta}{d^2}$ (iv) none of these

Answers

1. (iii)	2. (iv)	3. (ii)	4. (i)	5. (ii)	6. (iv)	7. (iv)
8. (i)	9. (i)	10. (iii)				

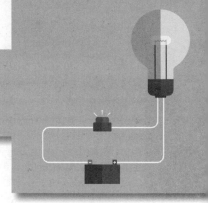

ADDITIONAL MULTIPLE CHOICE QUESTIONS

Chapter 1

1. The resistance of a conductor of diameter d and length l is R ohms. If the diameter of the conductor is halved and its length doubled, the resistance of the conductor is
   - (i) R ohms
   - (ii) $2R$ ohms
   - (iii) $4R$ ohms
   - (iv) $8R$ ohms

2. Which of the following does not represent the unit of power?
   - (i) VI
   - (ii) V/I
   - (iii) I^2R
   - (iv) J/sec

3. To obtain a high value of capacitance, the permittivity of the dielectric medium should be
   - (i) low
   - (ii) medium
   - (iii) high
   - (iv) that of air

4. Ten capacitors each of 10 μF are connected in series, the combined capacitance is
   - (i) 100 μF
   - (ii) 1 μF
   - (iii) 0.1 μF
   - (iv) none of these

5. Two resistors each of 4 Ω and 12 Ω are connected in parallel and the parallel combination is connected in series with a 2 Ω resistor. If this circuit is connected across a 100 V supply, the total current drawn is
   - (i) 50 A
   - (ii) 25 A
   - (iii) 20 A
   - (iv) 2 A

6. The energy stored in an inductor of inductance L henry is represented by
   - (i) i^2L
   - (ii) iL^2
   - (iii) $\dfrac{L^2}{i}$
   - (iv) $\dfrac{1}{2}Li^2$

7. The exact value of the absolute permittivity of vacuum is
   - (i) 1.0006
   - (ii) 1.0000
   - (iii) 8.85×10^{-12}
   - (iv) none of these

8. The unit of reluctance is
   - (i) weber
   - (ii) ampere turns/weber
   - (iii) ampere turns
   - (iv) tesla

9. The right-hand rule for determining the direction of the induced emf was enunciated by
   - (i) Faraday
   - (ii) Lenz
   - (iii) Fleming
   - (iv) Maxwell

10. What is the mmf required to produce a flux of 240 mWb in a coil of 200 turns and carrying a current of 2 A.
    - (i) 50 AT
    - (ii) 100 AT
    - (iii) 200 AT
    - (iv) 400 AT

11. The two laws that form the basis of circuit analysis were enunciated by
    - (i) Faraday
    - (ii) Ohm
    - (iii) Bohr
    - (iv) Kirchhoff

Chapter 2

1. Which of the element in the following is not bilateral?
   - (i) resistor
   - (ii) inductor
   - (iii) capacitor
   - (iv) transistor

2. A node in a network is defined as a
   - (i) closed path
   - (ii) junction point of two or more branches
   - (iii) group of interconnected elements
   - (iv) all of these

3. Which of the following characteristics are attributed to an ideal independent voltage source?
   - (i) independent of magnitude of current supplied
   - (ii) independent of the direction of flow of current
   - (iii) can absorb or deliver energy continuously at constant voltage
   - (iv) all of these

4. A voltage source v_s having an internal resistance of r_s can be mathematically represented by an equivalent current source by
   - (i) $i_s = \dfrac{v_s}{r_s}$ and r_s is connected in parallel with the

current source

(ii) $i_s = \dfrac{v_s}{r_s}$ and r_s is connected in series with the current source

(iii) $i_s = \dfrac{v_s}{r_s}$ and $r_s = 0$

(iv) none of these

5. Conversion of a voltage source into a current source and vice versa is undertaken to make numerical computations
   (i) more difficult (ii) more accurate
   (iii) quicker (iv) easier

6. When resistances are connected in series, the voltage across all the resistances is
   (i) equal
   (ii) unequal
   (iii) proportional to the square of the current
   (iv) none of these

7. The voltage drop across a resistance connected in series with a number of resistances is
   (i) inversely proportional to its magnitude
   (ii) directly proportional to its magnitude
   (iii) directly proportional to the ratio of its resistance to the total series resistance of the circuit
   (iv) all of these

8. Two resistances, each of 10 Ω and 20 Ω, are connected in parallel across a dc voltage source. The current through the 10 Ω resistance is
   (i) equal to the current supplied by the voltage source
   (ii) equal to the current through the 20 Ω resistance
   (iii) 50% of the current through the 20 Ω resistance
   (iv) 200% of the current through the 20 Ω resistance

9. Which of the following is not a unit of conductance?
   (i) mho (ii) siemens
   (iii) volt/ampere (iv) ampere/volt

10. When a number of inductances are connected in series, the equivalent inductance of the circuit is
    (i) the direct sum of the individual inductances
    (ii) sum of the reciprocal of the individual inductances
    (iii) the reciprocal of the direct sum of the individual inductances
    (iv) none of these

11. What is the equivalent inductance of two inductive coils of 5 H and 10 H connected in series?
    (i) 15 H (ii) $\dfrac{3}{10}$ H

(iii) $\dfrac{1}{15}$ H (iv) $\dfrac{10}{3}$ H

12. Star-delta transformation is employed to simplify circuit elements connected in
    (i) series
    (ii) parallel
    (iii) series parallel combinations
    (iv) none of these

13. Each branch of a star-connected load has a resistance of 10 Ω. The resistance of each branch of an equivalent delta connection will be
    (i) 30 Ω (ii) 100 Ω
    (iii) 1000 Ω (iv) none of these

14. When a delta-connected network, with all its branches having an equal resistance, is converted into an equivalent star-connected network, the magnitude of the resistance of each branch is
    (i) equal to the branch resistance in delta
    (ii) greater than the branch resistance in delta
    (iii) less than the branch resistance in delta
    (iv) none of these

15. In the nodal voltage method of analysis the independent variable is the
    (i) branch current (ii) node voltage
    (iii) branch voltage (iv) mesh current

16. Nodal voltage method of analysis is applicable to networks containing
    (i) only current sources
    (ii) only voltage sources
    (iii) current and voltage sources
    (iv) all of these

17. In a network if the number of nodes is n and the number of branches is b, the number of independent mesh equations required to solve the network is
    (i) $b - n + 1$ (ii) $b + n + 1$
    (iii) $b + n - 1$ (iv) $b - n - 1$

18. In applying superposition theorem to determine branch currents and voltages,
    (i) all current and voltages sources are removed
    (ii) only the current sources are removed
    (iii) only one source (current or voltage) is included at a time
    (iv) only the voltage sources are removed

19. The number of circuits required to solve a network using superposition theorem is equal to the number of
    (i) nodes
    (ii) branches
    (iii) voltage sources
    (iv) voltage plus current sources

20. Thevenin's theorem cannot be applied to networks that contain elements which are
    - (i) non-linear
    - (ii) linear
    - (iii) active
    - (iv) passive
21. In deriving the equivalent resistance at any pair of terminals of a network with the help of Thevenin's theorem
    - (i) all independent voltage sources are open-circuited
    - (ii) all independent current sources are short-circuited
    - (iii) the internal resistance of all independent sources is neglected
    - (iv) none of these
22. For the maximum power to be transferred to the load
    - (i) load resistance should be twice the internal resistance of the voltage source
    - (ii) load resistance should be equal to the internal resistance of the voltage source
    - (iii) load resistance should be half the internal resistance of the voltage source
    - (iv) none of these
23. When a circuit is suddenly switched on, transients will occur if it contains elements which are
    - (i) resistive and inductive
    - (ii) resistive and capacitive
    - (iii) resistive, inductive, and capacitive
    - (iv) all these
24. Which of the following is the unit of time constant of an RC network?
    - (i) second
    - (ii) $\dfrac{R \times ampere \times second}{V}$
    - (iii) $\dfrac{Rx \int idt}{V}$
    - (iv) all of these

25. Which of the following satisfies the condition for a critically damped *RLC* series circuit?
    - (i) $\left(\dfrac{R}{2L}\right) < \dfrac{1}{\sqrt{LC}}$
    - (ii) $\left(\dfrac{R}{2L}\right) > \dfrac{1}{\sqrt{LC}}$
    - (iii) $\left(\dfrac{R}{2L}\right) = \dfrac{1}{\sqrt{LC}}$
    - (iv) none of these
26. In an *RC* series circuit, the voltage is allowed to increase at its initial rate of rise. Which of the following gives the time at which its voltage becomes equal to the supply voltage?
    - (i) $t = 5\tau$
    - (ii) $t = 3\tau$
    - (iii) $t = \tau$
    - (iv) $1/\tau$

27. Which of the following governs the response of an unforced reactive circuit?
    - (i) Magnitude of the voltage source
    - (ii) Method of interconnection of the elements
    - (iii) Time constant of the circuit
    - (iv) All of these

Chapter 3

1. High permeability magnetic material helps
    - (i) to confine the flux within the magnetic circuit
    - (ii) to allow the flux to leak
    - (iii) in producing more current
    - (iv) none of these
2. The magnetomotive force set up in a magnetic circuit
    - (i) is another name for electromotive force
    - (ii) is analogous to potential difference
    - (iii) is analogous to resistance of an electric circuit
    - (iv) results in a flow of current in the magnetic circuit
3. The unit of mmf $\mathcal{F}$ is normally taken as
    - (i) weber
    - (ii) weber per metre^2
    - (iii) ampere turns
    - (iv) ampere turns per metre
4. If the direction of current flowing in a conductor is in the plane of the paper, the magnetic flux lines produced by it are
    - (i) concentric circles in the clockwise direction
    - (ii) concentric circles in the anticlockwise direction
    - (iii) straight lines parallel to the conductor and in the opposite direction of the current flow
    - (iv) straight lines parallel to the conductor and in the direction of current flow
5. Which of the following is representative of the unit of permeability?
    - (i) $\dfrac{newtons}{(ampere)^2}$
    - (ii) $\dfrac{volt \times second}{ampere \times metre}$
    - (iii) henry/metre
    - (iv) all of these
6. Which of the following does not apply to the reluctance $\mathcal{R}$ of a megnetic circuit?
    - (i) $\mathcal{R} \propto \dfrac{1}{\mu_0}$
    - (ii) $\mathcal{R} \propto \dfrac{1}{\mu_r}$
    - (iii) $\mathcal{R} \propto \dfrac{1}{l}$
    - (iv) $\mathcal{R} \propto \dfrac{1}{A}$
7. In a magnetic circuit, once a flux is set up,
    - (i) no further energy is required

(ii) energy is continuously required to maintain the flux

(iii) energy is released in the form of heat

(iv) none of these

8. The word 'hysteresis' means
    (i) to lead
    (ii) to be in step
    (iii) to lag
    (iv) none of these

9. The unit of $H \times B$ is
    (i) joule
    (ii) joule/metre
    (iii) joule/metre^2
    (iv) joule/metre^3

10. Self inductance of a coil is represented by
    (i) $\dfrac{e}{\dfrac{di}{dt}}$
    (ii) $\dfrac{\psi}{i}$
    (iii) $\dfrac{N^2}{\mathcal{R}}$
    (iv) all of these

11. Mutual inductance is defined as the ratio of change
    (i) of flux in coil 1 to the change of flux in coil 2
    (ii) of current in coil 1 to the change of current in coil 2
    (iii) of flux in coil 1 to the change of current in coil 1
    (iv) of flux in coil 2 to the change of current in coil 1

12. The law that states that when two coils are placed together in close proximity to each other, a change in flux in one coil induces an emf in the second coil was enunciated by
    (i) Lenz
    (ii) Faraday
    (iii) Hertz
    (iv) Ohm

13. Two coils of N_1 and N_2 turns are wound on a common core. If the current in coil 1 changes at 1 A/sec and it induces an emf of 1 V in coil 2, the magnitude of the mutua inductance is
    (i) $\dfrac{N_1}{N_2}$ henry
    (ii) $\dfrac{N_2}{N_1}$ henry
    (iii) $N_1 N_2$ henry
    (iv) 1 henry

14. The expression for energy stored in two mutually coupled and excited coils is given by
    (i) $\dfrac{1}{2} L_1 i_2^2 + \dfrac{1}{2} L_2 i_1^2 + M i_1 i_2$
    (ii) $\dfrac{1}{2} L_1 i_1^2 - \dfrac{1}{2} L_2 i_2^2 + M i_1 i_2$
    (iii) $\dfrac{1}{2} L_1 i_1^2 + \dfrac{1}{2} L_2 i_2^2 + M i_1 i_2$
    (iv) $\dfrac{1}{2} L_2 i_1^2 - \dfrac{1}{2} L_1 i_2^2 + M i_1 i_2$

15. The force of attraction produced by an electromagnet is a function of
    (i) B^2
    (ii) A
    (iii) μ_0
    (iv) all of these

16. How does the flux vary when a steel ring is replaced by a wooden ring of the same dimension?
    (i) No change
    (ii) Increases
    (iii) Decreases
    (iv) None of these

17. Which of the following compares well with the field intensity set up in the air gap and the ferromagnetic medium of an electromagnet?
    (i) Much higher
    (ii) Equal
    (iii) Much less
    (iv) None of these

18. The energy stored in the air gap of an electromagnet is proportional to
    (i) A
    (ii) B^2
    (iii) l
    (iv) All of these

19. For which of the following arrangements, the self-inductance and mutual inductance will have the same sign?

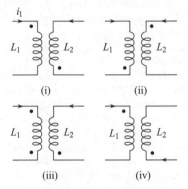

(i)   (ii)

(iii)   (iv)

20. Which of the following may not be used for magnetic circuit calculations?
    (i) Kirchhoff's laws
    (ii) Node method
    (iii) Mesh method
    (iv) None of these

Chapter 4

1. The alternating current system was put into practical use in
    (i) 1820
    (ii) 1880
    (iii) 1886
    (iv) 1896

2. The alternating current system found ready acceptance by the electric power industry because
    (i) voltages could be stepped up and down
    (ii) only real voltages and currents were involve
    (iii) long-distance transmission of power was not possible
    (iv) none of these

3. The alternating current transformer was invented by
    (i) Thomas Edison
    (ii) George Westinghouse

(iii) Nikola Tesla
(iv) Faraday
4. Which of the following do not represent an alternating quantity?
   (i) cosine wave   (ii) triangular wave
   (iii) rectangular wave   (iv) none of these
5. When a linear network is excited by an alternating waveform, the response in all parts of the network would have the same waveform and frequency if the excitation has a
   (i) sinusoidal waveform (ii) triangular waveform
   (iii) square waveform   (iv) sawtooth waveform
6. When two alternating waves attain their peak values simultaneously, the waves are
   (i) in quadrature
   (ii) in phase
   (iii) out of phase by 180°
   (iv) none of these
7. The root mean square value of an alternating quantity is another name for its
   (i) instantaneous value (ii) peak value
   (iii) effective value   (iv) average value
8. The highest speed of a 50-cycles/sec ac generator can be
   (i) 1000 rpm   (ii) 2000 rpm
   (iii) 3000 rpm   (iv) 6000 rpm
9. A sinusoidal voltage wave is represented by $e = 311 \sin(377t)$. The frequency of the voltage is
   (i) 50 cps   (ii) 60 cps
   (iii) 70 cps   (iv) 80 cps
10. An ac generator having six poles is required to generate voltage at 120 cps. What should be the speed, in rpm, of the generator?
    (i) 600   (ii) 1200
    (iii) 1800   (iv) 2400
11. The form factor of an alternating quantity is given by
    (i) $\dfrac{\text{average value}}{\text{rms value}}$ (ii) $\dfrac{\text{average value}}{\text{maximum value}}$
    (iii) $\dfrac{\text{rms value}}{\text{average value}}$ (iv) $\dfrac{\text{rms value}}{\text{maximum value}}$
12. The form factor of a rectangular wave is equal to
    (i) 1.0   (ii) 1.11
    (iii) 1.15   (iv) 1.414
13. The concept of representing a sinusoidal quantity by a constant amplitude line rotating at a frequency was introduced by
    (i) Euler   (ii) De Moivre
    (iii) Lagrange   (iv) Steinmetz
14. In a circuit, the applied voltage and the consequent current through the circuit is represented by $v =$ $V_m \sin \omega t$ and $i = I_m \sin(\omega t - \varphi)$. It can be stated that
    (i) the voltage and current are in phase
    (ii) the current leads the voltage by an angle φ
    (iii) the current lags the voltage by an angle φ
    (iv) none of these
15. A phasor can be denoted by
    (i) amplitude   (ii) angle
    (iii) amplitude and angle (iv) none of these
16. When two vectors are in quadrature, it means that
    (i) both vectors attain their maximum values simultaneously
    (ii) both vectors have the same amplitude
    (iii) both vectors reach their zero values at the same time
    (iv) one vector attains it maximum value $\pi/2$ rad prior to the other
17. Two voltage vectors are in quadrature and have effective values of 3 and 4 V. The sum of the two vectors is
    (i) 5 V   (ii) 7 V
    (iii) 1 V   (iv) 12 V
18. Every time a phasor is multiplied by the operator j, it causes the phasor to rotate in the anticlockwise direction through
    (i) 90°   (ii) 120°
    (iii) 180°   (iv) 270°
19. Phasor quantities can be represented in
    (i) rectangular form   (ii) polar form
    (iii) trigonometric form (iv) all of these
20. For dividing vector quantities, which form of vector representation is not convenient?
    (i) rectangular form   (ii) polar form
    (iii) trigonometric form (iv) none of these
21. When a sinusoidal voltage is applied to a purely inductive circuit, the current in the circuit
    (i) is in phase with the applied voltage
    (ii) leads the voltage by 90°
    (iii) lags the voltage by 90°
    (iv) none of these
22. With an increase in frequency, inductive reactance increases linearly and the current
    (i) remains constant   (ii) decreases inversely
    (iii) increases inversely (iv) increases linearly
23. If a 50-Hz sinusoidal voltage having an rms value of V volt is applied to a capacitance of C farad, the current in the circuit, with the applied voltage as the reference phasor, is given by
    (i) $jV\omega C$ (ii) $-jV\omega C$
    (iii) $jV/\omega C$ (iv) $V/j\omega C$
24. When the frequency of the applied voltage is varied in a capacitive circuit, the following conditions apply:

(i) when $f = 0$, current $= \infty$

(ii) when $f = \infty$, current $= 0$

(iii) capacitive reactance is inversely proportional to the frequency

(iv) all of these

25. An inductive impedance is represented by a resistance R and an inductive reactance X_L. Which of the following forms is not representative of the inductive impedance $\mathbf{Z}_L$?

(i) $Z_L \angle \varphi$ (ii) $Z_L(\cos \varphi + j \sin \varphi)$

(iii) $R - jX_L$ (iv) $R + jX_L$

26. The admittance of an inductive impedance $Z = R + jX_L$ is not denoted by

(i) $\dfrac{1 \angle \varphi}{Z_L}$

(ii) $\dfrac{1}{R + jX_L}$

(iii) $\dfrac{1}{Z_L(\cos \varphi + j \sin \varphi)}$

(iv) $G_L - jB_L$

27. The susceptance of an inductive admittance is represented by

(i) $R/(R^2 + X_L^2)$ (ii) $X_L/(R^2 + X_L^2)$

(iii) $X_L/(R^2 - X_L^2)$ (iv) none of these

28. In a series circuit containing a resistance of R ohm and a capacitance of C farad, the power factor angle is given by

(i) $\cos^{-1} \dfrac{R}{\sqrt{R^2 + \dfrac{1}{\omega^2 C^2}}}$ (ii) $\tan^{-1} \dfrac{X_C}{R}$

(iii) $\tan^{-1} \dfrac{1}{\omega CR}$ (iv) all of these

29. The complex impedance Z_C of the circuit in Question 28 is written as

(i) $R + \dfrac{1}{j\omega C}$ (ii) $Z_C \angle -\varphi$

(iii) $Z_C(\cos \varphi - j \sin \varphi)$ (iv) all of these

30. It is convenient to draw a phasor diagram of a series RLC circuit if the reference phasor is the

(i) current

(ii) applied voltage

(iii) voltage drop across the resistance

(iv) voltage drop across the inductance

31. The power factor in a RLC series circuit will be lagging if

(i) inductive drop < capacitive drop

(ii) inductive drop > capacitive drop

(iii) inductive drop = capacitive drop

(iv) none of these

32. The conductance G in a series RLC circuit can be written as

(i) $\dfrac{R}{R^2 + (X_L - X_C)^2}$ (ii) $\dfrac{X_L - X_C}{R^2 + (X_L - X_C)^2}$

(iii) $\dfrac{X_L}{R^2 + (X_L - X_C)^2}$ (iv) $\dfrac{X_C}{R^2 + (X_L - X_C)^2}$

33. In analysing single-phase parallel circuits connected to a common voltage source, which of the following apply?

(i) each branch can be treated independently as a series circuit

(ii) the current in the circuit can be determined by applying KCL to the node connecting all the parallel branches

(iii) the circuit current is obtained by the phasor sum of all branch currents

(iv) all of these

34. A resistance of R ohm is connected in parallel with a pure inductance of L henry. If the impedance of the parallel combination is Z, which of the following does not represent the power factor angle of the circuit?

(i) $\cos^{-1} \dfrac{R}{Z}$ (ii) $\tan^{-1} \dfrac{R}{\omega L}$

(iii) $\tan^{-1} \dfrac{I_R}{I_L}$ (iv) $\sin^{-1} \dfrac{\omega L}{Z}$

35. The magnitude of current in a parallel RL circuit, when the applied voltage has an rms value of V volt, is given by

(i) $\dfrac{V}{\sqrt{R^2 + X_L^2}}$ (ii) $V\left(\sqrt{R^2 + X_L^2}\right)$

(iii) $V\sqrt{\dfrac{1}{R^2} + \dfrac{1}{X_L^2}}$ (iv) none of these

36. The power factor of the circuit will be unity when

(i) $I_L > I_C$ (ii) $I_L = I_C$

(iii) $I_L < I_C$ (iv) none of these

37. In Question 36, if $R = X_L = X_C = 10\ \Omega$ and the applied voltage is equal to 100 V, the current flowing in the circuit is

(i) 30 A (ii) 10 A

(iii) 3.33 A (iv) none of these

38. The power factor angle in Question 36 can be expressed as

(i) $\tan^{-1} \dfrac{I_L - I_C}{I_R}$

(ii) $\tan^{-1} R\left(\dfrac{1}{X_L} - \dfrac{1}{X_C}\right)$

(iii) $\tan^{-1} \dfrac{B_L - B_C}{G}$

(iv) all of these

39. When a number of impedances are connected in parallel, the most convenient form to analyse a network is to express
    (i) impedances in Cartesian form
    (ii) admittances in Cartesian form
    (iii) admittances in polar form
    (iv) none of these

40. When a resistance is connected across an ac voltage source, the frequency of the power wave is
    (i) equal to the frequency of the voltage wave
    (ii) equal to the frequency of the current wave
    (iii) twice that of the voltage and current waves
    (iv) none of these

41. The power consumed in a pure inductive circuit is
    (i) zero
    (ii) $\dfrac{V_m I_m}{2}$
    (iii) VI
    (iv) $I^2 L$

42. Which of the following apply to the power curve of a pure capacitance when it is connected across a voltage source?
    (i) positive power means that the capacitance is absorbing power
    (ii) negative power means that the capacitance is returning power to the source
    (iii) there is a continuous exchange of power between the capacitance and the source
    (iv) all of these

43. Power in a reactive circuit having resistance is constituted of two components. The two components of power are called
    (i) real power
    (ii) reactive power
    (iii) quadrature power
    (iv) all these

44. The unit of reactive power is
    (i) watt
    (ii) VAR
    (iii) volt ampere
    (iv) none of these

45. Reactive power in a circuit signifies
    (i) energy consumed by the magnetic/electric field
    (ii) energy consumed by the resistance of the inductance/capacitance
    (iii) energy exchanged between the magnetic/electric field and the source
    (iv) energy consumed by the resistance in the circuit

46. The product of voltage and current is called
    (i) true power
    (ii) reactive power

(iii) apparent power
(iv) none of these

47. The unit of apparent power in an ac circuit is
    (i) watt
    (ii) volt-ampere
    (iii) volt-ampere reactive
    (iv) all of these

48. Which of the following do not represent reactive power
    (i) VI^*
    (ii) $S \cos \varphi$
    (iii) $VI \cos \varphi$
    (iv) all of these

49. The power factor of a circuit can be obtained by
    (i) $\dfrac{\text{volt-ampere reactive}}{\text{apparent power}}$
    (ii) $\dfrac{\text{real power}}{\text{apparent power}}$
    (iii) $\dfrac{\text{volt-ampere reactive}}{\text{real power}}$
    (iv) $\dfrac{\text{real power}}{\text{volt-ampere reactive}}$

50. From the point of view of electrical utilities, which of the following are of economic importance?
    (i) power
    (ii) reactive power
    (iii) apparent power
    (iv) all of these

51. Which of the following forms of tariff will encourage the consumers to have a high power factor?
    (i) high charge for kVA and low charge for kW
    (ii) low charge for kVA and low charge for kW
    (iii) low charge for kVA and high charge for kW
    (iv) high charge for kVA and high charge for kW

52. Which of the following is considered a serious disadvantage as a consequence of a low power factor?
    (i) high generating equipment cost
    (ii) operational problems in maintaining voltage at the consumer terminals
    (iii) financial losses to the supply company
    (iv) all of these

53. Static shunt capacitors are employed to improve power factor. The power factor is improved because leading current
    (i) reduces the heat loss
    (ii) increases the lagging current
    (iii) neutralizes the lagging current
    (iv) none of these

54. When an RLC series circuit is in resonance, the current through the circuit is determined by
    (i) $\dfrac{V}{X_L}$
    (ii) $\dfrac{V}{R}$
    (iii) $\dfrac{V}{X_C}$
    (iv) none of these

55. An *RLC* series circuit is connected across a sinusoidal voltage source with varying frequency. If the circuit is in resonance, the frequency f_0 at resonance is given by the relation

(i) $\dfrac{1}{2\pi\sqrt{LC}}$ (ii) $\dfrac{1}{2\pi LC}$

(iii) $\dfrac{L}{2\pi C}$ (iv) $\dfrac{C}{2\pi L}$

56. In an *RLC* series circuit, which of the following are correct when the frequency of the source is varied from zero to infinity?

(i) $X_L < X_C$ when $f < f_0$
(ii) $X_C < X_L$ when $f_0 < f$
(iii) $X_L = X_C$ when $f = f_0$
(iv) all of these

57. In Question 56 when the frequency of the source is kept below the resonance frequency, the power factor of the circuit is

(i) unity   (ii) lagging
(iii) leading   (iv) none of these

58. The *Q* factor of a circuit is defined as

(i) $\dfrac{\text{reactive power}}{\text{resistive power}}$ (ii) $\dfrac{1}{R}\sqrt{\dfrac{L}{C}}$

(iii) $\dfrac{2\pi f_0 L}{R}$ (iv) all these

59. Bandwidth is defined as the range of frequencies within which the power delivered to *R* is

(i) equal to half the power at resonance
(ii) less than half the power at resonance
(iii) greater than half the power at resonance
(iv) none of these

60. The bandwidth of a series *RLC* resonating circuit is represented by

(i) $\dfrac{R}{2\pi L}$ (ii) $\dfrac{C}{2\pi L}$

(iii) $\dfrac{C}{2\pi R}$ (iv) $\dfrac{L}{2\pi R}$

61. In a parallel *RLC* circuit, under the condition of resonance, which of the following are applicable?

(i) the conductance is maximum
(ii) the current is maximum
(iii) the current is in phase with the voltage
(iv) none of these

62. A circuit has an inductive coil of *L* henry and a resistance of *r* ohm ($r \ll \omega L$) connected in parallel with a capacitance of *C* farad. Under the condition of resonance, the impedance of the circuit is

(i) infinity   (ii) zero

(iii) $\dfrac{L}{Cr}$ (iv) *r*

63. Which of the following conditions is common to both series and parallel resonance?

(i) impedance is minimum
(ii) power factor is unity
(iii) power is low
(iv) *Q* factor depends on voltage amplification

64. Which of the following theorems can be modified to analyse ac networks?

(i) superposition
(ii) Thevenin
(iii) maximum power transfer
(iv) all of these

Chapter 5

1. Which of the following is not true of a three-phase power system?

(i) a non-zero starting torque for three-phase motors
(ii) the instantaneous sum of power delivered is always constant
(iii) the output of a three-phase machine is less than that of a single phase machine for the same frame size
(iv) for the same voltage the weight of conductor is 25% less than in a single-phase system

2. A wattmeter is required to measure power in an alternating circuit because power is given by the

(i) product of the rms voltage and the rms current
(ii) product of the rms voltage, the rms current, and power factor
(iii) product of the maximum voltage and the maximum current
(iv) product of the maximum voltage, the maximum current and power factor

3. In the two-wattmeter method of measuring power in a three-phase circuit, the two wattmeters will show equal readings when the power factor angle φ is equal to

(i) 90°   (ii) 60°
(iii) 30°   (iv) 0°

4. In Question 3, one of the wattmeters shows a negative reading. It implies that the

(i) connections are wrong
(ii) $\phi > 60°$

(iii) $\phi < 60°$

(iv) none of these

5. While measuring power in a three-phase circuit by the two-wattmeter method, one of the wattmeters indicates a negative reading. What should be done to read the wattmeter reading correctly?

    (i) reverse the connections of the pressure coil of the wattmeter showing negative reading

    (ii) reverse the connections of the pressure coil of the wattmeter showing positive reading

    (iii) reverse the connections of the current coil of the wattmeter showing negative reading

    (iv) reverse the connections of the current coil of the wattmeter showing positive reading

Chapter 6

1. For which of the following purposes are transformers used?

    (i) impedance matching

    (ii) isolating a part of the network

    (iii) heating filaments of electronic and display devices

    (iv) all of these

2. Which of the following describes a transformer?

    (i) mechanical device

    (ii) static device

    (iii) electrostatic device

    (iv) rotating device

3. Which of the following functions is a power transformer able to perform?

    (i) step down a voltage

    (ii) step up a voltage

    (iii) neither step down nor step up a voltage

    (iv) all of these

4. The voltage induced due to the varying flux

    (i) leads the flux by $\pi/2$

    (ii) lags the flux by $\pi/2$

    (iii) is in phase with the flux

    (iv) is in phase opposition with the flux

5. The effective value of the induced voltage is represented by

    (i) $4.44\,fN\phi$ (ii) $4.44\,fN^2\phi_m$

    (iii) $4.44\,f^2N\phi_m$ (iv) $4.44\,fN\phi_m$

6. Which of the following does not help in reducing the eddy-current losses in a transformer?

    (i) solid iron core

    (ii) laminated core

    (iii) silicon steel core

    (iv) laminated sheets of small thickness

7. In the shell-type construction of a transformer which of the following is not correct?

    (i) each limb carries the primary and secondary

    (ii) windings are sandwiched

    (iii) the low-voltage winding is placed nearer to the core

    (iv) all of these

8. Which of the following is employed in the construction of a core-type single-phase transformer?

    (i) both windings on the central limb

    (ii) concentric windings

    (iii) low-voltage winding placed nearer to the core

    (iv) all of these

9. Which of the following is not a function of a transformer tank?

    (i) absorbing heat

    (ii) providing a rigid support to fitments and accessories

    (iii) providing protection to the core and windings

    (iv) storing transformer oil

10. Which of the following is not a turns ratio of a transformer?

    (i) $\dfrac{E_1}{E_2}$ (ii) $\dfrac{N_1}{N_2}$

    (iii) $\dfrac{I_2}{I_1}$ (iv) $\dfrac{E_2}{E_1}$

11. Which of the following is not true of leakage flux?

    (i) It links both the windings through air.

    (ii) It links the primary winding through air.

    (iii) It links the secondary winding through air.

    (iv) It does not link both the windings.

12. Which of the following relationships applies for a transformer while supplying full load?

    (i) $I_1N_1 > I_2N_2$ (ii) $I_1N_1 < I_2N_2$

    (iii) $I_1N_1 = I_2N_2$ (iv) $I_1N_1 \cong I_2N_2$

13. Which of the following assumptions apply to an ideal transformer?

    (i) coefficient of coupling is unity

    (ii) permeability of the core is infinite

    (iii) copper and core losses are zero

    (iv) all of these

14. Which of the following applies to the phasor diagram of a transformer on no load?

    (i) The primary voltage is in phase with the primary current

    (ii) The primary voltage is in phase with the main flux.

    (iii) The primary current is in phase with the main flux.

    (iv) The secondary voltage is in phase with the main flux.

15. The transformation ratio of a transformer is a. Which of the following does not represent the correct trans-

formation for an ideal transformer supplying a load of current I_2 at voltage V_2?

(i) $I_2 = \dfrac{I_2}{a}$

(ii) $I_2 = a_1$

(iii) $V_1 = aV_2$

(iv) $V_2 = \dfrac{V_1}{a}$

16. What is done to balance the mmf setup due to the secondary current?
    (i) The primary voltage is increased .
    (ii) The core flux increases immediately.
    (iii) The current in the primary increases.
    (iv) All of these.

17. Which of the following is true for an ideal loss-free transformer under a load?
    (i) The complex power is balanced.
    (ii) The complex power is greater in the secondary than in the primary.
    (iii) The real power in the secondary is greater than in the primary.
    (iv) The real power in the primary is greater than in the secondary.

18. Which of the following represent the no-load current?
    (i) the algebraic sum of I_m and I_C
    (ii) the phasor sum of I_m and I_C
    (iii) the algebraic difference of I_m and I_C
    (iv) none of these

19. The no-load power factor angle is close to 90° because
    (i) $I_0 < I_C$
    (ii) $I_C > I_\phi$
    (iii) $I_C < I_\phi$
    (iv) $I_0 < I_\phi$

20. The effect of the core in the equivalent circuit of a transformer is represented by
    (i) R_C connected in parallel with the primary voltage source
    (ii) X_m connected in parallel with the primary voltage source
    (iii) R_C and X_m connected in series with the primary voltage source
    (iv) R_C and X_m connected in parallel with the primary voltage source

21. The secondary winding resistance and reactance can be, respectively, represented by
    (i) $a^2 r_2$ and $a^2 x_2$
    (ii) ar_2 an ax_2
    (iii) r_2 and x_2
    (iv) $\dfrac{r_2}{a}$ and $\dfrac{x_2}{a}$

22. Which of the following parameters represent the equivalent circuit of a transformer when all secondary quantities are referred to the primary?
    (i) $R_{eq} = r_1 + r_2/a^2$
    (ii) $X_{eq} = x_1 + x'_2 = x_1 + x_2/a^2$

(iii) $G_C + jB_m$

(iv) all of these

23. Which of the following prohibit testing of a transformer by actually loading it?
    (i) difficulty in simulating the load
    (ii) loss of energy
    (iii) inaccuracy in measurement
    (iv) all of these

24. Which of the following is the main reason for inaccuracy in measurement while testing a transformer by actually loading it?
    (i) instruments are not accurate
    (ii) sparking at the terminals
    (iii) the input and output power are almost equal
    (iv) human error

25. What is the purpose of testing a transformer?
    (i) determine its parameters
    (ii) determine its efficiency
    (iii) determine its regulation
    (iv) all of these

26. In the short-circuit test, which of the following is the reason for applying a reduced voltage?
    (i) transformer is not loaded
    (ii) test is conducted in the laboratory
    (iii) current is limited by the winding impedances
    (iv) full voltage is not available

27. The purpose of voltage regulation of a transformer is to
    (i) maintain the voltage within the prescribed limits
    (ii) control the power consumption
    (iii) control power theft
    (iv) all of these

28. Which of the following is a correct representation of the regulation of a transformer?
    (i) $\dfrac{V_{2,\text{full-load}} - V_{2,\text{no-load}}}{V_{2,\text{full-load}}} \times 100$

    (ii) $\dfrac{V_{2,\text{full-load}} - V_{2,\text{no-load}}}{V_{2,\text{no-load}}} \times 100$

    (iii) $\dfrac{V_{2,\text{no-load}} - V_{2,\text{full-load}}}{V_{2,\text{full-load}}} \times 100$

    (iv) $\dfrac{V_{2,\text{no-load}} - V_{2,\text{full-load}}}{V_{2,\text{no-load}}} \times 100$

29. When the power factor of the load is leading, the regulation of a transformer is negative because
    (i) $V_{2,\text{full-load}} > V_{2,\text{no-load}}$
    (ii) $V_{2,\text{full-load}} < V_{2,\text{no-load}}$
    (iii) $V_{2,\text{full-load}} = V_{2,\text{no-load}}$
    (iv) none of these

30. Which of the following do not represent the efficiency of a transformer?

   (i) $\dfrac{\text{output power}}{\text{input power}}$

   (ii) $1 - \dfrac{\text{losses}}{\text{input power}}$

   (iii) $1 - \dfrac{\text{losses}}{\text{output power} + \text{losses}}$

   (iv) none of these

31. The output of a transformer at full load and unity power factor is 400 kW. What is the output at 0.8 power factor and half load?

   (i)   200 W   (ii) 320 W
   (iii) 160 W   (iv) none of these

32. In Question 31 if the copper and core losses of the transformer are 4 W each, the efficiency of the transformer is

   (i)   100%   (ii) 97.57%
   (iii) 96.97%   (iv) 95.24%

33. A transformer has its maximum efficiency at full load and the core loss is 600 W. The transformer supplies full load and half load for 10 hours each, and for the rest of the day it is on no load. Energy loss during the day is equal to

   (i)   26.4 kWh   (ii) 21.9 kWh
   (iii) 12.0 kWh   (iv) 7.5 kWh

34. On no-load, the efficiency of a distribution transformer is governed by

   (i)   the relative magnitudes of the copper loss and the core loss
   (ii) the magnitude of the core loss
   (iii) the magnitude of the primary voltage
   (iv) the magnitude of the secondary voltage

35. Which of the following is not categorized as a special application transformer?

   (i)   current transformer
   (ii) audio-frequency transformer
   (iii) three-phase transformer
   (iv) all of these

36. Which of the following is applicable to an auto-transformer?

   (i)   It is a two-winding transformer.
   (ii) It is a transformer with two magnetic cores.
   (iii) It is a transformer with a part of the winding common to the primary and secondary windings.
   (iv) It is a transformer which has separate frequencies for the primary and secondary windings.

37 The saving in copper in an auto-transformer having the same input–output specification as a two-winding transformer is not represented by

   (i) $\dfrac{a-1}{a}$

   (ii) $1 - \dfrac{I_1}{I_2}$

   (iii) $\dfrac{I_1\,(N_1 - N_2) + (I_2 - I_1)\,N_2}{I_1 N_1 + I_2 N_2}$

   (iv) a

38. For which of the following applications it is not advisable to use an auto-transformer?

   (i)   as a regulating transformer
   (ii) as a step-down transformer
   (iii) for starting induction motors
   (iv) as a booster in ac feeders

39. Power transfer in an auto-transformer takes place

   (i)   partially inductively and partially conductively
   (ii) wholly conductively
   (iii) wholly inductively
   (iv) none of these

40. An auto-transformer has an output current of I_2 at a voltage V_2. The corresponding input quantities are I_1 and V_1, respectively. Which of the following expression gives the power transferred conductively?

   (i) $V_2 I_2$ (ii) $V_1 I_2$
   (iii) $V_1 I_1$ (iv) $V_1(I_1 - I_2)$

41. Which of the following is a disadvantage in a single-unit three-phase transformer?

   (i)   less space
   (ii) less weight
   (iii) in case of a fault the unit has to be replaced
   (iv) higher efficiency

42. The following forms of connections are used for various applications. Which of the forms below are correctly applied?

   (i)   star–delta for stepping up voltages
   (iii) delta–star for stepping down voltages
   (iii) star–star connection rarely used
   (iv) all correctly applied

43. Which of the following is the main advantage of using a delta–delta connection?

   (i)   magnetizing current inrush
   (ii) can be used with a reduced capacity when one transformer is under maintenance
   (iii) line and phase quantities on the primary side can be conveniently transformed with the help of transformation ratio
   (iv) all of these

44. Which of the following does not characterize a current transformer?

   (i)   It has a low power rating.

(ii) It is used to step down current.

(iii) It is used to step down the voltage.

(iv) It has a single turn primary.

45. If the secondary of a current transformer is open-circuited when the primary is loaded, which of the following is likely to happen?

(i) The core and windings get damaged due to excessive heat.

(ii) A high voltage is induced in the secondary.

(iii) An error in ratio of transformation.

(iv) All of these.

46. Which of the following characterize an audio transformer?

(i) a high mutual inductive reactance

(ii) a constant response over the audio-frequency range

(iii) output impedance matching of the amplifier with the load

(iv) all of these

Chapter 7

1. Which of the following is not true of electrical energy?

(i) It can be easily converted.

(ii) It can be transmitted.

(iii) It cannot be controlled

(iv) It is flexible and clean.

2. Which of the following characterize an electromechanical energy conversion process?

(i) A generator can operate as a motor.

(ii) A motor can operate as a generator.

(iii) The conversion process is reversible.

(iv) All of these.

3. Which of the following phenomenon is mainly used in analysing the behaviour of a rotating machine?

(i) A voltage is induced in the conductor moving in a magnetic field.

(ii) A voltage is induced in a stationary coil when there is a change in the flux linking the other stationary coil.

(iii) A mechanical force is exerted on a ferromagnetic material when it is brought near a magnetic field.

(iv) None of these.

4. The length of an air gap is kept as small and uniform as possible to

(i) increase the reluctance of the air gap

(ii) minimize the flux required in the air gap

(iii) reduce the dimensions of the rotary machine

(iv) none of these

5. Which of the following is the normal arrangement of the armature windings and field poles in a three-phase alternator?

(i) stationary field poles and rotating armature

(ii) stationary armature and rotating field poles

(iii) stationary field poles and stationary armature

(iv) rotating field poles and rotating armature

6. Which of the following advantages accrue from placing the three-phase winding on the stator of an alternator?

(i) high output machines possible

(ii) high voltage rating possible

(iii) robust construction of the rotor possible

(iv) all of these

7. Which of the following is a feature of the stator core?

(i) high permeability steel stampings

(ii) solid high permeability steel

(iii) slotted high permeability steel stampings

(iv) ventilated and slotted high permeability steel stampings

8. Which of the following is responsible for producing the required flux in a synchronous machine?

(i) three-phase stator winding

(ii) dc winding on the rotor

(iii) ac winding on the rotor

(ii) none of these

9. Which of the following is a characteristic of the exciter to produce the necessary flux?

(i) It is a dc shunt or compound generator.

(ii) Current is supplied to the rotor through two slip rings and brushes.

(iii) Its power rating is 0.3% to 1% of the power rating of the synchronous generator.

(iv) All of these.

10. The flux produced in a synchronous machine is

(i) close to a sine wave in a cylindrical rotor

(ii) close to a sine wave in a non-cylindrical rotor

(iii) not a sine wave in either rotor

(iv) none of these

11. Which of the following is a characteristic of a non-salient-pole synchronous machine?

(i) uniform air gap

(ii) uniformly distributed armature windings

(iii) harmonic content in the generated emf wave can be neglected

(iv) all of these

12. Which of the following is required in order to generate three-phase voltages?

(i) at least two coils displaced 180 mechanical degrees in space

(ii) at least two coils phase displaced 180 electrical degrees in space

(iii) at least three coils displaced 120 mechanical degrees in space

(iv) at least three coils phase displaced 120 electrical degrees in space

13. The unit for specifying the span of a coil is
    (i) mechanical degrees
    (ii) electrical degrees
    (iii) centimetres
    (iv) inches

14. Which of the following is an advantage of using distributed winding in a three-phase alternator?
    (i) the generated emf approximates more closely to a sine wave.
    (ii) efficient utilization of the stator periphery
    (iii) generation of high voltages
    (iv) all of these

15. The phase voltage in a three-phase alternator with distributed windings is obtained by
    (i) the algebraic sum of the voltages in each coil
    (ii) the algebraic difference of the voltages in each coil
    (iii) the phasor sum of the voltages in each coil
    (iv) the phasor difference of the voltages in each coil

16. Which of the following is a disadvantage of using short-pitched coils?
    (i) reduces high frequency harmonics in the generated voltage
    (ii) improves the voltage waveform of the phase voltage
    (iii) results in a saving in copper
    (iv) none of these

17. The stator of a three-phase alternator is wound with short-pitch coils. The pitch of the coil is $(\pi - \beta)$ radians electrical. Which of the following is correct for the voltages generated in the coil sides?
    (i) voltages are equal and in phase
    (ii) voltages are equal but are in phase opposition
    (iii) voltages are equal and differ in phase by angle β
    (iv) voltages are unequal and differ in phase by an angle β

18. Which of the following defines the pitch factor?
    (i) ratio of the algebraic sum of the voltage generated in the sides of a short-pitch coil to the algebraic sum of the voltage generated in the sides of a full-pitch coil
    (ii) ratio of the algebraic sum of the voltage generated in the sides of a short-pitch coil to the phasor sum of the voltage generated in the sides of a full-pitch coil
    (iii) ratio of the phasor sum of the voltage generated in the sides of a short-pitch coil to the sum of the voltage generated in the sides of a full-pitch coil
    (iv) none of these

19. Which of the following happens when the coils in a phase of a three-phase alternator are connected in parallel?
    (i) nothing happens
    (ii) phase voltage is doubled
    (iii) current is doubled
    (iv) current is halved

20. When a three-phase alternator supplies a balanced load, the armature current sets up a magnetic field. Which of the following is true of the magnetic field?
    (i) the field is stationary
    (ii) the field is rotating at twice the synchronous speed
    (iii) the field is rotating at the synchronous speed
    (iv) none of these

21. In Question 20, if the maximum value of the flux density set up by the phase currents is B_m, the maximum value of the resultant field is equal to
    (i) $1.5B_m$ (ii) B_m
    (iii) $1.06B_m$ (iv) $0.707B_m$

22. What happens to the resultant rotating magnetic field when the connections to any of the two phases are interchanged?
    (i) There is no magnetic field set up.
    (ii) It does not rotate.
    (iii) The direction of rotation is reversed.
    (iv) The direction of rotation is not reversed.

23. The interaction between the armature flux and the flux produced by the rotor direct current is called
    (i) commutation   (ii) armature reaction
    (iii) synchronization   (iv) phase reversal

24. A three-phase alternator supplies a constant load. How does the regulation vary when the power factor is changed from lagging to leading?
    (i) remains negative
    (ii) changes from negative to positive
    (iii) remains positive
    (iv) changes from positive to negative

25. Which of the following applies in conducting the open-circuit test?
    (i) A reduced voltage is applied to the armature with the rotor at standstill and rotor winding on open circuit.
    (ii) A reduced voltage is applied to the rotor winding with the rotor at standstill and the armature on open circuit.
    (iii) A varying voltage is applied to the armature with the rotor running at synchronous speed and rotor winding on open circuit.
    (iv) A varying voltage is applied to the rotor with the rotor running at synchronous speed and the armature winding on open circuit.

26. Two alternators *A* and *B* are operating in parallel. What happens when the input to generator *B* is reduced?
    (i) The generated voltages are no longer in phase opposition.
    (ii) A circulating current is set up in the closed loop of the alternators.
    (iii) The power angle δ increases.
    (iv) All of these.

27. What is the effect of varying the excitation on the operation of generators in parallel?
    (i) Load sharing between the generators can be changed.
    (ii) The terminal voltage can be controlled.
    (iii) The driving torque can be controlled.
    (iv) None of these.

28. Which of the following is essential for a synchronous machine to act as a motor?
    (i) Armature currents must set up a constant magnetic field.
    (ii) Armature currents must set up a constant synchronously rotating magnetic field.
    (iii) Armature and rotor magnetic fields must be stationary with respect to each other.
    (iv) None of these.

29. Which of the following is required in order to operate a synchronous machine as a motor?
    (i) Energy must be supplied to the prime mover.
    (ii) Energy must be supplied to the rotor.
    (iii) Energy must be supplied to the armature.
    (iv) Energy must be supplied to both the rotor and the armature.

30. When the excitation of the rotor field of a synchronous motor supplying a constant load is varied, the locus of the armature current phasor is
    (i) a circle with the magnitude of the armature current as its radius
    (ii) a vertical line perpendicular to the terminal voltage phasor and at a distance given by the horizontal component of the armature current phasor
    (iii) a horizontal line parallel to the terminal voltage phasor and at a distance given by the vertical component of the armature current phasor
    (iv) none of these

31. Which of the following is applicable to a synchronous motor supplying a constant load?
    (i) When the excitation is more than the normal excitation the motor is underexcited.
    (ii) When the excitation is more than the normal excitation the generated voltage is less than the voltage at normal excitation.

(iii) When the excitation is more than the normal excitation the motor has a leading power factor.
    (iv) When the excitation is more than the normal excitation the motor has a lagging power factor.

32. When the value of the field excitation current is varied from a low value to the normal value, which of the following depicts the variation of the armature current?
    (i) It increases from zero to normal value.
    (ii) It decreases from normal value to zero.
    (iii) It decreases from a large value to a minimum value.
    (iv) It increases from a minimum value to a large value.

33. Which of the following is likely to occur when an alternator connected to an infinite bus is disconnected?
    (i) The frequency drops.
    (ii) The bus voltage changes.
    (iii) The power supplied by the bus reduces.
    (iv) None of these.

34. Which of the following applies for the voltages generated in the two sides of a coil?
    (i) Voltages are in quadrature.
    (ii) Voltages are in phase.
    (iii) Voltages are in phase opposition.
    (iv) None of these.

35. On which of the following power factor is it possible to operate a synchronous motor?
    (i) lagging
    (ii) leading
    (iii) unity
    (iv) from lagging through unity to leading

Chapter 8

1. A three-phase induction motor is considered to be the workhorse of the industry. Which of the following is the reason for its popularity?
    (i) self-starting
    (ii) almost constant speed from no load to full load
    (iii) simple and rugged construction
    (iv) all of these

2. The first presentation of a poly-phase induction motor was made in
    (i) 1805   (ii) 1850
    (iii) 1891   (iv) 1905

3. The first presentation of a polyphase induction motor was made by
    (i) Tesla   (ii) Kirchhoff
    (iii) Faraday   (iv) Lenz

4. Which of the following types of winding is used for the stator of a three-phase induction motor?

(i) dc lap winding
(ii) dc wave winding
(iii) distributed ac star/delta connected
(iv) none of these

5. Which of the following is a justification for using high permeability laminated steel for the stator core?
    (i) ease in assembling
    (ii) to make the core flexible
    (iii) for heat dissipation
    (iv) for reducing core loss

6. Due to which of the following it is possible to use thicker laminations in the construction of a rotor of an induction motor?
    (i) rotor frequency is equal to the stator frequency
    (ii) rotor frequency is less than the stator frequency
    (iii) rotor frequency is greater than the stator frequency
    (iv) none of these

7. Which of the following is an appropriate classification of induction motors?
    (i) slip ring   (ii) squirrel cage
    (iii) wound rotor   (iv) all of these

8. Which of the following requires that the slots in a squirrel cage rotor be skewed?
    (i) need to reduce hum during running
    (ii) need to avoid magnetic locking at starting
    (iii) need to produce more uniform torque
    (iv) all of these

9. The winding of the rotor of a slip-ring induction motor is
    (i) similar to the rotor of a squirrel cage induction motor
    (ii) wound for the same number of poles as the stator
    (iii) short-circuited at the two ends
    (iv) none of these

10. When an induction motor is supplying a load, the relative speed between the rotor and the rotating magnetic field is
    (i) zero
    (ii) equal to the synchronous speed
    (iii) more than the synchronous speed
    (iv) less than the synchronous speed

11. When the relative speed of the rotor and the synchronously rotating magnetic field is small, the magnitudes of the rotor frequency and the rotor induced emf are
    (i) equal to supply frequency and voltage
    (ii) greater than the supply frequency and voltage
    (iii) small
    (iv) none of these

12. If the speed of the rotating magnetic field is N_S rpm and that of the induction motor is N rpm, which of the following is valid?
    (i) $N > N_S$
    (ii) $N = N_S$
    (iii) $N < NS$
    (iv) N is independent of NS

13. An induction motor produces a uniform torque at all possible rotor speeds. This is made possible because the relative speed between the rotor mmf and stator mmf is
    (i) more   (ii) less
    (iii) zero   (iv) none of these

14. Which of the following determines the direction of the induced emf in the stator just before it begins to rotate?
    (i) Fleming's right hand rule
    (ii) Kirchhoff's laws
    (iii) Lenz's law
    (iv) Faraday's laws of electromagnetic induction

15. Which of the following depicts the speed of an induction motor in terms of the slip and synchronous speed?
    (i) $N = sN_S$ (ii) $(1 - s)N_S$
    (iii) $(1 + s)NS$ (iv) $NS/(1 - S)$

16. The advantage of a slip-ring motor is that an external resistance can be connected in the rotor circuit. An external resistance is added
    (i) to improve its power factor
    (ii) for the purpose of starting
    (iii) to obtain higher speeds
    (iv) for higher efficiency

17. Which of the following does not apply when the mechanical load increases on the motor?
    (i) slip reduces
    (ii) rotor emf increases
    (iii) rotor current increases
    (iv) torque increases

18. Which of the following parameters are applicable to the rotor when the motor is running at a slip s?
    (i) $Z_2 = r_2 + jsx_2$
    (ii) $I_2 = (sE)/(r_2 + jsx_2)$
    (iii) $q_2 = \tan^{-1}[(sx_2)/r_2]$
    (iv) all of these

19. Which of the following represents the air gap power?
    (i) power input to the stator
    (ii) power input minus resistance loss in the stator winding
    (iii) power input minus resistance loss in the stator and rotor windings
    (iv) none of these

20. The total resistance loss in the rotor is given by
    (i) $s \times$ input power to the motor
    (ii) $s \times$ air gap power
    (iii) (air gap power)/s
    (iv) none of these

21. The power developed at the shaft is equal to
    (i) input power – gross mechanical power
    (ii) input power – mechanical losses
    (iii) gross mechanical power – mechanical losses
    (iv) none of these

22. Which of the following is applicable for the torque developed by an induction motor?

    (i) $T \propto \dfrac{sE_2^2 r_2}{r_2^2 + (sx_2)^2}$ (ii) $T \propto \dfrac{\phi^2 \, (r_2/s)}{(r_2/s)^2 + x_2^2}$

    (iii) $T \propto \dfrac{(r_2/s)}{(r_2/s)^2 + x_2^2}$ (iv) all of these

23. Which of the following is a correct expression for slip at maximum torque?
    (i) $s = r_2 x_2$ (ii) $s = x_2/r_2$
    (iii) $s = r_2/x_2$ (iv) none of these

24. The direction of rotation of a three-phase induction motor can be changed by
    (i) disconnecting it from the supply and re-connecting it
    (ii) re-designing the stator winding
    (iii) inter changing the connections between any two stator terminals
    (iv) none of these

25. Which of the following is not a mechanical loss in an induction motor?
    (i) rotor core loss   (ii) windage loss
    (iii) brush contact loss   (iv) friction loss

26. When in an induction motor $s = r_2/x_2$, which of the following is not true for torque T.
    (i) T is maximum   (ii) $T = 1/2x_2$
    (iii) T is minimum   (iv) all of these

27. In a three-phase squirrel cage motor, short circuit is applied by
    (i) short-circuiting the terminals of the stator
    (ii) short-circuiting the terminals of the rotor
    (iii) by preventing the rotor from rotating
    (iv) none of these

28. The starting current drawn by a three-phase induction motor is
    (i) 2 to 4 times the full-load current
    (ii) 4 to 7 times the full-load current
    (iii) 7 to 10 times the full-load current
    (iv) 10 to 15 times the full-load current

29. Which of the following is likely to happen when the motor draws a high secondary current?
    (i) disturb the voltage in the neighbourhood
    (ii) stator may get burnt
    (iii) rotor may get burnt
    (iv) the motor may attain a high speed

30. The star–delta starter method is used in
    (i) flour mills
    (ii) pumps
    (iii) machine tool drives
    (iv) all of these

31. Rotor resistance starting can be applied to which of the following motors?
    (i) heavy duty squirrel cage
    (ii) medium duty squirrel cage
    (iii) squirrel cage and slip ring
    (iv) slip ring

32. Which of the following is a disadvantage of a squirrel cage induction motor when compared to a slip ring motor?
    (i) low starting torque
    (ii) the rotor consists of bare conductors
    (iii) no brushes
    (iv) all of these

33. If, through an autotransformer, the voltage applied to the stator of a three-phase induction motor is a fraction of the supply voltage given by kV volts, the ratio of the autotransformer starting torque to the DOL starting torque is given by
    (i) k^2 (ii) k
    (iii) $1/k$ (iv) $1/k^2$

34. What happens when the phase sequence of the voltage applied to the stator of a three-phase induction motor is changed?
    (i) motor does not run
    (ii) slip changes
    (iii) direction of rotation is reversed
    (iv) motor gets heated

Chapter 9

1. Which of the following is a function of the brushes?
    (i) to convert ac to dc
    (ii) to convert dc to ac
    (iii) to collect current and deliver to the load
    (iv) none of these

5. In which of the following years were dc machines first used to drive a locomotive?
    (i) 1849   (ii) 1839
    (iii) 1829   (iv) 1819

3. It is essential for the armature coils in a dc machine to have a coil span of one pole pitch so that the emf generated in the sides of the coils is
    (i) additive   (ii) subtractive
    (iii) multiplicative   (iv) divisive

4. In a dc machine, the number of poles is P and the total number of armature slots is S. Which of the following expression is applicable to the span of an armature coil?

(i) $\dfrac{2P}{S}$ (ii) $\dfrac{2S}{P}$

(iii) $\dfrac{P}{S}$ (iv) $\dfrac{S}{P}$

5. Compared to a lap-wound machine, which of the following is true for a wave-wound machine?
   (i) voltage output is higher
   (ii) current output is higher
   (iii) both voltage and current outputs are higher
   (iv) none of these

6. Which of the following applies when a current flows through the commutator of a dc machine?
   (i) no armature flux is produced
   (ii) armature flux, which rotates at synchronous speed, is produced
   (iii) armature flux, which is fixed in space and perpendicular to the axis of the field flux, is produced
   (iv) armature flux, which is fixed in space and is parallel to the axis of the field flux, is produced

7. The direction of generated emf is determined by
   (i) Lenz's law
   (ii) Faraday's laws of electromagnetic induction
   (iii) Fleming's left-hand rule
   (iv) Fleming's right-hand rule

8. Which of the following applies to a dc machine?
   (i) it can operate only as a generator
   (ii) it can operate only as a motor
   (iii) it can operate both as a generator and a motor
   (iv) none of these

9. Which of the following can be categorized as self-excited dc generators?
   (i) series excited   (ii) shunt excited
   (iii) compound   (iv) all of these

10. In order that a dc generator be able to excite and generate voltage, the value of the field winding resistance should be
    (i) of any value
    (ii) less than the critical resistance
    (iii) equal to the critical resistance
    (iv) greater than the critical resistance

11. In a dc series generator, which of the following characteristics is not possible to plot?
    (i) internal voltage drop
    (ii) internal characteristic
    (iii) external characteristic
    (iv) all of these

12. Which of the following represents the effect of the armature current when a dc generator is supplying a load?

(i) armature mmf is produced
(ii) a cross-magnetizing effect is produced on the main flux
(iii) a demagnetizing effect is produced on the main flux
(iv) all of these

13. Which of the following is the cause of production of torque in a dc motor?
    (i) the resultant flux due to the field and armature currents
    (ii) the flux due to field current
    (iii) the flux due to armature current
    (iv) none of these

14. The emf generated in a dc motor is called back emf because
    (i) it is generated in the armature
    (ii) it opposes the direction of rotation of the motor
    (iii) it is in a direction opposite to the applied voltage
    (iv) none of these

15. Which of the following is a cause for the production of back emf?
    (i) generator action   (ii) motor action
    (iii) armature reaction   (iv) none of these

16. Which of the following is applicable to the speed of a dc series motor?

   (i) $N \propto \dfrac{V - I_a R_a}{\Phi}$ (ii) $N \propto \dfrac{E_b}{\Phi}$

   (iii) $N \propto \dfrac{1}{\Phi}$ (iv) all of these

17. For which of the following reasons is a dc motor never started with no load?
    (i) with a decrease in load, the motor is likely to develop a high speed
    (ii) when the armature current is of low value, the flux is very small
    (iii) at a high speed, the motor develops large centrifugal forces
    (iv) all of these

18. It is desired to select a dc motor for mining hoists. Which of the following motors is appropriate?
    (i) series motor
    (ii) cumulative compound motor
    (iii) differential compound motor
    (iv) shunt motor

19. Which of the following is an advantage of the armature control of the speed of a shunt motor?
    (i) continuous control of speed
    (ii) speed can only be reduced below its normal value
    (iii) the armature circuit rheostat needs to be of continuous rating
    (iv) excessive heat loss in the armature

20. It is desired to increase the speed of a dc shunt motor above its normal speed with a variable external resistance in the armature circuit. Which of the following will accomplish this?
    - (i) increasing the variable resistance to its maximum value
    - (ii) uniformly reducing the variable resistance to zero
    - (iii) decreasing the load
    - (iv) none of these

Chapter 10

1. Which of the following is a reason for using single-phase motors?
    - (i) economic operation.
    - (ii) domestic appliances have fractional kilowatt ratings.
    - (iii) it is a requirement in robotic applications.
    - (iv) all of these.

2. Which of the following cannot be classified in the fractional kilowatt range?
    - (i) servo motors
    - (ii) stepper motors
    - (iii) hysteresis motors
    - (iv) none of these

3. Single-phase induction motors are not self-starting because
    - (i) the magnitude of the flux produced in the stator is low
    - (ii) no voltage is induced in the rotor circuit
    - (iii) there is no relative motion between the stator and rotor magnetic fields
    - (iv) none of these

4. The principle that a single-phase pulsating stationary magnetic field, set up by a single-phase ac voltage, can be resolved into two equal rotating magnetic fields was enunciated by
    - (i) Fraday
    - ii) Ferrri
    - (iii) Tesla
    - (iv) Lenz

5. In a split-phase induction motor, the auxiliary winding is cut out when the motor attains
    - (i) 30% of the synchronous speed
    - (ii) 50% of the synchronous speed
    - (iii) 75% of the synchronous speed
    - (iv) 85% of the synchronous speed

6. When a servo motor transforms an electrical signal, which of the following is not its output?
    - (i) mechanical
    - (ii) electrical
    - (iii) torque
    - (iv) velocity

7. In a field-controlled dc servo motor, the field winding is
    - (i) separately excited
    - (ii) shunt excited

    - (iii) series excited
    - (iv) any one of these

8. Which of the following is not a characteristic of a shunt field controlled dc servo motor?
    - (i) large armature current for higher ratings
    - (ii) sluggish response to error signals
    - (iii) fast response to error signals
    - (iv) all of these

9. In which of the following types of controls, a dc servo motor is operated above the knee of the magnetization curve?
    - (i) field control
    - (ii) armature control
    - (iii) fixed field
    - (iv) none of these

10. Which of the following dc servo motors are also called torque motors?
    - (i) shunt motors with field control
    - (ii) shunt motors with armature control
    - (iii) shunt motors with fixed field
    - (iv) series motors

11. Which of the following represents the appropriate connection for an ac servo motor?
    - (i) reference phase connected to a constant voltage ac source
    - (ii) reference phase with a capacitor in series connected to a constant ac voltage source
    - (iii) reference phase connected to an ac amplifier
    - (iv) reference phase with a capacitor in series connected to an ac amplifier

12. Which of the following is not a characteristic of ac servo motors?
    - (i) low inertia
    - (ii) causes radiation and radio interference
    - (iii) no commutator arcing
    - (iv) can withstand high temperatures

13. The full-step angle of a stepper motor is a function of
    - (i) the number of phases
    - (ii) the number of rotor poles
    - (iii) the number of phases and the number of rotor poles
    - (iv) none of these

14. Which of the following is not a feature of a PM stepper motor?
    - (i) odd number of rotor poles
    - (ii) even number of rotor poles
    - (iii) salient pole rotor
    - (iv) non-salient pole rotor

15. The expanded form of VR in a VR stepper motor is
    - (i) very rugged
    - (ii) very round
    - (iii) variable reluctance
    - (iv) variable resistance

16. A HB stepper motor has a rating of 12 V and a current rating of 1 A/phase. If both the phases are excited, the rating of the motor is

(i) 24 W   (ii) 12 W

(iii) 6 W   (iv) none of these

17. Which of the following is not a feature of hysteresis motors?

    (i) cylindrical rotor

    (ii) non-cylindrical rotor

    (iii) no dc field excitation

    (iv) none of these

18. The torque produced by a hysteresis motor is proportional to

    (i) the fundamental component of the stator mmf

    (ii) the rotor flux

    (iii) sine of the hysteresis angle

    (iv) all of these

19. Which of the following is true for the magnitude of the torque due to eddy current loss?

    (i) maximum at $s = 1$

    (ii) varies linearly with s

    (iii) minimum at $s = 0$

    (iv) all of these

20. In which of the following are single-phase motors used?

    (i) washing machines   (ii) blowers

    (iii) office machines   (iv) all of these

Chapter 11

1. As per the Indian Electricity Rules, which of the following is a permissible range for frequency variation?

    (i) ±3%   (ii) ±5%

    (iii) ±7%   (iv) ±9%

2. For which of the following purpose is it necessary to measure the energy supplied?

    (i) to ensure there is no fault on the system

    (ii) for maintenance of the system

    (iii) for billing the consumer

    (iv) for protecting the system

3. Which of the following is an acceptable classification of measuring instruments?

    (i) absolute   (ii) digital

    (iii) secondary   (iv) all of these

4. Which of the following measuring instruments are not categorized as secondary instruments?

    (i) absolute   (ii) analogue

    (iii) digital   (iv) all of these

5. In a measuring instrument, variation in resistance is used to measure the magnitude of current. In which of the following categories will such a type of instrument be classified?

    (i) digital   (ii) absolute

    (iii) analogue   (iv) none of these

6. Which of the following effects are commonly employed in most measuring instruments?

    (i) electrostatic   (ii) electromagnetic

    (iii) heating   (iv) chemical

7. Which of the following is a function of the controlling system?

    (i) deflect the pointer

    (ii) prevent oscillations of the pointer

    (iii) bring the pointer to rest

    (iv) none of these

8. Which of the following is not employed for generating a controlling torque?

    (i) electromagnetic effect

    (ii) spring system

    (iii) gravity control

    (iv) all of these

9. Which of the following has a uniform scale?

    (i) spring control

    (ii) gravity control

    (iii) air damping

    (iv) eddy current damping

10. Which of the following is not applicable to a gravity control system?

    (i) scale cramped for low values of currents

    (ii) extended scale for higher values of current

    (iii) uniform scale for the whole range of currents

    (iv) all of these

11. Which of the following is a function of the damping system?

    (i) balance the deflecting and controlling torques

    (ii) increase the deflecting torque

    (iii) produce oscillations in the controlling torque

    (iv) bring the pointer quickly to its steady-state position

12. Measuring instruments employ dead-beat type of damping. Which of the following describes the type of system damping?

    (i) under damped   (ii) critically damped

    (iii) over damped   (iv) none of these

13. Which of the following method can be employed to use a voltmeter as an ammeter?

    (i) a low resistance across the terminals of the voltmeter

    (ii) a low resistance in series with the voltmeter

    (iii) a high resistance across the terminals of the voltmeter

    (iv) a high resistance in series with the voltmeter

14. Which of the following can be used for measuring alternating quantities only?

    (i) moving-coil dynamometer type

(ii) induction type

(iii) moving iron type

(iv) all of these

15. Which of the following can be used to extend the range of an ammeter?

    (i) a low resistance in series with the ammeter

    (ii) a high resistance in series with the ammeter

    (iii) a low resistance in parallel with the ammeter

    (iv) a high resistance in parallel with the ammeter

16. Which of the following cannot be used to measure both alternating and direct quantities?

    (i) moving-coil permanent-magnet type

    (ii) moving-coil dynamometer type

    (iii) moving iron type

    (iv) hot wire type

17. What happens when an alternating quantity is applied to a moving-coil instrument?

    (i) pointer shows a deflection

    (ii) pointer shows a negative reading

    (iii) pointer oscillates around the zero position

    (iv) none of these

18. Moving-coil dynamometer type instruments use air-cored coils. Which of the following reasons applies?

    (i) keep the weight of the coil low

    (ii) keep the construction of the coil simple

    (iii) eliminate errors due to hysteresis and eddy currents

    (iv) none of these

19. Which of the following is an advantage of moving-coil dynamometer instruments?

    (i) high cost

    (ii) low sensitivity

    (iii) measure both alternating and direct quantities

    (iv) none of these

20. Which of the following is not an advantage of a moving iron instrument?

    (i) low cost

    (ii) low precision in calibrating direct quantities

    (iii) sturdiness

    (iv) degree of reliability is good

21. Which of the following is not a type of wattmeter?

    (i) moving-coil permanent magnet

    (ii) electrostatic

    (iii) induction

    (iv) dynamometer

22. Which of the following makes a dynamometer type wattmeter useful for calibration work?

    (i) can be used to measure ac and dc power

    (ii) high degree of accuracy

    (iii) high cost

    (iv) all of these

23. Which of the following is a function of the copper band on the shunt magnet?

    (i) measure the power factor

    (ii) produce the required magnitude of flux

    (iii) make the resultant flux lag the applied voltage by 90°

    (iv) make the resultant flux lag the current in the series magnet by 90°

24. Which of the following applies to a wattmeter?

    (i) it has a voltage coil

    (ii) it has current coil

    (iii) the voltage coil has a high resistance

    (iv) all of these

25. Which of the following interactions produces the deflecting torque in an induction wattmeter?

    (i) between the flux produced by the shunt electromagnet and the flux produced by the series electromagnet

    (ii) between the flux produced by the shunt electromagnet and the current in series coil

    (iii) between the flux produced by the series electromagnet and the voltage across the series coil

    (iv) between the fluxes and the eddy currents

26. Which of the following finds application as a standard energy meter?

    (i) clock type   (ii) motor type

    (iii) electrolytic type   (iv) none of these

27. In which of the following fields an induction type of an energy meter finds application?

    (i) industry   (ii) domestic

    (iii) laboratories   (iv) all of these

28. Which of the following indicating system is used in an energy meter?

    (i) pointer type

    (ii) continuously rotating type

    (iii) digital type

    (iv) none of these

29. The principle of operation of a three-phase energy meter is similar to the measurement of power in a three-phase circuit by

    (i) one-wattmeter method

    (ii) two-wattmeter method

    (iii) three-wattmeter method

    (iv) none of these

30. Which of the following is a function of the adjustable magnetic field in a three-phase energy meter?

    (i) make the two torques unequal

    (ii) make the two torques exactly equal and opposite

    (iii) make the two torques exactly equal and in the same direction

(iv) none of these

31. In domestic or industrial networks the insulation resistance is of the order of
    (i) megohms   (ii) kilo-ohms
    (iii) hundred ohms   (iv) ohms

32. In a megger the most appropriate voltage source is
    (i) normal domestic supply
    (ii) a Ni–Cd battery
    (iii) a storage battery
    (iv) hand-driven generator

33. Which of the following is a feature of a megger?
    (i) it is sturdy   (ii) it is light
    (iii) it is portable   (iv) all of these

34. Which of the following produces the deflecting torque in a megger?
    (i) permanent magnet
    (ii) potential coil and current coil
    (iii) current coil
    (iv) potential coil

35. Which of the following is a requirement of measuring instruments?
    (i) sturdiness
    (ii) aesthetic appearance
    (iii) accuracy and repeatability
    (iv) none of these

36. Which of the following is the purpose of providing a mirror on the scale?
    (i) for illumination
    (ii) to see your reflection
    (iii) to remove the parallax between the pointer and the image
    (iv) to improve the aesthetic appearance

37. Which of the following is an appropriate description of a shunt resistance?
    (i) low resistance connected in parallel with the ammeter
    (ii) high resistance connected in parallel with the ammeter
    (iii) low resistance connected in series with the ammeter
    (iv) high resistance connected in series with the ammeter

38. Which of the following is a purpose of connecting a high resistance in series with a voltmeter?
    (i) improve accuracy
    (ii) make the readings repeatable
    (iii) extend the range
    (iv) none of these

Chapter 12

1. Which of the following is not a subsystem of an electrical energy system?
   (i) distribution subsystem
   (ii) protection and control subsystem
   (iii) transmission subsystem
   (iv) none of these

2. In a power station subsystem,
   (i) power is transmitted
   (ii) power is distributed
   (iii) electrical energy is obtained from a prime source of energy
   (iv) none of these

3. Which of the following is not a primary source of energy in alternative/unconventional power stations?
   (i) wind   (ii) plutonium
   (iii) geothermal   (iv) tides

4. In a hydroelectric generating system, the primary source of energy is
   (i) water
   (ii) wind
   (iii) head and volume of water
   (iv) all of these

5. Which of the following is a constituent of a generating station?
   (i) fuel   (ii) prime mover
   (iii) generator   (iv) all of these

6. The earliest generating station established was a
   (i) hydro-station   (ii) thermal
   (iii) nuclear   (iv) unconventional

7. Which of the following is a characteristic of the electric power generating systems set up in the earlier times?
   (i) individual power stations
   (ii) power supplied to individual consumers
   (iii) located at pitheads
   (iv) all of these

8. Which of the following describes the process of electrical energy generation in a coal-fired thermal station?
   (i) chemical energy in coal is converted into electrical energy
   (ii) chemical energy in coal is converted into thermal energy, which is converted into electrical energy
   (iii) chemical energy in coal is converted into thermal energy, which is converted into mechanical energy and then into electrical energy
   (iv) none of these

9. In a super-thermal plant in India, what is the unit capacity in use?
   - (i) 300 mW
   - (ii) 400 mW
   - (iii) 500 mW
   - (iv) 600 mW

10. Which of the following is a reason for designing and building units having a large mW rating?
    - (i) higher capital cost
    - (ii) possibility of generating more power
    - (iii) low maintenance
    - (iv) less cost per kilowatt

11. What is the capacity of the turbo-generator sets being produced today?
    - (i) 1200 mW
    - (ii) 1000 mW
    - (iii) 800 mW
    - (iv) 500 mW

12. Which of the following describes cogeneration?
    - (i) using coal-fired and hydro-generating stations in combination
    - (ii) using coal-fired and diesel-oil-fired generating stations in combination
    - (iii) using large capacity super-thermal units
    - (iv) using electricity and steam or hot water for industrial purposes

13. Which type of pollution occurs when a large amount of heat is released in a thermal generating station?
    - (i) thermal
    - (ii) environmental
    - (iii) noise
    - (iv) all of these

14. Which of the following is the cheapest method of transporting oil?
    - (i) rail
    - (ii) road
    - (iii) pipeline
    - (iv) ship

15. Which of the following is employed as a prime mover in a gas-fired station?
    - (i) the turboprop engine used in aircraft
    - (ii) the LPG engine
    - (iii) the steam turbine
    - (iv) the CNG engine

16. In a gas-fired power station, the combined-cycle method is used to generate electrical energy. Which of the following applies?
    - (i) the gas-fired power station is used with a diesel-oil-fired power station
    - (ii) gas turbines first generate power; the hot exhaust gases then produce steam, which is used to generate power in a conventional steam station
    - (iii) gas turbines first generate power; the hot exhaust gases are then used to run another gas station
    - (iv) none of these

17. Which of the following applies to a gas station when compared to a coal-fired steam station?
    - (i) low installation and operational costs
    - (ii) high installation and low operational costs
    - (iii) low installation and high operational costs
    - (iv) high installation and operational costs

18. In a diesel-oil-fired power station, the internal combustion engine is used as the prime mover. Which of the following types is used?
    - (i) turboprop engine
    - (ii) truck engine
    - (iii) ship engine
    - (iv) tube well engine

19. For which of the following reasons are hydro-generating stations sited in mountainous regions?
    - (i) large land area is required
    - (ii) safety
    - (iii) earthquake-free region
    - (iv) availability of the required head of water

20. Which of the following is employed to generate power in a hydro-station?
    - (i) potential energy of water
    - (ii) kinetic energy of water
    - (iii) potential energy and kinetic energy of water
    - (iv) all of these

21. Which of the following is an advantage of a hydro-power station?
    - (i) long gestation period
    - (ii) no air pollution
    - (iii) causes ecological imbalance
    - (iv) capital-intensive

22. Which of the following is not an advantage of a hydro-generating station?
    - (i) no waste products
    - (ii) long life
    - (iii) quick start-up time
    - (iv) none of these

23. The fuel used in a nuclear generating station is
    - (i) isotopes of uranium
    - (ii) crude oil
    - (iii) liquefied hydrogen
    - (iv) none of these

24. Nuclear fuel rods are embedded in heavy water. The role of heavy water is to act as a
    - (i) catalyst
    - (ii) lubricant
    - (iii) accelerating agent
    - (iv) decelerating agent

25. Which of the following functions as a prime mover in a nuclear power station?
    - (i) steam turbine
    - (ii) turboprop engine used in aircraft
    - (iii) diesel engine used in ships
    - (iv) Pelton wheel

26. Theoretically, a nuclear power station can be sited close to the load centres. Which of the following reasons applies?
    - (i) no air pollution
    - (ii) low operational cost
    - (iii) cost of transporting nuclear fuel is negligible
    - (iv) all of these

27. In India, which of the following states does not have wind power stations?
    - (i) Punjab
    - (ii) Tamil Nadu
    - (iii) Maharashtra
    - (iv) Orissa
28. Which of the following is an advantage of wind as a primary source of energy?
    - (i) unpredictable
    - (ii) unsteady
    - (iii) non-renewable
    - (iv) none of these
29. The potential of wind power in India is estimated at
    - (i) 10,000 MW
    - (ii) 20,000 MW
    - (iii) 30,000 MW
    - (iv) 40,000 MW
30. The potential of the available geothermal energy in India is
    - (i) poor
    - (ii) good
    - (iii) very good
    - (iv) excellent
31. Which of the following is not a site for tidal power generation?
    - (i) Bhavnagar
    - (ii) Diamond harbour
    - (iii) Ganga Sagar
    - (iv) Gobind Sagar
32. Which of the following is not a suitable location for siting a tidal power generating station?
    - (i) estuary
    - (ii) bay
    - (iii) lake
    - (iv) all of these
33. The potential for generating tidal power in India is
    - (i) excellent
    - (ii) very good
    - (iii) good
    - (iv) poor
34. Which of the following is a disadvantage of a tidal power station?
    - (i) high cost of civil works
    - (ii) power generation is intermittent
    - (iii) variation in low- and high-tide timings
    - (iv) all of these
35. Which of the following is a function of a transformer at the generator end?
    - (i) to transmit power by raising the voltage level
    - (ii) to transmit power by lowering the voltage level
    - (iii) to supply power
    - (iv) to consume power
36. The power transmitted over a transmission line is a function of the transmission voltage. It is proportional to
    - (i) the transmission voltage
    - (ii) the square of the transmission voltage
    - (iii) the cube of the transmission voltage
    - (iv) none of these
37. Which of the following factors goes in the favour of dc transmission of bulk power over long distances?
    - (i) easier to convert ac power into dc power
    - (ii) consumers prefer dc power
    - (iii) it is more economical

    - (iv) it is safe
38. The feeder voltage at which distribution of power is undertaken is
    - (i) 66 kV
    - (ii) 33 kV
    - (iii) 22 kV
    - (iv) 11 kV
39. Which of the following is a characteristic of a good distribution subsystem?
    - (i) reliable
    - (ii) economical
    - (iii) uninterrupted
    - (iv) all of these
40. Which of the following is a disadvantage of the parallel distribution system?
    - (i) reliable
    - (ii) single feeding point
    - (iii) alternate circuit available in case of a fault
    - (iv) none of these
41. It is desired to supply a large load reliably over a large area. Which of the following distribution systems is recommended?
    - (i) direct distribution from the grid
    - (ii) radial distribution
    - (iii) loop distribution
    - (iv) network distribution
42. Which of the following is not a feature of a good wiring design?
    - (i) equal voltage distribution
    - (ii) varying load current in each circuit
    - (iii) easy repair and maintenance
    - (iv) safety to consumer
43. Which of the following gives the expanded form of an MCB?
    - (i) minimum circuit-breaker
    - (ii) maximum current-breaker
    - (iii) modern circuit-breaker
    - (iv) miniature circuit-breaker
44. Which of the following gives the maximum load per wiring circuit and the maximum number of points in a circuit as specified by the BIS?
    - (i) 1000 W, 8
    - (ii) 800 W, 10
    - (iii) 600 W, 12
    - (iv) 400 W, 14
45. In the distribution board system, the number of sub-distribution systems is governed by
    - (i) the current-carrying capacity of each sub-distribution board system
    - (ii) the current-carrying capacity of the distribution board system
    - (iii) the current-carrying capacity of the supplier's feeder
    - (iv) the voltage level
46. Which of the following gives the expanded form of DPIC?
    - (i) double-phase, iron-clad

(ii) double-pole, insulated cover

(iii) double-pole iron clad

(iv) none of these

47. Which of the following is not a disadvantage of the loop-in system of wiring?

   (i) more material used

   (ii) no joint boxes required

   (iii) more expensive

   (iv) high copper loss

48. Which of the following is the cause for a human body to experience an electric shock?

   (i) static charge discharging through the human body

   (ii) charge being induced in the human body

   (iii) voltage being induced in the human body according to Faraday's laws of induction

   (iv) none of these

49. Which of the following is the purpose of earthing an electrical installation?

   (i) safety to personnel

   (ii) protection against electric shock

   (iii) fire protection

   (iv) all of these

50. In India, which of the following bodies specifies the standards for earthing of electrical installations?

   (i) CSIR   (ii) ISO

   (iii) BIS   (iv) CII

51. Which of the following gives the maximum permissible earth resistance for all installations except power stations and substations?

   (i) $16\,\Omega$ (ii) $12\,\Omega$

   (iii) $8\,\Omega$ (iv) $4\,\Omega$

52. For which of the following types of soils is strip earthing adopted?

   (i) clayey   (ii) rocky

   (iii) sandy   (iv) all of these

53. Which of the following method is most commonly adopted in earthing installations?

   (i) water mains   (ii) plate

   (iii) wire   (iv) rod

54. Which of the following gives the maximum specified resistance between the earth plate and any point inside an installation?

   (i) $1.0\,\Omega$ (ii) $2.0\,\Omega$

   (iii) $3.0\,\Omega$ (iv) $4.0\,\Omega$

55. Which of the following is the real purpose of filling the earthing pit with alternate layers of charcoal and salt?

   (i) easy availability

   (ii) cheap

   (iii) reduction of the earth's resistance

   (iv) none of these

56. Which of the following are the minimum specified dimensions of the earthing plate?

   (i) $90\ cm \times 90\ cm \times 12.5\ mm$

   (ii) $90\ cm \times 90\ cm \times 6.36\ mm$

   (iii) $60\ cm \times 60\ cm \times 12.5\ mm$

   (iv) $60\ cm \times 60\ cm \times 6.36\ mm$

Answers

Chapter 1

1. (iv)	2. (ii)	3. (iii)	4. (ii)	5. (iii)	6. (iv)	7. (i)
8. (ii)	9. (iii)	10. (iv)	11. (iv)			

Chapter 2

1. (iv)	2. (ii)	3. (iv)	4. (i)	5. (iv)	6. (ii)	7. (iv)
8. (iii)	9. (i)	10. (i)	11. (iv)	12. (i)	13. (iii)	14. (ii)
15. (iv)	16. (i)	17. (iii)	18. (iv)	19. (i)	20. (iv)	21. (ii)
22. (iv)	23. (iii)	24. (iv)	25. (iii)	26. (iii)	27. (iii)	

Chapter 3

1. (i)	2. (ii)	3. (iii)	4. (i)	5. (iv)	6. (iii)	7. (i)
8. (iii)	9. (iv)	10. (iv)	11. (iv)	12. (ii)	13. (iv)	14. (iii)
15. (iv)	16. (iii)	17. (i)	18. (iv)	19. (iii)	20. (iv)	

Chapter 4

1. (iii)	2. (i)	3. (iii)	4. (iv)	5. (i)	6. (ii)	7. (iii)
8. (iii)	9. (ii)	10. (iv)	11. (iii)	12. (i)	13. (iv)	14. (iii)
15. (iii)	16. (iv)	17. (i)	18. (i)	19. (iv)	20. (i)	21. (iii)
22. (ii)	23. (i)	24. (iv)	25. (iii)	26. (i)	27. (ii)	28. (iv)
29. (iv)	30. (i)	31. (ii)	32. (i)	33. (iv)	34. (iii)	35. (iii)
36. (ii)	37. (ii)	38. (iv)	39. (iii)	40. (iii)	41. (i)	42. (iv)
43. (iv)	44. (ii)	45. (iii)	46. (iii)	47. (ii)	48. (iv)	49. (ii)
50. (iv)	51. (i)	52. (iv)	53. (iii)	54. (ii)	55. (i)	56. (iv)
57. (iii)	58. (iv)	59. (iii)	60. (i)	61. (iii)	62. (iii)	63. (ii)
64. (iv)						

Chapter 5

1. (iii)	2. (ii)	3. (iv)	4. (ii)	5. (i)

Chapter 6

1. (iv)	2. (ii)	3. (iv)	4. (i)	5. (iv)	6. (i)	7. (iii)
8. (iv)	9. (i)	10. (iv)	11. (i)	12. (iv)	13. (iv)	14. (iii)
15. (i)	16. (iii)	17. (i)	18. (ii)	19. (iii)	20. (iv)	21. (i)
22. (iv)	23. (iv)	24. (iii)	25. (iv)	26. (iii)	27. (i)	28. (iii)
29. (i)	30. (iv)	31. (iii)	32. (iii)	33. (ii)	34. (ii)	35. (iii)
36. (iii)	37. (iv)	38. (ii)	39. (i)	40. (ii)	41. (iii)	42. (iv)
43. (ii)	44. (iii)	45. (iv)	46. (iv)			

Chapter 7

1. (iii)	2. (iv)	3. (i)	4. (ii)	5. (ii)	6. (iv)	7. (iv)
8. (ii)	9. (iv)	10. (i)	11. (iv)	12. (iv)	13. (ii)	14. (iv)
15. (iii)	16. (iv)	17. (iii)	18. (iii)	19. (iii)	20. (iii)	21. (i)
22. (iii)	23. (ii)	24. (iv)	25. (iv)	26. (iv)	27. (ii)	28. (iii)
29. (iv)	30. (ii)	31. (iii)	32. (iii)	33. (iv)	34. (ii)	35. (iv)

Chapter 8

1. (iv)	2. (iii)	3. (i)	4. (iii)	5. (iv)	6. (ii)	7. (iv)
8. (iv)	9. (ii)	10. (iv)	11. (iii)	12. (iii)	13. (iii)	14. (i)
15. (ii)	16. (ii)	17. (i)	18. (iv)	19. (iv)	20. (ii)	21. (iii)
22. (iv)	23. (iii)	24. (iii)	25. (i)	26. (iii)	27. (iii)	28. (ii)
29. (i)	30. (iv)	31. (iv)	32. (i)	33. (i)	34. (iii)	

Chapter 9

1. (iii)	2. (ii)	3. (i)	4. (iv)	5. (i)	6. (iii)	7. (iv)
8. (iii)	9. (iv)	10. (ii)	11. (iv)	12. (ii)	13. (i)	14. (iii)
15. (i)	16. (iv)	17. (iv)	18. (ii)	19. (i)	20. (iv)	

Chapter 10

1. (iv)	2. (iv)	3. (iii)	4. (ii)	5. (iii)	6. (ii)	7. (i)
8. (iii)	9. (ii)	10. (iii)	11. (i)	12. (ii)	13. (iii)	14. (i)
15. (iii)	16. (i)	17. (ii)	18. (iv)	19. (iv)	20. (iv)	

Chapter 11

1. (i)	2. (iii)	3. (iv)	4. (i)	5. (iii)	6. (ii)	7. (iii)
8. (i)	9. (i)	10. (iii)	11. (iv)	12. (ii)	13. (i)	14. (ii)
15. (iii)	16. (i)	17. (iii)	18. (iii)	19. (iii)	20. (ii)	21. (i)
22. (ii)	23. (iii)	24. (iv)	25. (iv)	26. (ii)	27. (iv)	28. (ii)
29. (ii)	30. (ii)	31. (i)	32. (iv)	33. (iv)	34. (ii)	35. (iii)
36. (iii)	37. (i)	38. (iii)				

Chapter 12

1. (iv)	2. (iii)	3. (ii)	4. (iii)	5. (iv)	6. (ii)	7. (iv)
8. (iii)	9. (iii)	10. (iv)	11. (i)	12. (iv)	13. (i)	14. (iii)
15. (i)	16. (ii)	17. (iii)	18. (iii)	19. (iv)	20. (iv)	21. (ii)
22. (iv)	23. (i)	24. (iv)	25. (i)	26. (iii)	27. (i)	28. (iii)
29. (ii)	30. (i)	31. (iv)	32. (iii)	33. (i)	34. (iv)	35. (i)
36. (ii)	37. (iii)	38. (iv)	39. (iv)	40. (iv)	41. (iv)	42. (ii)
43. (iv)	44. (ii)	45. (iii)	46. (iii)	47. (ii)	48. (i)	49. (iv)
50. (iii)	51. (iii)	52. (ii)	53. (ii)	54. (i)	54. (iii)	55. (iv)
56. (iv)						

BIBLIOGRAPHY

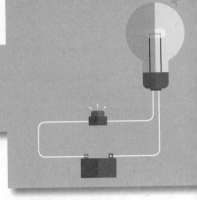

Bakshi, R., 2002 'Wind Energy in India', *IEEE Power Engineering Review*, vol. 22, no. 9, pp. 16–17.

Begamudre, E.R.D., 2000 *Energy Conversion Systems*, New Age International, New Delhi.

Bergen, A.R. and V. Vittal, 2001 *Power System Analysis*, 1st Indian reprint, Pearson Education Asia, Delhi.

Central Station Engineers of the Westinghouse Electric Corporation, Pennsylvania, *Electrical Transmission and Distribution Reference Book*, 1st Indian edn, Oxford Book Company, Calcutta, 1950.

Choudhuri, S., R. Chakraborty, and P.K. Chattopadhyay, 1999 *Electrical Science*, Narosa Publishing, New Delhi.

Clayton, A.E. and N.N. Hancock, 1961 *The Performance and Design of Direct Current Machines*, 3rd edn, Sir Issac Pitman & Sons, London.

Cotton, H., 1967 *Advanced Electrical Technology*, ELBS Publications, London.

Del Toro Vincent, 1994 *Electrical Engineering Fundamentals*, 2nd edn, Prentice Hall India, New Delhi.

Elgerd, O.I., 1985 *Electric Energy System Theory—An Introduction*, 2nd edn, Tata McGraw Hill Publishing, New Delhi.

Fitzgerald, A.E., D.E. Higginbotham, and A. Grabel, 1981 *Basic Electrical Engineering*, 5th edn, McGraw Hill International.

Fitzgerald, A.E., C. Kingsley, Jr, and S.D. Umans, 1992 *Electric Machinery*, 5th edn, McGraw Hill Company.

Golding, E.W., 1959 *Electrical Measurements and Measuring Instruments*, 4th edn, Sir Issac Pitman & Sons, London.

Hughes, E., 1995 *Electrical Technology*, 7th edn, revised by I.M. Smith, Addison Wesley Longman, England.

Kowsow, I.L., 1992 *Electric Machinery and Transformer*, 2nd edn, Prentice Hall India, India.

Nagrath, I.J., 1991 *Basic Electrical Engineering*, Tata McGraw Hill Publishing, New Delhi.

Nasar, S.A., 2002 *Electric Machines and Power Systems*: *Electrical Machines*, vol. 1, Tata McGraw Hill edn.

Rahman, S., 2003 'Green Power: What is It and Where Can We Find it', *IEEE Power and Energy*, vol. 1, no. 1, pp. 30–37.

Rizzoni, G., 2000 *Principles and Applications of Electrical Engineering*, 3rd edn, McGraw Hill Higher Education.

Say, M.G., 1983 *Design of Alternating Current Machinery*, 5th edn, ELBS, London.

Tiwari, S.N. and A.S. Bin Saroor, 1993 *A First Course in Electrical Engineering*, Wheeler Publishing, Allahabad.

Turan, G., 1986 *Electric Power Distribution System Engineering*, McGraw Hill Book Company, Singapore.

INDEX

About the Authors

Dr T.K. Nagsarkar obtained his BSc from Institute of Science, Nagpur (previously known as 'Government Science College') and BE from Visvesaraya Regional College of Engineering, Nagpur; he further pursued M Tech from IIT Kharagpur, and subsequently obtained his PhD from Panjab University, Chandigarh. He is a veteran in the teaching profession with an extensive career of 34 years until he retired as Professor and Head, Department of Electrical Engineering, Punjab Engineering College. Dr Nagsakar has guided many students for their PhD and ME thesis, and has published many technical papers in both national and international journals and conferences. He is a fellow of Institution of Engineers (India) and Life Senior Member IEEE (USA).

Dr Nagsarkar and Dr Sukhija have also jointly authored *Basic Electrical and Electronics Engineering* (OUP, 2012), *Power System Analysis 2e* (OUP, 2014), and *Circuits and Networks: Analysis, Design, and Synthesis 2e* (OUP, 2016).

Dr M.S. Sukhija, a PhD from Panjab University, was the founder principal of Guru Nanak Dev Engineering College, Bidar, Karnataka. He has several years of experience teaching both undergraduate and postgraduate students. Dr Sukhija has served the power cable industry for nearly ten years and was also Director (Technical) with Educational Consultants India Ltd., where he handled innumerable projects at the national as well as the international levels in the field of education.

Related Titles

Circuits and Networks Analysis, Design, and Synthesis 2e (9780199460922)

M.S. Sukhija & T.K. Nagsarkar

This second edition serves as a textbook for the undergraduate students of electrical, electronics, and instrumentation engineering. The new approach using MATLAB-based problem solving enhances the book's utility among professionals and practitioners as well as for laboratory-based learning.

Key Features
- Approximately 400 new solved examples
- MATLAB-based solved and unsolved problems
- Addition of new topics: sawtooth signals, doublet, four-wire system, terminating half section, composite filters, Bode plots, etc.

Microprocessors and Microcontrollers 2e (9780199466597)

N. Senthil Kumar, M. Saravanan & S. Jeevananthan

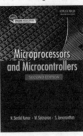

Microprocessors and Microcontrollers is an established textbook for engineering students pursuing a course in electrical and electronics, electronics and communication, computer science, and information technology. It is also a useful resource for practising professionals. This second edition of the book goes one step further in providing a comprehensive coverage of topics and an application-oriented approach.

Key Features
- New case studies on microprocessor-based temperature control system and thyristor triggering control
- Additional timing diagrams for 8085, debugging of assembly language programs, and ARM microcontrollers
- Numerous additional solved programming examples for 8051 to aid in better understanding of the theory.

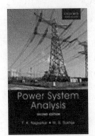

Power System Analysis 2e (9780198096337)

T.K. Nagsarkar & M.S. Sukhija

Power System Analysis serves as a basic text for undergraduate students of electrical engineering. It provides a thorough understanding of the basic principles and techniques of power system analysis as well as their application to real-world problems.

Key Features
- A large number of illustrative problems that use MATLAB in the analysis of power systems
- Includes Advanced topics such as contingency analysis and state estimation
- An introduction to HVDC and FACTS
- Numerous examples with step-by-step procedures and a variety of chapter- end problems

Protection and Switchgear

Bhavesh Bhalja, Nilesh Chothani & R.P. Maheshwari (9780198075509)

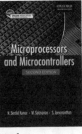

Protection and Switchgear is designed as a textbook for undergraduate students of electrical and electronics engineering. The book aims at introducing students to the various abnormal operating conditions in power systems and the apparatus, system protection schemes, and the phenomena of current interruption, to study various switchgears

Key Features
- In-depth coverage of apparatus protection, circuit breaking fundamentals, and ion and testing of circuit breakers using actual field data
- Solved examples and numerical exercises in each chapter along with review exercises and MCQs
- Chapters on recent developments in protective relays and power system computer-aided design (PSCAD)